QED and the Men Who Made It

PRINCETON SERIES IN PHYSICS

Edited by Philip W. Anderson, Arthur S. Wightman, and Sam B. Treiman
(published since 1976)

QED and the Men Who Made It: Dyson, Feynman, Schwinger, and Tomonaga

Silvan S. Schweber

Princeton University Press
Princeton, New Jersey

Published by Princeton University Press, 41 William Street, Princeton,
New Jersey 08540
In the United Kingdom: Princeton University Press, Chichester, West Sussex

Library of Congress Cataloging-in-Publication Data
Schweber, S. S. (Silvan S.)
   QED and the men who made it : Dyson, Feynman, Schwinger, and
Tomonaga / Silvan S. Schweber.
       p.   cm. – (Princeton series in physics)
   Includes bibliographical references and index.
   ISBN 0-691-03685-3; (pbk) 0-691-03327-7
   1. Quantum electrodynamics—History.   2. Physicists—Biography.
I. Title.   II. Series.
QC680.S34   1944   93-33550
537.6'7'09—dc20

This book has been composed in TIMES ROMAN and KABEL

Princeton University Press books are printed on acid-free paper, and meet
the guidelines for permanence and durability of the Committee on Production
Guidelines for Book Longevity of the Council on Library Resources

Printed in the United States of America

10 9 8 7 6

For my mother,
for Frank,
and
for Miriam

CONTENTS

8. Richard Feynman and the Visualization of Space-Time Processes

9. Freeman Dyson and the Structure of Quantum Field Theory

10. QED in Switzerland   576

PREFACE

During the middle 1970s my research interests shifted from physics to the history of science. I became fascinated with trying to understand how probabilistic concepts entered the sciences during the nineteenth century. With the characteristic hubris of the physicist, I had believed that probability was first introduced into the *physical* sciences by Maxwell and Boltzmann, only to discover that much earlier probabilistic concepts had been used by political economists and also by Darwin. I spent the academic year 1976/77 in the Department of History of Science at Harvard University exploring this topic further. I delved deeper into Darwin, random variations, and natural selection, and for the next several years I found myself happily immersed in nineteenth-century evolutionary and probabilistic thought.

It was Frank Manuel who urged me to make greater use of my past training as a physicist in my historical studies. We had often discussed creativity and novelty in the sciences and invariably the subject of the post-World War II advances in theoretical physics came up. He convinced me that others could write about Darwin, but not very many had the technical skills to convey what had been accomplished in quantum field theory.

In 1982 I started on the project that culminated with the present volume. Steve Weinberg's insightful article, "Some Notes for a History of Quantum Field Theory," which had appeared in *Daedalus* in 1979, mapped out what should be done. On several occasions in the intervening years he has suggested that the history of the theory of elementary particles could be understood as oscillations between quantum field theory and what has become known in modern parlance as S-matrix theory. My book constitutes one chapter of that story. It covers the period from 1927 till about 1950, with most of the emphasis on the post-World War II developments. It ends with what was the first decisive victory of the quantum field-theoretical viewpoint. Although the fruitfulness of quantum field theory had been recognized during the thirties—the successes of quantum electrodynamics (QED), Fermi's theory of beta decay, Yukawa's meson theoretic explanation of the nuclear forces, Pauli's proof of the connection between spin and statistics all attested to the power of this form of theorizing—the inability of these theories to make quantitative predictions in agreement with experimental data—except for quantum electrodynamics in the lowest order of perturbation theory—and their universal failure in higher orders of perturbation theory due to the divergence difficulties had raised serious doubts about the validity of the approach. The success of renormalization theory in quantum electrodynamics, and its extension to other field theories, made plausible the assertion that quantum field theory was the natural framework for the synthesis of the quantum theory and the theory of special relativity.

The first two chapters of this book review the developments of quantum field theory in the late twenties and during the thirties. The contrast between Dirac and Jordan sets the stage for understanding the subsequent oscillations between the "particle" and "field" points of view.

Many of the advances during the 1945–1950 period, especially on the experimental side, were made in the United States. One of the principal reasons for this was that the United States did not suffer the devastation that World War II had inflicted on Great Britain, continental Europe, and Japan. But there were other factors at play, reflecting the institutional settings in which physics was practiced in the United States and the great technical advances made during World War II. In the United States, theory and experiments were always housed under the same roof in one department. And perhaps more so than everywhere else, in the United States physics was about numbers, and theories were deemed to be algorithms for getting the numbers out. In the wartime laboratories an entire generation of young theorists was raised with this attitude. World War II played an important role in the maturing of the American physics community. It is thus not surprising that the physicists' wartime efforts and the influence of these experiences on the subsequent developments should receive attention here.

The gathering of many of the leading wartime American theorists at Shelter Island in June 1947 can be taken as the opening bell for the postwar developments. Lamb's landmark experiment on the fine structure of hydrogen was first reported at that meeting. It was the main impetus for reanalyzing the formulation of quantum electrodynamics and for making the notion of mass renormalization that had been advanced by Kramers central to the calculations of the Lamb shift by Bethe and others. Chapter 5 is devoted to the Shelter Island conference and to the two subsequent ones that were held at Pocono and Oldstone. Chapter 6 analyzes the Lamb shift experiment and experiments indicating that the magnetic moment of the electron differs from the value predicted by the Dirac equation.

The advances during the 1946–1949 period were principally the work of four remarkable individuals: Sin-Itiro Tomonaga, Julian Schwinger, Richard Feynman, and Freeman Dyson. My book is partly the story of these four physicists and of their community. The history of QED in the period from 1946 to 1949 has many similarities with the history of the developments of quantum mechanics from 1925 to 1927, when Schrödinger and Heisenberg had propounded two different approaches to quantum mechanics. The correspondence can be taken to be: Feynman is to Heisenberg what Tomonaga-Schwinger is to Schrödinger, with Dyson initially playing the role of Pauli, Schrödinger, and Eckart in proving the equivalence of the two approaches. But Dyson did much more, and his contributions in the period from 1947 to 1949 stand on par with those of the other three. It is in fact my view that his accomplishments were not duly rewarded. Dyson's contributions to renormalization theory and his suggestion that the criterion of renormalizability be adopted as a selection principle for acceptable theories were very influential in shaping the theoretical research programs of the next several decades.

I am aware of the imbalance that is created by the far lengthier accounts I have given of the life and work of Schwinger, Feynman, and Dyson as compared to those of Tomonaga. This should not be taken in any way as minimizing Tomonaga's accomplishments or passing judgment on his stature as a physicist. In fact, Tomonaga's story may perhaps be the most interesting of the four. My inability to read Japanese, my inadequate knowledge of Japanese history, and my lack of contact with Japanese culture are part of the explanation though not an excuse.

I see as the outcome of the episode that I narrate the establishment of quantum field theory as the appropriate mode of description of the atomic and subatomic realms. The victory was a conservative one. Feynman was the revolutionary among those responsible for the advance. He had hoped to formulate a new electrodynamics that would be self-consistent and intrinsically finite. In the end, however, the new theory was built on the old foundations and was rendered finite through "renormalization" procedures. But the new formulation also incorporated Feynman's impressive calculational techniques and the simplicity afforded by his intuitive visualization of processes by means of "Feynman" diagrams. As Schwinger has noted, Feynman brought field theory to the masses.

I have not tried to give a "full" history of QED during the thirties and forties. Fortunately there exist other accounts of the history I tell. The reader should consult Darrigol's (1982) impressive dissertation, and Pais's *Inward Bound*. In my telling of the story, the works of Tomonaga, Stueckelberg, Pais, Sakata, Pauli, Heisenberg, Heitler, Kramers, and others have been slighted to some extent. I believe that the history of science cannot escape some form of whiggism. The data are so rich that some selection must be made. Although part of the imbalance of my presentation reflects my access to primary source materials regarding Schwinger, Feynman, Bethe, Dyson, and other "Western" physicists, it did seem to me that the remarkable contributions of Tomonaga and his coworkers to the developments in the 1947–1950 period—in many ways the culmination of Yukawa's landmark contributions to quantum field theory during the thirties and early forties—were not as influential as those of Schwinger, Feynman, and Dyson. My account reflects an emphasis on what I assessed to be the influential contributions *at the time*. The criterion I used was to try to answer for myself the question: "How would the developments have been different had Tomonaga's work been totally unknown in the United States and Europe?" What I found striking was the remarkable similarity between the formulation of Tomonaga and his associates and that of Schwinger. The congruence reflects a common source in the earlier works of Pauli and Fierz, and of Wentzel; also, their researches in strong coupling meson theory, upon which their later work drew, had much in common. I was deeply moved by the conditions under which Tomonaga's contributions were made, and I came to appreciate their importance to the training of a generation of outstanding Japanese theorists that includes Nambu, Kinoshita, Fukuda, Hayakawa, and Nishijima, among others. I hope that an account that does justice to Tomonaga and to the Japanese context will be forthcoming.

I have not tried to fit my presentation of the history of quantum field theory into a preconceived pattern, whether that of Kuhn or that of Lakatos. My concern has been with the telling of the story, in order to understand better the dynamics of change. One could easily cast the history into a Lakatosian mold of research programs—with S-matrix and field theory the two competing modes (Weinberg 1980a, 1986a; Cushing 1990). Similarly, one could pick from that history examples that would instantiate both of Kuhn's notions of paradigm: namely, paradigm as achievement (the body of work that emerges from a scientific crisis and sets the standard for addressing problems in the subsequent period of normal science), and paradigm as a set of shared values (the methods and standards shared by the core of workers who decide what the interesting problems are and what counts as solutions, and who determine who shall be admitted to the discipline and what shall be taught to them). Furthermore, one could readily give examples of Kuhnian revolutions. Renormalization theory as formulated in the period from 1947 to 1949, culminating with the work of Dyson, is surely one such revolution, and broken symmetry (Goldstone 1961; Nambu 1960; Goldstone, Salam, and Weinberg 1962) is another. One probably could constrain the history of quantum field theory into a Kuhnian mold. But I believe that much would be lost in doing so, in particular a perspective on the cumulative and continuous, yet novel, components of the developments. It seems to me that Kuhn's recent emphasis on "lexicons"—the learnable language, laws, and facts of a given tribe of scientific workers—constitutes a more useful approach to the growth of our knowledge of elementary particles (Kuhn 1989). Equally helpful, I believe, is Hacking's notion of style of scientific reasoning: "A style of reasoning makes it possible to reason toward certain kinds of propositions, but does not of itself determine their truth value." It determines what may be true or false. Similarly, it indicates what has the status of evidence. Styles of reasoning, as Hacking has noted, tend to be slow in evolution and are vastly more widespread than paradigms. Furthermore, they are not the exclusive property of a single disciplinary matrix (Hacking 1985). Thus Feynman's (1948c) space-time approach to nonrelativistic quantum mechanics encapsulates a style of reasoning that can be traced back to Maxwell, Boltzmann, and especially to Rutherford: all physical measurements and interactions can be considered as scattering processes. I believe Hacking's notion of a style of reasoning captures something right about the history of quantum field theory. The uses of symmetry, of renormalization methods, are examples of styles of reasoning. Moreover, the fact that these styles of reasoning are useful in both particle physics and in condensed matter physics—and in point of fact cross-fertilized these fields—illustrates the (nonlinear) additive properties of styles of reasoning. Since a style of reasoning can accommodate many different paradigms, it is not surprising that one should discern Kuhnian revolutionary episodes. The delineations of such revolutions are helpful guidelines and periodizations of the history of the field. But it is the identification of the different styles of reasoning that is, I believe, the important task for the intellectual historian attempting to relate that history.

I have attempted to function as an "objective" intellectual historian. Nonetheless, my emotional ties to the theoretical physics community will be apparent to the reader: I have given loving biographies of the principals involved and an admiring account of the community of theoretical physicists. As Buber (1958, p. 53) observed: "Common reverence and common joy of soul are the foundations of genuine human community." Theoretical physicists share a common reverence for the accomplishments and the capabilities of the leading practitioners—for example, Newton, Maxwell, Boltzmann, Einstein, Bohr, Dirac, Landau, Feynman—and derive a common joy of soul from the understanding gained by the representations and interpretations of the empirical data put forth by the great theorists. I have attempted to convey some of that common exhilaration. The presentation in the book is at times technical. Nonetheless, I hope it will not discourage those without the technical background from reading it. In general, I have included technical details when these came from sources not generally available, for example, the private papers of the individuals involved or the unpublished records of conferences such as Shelter Island, Pocono, and Oldstone.

ACKNOWLEDGMENTS

Perhaps the most pleasant aspect of completing the writing of this book is that it gives me the opportunity to thank publicly the people who in one way or another made it possible. All my colleagues in the Department of Physics at Brandeis University—Steve Berko and Eugene Gross in particular—encouraged my historical activities even while I was doing the history of biology. Their helpfulness and friendly support made it possible to carry out my work in surroundings that we all look for but rarely find. I would like to believe that what I was able to do justifies the tenure system at American universities.

I still vividly remember the helpful suggestions and warm encouragement that Erwin Hiebert and Everett Mendelsohn gave me when I first entertained the idea of doing historical research. The Department of History of Science at Harvard has been the source of much intellectual stimulation over the past decade. I deeply appreciate the hospitality and the collegiality its faculty have tendered me. An outstanding asset of that department is its graduate student body. I was fortunate to begin my historical studies at the time that Raine Daston and Peter Galison were studying there. I value greatly what I have learned from them, and I especially value their friendship. Over the years I have been the beneficiary of extended discussions with the graduate students and visiting scholars of that department.

I have found the community of historians of science an open, friendly, and stimulating one. I have discussed with profit various aspects of my work with Joan Bromberg, Tuan Yu Cao, Nancy Cartwright, Jim Cushing, Raine Daston, Paul Forman, Peter Galison, Kostas Gavrolu, Yves Gingras, Ian Hacking, Steve Heims, Erwin Hiebert, John Heilbron, Gerry Holton, Evelyn Fox Keller, Martin Klein, Everett Mendelsohn, Tom Kuhn, Andy Pickering, Jürgen Renn, Skuli Sigurdsson, Roger Stuewer, Sharon Traweek, and Norton Wise. I thank them all.

I could not have written this book without the cooperation and helpfulness of many of the contributors to the developments I have related. Although I had studied their work, I did not know Feynman, Schwinger, or Tomonaga during the late forties or early fifties. For the most part I have relied on archival materials in the telling of my story. I did interview Schwinger, Feynman, Dyson, Bethe, Lamb, Rabi, and others but only after I had studied the materials they had deposited in their archives, or the letters, papers, and notes they had in their possession. I am deeply appreciative of the time and thoughtfulness of all the people I interviewed: Valia Bargmann, Hans Bethe, Robert Dicke, Stanley Deser, Bryce DeWitt, Freeman Dyson, John Edsall, Bernie Feld, Herman Feshbach, Richard Feynman, Bob Finkelstein, Markus Fierz, Roy Glauber, Murph Goldberger, Res Jost, Willis Lamb, Quin Luttinger, Francis Low, Hal Lewis, Phil Morse, Yoichiro Nambu, Bram Pais, Rudolf Peierls, Norman Ramsey, Dominique Rivier, Isadore Rabi, Ed Salpeter, Julian Schwinger, Felix Villars, Vicki Weisskopf, John A. Wheeler, Arthur Wightman, and Eugene Wigner.

I would like to thank the archivists of the Bodleian Library of Oxford University, the California Institute of Technology, the Carnegie Institution of Washington, Churchill College (Cambridge University), Cornell University, the Eidgenössische Technische Hochschule (ETH) in Zurich, Harvard University Archives, the MIT Archives, the Seeley Mudd Library at Princeton University, the Niels Bohr Library at the American Institute of Physics in New York, the National Academy of Sciences, the Pauli Archives in CERN, the Bancroft Library of the University of California/Berkeley, the University of California/Los Angeles, and the University of Washington for making available materials in their archives and for permission to quote from these documents. I also want to express my gratitude to the staff of the Brandeis University Library: to its circulation librarians, who stretched rules and battled computers to allow me to keep books out—both in quantity and in length of time—beyond the norms of reasonableness; and to its reference librarians, who helped me get countless books from other libraries.

I am grateful to the National Science Foundation and the Council of Learned Societies for support during the initial phases of this project. The office of the dean of the faculty at Brandeis made available funds to help defray part of the cost of preparing the manuscript. I thank Ann Carter, Jim Lackner, and Greg Shesko for obtaining these grants.

I would also like to express my appreciation and thanks to Mimi Fricks, Jane Jordan, Joan Thorne, and particularly to Maureen Meyer for their caring, friendly, and most helpful assistance in typing the several drafts and revisions. Hal Riggs is responsible for the computer generated diagrams and in doing so exhibited another facet of his many talents. Sara Van Rheenen at Princeton University Press was always supportive and invariably helpful. Alice Calaprice carefully edited the manuscript, made sure that my sentences made sense, and kept me honest in matching references in the text with those in the bibliography. The book is much the better due to her skills, and I am much indebted to her.

Larry Abbott, Hugh Pendleton, and Howard Schnitzer read various chapters and gave me constructive criticism and made many very helpful suggestions. I thank them. I am particularly indebted to Howard Schnitzer for sharing with me his valuable insights into the history of particle physics during the sixties and seventies. I have greatly benefited from these discussions with him.

Arthur Wightman, Freeman Dyson, Laurie Brown, and Paul Forman read carefully through a first draft of the book and made trenchant criticisms and valuable suggestions for improving the presentation. I have attempted to follow their suggestions as best I could, and the book is undoubtedly much the better for it. I am indebted to them.

My greatest intellectual debt is to Frank Manuel. It is one that I cannot repay adequately. Over the past fifteen years he has been my teacher and mentor. He is my model for what it means to be a historian. I have benefited immensely from his questions, suggestions, and criticisms. This book is dedicated to him as an expression of my affection, esteem, and respect. He and his equally remarkable wife, Fritzie, have become dear friends, and my visits and discussions with them have been deeply rewarding and occasions of great pleasure.

My wife Miriam bore the brunt of this undertaking. I thank her for her good cheer in the face of the demands of the task, and I deeply appreciate her adjustments to the vagaries of my scholarship.

Despite all the help I have received, responsibility for the content of the book rests with me. I apologize for the errors that are undoubtedly still present, for the omission of references that might be relevant, and for inadvertently not giving appropiate credit for some of the materials. I have tried meticulously to ackowledge the sources of my indebtedness. I do not doubt that I have missed some.

As is often the case with the writing of books, a point was reached where the project was abandoned rather than completed. I took as my motto Alexander von Humboldt's assertion: "Works are of value only if they give rise to better ones." My task will have been worthwhile and I will judge it successful if I have achieved this.

QED and the Men Who Made It

INTRODUCTION

In this introduction I sketch the history of quantum electrodynamics from 1927 to the late 1940s in order to delineate the main trends.

The foundational aspects of physics during the first half of the twentieth century have been principally concerned with the characterization of the "elementary" constituents of matter and the elucidation of the nature of the space-time framework in which their interactions take place. The discovery of the electron by Thomson, the precise characterization of its charge by Millikan, the demonstration of the nuclear atom by Rutherford, the photon hypothesis of Planck and Einstein, and Bohr's explanation of the spectrum of hydrogen were some of the landmarks of that history. These early efforts culminated in the mid-twenties with the formulation of quantum mechanics by Heisenberg, Dirac, and Schrödinger (Kuhn 1978; Heilbron 1975; Segrè 1980; Pais 1986; Mehra and Rechenberg 1982–1988).

The revolutionary achievements in the period from 1925 to 1927 stemmed from the confluence of a theoretical understanding (the description of the dynamics of microscopic particles by quantum mechanics), and the apperception of an approximately stable ontology (electrons and nuclei). Approximately stable meant that these particles (electrons, nuclei), the building blocks of the entities (atoms, molecules, simple solids) that populated the domain that was being carved out, could be treated as ahistoric objects (whose physical characteristics were seemingly independent of their mode of production and whose lifetimes could be considered as essentially infinite). These entities could be assumed to be "elementary" pointlike objects that were specified by their mass, spin, and statistics (whether bosons or fermions), and by electromagnetic properties such as their charge and magnetic moment.

Quantum mechanics came to be seen as correctly describing that domain of nature delineated by Planck's constant (h) : any system whose characteristic length (l), mass (m), and time (t) were such that the product ml^2/t was of the order of h, and such that l/t was much smaller than c, the velocity of light, was quantum mechanical and was to be described by the new nonrelativistic quantum mechanics.

Quantum mechanics reasserted that the physical world presented itself hierarchically. The world was not carved up into terrestrial, planetary, and celestial spheres, but was layered by virtue of certain constants of nature. As Dirac (1930e) emphasized in the first edition of his *The Principles of Quantum Mechanics*, Planck's constant allows the world to be parsed into microscopic and macroscopic realms. It is to be stressed that it is constants of nature—Planck's constant, the velocity of light, the masses of "elementary" particles—that demarcate the domains. In the initial flush of success, quantum mechanics was believed to explain most of physics and all of chemistry. All that remained to be done was "fitting of the theory with relativity ideas." Dirac's (1929b) famous assertion reflected the confidence and hubris of the community:

> The general theory of quantum mechanics is now almost complete, the imperfections that still remain being in connection with the fitting of the theory with relativity ideas. These give rise to difficulties only when high-speed particles are involved, and are therefore of no importance in the consideration of atomic and molecular structure and ordinary chemical reactions. . . . The underlying physical laws necessary for the mathematical theory of a large part of physics and the whole of chemistry are thus completely known, and the difficulty is only that the exact application of these laws lead to equations much too complicated to be soluble.

However, fitting the theory with relativity ideas proved to be a much more difficult problem than had been anticipated. How to synthesize the quantum theory with the theory of special relativity was—and has remained—the basic problem confronting "elementary" particle theorists since 1925–1927.

By the early thirties it had become clear that particle creation and annihilation were the genuinely novel features emerging from that synthesis. It should be recalled that the theoretical apparatus for the description of microscopic phenomena up to that time had been predicated on a metaphysics that assumed conservation of "particles." Dirac's (1927b,c) quantum electrodynamics was the first step in the elimination of that preconception (Bromberg 1976, 1977). Dirac's (1931b) hole theory was the first instance of a relativistic quantum theory in which the creation and annihilation of matter was an intrinsic feature. Hole theory was the first insight into what was entailed by a quantum mechanical description of particles that conformed with the requirements of special relativity.

The history of elementary particle physics can be analyzed in terms of oscillations between two viewpoints: one which takes fields as fundamental, in which particles are the quanta of the fields; and the other which takes particles as fundamental, and in which fields are macroscopic coherent states (Weinberg 1977, 1985a, 1986a). Figure I.1 outlines the history of relativistic quantum mechanics during the 1930s from this perspective. One research tradition (de Broglie → Schrödinger → Jordan → Pauli-Heisenberg) laid the foundation of the quantum theory of fields. The other, predicated on a metaphysics that took particles as the fundamental entities, has Dirac as its founding father and guiding spirit. It is exemplified by the hole theoretic formalism. In hole theory electrons were described by the Dirac equation. To this was appended the postulate that the vacuum was the state in which all the negative energy states were occupied. Pair creation was regarded as a transition of a "negative energy" electron to an unoccupied positive energy state—rather than the creation de novo of a positron and an electron (Dirac 1931b; Pais 1947, 1948). Although essentially equivalent to the field-theoretic

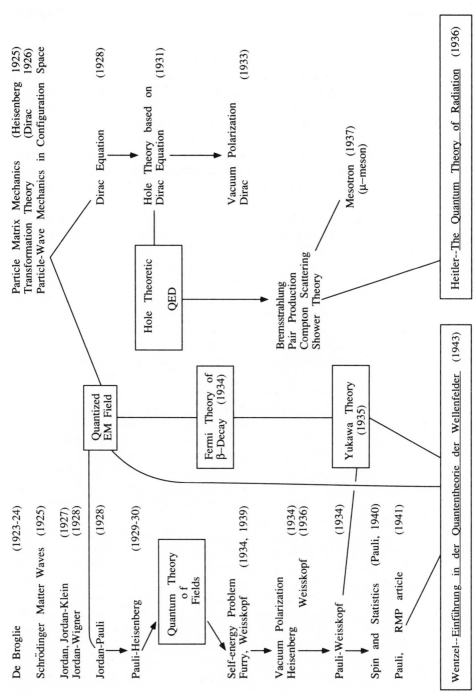

Figure I.1

approach for problems dealing with a single electron, the hole-theoretic formulation was ambiguous in its treatment of multi-electron systems. Hole theory was a "quasi" field theory in that it accepted the fact that a one-particle interpretation of the Dirac equation was impossible and recognized that it dealt with a (denumerable) infinity of particles. It was a particle theory in that particles were the primary entities and no reference was made to a quantized field. Hole theoretic QED was a particle theory as far as matter was concerned but a quantum field theory in its treatment of radiation (Heitler 1936). The electromagnetic field had a privileged role by virtue of the fact that it exhibited a classical limit. Both research traditions agreed that it should be quantized.

In his Nobel Prize speech, Feynman (1966a) made this insightful observation: "Theories of the known, which are described by different physical ideas, may be equivalent in all their predictions and hence scientifically indistinguishable. However, they are not psychologically identical when trying to move from that base into the unknown. For different views suggest different kinds of modifications which might be made and hence are not equivalent in the hypotheses one generates from them in one's attempt to understand what is not yet understood."

That the quantum field-theoretic approach was richer in potentialities and possibilities is made evident by the field-theoretic developments of the thirties. All these advances took as their point of departure insights gained from the quantum theory of the electromagnetic field, and in particular from the centrality of the concept of emission and absorption of quanta.

Fermi's theory of beta decay and Yukawa's theory of nuclear forces suggested that quantum field theory was the natural framework in which to attempt to understand what we now call the weak and strong interactions. By the late 1930s the formalism of quantum field theory was fairly well understood and the state of affairs can be inferred from Pauli's article in the *Reviews of Modern Physics* (Pauli 1941). But it was Wentzel's *Einführung in der Quantentheorie der Wellenfelder* (Wentzel 1943) in which a full account of relativistic quantum field theories was presented and which disseminated this approach to a wide audience after World War II.

In many ways, hole theory was equally successful. Most of the predictions of quantum electrodynamics during the 1930s—such as the cross sections for electron-positron pair production and annihilation, bremsstrahlung, Compton scattering—as well as the verification of the validity of the theory up to energies of the order of $137 \, mc^2$ and even greater, were based on hole-theoretic QED calculations. Incidentally, it was confidence in this theory that was responsible for the discovery and postulation of a new particle in the cosmic radiation, the "mesotron," the particle now identified as the muon (Cassidy 1981; Galison 1983). Heitler (1936) summarized the hole theoretic approach in his *The Quantum Theory of Radiation*, with a slightly revised edition appearing in 1944. Heitler's books were the primary and standard sources for learning how to "calculate" quantum electrodynamic processes.

But both approaches—field theory and hole-theoretic QED—were beset by overwhelming divergence difficulties that manifested themselves in higher-order calculations (Weinberg 1977; Pais 1986). These difficulties impeded progress and gave rise to a deep pessimism about the formalisms at hand (Rueger 1991). Numerous proposals to overcome the problems of the divergences were advanced. They can be classified as follows (Aramaki 1987):

1. *Attempts at eliminating the divergences.* Born and Infeld's (1935) non-linear theory of the electromagnetic field is one example of an attempt to remove the divergences (see also Pauli 1936). Wentzel's (1933, 1934) λ-limiting procedure, which reinterpreted the meaning of a local interaction is another. Several investigators tried to remove the divergences by introducing new interactions that "compensate" (i.e., cancel) the divergences of the original theory. Sakata's (1947) C-meson field and Pais's (1945, 1946, 1947) f-field are representative examples. This trick, however, only works for the self-energy divergences in lowest order of perturbation theory (Kinoshita 1951).

2. *Attempts at circumventing difficulties.* The procedure of redefining the charge-current operator in the problem of vacuum polarization so as to absorb the divergences (Dirac 1934a,d; Heisenberg 1934b; Weisskopf 1936) is the first example of the circumvention of the divergence difficulties by a process that would later be called "renormalization." Pauli and Fierz (1937) in the quantum case, and Kramers (1938a,b) in the classical case, similarly removed the self-energy divergence of a charged particle in interaction with the radiation field by a redefinition of the mass parameter in terms of which the theory was originally formulated. Dancoff (1939) made an attempt to obtain a divergence-free formulation of hole-theoretic quantum electrodynamics to lowest order in perturbation theory by renormalizing both the charge and the mass of the electron. His failure to include all the contributions to that order of perturbation theory doomed the effort at the time.

3. *Attempts to understand the structure of the theory and the nature of divergences.* Weisskopf (1939) computed the self-energy of an electron in QED in higher orders and concluded that the self-energy divergences are logarithmic to all orders of perturbation theory.

Almost all the proposals to eliminate the divergences that were made during the 1930s ended in failure. The pessimism of the leaders of the discipline—Bohr, Pauli, Heisenberg, Dirac—was partly responsible for the lack of progress. They had witnessed the overthrow of the classical concepts of space-time and were responsible for the rejection of the classical concept of determinism in the description of atomic phenomena. They had brought about the quantum-mechanical revolution and they were convinced that only further conceptual revolutions would solve the divergence problem in quantum field theory. Heisenberg in 1938 noted that the revolutions of special relativity and quantum mechanics were associated with fundamental dimensional parameters: the speed of light, c, and Planck's constant, h. These delineated the domain of classical physics. He proposed that the next revolution be associated with the introduction of a fundamental unit of length, which would delineate the domain in which the concept of fields and local interactions would be applicable (Heisenberg 1938a,b,c).

The circumvention of the divergence difficulties in the 1945–1950 period was the work of a handful of individuals—principally Kramers, Bethe, Schwinger, Tomonaga, Feynman, and Dyson—and the solution advanced was conservative and technical. It asked to take seriously the received dogma of quantum field theory and special relativity and to explore the limits of that synthesis. Renormalization theory—the technical name for the proposed solution of the 1947–1950 period— revived the faith in quantum field theory.

The history of the developments of quantum field theory in the period from 1943 to the early 1950s is summarized in figure I.2 in a form which again highlights the distinction between the field and the particle approach. Tomonaga, Schwinger, and Dyson were all field theorists. On the other hand, particles were the fundamental building blocks for Feynman, as had been the case for Dirac. In fact, Feynman can be said to have inherited Dirac's mantle. Feynman diagrams visualize the fundamental processes in terms of space-time trajectories of particles. Schwinger (1948b,c) and Tomonaga (1943b, 1946, 1948), by exhibiting a field-theoretic formalism that identified and eliminated the divergences in low orders of perturbation theory in a relativistically and gauge-invariant fashion, established the validity of relativistic quantum field theories. In Schwinger's and Tomonaga's works, the elimination was accomplished by a renormalization procedure that identified the divergent terms according to their relativistic and gauge transformation properties, and then showing that these divergent contributions could be absorbed in a redefinition of the mass and charge parameters entering the original Lagrangian. Most importantly, the formalism could make predictions about observable phenomena (e.g., the magnetic moment of the electron, the Lamb shift, radiative corrections to Coulomb scattering). Feynman's genius was such that his idiosyncratic approach (stemming from his work with Wheeler on an action-at-a-distance formulation of classical electrodynamics from which all reference to the electromagnetic field had been eliminated!) resulted in a highly effective computational scheme (Feynman 1949a,b). Dyson (1949a) then demonstrated that Feynman's results and insights were derivable from Schwinger's and Tomonaga's formulation of QED. Furthermore, Dyson (1949b) was able to exhibit a proof that mass and charge renormalization removed all the divergences from the S-matrix

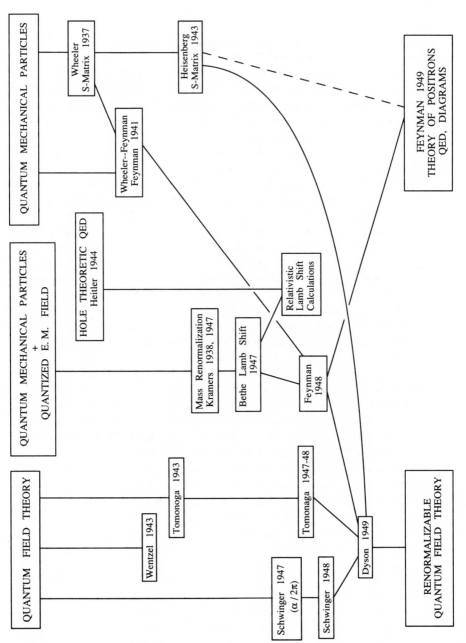

Figure I.2

of QED to all orders of perturbation theory. It suggested that renormalized QED was a consistent quantum field theory.

Let me briefly indicate some of the accomplishments of the 1947–1951 period. Foremost was the establishment of local quantum field theory as the best suited framework and formalism for the unification of quantum theory and special relativity. Furthermore, quantum electrodynamics, with the addition of renormalization rules, yielded calculated values for the fine and the hyperfine structure of hydrogen and positronium in remarkable agreement with experiments (Baranger 1951; Baranger et al. 1953; Karplus et al. 1952; Karplus and Klein 1952); and the same was true for the magnetic moment of the electron and muon.

In addition, a deeper understanding of the consequences of synthesizing quantum field theory and special relativity was obtained. Thus, it was shown that a relativistically invariant quantum field theory of charged fields automatically contains in its description oppositely charged particles. Also, a deeper insight was obtained into the necessity of quantizing spin zero and integer spin field theories with commutation rules and odd half-integer spin theories with anticommutation rules (Feynman 1949b, 1987; Pauli 1950; Schwinger 1950). Finally, the most perspicacious theorists—for example, Gell-Mann—noted the ease with which other symmetries besides the space-time ones could be incorporated into a local quantum field theory, and made field theory the framework in terms of which to describe the plethora of elementary particles then being discovered.

In January 1950, J. Robert Oppenheimer gave a series of lectures on the "Problems in the Interaction of Elementary Particles" at the California Institute of Technology. He reported on "the very great effort" on the part of theoretical physicists that had been devoted "during the last few years" to these problems and took it as his task to make clear "to what this effort has led and where we stand today." In his opening remarks Oppenheimer remarked that while preparing his lectures he was "appalled by how much we have learned and how little we knew [in the thirties]." He indicated that physics was "in the middle of a great and ... extremely deep advance." That "great advance" is the subject of this book.

1. The Birth of Quantum Field Theory

1.1 Introduction

In the next two chapters the history of quantum electrodynamics from 1927 to 1939 is adumbrated. Chapter 1 focuses on the early developments, chapter 2 on the 1930s.[1] In chapter 1, I concentrate on Jordan's and Dirac's work in the late 1920s. Jordan and Dirac are representative of two alternative points of view: Jordan assumed that fields were the fundamental entities, Dirac took particles as fundamental. The oscillation between these themata (Holton 1973) charts the subsequent history of the theory of "elementary" particles. The great successes of the 1945–1950 period can be interpreted as establishing quantum field theory as the most appropriate formalism for the unification of quantum mechanics and special relativity. The failure of meson theories in the 1950s to explain pion-nucleon interactions, nuclear forces, and the newly discovered particles led to a revival of the particle viewpoint: Chew's S-matrix program was its most ambitious expression (see, for example, Chew 1966; Cushing 1982, 1986, 1990; Cao 1991). The revival of the field viewpoint in the 1970s was sparked by the Weinberg-Salam theory of weak interaction and t'Hooft's proof that gauge theories of the Yang-Mills type were renormalizable. This rebirth stemmed largely from the fact that the physics of dynamical symmetry breaking is most easily and readily expressed in terms of quantum *field* theory (see, for example, Frampton 1987).

The developments of the quantum theory during the period from 1925 to 1930 harbored a deep ambiguity. Heisenberg and Dirac initially viewed quantum mechanics as the correct dynamics of microscopic *particles*, now described in terms of noncommuting variables p_i, q_i. Dirac's first paper on quantum electrodynamics (Dirac 1927b) similarly conceived of photons as "bosons," that is, as particles of zero rest mass obeying Bose-Einstein statistics. The vacuum was characterized by Dirac as a state with an infinite number of zero energy and zero momentum quanta. Photon emission was considered as a transition from this vacuum state to a state of a single photon with finite momentum and energy; photon absorption consisted of the reversed transition. Schrödinger, on the other hand, in his formulation of quantum mechanics, conceived of the "wave function" ψ as representing some kind of *wave*, and he interpreted the squared modulus $| \psi |^2$ as the density of electronic matter. For him, following de Broglie, *waves* were the fundamental entities (Raman and Forman 1969; Wessels 1977; Hanle 1975, 1977; Mehra and Rechenberg 1987, vol. 5).

It was Pascual Jordan who, more clearly than anyone else, saw the relation between the various approaches. Already in his initial work with Born in which the matrix formulation of quantum mechanics was outlined (Born and

Jordan 1926) and in the *Dreimännerarbeit* (Born et al. 1926), Jordan had suggested that the quantization rules $[q_i, p_j] = i\hbar\delta_{ij}$ that effected the transition from the classical to the quantum description of a system of particles also be applied to systems with an infinite number of degrees of freedom, that is, to field systems. Later, it was his view that Dirac's quantum electrodynamics was to be understood as the result of "quantizing" the electromagnetic field (Jordan 1929).

Influenced by the earlier work of Gustav Mie (1912), Jordan became committed to a unitary view of nature in which both matter and radiation were described by wave fields: the Schrödinger equation described the matter wave field and Maxwell's equation the electromagnetic field. He hoped to be able to understand the existence of particles as a consequence of the quantization of these field theories. The "classical" theory of matter waves was governed by the field equation

$$ih\frac{\partial\psi}{\partial t} = -\frac{\hbar^2}{2m}\nabla^2\psi + V\psi, \tag{1.1.1}$$

where the matter field amplitude $\psi(x, y, z, t)$ was to be regarded as analogous to the field quantities $E(x, y, z, t)$ and $H(x, y, z, t)$ entering Maxwell's equations. As Heisenberg (1930a) emphasized in his Chicago lectures, "This wave theory does not consider electrons, and e and m are merely universal constants of the wave equation." In eq. (1.1.1), V "no longer simply represents the potential of the external forces but also includes the potential of the matter waves themselves, that is, it takes account of the reaction of one part of the charge distribution upon another part." An interpretation of this classical theory was obtained by making the following identifications:

Charge density: $$\rho = -e\psi^*\psi \tag{1.1.2a}$$

Current density: $$j = -\frac{e\hbar}{2mi}(\psi^*\nabla\psi - \psi\nabla\psi^*) \tag{1.1.2b}$$

Energy density: $$u = \frac{\hbar^2}{2m}\nabla\psi^* \cdot \nabla\psi, \tag{1.1.2c}$$

for which the following conservation laws held by virtue of the field equation (1.1.1):

Conservation of charge: $$\frac{d}{dt}\int p\,dv = 0 \tag{1.1.3a}$$

Conservation of momentum: $$\frac{d}{dt}\int j\,dv = -e\int \nabla V\psi^*\psi\,dv \tag{1.1.3b}$$

Conservation of energy: $$\frac{d}{dt}\int u\,dv = e\int V\frac{\partial}{\partial t}(\psi^*\psi)dv. \tag{1.1.3c}$$

Small wave amplitudes, which were assumed to describe situations in which very low density matter waves were involved, obeyed the equation

$$-\frac{\hbar^2}{2m}\nabla^2\psi - i\hbar\frac{\partial\psi}{\partial t} = 0 \qquad (1.1.4)$$

from which the de Broglie theory was recovered. This theory gave a simple qualitative account of the diffraction experiments of Davisson and Germer, Thomson, Rupp, and others (Heisenberg 1930a).

In the case of matter waves interacting with other charges—such as atomic nuclei—and with themselves, the potential which enters eq. (1.1.1) satisfies the Poisson equation

$$\nabla^2 V = -4\pi(\rho + \rho_0), \qquad (1.1.5)$$

where ρ_0 is the density of the "external" charges and ρ is given by (1.1.2a). This theory does not contain any corpuscular elements: the total charge of the system

$$\int p\,dv = -e\int \psi^*\psi\,dv \qquad (1.1.6)$$

can take on any desired value, not merely integral multiples of e: $-e, -2e, \dots$. Similarly, the total energy can have any value. Nonetheless, the theory can be used to describe atomic phenomena in a manner analogous to that used by Bohr and Sommerfeld. Just as these authors introduced the quantum condition $\int pdq = nh$ into classical mechanics, similarly one can give an approximate account of atomic spectra by imposing the quantum condition

$$\int \psi^*\psi\,dv = n, \qquad (1.1.7)$$

which corresponds to the Hartree approximation.

In a series of seminal papers, Jordan, with Klein and Wigner as his collaborators, indicated how to quantize this "classical" field theory and recover from this procedure the "ordinary" Schrödinger configuration space description of a system of n bosons or fermions. In particular, Jordan and Klein (1927) showed that the quantum mechanics of a system of n identical particle systems obeying Bose-Einstein statistics (symmetric wave functions) is recovered if eq. (1.1.1) is interpreted as the equation of motion for an *operator* $\psi(x, t)$ that satisfies the equal-time commutation rules

$$[\psi(\mathbf{x}, t), \psi^*(\mathbf{x}', t)] = \psi(\mathbf{x}, t)\psi^*(\mathbf{x}', t) - \psi^*(\mathbf{x}', t)\psi(\mathbf{x}, t) = \delta(\mathbf{x} - \mathbf{x}'). \qquad (1.1.8)$$

If anticommutation rules are imposed,

$$[\psi(\mathbf{x}, t), \psi^*(\mathbf{x}', t)] = \psi(\mathbf{x}, t)\psi^*(\mathbf{x}', t) + \psi^*(\mathbf{x}', t)\psi(\mathbf{x}, t) = \delta(\mathbf{x} - \mathbf{x}'), \quad (1.1.9)$$

one similarly recovers the quantum mechanics of Fermi-Dirac particles described by antisymmetric wave functions (Jordan and Wigner 1928). Heisenberg and Pauli (1929, 1930) later extended the method to apply to any "wave field." Just as in the case of particle mechanics, Pauli and Heisenberg introduced a Lagrangian from which the equations of motion of the field variables could be derived, and in terms of which canonical momenta and a Hamiltonian could be defined. The quantum conditions $[q, p] = ih$ then translate into commutation relations for the field variables and their conjugate momenta.

Jordan's approach, however, did not become the accepted interpretation of nonrelativistic quantum mechanics during the 1930s. With the adoption of Schrödinger's wave mechanics after the demonstration of the equivalence of this approach with Heisenberg's matrix mechanics, and the acceptance of Born's probabilistic interpretation of the wave function, the huge success of this formalism in many applications made the description of electrons, protons, nuclei by particle variables, and associated Schrödinger wave functions the standard practice.

The tenacity of this approach was further reinforced by Dirac's success in deriving a relativistic wave equation for the electron (Dirac 1928a), which initially seemed to be interpretable in a way similar to Schrödinger's wave equation for a nonrelativistic particle. Dirac's equation not only accounted for the spin of the electron and its observed magnetic moment, but also correctly explained the fine structure of the hydrogen atom.

If the derivation of the Sommerfeld-like formula for the spectrum of the hydrogen atom was one of the striking successes of the Dirac equation, some of its other features were very troublesome. Besides the states of positive energy which seemed to agree closely with experiments, the Dirac equation also possessed solutions of negative energy. In the quantum theory—in contradiction to the classical situation—these states could not be ignored, for transitions to such states could not be ruled out. Dirac (1931b) found a partial solution to this difficulty by postulating that all the negative energy states were filled with electrons. The Pauli exclusion principle then prevented a positive energy electron from making a transition to a negative energy state. The vacuum—the state of lowest energy—corresponded to the state in which all the negative energy states were occupied and all the positive energy states were vacant. By assumption, no observable charge or current density resulted from the occupancy of the negative energy states. A one-electron state corresponded to an electron in a positive energy state and all negative energy states occupied. The removal of an electron from a negative energy state resulted in a hole which was identified as a particle of positive charge and positive energy.

Although initially Dirac had hoped that such a hole might correspond to a proton, he soon became convinced that the "antiparticle" had to have the same mass as the electron. The discovery of the positron in 1932 gave strong evidence for the correctness of the thus interpreted hole theory.

Heisenberg (1963) characterized the discovery of antimatter by Dirac "as the most decisive discovery in connection with the properties or the nature of the elementary particles.... This discovery of particles and anti-particles by Dirac... changed our whole outlook on atomic physics completely." Nothing had prepared the community for this new "outlook" in which matter—particles—could be created and annihilated. It was counterintuitive and had not been anticipated. It ran against the grain of the classical description of matter, and upset the intuitions that had just been achieved in *nonrelativistic* quantum mechanics.

It should be noted that for the most part the leading theorists resisted the introduction of new "elementary particles." When Dirac first advanced the suggestion that protons were "holes," he conceived his hole theory as a *unitary* theory of matter with "all matter built out of one fundamental kind of particle, instead of two, electrons and protons" (Dirac 1930d). He recalled that "I didn't dare to postulate a new particle at that stage, because the whole climate of opinion at that time was against new particles" (Dirac 1978b). The revolutionary stand of Bohr, Heisenberg, and Dirac with respect to theory was in marked contrast to their conservatism with respect to ontology.

The radical reinterpretation of quantum mechanics necessitated by relativity was made manifest by Furry and Oppenheimer (1934). They emphasized that, for example, the description of the interaction of an electron with a Coulomb field required a state vector with an infinite number of components, the various components giving the probability amplitude of finding an electron with no pairs (of electron-positrons) present, an electron and one pair (of electron-positrons) present, an electron and two pairs present, and so on. Heisenberg has with feeling described the change entailed by this new perspective: "Until that time [the introduction of the Dirac equation] I had the impression that in quantum theory we had come back into the harbor, into the port. Dirac's paper (on the spinning electron) threw us out into the open sea again. Everything got loose again and we got into new difficulties. Of course at the same time, I saw that we had to go that way. There was no escape from it because relativity was true" (Heisenberg 1963).

In the present chapter we review these developments.

1.2 Pascual Jordan

Pascual Jordan is the unsung hero among the creators of quantum mechanics. Major portions of the two papers he coauthored with Born and Heisenberg that elaborated matrix mechanics (Born and Jordan 1926; Born et al. 1926) following Heisenberg's initial insight (Heisenberg 1925) were Jordan's contribution. Similarly, he was responsible for laying the foundations of quantum field theory (Born and Jordan 1926; Born et al. 1926; Jordan 1927g; Jordan and Klein 1927; Jordan and Wigner 1928; Jordan and Pauli 1928). The fact that in the early 1930s his interests shifted to problems in biology, psychology, mathematics, geology, and cosmology and that his contributions thereafter were of a more speculative, and at times polemic, nature and never achieved the importance of his earlier work in quantum mechanics was very probably a factor in the subsequent assessment by the community. His difficulty in communicating with others—he stuttered badly—was another. He was the only one of the major contributors to the development of the quantum theory who did not attend the 1927 Solvay conference. His cooperation with the Nazis in the 1930s and 1940s undoubtedly also contributed to his lack of recognition. During that period he wrote approvingly of Nazi Germany's "work of renewal in domestic politics" and of war as "the normal way to accomplish something new in history."[2] However, as Dirac once remarked: "One must not judge a man's worth from his poorer work; one must always judge him by the best he has done" (Salam 1985).

Jordan was born in Hannover, Germany, in 1902.[3] He was the younger of the two children in the family; a sister some ten years older than Pascual was the older sibling. Both his parents were well read in the natural sciences. His father was a painter and he got the young Pascual interested in the geometrical concepts involved in the "perspective" of drawing at an early age. In his interview with T. S. Kuhn, Jordan recalled that as a young boy his father read him books from the *Kosmos* series that acquainted him with the writings of Darwin and Haeckel. His mother introduced him to the world of plants, animals, and stars. "From her I . . . learned that . . . light has to go eight minutes from the sun to here. She was also very interested in calculation, in numbers and so on and from her I learned the first steps in arithmetic and so on" (Jordan 1963, p. 1). She often took him to visit the local zoo and he remembered collecting pictures of extinct animals, particularly those of dinosaurs. In his early teens he thought of becoming a painter or an architect, but gradually his interests shifted to natural history and biology, and eventually to physics and mathematics. He was clearly quite gifted and ambitious: "At fourteen, I had a plan of writing a big book on all the fields of science linking them all together" (Jordan 1963, p. 5). He had by then read and absorbed such books as Pauly's *Darwinismus und Lamarkismus* and F. A. Lange's *Geschichte des Materialismus*. He had also studied by himself classical physics and a great deal of mathematics. While in Gymnasium he taught himself the differential and integral calculus from Nernst and Schoenfliess's *Kurzgefasstes Lehrbuch der Differential- und Integralrechnung*, and the theory of complex variables from Knoff's *Funktionentheorie*. During his last year in the Gymnasium he began to study physics in depth and carefully read Mach's *Mechanik* and *Prinzipien der Wärmelehre*. Mach's views influenced Jordan deeply and he became an ardent positivist. He later declared that he took up physics in order to help resolve the discrepancy he felt existed between Mach's teachings and the old quantum theory (Jordan 1936). He adopted as the central tenet of his philosophical outlook what he considered to be the essential and decisive principle of the positivistic theory of knowledge: that scientifically sound propositions are limited to those that can be proved experimentally.

Jordan entered the Technische Hochschule of the University of Hannover in 1921 intending to study physics. He had by that time learned some special and general relativity from Moritz Schlick's *Raum und Zeit in der gegenwärtigen Physik*, had mastered electromagnetic theory, and had carefully studied Sommerfeld's *Atombau und Spektrallinien*. Jordan found that physics was not taught well at the Technische Hochschule and he transferred to Göttingen in 1922. However, he had made good use of his year at the Hochschule taking courses in mathematics, electrical engineering, and physical chemistry. In Göttingen he attended Courant's course on mathematical methods for physicists and became the official note taker for the course. For a while he toyed with the idea of becoming a mathematician. But he came into Born's orbit, and under his influence and with his guidance became more and more committed to physics. When Born died, Jordan, in a brief eulogy for him, wrote: "He was not only my teacher, who in my student days introduced me to the wide world of physics—his lectures were a wonderful combination of intellectual clarity and horizon widening overview. But he was also, I want to assert, the person who next to my parents, exerted the deepest, longest lasting influence on my life" (Jordan 1971, p. 430).

The Bohr *Festspiel* of June 1922 gave Jordan a taste of the drama of physics. Heisenberg, Fermi, Pauli, and Hund were also in Göttingen at the time. Fermi was evidently left out of the Göttingen intellectual community (Segrè 1970, p. 32) but Jordan got to know the other three well, particularly Heisenberg and Pauli, and came to appreciate the company of these brilliant young men. But he was overshadowed by their brash and confident ways. Jordan was rather short, and his presentation of self reflected his physical stature. He gave the impression of being insecure, an impression that was reinforced by his stuttering (he in fact suffered a breakdown in the early 1930s).

Although Jordan did not enjoy his courses in experimental physics—he actually stopped attending them—he found the laboratory course in zoology in which he had enrolled very satisfying; he also faithfully attended Alfred Kuhn's lectures on heredity. In fact, he chose zoology as one of the minor subjects for his doctorate. Most of his energies, however, were spent on theoretical physics and mathematics. He helped Courant in the preparation of the book he was then writing with Hilbert, the famous *Mathematische Methoden der Physik* (Jordan 1963, p. 12). He also assisted Born with an article on crystal dynamics and became quite close to him.

For his dissertation under Born, Jordan worked on a problem in the theory of light quanta (Jordan 1925) that dealt with the interaction of electrons and radiation. In it he tried to disprove Einstein's hypothesis that in the process of absorption or emission of a photon of energy $h\nu$ by an atom an amount of momentum $h\nu/c$ is transferred by or to the photon. Einstein, in a brief note to the *Zeitschrift für Physik,* took exception with Jordan's work. He pointed out that Jordan's paper was based on a hypothesis that implied "that the amounts of radiation taken (by an atom exposed to blackbody radiation) from rays of different directions were treated as not being independent of each other" and that this would result in consequences contrary to observation (Einstein 1925). After finishing his thesis in the fall of 1924, Jordan worked with James Franck on problems connected with spectroscopy. He helped him write volume 3 of the series *Struktur der Materie* which Born and Franck edited. The book was published with Franck and Jordan as coauthors, with the title *Anregung von Quantensprüngen der Stosse.* During that same year Jordan wrote several papers dealing with problems in atomic structure and spectroscopy. He also collaborated with Born on a paper in the quantum theory of aperiodic processes. By generalizing Kramers and Heisenberg's dispersion-theoretic approach, they calculated the effect on an atom of an electric field whose time dependence is arbitrary. The arbitrary time dependence was to allow them to simulate the effect of a charged projectile particle during an atomic collision.

In the summer of 1925, Born received from Heisenberg his paper on a quantum-mechanical reinterpretation of kinematic and dynamical relations. Upon reflecting about the meaning of the rule that Heisenberg had given for the multiplication of two quantum-mechanical amplitudes, Born came to the conclusion that "Heisenberg's symbolic multiplication was nothing but the matrix calculus, well known to me since my student days from the lectures of Rosanes in Breslau" (1977, p. 217). In fact, Born and Jordan had considered such symbolic multiplications of transition amplitudes during their discussions when writing their paper on the absorption of radiation (Born and Jordan 1925a). However, they had not realized the implications of such symbolic multiplications—whereas Heisenberg had.

Born reformulated the quantum condition that appeared in Heisenberg's paper as an equation for the diagonal elements of the commutator pq-qp of the matrices p and q that represented the momentum and position of the oscillator Heisenberg had analyzed, that is, $(pq-qp)_{nn} = h/2\pi i$. Born had found this result just before going to a meeting of the German Physical Society in Hannover.

On the train ride to Hannover he met Pauli, his former assistant, and invited him to join him in the further exploration of his finding. But Pauli gave him a "cold and sarcastic refusal." "You are fond of tedious and complicated formalism," Pauli told Born. "You are only going to spoil Heisenberg's physical ideas by your futile mathematics" (Born 1977, p. 218).[4] Upon his return to Göttingen, Born asked Jordan to collaborate with him, and within two months Born and Jordan had laid the foundations of the new matrix mechanics (Born and Jordan 1925b). In their paper they showed that, starting with the basic premises given by Heisenberg, it is possible to build a closed mathematical theory of quantum mechanics that displays strikingly close analogies with classical mechanics, but at the same time preserves the characteristic features of quantum phenomena. Most of that paper is the work of Jordan.[5] Originally Born had told Jordan of his interpretation of Heisenberg's symbolic multiplication as matrix multiplication and had shown him his result that the matrix elements of $pq-qp$ along the main diagonal were all equal to $\hbar$. He had also given him his notes on Heisenberg's paper. Within a few days Jordan had derived the result that the commutator $pq-qp$ is equal to $\frac{h}{2\pi i}$ by computing the time derivative of $pq-qp$. He had also obtained a proof of the conservation of energy for Hamiltonians of the form $H = p^2/2m + U(q)$, and had derived the Bohr frequency condition. He then went on to justify Heisenberg's assumption that $|q_{mn}|^2$ determines the probability for the transition $m \rightarrow n$. To do so, the quantum-mechanical description of cavity radiation was adumbrated. Jordan represented this system as an (infinite) set of uncoupled harmonic oscillators and considered the p_ℓ and q_ℓ's describing these oscillators as matrices satisfying $[q_\ell, p_\ell] = \hbar$.

Jordan's work was done without knowing how to impose quantum rules for systems with more than one degree of freedom. It was Heisenberg, in early September 1925, who formulated such rules after becoming acquainted with Born and Jordan's matrix formulation of quantum mechanics. The famous *Dreimännerarbeit* (Born et al. 1926) extended Born and Jordan's work to systems with an arbitrary (but finite) number of degrees of freedom. For the paper Born and Heisenberg developed a perturbation theory which took into account the possible degeneracy of the energy spectrum of such a system. The paper also included a derivation of the conservation laws for energy and angular momentum. In a chapter entitled "Physical Applications of the Theory," due to Jordan and Heisenberg, the commutation rules of the angular momentum operators $[M_x, M_y] = M_x M_y - M_y M_x = (h/2\pi i)M_z$ were derived and a proof was given that in the representation in which M_z and M^2 are diagonal, the eigenvalues of M_z are $m(h/2\pi)$, with m integer or half-integer, and those of M^2 are $j(j+1)(h/2\pi)^2$, with j integer or half-integer ($j = \max m; -j \le m \le +j$). Using these results, formulas for the intensities and polarization of atomic transitions were derived. The final section of the paper entitled "Coupled Harmonic Oscillators: Statistics of Wavefields" analyzed the fluctuations of the radiation field, and was entirely Jordan's work. That it was Jordan's work we know from a letter Heisenberg wrote Pauli:

> A third thing that Jordan did for our work is a calculation of the statistical behavior of the natural oscillations of something like a membrane in the new theory. J[ordan] claims that the interference fluctuations come out right, i.e. both the classical and the Einsteinian terms, and he believes that he sees an analogy between our calculations and Bose's statistics. I am somewhat unhappy that I do not understand enough statistics to be able to judge how much sense it makes; but I cannot criticize either, because the problem itself and the subsequent calculations appear meaningful. (Pauli 1979 [102], p. 252)

The problem Jordan addressed in the last section of the *Dreimännerarbeit* was to give a dynamical derivation of Einstein's formula for the mean squared energy fluctuations in blackbody radiation,

$$< \Delta E^2 > = h\nu < E > + \frac{< E >^2}{(8\pi \nu^2 / c^3)\nu d\nu}. \tag{1.2.1}$$

Einstein had derived this important formula using only Boltzmann's principle and Planck's law, and had interpreted it as stating that if one accepted Planck's formula one also had to accept the corpuscular nature of light (Einstein 1909; Pais 1979, p. 69). The derivation is quite straightforward: Let there be blackbody radiation at temperature T within a cavity of volume V. Consider a small chamber of volume v within this cavity which is assumed to be in thermal equilibrium with the cavity. The average energy of the radiation with frequency between v and $v + dv$ contained in the chamber is

$$< E > = \frac{8\pi h\nu}{c^3} \frac{1}{e^{h\nu / kT} - 1} \nu^3 dv. \tag{1.2.2}$$

From Boltzmann's principle

$$< E > = \frac{\int E e^{-E/kT} d(\text{phase space})}{\int e^{-E/kT} d(\text{phase space})}, \tag{1.2.3}$$

it follows that the root mean square value of the energy is given by

$$< \Delta E^2 > = \langle (E - < E >)^2 \rangle$$
$$= \frac{\partial < E >}{\partial(-1/kT)}. \tag{1.2.4}$$

Eq. (1.2.1) is then readily obtained by performing the appropriate differentiation of (1.2.2). The contribution to $< \Delta E^2 >$ from the $h\nu < E >$ term stems from the "corpuscular" behavior of the radiation within the cavity, whereas that of the $< E >^2 / \nu^3 d\nu$ term derives from its "wavelike" behavior.[6] The radiation thus has both characteristics if Planck's formula holds. Note incidentally that when $h \rightarrow 0$, the classical "wave" picture is recovered. Einstein in 1925 demonstrated that a similar formula held for a gas of indistinguishable *material particles* (counting complexions as Bose had done for photons). The mean square deviation of the energy was again given by a sum of two terms, the first one characteristic of a system of independent particles, and the second of a system of standing waves. Einstein's derivation was again based in the thermodynamic formula (1.2.4). A dynamical treatment in which the fluctuations were calculated as time averages was unable to give *both* terms simultaneously. The first term by itself could be derived using the equations of classical dynamics for particles; similarly, the second term by itself could be derived from a wave equation. Starting from a description of the cavity radiation as a set of independent, uncoupled, harmonic oscillators, and imposing commutation rules $[q, p] = i\hbar$ on the variables describing these oscillators, Jordan derived eq. (1.2.1). He showed that in a quantum description the corpuscular properties of the electromagnetic waves were a consequence of the noncommutativity of the dynamical variables describing the electromagnetic field.

Actually, rather than analyze the more difficult situation of electromagnetic radiation in a cavity, Jordan investigated the fluctuations in the energy of a small portion ℓ of a stretched string of length L. Jordan derived the fluctuation formula

$$< \Delta E^2 > = h\nu < E > + \frac{< E^2 >}{Z_\ell(\nu)d\nu}, \qquad (1.2.5)$$

where $Z_\ell(\nu)$ is the number of proper oscillations of frequency between ν and $\nu + d\nu$ in the length ℓ. Jordan had proceeded by decomposing the motion of the string into its normal modes Q_s, imposing the quantum conditions $[Q_s, P_s] = i\hbar$ and calculating the time average of $< \Delta E^2 >$. The result was not without ambiguity, but this was not noted at the time.[7] The fact that $< \Delta E^2 >$ reflected the "corpuscular and discontinuous" character of radiation was later used by Jordan to argue against Schrödinger's hope of reestablishing a continuous description of atomic phenomena with his wave mechanics (Jordan 1927g).

In his interview with T. S. Kuhn, Jordan commented that no one read the last section of the *Dreimännerarbeit,* no one took notice of it, and no one wanted to believe it (Jordan 1963, p. 8). In fact, the quantum-theoretic description of the electromagnetic field was not addressed again until Dirac did so in the fall of 1926. Although much work was done—by Schrödinger, Dirac, and others[8]—on the interaction of charged particles with the electromagnetic field, with the latter treated semiclassically, it was Dirac who first attempted to give a fully quantum-mechanical treatment of this interaction. It is to this work that we turn next.

1.3 P.A.M. Dirac and the Birth of Quantum Electrodynamics

Salam and Wigner, in their preface to the Festschrift that honored Dirac on his seventieth birthday and commemorated his contributions to quantum mechanics, succinctly assessed the man: "Dirac is one of the chief creators of quantum mechanics. . . . Posterity will rate Dirac as one of the greatest physicists of all time. The present generation values him as one of its greatest teachers. . . . Of those privileged to know him, Dirac has left his mark . . . by his human greatness. He is modest, affectionate and sets the highest possible standards of personal and scientific integrity. He is a legend in his own lifetime and rightly so" (Salam and Wigner 1972, p. ix).

Dirac was not only one of the chief authors of quantum mechanics, but he was also the creator of quantum electrodynamics and one of the principal architects of quantum field theory. All the major developments in quantum field theory in the thirties and forties have as their point of departure some work of Dirac's. Let me list them:

> **1.** The hole theory and the prediction of anti-matter (Dirac 1930a). Dirac had advanced his hole theory in order to give a consistent quantum mechanical interpretation to the relativistic equation he had proposed for the electron—the equation which now bears his name (Dirac 1928a,b).

> **2.** The many-time formalism based on the "interaction" picture (Dirac et al. 1932a,b).

> **3.** The recognition that in the presence of an external electromagnetic field, hole theory implied the phenomenon of vacuum polarization. Dirac furthermore indicated how the associated divergence could be eliminated by a process of charge renormalization (Dirac 1934a,d).

> **4.** A relativistic formulation of the Lorentz electron including a classical version of mass renormalization (Dirac 1938b).

Heisenberg (1963) characterized the postulation of antimatter by Dirac "as the most decisive discovery in connection with the properties or the nature of elementary particles." Dirac's many-time formalism (Dirac et al. 1932a,b) was the basis of Tomonaga's and Schwinger's covariant formulation of quantum field theory. I might add that Dirac's work on the role of the Lagrangian in quantum mechanics, done roughly at the same time (Dirac 1933a), was the insight Feynman needed to elaborate his own space-time, integral-over-paths formulation of quantum mechanics. To the above list should be added Dirac's paper on magnetic monopoles (1931b), his researches on relativistic wave equations (1928a,b; 1936b), as well as his paper on Hilbert spaces with indefinite metric (1942), a work that motivated Heisenberg's introduction of the S-matrix in quantum field theory in 1942. And all this does not take into account his justly famous presentation of *The Principles of Quantum Mechanics*, which in its several editions was (and is) a bible to the post-1930 generations of physicists.[9]

In this section I will review Dirac's initial formulation of quantum electrodynamics. Before doing so let me relate his background.[10] Paul Adrien Maurice

Dirac was born on August 8, 1902, in Bristol, England, the son of a Swiss, French-speaking father, Charles Adrien Ladislas Dirac, and an English mother, Florence Hannah, née Holten.

Charles Dirac had obtained a Baccalauréat-ès-Lettres degree from the University of Geneva in 1887. After attending lectures in the Faculté des Lèttres for a year he left for England, where he supported himself by giving French lessons. Paul's mother grew up in Bristol. She met Charles while working in a library. They were married in 1899. In 1902 they bought a house in Bishopton, Bristol, which they named "Monthey" after Charles's birthplace. Paul was the second of three children. An older brother, Reginald Charles Felix, was born in 1900, and a younger sister, Beatrice, in 1906. All three were registered at birth as Swiss citizens in the canton of Valais. Only in 1919 did Charles give up his Swiss citizenship—and that of his children—and acquire British nationality.

In 1896 Charles obtained a position to teach French in the Merchant Venturers' Technical College (M.V.) at Bristol. When in 1909 Bristol University College and Venturers' College merged into Bristol University, Charles taught French at the university and continued teaching that same subject in the secondary school of Venturers' College. One of Charles's students, who became a staff member of M.V. after Charles had retired, described him as follows: "He had, next to Archbishop Temple, the largest cranium in Christendom. . . . He was a brilliant linguist, being able to speak eight or nine languages—it was said he learned a new language every summer holiday" (Dalitz and Peierls 1986, p. 145). Late in life Paul gave the following account of his father as a teacher:

> He was somewhat strict, and would frequently give the boys a test which was not announced beforehand, so that they were unable to prepare for it. He expected them to be ready for any sort of test.
>
> He was thus not very popular with the boys, but he was very successful in getting them through their exams, for which they were glad. He was nicknamed "Dedder." (Dirac 1980)

Dirac remembers his mother as a very simple, kindly woman. Paul's father, on the other hand, was a strong-willed, dominating personality who assertively influenced his son's early developments. Charles's own father evidently had been a rather difficult, highly emotional man whose life had not been easy. Charles left Switzerland without informing his family where he was going; nor did he tell them of his marriage until he visited his mother in Geneva in 1905, bringing with him his wife and two sons. Paul remembered the visit all his life, a memory probably nurtured by his mother.

For all his ambiguity about his own family, Charles did not reject his Genevan heritage. He asked his children to speak French to him as much as possible. At dinner he insisted they speak French—in fact, grammatically correct French—or they would be punished. Paul remarked that his own taciturnity was due to this interaction with his father: "My father made the rule that I should only talk to him in French. He thought it would be good for me to learn French in that way. Since I found that I couldn't express myself in French, it was better for me to stay silent than to talk in English. So I became very silent at that time—that started very early." It was his wife's opinion that "having been forced to remain silent may have been the traumatic experience that made him a very silent man for life" (Margit Dirac 1987, p. 5).

Because Paul's mother, sister, and older brother were unable to meet Charles's standards, they ate dinner in the kitchen, while Paul—who indeed learned to speak French fluently and correctly—ate in the dining room with his father. In fact, Paul's mother did not speak French, and Paul reported that he never saw his parents eat a meal together (Salaman and Salaman 1986); nor could he recall anyone ever making a social call at the house: "I had no social life at all as a child.... No one ever came [to our house] for social purposes" (Dirac 1963a).

Reginald, Paul's older brother, had wanted to become a physician, but Charles forced him to study mechanical engineering at Bristol. He obtained only a third-class degree upon graduating and accepted a position as a draftsman with an engineering firm in Wolverhampton. He committed suicide when he was twenty-four years old. The death of his oldest son deeply disturbed Charles, and for a while Paul feared that his father might lose his sanity—and he resolved that he would never take his own life no matter what the circumstances. Thereafter Paul's relationship with his father became chilled and they had very little interaction with one another. One manifestation of Paul's feeling toward his father was that throughout his life he avoided going to Switzerland, a country he associated with his father (Mehra and Rechenberg 1982). Paul invited only his mother to attend the ceremonies in Stockholm honoring him with the Nobel Prize in 1933. Dalitz and Peierls report that when Professor Tyndall, who had headed the physics department at Bristol University for three decades, gave a set of public evening lectures on modern physics in the early 1930s, he noticed a regular listener in the front row, a man much older than the others there, who was taking careful note of all that he said. At the end of the last lecture of the series, this old man came up to him to thank him, saying: "I am glad to have heard all this. My son does physics but he never tells me anything about it." The old man was Charles Dirac. Charles Dirac died in 1935. Paul was in Russia at the time to watch an eclipse of the sun; when informed of the seriousness of his father's illness, Paul flew back to England, but it was too late. The first letter he wrote his wife after his father's death was to say, "I feel much freer now" (Margit Dirac 1987, p. 5).

I do not know when Dirac became an atheist, but Heisenberg (1971) recalled that in 1927 during a discussion on religion among the younger members who were attending the Solvay Congress, the twenty-five-year-old Dirac asserted:

> If we are honest—and scientists have to be—we must admit that religion is a jumble of false assertions, with no basis in reality. The very idea of God is a product of the human imagination.... I can't for the life of me see how the postulate of an Almighty God helps us in any way. What I do see is that this assumption leads to such unproductive questions as why God allows so much misery and injustice, the exploitation of the poor by the rich and all other horrors He might have prevented. If religion is still being taught, it is by no means because its ideas still convince us, but simply because some of us want to keep the lower classes quiet. Quiet people are much easier to govern than clamorous and dissatisfied ones. They are also easier to exploit.... Hence the close alliance between those two great public forces, the State and the Church.

The ensuing heated discussion was capped by Pauli's observation that actually "our friend Dirac, too, has a religion, and its guiding principle is 'There is no God, and Dirac is His prophet.'"

Paul attended the secondary school of the Merchant Venturers' Technical College (M.V.), the public school where his father taught, and while there displayed great mathematical abilities. The school's academic standards were quite high but, being part of a technical college, its orientation was practical. Late in life Dirac recalled the following:

> The M.V. was an excellent school for science and modern languages. There was no Latin or Greek, something of which I was rather glad, because I did not appreciate the value of old cultures. I consider myself very lucky in having been able to attend the School.
>
> I was at the M.V. during the period 1914–18, just the period of the First World War. Many of the boys then left the School for National Service. As a result, the upper classes were rather empty; and to fill the gaps the younger boys were pressed ahead, as far as they were able to follow the more advanced work. This was very beneficial to me: I was rushed through the lower forms, and was introduced at an especially early age to the basis of mathematics, physics and chemistry in the higher forms. In mathematics I was studying from books which mostly were ahead of the class. This rapid advancement was a great help to me in my later career.
>
> The rapid pushing-ahead was a disadvantage from the point of view of Games—which we had on Wednesday after-

noons. I played soccer and cricket, mostly with boys older and bigger than myself, and never had much success. But all through my schooldays, my interest in science was encouraged and stimulated.

It was a great advantage, that the School was situated in the same building as the Merchant Venturers' Technical College. The College "took over" in the evenings, after the School was finished. The College had excellent laboratories, which were available to the School during the daytime. Furthermore, some of the staff combined teaching in the School in the daytime with teaching in the College in the evenings. (Dirac 1980, p. 9)

Dirac's schoolmates recall him as introverted, reticent, and aloof. A fellow student at the school remembered him as "a slim, tall, un-English looking boy in knickerbockers, with curly hair. He haunted the library and did not take part in games. On the one isolated occasion I saw him handle a cricket bat, he was curiously inept" (Dalitz and Peierls 1986, p. 142).

Upon graduation from M.V., Dirac followed in his older brother's footsteps and went to Bristol University to study electrical engineering. Although his favorite subject was mathematics, he thought that the only way that one could earn one's living as a mathematician was as a school teacher, and this prospect did not appeal to him.

In his interview with T. S. Kuhn (Dirac 1963a), Dirac reported that during a summer vacation he worked as a student apprentice in the engineering works of Thomson-Houston in Rugby, where his older brother was employed as an engineer. Although they would often see one another on the street, they "didn't exchange a word." Dirac did not find the work at Thomson-Houston challenging, and his employers judged his performance unfavorably. He himself observed that he "lacked keenness and was slovenly" when turning things on lathes and doing metal work.

Dirac was nineteen when he graduated Bristol University with first-class honors in 1921. Unable to find a suitable engineering position due to the economic recession that gripped England after World War I, Dirac, with the encouragement of his father, accepted a free tuition at Bristol University to study mathematics there. An 1851 Exhibition Studentship that Dirac had won in 1921 together with a grant from the Department of Scientific and Industrial Research, awarded because of his outstanding performance in the mathematics examination at Bristol University, made it possible for him to go to Cambridge as a research student in 1923. He had hoped to have Ebenezer Cunningham as his adviser, because Cunningham was working on relativity, and the special and general theory of relativity were the focal point of Dirac's interests at the time. But Cunningham no longer accepted any students, and Dirac therefore came under the influence of R. H. Fowler and A. Eddington. He attended Fowler's lectures on the old quantum theory and became his research student. Fowler made Dirac conscious of the importance of experimental results in advancing theoretical understanding. Fowler, in his quantum theory course, stressed the spectroscopic data and its relevance to the theoretical advances from 1913 to the mid 1920s. At Cambridge, Dirac was exposed to the experimental activities of the Cavendish Laboratory and he became a member of the intellectual circle over which Rutherford and Fowler presided. Fowler, incidentally, was Rutherford's son-in-law.

Within six months of his arrival at Cambridge, Dirac wrote two papers on statistical mechanics (Dirac 1924a,d) and in May 1924 he submitted his first paper dealing with quantum problems (Dirac 1924c).

In the spring of 1925 Bohr visited Cambridge and lectured on the problems facing the quantum theory. Dirac's reaction to Bohr was mixed: "While I was very much impressed by [him], his arguments were mainly of a qualitative nature, and I was not able to really pinpoint the facts behind them. What I wanted was statements which could be expressed in terms of equations, and Bohr's work very seldom provided such statements. I am not really sure how much my later work was influenced by these lectures of Bohr" (Dirac 1977c, p. 116). Later that year Heisenberg visited Cambridge and on July 28 gave a talk to the Kapitza Club entitled "Term Zoologie und Zeeman—Botanik" in which he mentioned his new ideas briefly. It is not clear whether Dirac attended the lecture; if he did he evidently did not perceive the importance of Heisenberg's remarks at the time (Dirac 1977c). Heisenberg sent the page proofs of his paper to Fowler, who gave them to Dirac for his comments. Dirac failed to see the significance of the paper when he first read it in September 1925. A week later, when he studied the paper again, he came to appreciate its content. He then realized that the appearance of noncommuting quantities was the essence of Heisenberg's new approach and came to believe that "Heisenberg's idea provided the key to the whole mystery." It occurred to him to try to connect Heisenberg's noncommutative products, in particular the commutator of two noncommuting dynamical variables, with their Poisson brackets. His success in linking commutors and Poisson brackets convinced him that the new quantum mechanics represented an extension of classical physics rather than—as Heisenberg had argued—a break with it. He stated his viewpoint in the opening sentences of his first paper on the new quantum mechanics: "In a recent paper Heisenberg puts forward a new theory which suggests that it is not the equations of classical mechanics that are in any way at fault, but that the mathematical operations on which physical results are deduced require modification. All the information supplied by the classical theory can thus be made use of in the new theory" (Dirac 1925d). Quantum mechanics differs from classical mechanics in that the dynamical variables representing the position and momentum of a particle do not commute with one another, but instead satisfy the commutation relation

$$qp - pq = [q, p] = i\hbar.$$

The correspondence between Poisson brackets and commutators that Dirac had established after reading Heisenberg's manuscript had taken him back to the familiar grounds of the Hamiltonian formalism describing the dynamics of *particles*.[11] When he first had come to Cambridge in the fall of 1923, Dirac had studied Hamiltonian mechanics as presented in Whittaker's *Treatise on the Analytic Dynamics of Particles and Rigid Bodies*. The following year he had worked through Sommerfeld's *Atomic Structure and Spectral Lines*, which relied on the Hamiltonian method of action and angle variables.

Dirac's position was stated succinctly in the paper he submitted to the *Proceedings of the Royal Society* in late January 1926: "Only one basic assumption of classical theory is false ... the laws of classical mechanics must be generalized when applied to atomic systems, the generalization being that the commutative law of multiplication as applied to dynamical variables, is to be replaced by certain quantum conditions" (Dirac 1926a).

Dirac's brilliance was recognized early. His contributions to the development of quantum mechanics were immediately acknowledged as central, and by 1927 he was a member of the core set that was passing judgment on theoretical advances. The deferential tone of the letters that Bohr, Heisenberg, Pauli, and Jordan wrote him is indicative of the high respect and esteem in which he was held by them. He was invited to visit all the leading European and American universities.[12] In the spring of 1928 Arthur Compton invited him to visit the University of Chicago at the considerable salary of $4,000 for the autumn and winter quarters. In September of that year, Compton offered Dirac the newly established chair in theoretical physics at Chicago, at an annual salary of $6,000, with the freedom to do anything he wanted for two of the four quarters. Although Dirac refused these offers, he accepted an invitation to visit the University of Wisconsin in the spring of 1929, and that summer he lectured at the Summer School in Theoretical Physics at the University of Michigan in Ann Arbor.[13] Van Vleck (1972) has written a charming recollection of Dirac's stay in Madison. A less well known but equally revealing account of that visit is an interview that Dirac granted the *Wisconsin State Journal*. Let me quote it in its entirety, because it gives an insight into Dirac's personality[14] and testifies to the fame that already accompanied him then.

ROUNDY INTERVIEWS PROFESSOR DIRAC

AN ENJOYABLE TIME IS HAD BY ALL

(Copied from Wis. State Journal of Apr. 31 [*sic*], 1929. P.A.M. issue)
By Roundy.

I been hearing about a fellow they have up at the U. this spring—a mathematical physist, or something, they call him—who is pushing Sir Isaac Newton, Einstein and all the others off the front page. So I thought I better go up and interview him for the benefit of State Journal readers, same as I do all other top notchers. His name is Dirac and he is an Englishman. He has been giving lectures for the intelligentsia of the math and physics departments—and a few other guys who got in by mistake.

So the other afternoon I knocks at the door of Dr. Dirac's office in Sterling Hall and a pleasant voice says "Come in." And I want to say here and now that this sentence "come in" was about the longest one emitted by the doctor during our interview. He sure is all for efficiency in conversation. It suits me. I hate a talkative guy.

I found the doctor a tall youngish-looking man, and the minute I seen the twinkle in his eye I knew I was going to like him. His friends at the U. say he is a real fellow too and good company on a hike—if you can keep him in sight, that is.

The thing that hit me in the eye about him was that he did not seem to be at all busy. Why if I went to interview an American scientist of his class—supposing I could find one—I would have to stick around an hour first. Then he would blow in carrying a big briefcase, and while he talked he would be pulling lecture notes, proof, reprints, books, manuscript, or what have you out of his bag. But Dirac is different. He seems to have all the time there is in the world and his heaviest work is looking out the window. If he is a typical Englishman it's me for England on my next vacation!

Then we sat down and the interview began.

"Professor," says I, "I notice you have quite a few letters in front of your last name. Do they stand for anything in particular?"

"No," says he.

"You mean I can write my own ticket?"

"Yes," says he.

"Will it be all right if I say that P.A.M. stands for Poincaré Aloysius Mussolini?"

"Yes," says he.

"Fine," says I, "We are getting along great! Now doctor will you give me in a few words the low-down on all your investigations?"

"No," says he.

"Good," says I. "Will it be all right if I put it this way— 'Professor Dirac solves all the problems of mathematical physics, but is unable to find a better way of figuring out Babe Ruth's batting average'?"

"Yes," says he.

"What do you like best in America?", says I.

"Potatoes," says he.

"Same here," says I. "What is your favorite sport?"

"Chinese chess," says he.

That knocked me cold! It was sure a new one on me! Then I went on: "Do you go to the movies?"

"Yes," says he.

"When?" says I.

"In 1920—perhaps also 1930," says he.

"Do you like to read the Sunday comics?"

"Yes," says he, warming up a bit more than usual.

"This is the most important thing yet, doctor," says I. "It shows that me and you are more alike than I thought. And now I want to ask you something more: They tell me that you and Einstein are the only two real sure-enough high-brows and the only ones who can really understand each other. I wont ask you if this is straight stuff for I know you are too modest to admit it. But I want to know this—Do you ever run across a fellow that even you cant understand?"

"Yes," says he.

"This well make a great reading for the boys down at the office," says I. "Do you mind releasing to me who he is?"

"Weyl," says he.

The interview came to a sudden end just then, for the doctor pulled out his watch and I dodged and jumped for the door. But he let loose a smile as we parted and I knew that all the time he had been talking to me he was solving some problem that no one else could touch.

But if that fellow Professor Weyl ever lectures in this town again I sure am going to take a try at understanding him! A fellow ought to test his intelligence once in a while.

Just two points about the story to corroborate its veracity. Van Vleck, in his account of his "Travels with Dirac," recalled that one of the high points of Dirac's visit to Madison in 1929 was deciphering what the initials P.A.M. stood for. I refer the reader to Van Vleck's article to see how that mystery was solved. The other remark concerns Dirac and chess.

During the very period that Dirac was visiting Madison, Heisenberg was in Chicago delivering the lectures upon which his book *The Physical Principles of the Quantum Theory* is based. At the end of the summer of 1929 the two of them traveled together to the West Coast, and from there to Hawaii and Japan on a Japanese steamer, the *Shinyo Maru*. There they parted company, Dirac traveling on his way home through Siberia and Russia on a trip arranged by Yoshio Nishina and Igor Tamm, whereas Heisenberg went through China, India, and the Red Sea to return to Germany.[15] Back in Leipzig, Heisenberg wrote Dirac on December 7, 1929, about his travels, informing him that the weather in China had been excellent and that the Himalayas were the best part of the trip. He went on:

> In India itself it was very hot and rather rainy. Once our train went off the rails in the middle of the Jungle and people were very afraid of tigers; the tigers probably pretty afraid too. In the Red Sea the temperature at night was 95° F. . . . I did not see the Southern Cross, so you are right again. You are wrong however in the question of mating a king and knight with King and Castle; this is *not* possible according to the edition of 1926 of Dufresne's Handbook of Chess (the best book about theory of Chess).
>
> With best wishes
> Yours Sincerely
>
> W. Heisenberg[16]

Brown and Rechenberg (1985) have given details of the trip. In introducing Heisenberg to an audience of physicists in Trieste in 1971, Dirac described an episode in Japan during that trip. In a few deft sentences Dirac not only painted a perspicacious portrait of Heisenberg, but he also indicated the qualities he admired and revealed the emotions that attend one's ascent to great heights of accomplishments. Dirac told how he and Heisenberg "had to climb" a high tower with a platform at the top, surrounded by a stone balustrade. "Heisenberg climbed up on the balustrade and then on to the stone-work at one of the corners and stood there, entirely unsupported, standing on about six inches square of stone-work. Quite undisturbed by the great height, he just surveyed all the scenery around him. I couldn't help feeling anxious. If a wind had come along then it might have had a tragic result" (Dirac 1971).

On several occasions Dirac indicated some of the influences that shaped his scientific outlook. Both in his interviews with Thomas Kuhn in 1963, and in his lecture in Varenna in 1972, Dirac commented that it was his engineering training that changed his outlook "to a very large extent." Whereas before that he was only interested in exact equations and it seemed to him that "if one worked with approximations there was an intolerable ugliness in one's work," thereafter he was able to see that "even theories based on approximations could sometimes have a considerable amount of beauty in them" (Dirac 1977c, p. 112). His becoming acquainted with the theory of relativity reinforced this change in outlook. Special relativity taught him that Newton's laws of motion were not exact, "only approximations," and he began to infer from that "maybe all laws of Nature are only approximations." "I was quite prepared at that time to consider all our equations as only approximations representing our present state of knowledge, and to take it as a task to try to improve on them" (Dirac 1977c, p. 113).

In his "Recollections of an Exciting Era" (1977c) Dirac stressed the impact of relativity upon him. In that lecture he dated the beginning of the "exciting era" in physics during the present century as occurring in 1919, with the "bursting of general relativity upon the world *at the end of a long and difficult war.*" He remembered the newspapers being full of articles on general relativity and of reports of Eddington's dramatic findings corroborating Einstein's prediction on the deflection of light by the sun. Moreover, he recalls that relativity was understood "in a very wide sense, and was taken up by philosophers and by people in all walks of life." Everyone wanted to get away from the strain of war and "eagerly seized on the new mode of thought.... The excitement was quite unprecedented in the history of science."

"I was caught up in this excitement produced by relativity" (Dirac 1977c, p. 110; see also Dirac 1982b). While an enginering student at Bristol, he attended Broad's lectures on relativity. Although Broad was a philosopher and lectured on the philosophical implications of relativity, he also imparted some exact information. Dirac vividly remembered Broad writing the formula

$$ds^2 = dx^2 + dy^2 + dz^2 - c^2dt^2$$

on the blackboard and the "tremendous effect" the minus sign had on him. Already as a young schoolboy Dirac had been interested in the relations of space and time and had thought about them a great deal. He had, in fact, considered the possibility of considering space-time from a general 4-dimensional point of view. The minus sign that Broad had introduced him to allowed him to "figure out by [himself] the basic relations of special relativity."

Relativistic considerations were central in all of Dirac's researches. One of his earliest manuscripts dating from when he first came to Cambridge in 1924 dealt with "The validity of Liouville theorem in all frames of reference." One of his first talks, given at Professor Baker's mathematical tea parties, was on general relativity. In one of his first published papers, Dirac generalized Einstein and Ehrenfest's (1923) work on the radiative equilibrium of charged particles and radiation using "detailed balancing" and made their arguments compatible with relativistic covariance (Dirac 1924d). In his Varenna lectures Dirac recalled: "My work during the first two years at Cambridge, which was before Heisenberg's theory appeared, was very much concerned with relativity ... there was a sort of general problem which one could take, whenever one saw a bit of physics expressed in a nonrelativistic form, to transcribe it to make it fit in with special relativity. It was rather like a game, which I indulged in at every opportunity, and sometimes the result was sufficiently interesting for me to write up a little paper about it" (Dirac 1977c, p. 120).

In October 1925, after reading Heisenberg's paper on matrix mechanics, he started a draft of a paper entitled "Heisenberg's Quantum Mechanics and the Principle of Relativity" which tried to overcome the limitation that the quantum rules of Heisenberg "can be applied only (if at all) to systems at rest."

The role of invariance was the lesson that relativity had taught him, and that lesson was applied with impressive results to the development of transformation theory in quantum mechanics. In his introduction to the first edition of *The Principles of Quantum Mechanics,* Dirac noted: "The formulation of the[se] laws [of nature] requires the use of the mathematics of transformations. The important things in the world appear as the invariants (or more generally the nearly invariants, or quantities with simple transformation properties) of these transformations. ... The growth of the use of transformation theory, as applied first to relativity and later to quantum theory, is the essence of the new method in theoretical physics" (Dirac 1930, p. v).

Making interactions satisfy the requirements of special relativity was one of the motivations for Dirac to study "The Quantum Theory of Emission and Absorption of Radiation" (1927b). This paper marks the birth of quantum electrodynamics. In his "Introduction and Summary," Dirac noted that the new quantum theory, based on noncommuting dynamical variables, was by then sufficiently developed to form a "fairly complete theory of any 'dynamical system' composed of a number of particles with instantaneous forces acting between them, provided it is describable by a Hamiltonian function." But hardly anything had been done "up to the present on quantum electrodynamics." "The questions of the correct treatment of a system in which the forces are propagated with the velocity of light instead of instantaneously, of the production of an electromagnetic field by a moving electron, and of the reaction of this field on the electron have not yet been touched. In addition there is a serious difficulty in making the theory satisfy all the requirements of the restricted principle of relativity, since a Hamiltonian function can no longer be used. The relativity question is, of course, connected with the previous ones and it will be impossible to answer any one question completely without at the same time answering them all."

In his interview with T. S. Kuhn, Dirac indicated that in fact the stated objectives of the introduction were not what initially motivated him: "I remember the origin of that work was just playing about with equations. I was intending to get a theory of radiation at the time. I was just playing about with the Schrödinger equation. I got the idea of applying the quantization to it and worked out what it gave and found out it just gave the Bose statistics" (Dirac 1963a, session 5, p. 20).

Res Jost (1972) has given a masterful technical analysis, and Joan Bromberg (1977) has presented an insightful historical analysis of Dirac's paper. I shall therefore touch only on its salient features. This seminal paper, which

Dirac wrote in Copenhagen during his four-month stay in Bohr's institute during the fall of 1926, was communicated by N. Bohr, F.R.S., to the *Proceedings of the Royal Society* [17] on February 2, 1927, and published shortly thereafter. Gregor Wentzel, who contributed significantly to the developments of quantum electrodynamics during the 1920s, commented in 1959:

> Today, the novelty and boldness of Dirac's approach to the radiation problem may be hard to appreciate. During the preceding decade it had become a tradition to think of Bohr's correspondence principle as the supreme guide in such questions, and, indeed, the efforts to formulate this principle in a quantitative fashion had led to the essential ideas preparing the eventual discovery of matrix mechanics by Heisenberg.
>
> Although some success had been achieved in the first half of the 1920s in the description of such processes as the photoelectric effect, there had been no possibility within the correspondence principle framework to understand the process of spontaneous emission or the disappearance of a photon. Dirac's explanation . . . came as a revelation. (Wentzel 1960)

In his paper, Dirac dealt with the problem of an atom interacting with the radiation field in two distinct ways that can be characterized as the "corpuscular" and the "wave" approaches. In the corpuscular approach, the light quanta are described as an assembly of "non-interactive particles moving with the speed of light and satisfying the Einstein-Bose statistics."

To describe a system of N noninteracting particles, Dirac proceeded as follows: Let H_0 be the Hamiltonian of an unperturbed atomic system and $H = H_0 + V$ the Hamiltonian when the system is under the influence of a perturbation V. The eigenfunctions of the perturbed system satisfy

$$i\hbar\partial_t\psi = (H_0 + V)\psi. \qquad (1.3.1)$$

Let ψ_r be a time-dependent solution of the unperturbed Hamiltonian

$$i\hbar\partial_t\psi_r = H_0\psi_r = W_r\psi_r \qquad (r \text{ labels the stationary states}). \qquad (1.3.2)$$

If $\psi = \sum_r a_r(t)\psi_r$ is a solution of (1.3.1) that satisfies the proper initial conditions, then $|a_r|^2$ is the probability of the system being in the state r at time t. The a_r's must be normalized initially so that $\sum |a_r|^2 = 1$; they will then remain normalized by virtue of (1.3.1).

"The theory will apply directly to assembly of N similar independent systems if we multiply each of these a_r by $N^{1/2}$ so as to make $\sum_r |a_r|^2 = N$. We shall now have that $|a_r|^2$ is the probable number of systems in the state r." Note that $|a_r|^2$ must then be an integer. The fulfillment of this requirement is the motivation for Dirac's introduction of "second quantization" (Kojevnikov 1990). The equation that determines the rate of change of the a_r is

$$i\hbar \dot{a}_r = \sum_s V_{rs} a_s, \tag{1.3.3a}$$

and that of the complex conjugate a_r^*

$$-i\hbar \dot{a}_r^* = \sum_s V_{rs}^* a_s^* = \sum_s a_s^* V_{sr}. \tag{1.3.3b}$$

Note we are in the interaction picture! If now the a_r and a_r^* are considered as canonical variables, then these equations, (1.3.3a) and (1.3.3b), assume the Hamiltonian form if the Hamiltonian function is taken to be

$$F_1 = \sum_{r,s} a_r^* V_{rs} a_s, \tag{1.3.4}$$

that is,

$$\frac{da_r}{dt} = \frac{1}{a\hbar} \frac{\partial F_1}{\partial a_r^*} \qquad i\hbar \frac{da_r^*}{dt} = -\frac{\partial F_1}{\partial a_r}. \tag{1.3.5}$$

Instead of the complex canonical variables a_r and a_r^*, Dirac then introduced the real canonical variables N_r, ϕ_r by the transformation

$$a_r = N_r^{1/2} e^{-i\phi_r} \tag{1.3.6a}$$

$$a_r^* = e^{i\phi_r} N_r^{1/2}. \tag{1.3.6b}$$

In terms of these variables,

$$F_1 = \sum_{r,s} V_{rs} N_r^{1/2} N_s^{1/2} e^{i(\phi_r - \phi_s)/\hbar} \tag{1.3.7}$$

and

$$N_r = -\frac{\partial F_1}{\partial \phi_r} \qquad \dot{\phi}_r = \frac{\partial F_1}{\partial N_r}. \tag{1.3.8}$$

Actually, Dirac found it more convenient to deal with the canonical variables

$$b_r = a_r e^{-iW_r t/h} \qquad b_r^* = a_r^* e^{iW_r t/h} \tag{1.3.9}$$

that satisfy

$$ih\dot{b}_r = \sum_{r,s} H_{rs} b_s$$

$$= W_r b_r + \Sigma_s v_{rs} b_s, \tag{1.3.10}$$

where

$$V_{rs} = v_{rs} e^{i(W_r - W_s)t/h}. \tag{1.3.11}$$

These equations can be considered to have been derived from a Hamiltonian

$$F = \sum_{r,s} b_r^* H_{rs} b_s, \tag{1.3.12}$$

where

$$b_r = N_r^{1/2} e^{-i\theta_r/h}. \tag{1.3.13}$$

This Hamiltonian can be written in the form

$$F = \sum_r W_r N_r + \Sigma_{rs} v_{rs} N_r^{1/2} N_s^{1/2} e^{i(\theta_r - \theta_s)/h}, \tag{1.3.14}$$

so that the Hamiltonian equations of motion become

$$\dot{N}_r = -\frac{\partial F}{\partial \theta_r} \qquad \dot{\Theta}_r = \frac{\partial F}{\partial N_r}. \tag{1.3.15}$$

Note if $v_{rs} = 0$, the phases θ_r increase linearly in time (whereas the ϕ_r are constants) and, as is to be expected, the N_r are then constants and do not depend on time.

Since one has a Hamiltonian and a set of canonical variables, "the development of the theory which naturally suggests itself is to make these canonical variables q-numbers satisfying the usual quantum conditions instead of c-numbers so that their Hamiltonian equations of motion become true quantum equations."

And so Dirac introduced the quantum conditions

$$b_r b_s^* - b_s^* b_r = \delta_{rs} \tag{1.3.16a}$$

$$[b_r, b_s] = 0 \tag{1.3.16b}$$

that will guarantee that $b_r^* b_r = N_r$ when diagonal has integer eigenvalues. The previous transformation equations must now be written in the form

$$b_r = e^{-i\theta_r/h} N_r^{1/2} \tag{1.3.17a}$$

$$b_r^* = N_r^{1/2} e^{i\theta_r/h} = e^{i\theta_r/h}(N_r + 1)^{1/2} \tag{1.3.17b}$$

in order that N_r, θ_r be canonical variables satisfying the commutation rules[18]

$$[\theta_r, N_r] = ih, \tag{1.3.18a}$$

or equivalently

$$[e^{i\theta_r/h}, N_r] = e^{i\theta_r/h}. \tag{1.3.18b}$$

The Hamiltonian becomes

$$F = \sum_{r,s} b_r^* H_{rs} b_s = \sum_{r,s} N^{1/2} e^{i\theta_r/h} H_{rs}(N_s + 1)^{1/2} e^{-i\theta s/h}$$

$$= \sum_r W_r N_r + \sum_r v_{rs} N^{1/2}(N_s + 1 - \delta_{rs})^{1/2} e^{i(\theta_r - \theta_s)/h}, \tag{1.3.19}$$

and the Schrödinger equation can readily be written in terms of the variables N_r', the eigenvalues of the N_r operators, by noting that by virtue of the r commutation rules (3.2.18) $e^{i\theta}$ and $e^{-i\theta}$ play the role of creation and annihilation operators[19]

$$e^{-i\theta_r/h}\psi(N_1', \ldots N_r', \ldots) = \psi(N_r', \ldots N_r' - 1, \ldots) \tag{1.3.20a}$$

$$e^{+i\theta_r/h}\psi(N_1', \ldots N_r', \ldots) = \psi(N_1', \ldots N_r' + 1, \ldots). \tag{1.3.20b}$$

One thus obtains

$$i\hbar \partial_t \psi(N_1', N_2', N_3', \ldots) = F\psi(N_1', N_2', N_3', \ldots)$$

$$= \sum_r W_1' N_r' \psi(N_1', N_2', N_3', \ldots)$$

$$+ \sum_{rs} v_{rs} \sqrt{N_r'} \sqrt{N_s' + 1 - \delta_{rs}}$$

$$\psi(N_1', \ldots N_r' - 1, \ldots, N_s + 1, \ldots). \quad (1.3.21)$$

Although Dirac referred to this procedure as "second quantization," it is not to be considered as the quantization of a classical matter wave field described by the Schrödinger equation. Rather, it should be regarded as an application of Dirac's transformation theory by which the state vector is expressed in terms of the variables $N_1', N_2', N_3', \ldots$. It guarantees that the particles that are being described satisfy Bose-Einstein statistics—since only the occupation number in each state r is specified and each N_r' can take on the values $0, 1, 2, \ldots$. "Second quantization" in its modern sense of quantizing the Schrödinger equation considered as a classical wave describing charged matter is Jordan's contribution.

Dirac next derived the transition rate from one state ψ_r to others of the same energy using perturbation theory, and incidentally corrected Born's original derivation of the transition rate for the scattering of a particle by a potential (Born approximation). He then turned to the application of this formalism to the description of the absorption and emission of light. The interaction of the light quanta with an atomic system was assumed by Dirac to be described by a Hamiltonian,

$$F = H_p + \sum_r W_r N_r + \sum_{r,s} v_{rs} N_r^{1/2} (N_s + 1 - \delta_{rs})^{1/2} e^{i(\theta_r - \theta_s)/h} \quad (1.3.22)$$

H_p is the Hamiltonian of the atomic system (considered the *perturbing system*), $\sum_r W_r N_r = \sum_r W_r b_r^* b_r$ is the energy of the light quanta, the *perturbed system*, and

$$\sum_{r,s} v_{rs} N_r^{1/2} (N_s + 1 - \delta_{rs})^{1/2} e^{i(\theta_r - \theta_s)/h} = \sum v_{rs} b_r^* b_s \quad (1.3.23)$$

is the perturbation energy. To start with, the v_{rs} are unknown. Note that the Hamiltonian

$$F_B = \sum_r W_r N_r + \sum_{r,s} v_{rs} N_r^{1/2} N_s^{1/2} e^{i(\theta_r - \theta_s)/h}$$

$$= \sum_r W_r b_r^* b_r + \sum_{r,s} b_r^* v_{rs} b_s \quad (1.3.24)$$

commutes with $\Sigma_r N_r$, the total number of boson (photon) operators, and hence seemingly this Hamiltonian is not applicable to the description of photons interacting with matter in which emission and absorption of photons occurs. Nonetheless, Dirac showed that the above Hamiltonian leads to Einstein's laws for the emission and absorption of radiation. How Dirac accomplished this is testimony to his uncanny physical intuition. Let me quote him:

> The light-quantum has the peculiarity that it apparently ceases to exist when it is in one of its stationary states, namely, the zero state, in which its momentum, and therefore also its energy, are zero. When a light-quantum is absorbed it can be considered to jump into this zero state, and when one is emitted it can be considered to jump from the zero state to one in which it is physically in evidence, so that it appears to have been created. Since there is no limit to the number of light-quanta that may be created in this way, we must suppose that there are an infinite number of light-quanta in the zero state, so that the N_0 of the Hamiltonian (1.3.24) is infinite. We must now have θ_0, the variable canonically conjugate to N_0, a constant, since
>
> $$\dot{\theta}_0 = \frac{\partial F}{\partial N_0} = W_0 + \text{ terms involving } N_0^{-1/2} \text{ or } (N_0 + 1)^{-1/2}$$
>
> and W_0 is zero. In order that the Hamiltonian (1.3.24) may remain finite it is necessary for the coefficients v_{r0}, v_{0r} to be infinitely small. We shall suppose that they are infinitely small in such a way as to make $v_{r0}N_0^{1/2}$ and $v_{0r}N_0^{1/2}$ finite, in order that the transition probability coefficients may be finite. Thus we put
>
> $$v_{r0}(N_0 + 1)^{1/2}e^{-i\theta_0/h} = v_r, \qquad v_{0r}N_0^{1/2}e^{i\theta_0/h} = v_r^*,$$
>
> where v_r and v_r^* are finite and conjugate imaginaries. We may consider the v_r and v_r^* to be functions only of the J's and w's of the atomic system, since their factors $(N_0 + 1)^{1/2}e^{-i\theta_0/h}$ and $N_0^{1/2}e^{i\theta_0/h}$ are practically constants, the rate of change of N_0 being very small compared with N_0. The Hamiltonian now becomes
>
> $$F = H_P(J) + \Sigma_r W_r N_r + \Sigma_{r \neq 0}\left[v_r N_r^{1/2}e^{i\theta_r/h} + v_r^*(N_r + 1)^{1/2}e^{-i\theta_r/h}\right]$$
>
> $$+ \Sigma_{r \neq 0}\Sigma_{s \neq 0}v_{rs}N_r^{1/2}(N_s + 1 - \delta_{rs})^{1/2}e^{i(\theta_r - \theta_s)/h}. \qquad \textbf{(1.3.25)}$$

The probability of a transition in which a light-quantum in the state r is absorbed is proportional to the square of the modulus of that matrix element of the Hamiltonian which refers to this transition. This matrix element must come from the term $v_r N_r^{1/2} e^{i\theta_r/h}$ in the Hamiltonian, and must therefore be proportional to $N_r'^{1/2}$ where N_r' is the number of light-quanta in state r before the process. The probability of the absorption process is thus proportional to N_r'. In the same way the probability of a light quantum in state r being scattered into state s is proportional to $N_r'(N_s' + 1)$. Radiative processes of the more general type considered by Einstein and Ehrenfest, in which more than one light-quantum take part simultaneously, are not allowed on the present theory.

Let me restate what Dirac did. He assumed the existence of a "zero state" for the light quanta—the vacuum state—in which an infinite number of them may be assumed to exist but in which *they are not observed*. That assumption allowed him to transform his original Hamiltonian to one of the form (1.3.25) that describes the creation and annihilation of photons. I have stressed Dirac's analysis of the vacuum state, because in 1930 he again made a striking assumption concerning the vacuum in the case of electrons when they are described by the Dirac equation, which assumption constituted the essence of the hole theory. Note further that by virtue of the $(N_r + 1)^{1/2}$ factor for the emission of a photon, Dirac could explain *spontaneous emission*: even when $N_r' = 0$ emission could still take place. In fact, by taking into account the relation between the energy density of photon of frequency ν and the number of photons in the modes r of frequency ν, Dirac could correctly derive the relation between the Einstein coefficients of induced and spontaneous emission.

In the last brief section of his paper, Dirac turned to the interaction of an atom with the electromagnetic field as described from the wave point of view. By this he meant considering the radiation field as described by a vector potential, which is resolved into its Fourier components κ_r in the Coulomb gauge.

The perturbation term to be added to the Hamiltonian according to classical theory (in the dipole approximation) is of the form $\frac{1}{c}\Sigma_r \kappa_r \dot{X}_r$, where X_r is the dipole moment of the atom in the direction κ_r. Dirac established that

$$\kappa_r = 2\left(\frac{h\nu_r}{c\sigma_r}\right)^{1/2} N_r^{1/2} \cos\theta_r/h, \tag{1.3.26}$$

and that the Hamiltonian describing an atom interacting with the radiation field is given by

$$F = H_P + \sum_r hW_r N_r + \sqrt{\frac{h}{c^3}} \sum_r \left(\frac{\nu_r}{\sigma_r}\right)^{1/2} \dot{X}_r \left\{ N_r^{1/2} e^{i\theta_r/h} + (N_r + 1)^{1/2} e^{-i\theta_r/h} \right\}.$$

$$\tag{1.3.27}$$

Dirac then concluded that (1.3.27) and (1.3.25) are consistent and in fact "the wave point of view . . . (1.3.27) gives values for the unknown interaction coefficients v_{rs} in the light quantum theory (1.3.25)."

In a lecture on the origin of quantum field theory in 1982, Dirac characterized the two approaches as follows:

> Instead of working with a picture of the photons as particles, one can use instead the components of the electromagnetic field. One thus gets a complete harmonizing of the wave and corpuscular theories of light. One can treat light as composed of electromagnetic waves, each wave to be treated like an oscillator; alternatively, one can treat light as composed of photons, the photons being bosons and each photon state corresponding to one of the oscillators of the electromagnetic field. One then has the reconciliation of the wave and corpuscular theories of light. They are just two mathematical descriptions of the same physical reality. (Dirac 1983, p. 49)

However, there seemed to exist a slight difficulty in Dirac's approach: "Since the wave theory gives $v_{rs} = 0$ for $r, s \neq 0$ it would seem to show that there are no direct scattering processes, but," Dirac noted, "this may be due to an incompleteness in the present wave theory." In a second paper on QED entitled "The Quantum Theory of Dispersion," Dirac showed that these scattering processes are in fact readily described in perturbation theory. That paper gave a quantum electrodynamic derivation of the Heisenberg-Kramers dispersion formula (Dirac 1927c). In it Dirac once again reiterated the view that "the Hamiltonian for the interaction of the field with an atom is of the same form as that for the interaction of an assembly of light quanta with the atom. There is thus a complete formal reconciliation between the wave and the light-quantum points of view" (Dirac 1927c).

Let me close this section on Dirac's initial formulation of QED by indicating that on February 19, 1927—that is, a few weeks after he had submitted his paper to the *Proceedings of the Royal Society*—he wrote a brief letter to Bohr telling him that he has "been able to integrate the equations of motion for the interaction of an atom and a field of radiation in a certain simple case, and thus obtain an expression for the breadth of a spectral line on the quantum theory."

This derivation is simple and elegant and predates the Weisskopf and Wigner paper on this subject by two years! The full content of this letter is as follows:

Geismarlandst. 1
Göttingen
19–2–27

Dear Professor Bohr,

      I have been able to integrate the equations of motion for the interaction of an atom and the field of radiation in a certain simple case, and thus obtain an expression for the breadth of a spectral line on the quantum theory.

We consider an atom in its lowest excited state ($J = J_1$, say) with no incident radiation. We must now find a solution of the wave equation of the form $\Sigma_{J'N_0'N_1'}a(J'N_1'N_2')\,\psi(J'N_1'N_2')$ [where $\psi(J'N_1'N_2')$ is the solution for no interaction between atom and field and $a(J'N_1'N_2')$ is a function of the time only] in which initially all the a's vanish except $a(J_1000)$. Only eigenfunctions that get excited to an appreciable extent by the interaction are those that refer to the atom being in the normal state and one light quantum of frequency aproximately equal to $\nu_0 \left[= \frac{H(J_1)-H(J_0)}{h} \right]$ in existence. We can without serious error suppose that all the other a's remain zero all the time. If for brevity we denote $a(J_1000\ldots)$ by a_1 and the a that refers to the atom being in the normal state with a single light-quantum of frequency ν by a_ν, then the equations for the rates of change of a_1 and a_ν are of the form

$$\left.\begin{aligned} ih\dot{a}_\nu &= \alpha a_1 e^{2\pi i(\nu-\nu_0)t} \\ ih\dot{a}_1 &= \alpha \int a_\nu e^{-2\pi i(\nu-\nu_0)t} \end{aligned}\right\} \tag{1}$$

where α is a constant. [These equations come from the general equation $ih\dot{a}_n = \Sigma_n a_n v_{mn} e^{i(W_m-W_n)t/h}$.]
It is easily verified that the solution of (1) is

$$a_1 = e^{-\gamma t}$$

$$a_\nu = \frac{a}{h}\frac{e^{[-\gamma+2\pi i(\gamma-\gamma_0)]t} - 1}{i\gamma + 2\pi(\nu - \nu_0)}$$

where γ is a real constant, equal to $\frac{|\alpha|^2}{2R^2}$, which may be shown to be equal to $\frac{1}{2T}$, where T is the mean life calculated from Einstein's A coefficient. The amplitude of the eigenfunction that refers to the initial state then falls off exponentially.

When $t = \infty$, a_ν has the value $\frac{\alpha}{h}\frac{1}{i\gamma+2\pi(\nu-\nu_0)} \cdot |a_\nu|^2$ is the probability of a light-quantum of frequency ν being in existence. Thus the probable distribution in frequency of the emitted radiation is according to the law

$$\frac{|\alpha|^2}{h^2}\frac{1}{\gamma^2 + 4\pi^2(\gamma - \gamma_0)^2}\cdot$$

With best wishes
Yours sincerely,

P.A.M. Dirac[20]

1.4 Jordan and the Quantization of Matter Waves

For Dirac the fact that the "Hamiltonian for the interaction of the [radiation] field with an atom is of the same form as that for the interaction of an assembly of light-quanta with the atom" meant that there was a complete reconciliation between the wave and light quantum point of view (Dirac 1927b). Waves and light quanta constituted complementary descriptions of the electromagnetic field. The Bose character of photons was the essential feature elicited by Dirac's second quantization. In fact, second quantization could be interpreted as a formal way to guarantee this property. In 1982 Dirac characterized the formalism he had developed in 1927 as follows:

> In the further development of the quantum mechanics of Heisenberg and Schrödinger, it occurred to me to apply a process that has been called second quantization. One simply takes the wave function $\psi(q)$ (q is a variable specifying a point in the domain of the wave function), and supposes that all the $\psi(q)$'s for different values of q are made into operators (Heisenberg operators) instead of just being numbers. Then one takes the conjugate quantities $\bar{\psi}(q)$ and makes them into operators. The $\psi(q)$'s are supposed to commute with each other, and likewise the $\bar{\psi}(q)$'s, but the $\bar{\psi}(q)$'s are supposed not to commute with the $\psi(q)$'s. It seemed to me that this was a very interesting idea, and I wondered where it would lead. I worked it out and found that it just led to an assembly of similar systems, all satisfying Bose statistics. It was a bit of a disappointment to find that nothing really new came out of the idea. I thought at first it was a wonderful idea and was very much looking forward to getting something really new out of it, but it turned out to be just a new way of going back to the idea of an assembly satisfying Bose statistics. (Dirac 1983, p. 48)

For Jordan, on the other hand, the "second quantization" that Dirac had introduced was to be viewed as the quantization of a classical field. Jordan had been very much drawn to the de Broglie-Einstein-Schrödinger view that fields were the primary entities. Moreover, he believed that this procedure should be applied to matter fields in the same way as it had been applied by him to the electromagnetic field in his paper with Born (1926a) and in the *Dreimännerarbeit* (Born et al. 1926). In a series of seminal papers stimulated by Dirac's formulation of QED, Jordan established that the "second quantization" procedure could in fact be thus interpreted. These papers laid the foundation of quantum field theory. He rightly considered them as his most important contribution to theoretical physics.

In the summer of 1927, Jordan wrote Schrödinger, telling him the path that had led to his interpretation of field quantization:

> [In 1924/5] I had given a lot of thought to Einstein's gas theory and I had specified the representation in a way similar to your work in the *Phys. Zeits.* [on Einstein's gas theory]: The number of atoms in a cell corresponds to the quantum number of a cavity-mode oscillation (we also made this point in the "Dreimännerarbeit"). Then your hydrogen paper gave hope that by following up this correspondence also the non-ideal gas could be represented by quantized waves—that therefore a complete theory of light and matter could be derived in which, as an essential ingredient, this wave field itself operates in a quantum non-classical way; the need to represent the light field as a quantum mechanical wave field was obvious to me after the result of the analysis of the fluctuation properties of quantized waves (cf. "Dreimännerarbeit"). The difficulty on which this hope seemed to founder at that time was just the validity of Pauli's statistics instead of Einstein's. Since this difficulty seemed insuperable, I gradually came to doubt the correctness of the whole representation. Furthermore, Pauli and Heisenberg did not want to hear much about it, while Born was initially very favourable, but later completely withdrew his support.[21]

The problem was thus to show how the formalism could yield particles that—like electrons—obeyed Fermi statistics. Reading Dirac's QED paper gave Jordan the insight on how to "second quantize" so that the quanta obeyed the Pauli exclusion principle. In Dirac's representation in which the N_r were diagonal, particles that obeyed Bose-Einstein statistics had eigenvalues for the number operator N_r equal to $N_r' = 0, 1, 2, 3, \ldots$, that is, the cells could be occupied by an arbitrary number of particles. For particles obeying Fermi-Dirac statistics, the eigenvalues of the occupation number operators, N_r', could only take on the values 0 or 1 and the operators N_r could thus be represented as 2×2 matrices

$$N_r = \begin{pmatrix} 0 & 0 \\ 0 & 1 \end{pmatrix}_r .$$ (1.4.1)

Since

$$[N_r, e^{i\theta_r}] = e^{i\theta_r},$$ (1.4.2)

Jordan suggested that

$$\theta_r = \frac{\pi}{2} \begin{pmatrix} 0 & 1 \\ 1 & 0 \end{pmatrix}_r ,$$ (1.4.3)

which can readily be shown to satisfy the commutation rules (1.4.2). One also easily verifies that the b_rs are then represented by

$$b_r = e^{-i\theta_r} N_r^{1/2} = -i\begin{pmatrix} 0 & 1 \\ 0 & 0 \end{pmatrix}_r. \qquad (1.4.4)$$

Jordan then showed that with the operators b_r and b_r^* thus represented, the Schrödinger equation

$$i\hbar\, \partial_t \psi = \left\{ \sum_r W_r b_r^* b_r + \sum_{r,s} v_{rs} b_r^* b_s \right\} \psi \qquad (1.4.5)$$

corresponded to a description of a system of fermions which was identical to the one given by antisymmetrical wave functions in configuration space. Jordan further computed the fluctuations in the number of particles (of energy between E and $E + dE$) inside a partial volume v, N_E, and rederived with his operator formalism the value for $< \Delta N_E^2 > /N_E^2$ that Pauli had derived in his paper on the paramagnetism of a Fermi gas (Pauli 1926). Jordan submitted his findings to the *Zeitschrift der Physik* at the beginning of July 1927 in a paper entitled "On the Quantum Mechanics of the Degeneracy of Gases" (Jordan 1927f). In the concluding section of his paper Jordan noted that "the results that have been obtained leave little doubt that one can formulate a quantum field theory of matter wherein electrons are represented by quantized waves in the usual three dimensional space—even though electrons satisfy the statistics of Pauli rather than those of Bose—and that the natural formulation of the quantum theory of electrons is obtained by simultaneously conceiving radiation and matter as waves in interaction in three dimensional space."

Jordan had written this paper during his three-month stay in Copenhagen during the spring of 1927. Bohr sent a copy of Jordan's ms to Pauli asking him for his opinion (Pauli 1979 [167], pp. 401–402). Pauli found Jordan's article quite stimulating, but he delayed giving Bohr an answer for over a month. Although he agreed with the main content of the paper ("which is correct and harmless"), he took exception with the viewpoint that Jordan had advanced in the last section of his article (Pauli 1979 [168], pp. 402–405).

Jordan's paper on the Fermi quantization of matter fields was the first of a series of five papers in which he developed his views on field quantization. Two of the papers were written by himself (1927f,h) and of the other three, one was written with Klein (Jordan and Klein 1927), one with Pauli (Jordan and Pauli 1928), and one with Wigner (Jordan and Wigner 1928). In the paper with Klein, written while the two of them were in Copenhagen in the spring of 1927, a generalization of Dirac's treatment of bosons was given to allow for the interaction of the bosons with one another. Their point of departure was a Schrödinger equation for the field operator containing a nonlinear term to account for the interaction of the field with itself:

$$i\hbar\frac{\partial\psi}{\partial t} = -\frac{\hbar^2}{2m}\nabla^2\psi + V^{ext}(\mathbf{x})\psi + e^2 \int \frac{\psi^*(x')\psi(x')}{|\,\mathbf{x} - \mathbf{x}'\,|}d^3x'\psi(x).$$

The equivalence of this description with that using symmetric wave functions in configuration space was established. The "particles" that emerged from the imposition of the quantum conditions (commutation rules) on the field variables thus obeyed Bose statistics. Heisenberg found the results of Jordan and Klein very attractive. In his interview with Kuhn and Heilbron in 1963, he recalled: "I like[d it] very much because now I could see, 'All right. There is an entirely different picture to start with (the wave picture), and if I quantize that picture—that is if I make this picture open to the same restriction as the particle picture—then the two pictures become equivalent.' That was exactly what I wanted" (Heisenberg 1963, session 8, p. 21).

Bohr at the Solvay meeting of 1927 saw Jordan and Klein's work as supporting his views of complementarity. Pauli at that same congress welcomed the formulation since it allowed to formulate the quantum theoretic description of an assembly of bosons *entirely in 3-dimensional space.* Jordan in all his papers emphasized this point. In his book, *Physics in the 20th Century,* Jordan elaborated on what was involved:

> For atoms with more than one electron ... Schrödinger wave mechanics assumes a very abstract form; in these cases Schrödinger waves are no longer waves in customary three-dimensional space but are simply a mathematical construction which mathematicians can "illustrate" to themselves as waves in space of more than three dimensions. This abstract, multidimensional space can be avoided through another (mathematically equivalent) method of representation, the construction of which was a special hobby of the author's. In this method of representation ("second quantization"), which clings especially closely to the fundamental dualism of waves and corpuscles, the waves dealt with are spread out in ordinary three dimensional space, but can only be described by means of the ideas of quantum mechanics. (Jordan 1944, p. 118)

In a letter to Kronig in November 1927, Pauli described the work of Jordan and Klein as "wirklich schön" (really beautiful) (Pauli 1979, [175] p. 416). In fact, the Jordan and Klein paper converted Pauli to the Jordan viewpoint about the quantization of matter fields. The article by Jordan and Klein made clear to both

Heisenberg and Pauli, who were then collaborating on a general theory of relativistic quantized fields, how to proceed in describing the interaction between the electromagnetic field and charges. Pauli, who up to that time had been reluctant to accept Jordan's views on the quantization of matter fields, embraced Jordan's viewpoint. After the publication of the Jordan-Klein article, Pauli and Heisenberg agreed that the quantization of matter fields was the correct approach. In December 1927 Heisenberg could write Bohr that the important work of Jordan and Klein had been the stimulus of his thinking long and hard on the formulation of relativistic quantum mechanics and that he and Pauli were making good progress. In February 1928 Heisenberg informed Dirac:

> I am writing a paper together with Pauli. We tried to change the Schrödinger-theory of de Broglie-waves + Maxwell waves ("EnergieImpuls-tensor" etc) into a theory of quantized waves (Bose or Fermistatistics does not make much difference); in so far the theory is a generalization of Klein-Jordans work as well as of Pauli-Jordans last paper (relativistically invariant Vertauschungs-relations between the F_{ik}). On[e] gets new terms in the Energy-Impuls-tensor, which compensates the interaction of the electron with itself, like in Klein-Jordan work. I think, one gets a definite idea, how retarded potentials etc. have to be treated in quantum mechanics. Of course the theory is to be extended for the spin in your way. — You will get proofs of the papers, as soon as possible.[22]

Although the basic idea in Jordan's paper formulating the quantization rules to yield Fermi-Dirac statistics had been correct (Jordan 1927g), Jordan had not correctly taken into account the phase factors necessary to guarantee that the creation operators for different energy states anticommute with one another. When Jordan discovered this shortcoming, he asked Wigner "to help him straighten out the mathematics of it" (Wigner 1963, p. 7). A joint paper ensued in which the Jordan-Wigner anticommutation rules

$$\left[a_r, a_s^*\right]_+ = a_r a_s^* + a_s^* a_r = \delta_{rs} \qquad\qquad \textbf{(1.4.6a)}$$

$$\left[a_r, a_s\right]_+ = \left[a_r^*, a_s^*\right]_+ = 0 \qquad\qquad \textbf{(1.4.6b)}$$

were first written down and the action of these operators on a state vector $\psi^{N'}(N_1',\ N_2'\dots\ N')$ specified by the eigenvalues N_r' of the $N_r(N' = \Sigma N_i')$ was exhibited (Jordan and Wigner 1928). Furthermore, the uniqueness (up to unitary transformation) of the representation they had obtained (for the case $N' < \infty$) was proven.

The abstract of their paper succinctly outlined their aim:

> The problem at hand is to describe an ideal or nonideal gas
> that satisfies the Pauli exclusion principle with the idea of not
> using any relation in the abstract (3N-dimensional) configu-
> ration space of the atoms of the gas, but of using only three-
> dimensional space. This is made possible by representing the
> gas by a three-dimensional quantized wave field, for which the
> particular non-commutative properties for multiplying wave
> amplitudes are simultaneously responsible for the existence of
> corpuscular atoms of the gas and for the validity of Pauli's ex-
> clusion principle. The features of the theory bear a close analogy
> to the corresponding theory for the ideal or non-ideal Einstein
> gas, formulated by Dirac, Klein and Jordan.

The reaction to the Jordan-Wigner paper was mixed. Heisenberg was enthusiastic (see de Broglie 1927). Most physicists, however, had difficulty with the
paper: the mathematics seemed recondite, overly rigorous, and perhaps too novel.
Dirac, who studied it carefully, objected to the paper on physical grounds. At the
Solvay Congress of 1927 he complained that the Fermi statistics did not arise naturally but only came about by a singular method of quantization of the waves,
specially chosen (by Jordan) in order to obtain the desired result. In Varenna in
1972, Dirac commented:

> At first I did not like this work of Jordan and Wigner, and I think
> I can attribute this dislike to my mind being essentially a geo-
> metrical one and not an algebraic one. In the case of the Bose
> statistics and the quantization which was connected with it, one
> has a definite picture underlying the basic equations, namely the
> picture that the theory would be applied to an assembly of os-
> cillators. There was no such picture available with the Fermi
> statistics, and I felt that was a serious drawback. I did not ap-
> preciate therefore the importance of this other kind of second
> quantization. (Dirac 1977, p. 140)

However, it is also likely that Dirac had disliked Jordan and Wigner's work because
he did not believe in "quantizing" matter fields in the late 1920s. In 1982 Dirac
expressed his initial objection to Jordan and Wigner's approach thus: In the case
of the bosons we had ... operators that were closely connected with the dynamical
variables that describe oscillators. We had operators that had classical analogues.
In the case of Jordan-Wigner operators, they had no classical analogues at all and
were very strange from the classical point of view. The square of each of them was
zero. I did not like that situation" (Dirac 1983, pp. 49–50).

1.5 Heisenberg and Pauli: The Quantum Theory of Wave Fields

It is somewhat ironic that Dirac, who was so concerned with the relativistic invariance of any formalism and considered quantum mechanics as an extension of classical mechanics, should have been the physicist who developed a formulation of quantum electrodynamics whose correspondence with Maxwell's equations is not immediately apparent, a QED based on a Hamiltonian formalism that made the relativistic invariance difficult to discern.

Soon after reading the manuscript of Dirac's QED paper, Pauli embarked on a program to construct his own version of quantum electrodynamics, one in which the relativistic invariance-covariance would be apparent and the relation to Maxwell theory manifest. He evidently outlined his proposal in a letter to Heisenberg that is no longer extant. In February 1927 Heisenberg countered: "I agree very much with your program concerning electrodynamics, but not quite concerning the analogy: quantum-wave-mechanics: classical mechanics = quantum electrodynamics: classical Maxwell theory. That one must quantize the Maxwell equations to get light quanta and so on à la Dirac, I believe already; but perhaps the de Broglie waves will later also have to be quantized in order to obtain charge and mass and statistics (! !) of electrons and nuclei" (Pauli 1979 [154], p. 376). Pauli and Heisenberg evidently disagreed about what had to be quantized. Heisenberg accepted Jordan's viewpoint and was prepared to quantize all wave fields— including matter waves. Pauli, on the other hand, was ready to quantize only the electromagnetic field. With that in mind he studied the mathematics of functionals that Vito Volterra had elaborated (Volterra 1913). On March 12, 1927, Pauli wrote Jordan: "I believe that I now have the essential understanding of the Hamilton-Jacobi theory of Maxwell's equations. My principal source is a (french) book by P. Levy. Lecons d'analyse fonctionelle. Paris 1922. We will thus see whether I can erect a quantum electrodynamic. For the present I am in good spirits" (Pauli 1979 [157], p. 386).[23] In late March 1927 Pauli sent Bohr a note to inform him that "at the moment I am much occupied with quantum electrodynamics. . . . I have written briefly to Heisenberg about my general foundational standpoint about quantum electrodynamics and would very much like to hear from him. . . . (I *dare* not ask you what your opinion is)" (Pauli 1979 [160], p. 389). A few days later Pauli received a letter from Heisenberg asking him a couple of things about his "Program": "I am in full agreement with the foundations of your program that $\mathscr{E}$ and $\mathscr{H}$ (the electromagnetic field variables) are not c[ommuting] fields, but are q fields, and that they must satisfy commutation rules that express this fact. But . . . " (Pauli 1979, [161], pp. 390–391). And so began the collaboration between Heisenberg and Pauli that eventually resulted in two important papers, "On the Quantum Dynamics of Wave Fields," that were published in February and September 1929. In them a general method for quantizing any field is presented. The starting point is a Lagrangian formulation of the classical equations, expressed in terms of a least-action principle for the action of the field system: $\delta S = 0$ with

$$S = \int L\left(Q_\alpha, \frac{\partial Q_\alpha}{\partial x_i}, \dot{Q}_\alpha\right) dV dt. \tag{1.5.1}$$

The Lagrangian coordinates of the field system are the field amplitudes at each point of space. The state of the system at a given time t is specified by the values of the field components $Q_\alpha(x_1, x_2, x_3, t)$ for every point of space. The Lagrangian allows one to define at every point of space the momentum P_α canonically conjugate to the Q_α at that point.

The quantization is then effected by imposing the usual commutation rules between coordinates and canonical momenta, which now read

$$[Q_\alpha(\mathbf{r}, t), Q_\beta(\mathbf{r}', t)] = 0 \tag{1.5.2a}$$

$$[P_\alpha(\mathbf{r}, t), P_\beta(\mathbf{r}', t)] = 0 \tag{1.5.2b}$$

$$[Q_\alpha(\mathbf{r}, t), P_\beta(\mathbf{r}', t)] = \frac{h}{2\pi i} \delta_{\alpha\beta}\, \delta(\mathbf{r} - \mathbf{r}'), \tag{1.5.2c}$$

where $\delta(\mathbf{r} - \mathbf{r}')$ is the Dirac delta function.

Unfortunately the usual Lagrangian for the free electromagnetic field

$$L = \frac{1}{4} F_{\mu\nu} F^{\mu\nu} \tag{1.5.3a}$$

$$F_{\mu\nu} = \partial_\mu A_\nu - \partial_\nu A_\mu \tag{1.5.3b}$$

yields for the momentum canonically conjugate to A_μ,

$$\pi_\mu = \partial L / \partial_0 A_\mu = F_{0\mu}. \tag{1.5.4}$$

Hence $\pi_0 = F_{00} = 0$ and cannot be introduced as an independent variable in a Hamiltonian. Moreover, $\pi_i = F_{0i} = -\mathscr{E}_i$. Since one of Maxwell's equations is div $\mathscr{E} = 0$, one deduces that div $\boldsymbol{\pi} = 0$, which is incompatible with the commutation relation

$$[A_j(\mathbf{x}), \pi_k(\mathbf{x}')] = i\delta(\mathbf{x} - \mathbf{x}')\delta_{jk}. \tag{1.5.5}$$

These problems indicated to Heisenberg that the formalism developed by Pauli must be amended. Since $\pi(\mathbf{x})$ and $A(\mathbf{x})$ are not independent variables, consideration should be given to variational problems with constraints (Pauli 1979, p. 270).

Pauli on the other hand, with Jordan, proceeded to show how to quantize the theory in the *free field* case (i.e., in the absence of charges) by dealing only with the field variables $F_{\mu\nu}(x)$: Jordan and Pauli (1928) derived relativistically invariant commutation rules for the Heisenberg operators $F_{\mu\nu}$, expressing $[F_{\mu\nu}(x), F_{\rho\sigma}(x')]$ in terms of the singular function

$$D(x - x') = -i \int \frac{d^4k}{(2\pi)^3} \delta(k^2)\epsilon(k_0)e^{-ik(x-x')}. \qquad (1.5.6)$$

The difficulties concerned with applying the canonical formalism to the electromagnetic field continued to plague Heisenberg and Pauli. They explored the formalism further and found further reasons for concern. They discovered that the energy-momentum tensor of a matter-field described by a Klein-Gordon equation interacting with the electromagnetic field is infinite (Pauli 1979 [187], pp. 435–438). By mid-1928 both Heisenberg and Pauli were pessimistic about the prospects of obtaining a satisfactory QED. Heisenberg began working on ferromagnetism, and Pauli expressed the opinion that no progress could be made without a fundamentally new idea.

In the fall of 1928 Heisenberg discovered a way to bypass the difficulties engendered by the fact that $\pi_0 = 0$ in the $L = \frac{1}{4}F_{\mu\nu}F^{\mu\nu}$ formulation of the action principle for classical electrodynamics. He suggested adding a term $-\frac{1}{2}\epsilon(\partial_\mu A^\mu)^2$ to the Lagrangian, in which case $\pi_0 = -\epsilon(\partial_\mu A^\mu)$ and the usual method of the canonical quantization scheme became applicable. The limit $\epsilon \to 0$ was to be taken at the end of all calculations. Pauli outlined the scheme in a letter to Klein in mid-February 1929 (Pauli 1979, [216], pp. 488–492) and included in his letter some of the conclusions Heisenberg and he had reached:

1. The theory contains divergences stemming from the self energy of the charged particles.

2. The matter field can seemingly be quantized so as to obey either Fermi or Bose statistics.

3. The theory introduces three kinds of fields: the electromagnetic field, the matter field describing electrons, and the matter field describing protons.

The first of the two lengthy papers Heisenberg and Pauli wrote on the quantum theory of wave fields was received by the *Zeitschrift für Physik* on March 19, 1929. Although their correspondence reflects a pessimistic assessment of their approach, in their paper Pauli and Heisenberg stated that although their theory was provisional they nonetheless believed that the "future correct theory" would exhibit many of the features of the present formulation. Jordan, upon reading their paper, found that Heisenberg and Pauli's work had not contributed much that was new: "Heisenberg and Pauli's work has now cleared up—and this seems to me the most important point of the whole paper—that together with the validity of the three-dimensional commutation relations, it is possible to maintain and demonstrate their relativistic invariance despite their manifest space-time asymmetry" (Jordan 1929).

For Jordan, the fact Heisenberg and Pauli had stressed that "the solution to the problem of understanding the existence of elementary particles—light quanta, electrons, protons—can be found by using quantization" was certainly not novel. He had emphasized this very point repeatedly. For him this was *the* decisive feature of matter wave quantization. In contrast to Pauli and Heisenberg's somewhat optimistic attitude toward the infinities that appeared in the expression for the energy of the interacting field system stemming from the zero-point energy of the radiation field and the Coulomb self-energy of the charged particles, Jordan was pessimistic. Pauli and Heisenberg's optimism was based on the fact that they believed these divergences to be infinite constants that would disappear when one evaluated the difference between two-energy eigenvalues of the system—the only quantities that were observable. Jordan, on the other hand, believed that

> the electron's self-energy is a variable infinite quantity and thus represents such a serious and worrying obstacle as to make it hardly possible in practical applications to arrive at the same point with Heisenberg and Pauli's theory as with the more primitive methods of Dirac.
>
> The difficulties we are now faced with have a totally different nature from those of the pre-quantum mechanics period, when there was no reliable firm basis for answering the question "Why do electrons exist?": it is no longer the existence of electrons, but it is now the mode and nature of their interactions with the electromagnetic field that still remain without solution and undoubtedly pose for us some difficult enigmas.

His paper concluded with the remark that "we can therefore say that we now have reached a theoretical understanding of the existence of the electron, but in no way that of its constitution."

Dirac, too, studied Heisenberg and Pauli's first paper carefully and took notes on it. These notes are extant and constitute a concise abstract of the paper. Let me merely point out two comments Dirac makes. Section 3 of Heisenberg and Pauli's first paper was concerned with whether the "quantum conditions are invariant under Lorentz transformation," since "these quantum conditions give commutation relations between p's and q's at two different points at the same time and are therefore very unrelativistic in form."

Commenting on the equation that Heisenberg and Pauli had derived as a sufficient condition for relativistic invariance

$$\Sigma_i \frac{\partial}{\partial x_i} \left(\frac{\partial^2 H}{\partial \frac{\partial Q_\alpha}{\partial x_k} \partial \frac{\partial Q_\beta}{\partial x_i}} \right) = 0, \tag{1.5.7}$$

Dirac wrote down the condition under which (1.5.7) is satisfied[24] and noted, as Heisenberg and Pauli had done in their paper, that "quantities referring to points connected by a finite spatial displacement $[\Sigma(\nabla x_i)^2 - c^2(\nabla t^2) > 0]$ always commute." Dirac then added the statement, "This one would expect physically as one can measure one of them without disturbing the other," thus making clear that he had fully appreciated the connection between commutation rules, measurability, and relativistic invariance already in 1929.

In comments on section 2 of Heisenberg and Pauli's paper, concerned with the application of the theory to the electromagnetic field, Dirac expressed his unhappiness with Heisenberg and Pauli's approach based on the Lagrangian

$$\mathcal{L} = \frac{1}{2} \left(\mathcal{E}^2 - \mathcal{H}^2 \right) \tag{1.5.8}$$

and the potentials ϕ_μ as the field quantities q.

They then form the momenta conjugate to the ϕ's and the trouble starts

$$p_4 = \frac{\partial \mathcal{L}}{\partial \dot{\phi}_4} = 0 \qquad p_1, p_2, p_3 = -\mathcal{E}. \tag{1.5.9}$$

The p's cannot be chosen arbitrarily at $t = 0$, on account of div $\mathcal{E} = 0$. The vanishing of p_4 is a very serious difficulty since it contradicts the Q condition

$$q_4(x)p_4(x') - p_4(x')q_4(x) = ih\delta(x - x') \tag{1.5.10}$$

and the previous theory is not applicable. However the commutability relations are quite all right so far as they concern the field $\mathcal{E}$, $\mathcal{H}$ and not the potential ϕ. They become

$$[\mathcal{H}_i, \mathcal{H}'_k] = 0 \qquad [\mathcal{E}_i, \mathcal{E}'_k] = 0 \qquad i, k = 1, 2, 3$$

$$[\mathcal{E}_1, \mathcal{H}'_2] = -[\mathcal{E}_2, \mathcal{H}'_1]$$
$$= c\delta(x_1 - x'_1)\,\delta(x_2 - x'_2)\,\delta(x_3 - x'_3) \tag{1.5.11}$$

It seems to me that the failure of the Q condition that refers to the potentials should be taken as meaning that one must work always with $\mathscr{E}$ and $\mathscr{H}$, which are the only physically important things.[25]

When Dirac was reading the Heisenberg and Pauli paper in 1929 he had already published his paper on the Dirac equation, and at the time he was concerned with the interpretation of the negative states. In his notes on the Pauli-Heisenberg treatment of the quantization of matter interacting with the electromagnetic field based on the Lagrangian

$$\mathscr{L} = +\Sigma_{\mu\sigma\rho}\left[\overline{\psi}_{\rho}\gamma^{\mu}_{\rho\sigma}\left(ihc\frac{\partial}{\partial x_{\mu}} + e\phi_{\mu}\right)\psi_{\sigma} - imc^{2}\overline{\psi}_{\rho}\psi_{\rho}\right], \qquad (1.5.12)$$

Dirac recorded:

To get the Fermi statistics we require these oscillators to be able to oscillate with only 0 or 1 quanta, which requires us to take difft. quantum conditions

$$q_{r}p_{s} + p_{s}q_{r} = ih$$
$$q_{r}q_{o} + q_{s}q_{r} = 0$$

This gives

$$\psi_{\rho}\psi'_{\sigma} + \psi'_{\sigma}\psi_{\rho} = 0 \qquad \psi_{\rho}\overline{\psi}'_{\sigma} + \overline{\psi}'_{\sigma}\psi_{\rho} = \delta_{\rho\sigma}\delta(x - x')$$

Invariance under Lorentz transf. still holds with this kind of Q condition.

Dirac then noted that "they remark that different ψ's are necessary for protons and electrons (but this is probably not necessary)." Commenting on the section concerned with approximations in the Heisenberg-Pauli scheme, Dirac observed: "They find the coulomb law. Also infinite zero point energy due to interaction of electron with itself. *This is rather remarkable in view of the fact that Coulomb forces are nowhere directly assumed* [italics mine]. They assumed only field equs. of Maxwell for continuous distribution of electricity and then applied quantum rules, and the appearance of particles with Coulomb forces between them justifies this procedure." Dirac's final observations were concerned with practical applications: "These are rather disappointing since they make so many approximations that all the special relativity features of the present paper disappear and they get results which could have been obtained from a much simpler non-relativity theory."

The full extent of Dirac's disagreement with Heisenberg and Pauli's approach emerged in a remarkable paper Dirac published in 1932 in the *Proceedings of the Royal Society* entitled "Relativistic Quantum Mechanics." In it, Dirac (1932) intuitively prefigured many of the subsequent developments of relativistic quantum field theory. In his introduction, Dirac commented that the development of the quantum theory was made possible by continually referring to Bohr's correspondence principle "according to which classical theory can give valuable information about quantum phenomena in spite of the essential differences in the fundamental ideas of the two theories." To Dirac, the masterful advance of Heisenberg in 1925 consisted in showing "how the equations of classical physics could be taken over in a formal way and made to apply to quantities of importance in quantum theory." Heisenberg's scheme was found "to fit wonderfully well with the Hamiltonian theory of classical mechanics ... and enabled one to apply to quantum theory all the information that classical theory supplies, in so far as this information is consistent with the Hamiltonian form."

Since classical mechanics gave a Hamiltonian description of the dynamics of a system of interacting particles, one could now formulate a satisfactory quantum mechanics for such an assembly provided the interaction between the particles could be expressed by means of an energy term in the Hamiltonian function.

To Dirac the new quantum mechanics "established the Correspondence Principle on a quantitative basis," and what Dirac meant by this is that the new rules allowed one to dispense with it.[26] But, Dirac went on, this "does not exhaust the sphere of usefulness of the classical theory. Classical electrodynamics, in its accurate (restricted) relativistic form, provides us with the insight that the idea of an interaction energy between particles is only an approximation and should be replaced by the idea of each particle emitting waves, which travel outward with a finite velocity and influence the other particles in passing over them." The challenge was how to incorporate this new information into the quantum theory and set up a relativistic quantum mechanics, "before we can dispense [completely] with the Correspondence Principle."

The specific problem Dirac wanted to address was to give a relativistic wave mechanical description of the motion of two or more charged particles. Dirac noted that Pauli and Heisenberg had attempted a "comprehensive theory" by regarding the electromagnetic field as a dynamical system amenable to Hamiltonian treatment and its interaction with the particles as describable by an interaction energy, "so that the usual methods of Hamiltonian quantum mechanics may be applied." But, Dirac went on, there are serious *physical* objections to these views, quite apart from the purely mathematical difficulties to which they lead:

> If we wish to make an observation on a system of interacting
> particles, the only effective method of procedure is to subject

them to a field of elecromagnetic radiation and see how they react. Thus the rôle of the field is to provide a means for making observations. *The very nature of an observation requires an interplay between the field and the particles.* We cannot therefore suppose the field to be a dynamical system on the same footing as the particles and thus something to be observed in the same way as the particles. The field should appear in the theory as something more elementary and fundamental.

How to achieve the desired ends was indicated by Dirac: "One ought to proceed ... by following the methods introduced by Heisenberg in 1925, which have already met with such great success for non-relativistic quantum mechanics. Heisenberg put forward the principle that one should confine one's attention to observable quantities, and set up an algebraic scheme in which only these observable quantities appear." More precisely: only *classically* observable quantities could constitute the basis for such a scheme. The impact on Dirac of Heisenberg's opening statements in his 1925 paper—"The present paper seeks to establish a basis for theoretical quantum mechanics founded exclusively upon relationships between quantities which in principle are observable"—cannot be overestimated. In his introductory remarks to the 1928 Cambridge "Lectures on Modern Quantum Mechanics" that formed the basis for *The Principles of Quantum Mechanics* published in 1930, Dirac commented:

> The main feature of the new theory is that it deals essentially only with observable quantities, a very satisfactory feature. One may introduce auxiliary quantities not directly observable for the purpose of mathematical calculation; but variables not observable should not be introduced merely because they are required for the description of the phenomena according to ordinary classical notions, e.g. orbital frequencies in Bohr's theory. The proper understanding of this point by Heisenberg formed the basis of modern theory. The theory enables one to calculate only observable quantities (e.g. probability coefficients for scattering processes, and not detailed connection between angle of scattering and incident path, as in classical theory, and any theories which try to give a more detailed description of the phenomena are useless. (Dirac 1928f)

He also emphasized that "the general quantum mechanics of l[ight]-quanta would only extend the [classical wave theory of light] domain of problem[s] that can be answered; not give more detailed answers (than can be experimentally verified)." In his 1932 paper Dirac reiterated these observations[27] and further noted:

Strictly speaking, it is not the observable quantities themselves (the Einstein A's and B's) that formed the building stones of Heisenberg's algebraic scheme, but rather certain more elementary quantities, the matrix elements, having the observable quantities as the squares of their moduli. The extra phase quantities introduced in this way are essential.

Let us see what are the corresponding quantities in relativistic theory. To make a relativistic observation on a system of particles we must, as mentioned in the introduction, send in some incident electromagnetic radiation and examine the scattered radiation. The numerical quantity that we observe is thus the probability of occurrence of a certain radiative transition process. This process may be specified by the intensities of the various monochromatic components of the ingoing and of the outgoing fields of radiation. (We shall ignore the purely mathematical difficulty that the total number of these components is an infinity of a high order).

In non-relativistic quantum mechanics the probability of occurrence of any transition process is always given as the square of the modulus of a certain quantity, of the nature of a matrix element or simply a transformation function, referring to the initial and final states. It appears reasonable to assume that this will still be the case in relativistic quantum mechanics. Thus the relativistic observable quantities, which are always transition probabilities, will all appear as the squares of the moduli of certain quantities. These quantities, which we shall refer to as probability amplitudes, will then be the building stones analogous to Heisenberg's matrix elements. *We should expect to be able to set up an algebraic scheme involving only the probability amplitudes and to translate the equations of motion of relativistic classical theory directly into exact equations expressible entirely in terms of these quantities* [Dirac's italics].

A point of special importance about the building stones of the new theory is that each of them refers to one field of ingoing waves and one field of outgoing waves, or to one initial field of a transition process and one final field. Quantities referring to two initial fields, or to two final fields, are not allowed. *This shows a departure from the theory of Heisenberg and Pauli*, according to which, if one is given any quantity referring to one initial field and one final field, one can obtain from it a quantity referring to two initial fields or to two final fields, by a straightforward application of the transformation theory of quantum mechanics. *The Heisenberg-Pauli theory thus involves many quantities which are unconnected with results of observations and which must be removed from consideration if one is to obtain a clear insight into the underlying physical relations* (italics mine).

These statements essentially outline what later will be called an S-matrix philosophy![28]

Dirac then turned his attention to the problem of how to translate these insights into a mathematical scheme. Since the classical theory itself is not free from ambiguity, Dirac first considered the one-electron problem and the problems encountered *at the classical level* in that situation. The classical equations that describe this situation are of two kinds, "(i) those that determine the field produced by the electron (which field is just the difference of the ingoing and outgoing fields) in terms of the variables describing the motion of the electron and (ii) those that determine the motion of the electron."

Although equations (i) are definite and unambiguous the latter, (ii), "express the acceleration in terms of field quantities at the point where the electron is situated and these field quantities in the complete classical picture are indefinite and undefined." Dirac expected that in an "accurate treatment—one that takes into account the reaction of the electron on the waves it emits"—the field determining the acceleration of the electron would be associated with both the ingoing and outgoing waves. Although classical attempts—by Lorentz and Abraham—had been made to improve the theory by assuming a definite structure for the electron and calculating the effect on one part of it of the field produced by the rest, "such methods are not permissible in modern physics" since these procedures are not relativistically covariant. Thus Dirac concluded that the limit of classical electromagnetic theory had been reached. Even though one had definite equations for determining the motion of the charged particle in terms of field quantities, the latter could not be easily interpreted since there was no reliable classical base, "the most that can be said about them is that they are related in some non-classical way to two fields, namely those of the incoming and of the outgoing field. . . . Further advance can only be made by introducing quantum ideas."

Dirac in the mid-1930s abandoned this philosophic stance of making the quantum theory primary, and devoted considerable efforts at first formulating a consistent relativistically invariant *classical* theory of charged particles which thereafter was quantized (Dirac 1938b, 1948a, 1951c). But in 1932 it was the quantum theory that motivated the approach. And so Dirac—to be consistent with his philosophy that "all quantities in relativistic quantum mechanics are of the nature of probability amplitudes referring to one ingoing and one outgoing field"—made the assumption that *"the passage from the field of ingoing waves to the field of outgoing waves is just a quantum jump performed by one field"* (Dirac's italics).

The significance of the new assumption lies in the fact that *"the classical picture from which we derive our equations must contain no reference to quantum jumps"* (Dirac's italics). Hence the classical picture must only involve one field, "a field composed of waves passing undisturbed through the electron and satisfying every-where Maxwell's equation for empty space."

Thus Dirac's physical intuition led him to a description of charged particles interacting with the electromagnetic field in terms of *interaction picture* wave functions. For a single charged particle this takes the form

$$F\psi = 0, \qquad (1.5.13a)$$

where the operator F "neglecting spin" is given by

$$F = \left(ih\frac{\partial}{\partial t} + eA_o\right)^2 - \left(ihc\frac{\partial}{\partial x} - eA_x\right)^2 - m^2c^4, \qquad (1.5.13b)$$

where A_0, A_1, A_2, A_3 are "not numerical, *but are operators satisfying the usual quantum conditions governing ... the electromagnetic field in empty space*" (italics mine) evaluated at the position $xyzt$ of the particle.

For the problem of the interaction of two charged particles, the wave function ψ is a function of the variables $x_1 y_1 z_1 t_1$ and $x_2 y_2 z_2 t_2$ describing the two particles, and of *one* set of A_μ's describing one field. (Note that the charged particles are treated as quantum mechanical particles, whereas the electromagnetic field is a continuum that is quantized.) This ψ satisfies the two equations

$$F_1\psi = 0 \qquad F_2\psi = 0, \qquad (1.5.14)$$

where F_1 is the operator

$$F_1 = \left(ih\frac{\partial}{\partial t_1} - eA_0(x_1 y_1 z_1 t_1)\right)^2$$
$$- \left(ihc\frac{\partial}{\partial x_1} - eA_x(x_1 y_1 z_1 t_1)\right)^2 - \ldots - m^2c^4, \qquad (1.5.15)$$

and similarly for F_2. Dirac stressed that "no terms of the type of a Coulomb interaction are required," the interaction of the two charged particles being due "to the motions of both being connected with the same field." To illustrate the formalism in as simple a manner as possible, Dirac considered a "toy model": the case of one spatial dimension (see also Dirac 1935b). The field $V(x, t)$ then obeys the field equation

$$\frac{1}{c^2}\frac{\partial^2 V}{\partial t^2} - \frac{\partial^2 V}{\partial x^2} = 0, \tag{1.5.16}$$

which can be derived from a variational principle

$$\delta \int \mathscr{L}\, dx\, dt = 0 \tag{1.5.17}$$

with

$$\mathscr{L} = \frac{1}{2}\left\{\left(\frac{1}{c}\frac{\partial V}{\partial t}\right)^2 - \left(\frac{\partial V}{\partial x}\right)^2\right\}. \tag{1.5.18}$$

Dirac then considered the interaction of a particle—assumed to move nonrelativistically—with the field. The Lagrangian for the combined system is taken to be

$$\begin{aligned} L &= L_p + L_{\text{field}} \\ &= \frac{1}{2}m\dot{X}^2 - eV(X) + \int \mathscr{L}\, dx. \end{aligned} \tag{1.5.19}$$

The variation of L_p yields

$$\delta L_p = m\dot{X}\delta\dot{X} - e\delta[V(X)] \tag{1.5.20}$$

with

$$\delta[V(X)] = \delta V_{at\ x=X} + \left(\frac{\partial V}{\partial x}\right)_{x+X}\delta X, \tag{1.5.21}$$

so that the variational principle yields

$$\begin{aligned} 0 &= \int\left\{-m\ddot{X}(t) + \left(\frac{\partial V}{\partial x}\right)_{x=X}\right\}\delta X dt \\ &+ \int\int dx dt\left\{-\frac{1}{c^2}\frac{\partial^2 V}{\partial t^2} + \frac{\partial^2 V}{dx^2} - e\delta(x-X)\right\}\delta V, \end{aligned} \tag{1.5.22}$$

from which follow the equations of motion for the particle

$$-m\ddot{X} + e\left(\frac{\partial V}{\partial x}\right)_X = 0 \qquad (1.5.23)$$

and that for the field

$$-\frac{1}{c^2}\frac{\partial^2 V}{\partial t^2} + \frac{\partial^2 V}{dx^2} = e\delta(x - X). \qquad (1.5.24)$$

There exists a static solution, $\frac{\partial V}{\partial t} = 0$, for the case of a particle fixed at $X = 0$ which is given by

$$\frac{\partial V}{\partial x} = -\frac{e}{2} \text{ for } x < 0$$

$$\frac{\partial V}{\partial x} = \frac{e}{2} \text{ for } x > 0, \qquad (1.5.25)$$

and therefore

$$V = \frac{1}{2}e\,|\,x\,|\,. \qquad (1.5.26)$$

This gives an attraction between particles of like sign. Dirac indicated that the particle equation of motion gave an undetermined force, $(\frac{\partial V}{\partial x})_{x=X}$, at the position of the particles, which was unsatisfactory. The case of two particles was considered next. Since Dirac wanted a quantum description, he considered the system described by the Hamiltonian

$$H = \sum_{i=1,2}\left\{\frac{1}{2}m_i\dot{X}_i^2 + e_iV(X_i)\right\}$$
$$+ \frac{1}{2}\int dx\left\{\frac{1}{c^2}\left(\frac{\partial V}{\partial t}\right)^2 + \left(\frac{\partial V}{\partial x}\right)^2\right\}, \qquad (1.5.27)$$

from which he inferred that the wave function $\psi(x_1, x_2, t)$ (in the interaction picture!) describing the two particle system satisfies

$$\left\{ih\frac{\partial}{\partial t} + \frac{h^2}{2m_1}\frac{\partial^2}{\partial x_1^2} + \frac{h^2}{2m_2}\frac{\partial^2}{\partial x_2^2} - e_1V(x_1,t) - e_2V(x_2,t)\right\}\psi(x_1, x_2, t) = 0,$$
$$(1.5.28)$$

where $V(x,t)$ is a field *operator* satisfying the free field equations (1.5.16) and the commutation rules

$$[(V(x,t), V(x',t')] = ihcD(x - x', t - t'), \quad\quad \textbf{(1.5.29)}$$

where D is the Pauli-Jordan invariant function that vanishes for spacelike separation of the space-time points xt and $x't'$, eq. (1.5.6). These equations were then generalized by introducing a multiple time wave function $\psi(x_1, t_1, x_2, t_2)$ that satisfied

$$\left\{ ih\frac{\partial}{\partial t_1} + \frac{h^2}{2m_1}\frac{\partial}{\partial x_1^2} - e_1 V(x_1, t_1) \right\} \psi(x_1 t_1, x_2 t_2) = 0 \quad\quad \textbf{(1.5.30a)}$$

$$\left\{ ih\frac{\partial}{\partial t_1} + \frac{h^2}{2m_2}\frac{\partial}{\partial x_2^2} - e_2 V(x_2, t_2) \right\} \psi(x_1 t_1, x_2 t_2) = 0 \quad\quad \textbf{(1.5.30b)}$$

and reduced to $\psi(x_1, x_2, t)$ for equal times of the particles: $t_1 = t_2 = t$.

Dirac next considered the scattering of two particles initially moving with definite momenta and showed that the solution obtained from (1.5.28) to second order in e (corresponding to the exchange of a V quantum between the particles) was the same as "if we were to solve for the 1st order correction to the equation

$$\left\{ ih\frac{\partial}{\partial t} + \frac{h^2}{2m_1}\frac{\partial^2}{\partial x_1^2} + \frac{h^2}{2m_2}\frac{\partial^2}{\partial x_2^2} - 4\pi e_1 e_2 \mid x_1 - x_2 \mid +K \right\} \psi \quad\quad \textbf{(1.5.31)}$$

where K is infinite, but contains no x dependence, i.e., it is an infinite self-energy term."

In other words, Dirac proved that eq. (1.5.28), with V a quantized free field, gives rise to an interaction energy $2\pi e_1 e_2 \mid x_1 - x_2 \mid$ between the particles, the electrostatic result that "we should expect from a one dimensional theory." However, Dirac believed he had made a mistake in sign, as he had obtained "an attractive force between like charges." But in a note added on April 20, Dirac indicated that Heisenberg had pointed out to him that the sign of the interaction energy given by the above calculation was correct, since for a one-dimensional longitudinal wave field "classical theory also requires an attractive force between like charges."

The sarcastic tone of the letter Pauli wrote Dirac upon reading his article is worth noting:

> Your recently published remarks in the Proceedings of the
> Royal Society concerning Quantum electrodynamics were ...
> certainly no masterpiece. After a confused introduction, that

consisted of only half understandable, because only half understood, sentences you come finally to results in a simplified one
dimensional example that are identical with those that the formalism of Heisenberg and I gives for that example. (This identity is immediately recognizable and has since been calculated in
much too complicated a fashion by Rosenfeld). This conclusion
of your work stands in contrast to your more or less unambiguous assertion in the introduction that somehow you can construct
a better quantum electrodynamics than Heisenberg and I. (Pauli
1985, [292], p. 115)

Dirac's paper could not have been very transparent until Rosenfeld (1932)
showed the equivalence of Dirac's new form of relativistic quantum mechanics
with Heisenberg and Pauli's formulation. Dirac, however, felt that "Rosenfeld's
proof (Rosenfeld 1932) was obscure and did not bring out some features of the
relation of the two theories . . . . To assist in the further development of the theory"
he gave a new "simplified proof of the equivalence" (Dirac et al. 1932b). It is
worth noting that whereas Rosenfeld in his proof of the equivalence had second-
quantized the fermion field that described matter, Dirac chose to describe matter
in terms of the individual particles' Hamiltonians. Dirac, Fock, and Podolsky[29]
proceeded as follows:

The Schrödinger equation for a system containing N charged
particles interacting with the electromagnetic field is given by

$$\left\{ \bar{H}_d + \sum_{n}^{N} H_n(q_n, p_n, a(q_n)) + \frac{\hbar}{i} \frac{\partial}{\partial t} \right\} \psi = 0. \qquad (1.5.32)$$

We consider now the unitary operator

$$u = exp\left\{ \frac{i}{\hbar} \bar{H}_{el} t \right\} \qquad (1.5.33)$$

and introduce the unitary transformation

$$A = uau^{-1} \qquad (1.5.34)$$

and the corresponding transformation of ψ

$$\Phi = u\psi. \qquad (1.5.35)$$

Then Φ satisfies the equation

$$\left\{ \sum_n H_n(q_n, p_n, A(q_n, t)) + \frac{\hbar}{i} \frac{\partial}{\partial t} \right\} \Phi = 0 \qquad (1.5.36)$$

In contrast to a, which was independent of time (Schrödinger picture), A contains t through u ...

Equation (1.5.36) is the starting point of the many time theory. In that theory one introduces the function $\Phi(q_1 t_1, \ldots q_N t_N)$, containing as many time variables $t_1, t_2, \ldots t_N$ as the number of particles, in place of the function $\Phi(q_1, q_2, \ldots q_N, t)$ containing only one time variable and suppose that this $\phi(q_1 t_1, \ldots q_N t_N)$ satisfies simultaneously the following N equations

$$\left\{ H_n(q_n p_n, A(q_n, t_n)) + \frac{\hbar}{i} \frac{\partial}{\partial t_n} \right\} \Phi(q_1 t_1, \ldots q_N t_N) = 0$$

$$= 1, \ldots N. \qquad (1.5.37)$$

Thus $\Phi(t_1, \ldots t_N)$ is related to the ordinary probability amplitude $\Phi(t)$ by

$$\Phi(t) = \Phi(t, \ldots t). \qquad (1.5.38)$$

The simultaneous equations (1.5.38) can be solved when and only when the N^2 conditions

$$(H_n H'_n - H'_n H_n)\Phi(q_1 t_1, \ldots q_N t_N) = 0 \qquad (1.5.39)$$

are satisfied for all pairs n', n. If the world point $(q_n t_n)$ lies outside the light cone whose vertex is at the point $(q'_n t'_n)$, it can be proved that $H_n H_{n'} - H_{n'} H_n = 0$. Hence, the function $\Phi(t_1 \ldots t_N)$ satisfying (5.38) can exist in the region where

$$(q_n - q'_n)^2 - c^2(t_n - t'_n)^2 \geq 0 \qquad (1.5.40)$$

is satisfied simultaneously for all values n and n'.[30]

Actually, I have quoted not from Dirac's paper but from Tomonaga's 1943 paper in which the generalization of the Dirac many-time formalism to a field system is given, and (what later will be called) the Tomonaga-Schwinger equation is introduced.

Incidentally, in 1934, the then 16-year-old Julian Schwinger read the Dirac-Fock-Podolsky paper and generalized it to the case where the charged particles are described not by the Hamiltonian $\Sigma_n H_n(q_n p_n)$ but by (second-quantized) Dirac field operators. He wrote but did not publish a paper entitled "On the Interaction of Several Electrons" which generalized Dirac's calculation to that situation (including a $3 + 1$ space-time) and in which he derived the Møller interaction between the electrons. Like Dirac, Schwinger had to discard the "infinite self-energy of the charges" (Schwinger 1983b).

Dirac had believed that his new relativistic quantum mechanics was "a departure from the theory of Heisenberg and Pauli" (Dirac et al. 1932a). However, as shown by Rosenfeld, the two were in fact identical. This must have suggested to him that he was on the wrong track, for a new approach was quickly developed. It is expounded in his paper, "The Lagrangian in Quantum Mechanics" (Dirac 1933a), which was submitted to the *Physikalische Zeitschrift der Sowjetunion* shortly after the Dirac, Fock, and Podolsky paper had been sent to the same journal.

The metaphysical assumptions which underlay Dirac's new approach were still the same: classically observable quantities must form the point of departure, but now relativistic invariance was to be maintained at all stages. The introduction to this paper outlined Dirac's motivation:

> Quantum mechanics was built up on a foundation of analogy with the Hamiltonian theory of classical mechanics. This is because the classical notion of canonical coordinates and momenta was found to be one with a very simple quantum analogue, as a result of which the whole of the classical Hamiltonian theory, which is just a structure built up on this notion, could be taken over in all its details into quantum mechanics.
>
> Now there is an alternative formulation for classical dynamics, provided by the Lagrangian. This requires one to work in terms of coordinates and velocities instead of coordinates and momenta. The two formulations are, of course, closely related, but there are reasons for believing that the Lagrangian one is the more fundamental.
>
> In the first place the Lagrangian method allows one to collect together all the equations of motion and express them as the stationary property of a certain action function. (This action function is just the time-integral of the Lagrangian.) There is no corresponding action principle in terms of the coordinates and momenta of the Hamiltonian theory. Secondly the Lagrangian method can easily be expressed relativistically, on account of the action function being a relativistic invariant, while the Hamilto-

nian method is essentially non-relativistic in form, since it marks
out a particular time variable as the canonical conjugate of the
Hamiltonian function.

　　　For these reasons it would seem desirable to take up
the question of what corresponds in the quantum theory to the
Lagrangian method of the classical theory. A little consideration shows, however, that one cannot expect to be able to take
over the classical Lagrangian equations in any direct way. These
equations involve partial derivatives of the Lagrangian with respect to the coordinates and velocities and no meaning can be
given to such derivatives in quantum mechanics. The only differentiation process that can be carried out with respect to the
dynamical variables of quantum mechanics is that of forming
Poisson brackets and this process leads to a Hamiltonian theory.
*We must therefore seek our Lagrangian in an indirect way. We
must try to take over the ideas of the classical Lagrangian, not
the equations of the classical Lagrangian theory* [italics mine].
(Dirac 1933a)

As we shall see, it was Feynman who took up these ideas and developed
a new formulation of quantum mechanics based on the Lagrangian (chapter 8).

1.6 Hole Theory

　　　The other set of papers that make up Dirac's amazing contribution to the
development of a relativistically invariant quantum theory are the ones connected
with the development of the spin 1/2 equation that bears his name.

　　　There is little question (Dirac 1963a, 1978; Kragh 1981; Moyer 1981a,b,c)
that it was the attempt to formulate a relativistic quantum mechanics based on the
transformation theory which he had formulated in 1926–1927 that led Dirac to the
first-order equation:

$$(p_0 + \boldsymbol{\alpha} \cdot \boldsymbol{p} + a_4 m_0 c)\psi = 0$$
$$\alpha_\mu \alpha_\nu + \alpha_\nu \alpha_\mu = 0 \quad (\mu \neq \nu) \quad \mu, \nu = 1, 2, 3, 4$$
$$\alpha_\mu^2 = 1. \tag{1.6.1}$$

Transformation theory was the foundation of Dirac's physical interpretation of
quantum mechanics and the basis of his belief of its validity. Transformation
theory required the equation of motion for the wave function to be of first order
in time.

　　　In introductory remarks to a lecture on the early history of relativistic
quantum mechanics that Dirac delivered at Ann Arbor in 1978, he addressed the

question: How does a theorist make a prediction? The answer he gave was: "The basic requirement is that one should have a theory in which one has a great deal of confidence" (Dirac 1978).

The transformation theory of quantum mechanics was such a theory, one in which he had a "very great deal of confidence." Dirac's transformation theory solved the quantum mechanical problem of obtaining the probability for a set of dynamical variables of a given physical system having specified values. It allowed one to calculate the probability of any dynamical variable having a specified value or of several variables simultaneously having specified values, provided they commute with one another.

Dirac was indeed pleased with his general transformation theory. In 1977, at a conference celebrating both his seventy-fifth birthday and the fiftieth anniversary of the publication of his relativistic wave equation for the electron, Dirac indicated that he thought that of all the work that he had done in his life his transformation theory was the piece that most pleased him: "It pleased me because it did not come from some lucky accident; it came from logical thinking step by step, seeing each step giving rather more detailed knowledge and leading on to the next question to examine and resolve. And in this step by step way I was able to pass to a general theory" (Dirac 1977a).

In his interview with T. S. Kuhn, Dirac recalled that he "had more confidence in that transformation theory than other people did" at the time (Dirac 1963a). Transformation theory had structured his approach to the quantum mechanical description of the interaction of radiation and matter. It also set the requirements that a relativistic theory of the electron had to meet if it was to be acceptable as a quantum mechanical theory (Dirac 1928a).[31] A relativistic wave equation had been published in 1925 and 1926 by Klein, by Fock, and by Gordon, among others.[32] In fact, Schrödinger's initial approach to his *Wellenmechanik* was based on a relativistic wave equation, but he abandoned that formulation because it did not yield the correct formula for the fine structure of hydrogen (Mehra and Rechenberg 1978, vol. 5). All these formulations of the quantum mechanical description of the motion of a free relativistic equation were based on the second-order differential equation

$$\left(\frac{1}{c^2} \frac{\partial}{\partial t^2} - \nabla^2 + \frac{m^2 c^2}{\hbar^2} \right) \psi = 0, \tag{1.6.2}$$

now usually called the Klein-Gordon equation. However, the interpretation of this relativistic equation—and its generalization for a particle interacting with a prescribed external electromagnetic field—was not consistent with the requirements of the general transformation theory. The reason for this is that the Klein-Gordon equation is of second order in $\partial/\partial t$, whereas Dirac's transformation theory is based on the equation

$$i\hbar\frac{\partial}{\partial t}\psi = H\psi. \tag{1.6.3}$$

The latter is of first order in $\partial/\partial t$ and yields a probability density, $|\psi|^2$, which is always positive. In contrast, the conserved "probability" for the Klein-Gordon equation is proportional to $\psi^*\partial_t\psi - \partial_t\psi^*\psi$, because the equation is of second order in $\partial/\partial t$, and is not always positive. There was thus a real difficulty in making quantum mechanics agree with relativity. This difficulty bothered Dirac very much, but he observed that "it did not seem to bother other physicists." Dirac had been "so strongly impressed by the beauty and the power of the formalism [of his transformation theory] . . . that [he] felt that one *had* to keep to this formalism and it would not do to pass over to a different kind of equation where [one] had $\partial^2/\partial t^2$ instead of $\partial/\partial t$." Dirac has recounted that Bohr came up to him during one of the conferences he attended in 1927 and asked him, "What are you working on now?"[33] Dirac remembered telling Bohr that he was working on the problem of attempting to find a satisfactory relativistic quantum theory of the electron and that Bohr had commented, "But Klein has already solved that problem." Dirac then tried to explain to Bohr why he was not satisfied with the solution of Klein because it involved a second-order equation in the time, but he was not able to do so because the conference reconvened just then and their discussion was cut short. This incident brought home to Dirac the fact that "so many physicists were quite complacent with a theory which involved a radical departure from some of the basic laws of quantum mechanics," and that "they did not feel the necessity of keeping to these basic laws" in the way that he felt. He remembered being quite disturbed that Bohr was "so satisfied" with the Klein-Gordon equation, "because of the negative probabilities that it led to" (Dirac 1963, p. 15). Dirac "worried over this point for some months," and then ultimately found a solution to the problem of the negative probabilities by deriving a new wave equation which now bears his name, the Dirac equation:

$$\left\{i\hbar\left(\frac{\partial}{\partial ct} + \alpha_1\frac{\partial}{\partial x_1} + \alpha_2\frac{\partial}{\partial x_2} + \alpha_3\frac{\partial}{\partial x_3}\right) + \alpha_m mc\right\}\psi = 0. \tag{1.6.4}$$

The ψ function in this Dirac equation has four components instead of just one as in the case of the Schrödinger or the Klein-Gordon equation; and the α's are 4×4 matrices that operate on the four components of ψ and anticommute with one another.

Heisenberg's reaction upon reading Dirac's papers summarizes that of the community: "I admire your last work about the spin in the highest degree."[34]

Dirac has related how he came to his equation for the electron in his interview with T. S. Kuhn: "A great deal of my work is just playing with equations and seeing what they give" (Dirac 1963). He had observed that the 2×2 Pauli matrices[35] that satisfied

$$\sigma_i \sigma_j + \sigma_j \sigma_i = 2\delta_{ij} \qquad (1.6.5)$$

implied that

$$\begin{aligned}(\sigma \cdot p)^2 &= (\sigma_1 p_1 + a_2 p_2 + \sigma_3 p_3)^2 \\ &= p_1^2 + p_2^2 + p_3^2. \end{aligned} \qquad (1.6.6)$$

"That was a *pretty* mathematical result [italics mine]. I was quite excited over it" (Dirac 1977c, p. 109). It implied that in some sense the square root of p^2 was $\sigma \cdot \mathbf{p}$. Dirac tried to generalize this result to a sum of four squares, so that the relativistic relation

$$p_1^2 + p_2^2 + p_3^2 + m_0^2 c^2 = E^2 = p_0^2 \qquad (1.6.7)$$

could be linearized. "It took me quite a while ... before I suddenly realized that there was no need to stick to ... σs with just two rows and columns. Why not go to four rows and columns?" (Dirac 1963). Eqs. (1.6.1) or (1.6.4) are the linearized version of (1.6.7). Note incidentally that here too Dirac's point of departure is the *classical* expression $\mathbf{p}^2 c^2 + m_0^2 c^2 = p_0^2$ (no factors involving h appear, even though the final equation will describe a particle of spin with an angular momentum $\hbar/2!$).

Already in the first paper in which he introduced his relativistic quantum theory of an electron, Dirac (1928a) announced that "it appears that the simplest Hamiltonian for a point-charge electron satisfying the requirements of both relativity and the general transformation theory leads to an explanation of all duplexity phenomena without further assumption." Dirac's equation not only accounted for the spin of the electron, and its observed gyromagnetic ratio and magnetic moment, but also correctly explained the fine structure of the hydrogen atom to the order calculated by Dirac.[36] In his interview with T. S. Kuhn, Dirac recounted why he did not attempt to solve exactly the equation he had obtained for the case of an electron in a Coulomb field:

DIRAC: When I first got that equation, of course I was very anxious
           to know whether it would work for the hydrogen atom, and
           I just tried it by an approximation method. I thought that
           if I got it anywhere near right with an approximation method,
           I would be very happy about that. It needed someone else,
           namely Darwin (and Gordon), to tackle that equation as an
           exact equation and see what the exact solutions were; I think

I would have been too scared myself to consider it exactly. I would be too scared that it would get unfortunate results which would compel the whole theory to be abandoned.

T. S. KUHN: That's fascinating; does this mean that you had yourself not tried to handle exactly before going to an approximation method?

DIRAC: That is so, yes.

KUHN: You looked for the approximation method from the start?

DIRAC: Yes, yes. Of course I had the fear that the whole theory was nowhere near right, and if I could get it approximately right, well my confidence would already be substantially increased in that way. It's just that one has a lack of confidence when one introduces something quite new. (Dirac 1963, p. 14)

Dirac has also noted that

> a person first gets a new idea and he wonders very much whether this idea will be right or wrong. He is very anxious about it, and any feature in the new idea which differs from the old established ideas is a source of anxiety to him. Whereas someone else who hears about this work and takes it up doesn't have the same anxiety, an anxiety to preserve the correctness of the basic idea at all costs, and without having this anxiety he is not so disturbed by the contradiction and is able to face up to it and see what it really means. (Dirac 1963, p. 14)

Both Dirac's approximate calculation and the exact calculation of Darwin and Gordon gave results for the hydrogenic spectrum in remarkable agreement with the then available experimental data. The derivation of the Sommerfeld-like formula for the spectrum of the hydrogen atom was a striking success of the Dirac equation. However, some of its other features were very troublesome. Besides the states of positive energy which seemed to agree closely with experiments, the Dirac equation also possessed solutions of negative energy. In the quantum theory—in contradiction to the classical situation—such states could not be ignored, for transitions to such states could not be ruled out.

Already in his first two papers dealing with the properties of the Dirac equation, Dirac was very much aware of the problem associated with the negative energy solutions. In his discussion of the difficulties connected with Gordon's interpretation of the probability density for the Klein-Gordon equation, Dirac noted:

> The second difficulty in Gordon's interpretation arises from the fact that if one takes the conjugate imaginary of the [KG]

equation one obtains

$$\left[\left(-\frac{W}{c} - \frac{e}{c}A_0\right)^2 + \left(-\vec{p} + \frac{e}{c}\vec{A}\right)^2 + m^2c^2\right]\psi = 0$$

which is the same as one would get if one put $-e$ for e. The [KG] wave equation thus refers equally well to an electron with charge e as to one with charge $-e$. If one considers for definiteness the limiting case of large quantum numbers one would find that some of the solutions of the wave equations are wave packets moving in the way a particle of charge $-e$ would move on the classical theory, while others are wave packets moving in the way a particle of charge e would move classically. For this second class W has a negative value.

One gets over the difficulty on the classical theory by arbitrarily excluding those solutions that have a negative W. One cannot do this on the quantum theory, since in general a perturbation will cause transitions from states with W positive to states with W negative. Such a transition would appear experimentally as the electron suddenly changing its charge from $-e$ to e, a phenomenon which has not been observed. The true relativity wave equation should thus be such that its solutions, split up into two non-combining sets referring respectively to the charge $-e$ and the charge e. (Dirac 1928a)

In a paper entitled "A Theory of Electrons and Protons," Dirac (1930a) amplified his previous observation and remarked that in quantum electrodynamics, "transitions can take place in which the energy of the electron changes from a positive to a negative value even in the absence of any external field, the surplus energy, at least $2mc^2$ in amount, being spontaneously emitted in the form of radiation." In this paper, which Dirac submitted to the *Proceedings of the Royal Society* on December 6, 1929, he took the next decisive step. He there pointed out that the negative energy difficulty is common to all relativistic theories since

$$\frac{W}{c} + \frac{e}{c}A_0 = \pm\sqrt{m^2c^2 + \left(\mathbf{p} + \frac{e}{c}\mathbf{A}\right)^2}, \tag{1.6.8}$$

and noted that "an electron with negative energy moves in an external field as though it carries a positive charge," which result had suggested to some people "a connection between the negative energy electron and the proton." But Dirac indicated that this would lead to the following paradoxes:

> **i)** A transition of an electron from a state of positive to one of negative energy would be interpreted as a transition of an electron into a proton, which would violate the laws of conservation of electrical charge.

> **ii)** Although a negative-energy electron moves in an external field as though it has a positive charge, yet, ... the field it produces must correspond to its having a negative charge, e.g. the negative energy electron will repel an ordinary positive energy electron although it is itself attracted to the positive electron.

> **iii)** A negative-energy electron will have less energy the faster it moves and will have to absorb energy in order to be brought to rest. No particles of this nature have ever been observed.

Dirac then put forward his famous suggestion: "Let us assume that there are so many electrons in the world, *that all the states of negative energy are occupied except perhaps a few of small velocity. . . . Only the small departure from exact uniformity, brought about by some of the negative-energy states being unoccupied, can we hope to observe . . . .* We are therefore led to the assumption that the holes in the distribution of negative electrons are the protons."

In order to give meaning to Maxwell's equation

$$div\mathscr{E} = -4\pi\rho$$

in the presence of the infinite density of negative energy electrons, Dirac proposed to use $\rho - \rho_{vac}$ instead of ρ. To the objection that the proton is much heavier than the electron, Dirac advanced the possibility that the interactions among the negative energy particles might change their mass.

The idea of the hole theory was suggested to Dirac by the chemical theory of valency in which one is used to the idea of electrons in an atom forming closed shells which do not contribute at all to the valency. One gets a contribution from an electron outside closed shells and also a possible contribution coming from an incomplete shell or hole in a closed shell. "One could apply the same idea to the negative energy states and assume that normally all the negative energy states are filled up with electrons, in the same way in which the closed shells in the chemical atom are filled up" (Dirac 1978, p. 50).

In 1929–1930, while writing his book on quantum mechanics, Dirac was also working on the description of the "chemical atom."[37]

The realization that there was no reasonable way to avoid the transition from positive to negative energy states forced Dirac to accept the picture of the vacuum as the state in which all negative energy states are filled. The vacuum as conceived by Dirac was a region of space that is in its lowest possible energy state. As one fills up the negative energy states, a lower and lower total energy is obtained. When all the negative energy states are filled, the lowest possible energy is obtained. "So this picture of the vacuum gives the vacuum as the lowest possible energy state of a region of space. And it is very reasonable from that point of view" (Dirac 1978, p. 19). One thus spoke of a sea of negative energy electrons, one electron in each of these states.[38] Late in life Dirac added: "It is a bottomless sea, but we do not have to worry about that. The picture of the bottomless sea is not so disturbing, really. We just have to think of the situation near the surface, and there we have some electrons lying above the sea that cannot fall into it because there is no room for them" (Dirac 1983, p. 51).

When he advanced his hole picture, Dirac was aware that the theory was symmetrical between positive and negative energies, so that the hole should have the same mass as the electron. But, Dirac recalled,

> at that time the only positively charged particle that was known was the proton. People believed that the whole of matter was to be explained in terms of electrons and protons, just those two particles. One needed only two particles because there were only two kinds of electricity, negative and positive. There had to be electrons for the negative electricity and protons for the positive electricity, and that was all. I just didn't dare to postulate a new particle at that stage, because the whole climate of opinion at that time was against new particles. So I thought this hole would have to be a proton. I was very well aware that there was an enormous mass difference between the proton and the electron but I thought that in some way the Coulomb force between the electrons in the sea might lead to the appearance of a different rest mass for the proton. So I published my paper on this subject as a theory of electrons and protons. (Dirac 1978, p. 20)

These ideas were communicated to both Bohr and Heisenberg by Gamow, who had visited Dirac in Cambridge in the fall of 1929. Bohr wrote Dirac on November 24, 1929, having heard that he had "made progress with the mastering of the hitherto unsolved difficulties in your theory of the electron." Bohr, too had been

> very interested in these problems and [had] thought that the difficulties in relativistic quantum mechanics might perhaps be connected with the apparently fundamental difficulties as regards conservation of energy in β-ray disintegration and the interiors of stars. My view is that the difficulties in your theory might be

said to reveal a contrast between the claims of conservation of energy and momentum on one side and the conservation of the individual particles on the other side. The possibility of fulfilling both these claims in the usual correspondence treatment would thus depend on the possibility of neglecting the problem of the constitution of the electron in nonrelativistic classical mechanics .... Only in regions where electronic dimensions do not come into play, the classical concepts should present a reliable fundament [*sic*, i.e., foundation] for the correspondence treatment.[39]

Dirac replied to Bohr on November 26: "The question of the origin of the continuous β-spectrum is a very interesting one and may prove a serious difficulty in the theory of the atom .... My own opinion of this question is that I should prefer to keep rigorous conservation of energy at all costs and would rather abandon even the concept of matter consisting of separate atoms and electrons than the conservation of energy."[40] In this same letter Dirac gave Bohr a synopsis of his paper outlining his hole theory, identifying holes with protons:

So long as one neglects interaction one has complete symmetry between electrons and protons. One could regard the protons as the real particles and the electrons as the holes in the distribution of −ve energy. However, when the interaction between the electrons is taken into account this symmetry is spoilt. I have not yet worked out mathematically the consequences of this interaction. It's the "Austausch" effect that is important and I have not yet been able to get a relativistic formulation of this. One can hope however, that a proper theory of this will enable one to calculate the ratio of the masses of proton and electron.

Bohr's answer to Dirac's letter was dated 5 December 1929, and in it he again stressed his belief that

on the whole it appears that the circumstance that hc/e^2 is large compared to unity does not only indicate the actual limit of the applicability of the quantum theory in its present form, but at the same time ensures its consistency within these limits. In fact the radius r_0 of the electron estimated on classical theory is $e^2/mc^2 = h/mc \cdot e^2/hc$, and we can therefore never determine the position of an electron with an accuracy comparable with r_0 without allowing an uncertainty in its momentum larger than mc, thus entailing an uncertainty in the energy surpassing the critical value mc^2. The idea that the reach of quantum mechanics is bound up with the actual existence of the electron would also seem to be in harmony with the fact that the symbols e and m appear in the fundamental equations of the present theory.

As far as the problem of β-ray spectra was concerned, Bohr felt that "we may now be outside this natural limit for the consistent applicability of the concepts of energy and momentum, and in this sense we may regard the expulsion of a β-ray from a nucleus as the birth of an electron as a dynamical individual ... [in which process we may have to abandon the conservation of charge]."[41]

Once again, Dirac answered promptly. On December 9 he wrote Bohr back:

I am afraid I do not completely agree with your views. Although I believe that the quantum mechanics has its limitations and will untimately be replaced by something better, (and that this applies to all physical theories) I cannot see any reason for thinking that quantum mechanics has already reached the limit of its development. I think it will undergo a number of small changes, namely with regard to its method of application, and by these means most of the difficulties now confronting the theory will be removed. If any of the concepts now used (e.g. potentials at a point) are found to be incapable of having an exact meaning, one will have to replace them by something a little more general, rather than make some drastic alteration in the whole theory.[42]

In his answer to Dirac's letter, Bohr retreated: "I quite agree with you that we have no reason to think that we have already reached the final limit of quantum mechanics. I am only inclined to take the present difficulties as indications that we have not yet obtained the proper expression for the correspondence with the classical electrodynamics. It is just in this connection that Klein and I suspect that the existence of the elementary electronic charge may perhaps prove as fundamental as the existence of the quantum of action."[43]

Bohr's response is to be contrasted with Heisenberg's reaction to hearing about Dirac's hole theory. On December 7, 1929, Heisenberg wrote Dirac: "I think I understand the idea of your new paper; it is certainly a great progress. But I cannot see yet, how the ratio of the masses etc. will come out. It seems to me already very doubtful, whether the terms of the electron (i.e. Sommerfeld formula) will not be completely changed by the interaction with the negative cells. One may hope, that all these difficulties will be solved by straight calculation of the interaction."[44]

A few weeks later Heisenberg again wrote Dirac after receiving a reprint of his paper: "I have calculated a little about the effect of interaction of the electrons in your theory. . . . One can prove that, electron and proton get the *same* mass. So I feel, that your theory [in identifying holes with protons] goes very far away from any correspondence to classical laws and also from experimental facts."[45] Pauli also objected to the interpretation that Dirac had given to the holes. Tamm informed Dirac that "at the Physical Congress in Odessa, . . . Pauli told us that he rigorously proved, that the system consisting of *m* positive energy electrons & *n* "holes" in the distribution of negative energy electrons, has the same energy as the system consisting of *m* holes & *n* electrons, the electrons having the velocities which previously belonged to the holes & vice versa. Pauli concludes that on your theory of protons the interaction of electrons can't destroy the equality of the masses of an electron & a proton."[46]

These objections, together with those published by Oppenheimer (1930b) and by Tamm (1930) on the rate of annihilation of protons and electrons into γ-rays[47] as predicted by hole theory, and the telling argument of Weyl[48] that inversion symmetry demanded that the hole have the same mass as the electron forced Dirac to the position that the antiparticle should correspond to something that was not observed in physics at the time: "A hole, if there were one, would be a new kind of particle, unknown to experimental physics, having the same mass and opposite charge to an electron. We may call such a particle an anti-electron" (Dirac 1931b, p. 61).

Dirac's original suggestion that protons were holes in a nearly filled sea of negative energy electrons is the only time that his physical intuition failed him in the period from 1925 to 1933. Dirac's explanation for his failure to postulate the existence of an anti-electron earlier, namely the conservatism of the theoretical community which would not have accepted a new kind of particle, is probably not the only reason.

When he first advanced the suggestion that protons were "holes," Dirac conceived his hole theory as a *unitary* theory of matter with "all matter built out of one fundamental kind of particle, instead of two, electrons and protons."[49] A unitary theory of matter held a certain attraction for Dirac.

In a talk before the British Association for the Advancement of Science in Bristol on September 8, 1930, Dirac indicated that Oppenheimer had recently put forward an idea that does get over these difficulties. Oppenheimer supposed that all and not merely nearly all of the negative energy states for an electron are occupied, so that a positive energy electron can never make a transition to a negative energy state. "There being no holes we can call protons, we must assume that protons are independent particles. The proton will now also have negative energy states, which we must again assume to be all occupied. The independence of the electron and proton according to this view allow us to give them any masses we please, and further there will be no mutual annihilation of electron and protons." But the difficulties had been gotten over "only at the expense of the unitary theory of the nature of electrons and protons." This unitary conception was not to be given up lightly as Dirac added: "At the present it seems too early to decide what the ultimate theory of the proton will be. One would like, if possible, to preserve the connection between the protons and electrons, in spite of the difficulties it leads to." Dirac's rejection of his identification of protons with holes in the face of Oppenheimer's, Weyl's, and Pauli's objections was stated in his 1931 paper on "Quantized Singularities in the Electromagnetic Field":

> It thus appears that we must abandon the identification of the holes with protons and must find some other interpretation for them. Following Oppenheimer, we can assume that in the world as we know it, all, and not nearly all, of the negative energy states are filled. A hole, if there were one, would be a new kind of particle, unknown to experimental physics, having the same mass and opposite charge to an electron. We should not expect to find any of them in nature, on account of their rapid rate of recombination with electrons, but if they could be produced experimentally in high vacuum they would be quite stable and amenable to observation. (Dirac 1931b)

Blackett—who was working in Cambridge at the time—told Dirac that he and Occhialini had evidence for this new kind of particle (Dirac 1977c, p. 21). Blackett did not want to publish his results without further corroboration, and "while he was obtaining the corroboration, Anderson quite independently published his evidence to show that the positron really existed."[50]

Dirac was of course quite pleased with the experimental verification of his prediction. He became actively involved with the interpretation of the cloud-chamber pictures that Blackett and Occhialini had obtained. In their paper Blacket and Occhialini thanked Dirac for "most valuable discussions [and] also for allowing us to quote the result of a calculation made by him [for the annihilation rate of positrons with electrons]" (Blackett and Occhialini 1933).

However, the conclusion of their paper reflects the cautious attitude of the Cavendish experimentalists toward Dirac's identification of holes with anti-electrons: "When the behavior of the positive electrons have been investigated in more detail, it will be possible to test these predictions of Dirac's theory. There appears to be no evidence against its validity, and in its favour is the fact that it predicts a time of life for the positive electron that is long enough for it to be observed in the cloud chamber but short enough to explain why it had not been discovered by other methods" (Blackett and Occhialini 1933, pp. 714–716). The skepticism of these experimentalists may have reflected Rutherford's disdain for theorists emblazoned in his famous quip: "They play games with their symbols but we in the Cavendish turn out the real facts of nature" (Blackett 1972, p. 58). Rutherford's attitude clearly also affected the theoreticians working there. Massey wrote of the anti-electron as an "out-of-this-world concept" when recollecting about research at the Cavendish in 1932 (in Hendry 1984, p. 95). But the skepticism was neither confined to the Cavendish nor to England.

In the summer of 1933, the Cal Tech physicists involved in "positron" research[51] asserted that the Dirac theory as interpreted by Blackett and Occhialini was not tenable: "The simplest interpretation of the nature of the interaction of cosmic rays with the nuclei of atoms lies in the assumption that when a cosmic-ray photon impinges upon a heavy nucleus, electrons of both signs are ejected from that nucleus ... the nucleus plays a more active role than merely that of a catalyst" (Anderson et al. 1934). Similarly, Bohr at first did not believe in Anderson's discovery, not even after Blackett had found clear evidence for electron pair production in cloud chamber photographs of cosmic ray showers. When the evidence became more and more convincing, one of his final remarks was: "Even if all this turns out to be true, of one thing I am certain: that it has nothing to do with Dirac's theory of holes!"[52]

Pauli likewise objected to Dirac's new version of his hole theory. Writing to Blackett on April 19, 1933, Pauli informed him:

> Yours and Occhialinis paper about the positive electron, is very interesting and the existence of the positive electron is very supported now by the paper of Meitner and Phillip in Naturwissenschaften. In this moment I come back to my old idea of the existence of a "neutrino" {that means a neutral particle with mass comparable with that of the electron; the italian name (in contrast to neutron) is made by Fermi}. If the positive and the negative electron both exist, it is not so phantastic to assume a neutral particle, consisting of both together.
>
> Further the paper of Sargent with the sharp upper limits of the energies in the β-spectra suggested to me again my old idea, that at every β-disintegration event a neutrino could be emitted and could save the conservation-law of energy (and momentum).
>
> What think the experimental physicists of the Cavendish laboratory *now* about those possibilities? Besides, I don't believe on the Dirac–"holes," even if the positive electron exist. (Pauli 1985 [307], p. 158)

Pauli wrote two weeks later to Dirac: "I do not believe on your perception of 'holes,' even if the existence of the 'anti-electron' is proved" (Pauli 1985 [308], p. 159).

The history of the discovery of the positron provides a striking illustration of the struggles to overcome ingrained ways of seeing the world (Hanson 1963, and especially de Maria and Russo 1985). Although in late summer 1932 Blackett and Occhialini, working at the Cavendish, had conclusive evidence for positrons in their cloud chamber pictures, and Blackett had had the benefit of extensive discussions with Dirac (1984c, p. 61),[53] Blackett and Occhialini were unable to overcome their "great fears." Only after Anderson had announced his discovery (Anderson 1932a,b) were Blackett and Occhialini—after repeating some of Anderson's procedures and confirming his results —"absolutely forced, independently of any theoretical presupposition, to admit the existence of a positive electron" (Occhialini 1933, quoted in de Maria and Russo 1985, p. 268).

In February 1933, Blackett and Occhialini communicated their findings to the Royal Society (Blackett and Occhialini 1933). A major part of their paper was devoted to the study of the "astonishing variety and complexity of [the] multiple tracks" they had observed and of the particles that produced them. They confirmed Anderson's "remarkable conclusion" that "some of the tracks must be due to particles with a positive charge but whose mass is much less than that of a proton." The last part of their paper was devoted to speculations about possible mechanisms responsible for the showers they had observed. These showers, Blackett and Occhialini asserted, were "almost certainly due to some process that involves the interaction of particles or photons of high energy with atomic nuclei." Three hypotheses were suggested about the origin of the positive and negative electrons: "They may have existed previously in the struck nucleus, or they may have existed in the incident particle, or they may have been created during the process of collision." The third one was considered the most reasonable: "One can imagine that negative and positive electrons may be born in pairs during the disintegration of light nuclei" (Blackett and Occhialini 1933, p. 714).

Although this last mechanism was suggested by Dirac's hole theory, and the authors were willing to accept Dirac's hole theoretic explanation for why positrons had not been observed previously—they annihilated with electrons to produce two or three photons—they were not willing to assert the correctness of Dirac's hole theory. They ended their paper with the statement: "When the behavior of the positive electrons have been investigated in more detail, it will be possible to test these predictions of Dirac's theory. There appears to be no evidence as yet against its validity, and in its favour is the fact that it predicts a time of life for the positive electron that is long enough for it to be observed in the cloud chamber but short enough to explain why it had not been discovered by other methods" (Blackett 1933a, pp. 714–716). Only in December 1933, upon reviewing all the experimental data on positrons available by then, did Blackett come out in favor of Dirac's theory: "These conclusions as to the existence and the properties of positive electrons have been derived from the experimental data by the use of simple physical principles. That Dirac's theory of the electron predicts the existence of particles with just these properties, gives strong reason to believe in the essential correctness of his theory" (Blackett 1933b, p. 918).

1.7 Postscript: Dirac and Scientific Creativity

The initial skepticism of the scientific community regarding antimatter reflected not only the startling novelty of Dirac's prediction, but also a distrust of his way of doing theoretical physics.

The reaction of two of the leading American theoretical physicists to Dirac's approach, if not quite typical, is nonetheless interesting to note. Kemble in 1933 wrote Birkhoff: "[Dirac] has always seemed to me to be a good deal of a mystic and that is, I suppose, my way of saying that he thinks every formula has a meaning if properly understood—a point of view which is completely repugnant to me and is one of the reasons that I have never been able to adopt his methods, as many other physicists have."[54]

Slater, in his scientific autobiography, compared Dirac's work to the activity of a magician, who "waves his hands as if he were drawing a rabbit out of a hat and who is not satisfied unless he can mystify his readers or hearers" (Slater 1975, p. 42).

I do not think that these delineations of Dirac are accurate. Dirac's writings are characterized by an amalgam of beauty and simplicity, and this is true not only of his scientific works. Central to the magic quality of his theoretical contributions is the role that mathematics played in them. In the fall of 1955, lecturing at Moscow University in answer to a request to state briefly his philosophy of physics, he wrote on the blackboard:

PHYSICAL LAWS SHOULD HAVE MATHEMATICAL BEAUTY.

His motto became an epigraph: it has been preserved to this day (Dalitz and Peierls 1986, p. 159). Earlier he had emphasized that "it often happens that the requirements of simplicity and beauty are the same, but where they clash, the latter must take precedence" (Dirac 1939a). For Dirac, simplicity and beauty are essential attributes of scientific truth. Thus he asserted: "The formalism [of nonrelativistic quantum mechanics] is so natural and beautiful as to make one feel sure of its correctness as the foundations of the theory" (Dirac 1940a).

Numerous examples can be given of Dirac's insistence on simplicity and beauty.[55] For example, he introduced his famous "delta-function" in order "to express in a concise form certain relations which we could if necessary, rewrite in a form not involving improper functions, *but only in a cumbersome way which would tend to obscure the argument*" (Dirac 1947a; italics mine). For Dirac, beauty and simplicity were also important factors in evaluating other people's work. He remained unconvinced by the renormalization procedures of Feynman, Schwinger, and Dyson because the theory that ensued did not meet these criteria: "Recent work by Lamb, Schwinger, Feynman and others has been very successful in setting up rules for handling the infinities and subtracting them away, so as to leave finite residues which can be compared with experiments, but the resulting theory is an *ugly and incomplete one* [italics mine], and cannot be considered as a satisfactory solution of the problem of the electron" (Dirac 1951).[56]

Dirac's belief in simplicity and beauty in fact motivated his research activities. A few years before his death Dirac revealed that "a good deal of my research work in physics has consisted in not setting out to solve some particular problem, but simply examining mathematical quantities of a kind that physicists use and trying to fit them together in an interesting way regardless of any application that the work may have. It is simply a search for pretty mathematics" (Dirac 1982a).[57] For Dirac, mathematics was synonymous with logic and precision. In one of his last scientific articles he noted that quantum electrodynamics was a

> theory in which infinite factors appear when we try to solve the equations. These infinite factors are swept into renormalization procedures. The result is a theory which is not based on strict mathematics, but is rather a set of working rules.
>
> Many people are happy with this situation because it has a limited amount of success. But this is not good enough. *Physics must be based on strict mathematics* [Dirac's italics]. One can conclude that the fundamental ideas of the existing theory are wrong. A new mathematical basis is needed. (Dirac 1984b)

To Dirac, "pretty" mathematics meant transparent, lucid mathematics, but still "strict" mathematics. His work is characterized by a lucidity whose consistency is unmatched. His creations are marked by certain childlike qualities: simplicity, purity, playfulness. He himself indicated that he obtained his famous relativistic wave equation by "playing around with three 2×2 matrices" (Dirac 1982a).

On several occasions Dirac attributed the success of his mathematical inventiveness as stemming from his ability to think "geometrically." In his "Recollections of an Exciting Era," he writes that he thought the tea parties of Professor Baker that he had attended regularly on Saturday afternoons while a research student at Cambridge had helped sharpen his "geometrical" way of thinking:

> At the end of the tea party someone would give a talk on some subject of geometry.... It was always [using] the methods of projective geometry ... and I was much impressed by the power of their methods....
>
> These tea parties did very much to stimulate my interest in the beauty of mathematics. The all important thing there was to strive to express the relationships in a beautiful form, and they were very successful. (Dirac 1977c, p. 114)

In his interview with T. Kuhn, he remarked that "I could settle down to algebra (the algebra of q numbers) when I had the basic ideas given, but to get new basic ideas I worked geometrically.... Once ideas are established, one can put them in algebraic form and one can proceed to deduce their consequences. That's just a question of algebra. The more important part is the getting of new ideas and that requires a geometrical mind, I believe" (Dirac 1963, 3d Kuhn interview). Commenting on how he had derived his relativistic equation Dirac told Kuhn: "I'd be thinking of wave functions ... forming some kind of density which one would picture spread about in space." When asked by Kuhn if by "pictorial models" he meant "something that interacts with mathematics but that is not simply mathematics," Dirac replied: "Some way which enables you to understand the equations independently of the approximate methods of solving the equations. I suppose I pictured the q-numbers as some kind of mysterious numbers which represented physical things.... The delta function came in just from picturing the infinity" (Dirac 1963, 5th Kuhn interview, p. 23).

Dirac's contribution to the developments of physics in the period from 1925 to 1933 were remarkable. In particular, the prediction of antimatter was a turning point in twentieth-century physics. In his Varenna lectures, Dirac noted that in order to make a prediction of this magnitude, "One must be prepared to follow up the consequences of [the] theory, and feel that one just *has to* accept the consequences, no matter where they lead" (Dirac 1977c, p. 1). Dirac had that courage to follow the consequences, no matter where they led. In that same lecture Dirac gave further insights into why he was able to accomplish what he did: "The research worker is only human and, if he has great hopes, he also has great fears. (I do not suppose one can ever have great hopes without their being combined with great fears)" (Dirac 1977c, p. 14). Dirac had the strength to overcome his great fears. He had the courage to venture onto entirely new ground, and to question and challenge accepted ideas. And as he tells us, "It is when one is challenging the main ideas that one has the great excitements and the great fears that something will go wrong" (Dirac 1977c, p. 65).

1.8 Fermi and the Regaining of *Anschaulischkeit*

Cambridge, Göttingen, and Copenhagen were not the only places where the formulation of QED was actively being pursued. In Rome, Fermi embarked on a study of the quantum theory of radiation during the winter of 1928. His initial step was to study Dirac's two papers on the subject[58] (Dirac 1927a,b). According to Amaldi (Fermi 1962, p. 305), the method used by Dirac "did not appeal" to Fermi. Fermi came to the conclusion that a simpler treatment of the quantization of the electromagnetic field was possible. In two papers written in 1929, Fermi (1929, 1930) presented his formalism, which indeed yielded a more satisfactory procedure for handling the subsidiary condition and the quantization of the theory. In particular, it justified the Hamiltonian that Dirac had postulated for QED: namely, a Hamiltonian which described the electromagnetic field in terms of (transverse) field operators $\mathcal{A}$ (satisfying div $\mathcal{A} = 0$), and in which the Coulomb interaction is essentially put in "by hand." These results, together with important applications, were the content of a course Fermi gave in April 1929 at the Institut Poincaré in Paris.

In the summer of 1930 Fermi was invited to the Summer Symposium on Theoretical Physics at the University of Michigan where he delivered a series of lectures on the "Quantum Theory of Radiation." These lectures were subsequently published in the *Reviews of Modern Physics* (Fermi 1932). They were very influential. It was from Fermi's *RMP* article that most of the more pragmatically inclined and experimentally oriented theoretical physicists—theorists like Bethe—learned their quantum electrodynamics. Written in the characteristically "simple" Fermi style, it presented an intuitive, transparent formulation of QED. It included a derivation of time-dependent perturbation theory, and by the use of telling examples (e.g., the propagation of light in vacuum, theory of Doppler effect) shed light on the working of the theory. It also included a fairly thorough presentation of the properties of the solutions of the Dirac equation, and how calculations were to be carried out when the electron is described by the Dirac equation. Fermi's closing remarks were: "In conclusion we may therefore say that practically all the problems in radiation theory which do not involve the structure of the electron have their satisfactory explanation, while the problems connected with the internal properties of the electron are still far from their solution." In Fermi's conceptual scheme, electrons were particles. He took the variables describing the electromagnetic field to be the Fourier expansion coefficients of the scalar and vector potentials (V, U):

$$(A_0 =) V = \sqrt{\frac{8\pi}{\Omega}} \, c \sum_s Q_s \cos\left(\frac{2\pi\boldsymbol{\alpha}_s \cdot \mathbf{X}}{\lambda_s} + \beta_s\right) \tag{1.8.1a}$$

$$(\mathbf{A} =) \mathbf{U} = \sqrt{\frac{8\pi}{\Omega}} \, c \sum_s \mathbf{q}_s \sin\left(\frac{2\pi\boldsymbol{\alpha}_s \cdot \mathbf{X}}{\lambda_s} + \beta_s\right) \tag{1.8.1b}$$

> These coefficients (Q_s; $\mathbf{q}_s$), which are of course functions only of the time, describe the scalar and the vector potential at each moment throughout the whole field that is being considered. (Fermi 1932, 1962, p. 306)

Fermi assumed that the vector and scalar potential obeyed the wave equation $\Box A_\mu = j_\mu$. He then deduced the equation of motions satisfied by the expansion coefficients. He also derived the relation that holds between the longitudinal and the scalar components of a given frequency as a consequence of the Lorentz condition

$$(\partial_\mu A^\mu) = \text{div } \mathbf{U} + \frac{1}{c}\frac{\partial V}{\partial t} = 0, \tag{1.8.2}$$

which must hold in order for the equations $\Box A_\mu = j_\mu$ to be equivalent to Maxwell's equations $\partial^\mu F_{\mu\nu} = j_\mu$. A Hamiltonian was then written down that yielded the equations for both the charged particle variables and for the Q_ss and the q_ss. Fermi ended his first note with this statement:

> Having thus written down the equations of motion of electro-dynamics and of the charges in canonical form, it remains to translate them into equations of quantum mechanics. For this it is sufficient to consider the Hamiltonian expression as an operator in which, according to the usual rules, the momenta are equivalent to the operation of differentiation with respect to the corresponding coordinate and to multiplication by $-\frac{h}{2\pi i}$. Schrödinger's equation will then be the following

$$H\psi = \frac{h}{2\pi i}\frac{\delta\psi}{\delta t} \tag{1.8.3}$$

> We shall demonstrate the application of this equation in subsequent notes.

In his next note, "On the Quantum Theory of Interference Fringes," Fermi showed how his formulation of QED accounts for interference phenomena by considering the theory of Lippman fringes and concluded "that one can well say that it includes all the properties of radiation in it."

What is striking about Fermi's approach is its matter-of-fact, pragmatic approach. Time-dependent perturbation theory is straightforwardly applied to the Schrödinger equation obtained from the previously derived Hamiltonian. There is no mention of wave-particle duality, nor of any of the other conceptual difficulties that had concerned Jordan or Dirac. The ability of the formalism to explain key experiments in a simple, lucid fashion is the primary requirement Fermi imposed on his formalism. Thus, in the treatment of Lippman fringes, Fermi considered two atoms, A and B. Atom A at time $t = 0$ is in an excited state, and atom B, which is separated from atom A by a distance r, is initially in its ground state. Fermi then computed the probability that at time t atom A is in its ground state and atom B is in an excited state. He first of all demonstrated that no transition could occur before a time $t = r/c$ had elapsed—this is the causality condition—and showed that the excitation probability as a function of time depended on the position of atom B according to the classical theory of standing waves. "We thus completely rediscover the results given by the classical theory of interference," Fermi asserted.

Pauli studied the May 1929 QED paper of Fermi and commented in a letter to Jordan "that it was completely independent of the work that Heisenberg and I wrote and methodologically interesting (even though it doesn't bring new results)" (Pauli 1979 [238], p. 525).

In their second paper on the quantization of wave fields, Heisenberg and Pauli devoted a long note to Fermi's papers (Heisenberg and Pauli, 1930). They translated Fermi's approach into a Lagrangian formalism, and noted that the Lagrangian

$$L = -\frac{1}{4} F_{\mu\nu} F^{\mu\nu} - \frac{1}{2}(\partial_\mu A^\mu)^2 - j_\mu A^\mu \qquad (1.8.4)$$

yields the equations of motion $\Box A_\mu = j_\mu$. Moreover with this Lagrangian $\pi_0 = \partial L / \partial(\partial_0 A_0) = -\partial_\mu A^\mu \neq 0$ so that a Hamiltonian can be constructed (which explains why Fermi could write down a Hamiltonian for the equations $\Box A_\mu = j_\mu$). Heisenberg and Pauli then went on to stress that the Lorentz condition $\partial_\mu A^\mu = 0$ (necessary for $\Box A_\mu = j_\mu$ to be equivalent to Maxwell's equations) cannot be imposed as an operator identity but only as a supplementary condition selecting admissible state vectors; state vectors describing physically realizable states of the electromagnetic field must satisfy $(\partial_\mu A^\mu)\Psi = 0$. In a third communication on quantum electrodynamics written after he had read the two papers of Heisenberg and Pauli on the quantum theory of wave fields, Fermi (1930) commented that a recent paper of Heisenberg and Pauli had treated the problem of quantum electrodynamics: "However since the methods followed by these authors are essentially different from mine, it is not, I think, useless to publish my results also." Fermi noted that in his previous two communications he had treated charged particles nonrelativistically. This shortcoming was now remedied by describing each spin 1/2 charged particle by a Dirac equation, so that the Hamiltonian referring to the unperturbed motion of the charged particles was given by

$$H_0^m = -c \sum_i \gamma_i \cdot p_i - \sum_i \delta m_i c^2 \qquad (1.8.5)$$

[γ_i, δ are the Dirac matrices in Fermi's notation ($\gamma = \alpha; \delta = \beta = \alpha_4$)]. The basis of his method was then succinctly stated: the electromagnetic field is described by a vector potential $U(= A)$ and a scalar potential $V(= A_0)$. Maxwell's equations $\partial^\mu F_{\mu\nu} = j_\nu$, are equivalent to the wave equations $\Box A_\mu = j_\mu$ only if the A_μ satisfy the Lorentz gauge condition $\partial^\mu A_\mu = 0$ (again the more modern notation $U^i = A^i, V = A_0$ has been used). If one imposes the boundary condition that at $t = 0 \, \partial^\mu A_\mu = 0$ and $\partial_t(\partial^\mu A_\mu) = 0$, then $\partial^\mu A_\mu = 0$ for all $t > 0$ by virtue of the fact that $\Box \partial^\mu A_\mu = \partial_\mu j^\mu = 0$, the right-hand side of this equation being the statement of charge conservation.

Following Heisenberg and Pauli, the equation $\partial^\mu A_\mu = 0$ is weakened to a subsidiary condition on admissible state vectors. Once again the Hamiltonian from which the equations of motion follow was written down. Fermi then showed how to eliminate the scalar and longitudinal oscillators and derived an effective Hamiltonian that involved only transverse oscillators, but in addition contained a Coulomb interaction term between the charged particles.

2. The 1930s

2.1 Introduction

In this chapter I survey some of the developments in quantum field theory during the 1930s. My exposition is not meant to be a history of the field during that period.[1] The reader is referred to the accounts by Wentzel (1960), Darrigol (1982), and Pais (1986) for more complete and balanced presentations, and to the remarkable Pauli correspondence of the thirties (Pauli 1985) in order to appreciate the heroic efforts of the leading protagonists, particularly Heisenberg and Pauli. My aim is to give an overview of the problems that were encountered, to sketch some of the solutions that were suggested, and to delineate the directions charted by the state of knowledge at the end of the thirties. This in order to make more understandable the developments in the 1947–1950 period.

The chapter is organized as follows. Section 2.2 structures the history of field theory during the thirties in terms of the two approaches that guided many of the efforts: quantum field theory and hole theory. Here I point out some of the landmarks and raise the question of why the problems that were solved in the 1945–1950 period had not been solved during the thirties, even though the materials were at hand to do so. I then turn to three conferences at the end of the decade to present the problems the community was addressing: the Warsaw conference of 1938, the Solvay conference of 1939, and the Washington conference of 1941. In the remainder of the chapter I look more closely at the divergence difficulties—the main obstacles in the development of the theory.

2.2 QED during the 1930s

Broadly speaking, two research programs operated during the thirties (see table I.1 in the Introduction). The one I labeled "Quantum Theory of Fields" emerged from the de Broglie-Schrödinger-Jordan conceptualization of fields and waves as the primary entities. It received its relativistic generalization and its formalization in the Heisenberg-Pauli papers (Heisenberg and Pauli 1929, 1930). In this approach matter is given a field description: the (scalar) Klein-Gordon wave field is used to describe spin zero particles, the (spinor) Dirac wave field for 1/2 particles. Later in the decade higher spin wave equations were introduced (Dirac 1936b; Proca 1936; Wigner 1939; Fierz 1939). All these wave equations were assumed to describe matter fields in the same way as Maxwell equations described the electromagnetic field. These fields (including the electromagnetic) were then quantized, thereby exhibiting the quanta of energy and momentum—the particles—associated with these fields. A reinterpretation of the spin 1/2 theory that incorporated the Dirac definition of the vacuum and explicitly exhibited the symmetry between positive and negative charges was given at the beginning of the decade by Fock (1933), Furry and Oppenheimer (1934), and Heisenberg (1934b). Their approach allowed one to describe electrons and positrons without ever referring to "holes" or to a "sea." Yet their suggestion was not generally adopted by those physicists who were calculating the observable consequences of QED until much later.

Quantum field theory had many successes during the thirties. All these advances took as their point of departure insights gained from quantum electrodynamics. The concept of emission and absorption of photons became generalized to that of particle creation and annihilation, and became a central feature of quantum field theoretic descriptions. The discovery in 1932 of the positron by Anderson (1932a,b), the corroboration of this finding by Blackett and Occhialini (1933; see also Blackett 1933b) and the identification of this particle with the "holes" of Dirac's theory (de Maria and Russo 1985) was an important step in this process.

The discovery of the positron and the discovery earlier that year of the neutron (Chadwick 1932a,b) overcame some of the previous prejudices of the community against new particles. It is interesting to note that a generational conflict had been at work (de Maria and Russo 1985).

Bohr and Rutherford—the two towering figures in "fundamental" theory and experimentation—initially opposed the identification of positrons with Dirac's holes. It was the younger theorists and experimenters—in particular Dirac (1934a), Oppenheimer (Oppenheimer and Plesset 1933), and Blackett (1933b)—who were responsible for establishing the new outlook.[2] Oppenheimer early in 1933 asserted that "the experimental discovery of the positive electron gives us a striking confirmation of Dirac's theory of the electron, and of his most recent attempts to give a consistent interpretation of that theory" (Oppenheimer and Plassett 1933). After 1933 the creation and annihilation of electron-positron pairs became an accepted way of theorizing about the interaction of photons with matter. Thus Weisskopf in the mid-1930s emphasized that "one of the most important results in the recent developments of electron theory is the possibility of transforming electromagnetic field energy into matter. A light quantum, for example, in the presence of other electromagnetic fields in empty space, can be absorbed and transformed into matter, with the creation of a pair of electrons with opposite charge" (Weisskopf 1936, p. 3).

But another factor—of equal if not greater importance—was responsible for the acceptance of the new viewpoint. By the end of 1933 positrons were being copiously produced in the laboratory and were being observed at "the rate of thirty thousand a second" (Darrow 1934). They were being focused in beams, their charge-to-mass ratio was measured, and their interaction with matter was being investigated (Thibeau 1934). Positrons were being "manipulated" in the laboratory—and their "reality" thus established (Hacking 1983). Moreover, it was being established that the theoretical calculations using Dirac's hole theory (the "representing" of these new entities) agreed with the careful experiments (the "intervening" and manipulating of these particles) that were being carried out to test the predictions of hole theory (Hacking 1983). Positrons and anti-electrons thus became identified with one another, and the hole theoretical description—in which pair production and annihilation was a central motif—became the accepted way to describe the interaction of photons with electrons.

Let me list some of the landmarks in the developments of quantum field theory during the 1930s:

1. The description of the interaction between charged particles as mediated by the exchange of photons (Bethe and Fermi 1932). Quantum field theory gave rise to a new view of how forces are generated between particles. It led to a conceptualization of the interaction between two charged particles as arising not from the creation of (continuous) electromagnetic fields that act on one another as in the classical description, but from the exchange of photons, which continually are "passed" from one another. In absorbing or emitting a photon, the momentum of a charged particle is altered; it is precisely this change per unit time of the momentum of the charged particle that is called a force. Similarly, other kinds of forces can be produced by exchanging other kinds of "virtual" particles; thus the exchange of gravitons (spin 2, mass 0 quanta) account for the gravitational interaction.

2. The quantization of the charged Klein-Gordon field (Pauli and Weisskopf 1934) which demonstrated the possibility of pair production without "hole theory." Pauli called it the "anti-Dirac" theory (Pauli 1935–1936).

3. Fermi's theory of β-decay (Fermi 1933, 1934a,b).

4. Yukawa's theory of nuclear forces (Yukawa 1935) which suggested that the short range forces between nucleons could be explained by the exchange of finite rest mass (spin zero) bosons between them (Brown 1982, 1985).

5. Pauli's proof of the connection between spin and statistics, namely that particles with zero or integer spins must obey Bose statistics, whereas those with odd half-integer spin had to obey Fermi statistics (Pauli 1940). For his proof Pauli assumed that the particles were described by a relativistic quantized field. The requirement that the total energy of the field system have a bounded lower limit implied that odd-half integer spin fields must be quantized using anticommutation rules and therefore were governed by Fermi statistics, whereas integer spin fields must be quantized using commutators, and hence obeyed Bose statistics. Already in his paper with Weisskopf on the quantization of the charged scalar field (Pauli and Weisskopf 1934) and more explicitly in the detailed account of this work that he gave in lectures in 1936 at the Institut Henri Poincaré (Pauli

1936), Pauli addressed the question whether the Klein-Gordon spin zero field could be quantized using anticommutation rules, so that the resulting "particles" would obey Fermi statistics. In his 1936 lectures Pauli had reached the conclusion that it was impossible to simultaneously satisfy the requirements of relativistic invariance and the condition that the charge density operators commute for spacelike separations when the theory is quantized with anticommutators. In 1939 Pauli's assistant, Markus Fierz, gave a description of free fields of arbitrary spins, and noted that classically the energy of the field system is positive for integral spin and indefinite for half-integral spin and that only quantization with anticommutation rules could make the energy positive semidefinite in the half-integral spin case (Fierz 1939). Pauli, in his 1940 paper on the connection between spin and statistics, made use of Fierz's insights and of his work with Fierz (Fierz and Pauli 1939) and with Belifante (Pauli and Belifante 1940) in giving a general "proof of the indefinite character of the charge in the case of integral, and of the energy in the case of half-integral spin" (Pauli 1941). By postulating "that all physical quantities at finite distances exterior to the light cone . . . are commutable" (the causality requirement), Pauli proved that "for integral spin the quantization according to the exclusion principle is not possible," and that quantization with commutors would leave the energy indefinite for fields with half-integral spin, an unacceptable situation.

Fermi's theory of β-decay was an important milestone in the history of quantum field theory. His paper put into sharp focus the advantages of a field theoretic approach: quantum field theory could easily give a description of phenomena in which the number of particles was not conserved. The introduction of his paper was explicit on this point:

> Besides the difficulty of the continuous energy distribution, a theory of the β rays faces still another essential difficulty in the fact that the present theories of the light particles do not explain in a satisfactory manner how these particles could be bound in a stable or quasi-stable manner inside a nucleus, considering the smallness of its volume.
>
> The simplest way for the construction of a theory which permits a quantitative discussion of the phenomena involving nuclear electrons, seems then to examine the hypothesis that the electrons *do not exist as such in the nucleus before the β emission occurs, but that they, so to say, acquire their existence at the very moment when they are emitted* [italics mine]; in the same manner as a quantum of light, emitted by an atom in a quantum jump, can in no way be considered as pre-existing in

the atom prior to the emission process. In this theory, then, the total number of the electrons and of the neutrinos (like the total number of light quanta in the theory of radiation) will not necessarily be constant, since there might be processes of creation or destruction of those light particles.

According to the ideas of Heisenberg, we will consider the heavy particles, neutron and proton, as two quantum states connected with two possible values of an internal coordinate ρ of the heavy particle. We assign to it the value $+1$ if the particle is a neutron, and -1 if the particle is a proton.

We will then seek an expression for the energy of interaction between the light and heavy particles which allows transitions between the values $+1$ and -1 of the coordinate ρ, that is to say, transformations of neutrons into protons or vice-versa; in such a way, however, that the transformation of a neutron into a proton is necessarily connected with the creation of an electron which is observed as a β particle, and of a neutrino; whereas the inverse transformation of a proton into a neutron is connected with the disappearance of an electron and a neutrino. . . .

The simplest formulation for a theory in which the number of the particles (electrons and neutrinos) is not necessarily constant is available in the method of Dirac-Jordan-Klein of the 'quantized probability amplitudes'. In this formalism, the probability amplitudes ψ of the electrons and ϕ of the neutrinos, and their complex conjugates $\psi*$ and $\phi*$, are considered as non-commutative operators acting on functions of the occupation numbers of the quantum states of the electrons and neutrinos. (Fermi 1934b)

Fermi's β-decay theory made clear that quantum field theory allowed a particularly simple and elegant description of phenomena in which particles were created or annihilated. The simplicity of Fermi's account of the β-decay process in terms of an interaction Hamiltonian

$$g_i \psi_p^*(x)\, \Gamma_i(x)\, \psi_e^*(x)\, \Gamma_i \psi_\nu(x) + \text{hermitian conjugate} \qquad (2.2.1)$$

written in terms of the creation and annihilation operators for the various particles (neutron, proton, electron, anti-neutrino) that take part in the basic reaction[3]

$$n \rightarrow p + e^- + \bar{\nu} \qquad (2.2.2)$$

should be contrasted with the attempt to describe the process (2.2.2) "hole theoretically." In that language one would have to say that in a transition from a negative energy state to a positive energy state a neutrino "becomes" an electron (Brown 1978)![4] It is not surprising that as new particles were introduced or discovered during the thirties (e.g., the various kinds of mesons to account for properties of nuclear forces) field theory was used for their description.

Fermi's theory was important for another reason. It was a theory that described nuclear β-decay in terms of particles that were assumed to be "elementary": neutrons, protons, electrons, and neutrinos (and their antiparticles). Furthermore, the meaning of "elementarity" was explicit: a particle is "elementary" if the Lagrangian of the system is expressed in terms of its associated field operators.

By the late 1930s, many features of the formalism of quantum field theory were fairly well understood. Pauli reviewed the state of affairs in the *Reviews of Modern Physics* in 1941. But it was Wentzel's *Einführung in der Quantentheorie der Wellenfelder* (1943), in which a particularly lucid and elegant account of relativistic quantum field-theories was given, that disseminated the field-theoretic approach to a wide audience after World War II. In particular, Wentzel made clear the meaning of quantizing the Dirac field in a charge-symmetric fashion. As a result of his exposition it became standard to regard the *quantized* Dirac equation as describing particles and antiparticles: all references to the "sea" had disappeared. The ("bare") vacuum was the state with no particles and no antiparticles present.

Fermi's theory of β-decay, Pauli and Weisskopf's quantization of the Klein-Gordon field which yielded a description of positively and negatively charged spinless particles, and Yukawa's theory of nuclear forces were examples of the range of possibilities that could be encompassed by quantum field theory. In the case of spin 1/2 charged particles interacting with the electromagnetic field, the field theoretic description (in terms of a quantized Dirac field) competed with an alternative approach. This second approach—which I have labeled "hole-theoretic QED"—regarded electrons as "particles" to be described quantum-mechanically by wave functions in configuration space. A single electron was assumed to be described relativistically by a wave function obeying the Dirac equation, to which was added the Dirac postulate that the vacuum was that state in which all negative energy states were occupied. Hence a single electron was described by a positive energy solution of the Dirac equation. Pair creation corresponded to a transition of a "negative energy" electron to an unoccupied "positive energy" state—rather than the creation de novo of a positron and an electron. Although equivalent to the field-theoretic approach for problems dealing with a single electron, the hole-theoretic formulation was much more complicated for multielectron systems. A system of N-electrons was to be described by an antisymmetric wave function in a $3N$ dimensional configuration space. (How to deal with the occupied negative energy states in this case was somewhat ambiguous.) Hole theory was a "quasi" field theory in that it recognized that a one-particle interpretation of the Dirac equation was impossible and that a consistent quantum-mechanical interpretation required it to be viewed as dealing with a denumerable infinity of "particles" (by virtue of the "sea" of occupied negative energy states). It was a "particle" theory in that particles were the primary entities and no reference was made to a quantized matter held. The theory which treated the interaction between charged spin 1/2 particles and the electromagnetic field—hole-theoretic QED—was a "particle" theory as far as matter was concerned but a "field" theory in its description of radiation. The fact that the electromagnetic field has a classical limit in which everything is measurable and that its behavior in that limit could be accounted for by the formalism was the justification of the quantized field-theoretic approach in that case. Heitler in 1936 summarized the hole-theoretic QED approach in his book, *The Quantum Theory of Radiation*. (A slightly revised edition appeared in 1944.) Until the appearance of Feynman's articles (1949a,b) Heitler's books were the primary and standard sources for learning how to "calculate" quantum electrodynamics processes.

It should be stressed that in its limited domain—the electromagnetic interactions of charged spin 1/2 particles—hole-theoretic QED was quite a successful theory. Most of the predictions of quantum electrodynamics during the 1930s— pair production, bremsstrahlung, Compton scattering—as well as the verification of the validity of the theory up to energies of the order of 137 mc^2 (and even greater) were based on hole-theoretic QED.

But the successes did not come easily. Although the spectroscopic data on the hydrogen atom were initially seen as a rousing confirmation of Dirac's relativistic wave equation for the electron, other consequences of hole-theoretic QED were more difficult to verify. The history of the quantitative corroboration of the energy dependence of the cross section predicted by QED to lowest order of perturbation theory for Compton scattering, for bremsstrahlung, for pair production, and of the formulas for the energy loss of high-energy charged particles in their passage through matter is an involved one, intimately related to the history of the discovery of the positron and of the μ-meson (Cassidy 1981; Galison 1983; de Maria and Russo 1985; see also Brown and Hoddeson 1983). The case of the scattering of γ-rays by electrons was typical. The cross section for the scattering of γ-rays from free electrons was calculated by Klein and Nishina in the fall of 1928 (Klein and Nishina 1929), and they used Dirac's equation to represent the electrons. The calculated cross section gave results in good agreement with experiments for low-energy X rays. The success of this calculation played an important role in the acceptance of the Dirac equation to describe charged spin 1/2 particles.[5] However, for the scattering of the most energetic photons then known—the 2.62 Mev γ-rays emitted by Th C''—by electrons in high Z elements (e.g., lead), the data indicated a large discrepancy: the observed scattering was much larger than predicted by the Klein-Nishina formula (Gray 1929; Tarrant 1930; Chao 1930a,b).[6] This anomaly eventually became known as the Meitner-Hupfeld effect (Brown and Moyer 1984). Until 1933 it was thought that the electrons that were believed to be present in the nucleus were responsible for the discrepancy. It was only after the discovery of the positron, and the identification of the positron with Dirac's antielectron, that the effect was recognized as resulting from the combined effect of the productions of electron-positron pairs, the subsequent annihilation of the positron, and of bremsstrahlung.

The discovery of the positron and of associated phenomena such as pair production and showers marked a turning point in the history of quantum electrodynamics.[7] It also marked a turning point in the evolution of cosmic ray physics. Thereafter the number of workers in that subdiscipline, the numbers of papers published, and the amount of experimental data to be accounted for by ambitious theorists increased dramatically.[8] De Maria and Russo put it succinctly: "Cosmic-ray [physics] ceased to be cosmic physics and became the experimental side of elementary particle physics" (de Maria and Russo 1985, p. 282). Cosmic ray physics up to that time had been a rather marginal field. The problem of the origin of cosmic rays had allowed ample freedom for bold speculations and for some, in particular Millikan (Kargon 1981, 1982; Galison 1983), it was the source of inspired religious rhetoric.

But the successes of hole theory also highlighted its shortcomings.

Only effects predicted by lowest-order perturbation theory could be compared with experiments, since both field-theoretic and hole-theoretic QED were beset by overwhelming divergence difficulties that manifested themselves in higher-order calculations (Weinberg 1977; Pais 1986). These difficulties had been recognized from the outset. The divergences associated with the self-energy of the electron were pointed out in the first field-theoretic papers by Heisenberg and Pauli. Thus, after completing his first paper with Heisenberg on the quantum theory of wave fields, Pauli wrote Bohr in July 1929: "I am *not* very satisfied with my and Heisenberg's theory (although I believe that it exhibits certain traits 'of a future exact theory.' In particular, the self energy of the electron makes much bigger difficulties than Heisenberg had thought at the beginning. Also the *new* results to which our theory leads are very suspect and the risk is very great that the entire affair loses touch with physics and degenerates into pure mathematics" (Pauli 1979 [231], p. 513).

These divergence difficulties were responsible for the fact that most of the theorists working on quantum electrodynamic problems had little faith in QED. Most of them believed that quantum electrodynamics would break down at energies of the order of $137 \, mc^2$ (10^8 volts), that is, at wavelengths of the order of the classical electron radius. Thus Bohr wrote to Dirac in 1930: "I have been thinking a good deal of the relativity problems lately and believe firmly that the solution of the present troubles will not be reached without a revision of our general physical ideas still deeper than that contemplated in the present quantum mechanics."[9] Bohr made similar statements throughout the decade. His approach was conditioned by his earlier success with the correspondence principle. For Bohr the QED of the early thirties had the same validity as the old quantum theory during the twenties. It was a provisional theory with a limited domain of applicability. The smallness of $\alpha = \frac{e^2}{hc}$ was responsible for the validity of some of its result; the divergences gave proof that the theory could not be trusted beyond its limits of applicability. In the mid-thirties Bohr wrote Dirac that it would be reasonable to assume that "the relativistic wave equation fails for energies of order 137 mc^2."[10] At the close of the decade he concluded a lecture on causality with the statement that because of the divergence difficulties he believed physics was "confronted with the necessity of a still more radical departure from accustomed modes of description of natural phenomena" than had been the case with quantum mechanics (Bohr 1939).

By the mid-thirties Dirac was ready to give up QED "without regrets." Moreover he indicated that "because of its extreme complexity, most physicists will be glad to see the end of it" (Dirac 1937). Even such "pragmatic" physicists as Bethe and Heitler despaired. Upon finding that the formula they had calculated for the energy loss of fast electrons passing through matter failed to agree with Anderson's experiments, they asserted: "One should not expect that ordinary quantum mechanics which treats the electron as a point-charge could hold under these conditions [i.e., when the electron has energy greater than 137 mc^2, and its wave length is smaller than the classical radius of the electron]." In fact they went on to state that they believed that the energy loss of fast electrons "provides the first instance in which quantum mechanics apparently breaks down for a phenomenon outside the nucleus" (Bethe and Heitler 1934).

Heisenberg in 1935 wrote Pauli "with respect to QED we are still at the stage in which we were in 1922 with regard to quantum mechanics. We know that everything is wrong. But in order to find the direction in which we should depart from what is extant we must know the consequences of the existing formalism much better than we do" (Pauli 1985 [407], p. 386). Heisenberg in 1936 concluded his paper with Euler on higher-order effects in quantum electrodynamics with the remark that "the present theory of the positron and quantum electrodynamics must be considered provisional" (Heisenberg and Euler 1936).

During the academic year 1935/36, Pauli visited the Institute for Advanced Study in Princeton and ran a seminar on "The Theory of the Positron and Related Topics." He summarized his views in a lecture on "The Foundations of Quantum Theory." Commenting on Dirac's theory of the electron, he suggested that "in this attempt the success seems to have been on the side of Dirac rather than of logic." Pauli furthermore indicated that he believed "that quantum theory is always successful when describing systems with a finite number of degrees of freedom, but when dealing with systems possessing infinitely many degrees of freedom it causes divergent results to appear.... The theory of holes postulates an infinite number of electrons, and therefore comes into the same category." He went on:

> It seems to me that our present methods are not fundamental enough, and there are two possibilities for overcoming the difficulties. The first is to change our concept of space and time in small regions. The second, to change the concept of state for systems with an infinite number of degrees of freedom.... I believe that the development of the theory along the correct lines

will then lead to a numerical value of the fine-structure constant $\alpha = e^2/\hbar c = \frac{1}{137}$, [11] and to an explanation of the fact that arbitrarily high masses do not appear concentrated in a given space region in nature. It seems likely that the future theory will be unitary in the sense that the duality of light and matter will disappear. By this I do not claim that we shall necessarily explain one in terms of the other, but perhaps both in terms of some more fundamental concept." (Pauli 1935–1936, pp. x–xi)

An ambiguous attitude toward quantum electrodynamics and quantum field theory was not a uniquely European phenomenon.[12] Addressing the University of Pennsylvania Bicentennial Conference in 1940, Oppenheimer indicated that his aim was "to explain why the theory of the mesotron, and more generally the quantum theory of fields, has failed so completely to deepen our understanding of nuclear forces and processes ... ; and to try to explain, too, why in spite of this the quantum theory of fields still seems a subject worth reporting on at all." Oppenheimer adduced the success of field theory in explaining such electrodynamic processes as pair production and cosmic ray showers as one of the reasons. Another was the discovery of the mesotron. Oppenheimer noted that even though "the discovery of the mesotron has ... sharpened all the difficulties of field theory; it has also given us some confidence that the fundamental ideas of the theory are right; for the mesotron was a prediction, very general and qualitative it is true, of this theory" (Oppenheimer 1941, p. 39).

Wentzel, an active and important participant in many of the developments of QED from 1925 on, recalled that "in spite of all failures, the general confidence in quantum electrodynamics as a supreme, though as yet imperfect, tool of atomistic theory remained alive.... Nevertheless, the awareness of the basic difficulties weighed heavily on our minds" (Wentzel 1960). He summarized the situation in the late 1930s as follows:

> Apart from self-energy divergences and related difficulties the positron theory appeared to furnish well defined and plausible rules for a quantitative prediction of observable phenomena. However, experimental methods for testing the finer details of the theory (vacuum polarization) were not yet available. From the aesthetic point of view, the subtraction devices seemed too artificial to be generally appealing. Pauli, in spite of being actively interested in the theory, revealed his misgivings by using the deprecatory term "subtraction physics." (Wentzel 1960)

In the period from 1927 to 1940 most of the problems, including many of the intractable ones, had been clearly perceived. How to describe the two-body system (e.g., a hydrogen atom) was one of the first QED problems addressed (Breit 1929; Oppenheimer 1930a). However, in addition to the more pedestrian divergence connected with the zero point energy of the electromagnetic field that Jordan had already encountered in the *Dreimännerarbeit* (Born et al. 1926), the work of Waller (1930a,b), Oppenheimer (1930b), Heisenberg (1930) and others, had indicated that relativistic quantum field theories were plagued with troublesome divergences. In QED these divergences made it impossible to calculate any higher-order effects. Oppenheimer in 1930 raised the question whether the radiative corrections to the *differences* in the energy levels of an atom could be finite despite the fact that the proper energy of each level diverged. The energy difference of the energy levels calculated by him in these pre-hole-theoretic days was logarithmically divergent, hence the conclusion was negative:

> The theory is however wrong, since it gives a displacement of the spectral lines from the frequency predicted on the basis of the non-relativistic theory which is in general infinite. This displacement arises from the infinite interaction of the electron with itself; this interaction depends upon the state of the material system; *and the difference in the energy for two different states is not in general finite* [emphasis mine]. Thus the present theory gives no more than the non-relativistic theory of Jordan, Klein and Wigner. (Oppenheimer 1930b)

The calculation of the self-energy of an electron in hole theory was carried out by Weisskopf and corrected by Furry and Carlson in 1934 and found to be logarithmically divergent. This result was an important landmark—because a logarithmic divergence is a very weak divergence. It gave rise to the hope that perhaps with some further effort a finite theory might be constructed. But by 1934 Oppenheimer's query about differences in energy levels had seemingly been shelved, though as we shall see not totally forgotten. There were also no convincing experimental data at hand to suggest that deviations from the predictions of the Dirac equation for an electron in a Coulomb field existed, deviations which might have encouraged a QED calculation of the level shifts.

   During this initial phase it became clear that in a relativistic quantum field theory the vacuum is no longer a simple entity. Fluctuations in the charge-current densities as well as in the electromagnetic field strengths gave the vacuum a complex structure. These phenomena also destroyed the correspondence with classical theory. It was pointed out by Furry and Oppenheimer (1934), by Dirac (1934a), by Heisenberg (1934a), by Peierls (1934), and by Weisskopf (1936) that an external electromagnetic field will induce charge-current fluctuations that will alter the original external field, giving the vacuum the characteristics of a polarizable medium. Even though the calculated polarizability of the vacuum turned out to have an infinite part (the contribution of high-energy electron-positron pairs) it was recognized that this (infinite) contribution to the induced charge-current density merely reduced the inducing charge-current by a constant factor. Only the total charge-current density (induced plus inducing) is observable. To quote Furry and Oppenheimer (1934): "Because it is in practice impossible not to have pairs present, we redefine all dielectric constants, as is customarily done, by taking that of the vacuum to be unity." Serber (1936) called this procedure: "to renormalize" the charge. By the mid-1930s it was also clear that it was the *local* coupling of the charge-current density to the electromagnetic field—corresponding in classical theory to the coupling of *point* charges to the field—that was responsible for giving arbitrarily small wavelengths a divergent role in both the self-energy and the vacuum polarization calculations.

Locality is a legacy of the point model of particles and its description of interactions among them. At first sight, locality seems merely to be a statement of the rejection of the possibility of action-at-a-distance, and a means to keep the representation in compliance with special relativity. But an examination of the construction of the point model of the electron reveals that it is also an attempt to resolve a difficulty in Lorentz's theory of the electron (1904). According to J. J. Thomson (1881), the energy contained in the field of a spherical charge of radius a is proportional to $e^2/2a$. Thus, when the radius a of the Lorentz electron goes to zero, the energy diverges linearly. But if the electron is given a finite radius, then the repulsive Coulomb force within the sphere of the electron makes the configuration unstable. Poincaré's response (1906) to the difficulty was the suggestion that there might exist a nonelectromagnetic cohesive force inside the electron to balance the Coulomb force, so that the electron would not be unstable. Two elements of the model exercised great influence upon later generations: (1) the notion that the mass of the electron has, at least partly, a nonelectromagnetic origin, and (2), that the nonelectromagnetic compensative interaction, when combined with the electromagnetic interaction, would lead to the observable mass of the electron. Thus Stueckelberg (1938b), Bopp (1940), Pais (1946), Sakata (1947), and others obtained their inspiration from Poincaré's ideas in their study of the problem of the electron's self-energy. In 1922 Fermi pointed out that the equilibrium of the Poincaré electron is not stable against deformations (cf. Rohrlich 1973). This observation elicited a novel response to the difficulty, which was first stated by Frenkel (1925). He argued that the electron is "elementary" and therefore has no substructure. Thus, the internal equilibrium of an extended electron is a meaningless problem within the classical framework. The idea of looking for a structure of the electron was given up because, as Dirac suggested (1938b), "the electron is too simple a thing for the question of the laws governing its structure to arise." By adopting the point model, Frenkel eliminated the "self-interaction" between the parts of an electron—and thus the stability problem—but he could not eliminate the "self-interaction" between the point electron and the electromagnetic field it produces without abandoning Maxwell's theory.

Frenkel's conception of the point electron was adopted by most physicists. It was incorporated into the conceptual foundation of quantum field theory and is expressed by the locality assumption. But the problem Frenkel left open became more acute when quantum field theory came into being with Heisenberg and Pauli's papers. Note that the locality assumption conceals the acknowledgment of our ignorance of the structure of the electron and that of other elementary entities described by QFT. The justification given for the point model—and the consequent locality assumption—is that they constitute good approximate representations at the energies available in present experiments, energies that are too low for exploration of the inner structure of the particles.

It was fully recognized in the thirties that since the application of field operators on the vacuum results in strictly local (i.e., delta function) excitations, the local coupling among quanta implied that virtual processes involving arbitrarily high momenta (and energy) would be present in quantum field-theoretical calculations. Mathematically, the inclusion of these virtual processes at arbitrarily high energy results in infinite integrals. The divergence difficulties were therefore seen as constitutive within the canonical formulation of QFT. The occurrence of the divergences pointed to a deep inconsistency in the conceptual structure of QFT.

In addition to various proposals for radically altering the foundations of QFT, two different responses were advanced to overcome this inconsistency. The first one was developed independently by Pais and by Sakata in the mid-forties, and was in the spirit of Poincaré's solution to the stability problem of the Lorentz electron. It put forth the idea of compensation: fields of unknown particles were introduced in such a way as to cancel the divergences produced by the known interactions. The second response was the renormalization program. In addition to confronting the intrinsic inconsistencies of quantum field theory, a conceptual descriptive scheme was also formulated. The work of Furry and Oppenheimer in 1933 showed that in QED the state vector describing a single charged particle had amplitudes for finding an arbitrary number of electron-positron pairs. Uehling (1935) demonstrated that these "attached" electron-positron pairs altered the Coulomb field of a charged particle at small distances. He calculated that this would result in an electron being slightly more tightly bound in the s levels of hydrogen due to the polarization of the vacuum caused by the field of the proton. "But," as Breit commented in reviewing Uehling's paper for Pauli's seminar on "The Theory of the Positron," "the polarization effect leads to an increase in the doublet separation and experiment shows a decrease to be necessary" (Pauli 1935–1936, p. 70).

Besides these vacuum polarization divergences, and the previously encountered self-energy divergences, another infinity plagued QED. When one attempted to calculate the contribution of radiative effects to the scattering of electrons by an external field (e.g., the Coulomb field of a nucleus), "infrared" divergences were encountered. Mott (1931) had found that in an expansion in powers of $\alpha = \frac{e^2}{\hbar c}$, the probability of scattering with emission of a single quantum with momentum between k and $k + dk$ behaves as dk/k at low frequencies, resulting in an infinite cross section (see also Sommerfeld 1931). In an important and seminal paper, Bloch and Nordsieck (1937) showed that this infrared catastrophe arose from the illegitimate neglect, implied in the expansion in powers of α, of processes involving the simultaneous emission of many light quanta. By taking these into account, and considering only frequencies so low that the energy and momenta of the photons could be neglected in comparison with those of the electron, Bloch and Nordsieck (see also Braunbeck and Weinmann 1938) indicated that in complete analogy with the classical result, the scattering probability is just that obtained by neglecting radiative effects entirely.

Pauli and Fierz, in a classic paper in 1938, further clarified the connection between Bloch and Nordsieck's and the classical description of the "infrared" radiation. A freely moving electron is accompanied by its appropriate electromagnetic fields. In the classical theory these attached fields correspond to the Lienard-Wiechert potentials, that is, the Coulomb and Biot-Savart fields of the particle; in QED the "attached" photons give rise to these electric and magnetic fields. As an important by-product of their analysis, Pauli and Fierz, in their nonrelativistic model of 1938, recognized that the fields which accompany the charged particle react back on it to produce an electromagnetic mass. They identified the sum of the particle's mechanical mass and of its electromagnetic mass with the observed, experimental mass of the electron. Kramers (1938a,b) at precisely this same period attempted to give a formulation of the quantum electrodynamics of extended nonrelativistic charged particles in which the structure and finite extension of the particles would not appear explicitly. In Kramers' formulation the quantity which was introduced as the mass of the charged particle was from the very beginning its observable experimental mass. The concept of mass renormalization in QED has its origin in these researches of Pauli and Fierz and of Kramers. By 1939–1940 Kramers had essentially worked out the physics of mass renormalization, and two of his students, Serpe (1940, 1941) and Opechowski (1941), published detailed calculations for some simple nonrelativistic systems. In his 1940 paper, Serpe used Kramers' ideas to consistently remove the infinite level shift found by Weisskopf and Wigner in their 1930 calculation of the line widths.

The problem of the polarization of the vacuum had been "solved" by 1936 by Dirac, Heisenberg, and Weisskopf by using relativistically invariant subtraction procedures and the idea of charge renormalization. In 1939 the problem of the self-energy of the electron in hole-theoretic QED had been clearly delineated by Weisskopf, and Weisskopf in particular gave very plausible arguments indicating that the degree of divergence of the self-energy would remain logarithmic in higher orders in α (Weisskopf 1939). How to circumvent the self-energy difficulties had been indicated by Kramers and by Pauli and Fierz in 1937–1938. It is an interesting historical problem to understand why there was not a *concerted* effort to combine these insights at that time to give a divergent free formulation of hole theory, at least to order α. One reason was probably that Kramers' research program was based on a philosophy that was not acceptable to many workers in the field. Kramers in 1938 stressed that his approach was based on the premise that the difficulties at the classical level were to be solved first so as "to arrive at a theory in which the mass and charge would be enough to characterize the electron, wishing to avoid the dangerous and superfluous considerations of [its] structure, [or] the electromagnetic part of its mass, etc." (Kramers 1938a, pp. 116–118). Only after having obtained a satisfactory structure-independent *classical* theory would quantization be imposed. Nor was Kramers alone in this approach. Dirac in 1938 had reached somewhat similar conclusions and had formulated a *classical* relativistic version of the Lorentz electron that was free of divergences (but which did manifest other difficulties). I quote from his paper: "One may think that this difficulty [the divergence of the self-energy of a point mass] will be solved only by a better understanding of the structure of the electron according to the quantum laws. However, it seems more reasonable to suppose that the electron is too simple a thing for the questions of the laws governing its structure to arise, and thus quantum mechanics should not be needed for the solution of the difficulty—our easiest path of approach is to keep within the confines of the classical theory" (Dirac 1938b, p. 148).

But to many others the success of QED, particularly in phenomena involving pair production, suggested that the *quantum* features of any field theory were fundamental. The breakdown of relativistic quantum field theories implied deep structural difficulties that, perhaps, were connected with the assumed description of space-time in the usual formulation of QFT.

There was actually one important investigation that did attempt to amalgamate all the previous insights in order to obtain a divergence-free formulation of hole theory to order α. At the suggestion of Bloch and Oppenheimer, Dancoff in 1939 investigated "to what extent the inclusion of relativistic effects modify the conclusions of Pauli and Fierz." The latter had treated the motion of the charged particles nonrelativistically. Dancoff calculated the radiative corrections to the elastic scattering of an electron in an external field—the electron being described hole theoretically. Unfortunately Dancoff only included the effects of the transverse photons in calculating radiative corrections and omitted the contribution of the Coulomb interaction terms[13] (repeating the oversight of Carlson and Furry in their 1934 computation of the self-energy!). He thus obtained a divergent result and concluded that in hole theory a *new type* of divergence occurred in the radiative corrections to the elastic scattering of an electron by an external field. Dancoff was actually partially correct. He did in fact encounter a new type of divergence— divergences that would later be called "vertex function" divergences. When combined with pieces of the self-energy divergences—and properly taking into account a wave function renormalization, which Dancoff did[14]—these divergences cancel one another. Bloch and Oppenheimer had actually noted this cancelation for some of the diagrams[15] and this was the point of departure of Dancoff's investigation.

Dancoff's results were puzzling. His calculations yielded a self-energy for the electron that depended on the scattering potential and he himself noted the "fortuitous nature of the results." In 1946 Bethe and Oppenheimer stated that "it is not possible to believe that these results can have any relation to reality" and concluded that as presently formulated quantum electrodynamics "made no sense at all" (Bethe and Oppenheimer 1946).

The complete cancelation of divergences to order α requires taking into account both the transverse photons and the Coulomb interaction terms. Lewis (1948), who redid Dancoff's calculation after the Shelter Island Conference, discovered Dancoff's mistake (see also Epstein 1948). Since no one before World War II checked Dancoff's calculation, his result only added to the "awareness of the basic difficulties" that weighed so heavily on the mind of all field theorists.

Why did no one redo Dancoff's calculation at the time? There were certainly many able physicists who could have done so, for example, Nordsieck, Serber, or Weisskopf. Why did the community not invest talent and energy to carry out this task? Had it done so, the difficulties of QED might have been resolved much earlier.

One reason was that other fields, particularly nuclear physics, β-decay theory, and cosmic ray physics, offered greener pastures. The wealth of new experimental data in these fields and the success of the theoretical attacks by Heisenberg, Wigner, Peierls, Uhlenbeck, and Bethe, among others, are clear proof of this (see, for example, Bethe et al. 1986). Quantum electrodynamics played a crucial role in establishing the existence of the "mesotron" in 1937. This success, in fact, helped shift the focus. Yukawa's 1935 paper, in which he had proposed, in analogy with QED, that nuclear forces were mediated by the exchange of massive bosonic quanta (rather than pairs of particles as had been suggested by Tamm 1934 and Iwanenko 1934 on the heels of Fermi's β-decay theory) had not been received with "immediate consent or sympathy" (Wentzel 1960). But when the "mesotron" was detected in 1937, it became immediately identified with Yukawa's nuclear force meson and Yukawa's paper became "the focus of universal attention" since "Yukawa had predicted its existence" (Wentzel 1960). Nuclear forces and meson theory were pushed to the center of theoretical interest; and the explanation of the electromagnetic properties of the nucleons and those of the deuteron, such as their magnetic dipole moments and electric quadrupole moments, became the challenging problems (Kemmer 1938; Fröhlich et al. 1938; Møller and Rosenfeld 1940; Schwinger 1939, 1942).

That is not to say that there were no intriguing experimental results available to the theoretician interested in testing QED. By 1938, the experimental data on the spectrum of hydrogen suggested that the $2\,^2S_{1/2}$ and $2\,^2P_{1/2}$ levels were not degenerate, contrary to the predictions of the Dirac theory. Both Houston (1937) and Williams (1938) had found these departures from the Dirac theory. Pasternack (1938), in fact, indicated that these deviations could be interpreted as an upward shift of the $2\,^2S_{1/2}$ level by 0.03 cm^{-1} $\approx$ 1000 megacycles relative to the $2\,^2P_{1/2}$ level.[16] Stimulated by these findings, Fröhlich, Heitler, and Kahn (1939a,b) calculated (incorrectly, as it turned out) the form factor of the proton according to a version of meson theory and obtained an upward shift of about 0.03 cm^{-1} in agreement with Pasternack's requirements. It was, incidentally, Willis Lamb (1939) who pointed out the mistake in their calculations! Later, a detailed analysis by Blatt (1945) showed that the result Fröhlich, Heitler, and Kahn had obtained was a consequence of their particular method of handling the divergences they encountered in the calculation. But the ambiguous nature of the experimental data should not be underestimated. The data were at the limit of the resolution of the optical apparatus then available, and difficult questions concerning the assumptions on which it was based could not easily be resolved (e.g., was the H gas in the discharge tube in thermal equilibrium as was assumed in imposing the usual Doppler-broadened shape to the observed lines?) Thus it is not surprising that these deviations from the predictions of the Dirac equation should be taken *cum grano salis* by the theoretical community. Anyone faced with the staggering complexity a hole-theoretic computation entailed would have wanted a firmer experimental inducement to undertake such calculations.

But probably the most important factor in delaying the completion of the subtraction program and its application to render QED finite were the events surrounding the outbreak of World War II. With the ominous developments in Nazi Germany, after the discovery of the fission of uranium the leading theoretical physicists in England and the United States turned their attention to the problems implied by the possibility of a chain reaction. In particular, this was the case for Weisskopf, one of the most likely persons to have made a relativistic quantum electrodynamical calculation of the spectrum of hydrogen before World War II. With the fall of the Low Countries and France in the spring of 1940, scientific contact with European colleagues on the Continent—except for those in Switzerland—essentially ceased. Thus Kramers' insights had to await the 1947 Shelter Island conference to be appreciated by the theoretical community in the United States. That there is truth in this explanation is borne out by the following: In the fall of 1946, Weisskopf, who had just come to MIT, gave Bruce French, then a graduate student, the problem of calculating the $2S_{1/2} - 2P_{1/2}$ level splitting in hydrogen.[17] This was well *before* the experiment of Lamb had become known! Oppenheimer's original suggestion that differences in energy levels might prove finite was to be the basis of the calculation. Weisskopf's 1939 analysis of the self-energy of the electron in hole theory had made the chances for the success of such a calculation much greater.[18]

If the boundaries of knowledge lie between the possible and the unthinkable, between sense and nonsense,[19] part of the reason that the workers of the thirties did not overcome the divergence problems was that they were too ready to focus on the unthinkable—that is, on radical and revolutionary departures. Progress, it would turn out, could be achieved by a conservative stand: by taking seriously the successes achieved by charge renormalization procedure in the vacuum polarization problem and using the insights gleaned from the work of Kramers and Pauli and Fierz regarding mass renormalization to circumvent the self-energy divergences. In order to corroborate further the account of the thirties that I have given, I turn next to three conferences that had been planned to take place at the end of the 1930s. The agenda of these conferences reveals what was thought possible at the time and delineates the directions charted by the state of knowledge.

2.3 The Warsaw Conference of 1939

2.3.1 Background

The Paris Peace Conference, which convened in the spring of 1919, besides proposing the terms the victorious Allies would impose on Germany in the Treaty of Versailles, also recommended the establishment of the League of Nations to preserve the peace and to foster international economic and political cooperation (Holborn 1951; Walter 1952; Mayer 1967). Hymans, the delegate from Belgium to the conference, made the entreaty that "international intellectual relations" also be an important component of the work of the League (Institut 1947, p. 11). Although this suggestion was not implemented at the time, the Assembly of the League of Nations in 1921 did adopt a recommendation to establish a "commission to study questions relating to international intellectual cooperation" (Institut 1947, p. 14).

In January 1922, the Council of the League established such a commission and it met for the first time on August 1 of that year in Geneva.

Twelve distinguished scholars, scientists, and educators were appointed to the commission—among them Marie Curie-Sklodowska, George H. Hale, Gilbert Murray, Robert A. Millikan, and Hendrik A. Lorentz—and the aged Henri Bergson, France's most eminent and distinguished philosopher, was asked to be its chairman. The commission was enlarged in 1924, and Albert Einstein joined its ranks. That same year, as a result of Henri Bergson's efforts, the French government offered the commission the Palais-Royal in Paris to house the Institut International de Cooperation Intellectuélle, the organization the commission had recommended to implement the charge by the League. The Institut, which was to be supervised by the commission and accountable only to the League of Nations, was to be an autonomous body, supported by subventions from all member states. In fact, until 1940, France provided the major part of its annual budget. The outbreak of World War II in 1939 essentially terminated the activities of the League and those of the Institut. With the founding of the United Nations in 1945, UNESCO took over most of the functions of the International Organization for Intellectual Cooperation, the Institut's official English title.

The several divisions of the Institut started their operations in 1926. The Section d'Affaires Générales oversaw a wide range of activities: from inquiries into every aspect of public elementary and secondary education to the compiling of bibliographies on "great current social and intellectual problems" (Institut 1947, pp. 27–28). Other sections dealt with universities (faculty and student exchanges, student organization, rights of students, etc.); the legal aspects of intellectual life (rights of authors, ownership of intellectual and scientific productions, etc.); scientific relations (professional scientific organizations, plans for the establishment of an international lending library, scientific nomenclature, creation of an international committee on the history of science, etc.); literature (translations, etc.); art (photography, popular art, cinema, etc.); information and its dissemination (press, radio, etc.). By 1929 the Institut had about seventy functionaries administrating its various projects, and some thirty member states sent delegates to its annual meeting. The view that intellectual cooperation consisted in the "coordination" of the intellectual activities that were being carried out *throughout the world* determined policy (Institut 1947, p. 39). The Institut helped establish and maintain contact among many institutions, and it encouraged and fostered cooperation in their *ongoing* activities. This underlying assumption governing the activities of the Institut was challenged at the end of the 1920s by another conception—one that was characterized as "Anglo-Saxon" (Institut 1947, p. 41). On that view, the task of the organization was to elicit the effective cooperation of scholars and scientists from different countries in order to solve specific problems. The Institut was to be a sort of "international academy" where scholars would meet, confer, and address definite questions with the aim of resolving clear-cut issues and problems. In the first conception of the Institut, its role was passive: it did not initiate new projects nor set new directions for intellectual activities. This view had encouraged the bureaucratization of the agency. The second was recognized as "antibureaucratic" and more idealistic.

In September 1931 in a plenary session, the assembly of the League of Nations adopted a resolution stating that "intellectual cooperation has as its object international collaboration to assure the progress of civilization and knowledge, and more particularly, the development and diffusion of the sciences, literature and the arts. Its aim is the creation of a climate of opinion favorable to the peaceful solutions of international problems. Its framework is that of the League of Nations" (Institut 1947, p. 45).

Having thus defined intellectual cooperation, the resolution enjoined the Institut to

1. Develop the exchange of ideas and personal contact among intellectuals of all nations.

2. Encourage and facilitate the cooperation between institutions dealing with intellectual matters.

3. Assist in the diffusion of intellectual productions.

4. Study some of the great problems having international scope.

5. Contribute to the international protection of intellectual rights.

6. Disseminate the principles of the League of Nations. (Institut 1947, p. 45)

Even though the bureaucratic, more passive conception won out, the commission agreed to decentralize[20] and to curtail the size of its nonspecialized staff. It also agreed to have the Institut address specific problems and to establish committees of experts to advise it on the selection of problems and to help and find solutions. A number of conferences dealing with issues in the social sciences were organized during the early 1930s. Rivalry and disagreements between the Institut and the International Council of Scientific Unions prevented the sponsoring of meetings on subjects in the physical and biological sciences. However, an agreement that overcame these difficulties and allowed the two organizations to work together was reached in July 1937. Thereafter, the Institut organized a series of "Réunions d'Etudes," the subjects of which were decided by the scientific advisory committee to the executive committee of the Institut and the secretariat of the International Council of Scientific Unions.

Prior to each conference, the reports that were to be submitted were translated into English and French and circulated among those invited. Usually, they were not read at the meeting so that most of the time could be devoted to discussions. The latter were stenographed *in extenso* and then published with the reports.

The first of the *reunion d'études* was a conference on phytohormones held in Paris in October 1937. In December of that year, a *reunion* was held in Prague to select the titles for a series of publications of facsimile reproductions of early scientific manuscripts. A third meeting, held that same month in Neuchâtel, dealt with "physico-chemical methods for the determination of atomic and molecular weights of gases." The fourth conference, on "New Theories of Physics," was held in Warsaw in June 1938 to honor the memory of Mme Curie-Sklodowska, who had been an active member of the commission until her death in 1934.[21] This conference had been suggested by Professor Bialobrzeski, a member of the commission that oversaw the activities of the Institut. A great deal of effort was spent in its preparation; several meetings were held to formulate the program and to decide on the list of rapporteurs and participants (Institut 1947, p. 366). The reports and the subsequent discussions were published in 1939 in two editions, one in French and one in English. It was undoubtedly the most distinguished of the nine conferences sponsored by the Institut before World War II.[22] Charles Fabry, the chairman of the International Council of Scientific Union that had cosponsored the conference, emphasized that "the subject [of the conference] covers a field of speculations of unprecedented magnitude. These 'New theories of Physics' have caused a great upheaval not only in pure physics but throughout the domain of scientific thought" (*New Theories,* 1939, p. X).

Bialobrzeski, in his opening remarks to the conference, expressed the hope that should the conference prove successful, "more frequent meetings would be arranged to include, besides physicists, representatives of other sciences and philosophy" (*New Theories,* 1939, p. 2).

Niels Bohr set the tone of the conference with his report on "The Causality Problem in Atomic Physics" (Bohr 1939). The other reports were by John von Neumann (on the necessity of the indeterministic interpretation of the formal structure of quantum mechanics); Louis de Broglie (on the relation between the quantum theory and the theory of special relativity); Hendrik A. Kramers (on the difficulties connected with the quantum theory of the electromagnetic field and the limits of the applicability of the present system of theoretical physics); Oskar Klein (on the quantum theory of charged fields); Léon Brillouin (on the "individuality" of elementary particles and Pauli's statement of the relation between spin and statistics); Arthur S. Eddington and E. A. Milne (on the cosmological implications and applications of the theory of quanta); and by Pierre Langevin (on the positivistic and realistic trends in the philosophy of physics). Besides these rapporteurs, the other participants were E. Bauer, C. Darwin, R. H. Fowler, G. Gamow, S. Goudsmit, E. Hylleras, L. de Kronig, C. Møller, F. Perrin, L. Rosenfeld, and E. P. Wigner, and several Polish physicists including W. Rubinowicz. Dirac, Fermi, and Heisenberg had also been asked to attend and invitations had been sent to several Russian theoretical physicists. Dirac declined. Political factors were undoubtedly the reason for the absence of the German, Italian, and Russian physicists at a conference sponsored by a department of the League of Nations.[23]

The meeting opened on May 31, 1938, with great formality and was typical of the small, elitist physics conferences held before World War II.

The Solvay Congresses, which were initiated in 1911 and brought together the most eminent workers in the field, had become the model for such meetings (Mehra 1975). Starting in the late 1920s Bohr had organized similar, somewhat more informal conferences at his institute in Copenhagen, which assembled the younger and upcoming members of the theoretical physics community in Europe.[24]

After the conference, S. Goudsmit, who had been one of the participants, wrote Establier, the head of the Scientific Relations Service of the Institut who had been responsible for the arrangements of the conference, to express his "extreme satisfaction" with the meeting. He indicated that "I was sent to Europe for the special purpose of studying the most recent developments of modern theoretical physics. Months of travel did not yield as much valuable information as the one week at Warsaw."[25] To his friend Bacher, Goudsmit gave the following assessment of the conference: "At first it looked like a funny half philosophical program. But thanks to the presence of Bohr and Kramers, most of the time was used for discussion of real worthwhile things and only one session was spoiled by Eddington. It lasted five days and there were too many formal dinners and receptions, including a fairylike diplomatic ball at the American Ambassador's, where a few of us were invited."[26]

Bohr had been influential in organizing the conference, and it is therefore not surprising that the agenda reflected his interests and his inclination to address "fundamental" problems and philosophical issues. The discussions following the various reports corroborate Goudsmit's assertion that Bohr and Kramers were the dominant figures. Although the reports dealing with field theory were not by the leading experts in the subject—Pauli, Dirac, Heisenberg, Weisskopf, Oppenheimer—they nonetheless convey the dominant concerns of the day. *New Theories of Physics*, the record of the conference, gives a valuable overview of the state of "particle" physics just prior to World War II.

2.3.2 Bohr's Address

In his lectures on quantum mechanics during the 1930s, Bohr tended to repeat material he had delivered on previous occasions, refining and sharpening statements made in the earlier versions, and would then proceed to consider issues or problems raised by the most recent advances. His inaugural address to the Warsaw conference followed this pattern. He began his presentation by recalling the inadequacies of the classical description of atomic phenomena that led to the renunciation of "the unrestricted applicability of the causal mode of description to physical phenomena." The "extreme simplicity" of the structure of atoms that stemmed from Rutherford's discovery of the atomic nucleus had disclosed the inadequacy of the laws of classical mechanics and electrodynamics in their application to atomic phenomena since they were unable to account for the stability of atoms. The quantum conditions which he—Bohr—had introduced in 1913 were not only "totally foreign to classical ideas," but also had implied an explicit renunciation of any causal description of atomic processes. In his characteristically involuted style, Bohr stressed that "as regards its possible transition from a given stationary state to another stationary state, accompanied by the emission of photons of different energies, the atom may be said to be confronted with a choice for which, according to the whole character of the description, there is no determining circumstance" (*New Theories*, p. 13). Hence, as Einstein had made clear in 1917, all predictions can only relate to "the probabilities for the various courses of the atomic processes open to direct observation" (*New Theories*, p. 13). Moreover, far from being a temporary compromise, this "recourse to essentially statistical considerations [was] the only conceivable means of arriving at a generalization of the customary ways of description," which, on the one hand, was of sufficiently wide latitude to account "for the features of individuality expressed by the quantum postulates" and which, on the other hand, went over to a classical description in those situations in which all actions involved are large compared to h, Planck's constant (*New Theories*, p. 13). Bohr then pointed out that Einstein's success in formulating the laws of emission and absorption of photons in radiative processes—laws from which Planck's formula for blackbody radiation could be derived—stemmed from the fact that the dynamical description of pure radiative fields could be shown to be equivalent to that of a system of harmonic oscillators. The (linear) superposition principle which holds in electromagnetic (field) theory implied that the "equivalent" harmonic oscillators do not interact with each other and act independently of one another. Although any solution to the (classical) equations of motions of interacting material particles—such as the constituents of an atom—can be represented by a superposition of purely harmonic oscillations, these oscillations, by virtue of the nonlinearity of the equations, are not themselves independent solutions of the mechanical problem.

It was the absence "in classical mechanics of a superposition principle like that of field theory" that was responsible for the great difficulties encountered in applying Einstein's insights to atomic problems. Bohr went on to stress that "it was indeed the recognition of the necessity to avoid any explicit use of mechanical motion in connection with the quantum postulates which led Heisenberg to the establishments of a rational quantum mechanics, based on a suitable formal representation of kinematic and dynamical concepts" (*New Theories,* p. 14).

In this formalism as formulated by Dirac, the canonical equations of classical mechanics

$$\frac{dp_i}{dt} = -\frac{\partial H}{\partial q_i} \qquad \frac{dq_i}{dt} = \frac{\partial H}{\partial p_i}$$

remain formally the same; however, the dynamical variables p_i, q_i are now linear (noncommuting) operators and the quantum of action is only introduced in the commutation rules these operators satisfy:

$$p_\ell q_j - q_j p_\ell = \frac{h}{2\pi i}\delta_{\ell j}.$$

While the whole scheme reduces to classical mechanics when $h = 0$, "all the exigencies of the correspondence argument are fulfilled" in the general case "through the possibility of combining a purely classical description of the Hamiltonian function H, with a proper account of the quantum characteristics of its behavior." Bohr went on to stress that "the essentially statistical nature of this account is a direct consequence of the fact the commutation rules prevent us to identify at any instant more than a half of the symbols representing the canonical variables [i.e. the p's and q's] with definite values of the corresponding classical quantities" (p. 15). In the case of an isolated atomic system, quantum mechanics "thus allows us to predict the possible values of all action variables—which characterize the stationary states of the system—while the conjugate variables are left entirely undetermined." It would be difficult to find a more succinct and more insightful characterization of Dirac's and Heisenberg's formulation of quantum mechanics than this presentation by Bohr, which was sensitive to the historical evolution of the concepts involved. He pointed out that although this formalism could be applied to any conceivable atomic problem, "the intricate mathematical operations involved" made it imperative for both practical purposes and for the elucidation of the consequences of the formalism "that the treatment of any quantum mechanical problem be shown to be essentially reducible to the solution of a linear differential equation, allowing to formulate for atomic systems a principle of superposition of states, analogous to that of field theory" (p. 15). This is what was accomplished by the Schrödinger's equation

$$H\psi = \frac{ih}{2\pi}\frac{\partial\psi}{\partial t},$$

where H is a differential operator obtained by replacing in the Hamiltonian function $H(p_i, q_i)$ the p's by the operators

$$p_i = \frac{h}{2\pi_i} \frac{\partial}{\partial q_i}.$$

Bohr then reminded his audience of the well-known equivalence of the approach of Schrödinger with that of Dirac and Heisenberg, which had been demonstrated by many people and "most elegantly" by von Neumann's (1932) axiomatic exposition. Von Neumann's formulation had the further particular merit that it made "evident that the fundamental superposition principle of quantum mechanics logically excludes the possibility of avoiding the noncausal feature of the formalism by any conceivable introduction of additional [hidden] variables." "The necessity of renouncing a causal description of atomic phenomena," Bohr went on to say, "is in no way affected by the preliminary disregard of relativity exigencies in quantum mechanics." The success of Dirac's relativistic quantum theory lends support to the view that this situation will persist in any future elaboration of the theory. The difficulties encountered in quantum electrodynamics—"connected with the use of the idealization of point charge for the electron"—that have impeded the further development of the theory must be "imputed to our failure, already acute in classical theory, to grasp some deeper feature of the stability of the individual particles themselves," and do not derive from the noncausal features of the quantum theory. After reviewing "the observation problem in quantum theory" which the quantum of action entailed, Bohr concluded his address with these remarks: "Quite apart from any prospect of mastering hitherto unresolved problems in atomic theory, . . . there can be no question in further developments of returning to a description of atomic phenomena in closer conformity with the causality ideal. Rather we are here confronted *with the necessity of a still more radical departure* from accustomed modes of description of natural phenomena implying a further extension of the view point of complementarity" (p. 29; my italics).

Bohr was followed by von Neumann, who gave an updated version of his earlier proof (von Neumann 1932) that quantum mechanics is a "properly" statistical theory such that "no causal explanation of quantum mechanics by hidden parameters is possible without sacrificing some part of the theory existing today" (pp. 30–38). The ensuing heated discussion dealt primarily with the measurement process in quantum mechanics.

On the second day of the conference, Klein presented a visionary— though at the time unappreciated—paper on "The Theory of Charged Particles." In this important communication, Klein anticipated several of the developments of gauge theories of the 1970s. Klein noted that the discovery of the so-called heavy electron or mesotron, the particle which is supposed to be the "quantum" whose exchange generates the attractive forces at small distances between neutrons and protons—"a role suggested by Yukawa already before the discovery"—implied a considerable enlargement of the region of applicability of the field concept "which had hitherto been limited by the self-energy difficulties." Since the new particle's mass, μ, was two orders of magnitude smaller than the Compton wavelength of the electron, $\frac{h}{mc}$, the enlargement of the field concept "would seem to require the removal of the self-energy difficulty of the electron at least down to distance approaching radius of the new particle." Since the range of the nuclear forces, $\frac{h}{\mu c}$, is of the order of the classical electron radius, $\frac{e^2}{mc^2}$, "it would not seem unreasonable to assume that a theory explaining nuclear attractions would also account for the rest mass of the electron, the attractive forces required as a compensation of the Coulomb repulsion being of a similar nature as the nuclear forces" (p. 78). If this is to be accomplished, "the new forces belonging to the heavy electron field are determined by means of the elementary electric charge in a similar way as the electromagnetic forces, so that no other independent constant than the mass of the new particles will appear in the theory." Klein based his formulation on a five-dimensional representation of field theory that he had developed earlier in 1926 (Kaluza 1921; Klein 1926)—the famous Kaluza-Klein theory—one that lead "exactly" to the Einstein-Maxwell theory of gravitation and electromagnetism. In this theory the electric charge becomes quantized as a consequence of the requirement that the solutions of the theory be periodic in x_o, the "fifth" dimension that corresponded to a length $\ell_o = \sqrt{G\hbar/c^3}$, where $\hbar$ is Planck's constant, c the velocity of light, e the electronic charge, and G Newton's gravitational constant. By extending the theory to include charged and uncharged integer spin fields, and requiring that these new fields be likewise invariant under gauge transformation of x_o, a theory was obtained where no new coupling constant besides e^2 was introduced.

At the session on the "Limits of the Applicability of the Present System of Theoretical Physics," which Kramers chaired, Kronig gave an outline of Heisenberg's views on these matters based on the manuscript Heisenberg had recently communicated to the *Annalen der Physik* (Heisenberg 1938a). In his article Heisenberg had pointed out that often in the past, "refinement and development of methods of measurements have led to apparent paradoxes in the theoretical interpretation." Thus the Michelson-Morley experiment led to a revision of our concepts of space and time, concepts that were embodied in the theory of relativity. "The progress made may be characterized by saying that in pre-relativistic theories a certain constant c, the velocity of light, had been dealt with as if it were infinitely large while according to the new viewpoint its finite value is taken consistently into account." Similarly, the failure of pre-quantum theories can be attributed to the fact that Planck's constant, h, was "regarded as infinitely small." Heisenberg then noted that the divergence difficulties encountered in all field-theoretic calculations in higher orders of perturbation theory are closely connected with the point character of the charges that were presupposed in the calculations. These difficulties may be avoided if in the Fourier development of the field quantities, fields with wavelengths smaller than a certain length are left out or if one ascribes to the charges a finite extension. However, as is well known, this cannot be accomplished in a relativistically invariant fashion. Heisenberg then proposed that just as the paradoxes of space-time and quantum phenomena were overcome by taking into account the finite value of c and h, the solutions of the difficulties encountered in quantum field theory "will have to be sought by ascribing a finite value to another constant... [that has] the dimension of a length." As to the physical significance of this length, Heisenberg suggested:

> The passage from the pre-relativistic to the relativistic theory or from the pre-quantum to the quantum theory was not so much a correction of the older theories but the recognition that upon ascribing finite values to c and h to the possibility of visualizing physical phenomena in terms of the concepts of daily life must in part be abandoned. Thus in the relativity theory the finite velocity of light precludes the introduction of an absolute time independent of the state of motion of the observer while in quantum theory the limit reaction of the observer upon the objects makes illusory every attempt of measuring the position and velocity of a particle simultaneously with any desired accuracy. (p. 101)

The fundamental length r_0 which Heisenberg introduced was assumed to be of the order of the classical electron radius

$$r_0 = \frac{e^2}{mc^2} = 2.81 \times 10^{-13} cm$$

and was to represent "a sort of limit below which the concept of length loses its significance." He wanted to relate this length to the phenomena of "bursts" that experimenters believed occurred when cosmic ray particles of great energy enter the atmosphere. Besides the showers that had been observed in cloud chambers, and were accounted for by the theory advanced by Oppenheimer and Snyder and by Heitler and Bhabba (see Heitler 1936), Heisenberg believed that true multiple processes occurred in which an incident high-energy particle created a large number of particles "in one act." In fact, he was of the opinion that the empirical study of such bursts would yield clues about the features of the future correct theory. In any attempt to measure lengths on the order of smaller than r_0, one would have to resort to wave phenomena with wavelengths $\lambda < r_0$, "But upon employing such waves, explosions will occur which make it impossible to carry out the measurements in question" (Heisenberg 1938; see also Heisenberg 1939a).

Rosenfeld then proceeded to outline the content of a letter that Heisenberg had written Bohr, which detailed how a quantum field theory involving a fundamental length could be constructed. Starting from the Schrödinger equation describing the field system

$$i\hbar\frac{\partial\Psi}{\partial t} = \left\{H_0 + \int H_I d^3r\right\}\Psi, \tag{2.3.1}$$

where H_0 is the Hamiltonian of the noninteracting fields and H_I the interaction energy density, Heisenberg introduced the Dirac picture description of the system, by defining the vector χ

$$\chi = \exp\left[\frac{iH_0 t}{h}\right]\Psi, \tag{2.3.2}$$

which satisfies the equation

$$ih\frac{\partial\chi}{\partial t} = \int H_I(r,t)d^3r\chi, \tag{2.3.3}$$

where

$$H_I(r,t) = e^{iH_o t}H_I(\mathbf{r})e^{-iH_o t}. \tag{2.3.4}$$

Eq. (2.3.3) can also be written as

$$\chi(t+dt) = \left(1 - \frac{idt}{h}\int H_I(r,t)d^3r\right)\chi(t) \tag{2.3.5}$$

$$= \left(1 - \frac{i}{h}\int H_I(r,t)d\omega\right)\chi(t), \tag{2.3.6}$$

where $d\omega = d^3r dt$.

In this formulation the relativistic invariance of the theory is manifest, since "the density of interaction energy," that is, $H_I(r,t)$, is a scalar. Heisenberg speculated that if a fundamental length existed, only spatiotemporal elements of finite magnitude, rather than the infinitely small space-time $d\omega$, would appear in the theory. How to do so without breaking the relativistic invariance was, however, not obvious. Heisenberg was of the opinion that the form (2.3.6) might nonetheless be meaningful in a future theory in which dt will only be greater than $\frac{r_0}{c}$. It might, however, be necessary to give up the notion of an interaction-energy density defined at a point.

Part of the ensuing discussion focused on the experimental status of the bursts adduced by Heisenberg. The recent results by Fussell (1937) were discussed. In this experiment cosmic rays were made to cross in which three lead plates had been inserted, whose thickness had been so chosen as to make probable that in each of the plates a simple process would occur, namely either the creation of

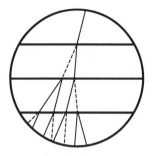

Figure 2.3.1 A typical event in Fussell's cloud chamber.

an electron-positron pair or the emission of a photon (see fig. 2.3.1). Some of Fussel's pictures—3 plates out of 900—seemed to indicate showers diverging from a well-defined point that were unaccountable by cascade theory. Heisenberg wanted to identify these as (nuclear) bursts (Euler and Heisenberg 1938).

The conference next turned to the report by Kramers. In it Kramers suggested that the difficulties encountered in all relativistic field theories stemmed from a difficulty already present at the classical level. Quantum electrodynamics is based on an interaction term that has the form

$$\frac{-e}{c} v \cdot \mathcal{A} + e\Phi, \qquad (2.3.7)$$

where v is the velocity of the particle, e is its charge, and $\mathcal{A}$ and Φ are the electromagnetic 4 potentials evaluated at the position of the charged particle. For the quasi-stationary motion of an electron in an external field, use of expression (2.3.7) is justified when it is realized that the potentials $\mathcal{A}$, Φ describe this external field and not the whole field (i.e., including the fields created by the charge particle which became singular at the location of the charged particle). Kramers then outlined his work in which he "attempted to represent the theory in such a way that the question of the structure and finite extension of the particles does not appear explicitly" and that the quantity which is introduced as "particle mass" is from the very beginning the *experimental mass* (Kramers 1938b; also 1956, p. 831). Kramers "hoped to arrive at a theory, in which the mass and the charge would be enough to characterize the electron, wishing to avoid dangerous and superfluous considerations of the structure, the electromagnetic part of its mass, etc."

Originally it had been hoped that on the fourth day Dirac would make a presentation on the question of the individuality of elementary particles, but at the last minute "he found it impossible . . . to send in this paper" and Louis Brillouin was called to replace him.

Brillouin presented a somewhat patchy review of the property and description then known as "indivisible elementary particles." These were the electron, the positron, the photon, the neutrino, the neutron, the proton, a semiheavy electron with an approximate mass of about 100 electronic masses—the proof of whose existence "does not seem very substantial"—and a "baryon" or "mesotron" with a mass of the order of two hundred times the electronic mass. Brillouin then reviewed the quantum-theoretical description of these particles, including de Broglie's method of fusion for constructing higher-spin objects (see de Broglie 1943). The statistics obeyed by assemblies of such "elementary particles," and

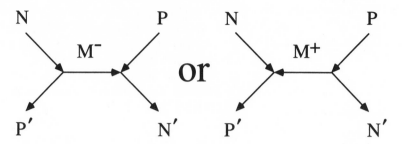

Figure 2.3.2 The neutron-proton interaction as mediated by mesotron exchange.

their description by quantized wave fields, was then discussed. Finally, Brillouin described the way interactions had been introduced within the field-theoretic formalism. All were based on the analogy to electrodynamics: the Fermi theory describing β-decay and most recently the neutron-proton interactions mediated through charged mesotron exchanges. The latter were illustrated by the two suggestive diagrams of figure 2.3.2 where the lines N and P represented the neutron and proton before interaction; N', P' the neutron and proton after interaction; and the lines marked M^-, M^+ a positive or negative mesotron. Brillouin also indicated how the basic processes that occurred at the vertices of the above diagrams

$$N \rightarrow P + M^-$$
$$P \rightarrow N + M^+$$

and reflected the basic interaction energy term in the Hamiltonian could account for the magnetic moments of the proton and neutron. A neutron during "a certain fraction of its existence finds itself in a virtual state "consisting of a proton and a negative mesotron." A magnetic field acts upon the "constituents" of the neutron so that one is led to attribute, "by an average," a magnetic moment to the neutron. Brillouin then reported on the calculations of Fröhlich and Heitler, who had obtained very acceptable values for the magnetic moment of the proton and neutron.

2.4 The Washington Conference of 1941

In order to get a fuller view of the problems that "elementary" particle physicists were confronting, let me briefly consider two other conferences that were to convene at the end of the 1930s to deal with the subject of elementary particles. The eighth Solvay conference had been planned to be held in Brussels October 22–29, 1939. It was canceled because France and Great Britain declared war on Germany on September 1, 1939, following Hitler's invasion of Poland. Some forty persons had been invited to attend. The list included H. A. Bethe, F. Bloch, N. Bohr, P.A.M. Dirac, A. Einstein, E. Fermi, G. Gamow, W. Heisenberg, W. Heitler, H. A. Kramers, W. Pauli, A. Proca, G. E. Uhlenbeck, C. F. von Weizsäcker,

E. J. Williams, and H. Yukawa among the theoreticians, and C. D. Anderson, P. Auger, P.M.S. Blackett, F. Joliot-Curie (M. et Mme.), L. Meitner, and B. Rossi among the experimenters. The list of requested papers that was appended to the letter inviting the participants was as follows:

INSTITUT INTERNATIONAL DE PHYSIQUE SOLVAY
Université Libre
Bruxelles

LISTE DES RAPPORTS DEMANDES

M. HEISENBERG.	Problèmes généraux—Equations des particules.
	Cas des électrons positifs et négatifs—Matérialisations.
	Limites d'applicabilité de la théorie des quanta.
M. Blackett.	L'électron lourd (mésoton) au point de vue expérimental.
	Rayons cosmiques pénétrants.
M. Heitler.	Le mésoton au point de vue théorique.
M. Fermi.	Le neutrino—Expérience et théorie.
M. Bethe.	Les interactions électron - proton - neutron.
	Indications tirées de l'étude des noyaux simples et des déviations—Forces d'Heisenberg et de Majorana.
M. P. Auger.	Les gerbes au point de vue expérimental.

RAPPORTS COMPLEMENTAIRES

M. L. de Broglie.	Le photon.
M. Weizsäcker.	Les indications astronomiques concernant les propriétés des particules.
M. F. Bloch.	Les moments magnétiques des protons et des neutrons.

The second conference that I want to look at is the sixth annual Washington conference of theoretical physics which was to be held in the winter of 1940, and which also had been planned to deal with the subject of elementary particles. These conferences had taken place annually in Washington since 1935 under the sponsorship of George Washington University and the Department of Terrestrial Magnetism of the Carnegie Institution of Washington. Its organizers were G. Gamow and E. Teller from George Washington University, and J. A. Fleming and Merle Tuve from the Department of Terrestrial Magnetism. Their avowed purpose was to evolve in the United States something similar to the Copenhagen conferences that Bohr had organized in his institute during the late twenties and early thirties, in which a small number of theoretical physicists working on related problems assembled to discuss in an informal way difficulties met in their researches. The total number of persons attending any one of the Washington conferences from 1935 to 1942 never exceeded sixty, and in the ones dealing with foundational problems in physics the number was much smaller, ranging from twenty-five to thirty participants. The persons invited and attending the conferences were the active workers in the field; and since the conferences dealt primarily with subjects that could be elucidated by quantum mechanics, the participants were primarily the younger members of the physics community. For example, at the third Washington conference in 1937 that was devoted to the problems of the properties and interactions of elementary particles and the related questions of nuclear structure, the elder statesmen were N. Bohr—who at the time was actively engaged in the study of nuclear models—and J. Frank. Most of the participants had started their researches after 1925—Breit, Teller, Wigner, Gamow, Bethe, Bloch, Rabi—and several of them were of an even younger generation—Furry, Plesset, Wheeler. Wheeler, after attending his third Washington conference, wrote his thanks to the organizers of the conference and added the following statement to the letter: "This conference was more like those fruitful ones of Bohr that I knew in Copenhagen than any I have ever joined elsewhere. Bohr's own summary of the status of physics, what things are established, where are the basic problems, on what lines can progress be made, was one of the valuable things I personally got from the Conference—the general outlook that helps one to formulate his later work, most effectively."[27]

The tentative program for the sixth conference as outlined in a handwritten note by M. Tuve,[28] written during discussions with Teller and Gamow, was to include the following subjects for discussion:

1. Mean calculations—nuclear forces

2. Weisskopf   self energy of electron
      Also Dirac two years ago

3. Quadruple moment of deuteron

4. Cosmic ray mesons—annihilation } cross sections
      creation

5. What experiments will yield essential
      information about role of electron?

6. [illegible] interaction of nuclei and
      electrons? [illegible] demonstrated by Rabi

The list of people that were to be invited included Fermi, Bethe, Breit, Rabi, Weisskopf, Wheeler, Feenberg, Furry, Wigner, Uehling, Nordsieck, Oppenheimer, Lamb, Critchfield, Nordheim, Van Vleck, Uhlenbeck. Carl Anderson? Compton? Rossi? Street? The question marks implied that there was some doubt whether experimentalists were to be invited. A subsequent memorandum issued by Fleming was more detailed. It revealed that the subjects, "The Elementary Particles" and "The Interior of the Earth," were both being considered as possible topics for the conference. In weighing the issue of which of the two topics the conference ought to address Fleming indicated that "a particular reason for holding a conference this winter on this subject [of elementary particles] is the fact that we need help in directing our experimental program toward a subject which is regarded as of the same importance as the problem of the forces between heavy particles; namely, what is the role of the electron in the nucleus, and what are the laws governing electron interactions at very close distances between particles."[29]

Fleming was referring to the recent work by Rabi on the electron-neutron interaction. The topics for discussion that Fleming outlined in his memorandum were expanded versions of the notes he had made in his discussions with Teller and Gamow:

> **1.** Calculations concerned with the heavy electron in relation to nuclear forces...
>
> **2.** Calculations concerned with the quadrupole moment of the deuteron in particular relation to the magnetic moments of proton, neutron, and deuteron and also the magnetic moments of more complex particles.
>
> **3.** Calculations concerned with the self-energy of the electron.
>
> Dirac published a paper two years ago which appears to be consistent with relativity, and Weisskopf has recently made calculations of another type showing that the divergence for an electron is no longer as $1/r$, but because of the positron, is divergent only as $\log r$.
>
> **4.** Calculations concerned with the cross sections for creation and annihilation of the heavy electron, having particular reference to the cosmic ray data.
>
> **5.** A discussion of possible non-electrical interactions between electrons and nuclei, and perhaps between electrons and electrons.

In fact, the sixth Washington conference which was held in March 1940 dealt with the subject of "The Interior of the Earth." It was the seventh annual Washington Conference of Theoretical Physics, which convened from May 22 to May 24, 1941, that dealt with the theory of elementary particles. Besides the sixteen physicists who had been invited as "convenors" of the conference, eighteen others took part. The report of the conference[30] gave a summary of the various sessions that were held. On the 22nd, Oppenheimer led a discussion on the theory of the meson. The main problems discussed were the magnitude of its spin and the value of its magnetic moment. During these discussions it was suggested that the meson may have spin zero but be a pseudoscalar particle, that is, that its wave function changes sign under a spatial inversion of the coordinate system. The difficulties in accounting for both the "strong" interaction of mesons with nuclei as manifested by multiple meson production in high-altitude cosmic ray phenomena and the "weak" scattering of mesons by nuclei—as seen at sea level—were stressed. On the second day, Wigner led the discussions that centered on the structure of complex nuclei and their interaction with electromagnetic radiation. Weisskopf was the discussion leader for the third session, which dealt with "the nature of the forces represented by various fields which have been used both in classical physics and in modern theory." The question of "artificial" meson production in the new cyclotrons that were then being built was also discussed. Weisskopf presented the results of his research in quantum electrodynamics on the interaction between charged particles at small distances, and Wheeler outlined his "radical suggestion" that would abandon completely the concept of field and would reintroduce instead the idea of interaction of particles at a distance, work which he had carried out with R. P. Feynman.

2.5 The Divergences

In this section the status of the divergence difficulties at the end of the 1930s is outlined.

2.5.1 Zero Point Energy, Fluctuations

The first divergence of the quantum theory of fields was encountered by Jordan in his quantum-mechanical treatment of the vibrating string in the *Dreimännerarbeit* (Born et al. 1926). Jordan discarded this infinity by dropping it, thereby performing the first infinite subtraction, or renormalization, in quantum field theory.

The origin of the divergence is quite simple. The transverse displacement, $U(x, t)$, of the string (of linear density ρ and tension T) fixed at $x = 0$ and $x = \ell$ was specified in terms of the Fourier coefficients $q_k(t)$,

$$U(x,t) = \sum_{k=1}^{\infty} q_k(t)\sqrt{\frac{2}{\ell}} \sin \frac{\pi k}{\ell}x. \tag{2.5.1}$$

The equation of motion of the string[31]

$$\rho\frac{\partial^2 u}{\partial t^2} = T\frac{\partial^2 u}{\partial x^2} \tag{2.5.2}$$

transcribes into the following equations for the $q_k(t)$:

$$\frac{d^2 q_k(t)}{dt^2} + \omega_k^2 q_k(t) = 0, \tag{2.5.3}$$

with

$$\omega_k = \sqrt{\frac{T}{\rho}} \frac{\pi k}{\ell}. \tag{2.5.4}$$

Similarly, the total energy

$$H = \frac{1}{2} \int dx \left\{ \rho \left(\frac{\partial u}{\partial t} \right)^2 + T \left(\frac{\partial u}{\partial x} \right)^2 \right\} \tag{2.5.5}$$

when expressed in terms of the q_k's and $\dot{q}_k$'s becomes

$$H = \frac{1}{2} \sum_{k=1}^{\infty} \left\{ \rho \dot{q}_k^2(t) + \left(\frac{\pi k}{\ell} \right)^2 T q_k^2 \right\}. \tag{2.5.6}$$

In terms of the canonical variables $P_k = \rho q_k, Q_k = \rho q_k$ the Hamiltonian is given by

$$H = \sum_{k=1}^{\infty} \frac{1}{2} (P_k^2 + \omega_k^2 Q_k^2). \tag{2.5.7}$$

H is thus a sum of independent harmonic oscillators. Each oscillator is quantized separately by imposing the commutation rules

$$[Q_k, P_k] = i \frac{h}{2\pi}. \tag{2.5.8}$$

The energy of the kth oscillator is then given by $\hbar \omega_k (n_k + \frac{1}{2})$, where n_k can take on the values $0,1,2,\ldots$ and the factor $1/2 \hbar \omega_k$ represents the zero point energy of that oscillator. Jordan's infinite subtraction consisted in dropping the sum $\sum_k \frac{1}{2} \hbar \omega_k$ (the zero point energy of all the oscillators) from H. Note that the time evolution of an observable, $O(t)$,

$$i \frac{dO(t)}{dt} = [H, O(t)], \tag{2.5.9}$$

is not affected by this subtraction since (presumably) any constant term commutes with $O(t)$. Equivalently, this infinite constant cancels out in the integrated form of (2.5.9),

$$O(t) \; = \; e^{\frac{i}{\hbar}Ht}O(0)e^{-\frac{i}{\hbar}Ht}. \tag{2.5.10}$$

The dropping of the zero point energy became transcribed into a formal rule: the Hamiltonian of the field system should be so defined that its expectation value in the vacuum state is equal to zero (Rosenfeld and Solomon 1931; Pauli 1933, p. 256).[32] For example, Pauli in his *Handbuch* article wrote the energy density of the electromagnetic field in such a way that the zero point energy automatically dropped out:

$$H(x) \; = \; \frac{1}{2}(\mathcal{E}^2(x) + \mathcal{H}^2(x)) + i\left[\mathcal{E}(x), \frac{1}{\sqrt{-\nabla^2}}\nabla \times \mathcal{H}(x)\right], \tag{2.5.11}$$

where $\mathcal{E}$ and $\mathcal{H}$ are the operators describing the electric and magnetic field, respectively.

The meaning of expressions such as (2.5.11) was clarified only after World War II with the introduction of Wick ordering (Houriet and Kind 1950; Wick 1950). In terms of Wick ordering, (2.5.11) can be written as

$$H(x) \; = \; \frac{1}{2}\{: \mathcal{E}^2(x) : + : \mathcal{H}^2(x) :\}, \tag{2.5.12}$$

where the Wick expression $\mathcal{E}^2(x):$ is to be understood as the following "point splitting" operation:

$$: \mathcal{E}^2(x) : \; = \; \lim_{x \to y}[\mathcal{E}(x) \cdot \mathcal{E}(y) - (\Phi_0, \mathcal{E}(x) \cdot \mathcal{E}(y)\Phi_0)]$$

$$\Phi_0 = \text{vacuum state}. \tag{2.5.13}$$

The efficacy of Wick's formula stems from the fact that $\mathcal{E}(x) \cdot \mathcal{E}(y)$ has a singularity at $x = y$ that is canceled by $(\Phi_0, \mathcal{E}(x) \cdot \mathcal{E}(y)\Phi_0)$. In fact, this subtraction cancels the singularity in all physically significant matrix elements. That the cancelation of the singularity in the vacuum expectation value is sufficient to cancel all singularities was also recognized only much later and forms the subject of operator product expansions developed by Wilson and others in the 1970s. (For a review, see Wightman 1977, 1978a,b, 1986.)

The fact that the vacuum expectation value of the energy could be renormalized to zero does not imply that the mean square value of the energy vanishes. These mean square fluctuations have observable consequences. They were essential in the derivation by Jordan of the Einstein mean square energy fluctuation formula in the *Dreimännerarbeit*. However, in 1931 Heisenberg showed that Jordan's derivation was incorrect: the energy fluctuation was in fact infinite. The infinity stemmed from the fact that the fluctuation had been calculated for a sharply defined region. The fluctuation in the energy in a volume v averaged over a time interval $(t, t + T)$

$$\frac{1}{T} \int_t^{t+T} dt \int_v d^3x\, H(x) \qquad (2.5.14)$$

is infinite if the volume v has sharp boundaries. Only by smoothing the boundary of the volume v will the fluctuations be finite. The recognition that in quantum field theory only smeared operators—that is, only operators suitably averaged over small regions of space-time—make sense was a central idea in Bohr and Rosenfeld's paper on the measurability of the electromagnetic field (Bohr and Rosenfeld 1933). The detailed history of how that important paper came to be written has not yet been told. In rough outline the story is as follows (Rosenfeld 1955; Peierls 1963; Kalckar 1971). The quantization of a field à la Heisenberg and Pauli implies that there will be a limit to the accuracy with which one can measure such a field. Just as the quantum rules $[q, p] = i\hbar$ for a particle imply that the uncertainties in the momentum and position coordinates of the particle must be such that $\delta p\, \delta q \geq h$, similarly there will be uncertainty relations limiting the accuracy of certain field measurements. There is, however, a further complication in the case of field measurements: the uncertainties in the field variables at a given *point* in space are infinite (a reflection of the δ function in the right-hand side of the commutator of two field operators defined at the points x, t and x', t). In his 1929 Chicago lectures Heisenberg demonstrated that the best one can hope to do is to measure an average over a small region. If one does this, then for a region of linear dimension L, the uncertainty relations for the electromagnetic field were found to be of the form

$$\delta \mathcal{E}_i\, \delta \mathcal{H}_j > \frac{ch}{L^4}(i \neq j, i, j = x, y, z) \qquad (2.5.15)$$

(Heisenberg 1930a). Hence as the region gets smaller the fluctuations become larger.

The uncertainty relations for the electromagnetic field connect uncertainties in simultaneous measurements of components of the electric and magnetic fields in different directions; there are no such limitations on the simultaneous measurements of components of the electric field by themselves or of the components of the magnetic field by themselves. However, in 1931 Landau and Peierls claimed that it was not actually possible in practice for any kind of device to measure components of one of the fields by themselves beyond a certain limit of accuracy:

the radiation emitted would interfere with the field of the test bodies that were being used to observe the original fields.

Bohr disagreed sharply with the view expressed by Landau and Peierls, and in a famous paper with Rosenfeld demonstrated that their claims were wrong (Bohr and Rosenfeld 1933). Bohr and Rosenfeld pointed out that measurements at a point are not possible because extended test bodies have to be used to make field measurements. The only observables are space averages of the field operators. They showed that one could in principle construct devices that would measure one component of the field averaged over a finite volume (or over a finite time) to any degree of accuracy. Bohr and Rosenfeld further demonstrated the consistency of the uncertainty relations implied by the commutation rules of the electromagnetic field components.[33]

2.5.2 Dirac's Hole Theory and Vacuum Polarization

Heisenberg thought that Dirac's "discovery of antimatter was perhaps the biggest jump of all the big jumps in physics of our century. It was a discovery of utmost importance because it changed our whole picture of matter" (Mehra 1973, p. 271). Hole theory implied that a light quantum could create an electron-positron pair, which formally meant that the number of particles with a given charge was no longer a good quantum number and that therefore there was no longer a conservation law for the number of particles. Another consequence of Dirac's hole theory was that the vacuum became a complex entity: it was a state filled with an infinite number of particles of negative energy that were not observable. In his interview with Kuhn, Heisenberg spoke of the mood of the early thirties:

> It is very difficult to describe that state because it was psychologically so different from the state in 1923 or 1924. In 23 and 24 we knew there were difficulties and we also had the feeling that we were quite close to the final solution of these difficulties. Just one step and we will be in the new field. It was as if we were just before entering the harbor, while in this later period we were just going out into the sea again, i.e., all kinds of difficulty coming up and really we didn't know where it would lead to. And even if new and good ideas came up, these ideas would work a short way and then again one had new difficulties. It was clearly seen that this was now an entirely new story. So nobody expected quick results at that time. (Heisenberg 1963)

Heisenberg, in 1929, became so frustrated with the difficulties presented by Dirac's equation and its negative energy states that he turned to the problem of ferromagnetism in order not to be "forever irritated by Dirac." Nonetheless, both he and Pauli worked hard throughout the thirties to understand the novel features introduced by the hole theory. Their correspondence between 1928 and 1939 (one of the great scientific correspondences) is an impressive testament to their intellectual power and vigor (Pauli 1985).

Once the problem of transitions to negative energy states had been solved by Dirac's "hole theory," the next "irritation" proved to be the problem of the polarization of the vacuum.

Dirac wrote a paper[34] on the "Theory of the Positron" for the seventh Solvay Congress that took place in Brussels October 22–29, 1933. In its opening sentence, Dirac noted: "The recent discovery of the $+ve$ charged electron or *positron* has revived interest in an old theory [*sic*] about the states of negative kinetic energy of an electron, as the experimental results that have been obtained so far are in agreement with the predictions of the theory."

Noting that in a quantum theory the negative energy states cannot be excluded and that it is not possible to assume that the energy of a particle is always positive "without getting inconsistencies in the theory," Dirac faced the two courses that were open: "Either we find some meaning for the negative-energy states *or* we must say that the relativistic quantum theory of the electron is inaccurate to the extent that it predicts transitions from positive to negative energy states."

Arguing that there does not seem to be any fundamental reason why the present quantum mechanics should not be applicable for processes involving energy changes of the order of $2 mc^2$ that involve transition to negative states—and which are necessary to account for Thomson scattering—Dirac suggested that "it would seem that the more reasonable course is to try to get some physical meaning for the negative-energy states." He left open the possibility that the theory might break down at energies of the order of $hc/e^2 mc^2$, since as Bohr had emphasized, "The present quantum mechanics cannot be expected to apply to phenomena in which distances of the order of the classical radius of the electron, e^2/mc^2, are important, since the present theory cannot give any account of the structure of the electron, but such distances . . . correspond to energies of the order $hc/e^2 mc^2$, which is much greater than the energy changes above mentioned."

Dirac then proceeded to review his formulation of the hole theory, sharpening his previous statements somewhat so that they could readily be applied to the problem at hand. "A hole can be described by a Schrödinger wave function like that describing the motion of an electron, and second, that the hole is deflected by a field in the same way as positive energy electron." Dirac also stated two important inferences: "The mass of the positron must exactly equal that of the electron and its charge must be exactly opposite to that of the electron." A further assumption Dirac made was that the distribution of electrons in which all positive energy states were unoccupied and all negative energy states occupied produce no electric field. Only *departures* from this distribution produced a field in accordance with Maxwell's equation

$$\text{div}\mathscr{E} = -4\pi(\rho - <\rho>_{\text{vacuum}}). \tag{2.5.16}$$

The question Dirac then addressed was: What happens when there is an external electric field present? In a region of space in which there is no field, the division of the states into positive and negative energy states is unambiguous and the above stipulations do not result in any difficulties. However, "when applied to space in which there is an electromagnetic field, ... one must specify just which distribution of electrons is assumed to produce no field and one must also give some rule for subtracting this distribution from the actually occurring distribution in any particular problem" to get a finite difference to be used in the right-hand side of the div $\mathscr{E} = -4\pi\rho$ equation, "since in general the mathematical process of subtracting one infinity from another is ambiguous."

Dirac went on to consider the case of a weak, time-independent electrostatic field, "in which the necessary assumptions seem fairly obvious." The field was assumed to be sufficiently weak "for a perturbation method to be applicable." He then found "that the distribution which produces no field does not satisfy the equations of motion. By subtracting this distribution from that distribution which does satisfy the equations of motion and which corresponds to a state with no electrons or positrons present we are left with a difference which can be interpreted physically as an effect of polarization of the distribution of negative-energy electron by the electric field."

In his calculation Dirac used a density matrix description in the Hartree-Fock approximation—a method he had developed to describe complex atoms (Dirac 1930c). The density matrix is defined as

$$(q'|R|q'') = \sum_r \bar{\psi}(q)\psi(q), \tag{2.5.17}$$

where the sum runs over all occupied (negative energy) states. Its equation of motion is

$$i\hbar\dot{R} = [H,R], \tag{2.5.18}$$

where

$$H = c\rho_1(\boldsymbol{\sigma} \cdot \mathbf{p}) + \rho_3 mc^2 - eV \tag{2.5.19}$$

is the Hamiltonian for an electron in the electric field $\mathscr{E} = -\nabla V$. Furthermore, since the distribution must satisfy the Pauli principle,

$$R^2 = R. \tag{2.5.20}$$

Let R_0 be the distribution that produces no field. Dirac assumed it to be given by the gauge-invariant expression

$$R_0 = \frac{1}{2}\left(1 - \frac{W}{|W|}\right), \qquad (2.5.21)$$

where

$$W = c\rho_1(\boldsymbol{\sigma} \cdot \mathbf{p}) + \rho_3 mc^2 \qquad (2.5.22)$$

is the free-particle Hamiltonian. In the representation in which W is diagonal, R_0 is diagonal and has eigenvalues 0 or 1 depending on whether W is positive or negative. (Note that $[H, R_0] \neq 0$ so that R_0 does not satisfy the equation of motion.) Dirac then solved the equations $R^2 = R$ and $[H, R] = 0$ by looking for solutions in a power series in V:

$$R = R_0 + R_1, \qquad (2.5.23)$$

where R_1 is assumed to be order V. The quantity of physical interest is the electric charge density corresponding to R_1; it is given by $(-e)$ times the diagonal sum with respect to the spin variables of R_1 evaluated at $q' = q'' = x$. Dirac denoted this quantity by $D(R_1)$.

"After a complicated integration" Dirac found that $(x|D(R_1)|x)$ is logarithmically divergent, but if the domain of integration is cut off at P, the final result is

$$-e(x|D(R_1)|x) = -\frac{e^2}{hc}\frac{2}{3\pi}\left(\ln\frac{2P}{mc} - \frac{5}{6}\right) - \frac{2}{15\pi}\frac{e^2}{hc}\left(\frac{h}{mc}\right)^2 \nabla^2\rho, \qquad (2.5.24)$$

where ρ is the charge density producing the potential V, so that $\nabla^2 V = 4\pi\rho$, and terms involving higher derivatives of ρ than the second have been neglected.[35]

For a cut-off $\frac{P}{mc} \approx 137$, that is, assuming that the theory is not valid for energies greater than $\frac{hc}{e^2}mc^2$, the first term is approximately equal to $\frac{e^2}{hc}\rho$, which Dirac explained meant "that there is no induced electric density except at the places where the electric density producing the field is situated, and at these places the induced electric density cancels a fraction of 1/137 of the electric density producing the field." The second term is a finite correction that arises when the charge density producing the field varies rapidly with position and changes appreciably through a distance of the order h/mc. Dirac concluded his report with the statement: "As a result of the foregoing calculation, it would seem that the electric charges which one ordinarily observes on electrons and protons and the other particles of physics are not the actual charges which these particles carry (appearing in the fundamental equations) but are all slightly smaller, in the ratio of about 136/137."

The physical meaning of these results was summarized in a letter Dirac wrote Bohr after he completed his Solvay report:

Dear Bohr,

Peierls and I have been looking into the question of the change in the distribution of negative-energy electrons produced by a static electric field. We find that this changed distribution causes a partial neutralization of the charge producing the field. If it is assumed that the relativistic wave equation is exact, for all energies of the electron, then the neutralisation would be complete and electric charges would never be observable. A more reasonable assumption to make is that the relativistic wave equation fails for energies of the order 137 mc^2. If we neglect altogether the disturbance that the field produces in negative-energy electrons with energies less than $-137\ mc^2$, then the neutralization of charge produced by the other negative-energy electrons is small and of the order 1/137. We then have a picture in which all the charged particles of physics electrons, atomic nuclei, etc. have effective charges slightly less than their real charges, the ratio being about 136/137. The effective charges are what one measures in *all* low energy experiments, and the experimentally determined value for e must be the effective charge on an electron, the real value being slightly bigger. In experiments involving energies of the order mc^2 it would be the real charge, or some intermediate value of that charge which comes into play, since the "polarisation" of the negative-energy distribution will not have then to take in its full value. Thus one would expect some small alterations in the Rutherford scattering formula, the Klein-Nishina formula, the Sommerfeld fine-structure formula, etc. when energies of the order mc^2 come into play. It should be possible to calculate these alterations approximately, since, although the ratio effective charge/real charge depends on the energy at which we assume the relativistic wave equation to break down, it does so only logarithmically, and varies by only about 12% when we double or halve this energy. If the experimenters could get sufficiently accurate data concerning these formulae, one would then have a means of verifying whether the theory of negative-energy electrons is valid for energies of the order mc^2.

I have not yet worked out the effect of magnetic fields on the negative-energy distribution. They seem to be rather more troublesome than electric fields.

With best wishes, and hoping to see you in September

With best wishes, and hoping to see you in September

Yours sincerely,

P.A.M. Dirac[36]

In his letter, Dirac had succinctly stated the physics entailed by vacuum polarization. The physical picture he sketched is still accepted at present. It was elaborated by Uehling (1935) and by Weisskopf (1936). In quantum electrodynamics the vacuum is not an empty medium but contains virtual electron-positron pairs. These are responsible for producing a dielectric constant, ϵ, in the vacuum. This dielectric constant is distance dependent: when a bare charge $e(0)$ is introduced into the vacuum the charge that is observed at a distance R is given by $e(0)/\epsilon(R)$, that is, the observed charge is reduced by an amount $\epsilon(R)$. Uehling determined that the larger R the more screening occurred: the effective charge at R increases as R decreases. (Coleman and Gross in 1973 proved that all quantum field theories—with the sole exception of non-Abelian gauge theories—have effective charges that decrease as the distance increases.) The picture can readily be generalized to the case of time-dependent electric fields, in which case the dielectric constant becomes frequency dependent. Dirac, in fact, indicated how this is to be done.

In the presence of a time-dependent electric field, the procedure Dirac had outlined in his Solvay report is ambiguous, because one cannot in general separate the solutions of the Dirac equation in the presence of the time-dependent field into positive and negative energy states in an unambiguous fashion. Nonetheless, Dirac found a way to resolve these ambiguities by exhibiting the characteristic singularities of the density matrix on the light cone. Shortly after the Solvay Congress, Dirac wrote Bohr:

> I have been working at the problem of the polarization of the distribution of negative-energy electrons, from a relativistic point of view. If I have not made a mistake, then there is just one relativistically invariant, gauge invariant treatment, which gets over all the difficulties connected with the infinites, to the accuracy with which the Hartree-Fock method applies.... I have not yet seen whether this relativistic treatment leads to any kind of compensation of charge arising from the vacuum polarization.... I shall write to Pauli about this and hope it will satisfy his objections to the theory of holes."[37]

In the paper which presented this new approach (Dirac 1934d), the off-diagonal density matrix R is again defined by

$$\sum_{\text{occupied}} \psi_n(x',t',k')\psi_n^\dagger(x'',t'',k'') = (k',x',t'|R|k'',x'',t''), \qquad (2.5.25)$$

where $(x', t', k'), (x'', t'', k'')$ are distinct sets of space-time-spin variables, and the summation is over all occupied states. The spinors ψ and its adjoint $\psi^\dagger$ are one-particle c-number Dirac wave functions in the presence of the external field, which are approximately determined by the Hartree-Fock self-consistent method. Dirac found that the characteristic singularities of $(k', x', t'|R|k'', x'', t'')$ depended only on $x^2 = (x' - x'')^2$, and that these occurred when $x^2 = 0$. Dirac proposed that the infinities to which the singularities give rise be dropped, a process we now call "charge renormalization."

Heisenberg in 1934 clarified and generalized Dirac's approach. Motivated by "the necessity to formulate the fundamental equations of the theory in a way that goes beyond the Hartree-Fock approximation method," Heisenberg introduced the *operator*

$$R(x', x'') = \psi(x')\psi^*(x''), \tag{2.5.26}$$

where $\psi(x')$ and $\psi^*(x'')$ are *quantized* field operators satisfying the Dirac equation in the presence of the external field. The vacuum expectation value of $R(x, x')$ reduces to the $(x't'|R|x''t'')$ considered by Dirac. Incidentally, Heisenberg was using a charge-symmetric formalism that treated electrons and positrons on an equal footing, so the removal of all singularities was then straightforwardly effected by considering the limit $x^2 = (x' - x'')^2 \rightarrow 0$ of the operator $R(x', x'')$. A "modern" version of Heisenberg's formulation can be stated as follows.

One first establishes that for a reasonable class of external fields, the solutions of the Dirac equation for the field operator $\psi(x)$ are well behaved and free of divergences. In terms of the operator $R(x, x')$, the current operator is defined as

$$j^\mu(x) = -e \lim_{x' \to x} Tr[(R(x, x') - \text{subtraction terms})\gamma^0\gamma_\mu], \tag{2.5.27}$$

where the trace is over the Dirac spinor indices. The choice of the subtraction terms is a matter of definition. The requirements that $j^\mu(x)$ exists as a well-defined operator on the state vectors of the system and that it be conserved, $\partial_\mu j^\mu(x) = 0$, impose constraints on the subtraction terms. That these constraints could be satisfied is essentially what Heisenberg showed in his 1934 paper.[38] In that same article Heisenberg also considered the extension of the formalism to the case that both $\psi(x)$ and $A(x)$ are quantized, that is, the full quantum electrodynamics. He noted that the subtraction terms for $R(x, x')$ could be determined order by order in perturbation theory, but that, seemingly, the higher-order terms contain divergences similar to that of the electron self-energy—divergences that the subtraction procedure could not remove. In particular he found that to order e^2, the photon self-energy diverged even before the limit $x_\mu \rightarrow 0$ was taken. Serber shortly thereafter noted that Heisenberg's result was a consequence of an incorrect canonical transformation: the gauge invariance of the theory guaranteed that the photon's mass was zero (Serber 1935).

Heisenberg's last investigation of positron theory during the 1930s was a paper he wrote jointly with his student Euler (Heisenberg and Euler 1936). In 1933 Halpern had noted that hole theory implied that a photon could scatter off another photon: the scattering of light by light (or "Halpern scattering," as it was called during the thirties). Heisenberg set his students Euler and Kockel to calculate the cross section for Halpern scattering, and after a lengthy and difficult calculation they reported that to order α^2 the scattering matrix element was finite (Euler and Kockel 1935; see also Akhieser 1937). Euler and Kockel had expressed their result in terms of an effective Lagrangian: the interaction between photons resulted from the presence in the Lagrangian of a gauge-invariant term proportional to $a(\mathscr{E} \cdot \mathscr{H})^2 + b(\mathscr{E}^2 - \mathscr{H}^2)$ (with a and b constants proportional to α^2) in addition to the "free field" Lagrangian $\frac{1}{8\pi}(\mathscr{E}^2 - \mathscr{H}^2)$.

Heisenberg and Euler's paper generalized Euler and Kockel's work. They found the effective Lagrangian to order α^3 induced by static, homogeneous, external fields when no real electron-positron pairs could be produced. Shortly thereafter, Weisskopf simplified their calculation and gave a thorough discussion of the physics involved in "charge renormalization" (Weisskopf 1936).[39]

In their paper Heisenberg and Euler also pointed out that the theory contained divergent vacuum self-energy contributions (nowadays called closed loop or "bubble" diagrams) that had to be subtracted (Fig. 2.5.1).

They also noted that in QED the fourth-order contribution to Compton scattering diverged, as did the sixth-order contribution to the scattering of light by light. The paper ended on the previously quoted pessimistic note: hole-theoretic QED must be considered provisional.

2.5.3 Hole Theory: The Self-Energy of the Electron

Before turning to the problem of the self-energy of an electron in hole-theoretic quantum electrodynamics, I want to look briefly at the problem of Thomson

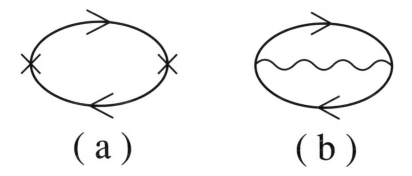

Figure 2.5.1 "Bubble" diagrams.

scattering, that is, the scattering of long-wave radiation by a charged particle. This problem has a long and venerable history in the development of the quantum theory. The relation of Thomson scattering to the weak-field Stark effect in hydrogen was the point of departure of Heisenberg's sharp criticism of Bohr's presentation at the "Bohr Festspiel" in Göttingen in 1922 (see, for example, Mehra and Rechenberg 1982, vol.1, part 2, chap. 3; Dresden 1987), pp. 124ff).

In Dirac's 1927 formulation of dispersion theory based on the nonrelativistic Hamiltonian,

$$H = \int \frac{E^2 + (\nabla \times A)^2}{2} d^3x + \frac{1}{2m}(p - eA(x))^2$$
$$= H_0 + H',$$
(2.5.28)

the only term that contributes to the Thomson scattering amplitude is the $\frac{e^2}{2m}A^2$ term (since the electron could be considered at rest so that $p = 0$), and yields in lowest order of perturbation theory the (correct) scattering amplitude[40]

$$R_{fi} \cong H'_{fi} = \frac{e^2}{2m}\frac{1}{2kV}\boldsymbol{\epsilon} \cdot \boldsymbol{\epsilon}',$$
(2.5.29)

$\boldsymbol{\epsilon}$ and $\boldsymbol{\epsilon}'$ being the polarization vector of the initial and final state photon.

Waller (1930a) considered the problem of Thomson scattering in Dirac's one-particle theory, based on the Hamiltonian

$$H = \sum \hbar k c\, a^*_{k\epsilon} a_{k\epsilon} + H_{\text{coulomb}} + \boldsymbol{\alpha} \cdot (p - eA) + \beta m.$$
(2.5.30)

Note that there is now no A^2 term, the interaction being linear in A; $H' = -e\boldsymbol{\alpha} \cdot A$. The scattering amplitude (in the limit $k = 0$) is

$$R_{fi} \cong \sum_\ell \frac{H_{f\ell}H_{\ell i}}{E_i - E_\ell}$$
$$= \frac{e^2}{2kV}\left\{ \sum_n \frac{\langle 0|\boldsymbol{\alpha} \cdot \boldsymbol{\epsilon}'|n\rangle\langle n|\boldsymbol{\alpha} \cdot \boldsymbol{\epsilon}|0\rangle}{E_0 - E_n} \right.$$
$$\left. + \sum_n \frac{\langle 0|\boldsymbol{\alpha} \cdot \boldsymbol{\epsilon}'|n\rangle\langle n|\boldsymbol{\alpha} \cdot \boldsymbol{\epsilon}|0\rangle}{E_0 - E_n} \right\},$$
(2.5.31)

where the sum $\sum_n$ runs over **all** electron states, E_+ and E_-. We can rewrite (2.5.31) as follows:[41]

$$R_{fi} = \frac{e^2}{2kV} \left\{ \langle 0|\boldsymbol{\alpha} \cdot \boldsymbol{\epsilon}' \frac{1}{E_0 - (\boldsymbol{\alpha} \cdot \boldsymbol{p} + \beta m)} \boldsymbol{\alpha} \cdot \boldsymbol{\epsilon}'|0\rangle \right.$$
$$\left. + \langle 0|\boldsymbol{\alpha} \cdot \boldsymbol{\epsilon}' \frac{1}{E_0 - (\boldsymbol{\alpha} \cdot \boldsymbol{p} + \beta m)} \boldsymbol{\alpha} \cdot \boldsymbol{\epsilon}'|0\rangle \right\} . \tag{2.5.32}$$

Upon expanding R_{fi} in powers of $\boldsymbol{p}$, the term linear in $\boldsymbol{p}$ gives zero. Hence

$$R_{fi} = \frac{e^2}{2kV} \left\{ \langle 0|\boldsymbol{\alpha} \cdot \boldsymbol{\epsilon}' \frac{1}{E_0 - \beta m} \boldsymbol{\alpha} \cdot \boldsymbol{\epsilon}|0\rangle = \langle 0|\boldsymbol{\alpha} \cdot \boldsymbol{\epsilon}' \frac{1}{E_0 - \beta m} \boldsymbol{\alpha} \cdot \boldsymbol{\epsilon}'|0\rangle \right\}$$
$$= \frac{e^2}{2kV} \left\{ \langle 0|\boldsymbol{\alpha} \cdot \boldsymbol{\epsilon}' \boldsymbol{\alpha} \cdot \boldsymbol{\epsilon} \frac{1}{E_0 + \beta m} |0\rangle = \langle 0|\boldsymbol{\alpha} \cdot \boldsymbol{\epsilon} \boldsymbol{\alpha} \cdot \boldsymbol{\epsilon}' \frac{1}{E_0 - \beta m} |0\rangle \right\}$$
$$= \frac{e^2}{2kV} \boldsymbol{\epsilon} \cdot \boldsymbol{\epsilon}' \frac{1}{m}, \tag{2.5.33}$$

since

$$\beta \, |0> \, = |0>$$
$$E \, |0> \, = m \, |0> \tag{2.5.34}$$

and

$$\alpha_i \alpha_j + \alpha_j \alpha_i = \delta_{ij}. \tag{2.5.35}$$

The Thomson amplitude is again recovered. Note, however, that to obtain it, it was necessary to include the negative energy states. In the paper in which Dirac introduced his hole theory he addressed the problem of the derivation of the Thomson amplitude in that theory. In hole theory, all the negative states are filled and the transitions to the negative energy states are prohibited by the Pauli principle. However, in the vacuum there are virtual transitions to positive energy states and back. These are unobservable, since they are always present, and transition rates for any physical process must be calculated relative to this vacuum transition rate. Hence the Thomson amplitude is now given by[42]

$$R_{fi} = \frac{e^2}{2kV} \left\{ \sum_{n+} \frac{\langle 0 \mid \boldsymbol{\alpha} \cdot \boldsymbol{\epsilon}' \mid n\rangle\langle n \mid \boldsymbol{\alpha} \cdot \boldsymbol{\epsilon} \mid 0\rangle}{E_0 - E_n} + \sum_{n+} \frac{\langle 0 \mid \boldsymbol{\alpha} \cdot \boldsymbol{\epsilon} \mid n\rangle\langle n \mid \boldsymbol{\alpha} \cdot \boldsymbol{\epsilon}' \mid 0\rangle}{E_0 - E_n} \right\}$$
$$- \left\{ \sum_{n-} \frac{\langle n \mid \boldsymbol{\alpha} \cdot \boldsymbol{\epsilon}' \mid 0\rangle\langle 0 \mid \boldsymbol{\alpha} \cdot \boldsymbol{\epsilon} \mid n\rangle}{E_0 - E_n} - \sum_{n-} \frac{\langle n \mid \boldsymbol{\alpha} \cdot \boldsymbol{\epsilon} \mid 0\rangle\langle 0 \mid \boldsymbol{\alpha} \cdot \boldsymbol{\epsilon}' \mid n\rangle}{E_0 - E_n} \right\} . \tag{2.5.36}$$

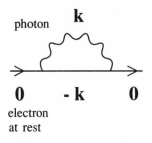

photon **k**

0   - k   0

electron
at rest

Figure 2.5.2 Self-energy diagram.

The last two terms in eq. (2.5.25), upon changing the denominators to read $-(E_o - E_n)$ and commuting the order of the (c-number) matrix elements in the numerators, are readily seen to be identical with the first two terms, except that the sum runs over all **negative** energy states. Hence combining $\sum_{n+} + \sum_{n-}$ yields the previous formula, eq. (2.5.31), obtained in one-particle Dirac theory, which in turn yields the Thomson amplitude.

The situation for the self-energy of an electron is more complicated. It was Waller (1930b) who first calculated the self-energy in Dirac one-particle theory. For an electron at rest, the expression for the self-energy is given by

$$\Delta E = \sum' \frac{|< n \,|\, H' \,|\, 0 >|^2}{E_0 - E_n},$$ (2.5.37)

which corresponds to the diagram of figure 2.5.2.

The denominator in (2.5.37) has the value

$$E_0 - E_n = \left\{ m \mp \sqrt{m^2 + k^2} - k \right\}.$$ (2.5.38)

If n belongs to a negative energy state, $E_0 - E_n \to m$ for large k. For large k the numerator in (2.5.26) goes like

$$|< n \,|\, H' \,|\, 0 >|^2 \sim \frac{e^2}{k},$$ (2.5.39)

and since

$$\sum_n \to \int d^3k = \int_0^\infty k^2 dk,$$ (2.5.40)

we obtain that

$$\delta E \sim \propto \int \frac{k\,dk}{m},$$ (2.5.41)

that is, ΔE diverges quadratically.

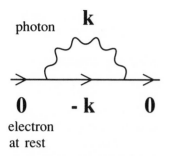 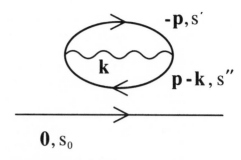

photon   **k**

0   **- k**   **0**

electron
at rest

0, s_0

Figure 2.5.3a Self-energy diagrams
in Hole theory.

Figure 2.5.3b

It would seem that in hole theory the denominators can only go like k for large k and hence the ΔE (hole theory) $\propto \int k\,dk\,\frac{1}{k}$, that is, it would be linearly divergent. But the contributions from those vacuum fluctuations which are forbidden by virtue of the presence of the electron give rise to similar terms, but of opposite sign, and the term turns out to be of order $\int \frac{dk}{k}$, that is, only logarithmically divergent (Weisskopf 1934, 1939). In hole theory, the transverse self-energy (coming from the term $-e\boldsymbol{\alpha} \cdot \boldsymbol{A}$) gets two contributions as indicated in figure 2.5.3a,b.

Since the true vacuum—the sea of negative energy electrons—is taken as the zero of energy, we must consider the processes that are forbidden in the vacuum that would have been possible if the electron were not present. These are indicated in the diagram of figure 2.5.4.

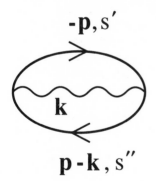

Figure 2.5.4 Vacuum Self-energy
diagram.

Now, the contributions coming from figures 2.5.3b and 2.5.4 have opposite signs and cancel, except for one term which is present in 2.5.4 but absent in 2.5.3b: the term when $p = 0$ and $s = s_0$.

Thus in hole theory the transverse self-energy is given by

$$\Delta E = \sum_{k_+} \sum_{\epsilon} \sum_{s} \frac{e^2}{2kV} \frac{\langle 0_1 s_0 \mid \boldsymbol{\alpha} \cdot \boldsymbol{\epsilon} \mid -\boldsymbol{k}, s \rangle \langle -\boldsymbol{k}, s \mid \boldsymbol{\alpha} \cdot \boldsymbol{\epsilon} \mid 0 s_0 \rangle}{m - k - E_k}$$
$$- \sum_{k_-} \sum_{\epsilon} \sum_{s} \frac{e^2}{2kV} \frac{\langle ks \mid \boldsymbol{\alpha} \cdot \boldsymbol{\epsilon} \mid 0, s_0 \rangle \langle 0, s_0 \mid \boldsymbol{\alpha} \cdot \boldsymbol{\epsilon} \mid 0s \rangle}{-E_k - m - k}. \qquad (2.5.42)$$

Weisskopf in his 1934 calculation of the self-energy of an electron in hole theory had made a mistake in sign. Upon reading his result, Furry wrote him:

> The *Zs f. Physik* containing your paper on the proper energy of the electron arrived here some days ago. Professor Oppenheimer and I were much interested in the method you followed and note that it is in complete agreement with the methods of "pair theory," as we formulated it in our *Physical Review* paper. The result for the electrostatic proper energy was new to us, as for some reason we had not previously realized the need of recalculating it. We are thoroughly convinced that your result is right, and think it quite interesting that it is of order $\int \frac{dk}{k}$. We are, however, not able to agree with your result for the magnetic proper energy. About a year ago, at Prof. Bohr's suggestion, Dr. Carlson and I made this calculation. Our result was

$$E^D = \frac{mc}{\sqrt{m^2c^2 + p^2}} \left(1 - \frac{4}{3}\frac{p^2}{m^2c^2} \right) \frac{e^2}{hc} \int \frac{dk}{k} + \text{finite terms}$$

> and is of the same order as your result for the electrostatic proper energy.
>
> The point at which we differ with you is found on page 38 of your paper. The assertion that the last term in your expression for E^D_{vak+1} is equal to J^k_- seems to us to be definitely false, and we believe it must be due to mistake in calculation. Similarly the earlier statement that $E^D_{vak} = \sum_{k=3,4} \int J^k_-(p)dp$ is incorrect. In both these cases we get for J^k_- an integral whose integrand has the denominator

$$\sqrt{m^2c^2 + (p+k)^2} + |k| + \sqrt{m^2c^2 + p^2}$$

> instead of the much smaller denominator

$$\sqrt{m^2c^2 + (p+k)^2} - |k| + \sqrt{m^2c^2 + p^2}.$$

> This means that the contribution to E^D from the negative K.E. states is much **smaller** than in the calculation with empty negative K.E. states, instead of merely being opposite in sign. This is

physically evident when we use the light quantum picture and treat the energy displacement as due to the transitions of the system from a given state to other states and back with corresponding emission and reabsorption of light quanta. For in the case of transition to a state of large negative K.E. and simultaneous emission of a quantum there is a high degree of resonance, and in the case of simultaneous pair production and light quantum emission such a resonance is lacking.

Our result which I quoted above has not been checked in detail and I cannot absolutely guarantee its accuracy; but I am quite sure that the above remarks about the form of the energy denominator and its influence on the order of magnitude must be right.

Sincerely Yours

W. H. Furry[43]

Acknowledging his mistake, Weisskopf wrote Furry, "You are *quite* right. The self-energy diverges only logarithmically and I have already sent a correction to the Zeitschrift für Physik. However my calculation yields $E^D = \frac{mc}{\sqrt{m^2c^2+p^2}}(2 - \frac{1}{3}\frac{p^2}{m^2c^2})\frac{e^2}{hc}\int \frac{dk}{k}$ + finite terms."[44] Heisenberg, upon becoming aware of the results of Weisskopf's calculation, repeated it and found

$$E^S = \frac{e^2}{h\sqrt{m^2c^2+p^2}}(2m^2c^2 + p^2)\int \frac{dk}{k}$$

$$E^D = \frac{mc}{h\sqrt{m^2c^2+p^2}}\left(m^2c^2 - \frac{4}{3}\right)p^2\int \frac{dk}{k}. \qquad (2.5.43)$$

He characterized these results in a letter to Weisskopf as "implausible and suspicious."[45] This because "one must expect on relativistic invariance grounds that

$$E^S + E^D = \text{constant}\frac{e^2}{h}\sqrt{m^2c^2+p^2}\int \frac{dk}{k}." \qquad (2.5.44)$$

The lack of proper relativistic covariance was to plague all self-energy calculations in the prewar period.

The most detailed analysis of the self energy problem and of the electromagnetic field of an electron was given by Weisskopf in an important paper published in 1939. He there evaluated the expression for the transverse self-energy.

Since

$$\sum_{\epsilon,s} |\langle -k, s \mid \boldsymbol{\alpha} \cdot \boldsymbol{\epsilon} - s_0 + \rangle|^2 = \frac{2k^2}{2E_k(m + E_k)}$$

$$\sum_{\epsilon,s} |\langle 0, s_0 \mid \boldsymbol{\alpha} \cdot \boldsymbol{\epsilon} \mid k, s - \rangle|^2 = \frac{E_k + m}{E_k}, \tag{2.5.45}$$

he obtained the result

$$\Delta E = \frac{\alpha}{\pi} m \int \frac{k \, dk}{E_k(k + (k + E_k))} \quad ; \quad \alpha = \frac{e^2}{4\pi}, \tag{2.5.46}$$

so that for large k, $\Delta E \sim \int \frac{dk}{k}$ and the integral is logarithmically divergent.
Weisskopf's analysis was based on the Hamiltonian

$$H = H_0^{(\text{em})} + H^{(\text{electron})} + H_{\text{int}}, \tag{2.5.47}$$

where

$$H_0^{(\text{em})} = \frac{1}{2} \int (\nabla \times A)^2 + (E)^2 d^3 x \tag{2.5.48}$$

is the Hamiltonian of the electromagnetic field,

$$H_0^{(\text{electron})} = \int \psi^+ (\boldsymbol{\alpha} \cdot \boldsymbol{p} + \beta m) \psi d^3 x \tag{2.5.49}$$

is the Hamiltonian of the electron-positron field, and

$$\begin{aligned} H_{int} = &- e \int \psi^+ \boldsymbol{\alpha} \cdot A \psi d^3 x \\ &+ e^2 \int d^3 x \int d^3 x' \frac{\psi^+(x)\psi(x)\psi^+(x')\psi(x')}{4\pi \mid x - x' \mid} \end{aligned} \tag{2.5.50}$$

describes the interaction with the radiation field and the Coulomb interactions. The electron-positron field operator, $\psi(x)$, has the usual representation

$$\psi(x) = \frac{1}{\sqrt{V}} \sum_{k,s} \left\{ u_{k,s}^{(+)} e^{ik \cdot x} b_{k,s} + u_{-k,s}^{(-)} e^{ik \cdot x} d_{-k,s}^* \right\}, \tag{2.5.51}$$

with $b_{k,s}$ an annihilation operator for an electron of momentum k; s, $d^*_{k,s}$ a creation operator for a positron of momentum k spin s; and $u^{\pm}_{k,s}$ are positive $(+)$ and negative $(-)$ energy Dirac spinors for momentum k, spin s. Using the Hamiltonian (2.5.47) there are two contributions to the self-energy of an electron to second order in e: (1) a first-order perturbation term from the Coulomb energy, (2) a second-order perturbation term from the $\int j \cdot A d^3 x$ term, the transverse self-energy. This "transverse" self-energy was further analyzed by Weisskopf into two parts: (1) a magnetic self-energy due to currents which give rise to magnetic moment, and (2) fluctuation self-energy. One readily verifies that the $j \cdot A$ term to second order gives the previously computed term, eq. (2.5.46).

The Coulomb self-energy is defined as

$$\Delta E^{\text{coulomb}} = < H_{\text{coul}} >_{1 \text{ electron at rest}} - < H_{\text{coul}} >_{\text{vac}}, \qquad (2.5.52)$$

where

$$H_{\text{coul}} = \frac{e^2}{2} \int \int d^3 x \, d^3 x' \frac{\psi^+(x)\psi(x)\psi^+(x')\psi(x')}{4\pi \mid x - x' \mid}. \qquad (2.5.53)$$

Let

$$\xi = < \pi(x)\rho(x') >_{1 \text{ electron at rest}} - < \rho(x)\rho(x') >_{\text{vacuum}} \qquad (2.5.54)$$

with

$$\rho(x) = e\psi^+(x)\psi(x). \qquad (2.5.55)$$

In hole theory, Weisskopf calculated that

$$\xi = \frac{1}{V} \frac{e^2}{(2\pi)^3} \int d^3 k e^{ik \cdot (x - x')} \frac{m}{E_k}. \qquad (2.5.56)$$

It is nonsingular, but nonlocal: the charge is smeared out. In one-particle theory, on the other hand,

$$\xi_{1 \text{ particle}} = \frac{e^2}{V} \delta(x - x') \qquad (2.5.57)$$

is local but singular. From these expressions Weisskopf further calculated that

$$< H_{\text{coulomb}} > = \frac{1}{8\pi} \int d^3 x \, d^3 x' \frac{\xi(x, x')}{\mid x - x' \mid} = \frac{\alpha}{\pi} \int_0^{\infty} dk \begin{cases} \frac{m}{E_k} & \text{(hole theory)} \\ 1 & \text{(1 particle theory)} \end{cases}$$

that is, the Coulomb self-energy is linearly divergent in the one-particle theory but in hole theory it is only logarithmically divergent (cf. Gell-Mann 1953).

2.5.4 Hole Theory: The Sea

In 1933 Fock (1933) and Furry and Oppenheimer (1934) introduced a formalism which dispensed talking of a sea, by introducing creation and annihilation operators for positrons and electrons. There is then no sea. There are only electrons and positrons. It is in fact the formalism that is used in the more modern treatments (see, e.g., Wentzel 1943). Yet throughout the thirties almost all calculations were done using the hole-theoretic formalism (see, e.g., Dancoff 1939). Questions associated with the gauge invariance of the Furry-Oppenheimer formalism (the decomposition into positive and negative energy states is not gauge invariant!) probably contributed to the continued use of the hole-theoretic formalism. The charge symmetric formulation, as outlined for example, in Heisenberg's (1934b) paper did make calculations less ambiguous. Thus Weisskopf in his 1939 paper on the self-energy of a free electron of momentum p, spin s adopted the charge-symmetric formalism. In positron theory the self-energy meant: "Self-energy of one electron in the state p, s plus the vacuum electron" minus "self-energy of the vacuum electrons alone." The highly divergent interaction of the extra electron with the infinite charge density of the vacuum electrons had to be removed. This was done by the process of symmetrization (Heisenberg 1934) "in which the calculation was also made using the equally justified picture that all the electrons in existence are positively charged, so that the observance of a negatively charged electron in the state $[ps]$ corresponds to a vacancy in the sea of negative energy states otherwise filled with positively charged particles. Then the results of the two methods of calculation are averaged" (Kroll 1948). The result was the avoidance of all singularities worse than logarithmic in hole-theoretic calculations.

The question is: Why did most theorists hold on to the hole-theoretic formalism so tenaciously? Undoubtedly, Dirac's authority made everyone take hole theory seriously. To some the physics that justified positron theory was more apparent in the hole-theoretic formulation. Thus as late as 1985 Peierls wrote:

> I may be old-fashioned, but I still regard the "hole theory" as a good way of describing positron theory. It seems to contain much of the essence of the situation, for example it makes it obvious that one could never have spin-one-half bosons, because the Pauli principle would not be there to prevent transitions to negative-energy states. Of course one realizes that the picture is symmetric, and one could just as easily take the electron as a hole in the sea of positrons with negative energy. And if one tries to make things quantitative one gets involved with infinite subtractions, but that one is used to anyway.

The consistency of the alternative way of looking at the situation always seemed to me to come out by a miracle, unless one knew it was equivalent to the "hole" description.

But of course it is a matter of taste, since the two ways of looking at the problem are equivalent.[46]

Heitler's book—one of the few comprehensive presentations of quantum electrodynamics—relied on the hole-theoretic formulation of QED and thus educated an entire generation of physicists based on this conceptualization. But there was another reason for the tenacity of the hole-theoretic viewpoint, or rather the resistance to adopting the field viewpoint with respect to particles. The fact is that there is "the great difference between the wave field describing a particle and the electromagnetic field describing radiation." The electromagnetic field is something measurable in principle as well as in practice, because a classical limit exists. However, the wave field representing the electron is never measurable, nor can one obtain a classical description for such waves (Peierls 1960).

3. The War and Its Aftermath

3.1 Introduction

It has been suggested that "in part at least," the explanation for the creativity of the theoretical physicists in Germany who were responsible for the development of quantum mechanics from 1920 to 1927 is to be found "in the very hostility of the Weimar intellectual milieu" (Forman 1971, p. 2), and that the resignation so characteristic of the Copenhagen interpretation of quantum mechanics during the 1930s[1] was "but the physical expression of that general, deep, cultural pessimism that is a basic theme in our time" (von Laue 1934). Similarly, part of the explanation for the creativity of American theoretical physicists in the period following World War II must surely include the prestige, confidence, and élan they acquired from the success and importance of their wartime activities. In the present chapter I explore briefly some of the factors responsible for the "atmospherics" of the American physics community as it emerged from World War II. That the hostilities affected the postwar intellectual production of the belligerents differentially is to be expected.

The upheaval created by the war is difficult to convey; in Holland, Denmark, Italy, France, and the USSR the disruption and the distress it caused were enormous. Even in Switzerland, which had stayed neutral, the impact was considerable. Most young men there spent a considerable part of the war years in military service, and even older persons served as much as three months each year in the army. The German and Italian physics community did not begin to recover from the war and the earlier intellectual migration until the 1950s. The situation in Japan is beginning to become known[2] and constitutes a most interesting story. How that community was able to contribute so significantly to the developments in theoretical physics in the post-World War II period in the face of the devastation and upheavals that the conflagration had produced constitutes an important historical problem that I am not competent to address.

Much has been written about the wartime activities of the physicists in the United States and Britain (see for example Kevles 1978; Gowing 1964), and I do not intend to repeat that story here. Rather, I focus on certain aspects of the wartime activities that were in part responsible for both the wartime and the subsequent successes. The wartime experience reinforced the close ties between experimentalists and theorists that had existed in the physics departments of American universities before the war. A pragmatic, utilitarian, instrumental outlook was de rigueur among theorists working in these laboratories. In large measure, physicists were responsible for the successes of these laboratories. Physicists had been at their helm—DuBridge at the MIT Radiation Laboratory, Compton at the Metallurgical Laboratory in Chicago, Oppenheimer at Los Alamos. Physicists discovered their powers at these laboratories, and they were quick to make use of their new status: by 1944 they were busy planning for the postwar expansion of the physics departments of the universities.

Although I shall not attempt a history of the *mentalité* of the wartime and postwar American physics community, I do want to suggest that a detailed study of its culture and its rhetoric during that period would be a valuable enterprise.

The tensions and polarities that were so evident after the war were the by-products of the wartime accomplishments. The cooperative spirit that was nurtured at these laboratories was a factor accounting for their impressive successes. The development of the atom bomb and of radar were examples on the largest scale and of the maximum intensity of scientific enterprise on the new model of *planned* research and development. As Bernal observed shortly after the war:

> [At Los Alamos] pure scientists, engineers, administrators and soldiers all worked together in a coordinated way according to a definite plan, flexible enough to allow for the necessarily unforeseen contingencies of scientific discovery. In some ways this organization was as important an invention as the liberation of atomic energy itself.
>
> The lesson that it teaches is that it is possible, given a definite end and a social will to achieve it, to attack any problem and to reach practical success in a small fraction of the time in which it might have been reached by the uncoordinated methods that ruled before the war. (Bernal 1946, p. 427)

Having experienced what it meant to work under such conditions, physicists after the war had to confront difficult questions: Should they carry out their research individually or in groups? What was the appropriate relationship between the individual scientist and the collective? How should they choose between collective activities and the tempo of progress and the planning that these implied, and the autonomy, freedom, and independence allowed by individual action? The war had made physicists aware of the dialectic between pure and applied research, between service to science for science's sake and service to the state. How to be a *pure* scientist yet a good citizen became an important issue for everyone who had been engaged in wartime research (see, for example, Bernal 1946; DuBridge 1947).

Section 3.2 takes stock of the mood of the physics community at the beginning of the war, as conveyed by remarks of K. T. Compton. The fierce single-mindedness with which American science mobilized and was mobilized for war is one of the striking facts that emerges from even a cursory study of the prewar and wartime activities of American scientists. In this and in much else physicists led the way. In section 3.3 I look at the Radiation Laboratory of MIT and its satellites.

It was in the wartime laboratories that many of the theoreticians who made their mark after World War II were molded: for example, Chew, Feynman, Goldberger, Kroll, Marshak. I detail the wartime experience of Norman Kroll at the Columbia Rad Lab in section 3.4.

The MIT Radiation Laboratory, the Metallurgical Laboratory, the Los Alamos lab, and the other wartime laboratories produced spectacular results. Because of their accomplishments, the armed services and the informed public came to view scientific research as a central activity of war. The strength of a nation was henceforth to be measured in terms of its scientific capabilities. Thereafter, everyone—the armed forces, Congress, the universities, the public at large— wanted to ensure the scientific well-being of the nation.

These wartime laboratories became the models of the large university laboratories that were to shape the institutional framework of physics after the war. However, it was not widely appreciated at the time that the research that had been carried out in the wartime laboratories had lost much of its character as fundamental research and had become programmatic research or development.[3] Philip Morrison insightfully noted in 1946: "War experience showed that the extraordinary pressures of war could bring great accomplishments from the large laboratory, and that from a heavily supported effort worthwhile results almost inevitably flowed. This was a well-learned lesson; I am afraid it is too well learned."[4] Everyone had been mesmerized by the great accomplishments.

Section 3.5 briefly looks at the transformation of physics in the United States brought about by the war. Section 3.6 highlights some of the physics problems that the community addressed in 1946, and section 3.7 considers some of the conferences held immediately after the war.

3.2 The Community in 1941

Bush, Conant, K. T. Compton—the men who assumed responsibility for the scientific war effort—had recognized early the threat posed by Hitler and fascism. By the end of the 1930s they were quietly and effectively mobilizing their own institutions and the scientific community at large to meet the challenge. Due to the efforts of its president, K. T. Compton, MIT in the summer of 1940 agreed "always to give first precedence to any important opportunity for service in the crisis; never to let this service be delayed by arguments over conditions or contracts; never to let the self-interest of the institution prevail over the interest of the nation."[5] After the fall of France, these men intensified their efforts. They also made their concern public. In a lead article in the *Technology Review* in the summer of 1941, K. T. Compton urged that nothing stand in the way of military preparedness. "Willingness to make sacrifices is a hallmark of the character essential to a high standard of civilization." It is not the mere "possession of scientific knowledge, engineering proficiency, and inventive skill" that characterized "a better man," Compton argued, but the readiness to "sacrifice the future for the present" under the threat of war.[6]

In the fall of 1941, a few months before Pearl Harbor, K. T. Compton made his fullest public statement on the occasion of the academic convocation celebrating the one hundred and seventy-fifth anniversary of the founding of Rutgers University. The subject of Compton's address was: "Scientists Face the World of 1942" (Compton 1942). "Except for one thing," Compton declared, "science approaches 1942 on the crest of a great wave of achievements" and with every expectation for "further triumphs of scientific discoveries [and] . . . pioneering applications to the arts of living." Before revealing what this impediment was and what should be done about it, Compton adumbrated a "suggestive" picture of the principal lines of activities and new developments in some of the important fields of science.

Physics, Compton claimed, was passing through a period of brilliant interpretation of the composition of the universe in terms of the basic structural elements of matter—electrons, protons, neutrons and photons—and their interactions. And he added: "Most recently and most importantly" physicists "have succeeded in analyzing and to a certain extent controlling the composition even of the atomic nucleus." Chemistry had produced an enormous number of new compounds, and chemists had studied many complex chemical substances that play a fundamental role in the structure and functioning of living organisms. "We are rapidly entering an age of plastics," Compton contended. In physical chemistry the factors that govern the rates and control chemical reactions were being elucidated. "In all these directions the principle of chemistry and of physics tend to be joined to give a unified interpretation of the properties of matter" (p. 5).

Biology was in a stage of transition from the phase "of extensive observation and classification of form and function" to that of using "physical and chemical techniques for penetrating the inner workings of biological processes." The facts of genetics were becoming understood and experimental techniques "for accelerating or modifying natural trends" were being developed. Compton declared that there is good reason to believe that biology was entering into an era of accelerating developments analogous to that into which physics passed with the discovery of the electron and radioactivity. Biochemistry was coming to the fore and was "leading the way to a field we may call biological engineering"[7] (p. 5).

Compton painted an equally glowing future for geology, a science that was being transformed by the application of quantitative physical, chemical and engineering principles. Moreover, the important progress that was being made in that field had enormous economic implications in locating and surveying oil and metal deposits.

All the sciences were being helped by the advances in mathematics, and "there are signs of the beginning of a new era in applied mathematics with the aid of new concepts of analysis and statistics and the invention of mechanical or electrical aids which permit the mathematical handling of problems hitherto inaccessible because of the magnitude of the labor involved in their solution" (p. 6). The "mechanical and electrical aids" Compton had in mind were, of course, the differential analyzer Vannevar Bush had built at MIT during the thirties and the other analog devices that were being developed for solving differential equations.

Similar impressive advances could be noted in the applied sciences, engineering, agriculture, and medicine. Compton gave as one of the reasons for this extraordinary rate of progress in the pure and applied sciences, the increasing number of able young men and women interested in careers in these fields. Another "exceedingly important" reason for the accelerated rate of scientific progress was "the fact that science is continually building on the accumulated foundation of all previous scientific discovery. Scientific progress is not a sudden burst or revolution, but it is an evolution in which nothing of the past is really lost and in which all current activity takes root and nourishment in what has gone before" (p. 6).

It was of course fitting that the president of one of the leading scientific institutions should so forcefully proclaim science as the engine of progress. The belief in progress driven by science and technology was quintessentially American and Compton, like most scientists of his day, regarded technology as the application of knowledge that had its origin in science: technologists apply the new knowledge that scientists discover. But equally important, Compton had great confidence in detailed *specialized* knowledge as a guide to decision making and valued it above general knowledge and experience. He had rebuilt MIT based on these premises.[8]

But, Compton contended, the war that was raging in Europe and Asia, and which was threatening to engulf the remaining portion of the earth still at peace, was so critical "for the welfare of the country and for the future welfare of science itself" that American scientists had no alternative but to lay aside their normal scientific work and to contribute their best efforts to bring about a speedy and satisfactory conclusion to the present emergency. "The conditions of thought and of individual activity which had been the favorable environment for scientific progress ever since the modern age of science began" (p. 89) were being menaced by the fascist totalitarian states.

Compton went on to inform his audience that since the outbreak of the war in Europe in September 1939, American scientists had in fact been deeply involved in responding to the emergency: over 50% of the chemists and over 70% of the physicists whose names were starred in *American Men of Science* were engaged in defense activities. Compton then outlined how this mobilization had been effected through the efforts of the National Academy of Sciences, the National Research Council, and particularly by the National Defense Research Committee (NDRC), which had been created in June 1940.

One of the first steps taken by the NDRC had been to establish a national roster of scientific and specialized personnel. Another was to implement a program for the emergency training of technologists. The magnitude of this Engineering Defense Training program could be inferred from the figures that Compton presented: In 1940 the total enrollment in the engineering schools in the United States was around 110,000, with about 12,000 students graduating each year. In the fall of 1942, as a result of the program, 120,000 *new* students were being enrolled. Compton informed his audience that in August 1941, the Office of Scientific Research and Development (OSRD) was established by an executive order, with Vannevar Bush as its director and NDRC as one of its operating divisions. It was directed "to coordinate, aid, and supplement scientific research activities relating to national defense carried on by the Departments of War, Navy and other agencies of the federal government" and "to develop broad and coordinated plans, to initiate and support scientific research and to perform such other duties as the President of the United States may from time to time assign to it all in the field of mechanisms, devices, instrumentalities and materials of warfare, required for national defense or of medical problems affecting national defense" (p. 19).[9]

In his address, Compton indicated that some five hundred contracts had already been approved "of which about two thirds are with educational institutions and about one third with commercial firms" involving between four thousand and five thousand people, half of these being scientists and engineers and the remainder technicians, mechanics, guards, secretaries, and other types of assistants. Should the scientists who have been engaged in these national defense projects be asked to describe their states of mind regarding their work, Compton believed they would be "practically unanimous" on the following points:

1. The work is exceedingly interesting and of a type to challenge first class scientific ability.

2. The objectives are obviously of great importance and for this reason the work is approached with enthusiastic conviction.

3. In many cases there has been some inclination to doubt whether the armed services give adequate recognition to the significance of the results which are being obtained and show as much enthusiasm as could be wished about putting the results into production and use. (Compton 1942, p. 23)

Compton argued for a partnership between scientists and the military as the most effective way of meeting the emergency. He suggested that this was in fact happening as "mutual acquaintance and understanding develop between the scientific groups and the officers with whom they work" (p. 25). He also stated his conviction "that the great program of educating the youth of the country, and specifically of developing a new generation of scientists, is so fundamentally important to the future of our country that every effort must be made to maintain the educational programs of our universities just as effectively as the conditions of the emergency permit" (p. 25).

Compton concluded his address by stressing that the emergency was greater than most people realized and "that we must be prepared to go much further than we have gone thus far." Scientists had already done so: "They were effectively at work, determined to do their full share for national service, and looking forward eagerly to the time when they can resume their important scientific work in an environment of freedom and peace" (p. 32).

The partnership that Compton envisioned flourished during the war. Scientists eagerly assumed their responsibilities. They in fact became the driving force in the partnership. The OSRD had been given broad powers and had put into *civilian* hands responsibility and control over the development of the weapons to be used by the army and navy—functions and powers that had been in the hands of the armed forces before the war. The relationship between science and the military in the post-World War II period was in part determined by the efforts of the armed forces to regain control over the planning and deployment of new weapons systems.

3.3 The MIT Radiation Laboratory

RADAR (radio detection and ranging) works by sending out pulsed radio waves from a transmitter powerful enough so that measurable amounts of radio energy will be reflected from the objects to be "seen." These reflected waves are then detected by a radio receiver that is usually located for convenience at the same site as the transmitter (Ridenour 1947, chap. 1).

Successful pulse radar systems had been developed independently in the United States, England, France, and Germany during the 1930s. As early as 1930 a French subsidiary of the ITT had established a telephone link across the English Channel that operated at a wavelength of 20 cm. In the face of the growing militancy of Nazi Germany, the Air Ministry of Great Britain during the winter of 1934/35 set up a committee for the Scientific Survey of Air Defense.

Among the suggestions it received was a carefully worked out plan for the detection of aircrafts by a pulsed radio signal method submitted by the Scottish physicist Robert Watson-Watt. By late spring of 1935, the first such experimental radar system was set up, and by March 1938 a set of five radar stations using 10-meter radio waves, designed to protect the Thames estuary, had been put in place. When the war broke out, work had been successfully carried out in England to develop airborne radar for both the detection of surface vessels by patrol aircraft and for the detection of enemy aircraft, using radio waves with a wavelength of 1.5 meters. The advantages of shorter wavelength radiowaves were clearly recognized—they would result in better resolution, smaller emitters and detectors, and afforded the possibility of higher power—and a push was therefore made to develop even shorter wavelength radar. These efforts succeeded, and early in 1940, 10-cm microwaves became readily available (Swords 1986).

Developments in the United States were less successful. Although by 1938 the navy had built a radar set to be placed on navy vessels for the detection of other ships, and the army had designed and built radar position-finding equipment intended for the control of antiaircraft guns and searchlights, further progress was hampered by the lack of adequate sources of microwave power. When the war broke out in Europe in 1939, efforts were intensified (Allison 1981; Gebhard 1979).

In early 1940 an agreement was reached by NDRC with the army and navy that it would concentrate on radar at 40 cm and shorter. Under the leadership of Alfred Loomis, the Microwave Committee of NDRC set about the task of developing a microwave technology that would be useful in the war. The obstacles were considerable because no adequate sources of radiation with 40 cm or less were available in the United States. A survey indicated that only two tubes seemed to hold any promise: a low-power klystron, already commercially available from the Sperry Corporation, and the resonatron, a multielement vacuum tube developed at the University of California. It was precisely at this junction, in the fall of 1940, that the British Technical Mission led by Sir Henry Tizard came to the United States carrying with them the cavity magnetron that had been developed in early 1940 by John T. Randal and Henry A. Boot in Oliphant's laboratory at Birmingham University. This cavity magnetron was a revolutionary source of microwaves: it was capable of producing one microsecond pulses with a peak power of 10 kilowatts at 3000 megacycles, that is, 10 cm. It proved to be one of the most important vacuum tubes ever developed. Its use in the early radar systems designed by the British saved Great Britain during the early phases of World War II when it faced Germany alone (Baxter 1948).

It became immediately clear to the Microwave Committee that further development ought to be based on the principles embodied in the cavity magnetron. The success that the British had obtained by concentrating their work on magnetrons in one location convinced the Microwave Committee that a single large laboratory, staffed by research physicists, and directed by a civilian administration, ought to be established in the United States. MIT was the site eventually chosen.

That MIT was chosen reflected not only K. T. Compton's commitment to meet the challenges that Hitler and fascism posed, but also MIT's long history of cooperation with industry and the government. The contract system by which OSRD carried out its activities was developed at MIT during the thirties to govern its dealings with industry. It had been devised to avert possible criticism that MIT's collaboration with industry would make it subservient to commercial interests (Greber 1987).

In October 1940, NDRC approved a contract with MIT to establish the Radiation Laboratory. It allocated to MIT the considerable sum of $455,000 for the first year's operation. By the end of the war, the Radiation Laboratory had grown from an initial staff of about fifty (including mechanics and secretaries) to a scientific and technical staff of 1200, supported by some 2700 technicians, assistants, mechanics, stenographers, business staff. Its operating budget for the fiscal year July 1944 to July 1945 was over eighty times the original budget. In 1945 it occupied over 15 acres of floor space in Cambridge, and operated satellite laboratories at Harvard and at Columbia and large sublaboratories in adjoining municipalities, as well as in France, Great Britain, and Australia. During the war its staff participated in every major military operation: in the sorties against the German submarines in the Atlantic, in the North Africa campaign, in the invasions of Sicily, Italy, D-day, in all the battles in the Pacific. It maintained extensive and intimate contact with all the major radar manufacturing concerns. Except for the atomic bomb project, the Radiation Laboratory was the largest civilian research and development agency (Burchard 1948). The director of the laboratory was Lee A. DuBridge, the chairman of the physics department at Rochester; the associate director was F. Wheeler Loomis, head of physics at the University of Illinois. I. I. Rabi of Columbia was in charge of overseeing research. In the beginning Rabi, L. N. Ridenour, K. T. Bainbridge, L. C. Marshall, L. W. Alvarez, R. F. Bacher, L. A. Turner, J. R. Zacharias, and J. T. Rump, all physicists, helped DuBridge and Wheeler direct the affairs of the laboratory.[10] Ernest Lawrence of the University of California played an important role during the earliest phase and was instrumental in persuading people to come to work at the MIT laboratory.

Incidentally, the name Radiation Laboratory was chosen in order to deliberately mislead the casual passer-by into believing that the laboratory was engaged in nuclear physics, a field that at the time was considered by most scientists to be a harmless activity.

Initially the Radiation Laboratory was charged with three missions. The first of these, urgently requested by the British, was to develop a 10 cm airborne radar for use by night fighters; the second was to develop an accurate precision gun laying radar; the third was to design a long-range navigational device. Each of these developments has a long history and is chronicled in Guerlac's (1987) account of the Radiation Laboratory. Each was a success. And in the process of achieving these successes, the physicists working on these problems made significant improvements in pulse modulators, antennas, receivers—the components making up the systems.

Within a few months of the establishment of the laboratory, much better magnetrons had been developed, operating at 10 cm (S-band), 3 cm (X-band), and 1.25 cm (K-band). (The 1.25-cm, K-band choice turned out to be a terrible mistake, because it corresponded almost precisely to an absorption line for the water molecule that strongly limited the usefulness of this system.) These new magnetrons required the reevaluation of all the other components and introduced some new ones, such as waveguides, whose properties were carefully studied.

There is little doubt that radar turned the tide of the war: it won the air war over Britain in 1940 and it helped eliminate the threat of German submarines to the Allied convoys in the Atlantic by locating and destroying them at night when they were recharging their batteries. Radar made possible blind bombing by and blind landing of aircrafts. It also revolutionized antiaircraft guns which up to that time had relied on visual tracking but could, with radar, function automatically and reliably irrespective of weather, day or night. The Rad Lab also built the huge and powerful Microwave Early Warning systems designed to detect the approach of enemy aircraft over considerable distances.

Perhaps the most striking feature of the Rad Lab was the speed with which most of these developments were carried out. By 1943 most of the fundamental and seminal research and development had been done to the point that many of the top scientists at the Rad Lab could be released to go to Los Alamos. Similarly impressive were the technical accomplishments. By the end of the war the power of magnetrons had been vastly increased, and sensitive mixers and amplifiers had been designed and built, to the point that pulses with several megawatt of power could be sent out and echoes of 10^{-13} watts detected (Collins 1948)!

Experimental physics after the war benefited greatly from the advances made at the Rad Lab at MIT and at similar installations at Harvard and Columbia. The first and most obvious benefit was the great technical skills it gave experimenters in manipulating microwaves: Bloch, Hansen, Purcell, Pound, Lamb, Dicke, Rabi all made use of the microwave techniques they had learned during the war in the classic experiments they performed after World War II. Incidentally, it was the Rad Lab that was responsible for making oscilloscopes *the* standard piece of equipment in any physics laboratory.

An equally important dividend accruing from the work at the Rad Lab— one that also stemmed from cooperation between theorists and experimentalists— was the research of George Uhlenbeck and his colleagues on the detection of very weak signals. Their analysis clarified the relation between signal and noise (Lawson and Uhlenbeck 1949), and resulted in techniques that allowed the extraction of extremely weak signals from background noise. Because noise consists of random fluctuations that average to zero, it was found that averaging a signal over a long period often reduced the apparent noise level significantly. Purcell has stated that "one of the most useful things we learned at the Rad Lab was the idea of signal to noise and how to calculate it" (Gerstein 1987). These lessons were thoroughly assimilated and many of the averaging techniques developed at the Rad Lab became standard components of an experimentalist's tool kit after the war.

The twenty-eight-volume Radiation Laboratory Series, edited by Ridenour, which dealt with every aspect of radar—volumes that Rabi had urged be written when it became clear in 1944 that most of the mission of the Rad Lab had been accomplished—were the vehicles whereby the lessons learned at the Rad Lab were disseminated to the American experimental physics community.

One further aspect of the MIT Rad Lab experience ought to be stressed. Not only had physicists displayed a mastery of the fundamental science governing the physical phenomena, but they also had proven themselves talented gadgeteers and engineers, as well as able production managers. Physicists had been concerned with every facet of the development of radar from the design of magnetrons and klystrons to the oversight of the manufacture of the completed sets including antennas and scopes. Shortly after the war, Lee DuBridge described the transformation of the physicists who had come to the Rad Lab at MIT:

What happened to that group of academic physicists who had
been so suddenly plucked from their ivory tower labs? Well, of
course, by 1943 there were several hundred of them—and they
were now being aided by hundreds more of engineers, mathematicians, astronomers, physiologists, lawyers, businessmen, to
say nothing of machinists, secretaries, janitors, plumbers, car-
penters and an unaccountable number of guards. But it was always that initial group of 75 physicists who supplied the leadership. Maybe the prejudices of the Director had something to do
with it. Maybe we were able to attract the best physicists and not
the best engineers. But the physicists were just *good*. They were
objective, they were imaginative. They jumped at new ideas,
they were enthusiastic, genial, cooperative. They had a happy,
exciting, wonderful time.

But they weren't really physicists any more. They
weren't doing research for the fun of learning new things. They
were developing weapons for war for use *tomorrow*. They had
become engineers, military strategists, salesmen, production
experts. And they were the most *travelling* group you ever saw.
We quickly found that the only satisfactory way to exchange information rapidly and effectively was to do it person to person.
Many of us seldom had a straight week at home. Some were
gone for weeks or months at a stretch. They visited air bases,
factories, the Pentagon and the Navy buildings, other Army,
Navy, and civilian laboratories in this country and in Canada,
in England, in Australia. In 1943 we established a branch laboratory in England to keep in closer touch with British radar
workers and to be of more direct help to our own fighting forces.
A few days after the liberation of Paris we had a field station
there. (DuBridge 1949)

Bethe attributed the physicists' resourcefulness to their training in "fundamental research." Arguing that after the war the physics department at Cornell
ought to stress "fundamental" rather than "applied" research, Bethe noted that

there is a good argument for fundamental research also from
the purely practical point of view. This is that the greatest contributions to the applied research developments at such places
as the Radiation Laboratory at M.I.T. were made by physicists who had done only pure research before the war. The
men who had been in the field of electronics before the war,
whether they had been at a scientific institution such as M.I.T.
or at an industrial laboratory such as G.E. or RCA, could not

keep pace with the newcomers after a very few months. I think
the reason is that fundamental research is intrinsically so much
more difficult and so much more varied that it keeps a scientist
flexible and able to take any problem, including applied physics
problems, with much more ideas and flexibility. Therefore, even
if it comes to the problem of solving applied physics questions
in a pinch, the pure research worker is generally superior.[11]

Much the same story can be told about physicists working on the development of the atomic bomb. Physicists not only constructed the first experimental pile, designed the first uranium and plutonium bomb, but they also were deeply involved in the construction of the Hanford reactor and the Oak Ridge separation plant.

3.4 Training a New Generation of Physicists: Norman Kroll

It was at Los Alamos, at the Rad Lab, and at the other wartime laboratories that an entire generation of new theoreticians were trained. The story of Norman Kroll and his experiences at the Radiation Laboratory at Columbia is similar to that of David Falkoff, David Saxon, Harold Levine, and others at the Rad Lab at MIT, or of Roy Glauber, Ted de Hoffman, Murph Goldberger, and many of their cohorts at Los Alamos. All these young men graduated college as physics majors in the early 1940s, and because of their recognized talents were sent to work at one of the wartime laboratories as research assistants. Who they worked with there often determined where they went on to do their graduate studies after the war. I relate Norman Kroll's experiences for two reasons. The first is that he played an important role in the development of quantum field theory after World War II (Kroll and Lamb 1949; Karplus and Kroll 1949). The second is that his story is taken from an interview with Joan Bromberg, who was interested in a very different aspect of Norman Kroll's career than the one of relevance to the subject of this book.[12]

The mission of the Rad Lab at MIT was to develop as quickly as possible effective radar systems that would be useful in the war. No "basic" research was ever conducted unless this criterion of "usefulness" could be satisfied. Rabi, the head of the Advanced Development Division of the Rad Lab "was one of the most forceful and insistent advocates of [this] policy of getting new radar systems into the field" (Guerlac 1987). The "usefulness" and value of radar with wavelengths shorter than 3 cm had always been recognized, and in late 1941 it was agreed to proceed with the development of such high-frequency power sources. But in order not to dilute further the activities of the scientific manpower at the Rad Lab at MIT, it was decided that this work should be undertaken elsewhere. In March 1942 a branch of the Rad Lab was established at Columbia University. Its mission was to build a magnetron that would generate 1.25 cm microwaves; Rabi was appointed the nonresident director, and J.M.B. Kellogg the associate director of the laboratory. Initially, the laboratory was housed in Pupin, in rooms previously occupied by Rabi's molecular beam experiments (Rigden 1987). Millman and Kusch, who had been members of Rabi's molecular beam group during the thirties, were among the first to be recruited for the project. Soon thereafter, Arnold Nordsieck, a young assistant professor in the physics department at Columbia, became a member of the staff. Willis Lamb, who at the time was a lecturer in the physics department at Columbia and had had some difficulties obtaining his clearance because his wife had been born in Germany and still had family there, only joined the laboratory in the winter of 1943.

The 21-year-old Norman Kroll was recruited as a junior staff member in the fall of 1943. A year earlier he had graduated from Columbia University. He had become a physics major there in his senior year upon transferring from Rice University, and he remained at Columbia for his graduate studies. In his first year as a graduate student he took the quantum mechanics course that was being taught that year by J. Keller as well as two courses with Willis Lamb: one in classical mechanics and one in electricity and magnetism. During the summer of 1942 he attended a course on microwave physics at NYU, and in summer 1943 he went to Brown, where he took several courses in its famous Applied Mathematics Program and remembers attending lectures by Leon Brillouin and J. D. Tamarkin.[13] When Kroll joined the Rad Lab at Columbia he was shown a microwave magnetron, the "MO," and he recalls that someone explained to him how it worked and that "it all sounded perfectly reasonable to me." He was then taken to a room where "there were big magnets and modulators" and given instructions on how magnetrons were tested. "So I started testing magnetrons, like all the other people in my situation" (Kroll 1987, p. 6).

The cavity magnetron that the British had invented in 1939–1940 was a traveling wave type magnetron with six internal resonators. These resonators present to the space charge a radio frequency field of the desired frequency to generate the microwave radiation and simultaneously transmit a portion of the generated radiation to the external load. Several ingenious schemes had been devised for separating the various modes of oscillation of the magnetron (e.g., strapping), but all of them became inefficient or impractical at higher frequencies (Collins 1948, chap. 4). In 1942 Nordsieck and Millman had the idea of making the size of the resonators of the magnetrons unequal in length: "Alternate resonators were alike, but adjacent resonators were not alike" (Kroll 1948, p. 8). This magnetron became known as the rising-sun magnetron because "it looked like the Japanese flag." Alternating the size of the resonators produced good mode separation at high frequencies.

Willis Lamb had done an equivalent circuit analysis of the symmetric magnetron, and had presented his methods in "the little talks that were given to educate the people in the laboratory." He asked Kroll, whom he knew as an outstanding graduate student, to do a similar analysis for the rising sun magnetron. "I looked at that for a day and I decided it would not work the way [Lamb] asked me to do it" (Kroll 1987, p. 8). Kroll had seen some papers by Goldstein and Clogston on an alternative analysis for the symmetric case—"a matching field theory"—and decided it would work:

> ... and so I in fact worked it out. I had trouble getting Lamb to listen to me at first because he wanted me to do what he told me to do. I said "it won't work," and I explained why it wouldn't work, and he listened to me. So then I showed him what I had done. I'd actually gotten the basic equations. So that was fine. *The notion that once you had a formula, that that wasn't the end of the problem, that you actually wanted to get numbers was something I hadn't yet learned* [emphasis mine]. So I said, "Well here it is. That's the formula". So then a little later, [Lamb] got some of the calculated results of the formula. He did it himself. ... It was a great big plot of frequency versus the ratio of the large to the small resonators, and so that was an important educational experience, because I learned that one was actually interested in the answers. ... Then it occurred to me that it would be nice to plot it in a way that might be more understandable, that one might see what was going on better than it seemed to me that that plot did, and I made the simple change of plotting the wavelength instead of the frequency, and then it all became clear, as to exactly what was going on. (Kroll 1987, p. 9)

That was the first research Kroll ever carried out: "It's the first time I had the slightest glimmer of what physicists did, actually," Kroll recalls. "[My calculations] turned out to be very useful, because it was immediately apparent what the whole physics of the rising sun magnetron [was]." Kroll's work was a major contribution to the understanding of how that magnetron worked. Kroll proceeded to write a paper on the subject, and a generalization of this approach to "the situation in general" formed the major part of the dissertation he later submitted for his doctorate. Kroll also wrote the chapter on rising sun magnetrons in the Radiation Laboratory Series volume on *Microwave Magnetrons* (Collins 1948, chap. 3) attesting to his expert status in the field.

Another project that Kroll became involved with while at the Rad Lab was the measurement of the absorption of the 1.25-cm radiation in its passage through the atmosphere. Willis Lamb had not only worked out the theory,

> but he really invented the method for doing the absorption experiment. There was a large copper echo box ... a big thing ...

like $12 \times 12 \times 12$, or maybe $10 \times 10 \times 10$ [feet]. You sent microwaves into it, and you compared the level of radiation in the box . . . sealed, and then with holes of various sizes opened in the sides, and the idea was to compare absorption, that is, the loss rate from the box through the holes in the side, which you can calculate very easily by a rather simple photon argument, and with the absorption caused by the gas.[14] . . . In fact I had some ideas [on] . . . another way one might do it. . . . I don't remember the details, but it did involve the use of silicon. I remember getting this sample of silicon . . . and making measurements on it. I recall the method I proposed had . . . more . . . theoretical appeal to me, but I think in actual fact the method used was much better. (Kroll 1987, p. 7)

Kroll's experience was paridigmatic. All the young theoreticians at the wartime laboratories learned that physics is about numbers and about the results of experiments. Good theories yield numbers, explain numbers, and help design good apparatus. They allow you to control and manipulate the objects they describe.

3.5 The Universities: 1945–1947

The war ended in August 1945. In the United States many of the physicists who had been working on war work at government laboratories, remained there to discharge their responsibilities, finishing projects and writing reports. Only in the spring of 1946 did most of them return to academic life—to faculty positions or to complete their Ph.D.'s. Once the war was over, all of them thought about the physics they would be doing now that they were free to pursue their own interests. At the wartime laboratories—where resources were seemingly unlimited— the more senior physicists had acquired a taste to do things on a large scale. Gone was their reluctance to think big, or to do with less. Most of them welcomed the new opportunities and pondered about the kind of research their department and universities ought to be undertaking.

The war had ended the Great Depression of the 1930s and the United States was entering a period of great economic expansion. Now—1946—clearly was the time to put forth plans for the growth of physics. Physics departments did expand; the increase in the number of theoreticians and the number of theoretical graduate students was particularly noticeable.

Commenting in 1949 on the effects of World War II on physicists, DuBridge noted that they had become more numerous, more affluent, and much more famous.

In spite of the fact that universities had practically stopped training physicists during the war, membership in the American Physical Society increased from

3341 in 1938, to 4120 in 1942, 5714 in 1946, 7238 in 1948, and 8100 in 1949. During the war "every person who ever had a course beyond Physics was dragged into a war laboratory," and this experience made many of these people decide to become physicists (DuBridge 1949). The war increased the demand for physicists, and this demand did not abate after the war since government and industrial laboratories continued to increase their staff of physicists. Similarly, physics departments at colleges and universities clamored for more personnel to meet the swollen enrollments brought about by the GI bill of rights—with its opening of educational opportunities for veterans—and the demands of the expanded research activities being supported by lavish government funding. In 1949 the gap between the number of vacant jobs and the number of physicists was widening.

DuBridge noted another effect of the war: "Physicists now seem to have more fun than they used to. They not only enjoy their work, but they also enjoy each other." The spirit of the physics community had evidently been buoyed by the lasting friendships that were formed at Los Alamos, at the Rad Lab, and at other such places. DuBridge suggested that many of these war laboratories had been rather like permanent meetings of the Physical Society. Physicists there developed an esprit de corps. It is of course not surprising that there should be a much greater pride in being a member of the physics community given the immense prestige that had been bestowed upon physicists during the war.

The status physicists had acquired made them influential on the campuses, and they helped transform American universities. It was the physicists, and in particular the theoretical physicists, who broke down the barriers that had existed before the war at many universities in the hiring of Jews and other minority members. A (more or less) tacit quota system had existed at many universities before World War II for the hiring of Jews; and many of these quotas had been reduced—or filled—by the influx of the German refugees in the mid-1930s (Synott 1979; Hoch 1983).

The situation changed radically after the war. For example, in 1945 Oppenheimer was offered an appointment at Harvard but declined;[15] Schwinger accepted an offer to go there, and in 1947 was made a full professor at age 29; Weisskopf went to MIT. The hiring at the junior levels was even more dramatic: Feynman and Morrison at Cornell, Bohm at Princeton, Weinberg at Minnesota, Schiff at Pennsylvania, Feshbach at MIT. These practices within physics departments helped bring about similar changes in the other science departments: in chemistry, where discriminatory practices had been even more intense than in physics during the thirties; in mathematics and in biology, in which physicists who had retooled themselves were being hired as molecular biologists. By the mid-1950s, faculty hiring was based on ability, with ethnic and religious background playing almost no role in the major universities on the east and west coasts.

3.6 The Conferences

A large number of conferences were organized in the United States after the war to gauge and utilize the technological and conceptual advances made during the war. Many of them turned out to be important and seminal. The Josiah Macy Conference of 1946, which was attended by von Neumann, Weiner, McCulloch, Pitts, Rosenblith, among others, was the first of a series of five conferences that were very influential in the development of cybernetics and neurobiology (Heims 1980, 1991). The postwar Cold Spring Harbor conferences stimulated the development of molecular biology. Similarly, the eleven small conferences initiated by Duncan McInnes and sponsored by the National Academy of Sciences had considerable impact on the research carried out in elementary particle physics, low temperature and solid state physics, statistical mechanics, and theoretical chemistry. These NAS conferences were small and usually focused on a particular area of research. I will consider three of these conferences—Shelter Island, Pocono, and Oldstone—in the next chapter. The function of all these small postwar conferences was primarily to help the elite in the field exchange ideas, assess directions, and restore some of the traditional ways of doing business that had been disrupted by the war.

But there were also other, much larger, conferences aimed at reasserting the continuity of the scientific enterprise, reestablishing ties and contacts between practitioners who had not been in touch with one another since the outbreak of the war in 1939, reaffirming communal values and restoring the sense of belonging to a scientific community whose goals were pacific. The autumn meetings of the National Academy of Sciences and the American Philosophical Society in 1945 and 1946 served all these purposes. In fact, in 1945 these two institutions held their meetings jointly. The major event at this joint meeting was a symposium on "Atomic Energy and Its Implications," whose content I shall return to later.

The war had ruptured the peace time links that had held the international scientific community together. Communication with the rest of the world had all but disappeared with the secrecy requirements and the other demands of wartime. By 1946 only little progress had been made toward reasserting an international dimension to the scientific enterprise. In order to reestablish ties with the international scientific and intellectual community, in 1946 the National Academy of Sciences invited twenty-three foreign delegates to attend its fall meeting, as representatives of sister academies and learned societies. These foreign delegates were the guests of the Academy with all their expenses paid, support having been obtained from the Rockefeller Foundation and the Carnegie Corporation. On October 17 and 18 they attended the American Philosophical Society meeting in Philadelphia, on the 19th they partook in the Bicentennial Convocation of Princeton University (C. Osgood 1951), and on the 20th they went on to Washington to participate in the NAS fall meetings. These gatherings were followed by trips to the West Coast. Upon the conclusion of their stay in the United States, Robin Allan, the representative of the Royal Society of New Zealand, conveyed the appreciation of the guests to the president of the National Academy of Sciences:

> I believe that you hoped that the contacts which your invita- tions were to make possible would act as a powerful flux, fa- cilitating and stimulating that international understanding and co-operation between scientific societies and individuals which has been so fruitful in the past, and on which the future of scien- tific endeavour depends. You were aware of the dangers which threaten the freedom of scientists; you hoped to forge interna- tional links which would offset those dangers. I think, too, that you saw clearly that the future of our civilization, and with it the American "way of life" (which, as your guests, we have been privileged to share) are also in danger. You realized, and we share the conviction, that a potent factor for good lies in a wider appreciation and deeper understanding of the attitude which a scientist brings to the solution of his problems. This attitude, I believe, derives from, and is implicit in, the proper use of the scientific method. The proper use of the scientific method requires freedom of speech, freedom to publish results, freedom from bureaucratic controls, and that freedom to inter- change ideas and techniques which can come only through per- sonal contacts and travel. A first step towards world sanity lies in the re-establishment of free intercourse between scientists pre- viously separated or isolated. This is not a selfish wish; it is an imperative pre-requisite to a civilization conceived on an inter- national scale.[16]

I want to turn next to the "Symposium on Atomic Energy," which was held on November 16, 1945, during a joint fall meeting of the APS and NAS, and which brought to the attention of the assembled academicians the dramatic events which had taken place during the war.[17] The roster of speakers included most of the major figures involved in the development of the atomic bomb. The program was as follows:

Henry D. Smyth	— Fifty Years of Atomic Physics
J. R. Oppenheimer	— Atomic Weapons
E. Fermi	— The Development of the First Chain Reacting Pile
E. P. Wigner	— Nuclear Reactions
H. C. Urey	— Methods and Objectives in the Separation of Isotopes
J. A. Wheeler	— Problems and Prospects in Elementary Physics Research

J. Willits	— Social Adjustments to Atomic Energy
Jacob Viner	— The Implications of the Atomic Bomb for International Relations
J. Shotwell	— The Control of Atomic Energy
I. Langmuir	— World Control of Atomic Energy
Arthur H. Compton	— Atomic Energy as a Human Asset

I shall focus on John A. Wheeler's paper for two reasons. The first is to compare it with an address by Felix Klein in Göttingen in 1918. Wheeler's speech in 1945 was a repeat performance, but with a vision that would become reality: to the victors belong the spoils. The second is that Wheeler's paper gives an overview of the state of the art in elementary particle research in the fall of 1945 by an imaginative and insightful scientist.[18]

In the summer of 1918 Felix Klein delivered a lecture to the Göttingen Society for the Advancement of Applied Mathematics and Physics. Klein, an outstanding mathematician, was addressing an audience that included leaders of industry and government as the representative of the scientific community that had contributed so importantly to the success of the German war machine. Self-confident and self-satisfied, he indicated that "the closer we appear to approach the victorious conclusion of the war, the more our thoughts are dominated by the question what, after peace is successfully won, ought then to come?" He answered his own question and listed his desiderata, which included a mathematical institute at Göttingen, an intensified harmonious collaboration of German science with industry and the military, and a reorientation of German education at all levels to achieve this "preestablished harmony" with the military-industrial complex.[19] By November 1918 all these dreams—including the academicians' hope for large funding increases for education and research—were shattered. The German empire was utterly defeated and had to accept an unconditional surrender. The vindictive terms of the peace treaty imposed on Germany by the victorious Allies made an economic recovery impossible, and much of the political unrest that was to engulf Germany until Hitler seized power can be attributed to the consequences of the harshness of the Versailles treaty.

The defeat brought about strong sentiments of antagonism toward science and the scientific establishment and great resentment toward the armed forces and the military-industrial complex.

In a challenging and influential paper published in the early 1971, Forman tried to relate these contextual components to the dramatic ideological changes that marked the development of quantum mechanics. He suggested that his researches gave

> overwhelming evidence that in the years after the end of the first World War but before the development of an acausal quantum

mechanics under the influence of "currents of thought" large numbers of German physicists, for reasons only incidentally related to developments in their own discipline, distanced themselves from, or explicitly repudiated, causality in physics.

Thus the most important of Jammer's theses—that extrinsic influences led physicists to ardently hope for, actively search for, and willingly embrace an acausal mechanics—is here demonstrated for, but only for, the German cultural sphere. (Forman 1971, p. 3)

Forman's analysis was a sociological one. It eschewed the psychological makeup of the individual scientists concerned, and assumed that their "mental posture" was the "socially determined response to the immediate environment."

Forman characterized the outlook of the Weimar milieu as neoromantic and existentialist. Totally insecure in its present, Weimar Germany built a future out of its past, returning to the romanticism of Schiller and Goethe. Weimar Germany rejected the utilitarian standards of its French and British conquerors and sought to reestablish its position of leadership in the cultural sphere. In Weimar Germany the previously embraced ideals of technology and science were considered suspect and vilified as being tainted with Anglo-French, Judeo-Marxist materialism.

The teaching of mathematics in the public school became noticeably reduced, and a new emphasis was placed on "cultural" education. It is not surprising that Spengler should have found in Weimar Germany such fertile soil among educated Germans for the dissemination of his views. Spengler's hostility toward the ideology of the exact sciences and the pessimistic outlook of his *Decline of the West* resonated with the general mood of gloom and doom, particularly with the psychological disposition of the exact scientists who had lost their confidence in the future of their discipline.[20] A measure of the impact of *Das Untergang des Abendlandes* is conveyed by the fact that it went through some sixty editions in the eight years after its publication in July 1918.

Forman in his article presented evidence that the Weimar physicists and mathematicians "accommodated" themselves to a generally Spenglerian point of view and argued that there existed a widespread movement to dispense with causality in physics—a movement that, according to him, could not be related to any internal development of physics, and particularly of quantum theory.[21] Both Hermann Weyl and Erwin Schrödinger came under the influence of Spengler's philosophy. Both became convinced that mathematics and physics had by 1920 reached a state of crisis that left radical revolution as the only open road. Even before 1918, Weyl had embraced Brouwer's intuitionism, which denied the validity of a large part of classical mathematics. Weyl attempted to give a place to *Anschauung* (intuition) rather than logic at the foundation level. In 1922 Schrödinger joined in the call for the restructuring of the laws of physics. Weyl and Schrödinger agreed with Spengler that the coming revolution would do away with the principle of causality. It is ironical—the cunning of reason—that Hilbert and Einstein, the former revolutionaries, found themselves in the position of defenders of the status quo, "Hilbert defending the primacy of formal logic in the foundations of mathematics, Einstein defending the primacy of causality in physics." In the short run, Hilbert and Einstein could not stem the tide and they were defeated. The Spenglerian ideology of revolution triumphed both in physics and mathematics: "Heisenberg discovered the limits of causality in atomic phenomena and Gödel the limits of formal deduction and proof in mathematics" (Dyson 1983).

But in the longer term, "the achievement of revolutionary goals destroyed the revolutionary ideology which gave them birth. The visions of Spengler became irrelevant: chemists who never heard of Spengler could use Quantum Mechanics to calculate accurately molecular binding energies. The discoveries of Gödel did not lead to a victory of intuitionism but rather to a recognition that no single scheme of mathematical foundations has a unique claim to legitimacy" (Dyson 1983).

Yet it cannot be doubted that external factors—the Weimar setting, the Spenglerian philosophy, and so forth—were important. They help explain, in part, why quantum mechanics was formulated where it was and help understand the reception of its formulation. There is little question that a formulation of quantum mechanics—with many if not most of its present features—would have been put forward at some later time by others if Heisenberg, Schrödinger, Pauli, and Born had not done so in the 1925–1927 period: a real crisis existed in physics and internal factors demanded a resolution.[22] The Weimar setting accelerated the process. It provided not only viewpoints that would prove helpful but also a setting in which the search for revolutionary solutions was the accepted practice.

The developments of theoretical physics in the United States in the post-World War II era can be analyzed similarly (Schweber 1989). Here it was the pragmatic, utilitarian outlook—which had been reinforced by the wartime experiences—that gave the philosophical and ideological underpinning. The kinds of problems tackled helped maintain that viewpoint. Furthermore, the sources of support and what was interpreted as the realities of the international situation helped reinforce that viewpoint and further polarized problem selection.

Thus at the NAS symposium on atomic energy in 1945 Wheeler opened his presentation with this exhortation:

> Discovery how to release the untapped energy [of the nucleus, were it possible to convert its mass completely into energy] on a reasonable scale might completely alter our economy and the basis of our military security. For this reason we owe special attention to the branches of ultranucleonics—cosmic-ray phenomena, the mechanism of energy production in special stars, field theory, and particle transformation physics—where

a single development may produce such far-reaching changes. Other nations have not neglected this work during the war. We must prepare to resume it vigorously. (Wheeler 1946)

And he concluded his lecture with these remarks:

From this survey of some of the problems of theory it is apparent that many lines of progress are open. Elementary particle physics will accomplish its task to reduce our experience to order just as it will get sources of energetic particles and use these sources to study the transformations of the elementary particles. The achievement of all these tasks together will carry us over the last mountain peak of the continent of ultranucleonics. Then we shall see the whole rich new land spread out beneath us.

Shall we ask at the end of the journey what is the good of this new continent? Shall we inquire what it will profit us to know how the universe is put together? Who will it be who can decide so soon the future uses of this new land? That will be the work of those who will come next.

But, if at the moment when the far shore shall be reached and before the days of colonization shall have begun the world shall ask what are the most precious immediate assets at its disposal from this exploration, we know now from the experience of the past war what the answer will be. We have only to recall the essential role played in all parts of the war effort, far from their own specialties, by men trained in fundamental science, to realize that we should look for the answer, not at the gold which will be brought back from the great adventure, but at the band of explorers themselves. On men like these searchers will depend our future in war and in peace. They will make for us new tools of defense in days of danger. They will leaven our applied science, our technology, our industry, and our intellectual life in the days of peace. Their qualities of mind and heart are the prize.

We must seek out our able young men, outfit them, and send them forward to work their way through the unknown, not only because the land is rich, but most of all because only participation in this great Odyssey will develop men of the kind on whom our future as a nation depends. There is no other way. (Wheeler 1946)

In his program for advancing ultranucleonics, Wheeler called for a close cooperation between theory and experiment:

The theoretical work must proceed hand in hand with experiment for the most effective progress . . . [and] a continual assessment of results of [experimental] investigations and of choice of further problems in the light of the outstanding unsolved [theoretical] questions is essential. . . .

The task of reducing our experience to order includes not only this close collaboration of theory with experiment but also the creative function to assimilate the fruits of such collaboration into a unified view of matter.

Wheeler's survey of the state of elementary particle physics at the NAS symposium was wide ranging and penetrating and is of great historical interest. He described the way nature was understood then in terms of four types of interactions—gravitation, electromagnetism, nuclear (strong) forces, weak-decay interactions—and suggested experiments and theoretical activities to gain further understanding into them.

For Wheeler the interesting and exciting areas of research were the investigations of the strong and weak interactions. He urged that every means to gain further insights into the properties of mesons and nucleons be explored. Although high-energy machines should certainly be built (by industry) he opted for an intensification of cosmic ray research because of the much higher energy available in these phenomena. Cosmic rays research is where the action is! This assessment came from a physicist who at the time was engaged in exploring reformulations of *electrodynamics* as an action-at-a-distance theory, and who had suggested to Dicke the possibility of experiments on *positronium* formation as a way of exploring the electromagnetic properties of a purely "electronic" system.

3.7 Physics in 1946

In 1941, *The Journal of Applied Physics* initiated the practice of publishing an article that reviewed the state of physics during the previous year (Osgood 1942, 1944). Although the articles' appearance was interrupted for several years during the war, the practice resumed in 1947 with Philip Morrison's article, "Physics in 1946." He characterized the year as one in which many plans, "often great and exciting ones," were being formulated, and noted that the "Physical Review" was being spoken of as the "Physical PREview." Much work was begun, but not much completed. A good deal of the article commented on what was to be expected.

Morrison made particular reference to the large high-energy accelerators that were being built or had just gone into operation. He reported that as of November 1, 1946, the giant Berkeley cyclotron that Lawrence had built in 1940 and whose 17-foot magnet was used during the war as a prototype for the electromagnetic separation plant at Oak Ridge, had produced 200 MeV deuterons. At

Illinois, Kerst had begun the construction of a 300 MeV betatron, a 100 MeV version of which had successfully operated at the General Electric Laboratories in Schenectady, New York. At the University of California in Berkeley, MacMillan was building a 300 MeV electron synchrotron, as were teams at Michigan, MIT, and General Electric. Also at Berkeley, Alvarez was engaged in building a large linear accelerator for protons, and it was expected that by February 1947 the first section delivering protons of more than 32 MeV would be operational. All these machines depended on the availability of new high-power radar oscillators—a technology that had been developed during the war.

In the second part of his article, in a section entitled "The Legacy of the War," Morrison reviewed some of the technical advances that the war had produced which undoubtedly would prove useful in experimental physics. He noted that during 1946 hundreds of war-experienced physicists had returned to their old laboratories "brimful of information about what had been done, and confident in their understanding of whole fields of technique which had been vague general possibilities in 1940." These new devices and techniques included microwave radiation ranging from a few millimeters to 30 cm, nuclear piles as sources for both fast neutrons and for well-collimated beams of thermal neutrons, and rockets for high-altitude cosmic ray work.

Morrison's article concluded with a report on two new methods by which nuclear magnetic moments had been detected: the nuclear magnetic induction techniques of Purcell, Torrey, and Pound and that of Bloch and Hansen.

The emphasis on the experimental practice in Morrison's article should be noted. Indeed, the dividends from the wartime activities manifested themselves there first. On the theoretical front, most of the papers on foundational problems in the 1945–1947 period were written by theorists outside the United States—for example, Bhabha (1945), Heitler (1947), Heisenberg (1944), Jost (1946, 1947), Møller (1945, 1946a,b), Pais (1946), Racah (1946), Stueckelberg (1945, 1946), Wentzel (1947)—or by theorists in the United States who had not taken part in the war effort, for example, Pauli (1947), Havas (1944, 1945).

The state of affairs in foundational physics can be inferred from the reports given at conferences held during 1946. In July 1946 an international conference on "Fundamental Particles and Low Temperature" took place at the Cavendish Laboratory in Cambridge, England. It was the first gathering of the European physics community after the war. Those who attended recall the meeting principally as the first opportunity since 1939 to greet colleagues and friends, exchange wartime experiences, mourn common losses, and re-cement friendships.[23] At the conference Bohr, Pauli, and Dirac gave their assessment of the physics of fundamental particles.

Their pessimistic outlook is conveyed by the titles of their talks. Bohr spoke on "*Problems* of Elementary-Particle Physics," Pauli on the "*Difficulties* of Field Theories and of Field Quantization," and Dirac on the "*Difficulties* in Quantum Electrodynamics."[24]

This same pessimism—primarily due to the divergence difficulties encountered in quantized *field* theories of the Hamiltonian form—had led Heisenberg in the late 1930s to suggest that the future correct (divergence-free) theory would contain in its foundation a fundamental length. This constant would play the same role in limiting the applications of the ordinary concepts of quantum mechanics, as does Planck's constant in restricting the unambiguous applications of the concepts of classical mechanics to quantum mechanics. If such a fundamental length is introduced into the theory, relativistic invariance then requires the introduction of a minimal time. This implies that the possibility of continuous time displacement of the state vector (as determined by the Schrödinger equation $i\hbar\partial_t\Psi = H\Psi$) must be given up. The problem then to be faced is how to describe experimental results when the system under consideration has no Hamiltonian and consequently no Schrödinger equation. In a series of papers published during the war, Heisenberg made important contributions toward the solution of this problem (Heisenberg 1943a,b, 1944; see Rechenberg 1989).

He noted that in a theory without a Hamiltonian, an atomic system must then be defined by other "fundamental functions," and set out to determine these functions and to investigate their properties. He translated this problem into the problem of specifying the quantities that will be observable in the theory. Although Heisenberg could not give a definitive answer to the question he had posed, he assumed that any quantity whose determination is unaffected by the existence of a minimal length may be considered as observable. He took the following to be such "observables":

1. The energy and momentum of a free particle.

2. The cross section of any collision process.

3. The discrete energy levels of atomic systems in closed stationary states.

4. The decay constants of radioactive systems.

These quantities can be calculated by means of the Schrödinger equation when the Hamiltonian of the system is known. However, collision cross sections can be obtained more directly in terms of the matrix elements of a certain unitary matrix, S, "the characteristic matrix" of the system. This matrix determines the asymptotic behavior of the wave function for large values of the mutual distance between the particles in ordinary quantum mechanics. Heisenberg assumed that his characteristic matrix would in the future take the role played by the Hamiltonian in quantum mechanics. In fact, it may be shown that all the observables listed under (2), (3), and (4) can be calculated if S is known as a function of the energy and the other observables that commute with S.[25] However, answers to such questions as the probabilities of the particles being separated by a small distance r cannot be answered unambiguously if only S is known. Heisenberg found the general conditions satisfied by S, and assumed that all conditions satisfied by S in quantum mechanics would also hold in the new theory. It was furthermore shown by Møller (1945, 1946a,b,c) that the S-matrix is an invariant matrix that has a meaning independent of the Lorentz frame used in the description of the system.

Important contributions to the elucidation of an S-matrix description were made by Kramers, Møller (1945, 1946), Stueckelberg (1944b,c,1945), Jost (1946, 1947), and Kronig (1946). But it was not apparent at the time how the S-matrix was to be constructed in a theory without a Hamiltonian, even though it was recognized that Lorentz invariance, unitarity, and the analyticity requirements imposed by causality (Kronig 1946) severely restricted the S-matrix elements. Heisenberg, in one of his papers, did give a concrete example of an S-matrix that predicted a nontrivial scattering matrix of a single kind of meson and another that predicted multiple particle production. His idea was to write $S = \exp(i\eta)$ and then express η as a polynomial function of a free field ϕ (which we would now regard as the in-field ϕ^{in}). Heisenberg's work was first reported in the United States in the lectures on meson theory that Pauli delivered at the MIT Rad Lab in the fall of 1944.[26]

Møller and Stueckelberg lectured on S-matrix theory during their visit to Great Britain in 1946,[27] and Wentzel devoted several sections of his widely read *Reviews of Modern Physics* report on "Recent Research in Meson Theory" to Heisenberg, Kramers, and Stueckelberg's work (Wentzel 1947).[28]

Most of the work on the S-matrix research program stopped after the field theoretical advances initiated by the Shelter Island conference. It is to this conference that we turn in the next chapter.

4. Three Conferences: Shelter Island, Pocono, and Oldstone

4.1 Introduction

The Shelter Island conference was a landmark. The first in a series of three small post-World War II conferences on theoretical physics sponsored by the National Academy of Sciences (NAS), it took place on 2–4 June 1947 at the Ram's Head Inn on Shelter Island at the tip of Long Island. The initial impetus for the conference came from Duncan MacInnes, a distinguished physical chemist at the Rockefeller Institute, who had suggested to Frank Jewett, then president of the NAS, that the Academy sponsor a series of small conferences. MacInnes enlisted the help of Karl K. Darrow, the secretary of the American Physical Society, and of John A. Wheeler, a young professor of theoretical physics at Princeton, in organizing a conference on the foundations of quantum mechanics. The Shelter Island conference grew out of these initial efforts. The participants were primarily theoretical physicists, most of whom had been leaders in the highly successful wartime laboratories: the MIT Radiation Laboratory, the Chicago Metallurgical Laboratory, Los Alamos, and Hanford (Kevles 1978). Also in attendance were several experimentalists: Lamb and Rabi, who reported on experiments on the spectrum of hydrogen recently performed at Columbia University; and Rossi, who reported on the results of research carried out in Rome on the absorption of cosmic rays in the atmosphere. The discussions at Shelter Island were primarily concerned with the interpretation of these experiments.

The significance of the conference was soon as evident to the participants as it is in retrospect. Six months after the conference, in January 1948, Darrow wrote MacInnes a brief postcard: "I must quote [you] the words of warm commendation used yesterday by I. I. Rabi anent your Shelter Island meeting—he said that it has proved much more important than it seemed even at the time, and would be remembered as the 1911 Solvay Congress is remembered, for having been the starting-point of remarkable new developments"[1] Similarly, Richard P. Feynman recalled many years later: "There have been many conferences in the world since, but I've never felt any to be as important as this."[2]

The meeting turned out to be one of the most seminal conferences to be held right after the end of World War II — a conference whose impact was indeed comparable to that of the Solvay Congress of 1911. Just as the Solvay Congress of 1911 set the stage for all the subsequent developments in quantum theory (see, for example, M. de Broglie 1951), similarly Shelter Island provided the initial stimulus for the post-World War II developments in quantum field theory: effective, relativistically invariant, computational methods; Feynman diagrams; renormalization theory. The conference was also responsible for the elucidation of the structure of the mesonic component of cosmic rays. Marshak and Bethe, in their paper on the two-meson hypothesis, acknowledge the role Shelter Island played in the genesis of their theory (Marshak and Bethe 1947, p. 509).

Section 4.2 deals with how the Shelter Island conference, and the Pocono and Oldstone conferences that followed it, came into being. Although these conferences are now remembered primarily as having been devoted to theoretical physics, the experimental results presented there were of crucial importance. At Shelter Island the findings of Lamb and R. C. Retherford and those of John E. Nafe, Edward B. Nelson, and Rabi on the spectrum of hydrogen, and those of Marcello Conversi, Ettore Pancini, and Oreste Piccioni on the absorption of cosmic rays at sea level stimulated the renewed interest in quantum electrodynamics and the clarification of the constitution of the mesonic components of cosmic rays. Section 4.3 presents the scientific content of the Shelter Island conference, and section 4.4 tells of the subsequent impact of these conferences.

4.2 The Genesis of the Conferences

The Shelter Island, Pocono, and Oldstone conferences were small, closed and elitist in spirit. In a sense they mark the postponed end of an era, that of the 1930s, and its characteristic style of doing physics: small groups and small budgets. Conferences—and small ones in particular—have ritual functions. Coming after World War II, these conferences reasserted the values of pure research and helped to purify and revitalize the theoretical physics community. They also asserted the new social reality implied by the newly acquired power of the theoreticians and helped integrate the most outstanding of the younger theoreticians—Richard Feynman, Julian Schwinger, Robert Marshak, and Abraham Pais at Shelter Island, and Freeman Dyson at Oldstone—into the elite.

These characteristics embodied the ideal of Duncan MacInnes, a member of the National Academy of Sciences and a past president of the New York Academy of Sciences. In the fall of 1945 he wrote Frank Jewett, the head of the NAS,[3] asking to see him when it would be "quite convenient for him" in order to discuss "a proposal for an additional activity of the National Academy of Sciences."[4] Jewett promptly answered MacInnes indicating that he would be delighted to talk with him,[5] and a meeting between them was held in mid-October. On October 24, 1945, MacInness sent Jewett "A Proposal for the Establishment of Conferences under the Auspices of the National Academy of Sciences," which read as follows:

> There appears to be a real need for the meeting, at intervals, of the relatively few men doing active research work in each field of science. To a very large extent the meetings of the larger scientific societies are failing to meet this need. The

meetings have become so large that it is hard for the key men to get together. The programs are so crowded that presentation of the papers takes most of the time and discussion is slighted if it takes place at all. The papers as presented and the discussion cannot be of high intellectual caliber or they will tend to be unintelligible to most of the large audiences assembled. A further difficulty is that the meetings usually deal with only one science, whereas many subjects under research involve several branches of sciences.

To consider a definite case: a conference on photosynthesis would certainly include mathematical physicists versed in quantum theory, experimental physicists, physical and organic chemists, and possibly others. Also to be really effective the conditions should be arranged so that only those who can definitely contribute to the discussion will be in attendance. This is important, in that informal incisive discussion does not occur in large gatherings.

It is proposed, therefore, to establish a series of conferences under the auspices of the National Academy of Sciences, to include the following ideas: (a) a topic for consideration on which active research is going on, (b) a small number of papers, e.g., six for a two-day meeting, (c) an allowance of more time for discussion than for presentation of papers, and (d) an invited group of participants, not more than twenty-five or thirty; (e) the papers, revised in the light of the discussion that has taken place, to be published as a monograph by the Academy. In some cases an additional conference might well take place before the appearance of the monograph.

It may be found desirable to prepare the papers in mimeograph form to be distributed some weeks before the conference. The need for meetings of this kind has been shown by the establishment of symposia and conferences at Cold Spring Harbor, at Gibson Island, by the New York Academy of Sciences, and elsewhere. There has been a tendency for all such meetings to get too large. It is in the matter of strict control of the size of the conferences so that effective discussion can take place that the National Academy may make a real contribution. The Academy would, of course, add prestige both to the conferences and the monographs.

It may be well to hold each conference at a center which involves the least travel for the participants, and is otherwise convenient.

Some topics on which conferences might be held follow:
Photosynthesis
The Assumptions of Quantum Mechanics
Bioelectric Potentials
Fluorescence
Nuclear Forces
There are, of course, many other topics that could be considered. The important point is that active workers in each field should meet under conditions which allow for effective discussion of the pressing research problem.[6]

For a number of years before becoming president of the New York Academy of Sciences in 1944, MacInnes had helped arrange a series of small conferences sponsored by that organization, "designed to promote active discussion of different scientific topics."[7] For the conferences to achieve their aim, he felt, it was essential that the topics "be in areas in which actual work was in progress and that participation be restricted to currently active investigators in the designated fields." Although the early conference sponsored by the New York Academy had achieved this objective, MacInnes believed that the effectiveness of the later ones had been impaired "by their success in attracting too large a crowd."[8] He felt very strongly that attendance should be limited. In October 1944 he declined to run again as president of the New York Academy when a decision on this matter by its council went against him. In fact, in January 1945[9] MacInnes resigned from the Academy over this issue.

At a subsequent meeting, MacInnes asked Jewett whether the NAS would be willing and in a position to sponsor a series of such conferences.[10] Jewett answered that the idea appealed to him provided the problems the conferences addressed were real, and that their solutions "would be aided by the concentrated consideration of a small highly qualified group, meeting in intimate association for discussion after study of the prepared papers"; he added that it seemed to him that "the idea of a two or three day meeting at some quiet place where the men [*sic*] could live together intimately seemed ... the best way to obtain the desired results."[11] Jewett felt confident that the Academy would be willing to sponsor the undertaking if it met the above criteria. He suggested that MacInnes pick out one or two problems that seemed promising and use these as "pilot plants," rather than address the problem of a broad program *ab initio*.

Given this encouragement, MacInnes, after discussion with colleagues, suggested two conferences, one on "The Nature of Biopotentials" and the other on "The Postulates of Quantum Mechanics." Biopotentials were the focus of research of MacInnes's esteemed colleague and friend W.J.V. Osterhout, and were of considerable importance in MacInnes's own work. Also, biopotentials represented an interesting field for involving outstanding physicists and physical chemists looking for challenging new problems after the war. The topic of the second conference was an area that intrigued MacInnes, who at the time was studying wave mechanics.[12] Darrow—the urbane, by then perennial, secretary of the American Physical Society, a theoretical physicist who had been at the Bell Laboratories in Murray Hill since 1925 and who was a popularizer of science of some stature— offered MacInnes his help in organizing the conference on quantum mechanics.[13] In late December 1945, MacInnes and Darrow consulted Darrow's friend León Brillouin because "he had a great deal of experience with Solvay Congresses."[14] Reporting on this meeting to Jewett, MacInnes indicated that Brillouin had made a number of worthwhile suggestions, one of which was that they consult with Pauli, who had just obtained his Nobel Prize. Pauli was then still at the Institute for Advanced Study in Princeton, where he had been since the outbreak of World War II. Darrow wrote Pauli indicating that MacInnes had thought of the idea "of convoking a congress similar to the excellent Solvay Congresses of old" to discuss problems connected with the foundations and philosophy of quantum mechanics and "subsequently to publish [their] discussions."[15] He also informed Pauli that the National Academy of Sciences was favorably disposed to sponsor the conference "so that it seems possible to solve the problem of financing."[16] Jewett had indeed explored further the matter of funding and had obtained indications from Raymond Fosdick and Warren Weaver of the Rockefeller Foundation, and from Devereux C. Josephs of the Carnegie Foundation that they would be interested in supporting the enterprise.[17]

Pauli met with Brillouin, Darrow, and MacInnes in New York on January 18, 1946. Pauli, who had just been invited to the international conference that was to convene at the Cavendish Laboratories[18] in the summer of 1946, was convinced that such conferences were important for reasserting the continuity of the scientific enterprise and reestablishing communal values after the terrible ordeal of World War II; he suggested a large conference rather than one patterned after the small and elitist Solvay Congresses. Moreover he "was a bit doubtful whether there would be anything very new to consider [in wave mechanics], but got more enthusiastic when the topic was changed to "Fundamental Problems of Quantum Theory." He also suggested quite a number of names from abroad for those who might attend "[promising] to make up a list of those who would present papers and may see some of the people when he goes abroad in connection with his Nobel Prize."[19]

MacInnes did not like Pauli's suggestions and became disturbed "at the way the plans for the quantum theory conference seemed to be going. Pauli was planning for too many of the older men, and [MacInnes thought] that the best results would be to get out the coming generation."[20] On January 21, MacInnes conferred with Jewett "who heartily agreed" with him on the matter, and "also suggested having the conference at an inn somewhere."[21]

After his consultation with Jewett, MacInnes wrote Pauli that "Dr. Jewett . . . is very much in favor of smaller conferences of the type that we outlined to you." Also, Dr. Fosdick of the Rockefeller Foundation, who "has been thinking along that line for some time," would give financial support only to a small gathering. Additionally, "Dr. Jewett thinks, as I do, that if at all possible we should try to discover and make prominent in such a conference the younger men who are coming along and give them, if possible, the job of preparing and presenting the papers.[22] MacInnes indicated that he has in mind "such men as J. A. Wheeler" and that there must be others. It is these men "who are most likely to do a good job and to bring in fresh ideas." Thanking Pauli for his help, MacInnes requested of him suggestions as to the names of "such coming men."

John Archibald Wheeler was indeed representative of the "younger [American] men." A brilliant young theoretical physicist, he had been educated at Johns Hopkins. As a National Research Council Fellow he had worked in 1933/34 with Gregory Breit at NYU on problems in quantum electrodynamics and had spent the academic year 1934/35 in Copenhagen at Niels Bohr's Institute.[23] He had made important contributions to nuclear physics, especially in his paper on the scattering matrix and his researches with Bohr on the fission process (Wheeler 1937; Bohr and Wheeler 1939). In 1938, at the age of 26, he had joined the Department of Physics of Princeton University as an assistant professor. During the war he distinguished himself at the Chicago Metallurgical Laboratory and bore major responsibility for the design and construction of the Hanford reactors.[24]

Wheeler's stature in the theoretical physics community was clearly recognized. In the fall of 1945 he had been chosen to present a paper on the "Problems and Prospects in Elementary Particle Research" at the Symposium on Atomic Energy and Its Implication. This symposium was the high point of the first postwar meeting of the NAS (held jointly with the American Philosophical Society). At that symposium he shared the limelight with Enrico Fermi, J. Robert Oppenheimer, Eugene Wigner, Arthur H. Compton, Harold Urey, and Irving Langmuir. His impressive address indicated that a new generation of American physicists was taking over the intellectual leadership of the field of "elementary particle physics." Very probably, Wheeler had come to MacInnes's and Darrow's attention at that symposium.[25]

On February 15, 1946, Darrow wrote Pauli that he and MacInnes were anxious to get his list of talented young physicists "worthy of invitation" to the proposed conference.[26] Soon thereafter MacInnes got a letter signed jointly by Pauli and Wheeler in which they indicated that "a proposal for a conference on Wave Mechanics had been made by both Bohr and [MacInnes], the former suggesting that one be held in Denmark early in 1947."[27] Pauli and Wheeler both remembered "the stimulating discussions which have taken place at Bohr's institute and the important consequences for science which have come out of them under his leadership." Since plans were then "underway" for a conference in New York in September on "Physics of the Elementary Particles," they suggested that the "second" conference which had been proposed both by MacInnes and by Bohr be held in Copenhagen. Furthermore, if MacInnes approved this suggestion, Pauli and Wheeler recommended that the funds "which would in any case have been set aside for travel should be used to send scientists from other countries to the conference in Copenhagen. This arrangement might be especially appropriate through the fact that Bohr's Institute has always enjoyed extremely close relations with the Rockefeller Foundation."[28]

Pauli and Wheeler preferred to promote Bohr's conference in Copenhagen rather than one to be held in the United States because of their respect and affection for Bohr and because this would be a "means to promote international collaboration in science." Although MacInnes was disposed to go along with them, Darrow convinced him otherwise and suggested that they "get Pauli and Wheeler to NYC."[29] Darrow called Pauli who, because he was leaving for Europe, delegated Wheeler to act for him in his absence. Darrow then wrote Wheeler to express his disagreement with Pauli that "Bohr's proposed conference should supersede the one which MacInnes is planning," noting that

(1) Few Americans are likely to be asked to Bohr's conference...

(2) Even among the Americans whom Bohr will invite... [few will come because of] the difficulty and the cost of the journey...

(3) Bohr did not publish the proceeding of his conferences, whereas MacInnes is in a position to do so."[30]

Darrow therefore suggested to Wheeler that he take over "the fulfillment of the request which MacInnes and I made of Pauli, viz. that you select say fifteen of the young and promising Americans in the field of quantum mechanics and send your list to MacInnes and me. We will then combine it with the list of established theoretical physicists which MacInnes, Pauli and I drew up some weeks ago, and we will seek an occasion to meet with you and discuss our plans."

Darrow in this letter to Wheeler had made explicit what had been implicit with MacInnes and Jewett: the conference was to be an American one, designed to "bring out" the young American theoretical physicists who had played such a large role in the successful war effort. Darrow, MacInnes, and Jewett wanted the conference to demonstrate that theoretical physics in the United States had come into its own with the "younger men" who had been born and trained there.[31] The conference was to prove the strength of American theoretical physics not only in wartime activities, but also in "pure" physics.

MacInnes also wrote to Wheeler, indicating that it seemed to him "that there is room for both the National Academy conference and Dr. Bohr's conference ... since somewhat different groups might be involved. In conversation with Dr. Jewett, he agreed with me that it would be well to include in the group, of say thirty, as many of the younger active producing physicists in the field as possible and have a relatively smaller number of the older men. It is the latter group, of course, that would be most likely to be invited to Dr. Bohr's conference." Moreover, "if the

National Academy conference could be held during the coming fall it might prove to be an excellent preliminary to the European meeting.[32]

Darrow and MacInnes met with Wheeler over lunch on Monday, March 11, 1946, in New York. At that meeting Wheeler suggested to them that their conference deal with "more advanced topics," namely electron-positron theory. He also identified for them some of the "younger men." On March 14 MacInnes wrote Wheeler a letter containing the names he had obtained from various sources, as well as those Wheeler had given him, for the conference to be tentatively called "Fundamental Problems of Quantum Mechanics." The names on MacInnes's list were as follows:

Saul T. Epstein	Fritz London	John G. Kirkwood
William V. Houston	Edward Teller	Albert Einstein
Darrow	Paul Dirac	Wolfgang Pauli
Edwin C. Kemble	Yakor Frenkel	William S. Kimball
Vladimir Rojanski	ErwinSchroedinger	Niels Bohr
Rabi	John Van Vleck	Max Born
Eugene Wigner	Robert Oppemheimer	Jean Louis Des
John Von Neumann	[sic]	Touches [sic]
Hans Bethe	Gregory Breit	Louis De Broglie
Henry Eyring	Peter Debye	Lev Landau
Linus Pauling	Richard Tolman	Lars Onsager
John C. Slater		

The list reflected various inputs. Some stemmed from MacInnes', reading in quantum mechanics, for example, Rojanski, Houston and Kemble; others from his professional activities in physical chemistry—Debye, Kirkwood, Tolman, Kimball, Pauling, Eyring, Onsager. Darrow, very likely with the advice of Rabi, had supplied the names of many of the physicists: Oppenheimer, Wigner, Bethe, Rabi, etc. The names of Destouches and de Broglie had probably been suggested by Brillouin. After noting that he had misplaced the memorandum containing the names of "Europeans" that Pauli had given him, MacInnes recorded Wheeler's suggestions:

E. B. Wilson	Robt. Christy
Richard Feyerman [sic]	Leonard Schiff
Chas. Critchfield	E. H. (?) Hill
Wendell Furry	Julian Schwinger
Victor Weisskopf	

He concluded his letter with the statement that his "present idea would be to get together a group consisting mostly of the younger men, who would understand each other's jargon."[33]

After his meeting with Wheeler, MacInnes also wrote to Jewett:

Dr. Darrow and I have come more or less to the conclusion that
it would be the most effective thing to get together a group of
the younger men who are actively at work or else slowly getting
back to their normal work after war research. There should also
be some of the older men present as well. Dr. Wheeler has been
of great assistance in pointing which of the younger men should
be invited and he has outlined the liv[eli]er portions of the topics
to be considered. The idea of a long weekend at some country
inn seems most desirable for this group.[34]

The emphasis on bringing together younger men who had been involved in war
work implied that other physicists, even though they had made contributions to
"fundamental problems in physics" but who had not been at one of the wartime
laboratories, were not invited to the conference. Peter Havas and Joseph Jauch are
two such cases.[35]

The format for the conference thus shifted from that of the Solvay to that
of the Washington Conferences on Theoretical Physics which had been patterned
after the ones Bohr had organized at the Institute of Theoretical Physics in Copenhagen during the 1920s and 1930s. Wheeler had in fact attended the Washington
conferences from 1937 on. The avowed purpose of these conferences had been
to evolve in the United States something similar to the Copenhagen conferences,
in which "a small number of theoretical physicists working on related problems
assemble to discuss in an informal way difficulties met in their researches."[36]

After MacInnes had requested a program for the contemplated theoretical
physics conference, Wheeler, in a letter dated March 19, outlined possible topics
dealing with "Problems on the Quantum Mechanics of the Electron," and drew up
the following list of suggested participants:[37]

Wigner	Breit	Feynman	Hill
Von Neumann	Pauli	Critchfield	Schwinger
Bethe	Bohr	Furry	Serber
Teller	Landau	Weisskopf	Darrow
Dirac	Kusaka	Christy	
Oppenheimer	E. B. Wilson	Schiff	

On March 28 MacInnes wrote Wheeler to thank him for his outline and indicated
to him that he was particularly pleased that he considered "a group as small as fifteen." In addition he informed Wheeler that Jewett had raised the question whether
the contemplated conference "could utilize some of the men who are to be brought
over from Europe during the coming October" to attend the autumn meeting of
the National Academy of Sciences and the American Philosophical Society in

Washington and Philadelphia. He also asked Wheeler to "sound out various of the workers as to their willingness to prepare papers as you describe," which were to be published under the auspices of the NAS "after revision in the light of the discussions at the conference."[38]

In the middle of April 1946, Wheeler answered him and detailed both possible problems to be addressed by a three-day conference and likely persons to present papers on these topics. In his letter, Wheeler noted:

> I believe that the final program should be developed as a result of extended discussions with the individual participants— discussions which in toto might properly consume as much time as a week. For such discussions, the following list can be regarded only as a start, to be modified in the course of the conversations. The men themselves would, I believe, be the best choosers of their subject matter.

Problems of the Quantum Mechanics of the Electron
Question:
Does there exist a logical, consistent and comprehensive theory of electrons and positrons, including their creation, interaction with each other, and annihilation? If so, how can this theory suitably be formulated? If not, what is the range of validity of existing theory? Are there any suggestions how this range of validity may be extended?

Day I
A. The fundamental principles of quantum theory developed from a few crucial experiments. Bohr (alternates: Teller, Furry).
B. Unified mathematical formulations of quantum theory. Von Neumann (alternates: Wigner, Schiff).
C. Interrelations between correspondence principle, complimentarity principle, indeterminism, matrix irreducibility, and spin-statistics relationships as guiding principles in the present and future development of quantum theory. Pauli (alternates: Landau, Serber).
D. Discussion and analysis of principles of quantum theory (general participation).

Day II
A. Analysis of measurability of the electromagnetic field quantities in quantum theory. Rosenfeld (alternates: Breit, Kusaka).

B. Present status of the theory of electron-positron pairs as a formalism. Dirac (alternates: Oppenheimer, Christy).

C. The measureability of the electron-positron field quantities. Bohr (alternates: Rosenfeld, Weisskopf).

D. General discussion and analysis of the characteristic difficulties in the unification of relativity theory and quantum theory.

Day III

A. Presentation of general points of view and methods of approach bearing on or related to the relativistic quantum theory of electrons and positrons. (The intention would be to provide maximum opportunity for presentation of original ideas.)

B. Concluding session to draft, if possible, a one-page statement summarizing progress made in answering the question originally posed to the conference.

Publication of papers and discussions in a widely accessible form, as you suggest, would do much to maximize the value of the conference. In addition, a simplified version of the conclusions of the conference would reach a different audience and would presumably have equally valuable results. For person to prepare such a report I can think of no one better suited than Dr. Darrow himself.

A meeting in New York or any other large city would, I am afraid, find several of the participants enmeshed in the distracting obligations which are such a bane to contemplative thinking in this first post-war year. Consequently I favor the idea you propose of a meeting in some comfortable country inn.

For date January 1, 1947 plus or minus a month or two might be suitable.

I am sorry that absence from Princeton on still another trip makes it necessary for me to write to you in long hand. If your secretary should find it possible to send me a copy of this letter, I should be most appreciative.[39]

Wheeler also estimated the probable cost of such an undertaking. MacInnes promptly thanked Wheeler,[40] and on the same day wrote to Jewett—enclosing Wheeler's letter—to give him his estimate of the cost of the two conferences. The estimate for the biopotential conference amounted to $1,500; that for the quantum mechanics conference was as follows:[41]

Two days at inn (30 people)	420.00
Train Fare for those coming from distance	750.00

Monograph	1500.00
Stenographic work (if used)	300.00
Mimeographing, postage, etc.	150.00
	$3120.00

Jewett in his reply to MacInnes indicated that there ought to be no difficulty in getting the $5,000 or $6,000 to cover the expenses of the two conferences. If necessary, Jewett as president would stretch his authority over the National Academy's Emergency Fund and cover the expenses of the conferences from it. "Further, if these first two prove really successful I think we will have no trouble financing others."[42]

With funding of the conference assured, Darrow and MacInnes instructed Wheeler on May 7, 1946, to select "20 or so participants for the conference" to be held in late January or early February 1947 and, in consultation with Darrow, choose the authors of the papers to be presented. They also recommended that N. Bohr ought to be gotten to attend "if possible." Wheeler was also informed that a decision had been made not to have a stenographic recording of the discussion "but to have careful notes made of such discussion by selected members of the conference"[43] and that the conference was to take place at a country inn to be selected by MacInnes.

In his reply to MacInnes, Wheeler noted that he had discussed the conference with Charles Critchfield of George Washington University and with his colleague Eugene Wigner at Princeton—both of whom were likely participants. They were very enthusiastic about the choice of subject. However, both felt that since George Washington University and the Carnegie Institute of Terrestrial Magnetism were considering sponsoring their first postwar Washington Conference on Theoretical Physics that October "to take advantage of the presence in this country of Bohr and Dirac" (who were to attend the joint NAS and APS meetings), and since Wheeler's proposed topics "would be suitable for consideration as a program for such a Fall Conference that it would be appropriate to unite the [MacInnes] conference with the Washington conference."[44] Wheeler stressed that the presence of Bohr at such a conference was essential. As Victor Weisskopf had recently put it to him, "There is no one in the world who possesses the same qualifications as Bohr to clear up conceptual difficulties such as he in this field [electron-positron theory]."[45] MacInnes forwarded Wheeler's recommendation to Darrow and requested his reaction. MacInnes's view was that "the conference has been developing into something quite different from what I originally intended and now shows signs of being captured by another group."[46] Although he professed himself willing to compromise on a combined conference if it were held, at least partly, under NAS auspices, he admitted that he did not like it. Moreover, the early date of October 1946 did not leave enough time for proper preparation. Finally, "Though the presence of Bohr and Dirac would help, it seems to me that it would be much more profitable to have the younger men in the field do the job and not rely too much on authority of big names."[47] Darrow's reaction was swift and to the point. On June 21 he sent a telegram to MacInnes stating: "Deprecate merger of your conference with other."[48]

MacInnes, with Jewett's backing, carried the day. With Wheeler as the principal consultant, planning for the conference proceeded on the assumption that it would be held in the spring of 1947 to allow proper preparations.[49] By November 1946, Darrow had prepared a first draft of an invitation to the conference.[50] The announcement stressed that there would be no formal papers—except for three or four brief ones "to invite and excite discussion"—and no agenda, and that, following the practice of the Copenhagen and Washington conferences, there was no longer any plan to make a record of the proceedings.

The conference on biopotentials, in honor of W. J. V. Osterhout, was held at the Rockefeller Institute under the auspices of the NAS on December 13 and 14, 1946. It turned out to be a great success,[51] and publication of its proceedings was expected. In contrast to the biopotential conference, no plans had been made to produce either a monograph or to make available to the scientific public a record of the proceedings of the conference on quantum mechanics.

This in Jewett's opinion detracted considerably from the value of the contemplated conference. To meet this objection MacInnes and Darrow decided to postpone the conference[52] on "The Quantum Mechanics of the Electron" to late May 1947 as "this will give time for the preparation of several papers to be distributed in mimeographed form as a basis of discussion and final publication."[53]

Darrow had recommended that Oppenheimer write a paper.[54] When asked, Oppenheimer, who would be traveling from California, indicated that he would be able to attend only if the conference were held "during the two days just before or the two days just after the Memorial Day weekend." Darrow advised MacInnes to accept the later date "because Oppenheimer is so great a man" and also "because it may facilitate greatly our task of getting accommodations"[55] (since many inns and hotels opened their premises only after Memorial Day). They set the date of the conference to be Monday, Tuesday, and Wednesday (June 2, 3, 4) and decided that Darrow would be its convener and chairman. The other persons asked to prepare papers for the conference were Edwin Kemble, who declined, Victor Weisskopf, and Hendrik Kramers.[56] In his reply to Darrow, Weisskopf stated that he thought "it is a good sign that somebody [i.e., MacInnes] is again interested in discussing the foundations of quantum mechanics instead of thinking only of high voltage machines and how to produce mesons. The whole idea of a few quiet days in the country together with Heisenberg's 'Uncertainty Relations' seems to me extremely attractive, and reminds me of wonderful days twenty years ago in Göttingen or Copenhagen."[57]

On February 28, 1947, Jewett gave MacInnes "the green light on the coming conference."[58] The matter of getting a place for the conference was attended to by MacInnes, and he turned to the Commerce Commission of the state of New York for assistance. John Deming, of the commission, recommended to him "an inn on

Ram's Island at the end of Long Island quite highly ... [that sounded] very good but possibly too expensive." On March 10, 1947, MacInnes visited the inn "to judge its fitness" and though it was closed deemed it "excellent for the conference."[59] On March 26 he informed Peter Katavalos, the manager of the inn, that "we would use his Rams Head Lodge for our conference."[60]

Although Oppenheimer and Weisskopf had initially agreed to prepare manuscripts of their papers in advance of the conference—which were then to be copied and circulated and were "to form the backbone of the published monograph which [was to be brought] out in due time"—by late March they proposed that the monograph be written wholly after the conference. Darrow and MacInnes were willing to accept the proposal but were unwilling to convene the conference without a written agenda. They therefore asked each man to prepare "an outline of 500 words or thereabouts, of the topics, problems and questions which you propose for consideration" to be delivered by April 30 to MacInnes. They also asked each of them to draw up a list of people "whom they propose for taking part in the conference."[61] On April 7 Darrow sent MacInnes a list of 25 proposed participants that Rabi and he had drawn up, "having before us Wheeler's list from which we made very few deletions."[62]

The names on that list were the following:

1. MacInnes	10. A. Einstein	19. E. Teller
2. Darrow	11. R. P. Feynman	20. Felix Bloch
3. H. A. Kramers	12. I. I. Rabi	21. G.E. Uhlenbeck
4. J. R. Oppenheimer	13. J. H. Van Vleck	22. David Bohm
5. V. F. Weisskopf	14. A. Pais	23. R.E. Marshak
6. E. P. Wigner	15. Hermann Weyl	24. W.E. Lamb, Jr.
7. John von Neumann	16. Julian Schwinger	25. Arnold Nordsieck
8. H. A. Bethe	17. J.A. Wheeler	
9. G. Breit	18. E. Fermi	

The notable additions to Wheeler's list were A. Einstein and Hermann Weyl. An invitation was sent out by MacInnes to the twenty-five proposed participants on April 11, 1947, detailing the plans of the conference:

> Under the auspices and with the support of the National Academy of Sciences an invitation conference on the foundations of quantum mechanics will be held at Ram's Head Inn, Shelter Island, Long Island, New York, on Monday, Tuesday and Wednesday, June 2, 3 and 4, 1947, under the chairmanship of Dr. Karl K. Darrow. Discussion will be led by Drs. H. A. Kramers, J. R. Oppenheimer and V. F. Weisskopf. The number of participants will be limited to twenty-five.
>
> It is planned that the group will gather at the American Institute of Physics, 57 East 55th Street, New York 22, on

Sunday, June 1, at 3 P.M., and will go by bus to the Inn. The Institute building will be open from 2 P.M. The return will also be by bus on Wednesday afternoon, June 4, arriving in New York about five o'clock. It is difficult, though not impossible, to reach the Inn by trains and taxis, so that it is highly desirable that participants avail themselves of this transportation. Instructions for reaching the Inn by private car will be given to those asking for them.

The three discussion leaders will prepare brief outlines of topics for consideration, which will be forwarded in advance of the conference, but the participants will not be bound to these subjects. It is expected that all participants will share in the interchange of ideas. A monograph based on the discussions will eventually be prepared by the chairman and the leaders of the discussions.

It is not intended that the discussions shall occupy all the time available. There will be opportunities for boating, swimming and golf.

A grant from the Academy arranged by its president, Dr. F. B. Jewett, has made it possible to pay the expenses of the sojourn at the Inn and the buses from and to New York City. The accommodations will be two in a room but with separate beds. An allowance may be arranged toward travelling expenses for those coming over two hundred and fifty miles. You are urgently requested to inform Dr. K. K. Darrow, Bell Telephone Laboratories, 463 West Street, New York 14, N.Y., **promptly** whether or not you will be able to come, as we wish to have the full complement of participants.[63]

By April 17 Darrow communicated to MacInnes that Pais, von Neumann, and Lamb had accepted but that Weyl had declined their invitation because he would be abroad at the time. He suggested that Hermann Feshbach, a young theoretical physicist at MIT, be invited as a substitute for Weyl.[64] On April 18 MacInnes informed Darrow that he had received a letter from Einstein declining the invitation because of ill health. He suggested that an invitation be sent to Pauling because this "might be tactful in connection with the National Academy politics" and "it might not be a bad idea to get a chemical point of view."[65] Bloch also declined because he had to undergo an operation,[66] this spot was filled by Pauling. Condon, a member of the NAS, was invited to fill the vacancy left by Einstein's refusal, but he likewise did not accept. An invitation was then sent to Rossi, a cosmic ray physicist from MIT, whose name had been suggested by Weisskopf.[67] From the original list only Bloch, Einstein, Weyl, and Wigner declined,[68] this last vacancy being filled by Robert Serber, a young theoretical physicist from

Berkeley, and the only other person besides Oppenheimer to come from the West Coast. Darrow had telegraphed Serber on May 21, 1947, inviting him to the conference. On that same day he wrote him a letter explaining "that our funds did not permit us the desired luxury of providing travel expenses from the coast, and this is why you were not asked until Oppenheimer told me that your fare could be paid from other sources."[69] The Shelter Island conference, as well as the Pocono and Oldstone conferences, acquired a distinctly East Coast character as a result of this decision to consider travel from the West Coast a luxury.[70]

The final announcement to the members of the conference detailing travel arrangements was sent out on May 24. It also included a list of the participants:

CHAIRMAN

K. K. Darrow, Bell Telephone Laboratories, New York 14, New York

DISCUSSION LEADERS

H. A. Kramers, Institute for Advanced Study, Princeton, N.J.

J. R. Oppenheimer, University of California, Berkeley, Calif.

V. F. Weisskopf, M.I.T., Cambridge 39, Massachusetts

H. A. Bethe, Cornell University, Ithaca, N.Y.

David Bohm, Princeton University, Princeton, N.J.

Gregory Breit, Yale University, New Haven, Conn.

Enrico Fermi, University of Chicago, Chicago 37, Ill.

Herman Feshbach, M. I. T., Cambridge 39, Mass.

R. P. Feynman, Cornell University, Ithaca, N.Y.

W. E. Lamb, Jr., Columbia University, New York 27, N.Y.

D. A. MacInnes, Rockefeller Institute, New York 21, N.Y.

R. E. Marshak, University of Rochester, Rochester, N.Y.

John von Neumann, Institute for Advanced Study, Princeton, N.J.

Arnold Nordsieck, Bell Telephone Laboratories, Murray Hill, N.J.

A. Pais, Institute for Advanced Study, Princeton, N.J.

L. Pauling, California Institute of Technology, Pasadena, Calif.

I. I. Rabi, Columbia University, New York 27, N.Y.

Julian Schwinger, Harvard University, Cambridge 38, Mass.

R. Serber, University of California, Berkeley 4, Calif.

Edward Teller, University of Chicago, Chicago 37, Ill.

G. E. Uhlenbeck, University of Michigan, Ann Arbor, Mich.

J. A. Wheeler, Princeton University, Princeton, N.J.

Bruno Rossi, M.I.T., Cambridge 39, Mass.

J. H. Van Vleck, Harvard University, Cambridge 38, Mass.

The conference arrangements proceeded without further incidents,[71] although MacInnes noted in his diary that some of those invited "have behaved like prima donnas at least, and are apparently badly spoiled."[72] The conferees gathered in New York at the American Institute of Physics, 55 East 55th Street, on the afternoon of Sunday, June 1, 1947. From there they were taken "on an old and shaky"[73] bus to Greenport at the tip of Long Island. On the final phase of the trip they were accompanied by a police motorcycle escort and their bus didn't stop at any traffic lights.[74] As they passed each county line a new police escort would meet them. The police escort had been secured by John F. Deming, the regional manager of the Department of Commerce of the state of New York, in charge of the Nassau-Suffolk office, and the prime agent in arranging the "happy sojourn at Ram's Head Inn."[75]

In Greenport the conferees were wined and dined at Mitchell's restaurant—a big oyster and steak dinner[76]—as guests of the Chamber of Commerce. The dinner, although officially given by the Chamber of Commerce, was actually paid for by John C. White,[77] its president. He had been deeply impressed by the physicists' war efforts and wanted to express his appreciation and indebtedness. In an after-dinner speech he told the audience that he had been a marine in the Pacific during the war and that he—and many like him—would not be alive were it not for the atomic bomb. Deeply touched, Darrow and Oppenheimer graciously replied on behalf of the conferees to express their thanks for the hospitality that Greenport was extending the conference, and for its generous gesture in tendering the dinner. After the dinner they were taken by ferry to Shelter Island, where "they were received by a siren-shrieking police escort to their Inn."[78]

In the late 1940s, Shelter Island was sparsely populated. A good deal of its land was still being farmed, and its coastline facing the Atlantic was dotted with great nests of ospreys. During the summer the island's population swelled to about two-thousand, as a fair number of summer homes and several summer resorts were located on it. Ram's Head Inn—an unpretentious but elegant and comfortable two-storied clapboard structure—had just "been painted and furnished with excellent furniture, and looked most inviting."[79] The conferees were its first guests for that season. In fact, it had opened ahead of schedule to host the conference.[80] The inn can accommodate about thirty people and was just large enough to hold all the participants. Located on a western cove of Shelter Island, it is perched on a promontory, surrounded by lush, green lawns that abut the dunes; and beyond the dunes lies a beach that can be reached in a leisurely few minutes' walk from the inn. It was an ideal place to come together to address once again the problem of pure science. Stephen White covered the conference for the *New York Herald Tribune.* His article, dated June 2, reported on the conference's first day:

> Twenty three of the country's best known theoretical physi-
> cists—the men who made the atomic bomb—gathered today in
> a rural inn to begin three days of discussion and study, during

which they hope to straighten out a few of the difficulties that beset modern physics.

It is doubtful if there has ever been a conference quite like this one. The physicists, backed by the National Academy of Science, have taken over Ram's Head Inn.... They roam throught the corridors mumbling mathematical equations, eat their meals amid the fury of technical discussions and gather regularly in the lobby for blackboard discussion....

The meeting is officially entitled "A conference on the Foundations of Quantum Physics." What it amounts to, is an attempt to clarify, now that war-time work can be set aside, the problems that lie before physicists in the realm of the most advanced portion of the science....

... Although quantum physics has yielded ... practical results—the atomic bomb among them—the physicists have always been conscious of the fact that the theories rest on insecure grounds. The science is far too approximate for their taste. In these three days they hope to find, if nothing else, what directions their work should take....

... The conference is taking place with almost complete informality, aided by the fact that the scientists have the inn all to themselves and feel that there is no one to mind if they take off their coats and get to work.

The article gave a list of the participants. Singled out from among them were Dr. J. Robert Oppenheimer, "who directed the construction of the first atomic bomb"; Dr. I. I. Rabi, Nobel prize winner; Dr. Hans Bethe, Dr. Julian Schwinger (identified as Schwainger), "who was recently made a full professor at the age of twenty nine" at Harvard University, and Dr. John von Neumann, "man of all science."[81]

The reporting was accurate: The conference was informal, the pace somewhat frenetic, the atmosphere intense. All the participants were aware that the experimental results being presented to them by Lamb and Rabi on the spectrum of hydrogen and by Rossi and others on recent cosmic ray findings were of exceptional significance.

Darrow chaired the conference, but Oppenheimer was the dominant personality and "in absolute charge."[82] MacInnes recorded in his diary that "it was immediately evident that Oppenheimer was the moving spirit of the affair."[83] When, at Darrow's insistence, the evening sessions were terminated at 9:00 P.M., the appointed hour, Oppenheimer adjourned the formal sessions to informal meetings that lasted late into the night.[84] Darrow himself, in his diary, gave a revealing account of Oppenheimer:

As the conference went on the ascendency of Oppenheimer became more evident—the analysis (often caustic) of nearly every argument, that magnificent English never marred by hesitation or groping for words (I never heard "catharsis" used in a discourse on [physics], or the clever word "mesoniferous" which is probably O's invention), the dry humor, the perpetually-recurring comment that one idea or another (incl. some of his own) was certainly wrong, and the respect with which he was heard. Next most impressive was Bethe, who on two or three occasions bore out his reputation for hard & thorough work, as in analysing data on cosmic rays variously obtained (An amusing interchange in which x—I've forgotten who, Teller I think—had put a math'l argument on the board; y said "is there not a logarithm?"; x replied "when I do it decently the logarithm will be there"; Bethe said, "When it is done *really* decently, there is no logarithm" (laughter!).[85]

Let me give one further indication of Oppenheimer's stature at the time of the meeting. The conference lasted through the morning of Wednesday, June 4. After the conference, Oppenheimer had to go to Harvard where he was to receive an honorary degree. Arrangements had been made to fly him, Rossi, Schwinger, and Weisskopf by seaplane from Shelter Island to Boston. However, bad weather forced the plane to come down at the New London coast guard station, which is not open to civilian aircraft. The pilot was very worried since they were not supposed to land there. They were met by a naval officer who was clearly furious and ready to read them the riot act. As they opened up and jumped out, Oppenheimer told his very nervous pilot, "Don't worry." Hand outstretched, he introduced himself to the ranting and raging officer with the statement: "My name is Oppenheimer." The bewildered officer queried: "The Oppenheimer?" To which came the reply: "An Oppenheimer!" After an "official" welcome in the officers' club, they were driven—with a military escort—to the New London railway station, where they boarded a train for Boston.[86]

Oppenheimer wrote Jewett from Cambridge:

We have just come away from the Ram Island Conference on the Foundation of Quantum Mechanics, that the Academy has sponsored and aided. On behalf of all participants, and for myself, I should like to thank you, as President of the Academy. The three days were a joy to us, and perhaps rather unexpectedly fruitful; we had a long needed chance for clarification and exchange of views, and came away a good deal more certain of the directions in which progress may lie. We recognized this so generally that we agreed that in our work of the next year, we would wish in publication to acknowledge our indebtedness to the conference and to the Academy explicitly, wherever it touched upon matters there discussed. Since these were indeed many and deep, I expect that the Physical Review will be full of praises of Ram Island. And that is only proper.

We worked rather hard, and could well have spent another day or two. It is my hope to ask the conference to reconvene next winter.

A similar impression was conveyed to MacInnes by Wheeler on June 6. "The conference may not have any great progress to report at the moment, but it has brought about the closest meeting of minds in physics in America which has taken place since 1941 and has brought theoretical physics to a starting point which, without the aid of the Ram's Head conference, might not have been reached for many months to come."[88] Writing to Jewett in July, Wheeler indicated that he knew of no meeting since the war which has done more than the Shelter Island conference "to restore to us all a common appreciation and assessment of the deeper problems of elementary particle physics."[89]

Similar sentiments were expressed by the other conferees. Breit called the conference "one of the most successful I ever attended" and expressed the hope that the NAS will find it possible to have similar conferences in the future.[90] Even MacInnes, for whom the conference "was all pretty much above my head,"[91] had been immediately aware that "there is some chance" that "the National Academy of Science Conference on Quantum Mechanics at Ram's Head . . . has made scientific history."[92] Writing to Richards later that year, Oppenheimer stated that "for most of us [Shelter Island] was the most successful conference we had ever attended. Out of it came a quite new understanding of the probable role of the meson in physical theory, and the beginnings of a resolution of the long outstanding paradoxes of the quantum electrodynamics. In the intervening months, many excellent papers have been prepared, all acknowledging the debt to this meeting and carrying further the progress which was there initiated."[93]

Jewett was elated by the results of the conference. Its cost had been minimal: Only $872 of the $1,500 appropriated had been spent. Both of the conferences that MacInnes had initiated and overseen "had paid dividends" and Jewett was gratified by the fact that they had been held under Academy auspices during his presidency.[94] Although his term expired on June 30, 1947, he recommended to A. N. Richards, the incoming president, that steps be taken to guarantee the continuance of such conferences under NAS sponsorship. Richards was favorably disposed and he requested MacInnes to prepare a memorandum assessing the two "experimental" conferences, so that a recommendation could be submitted to the Council of the NAS.

MacInnes did so in November 1947. In his report[95] he indicated that a monograph on the bioelectric potentials conference would appear soon and that it would emphasize the direction that future research in that field should take. He also reported that although it was decided not to publish a monograph on the Shelter Island Conference, the conferees agreed to give credit to the conference in papers published as a result of its occurrence. He noted that several papers making such acknowledgement had already appeared: Lamb and Retherford's paper on the fine structure of the hydrogen atom (*Phys. Rev.* 72: 241–243 [1947]), Bethe's nonrelativistic Lamb shift calculation (*Phys. Rev.* 72: 339–341 [1947]), Marshak and Bethe's paper on the two-meson hypothesis (*Phys. Rev.* 72: 506–509 [1947]), and Breit's paper on the hyperfine structure of the hydrogen atom, which was in press at the time of the writing of MacInnes's report. MacInnes's conclusion was that "it would appear that these trial conferences were sufficiently successful to justify the continuation."[96]

Richards presented MacInnes's memorandum to a business meeting of the NAS on November 17. Those attending endorsed the suggestion that the Academy encourage such conferences in the future. In December 1947 Richards wrote Oppenheimer, the now acknowledged leader of the previous conference, that the NAS would be prepared to underwrite a meeting of theoretical physicists similar to that held at Shelter Island—up to a sum of $3,000.

Oppenheimer, a few days earlier, had informed him of his plans to convene a second such conference and had requested $3,000, a sum that "would go a long way toward meeting all expenses."[97] MacInnes also wrote to Oppenheimer to express his satisfaction that his idea of the conferences is proceeding so well, and to pledge his support for continued NAS sponsorship. But he insisted "that the conference be not allowed to grow in size."[98] Although this might entail that some feelings might be hurt, he felt very strongly that this was an essential factor for the conferences to be "really successful." On December 10, Oppenheimer circulated a memorandum containing his plans for a second meeting tentatively scheduled to be held from March 30 to April 2, 1948—a date chosen so that Bohr, who would be visiting the United States at that time, could be invited. Because the Ram's Head Inn was not available for that period, he solicited suggestions for another meeting place. Pocono Manor was the site eventually agreed upon for this second conference. It is located approximately midway between Scranton and the Delaware Water Gap in Pennsylvania, and it afforded the same kind of setting as had Ram's Head Inn—a place where, as Oppenheimer put it, the conferees could "be together and undisturbed."

The conference was indeed held from March 30 to April 2, 1948. Its size was increased to twenty-eight, because of the presence of foreign guests. Of those who had attended Shelter Island, Kramers, who had suffered a heart attack, and MacInnes, Pauling, and Van Vleck, whose research interest did not overlap with the chosen topics, were absent. The new participants included Dirac, who was a member of the Institute for Advanced Study during 1947/48; Aage Bohr, Niels Bohr, and Walter Heitler, who were visiting Columbia that year; Gregor Wentzel, who had just come to the University of Chicago; and Eugene Wigner, who had been unable to attend the Shelter Island conference.[99]

At the conclusion of the conference, Oppenheimer wrote Richards:

> We have just come home from the second Academy conference on theoretical physics.... All of those invited attended and, as at the first conference, there was a unanimous and cordial expression of satisfaction and gratitude. We worked quite hard, meeting all day, and often in the evenings, and turned our attention primarily to the new evidences on the nature of the meson, on the interaction of nucleons with electromagnetic fields, and the magnificent new developments in modern electrodynamics which have so deeply increased our insight.[100]

Oppenheimer also informed Richards that it had been agreed to attempt another meeting "about a year from now," since both Shelter Island and Pocono had proved so useful, important, and inexpensive—the total cost of the Pocono conference amounting to only $1,550. He added in closing: "I believe that these conferences, quite without publicity, with a minimum of organization, and undertaken only for an exchange of views and for furthering our understanding of the foundations of physical theory, are singularly appropriate for Academy support."[101]

The plan for financing the conferences that MacInnes had conceived was discussed at a NAS Council meeting in Berkeley in November 1948. The Council decided to follow Jewett's original suggestion to approach private foundations for the funding of future conferences, and instructed Richards to do so. When Richards discovered that there still existed a balance of nearly $44,000 in the Academy Emergency Grant—monies originally given by the Carnegie Corporation "to help the Academy realize its usefulness during the early years of the war emergency"[102]—he asked the Council to be relieved of the duty of soliciting outside funds, and to be allowed to expend the balance on future conferences. Richards stated his reasons very explicitly:

> The unique character of the conferences—the fact that they are not expected to eventuate in a publishable report—makes me wish to keep them out of administrative machinery. I believe them to be a means of fostering scientific leadership in the truest sense—the kind of leadership for which our Academy membership frequently clamors without knowing what they are clamoring for.
>
> I think the Academy, more than the Research Council [and the Carnegie Foundation] need to be the sponsor of these meetings and to be known as such.[103]

From 1946 to 1951 eleven such small conferences[104] were held under the auspices of the NAS and financed at a total cost of approximately $48,000. Each of the first four had a budget of less than $1,200; none of the physics conferences on theoretical physics held between 1947 and 1949 expended more than $1,200. All eleven conferences proved to be important, seminal, and successful. They are a fitting tribute to MacInnes's vision and integrity. Incidentally, many of these NAS-sponsored conferences were held at Ram's Head Inn on Shelter Island.

In January 1949, Oppenheimer formally wrote Richards asking support from the NAS for a third meeting. He indicated that the subjects of this third conference "in so far as they can now be anticipated, will probably again be mesons, field theory, the relations of the elementary particles to nuclear forces, and cosmic rays." He added that "much of the progress of the last eventful years in this field was germinated at Shelter Island and at the Poconos."[105] Moreover, he noted that a sign of the seriousness and nature of these meetings is the mandate he had received to "arrange for four full days instead of the two and a half with which we originally started." At Shelter Island and Pocono the participants had put in "ten and eleven hour days" which may be "some indication of how much we cherish the privilege of these meetings."[106] The site for this third conference, which was held for four days from April 11 to April 14, 1949, under the chairmanship of J. Robert Oppenheimer, was Oldstone-on-the-Hudson in Peekskill, New York.[107] Its cost was $1,150. After the conclusion of the conference, Oppenheimer wrote to Richards:

> It is again my privilege and my pleasant duty to report to you on the conference on fundamental physics which the National Academy of Sciences has sponsored and supported.
>
> The conference was held for the four days: April 11, 12, 13 and 14; and was attended by some twenty-four scientists, many, but not all of whom had attended the conferences in 1947 and 1948. Again on behalf of all participants, I have been asked to express profound appreciation to the National Academy, and a real sense of satisfaction for the fruitfulness and value of the conference; this I gladly do.
>
> The two years since the first conference have marked some changes in the state of fundamental physics, in large part a consequence of our meetings. The problems of electrodynamics which appeared so insoluble at our first meeting, and which began to yield during the following year, have now reached a certain solution; and it is possible, though in these matters prediction is hazardous, that the subject will remain closed for some time, pending the accumulation of new physical experience from domains at present only barely accessible. The study of mesons and of nuclear structures has also made great strides; but in this domain we have learned more and more convincingly that we are still far from a description which is either logical, consistent or in accord with experience.

The members of the conference this year did not determine positively that they desired to reconvene next Spring, since determination on the fruitfulness of this will in part depend on the developments in physics in the year to come; but from the many expressions of gratitude and appreciation, it was clear that this conference needed to be held.[108]

In fact, no further NAS-sponsored conference on particle physics was held.

The Shelter Island, Pocono, and Oldstone conferences were the precursors of the Rochester conferences on high energy physics (Marshak 1970). But they differed from the later conferences in important ways. While the first three conferences reflected the style of an earlier era, the Rochester conferences were more professional and democratic in outlook and had the imprint of the new era: the large-group efforts and the large budgets involved in machine physics. Also, whereas Shelter Island and Pocono looked upon quantum electrodynamics (QED) as a self-contained discipline, the Rochester conferences saw "particle physics" come into its own, with QED as one of its subfields—albeit one with a privileged, paradigmatic position.

The theoretical success reported at these conferences, as impressive as they were, were undoubtedly further magnified by Oppenheimer, the charismatic leader of Los Alamos. Oppenheimer was the dominant figure at these conferences, and it was he who made the assignment of the relative merit of the various contributions reported there. These conferences gave further proof of the remarkable intellectual powers of the leading theoreticians. In many ways, they continued the heroic efforts that had been demanded by the war. In fact, most of the theoretical advances were carried out by the same Promethean figures who had solved the wartime problems.

4.3 The Scientific Content of the Conference

By May 1947 Darrow had obtained the "abstracts" that Kramers, Oppenheimer, and Weisskopf, the discussion leaders, had been asked to write.[109] Because they were to form the basis of the discussions at the conference, they were sent to the other participants. Weisskopf's paper outlined the problems in "elementary particle" physics in broad and general terms:

FOUNDATIONS OF QUANTUM MECHANICS
Outline of Topics for Discussion

Victor F. Weisskopf

The theory of elementary particles has reached an impasse. Certain well known attempts have been made in the last fifteen years to overcome a series of fundamental problems. All these attempts seem to have failed at an early stage. An agenda for a conference on these matters contains, necessarily, a list of these attempts. After returning from war work, most of us went through just these attempts and tried to analyze the reason of failure. Therefore, the list which follows will be well known to everyone and will probably invoke a feeling of knocking a sore head against the same old wall. The success of the conference can be measured by the extent it deviates from this agenda.

It may perhaps be useful to divide the discussion into three parts:

A. The difficulties of quantum electrodynamics.

B. The difficulties of nuclear and meson phenomena.

C. The planning of experiments with high energy particles.

These topics can be subdivided:

A. Quantum electrodynamics
   (1) Self Energies. Attempts to remove infinite self energies.
      (a) Classical attempts: theories using advanced minus retarded potentials. Non linear theories.
      (b) Attempts which change formalism after quantization: λ-process, Dirac negative light quanta, Heitler-Peng, Riesz-method. Why do logarithmic divergencies defy most of these methods?
   (2) How reliable are the "finite" results of quantum electrodynamics derived by means of a subtraction formalism? Polarization of the vacuum and related effects. Is there a high energy limit to quantum electrodynamics?
   (3) Infra-red catastrophe and related topics.

B. Nuclear forces and mesons
   (1) Present knowledge of nuclear forces, range, type, form, singularities, saturation.
   (2) Present knowledge of meson properties.
   (3) Resumé of meson theories and their representation of nuclear forces.
   (4) Discrepancies in the interaction of mesons with matter (Piccioni experiment). Shower formation by primary cosmic radiation (Bridge-Rossi experiment).
   (5) Discussion of new cosmic ray experiments.
   (6) Problems connected with the theory of β-decay. Inconsistencies. Connection with meson decay.

    C. Proposed experiments
Discussion of fundamental experiments to be done with the high energy machines.
(1) Electron accelerators.
(2) Proton accelerators.
(3) The relative importance of different energy regions and the planning of new machines.

    In view of the failure of the present theories to represent the facts, and the small probability that this conference may produce a new theoretical idea, Part C of this agenda could become the most useful part of the conference. A number of very powerful accelerators in the energy region of 200–300 Mev are near completion, and it is time to inaugurate a systematic program of research, with experiments which can be reliably interpreted. All too often much more thought is given to the production of the beam and too little to a reasonable instrumentation.

    Weisskopf's outline was a succinct statement of what he considered the outstanding problems of physics and an assessment of the methods available to tackle them. It was the abstract of a leading theoretician who had devoted all his energies to the solution of fundamental problems in pure physics since "returning from war work." The frontiers of physics were located in QED, nuclear and meson phenomena, and "high energy particle physics." The theoretical foundations were considered shaky and the mood conveyed was one of frustration: once again, Weisskopf had "been knocking his sore head" against the same old wall that he had confronted during the thirties. The proposals Weisskopf made for the topics of discussion centered on experiments that either had just been performed—the Conversi-Pancini-Piccioni and the Bridge-Rossi cosmic ray experiments—or ones that could be performed in the near future when the accelerators then being built at Berkeley, General Electric, Cornell, and Rochester would be running. His emphasis on experiments and what could be learned from them, and his advocacy of a much closer cooperation between experimentalists and theorists, were undoubtedly a legacy of his Los Alamos experience.

    Oppenheimer's outline was more narrowly focused and concentrated on cosmic ray phenomena.

THE FOUNDATIONS OF QUANTUM MECHANICS
Outline of Topics for Discussion

J. R. Oppenheimer

    It was long ago pointed out by Nordheim that there is an apparent difficulty in reconciling on the basis of usual quantum mechanical formalism the high rate of production of mesons in the upper atmosphere with the small interactions which these mesons subsequently manifest in traversing matter. To date no completely satisfactory understanding of this discrepancy exists, nor is it clear to what extent it indicates a breakdown in the customary formalism of quantum mechanics. It would appear profitable to discuss this and related questions in some detail.

We might start this discussion by an outline of the current status of theories of multiple production. Some illuminating suggestions about these phenomena can be worked out in a semiquantitative way, for instance on the basis of the neutral pseudoscalar theory of meson couplings. The suggested results appear to agree reasonably well as to energy dependence of multiplication, energy and angle distribution with the experimental evidence, which is admittedly sketchy. However, no reasonable formulation of theories along this line will satisfactorily account for the smallness of the subsequent interaction of mesons with nuclear matter. Similar difficulties appear when one attempts to make a theory involving couplings of meson pairs to nuclear matter. There are two reasons for these apparent difficulties. One is that in all current theory there is a formal correspondence between the creation of a particle and the absorption of an antiparticle. The other is that multiple processes are in these theories attributable to the higher order effects of coupling terms which are of quite low order, first or second, in the meson wave fields. The question that we should attempt to answer is whether, perhaps along the lines of an S matrix formulation, both these conditions must be abandoned to accord with the experimental facts.

It would be desirable to review the experimental situation with an eye to seeing how unambiguous current interpretations are.

The calculation of the multiple production of mesons is in some ways an extension of the treatment given by Bloch and Nordsieck of the radiation of electrons during scattering. The difficulties of a complete description of these phenomena appear in exaggerated form in the problem of meson production. It would therefore be profitable to review the present status of the theory of radiation reaction and of certain recent suggestions for improving the theory.

Oppenheimer's interests after his war work were less in physics than in statesmanship,[110] but his abilities were such that he could be both an outstanding spokesman for the scientific community and a most impressive theoretical physicist. His presentation reflected his current interests in physics. His students, H. A. Lewis and S. A. Wouthuysen, had just completed some calculations on multiple production of mesons (Lewis 1948),[111] and he had written a paper with Bethe on infrared divergences to prove that the prescriptions that Heitler had recently given for rendering field theories finite were inadmissible as they resulted in not allowing certain necessary processes (Bethe and Oppenheimer 1946). All these calculations were based on the Bloch-Nordsieck (Bloch and Nordsieck 1937) and Pauli-Fierz (Pauli and Fierz 1938) methods for handling multiple particle processes.

Both Weisskopf's and Oppenheimer's outline reflected their interest— and that of the other theorists—in cosmic rays. Since the early thirties experiments exploiting the interaction of cosmic ray particles with the air atoms in their passage through the atmosphere had been an important source of information on "elementary particles." In 1937 cosmic ray shower experiments and their theoretical interpretation had established the existence of a new (unstable) charged particle, the mesotron, which existed in both a positive and a negative variety (Cassidy 1981; Galison 1983). Its lifetime was about 10^{-6} second (Rossi 1983). This mesotron, or meson, was assumed to be the particle that Yukawa had postulated to account for the nuclear forces, and therefore it should interact strongly with nuclei. This identification led to the difficulty that Nordheim had pointed out in 1939 (Nordheim and Webb 1939) and that Oppenheimer emphasized in his outline: "If—as should be the case—these mesons are copiously produced by cosmic rays in nuclear interactions in the upper atmosphere, why were not more of them absorbed in their traversal through the atmosphere?"[112]

These difficulties had been dramatically exhibited in an important and striking experiment carried out by Conversi, Pancini, and Piccioni, the results of which were published early in 1947 in the *Physical Review* (Conversi, et al. 1947; Conversi 1983). Their experiment provided the first demonstration that a substantial fraction of the slow negatively charged mesons found at sea level *decayed* in a carbon plate—rather than being absorbed by the carbon nucleus ($Z = 6$). However, when these mesons were stopped in an iron plate, they were absorbed by the iron nuclei ($Z = 26$). Conversi and his collaborators also found that all positively charged mesons decayed in both the carbon and the iron plates. Now, according to theory (Araki and Tomonaga 1940), positively charged nuclear force mesons should always decay (in agreement with experiment), but negatively charged nuclear-force mesons should never decay (in contradiction with the experiment of Conversi et al.).[113]

Since the mesons are unstable, the time taken by the negatively charged mesons to slow down and be captured into Bohr orbits in an air nucleus (from which nuclear absorption can readily take place) had to be compared with the lifetime for decay. If the lifetime for decay is much shorter than the slowing-down time, the mesons will rarely be absorbed by the nucleus and will most often decay. Fermi, Teller, and Weisskopf (1947), in an analysis of the Conversi experiment carried out early in 1947, came to the startling conclusion that "the time of capture from the lowest orbit of carbon is not less than the time of natural decay, that is, about 10^{-6} second." This was in disagreement—by a factor of about 10^{12}—with the estimate of Araki and Tomonaga (1940), which assumed the meson to be responsible for nuclear forces. This striking discrepancy was behind Oppenheimer's and Weisskopf's call for a discussion of the Conversi experiment.

Kramers, for his part, chose to review the difficulties encountered in QED since its inception in 1927 and to indicate one way out of these problems. He outlined his own work (Kramers 1938b, 1944), and that of his students, Serpe and Opechowski, which had been carried out in 1940 (Serpe 1940; Opechowski 1941), and presented a theory in which all-structure effects had been eliminated. His abstract described "how an electron with *experimental mass* behaves in its interaction with the electromagnetic field":

THE FOUNDATIONS OF QUANTUM MECHANICS
Outline of Topics for Discussion

H. A. Kramers

At almost every important stage of the development of quantum mechanics, not only were new positive results added to what had already been achieved, but also certain "defects" revealed themselves. In some cases such defects were removed by the next step; in some cases they just stayed or even were emphasized more strongly.

When in 1926 the field of free Maxwell radiation was quantized, the infinite zero-point energy revealed itself. We may consider this as a defect which even now has by no means been removed in a satisfactory way.

When in 1927 Dirac gave his non-relativistic quantum theory of the interaction between radiation field and charged particles, the emission, absorption and scattering of photons were, on the whole, described in a satisfactory way. Several defects could be noticed, however, pertaining mainly to the divergencies involved in many perturbation-calculations of the second order. We mention the infinite shift of spectral lines, the impossibility of describing a steady state of an atomic system in the radiation field, and—connected therewith—the impossibility of arriving at an exact dispersion formula in which the phase of the scattered wave was duly accounted for, and finally the difficulty, if not impossibility, of describing the reaction of the radiation on the atomic particles.

The year 1928 brought the Dirac theory of the spin electron and its incorporation in a tentative relativistic quantum mechanics of electrons in the radiation field. Second order divergencies stayed, the negative energy-states gave new troubles, but the relativistic energy levels of the H-atom and the Klein-Nishina formula were significant achievements. They were both obtained, it should be remembered, without basing oneself explicitly on a quantization of the radiation field.

The hole theory of 1930, 1931 brought new significant results. Those which could immediately be connected up with experiments: existence of the positron, materialization and annihilation of electrons, did not—as far as I can see—require explicitly the introduction of the concept of the infinite sea of negative-energy states. Calculations which needed this concept led on the one hand to the not yet observed interaction of light with light; on the other hand they led, as regards the self energy, to results which threw a new light on the divergencies resulting from second and higher order perturbations (logarithmic divergency discussed by Weisskopf and by Pais).

The meson theory of nuclear forces showed encouraging features, but also brought new divergence sorrows.

With regard to this general situation two points seem to me worth closer attention. The first refers to the fact that the nonrelativistic Dirac theory of 1927 showed, in its fundaments, by no means the necessary correspondence with classical electron theory. If we ask for a system of formulae which, by means of a variational principle, describes how an electron with experimental mass behaves in its interaction with the electromagnetic field, we get a Lagrangian different from that on which Dirac's 1927 theory is based. The latter gives, even in purely classical problems, often divergent results where the former automatically leads to convergent results. As an example, it can be shown that the infinite shift of spectral lines, with the Dirac Lagrangian, is immediately connected with the divergence of the electromagnetic mass for a point-electron (Serpe 1941). As a second example we mention that the main interaction term in the Hamiltonian expressing the coupling of field and electron, which is Dirac's theory, took the form $-\frac{e}{c}(\overline{v}\overline{A})$, in the correspondence-theory referred to above takes the form $-\frac{e}{c}(\overline{v}\overline{Z})$, where Z is the Hertz potential of the electromagnetic radiation field $(\overline{A} = \frac{1}{c}\frac{\partial \overline{Z}}{\partial t})$. With the latter interaction term the ultraviolet catastrophy in 2nd order perturbation calculations disappears.

The second point refers to an—in my opinion—unsatisfactory feature which is already shown by the relativity-treatment of *free* particles. It consists in this, that with an arbitrary initial choice of the symbolic wave functions, which are supposed to describe the presence of such particles, the knowledge of those initial wave functions in a finite domain of space does not allow precise prediction of the probable existence of those particles within that finite domain, as was the case in nonrelativistic quantum theory. This situation appears unsatisfactory with respect to the exigencies of the general idea of "Nahewirkung" in physics.

It is proposed, if there is time, to go into more specific details as regards the first of the points mentioned.

Kramers' essential point was that the infinite shift of spectral lines that resulted from the Hamiltonian describing the interaction of a charged particle with the radiation field that Dirac had written down in 1927 (Dirac 1927b) could be "immediately connected with the divergence of the electromagnetic mass for a point electron." In Kramers' treatment the interaction energy, which in Dirac theory was $-\frac{e}{c}(\mathbf{v} \cdot \mathbf{A})$, took the form $-\frac{e}{c}(\mathbf{v} \cdot \mathbf{Z})$, where $\mathbf{Z}$ is the Hertz vector $\mathbf{A} = \frac{1}{c}\frac{\partial Z}{\partial t}$. And with "the latter interaction term the ultraviolet catastrophy in 2d order perturbation calculation disappears." Kramers' remarks proved to be very important at the conference.

Equally important was the report of Lamb and Retherford's experiment at Columbia on the fine structure of the $2S$ and $2P$ levels in hydrogen. Rabi, of course, had known of the progress of Lamb and Retherford's experiment on the fine structure of the $2S$-$2P$ levels of hydrogen being carried out at Columbia and had Lamb invited to the conference. On Saturday April 26, 1947, that experiment first succeeded and indicated "that, contrary to [Dirac] theory but in essential agreement with Pasternack's hypothesis, the $2^2S_{1/2}$ state is higher than the $2^2P_{1/2}$ by about 1000 mc/sec" (Lamb and Retherford 1947).

By early May rumors of the experimental findings in Rabi's laboratory were spreading. In a letter written in mid-May, Weisskopf informed Bethe that he had as yet no news about Rabi's hyperfine structure experiments and inquired from Bethe "whether he had heard rumors that he [Rabi] had measured the spread of the excited level ($2S^{1/2} - 2P^{3/2}$) and got a 10% deviation from theory?"[114] He also indicated that he had written his conference outline "only because I had no specific idea ready to discuss," adding that "since Kramers, and to some extent Oppenheimer, have something to say, let us give them all the time they need to do so. A few days earlier he had written Bethe that "he would like to hear from Kramers in great detail about his new theory."[115]

In mid-May, Bethe wrote Weisskopf:

A completely different subject:
  The Long Island Conference. I read your outline and got the impression that we should be careful not to try and do too much. I should like especially to hear from Kramers in great

detail about his new theory. Generally, I think we should try to hear from people who have actually got some results and avoid as much as possible discussions in the vacuum. For this reason I am not much in favor of extending discussions of experiments to be done in future accelerating equipment. If our discussion of concrete theoretical problems lead to desirable experiments, this is very good; but I believe we should not continue the war spirit of planned research too much.

Would it be possible to set one day aside for Kramers and have a really good discussion on that, then perhaps have half a day or one day for meson production starting with Oppy's theory, and then have another day devoted to the Piccioni experiment and its interpretation.[116]

Note Bethe's insistence that the conference "not continue the war spirit of planned research too much." The "not . . . too much" indicates that his view was moderated by the fact that the new high-energy accelerators that were being built did require some "planned research." By the end of May the preliminary results of Lamb and Retherford were confirmed. Schwinger and Weisskopf discussed the theoretical implication of the experiment on their train ride from Boston to New York to attend the Shelter Island conference and agreed that the effect was very likely quantum electrodynamic in origin. Schwinger recalls their discussing that the fact that the electron self-energy was logarithmically divergent in hole theory implied that the $2S_{1/2} - 2P_{1/2}$ energy difference would be finite when calculated with that theory.[117] In fact, the matter may have been discussed even earlier over the lunches Schwinger and Weisskopf took together periodically, since this hydrogenic level shift was one of Weisskopf's current research interests. In the fall of 1946, he had given the problem of a hole-theoretic computation of the $2S - 2P$ level shift in hydrogen to Bruce French, who had been working on it ever since. It is likely that Weisskopf would have told Schwinger, elicited his reaction, and sought his advice on effective computational approaches.[118]

This is how matters stood before the opening of the conference. It was clear that Lamb and Retherford's experiment on the fine structure of H and that of Conversi et al. on the absorption of sea level mesons would have an important place in the presentations at the conference. In fact, it had been arranged that Rossi would give a general report on recent cosmic ray experiments and also that Rabi would report on the other atomic beam experiments being carried out at Columbia investigating the hyperfine structure of hydrogen, deuterium, and more complex atoms.

The conference opened on Monday morning, June 2. Only the notes of one of the participants, Gregory Breit, are extant. They are, however, sufficiently detailed to allow an accurate reconstruction of the proceedings.[119] The first day of the conference was devoted to presentations of experimental results. The opening talk was by Lamb. He reported on experiments, carried out with Retherford, on the fine structure of hydrogen. He reviewed the history of previous measurements and recalled the suggestion by Pasternack (1938) that the observed structure of the $n = 2$ levels of hydrogen could be understood if, contrary to the prediction of Dirac's theory, the $2S_{1/2}$ and $2P_{1/2}$ were *not* degenerate, but were separated by .03 cm^{-1}, with the $2S_{1/2}$ level lying above the $2P_{1/2}$.

After a description of the techniques and apparatus used in the experiment, Lamb concluded his lecture with his finding: "The S level is $\sim$ .033 cm^{-1} up." In cycles per second, this level shift amounts to .033 cm$^{-1} \times 3 \times 10^{10}$ cm/sec $\cong 1000\ 10^6$ cycles/sec, or 1000 Mc/sec. The result of the Lamb-Retherford experiment became one of the central and dominant concerns of the conference. According to the article that appeared in the *Herald Tribune* on June 3, "the subsequent discussion was led by Oppenheimer and Weisskopf. The meeting discussed recent work on the simplest of all atoms, the hydrogen atom."[120]

Breit, in his notebook, abstracted the highlights of the discussion following Lamb's report: "Oppenheimer believes in Electrodynamic Term Shift as the least improbable explanation." Also mentioned are "Kramer's th. and Pais f-field."

The second lecture was by Rabi, who told the conference of his experiments with Nafe and Nelson on the "HFS (hyperfine structure) of H^1 and H^2." In hydrogen (H^1), the state in which the proton's spin and the electron's spin are parallel ($F = 1$) has a slightly different energy from that in which the spins are antiparallel ($F = 0$). The energy difference between these two states is known as the HFS. Since the spin of the deuteron is $1\hbar$, the corresponding states in deuterium are $F = 1/2$ and $F = 3/2$. Rabi reported that the measured values for the hyperfine structure were

$$\nu_D = 327.37 \pm 0.03\ 10^6 c/sec$$
$$\nu_H = 1421.3 \pm 0.2\ 10^6 c/sec,$$

at compared to computed values of

$$\nu_H = 1416.9 \pm 0.54$$

and

$$\nu_D = 326.53 \pm 0.16.$$

In his notes Breit recorded that the calculated values of ν_D and ν_H were ".22% off" and that the calculated ratio (ν_H/ν_D) gave a value of 4.3393, as compared to the measured ratio 4.3416. "Moreover the value 4.3416 for ν_H/ν_D had been measured and checked at Argonne." The discrepancy between measurement and theory was "way beyond error [the] disagreement [being] $\sim$1 in 2000."[121]

Commenting on the events of the morning, Darrow made the following entry in his diary: "Lamb and Rabi gave long papers on the terms of H and D, each embodying experiments of their own: these held to the end of the conf'ce their rank as outstanding contrib.^{ns}"[122]

In the afternoon, Rossi gave an extended account of recent cosmic ray experiments. The discussion of the Conversi experiment was "animated" (Marshak 1983, p. 381). That evening, in a session that lasted until 11:30 P.M., Oppenheimer reported on cosmic-ray experiments using rockets, and Bethe summarized work with Feynman that attempted to interpret all the new cosmic ray data.[123]

The second day of the conference was devoted to fundamental theory.

In the morning, Kramers gave a detailed account of the *classical* version of his theory of the electron in which only the "observed" mass of the charged particle enters the final result. Breit in his notes recorded that the point of departure of Kramers was the *classical* Hamiltonian

$$H = \frac{(\vec{P} - \frac{e}{c}\vec{A})^2}{2m_0} + e\phi + \frac{1}{8\pi} \int (\mathcal{E}_\perp^2 + H^2)d\tau,$$

where m_0 is the "intrinsic mass" of the charged particle. The "actual mass," m, of the particle is the sum of the "intrinsic mass," m_0, and of the electromagnetic mass of the particle, μ:

$$m = m_0 + \mu.$$

The intent of Kramers' theory was to recast the above Hamiltonian into a form in which only m appears. The content of Kramers' lecture (as outlined in Breit's note) was identical to the presentation of the *classical* theory that Kramers prepared for the Solvay conference of 1948 (Kramers 1948; see also sec. 3.2). In the afternoon, Weisskopf reviewed the divergence difficulties in the hole-theoretical calculations of the self-energy of an electron. He pointed out that in hole theory, where the self energy is logarithmically divergent, the level shift "ΔE due to electron mass for same $\overline{p^2}$ does not diverge" in the relativistic limit.[124]

During the ensuing discussion, Weisskopf and Schwinger indicated how a hole-theoretic calculation of the level shift in hydrogen might be attempted and suggested reasons why a finite answer might result fom the application of Kramers' ideas.

In his presentation, Weisskopf had made clear that Oppenheimer's hunch that the Lamb shift was a quantum electrodynamical radiative effect could be tested, and he also pointed out that a hole-theoretic calculation would probably yield a finite answer for ΔE. After the conference was over, Bethe performed a nonrelativistic calculation and proved that the level shift was indeed of quantum electrodynamic origin (Bethe 1947).

After Weisskopf's talk, Bethe presented the results of his recent work with Oppenheimer on the infrared catastrophe (Bethe and Oppenheimer 1946) and also reviewed earlier work on the subject by Bloch and Nordsieck (1937), Pauli and Fierz (1938), and Dancoff (1939). Von Neumann then discussed the recently published paper by Snyder (1946) on quantized space-time and indicated ways to generalize this work. The rest of the session was taken up by a talk by Teller on the capture of mesons by nuclei.

The morning of the last day of the conference was devoted to an extended discussion, led by Oppenheimer, of the "paradox" of the mesons: their copious production at the top of the atmosphere in nuclear interactions, yet their anomalously low scattering and absorption at sea level.

Oppenheimer suggested that one consider giving up microscopic reversibility. Weisskopf's suggestion for a way of overcoming the apparent lack of reversibility between production and absorption was a mechanism whereby a primary cosmic ray particle when interacting with a nucleus converted one of its nucleons into an "excited nucleon," capable subsequently of emitting mesons. The lifetime of the "pregnant" nucleon could be chosen sufficiently long to account for the subsequent weak interaction between mesons and nucleons. Marshak proposed his famous two-meson hypothesis: There exist in nature two kinds of mesons, possessing different masses. The heavier Yukawa meson (subsequently known as the π-meson) is responsible for the nuclear forces and is the one produced with a large cross section in the upper atmosphere; the lighter meson (the μ-meson) is a decay product of the heavier one and is the meson normally observed at sea level to interact weakly with matter.[125]

The history of the two-meson hypothesis, including an earlier Japanese version,[126] has been told by Marshak (1952, 1983) and by Brown (1981). Marshak's hypothesis was put forth before Blackett and his collaborators announced the discovery of the π-meson. The discussion of the long lifetime for meson absorption from the carbon K shell, 2×10^{-6} sec,[127] which had been estimated by Fermi, Teller, and Weisskopf on the basis of the results of the Conversi experiment, spawned the two-meson hypothesis at Shelter Island.

The final session on the last day of the conference was given over to Feynman for a presentation of his "space-time approach" to quantum mechanics (Feynman 1948c)[128] Darrow recorded in his diary: "At the morning session, Oppenheimer spoke with his usual quiet grace and Feynman with a clear voice, great rush of words and illustrative gestures sometimes ebullient."[129]

As is well known,[130] on the train ride from New York to Schenectady after the Shelter Island conference, Bethe made a nonrelativistic calculation of the $2S - 2P$ level shift using Kramers', Weisskopf's, Schwinger's and Oppenheimer's suggestions[131] and discovered that with suitable cutoffs such a calculation could account for a major part (1040 megacycles) of the observed level shift. This had not been obvious at Shelter Island. Bethe's calculation was crucial and essential and confirmed the feeling expressed at Shelter Island that the effect was a quantum electrodynamical one. Incidentally, until he received Bethe's paper in mid-June, Lamb believed that the effect was of nonelectromagnetic origin.[132]

Bethe's paper (1947) was completed by June 9 and he circulated it to the participants of Shelter Island. His accompanying letter to Oppenheimer was brief and to the point:

> Enclosed I am sending you a preliminary draft of a paper on the line shift. You see it does work out. Also, the second term already gives a finite result and is not zero as we thought during the conference. In fact, its logarithmic divergence makes the order of magnitude correct. It also seems that Vicky and Schwinger are correct that the hole theory is probably [handwritten insertion] important in order to obtain convergence. Finally, I think it shows that Kramers cannot get the right result by his method. It was good seeing so much of you during the conference. This should be repeated.

The accompanying letter to Lamb noted that

> the line shift...explains your results very beautifully. Moreover, it seems likely that from a more accurate theory and from precise experiments one could test the various relativistic theories of the electron, particularly the hole theory. I think all this is very exciting and quite important.
>
> Have you published your results or are you going to do so in the near future? I should like my note to appear together with, or soon after, yours. It has not been sent away yet.
>
> Please show this to Rabi...and to others interested. It was fine having that conference. Is there any news on the experiments? [133]

Bethe's remarks concerning Kramers in his letter to Oppenheimer are worth noting. Although Kramers' work was clearly very important—it indicated the importance of expressing observables in terms of the *experimental* mass of the electron—Bethe's letter corroborates that Kramers' presentation at the conference had only presented the *classical* version of his approach and that its extension to the quantum electrodynamic situation was not obvious to the participants. This is confirmed by Schwinger's letter to the *Physical Review* in which he reported his calculation of the quantum 2π electrodynamic correction to the magnetic moment of the electron (Schwinger 1948b). He indicated there that H. A. Kramers had presented a "classical non-relativistic theory" of mass renormalization at the Shelter Island conference. Bethe in his paper (1947) acknowledged that the comments by Weisskopf and Schwinger—that a hole-theoretic calculation of energy level *differences* would be finite—had a great and immediate impact on him. Bethe was interested in calculating the level shift quantum electrodynamically and the insight of Schwinger and Weisskopf gave him the justification for cutting off the divergent integrals he encountered in his nonrelativistic approach. Dresden (1987), in his biography of Kramers, has suggested that Kramers did not receive adequate credit for his contributions at Shelter Island. Perhaps so. The challenge was to get the numbers out and to explain the magnitude of the $2S - 2P$ shift in hydrogen and the new values of the hyperfine splitting in H and D that Nafe and Nelson had measured. Bethe did that for the Lamb shift. Moreover, I would suggest that if Kramers had *not* attended the Shelter Island conference, the course of developments would not have been very different. During the conference Oppenheimer, Schwinger, and Weisskopf had all suggested that the Lamb shift must be an electromagnetic radiative effect. Also, as noted above, Bethe had given a lecture on the infrared problem and had discussed the papers by Bloch and Nordsieck and by Pauli and Fierz.

In the latter, a fully quantum mechanical treatment of the interaction between a nonrelativistic charged particle and the electromagnetic field is given and an explicit mass renormalization is carried out to remove the electromagnetic self-energy. Kramers in his lecture at the conference had only dealt with the classical theory and had insisted that a consistent classical theory must be developed before the quantum theory could be addressed. Bethe's train-ride paper in which he calculated the nonrelativistic Lamb shift is much closer in spirit and style to the Pauli-Fierz paper than to Kramers' work and the content of Kramers' lecture at Shelter Island.[134]

The importance of Bethe's calculation is apparent from Weisskopf's reaction to it. Weisskopf received Bethe's manuscript on June 11 and after studying it wrote Bethe that he was

> quite enthusiastic about the result. It is a very nice way to estimate the effect and it is most encouraging that it comes out just right. I am very pleased to see that Schwinger's and my approach seems to be the right one after all. Your way of calculating is just an unrelativistic estimate of our effect, as far as I can see. "I am all the more pleased about the result since I tried myself unsuccessfully to estimate the order of magnitude of our expression. *I was unable to do this* [emphasis mine] but I got more and more convinced that the method was sound.... I would like to talk it over with you especially the "korrespondenz Deutung" of the effect.

And he added:

> I do not quite agree with your treatment of the history of the problem in your note. That the $2S_{1/2} - 2P_{1/2}$ split has some-

thing to do with radiation theory and hole theory was proposed by Schwinger and myself for quite some time. We did not do too much about it until shortly before the conference. We then proposed to split an infinite mass term from other terms and get a finite term shift, just as I demonstrated it at the conference. Isn't that exactly what you are doing? Your great and everlasting deed is your bright idea to treat this at first unrelativistically. "Es mochte doch schon sein" if this were indicated in some footnote or otherwise.

See you soon I hope
Yours

Vicki[135]

A few days later, Bethe answered him in his quiet and thoughtful way. Weisskopf soon acknowledged that Bethe's "abstract" was "*harmloser*" (much more harmless) than he initially thought and agreed, "Let's forget about patent claims."[136]

Kramers, who after the Shelter Island conference had gone to Ann Arbor to lecture at the University of Michigan Summer School, there indicated in his lectures another way of deriving the nonrelativistic Lamb shift: "You can calculate the self energy W of an oscillatory bound l and *free* electron by adding up $W = \Sigma h \Delta k_i$ where Δk_i are the changes in the eigen frequencies of a Hohlraum ... (due to the [presence of] the electron). This shifts Δk_i are, of course, different for the free and bound case, and the sum of the difference is finite.[137]

Bethe's work made a relativistic calculation of the Lamb shift the next desideratum and he himself and two of his students, Scalletar and Lennox, started on such a calculation in July. French and Weisskopf at MIT continued their calculation, but now made use of Kramers' ideas and thus greatly simplified the subtraction of infinities. Lamb at Columbia started on a hole-theoretic level shift calculation during the summer of 1947. He was soon joined by Kroll, then a graduate student at Columbia.[138] Schwinger, who got married right after Shelter Island and spent nearly two months traveling throughout the United States on his honeymoon, began work on such a calculation in late July. At Chicago, M. Goldberger started on such a calculation under E. Fermi's direction.[139] All of them used the noncovariant calculational methods developed in the 1930s, and which were to be found in Heitler's *Quantum Theory of Radiation*, whose second edition had appeared in 1944. Bethe, however, also felt strongly that what was needed was a relativistically invariant way of computing quantum electrodynamic effects. On his return to Cornell after Shelter Island, he got Feynman interested in this problem. Feynman, making use of his previous work with Wheeler on electrodynamics as an action-at-a-distance theory and of his Lagrangian formulation of quantum mechanics, worked on a formulation of quantum electrodynamics with a relativistic cut-off (Feynman 1948a,b, 1966b).

4.4 The Later Developments

Since the mid-1930s, nuclear and high-energy phenomena had been considered the exciting "frontier" fields within physics. The Columbia experiments indicated that much could yet be learned from quantum electrodynamics.

Lamb and Retherford's experiment on the fine structure of hydrogen and Nafe and Nelson's experiment on the hyperfine structure of hydrogen, both of which were performed at Columbia during the fall of 1946 and the spring of 1947, were undoubtedly the major stimuli for the renewed interest by theoretical physicists in quantum electrodynamical problems. By yielding accurate values for the electromagnetic properties of free and bound electrons, the Columbia experiments placed quantum electrodynamics once again at the center of interest of the theoretical community. As Oppenheimer (1948) indicated in his report to the Solvay Congress of 1948, in their application to level shifts "these developments [mass and charge renormalization] which could have been carried out at any time during the last fifteen years, required the impetus of experiments to stimulate and verify."

The experiments themselves, however—and not only the high precision experiments of Lamb, Foley, Kellogg, Kusch, Rabi, and others at Columbia, but also those of Dicke at Princeton, on the spectrum of hydrogen, the g-factor of the electron, and the search for positronium—depended upon the wartime developments of microwave technology.

The publication during the war of Wentzel's book on quantum field theory (Wentzel 1943) and its availability in the United States after the war (it was reprinted by Edwards Bros. in Ann Arbor in 1946) also proved helpful: this book detailed a coherent and unified presentation of relativistic quantum field theories incorporating most of the advances in the field until 1942. Thus the Shelter Island conference brought together members of the experimental and theoretical community after each had either assimilated important new advances or was ready to do so. How rapidly these advances were amalgamated can be gauged from the fact that in his first paper—the short note on the calculation of the $\frac{\alpha}{2\pi}$ quantum electrodynamic correction to the electron's magnetic moment submitted to the *Physical Review* in December 1947—Schwinger (1948b) was already quite explicit as to the meaning of the renormalization procedure:

> Electrodynamics unquestionably requires revision at ultra relativistic energies, but it is presumably accurate at moderate relativistic energies. It would be desirable, therefore to isolate those aspects of the current theory that involve high energies, and are subject to modification by a more satisfactory theory, from aspects that involve only moderate energies and are thus relatively trustworthy.
>
> This goal has been achieved by transforming the Hamiltonian of current hole theory electrodynamics to exhibit

explicitly the logarithmically divergent term proportional to the interaction energy of the electron in an external field [polarization of the vacuum] . . . .

The interaction between matter and radiation produces a renormalization of the electron charge and mass, all divergences being contained in the renormalization factors.

This is essentially also the view propounded by Feynman in the spring of 1948:

The philosophy behind these ideas [mass and charge renormalization] might be something like this: A future electrodynamics may show that at very high energy our theory is wrong. In fact we might expect it to be wrong because undoubtedly high energy gamma rays may be able to produce mesons in pairs, etc., phenomena with which we do not deal in the present formulation of the electron-positron electrodynamics. If the electrodynamics is altered at very short distances then the problem is how accurately can we compute things at relatively long distances. The result would seem to be this: the only thing which might depend sensitively on the modification at short distances is the mass and the charge. But that all observable processes will be relatively insensitive and we are now in a position to be able to compute these real processes fairly accurately without worrying about the modifications at high frequencies. Of course it is an experimental problem yet to determine to what extent the calculations we are now able to make are in agreement with experience.

In other words, it seems as though with these methods of mass and charge renormalization we have a consistent and definite electrodynamics for the calculation of all possible processes involving photons, electrons, and positrons.[140]

By the American Physical Society meeting of January 1948 in New York, it was apparent to Bethe (1948), and to Schwinger and Weisskopf (1948) that it was imperative to use relativistically invariant and gauge-invariant calculational techniques. To obtain finite answers all of them had made use of the ideas of Kramers on mass renormalization and the earlier ones of Heisenberg and Weisskopf on charge renormalization.[141] But these finite answers were plagued with inconsistencies. The ambiguities that Schwinger encountered in calculating the magnetic moment of a bound electron and the Lamb shift convinced him that a new fully covariant approach was needed.[142] By the Pocono conference in April 1948 both Schwinger and Feynman had developed relativistically invariant calculational schemes. Schwinger, going back to the early work of Dirac, Fock, and Podolsky (Dirac et al. 1932b) on the interaction picture for QED worked out a systematic approach based on a series of canonical transformations designed to make description of the vacuum and of the one-particle states simple. The vacuum was to be the relativistically invariant state of zero charge, energy, and momentum; the one-electron states had to describe correctly a particle of experimental charge e_{exp}, experimental mass m_{exp}, whose energy and momentum were related by $E_p = c\sqrt{p^2 + m_{exp}^2 c^2}$.

Feynman's computational scheme, on the other hand, was much more intuitive. It was based on his integral over path quantization procedure, applied to a cut-off version of Wheeler-Feynman theory, which had been altered, "mapped," and catalogued to give results in a perturbative expansion identical to those found in the perturbative expansion of hole theory (Feynman, 1949a,b, 1966). One root of Feynman's diagrammatic approach is to be found in this cataloguing procedure. Feynman's hurried presentation at Pocono, coming on the heels of Schwinger's day-long presentation, was met with skepticism and even some hostility. Bohr, in particular, objected to his viewing positrons as electrons going backward in time, and to his use of diagrams in general, insisting that the uncertainty principle precluded the assignment of trajectories to particles. He also objected to Feynman's belief "that it is possible to get a consistent theory without using loops," that is, without vacuum polarization terms. Bohr's remarks were abstracted in Wheeler's unpublished notes of the Pocono conference:

> Bohr: Discussion of Schwinger, Feynman, etc.
>
> It was a mistake in the older days to be discontented with field and charge fluctuations. They are necessary for the physical interpretation. Schwinger's treatment is an advance in bringing the theory into a regular order where effects can be interpreted. As regards the infinities he had other views than those just discussed. He objected to Feynman's view of the electron going backward in time. It is also unreasonable to cut things out.
>
> No sound argument yet exists as to where to cut off electron theory, if at all. It may however be an extravagant wish to hope to have electron theory stand alone.[143]

Despondent, though not despairing, Feynman realized that he would convince his audience only by publishing the details of his theory. After Pocono he embarked on the writing of his classic papers on "The Theory of Positrons" and "Space-Time Approach to QED" (Feynman 1949a,b).

The Pocono conference was the most "theoretical" of the three postwar NAS conferences. The table of contents of Wheeler's notes conveys succinctly the emphasis of the conference—and the central role Schwinger played in it:

3. Quantum electrodynamics	J. Schwinger	(40 pages)
4. Alternative formulation of quantum electrodynamics	R. Feynman	(12 pages)
5. Discussion	N. Bohr	(1 page)
6. Analysis of expected line shift	H. A. Bethe	(1 page)
7. Experiments on line shift	W. Lamb	(4 pages)
8. Neutron electron interaction	E. Fermi	(3 pages)
9. Neutron electron interaction	I. Rabi	(1 page)
10. Hyperfine structure of hydrogen	I. Rabi	(2 pages)
11. Theory hyperfine structure	A. Bohr, E. Teller, J. Schwinger	(3 pages)
12. Conditions for consistency of field theory	P. Dirac	(2 pages)
13. Implications of magnetic poles	P. Dirac	(4 pages)
14. Self energy	A. Pais	(2 pages)
15. Other discussions—Wheeler on μ-mesons (decay etc) Corrigenda		(5 pages)

Essentially the same impression is gained from Breit's notes on the conference. Out of twelve pages of notes, seven are devoted to Schwinger's presentation; less than one to Feynman's; one to Dirac; and the rest record brief summaries of the other presentations.

Upon his return to Princeton from the Pocono conference, Oppenheimer prepared the abstract of the paper on the "Two Academy Conferences on Theoretical Physics," which he was to present at the meeting of the NAS in Washington in late April 1948. In it, Oppenheimer summarized his assessment of the state of fundamental physics:

Among the many problems discussed, three were most intensively considered at both conferences:
(1)   The relation of mesons with nucleons.
(2)   The electromagnetic properties of nuclear matter.
(3)   The development of electrodynamics to form a consistent theory, and the comparison with experiment of its prediction for the reactive corrections in atomic phenomena.

The rapid progress of the last year tended to confirm the earlier conclusions (a) that the paradoxes of electrodynamics were far less serious and its range of validity far greater, than had long been supposed, (b) that far less was known of the essential experimental facts about mesons than imagined, and (c) that the unambiguous detection of specific interactions between nuclear matter on the one hand and electrons and radiation on the other may require a careful analysis of the finer points in the quantum theory of interacting fields. In the course of the discussion, there was speculation about the synthesis of these various areas of experience in the atomic world.[144]

Meanwhile, the Shelter Island conference had also inspired parallel work in Japan and Switzerland. In Japan an extensive research program in quantum electrodynamics was being carried out in response to the results of the Lamb experiment, which the Japanese had gleaned from a *Newsweek* article in the summer of 1947.[145] Right after the Pocono conference, Oppenheimer received a letter from Sin-itiro Tomonaga informing him of this work. Tomonaga's influential article on a relativistically invariant formulation of quantum field theory—written in 1943 and published in Japanese during the war—had been translated into English and had appeared in the first issue of the *Progress of Theoretical Physics* in the winter of 1946. In fact, Oppenheimer had called it to Schwinger's attention after the latter's talk at the Physical Society meeting in New York in January 1948. But no one in the United States had been aware of the recent work until the arrival of Tomonaga's letter.[146] Tomonaga's letter was sent to all the members of the Pocono conference by Oppenheimer. It is reprinted here in its entirety, as it details some of the other important developments in QED following the Shelter Island conference, in particular H. Lewis's work on radiative corrections to the scattering of an electron by a potential.

Institute for Advanced Study
Princeton, N.J.

April 5, 1948

To Members of the Pocono Conference
From Robert Oppenheimer

When I returned from the Pocono Conference, I found a letter from Tomonaga which seemed to me of such interest to us all that I am sending you a copy of it. Just because we were able to hear Schwinger's beautiful report, we may better be able to appreciate this independent development.

"I have taken the liberty of sending you copies of several papers and notes concerning the reaction of radiation field in scattering processes and related problems, which my collaborators and I have been investigating for last six months. I should be much obliged if you would be so kind to look them over. I should like to take this occasion to relate the circumstances of their formation and the reason why I have made up my mind to send you these manuscripts.

"During the wartime, when we were perfectly isolated from the progress of physics in the world, Dr. S. Sakata, one of the main research workers on the theory of elementary particles in our country, made an attempt to overcome the divergence difficulty of the self-energy of the electron by introducing a neutral scalar field (the so-called C-meson field) which interacts with electrons, a hypotheses which we afterwards found to be identical with Dr. A. Pais' theory of f-field [*Phys. Rev.* 63:227 (1946)]. In view of its promising feature my collaborator and I have applied this theory to the problem of elastic scattering of an electron in order to put it to a further test because I supposed, as you and Dr. H. A. Bethe have emphasized, that 'this simple problem may afford a useful test of future theories of radiation.' We have thus carried out a perturbation calculation following Dr. S. M. Dancoff [*Phys. Rev.* 55:959 (1939)] and, though we at first made the same oversight as Dr. Dancoff missing some intermediate states in the calculation (the first note of Ito Koba and Tomonaga), finally arrived at the conclusion that the new field is also capable of eliminating the divergence in the scattering cross-section in e^2-approximation (the second note of Ito Koba and Tomonaga and the two papers of the same authors).

"Shortly before we finished this work we received the striking report about the experimental evidence of the level shift of hydrogen atoms and its theoretical explanation by Dr. H. A. Bethe in terms of radiation reaction. Thereupon I proposed a formalism to express Dr. Bethe's fundamental assumption in a more closed and plausible—as I believe—form, in which the separation of terms to be subtracted is made by a canonical transformation (the note of Tati and Tomonaga). This formalism, which we called 'self-consistent subtraction method,' was then applied to the scattering problem mentioned above and it was confirmed that the non-diagonal part of the mass-correction term plays a decisive role in the divergent part of the cross-section and just cancels the infinity that appeared in the usual formalism (the note of Koba and Tomonaga and the paper of the same authors).

"After we had finished this work the January 15th issue of the Physical Review (1948) arrived, in which we found Dr. H. W. Lewis' and Dr. S. T. Epstein's works dealing with the same problem, which, as they wrote, had been undertaken by your suggestion. As the conclusion of these authors was identical with ours, we at first hesitated to make our paper public, but because it should play an introductory part of the series of papers to be published, we thought it not unreasonable to send our manuscript to the editor of our English journal 'Progress of Theoretical Physics.' I hope that you and Dr. Lewis and Dr. Epstein would be good enough to acknowledge our works too.

"Under the unfavorable condition after the wartime, however, it will take a long time—a year or so—before our papers will appear in print, and thus I have resolved on sending you our manuscripts. (The short notes will appear soon in the journal mentioned.)

"In succession to these works we are further developing the formulation of our subtraction method in a relativistically more elegant form according to the formalism proposed by me some years ago [*Prog. Theor. Phys.*, 1:27 (1946); 2:101 (1947)] along the line mentioned in the note of Tati and Tomonaga above, and, on the other hand, we are examining the reaction of the radiation field in the collision processes of elementary particles, among which the consideration about the simplest example, the e^2-correction for Klein-Nishina Formula, is in the course of publication (the note of Koba and Takeda). I should be much obliged if you would kindly take some interest in these works.

"In conclusion I wish to express my hearty thanks to you and other scientists of the United States for having bestowed so many favors upon us, such as presentation of journals and literatures, by which the scientific activity in our country will be much incited.

"With kindest regards,

>Very sincerely yours
>(s) Sin-itiro Tomonaga"
>Physics Institute, Tokyo Bunrika Daigaku,
>Hyakunincho 4-400
>Shinjuku-ku, Toyko, Japan
>(Letter of April 2, 1948)"[147]

On April 13, Oppenheimer sent Tomonaga a telegram "strongly" suggesting that he write "a summary account of present state and views for prompt publication [in the] *Physical Review*." He also indicated in it that he would "be glad to arrange" its publication.[148]

In mid-May, Tomonaga communicated to Oppenheimer that he had written a "summary account" of the Japanese investigations and that he had sent him the paper under separate cover. The manuscript was conveyed from Japan to Oppenheimer by the special staff of the War Department.[149] Oppenheimer arranged for its publication, and it appeared together with a note by Oppenheimer in the July 15, 1948, issue of the *Physical Review* (Tomonaga 1948; Oppenheimer 1948). Upon reading Oppenheimer's and Tomonaga's letters, Freeman Dyson perceptively noted:

> The reason that everyone is so enormously pleased with this work of Tomonaga is partly political. Long-sighted scientists are worried by the growing danger of nationalism in American science, and even more in the minds of the politicians and industrialists who finance science. In the public mind, experimental science at least is a thing only Americans know how to do, and the fact that some theorists have had to be imported from Europe is rather grudgingly admitted. In this atmosphere the new Schwinger theory tended to be acclaimed as a demonstration that now even in theoretical physics America has nothing to learn, now for the first time has produced her own Einstein. You can see that if the scientists can say that even in this chosen field of physics America was anticipated and indeed by a member of the much-despised race of Japanese, this will be a strong card to play against nationalistic politics.[150]

The history of the Shelter Island, Pocono, and Oldstone conferences encapulates the development of QED from 1947 to 1950 and thus includes the story of how Schwinger, Feynman, and Dyson worked out their respective formulations.

Shelter Island indicated the problems and possible solutions, in part by virtue of the focus that Kramers' insight had given. At Pocono, Schwinger and his methods took center stage. Oldstone was Feynman's show—and also Dyson's. By then the great power of Feynman's calculational schemes had become patently clear. By April 1949 Karplus and Kroll were well on their way to complete their calculation of the fourth-order α^2 corrections to the magnetic moment of the electron using Feynman-Dyson methods. The deeper differences between Schwinger and Feynman were also apparent at Oldstone. As the unpublished notes from the conference indicate,[151] Dyson, in his report, stressed that "Feynman [has] more interest in solutions than in equations. Most though not all problems calculable as scattering problems.... Largely from Feynman one has systematic technique for writing down solutions of scatter[ing] problems." By the time of the Oldstone conference it had also become clear that QED in its renormalized version could probably account for all the data available then on the properties of an electron— its anomalous magnetic moment, the energy level sprectrum of H and He$^+$, etc. In fact, by Oldstone the interest of the theoretical community had already shifted to "π-meson physics."

The extent of the shift can be gleaned from the table of contents of the Oldstone notes:

Schwinger's primary accomplishment in the period from 1947 to 1949 was the formulation of a somewhat unwieldy but coherent and systematic apparatus for doing relativistic field theoretical calculations and the proof that these methods could be successfully applied to interesting problems (e.g., computation of electron g factor, Lamb shift, radiative corrections to Coulomb scattering). Feynman provided deep new insights by visualizing the processes in a manner that translated these intuitive representations into simple, extremely efficient and effective calculational methods for computing observable quantities. For the first time one could conceive of doing higher-order calculations routinely. But it could be argued that after Kramers's suggestion for mass renormalization and the earlier one of Dirac, Heisenberg, and Weisskopf for charge renormalization had been adopted, both Feynman's and Schwinger's formulations, although technically much more manageable and much superior, were not major advances in terms of physical content from the quantum field theory of the thirties. In fact the first correct calculation of the Lamb shift by French and Weisskopf (1949) and by Kroll and Lamb (1949) still used the computational apparatus of the 1930s. The question to be answered, therefore, is why—in contrast to the generation of the 1930s—was the post-World War II generation successful in bypassing the obstacles posed by a relativistic quantum field theory. For answers we must look to the broader context in which these developments took place. The workers of the thirties, particularly Bohr and Dirac, had sought solutions of the problems in terms of revolutionary departures (e.g., nonpositive metrics in Hilbert space). Special relativity and quantum mechanics had been brought forth by such revolutionary steps, and Bohr in the 1930s constantly advocated and encouraged such attitudes for the solution of the field theoretic problems.[152] The solution advanced by Feynman, Schwinger, and Dyson was in its core conservative: it asked to take seriously the received formulation of quantum mechanics and special relativity and to explore the content of this synthesis. A generational conflict manifested itself in this contrast between the revolutionary and conservative stance of the pre- and post-World War II generations (Dyson 1965).

Oppenheimer's observation should also be recalled:

> When quantum theory was first taught in the universities and institutes, it was taught by those who had participated, or had been engaged spectators, in its discovery. Some of the excitement and wonder of the discoverer was in their teaching; now, after two or three decades, it is taught not by the creators but by those who have learned from others who have learned from those creators.... It is taught not as history, not as a great adventure in human understanding, but as a piece of knowledge, as a set of techniques, as a scientific discipline to be used by the student in understanding and exploring new phenomena in the vast work of the advance of science.... It has become ... an instrument of the scientist to be taken for granted by him, to be used by him, to be taught as a mode of action, as we teach our children to spell and to add. (Oppenheimer 1953, pp. 36–37)

Schwinger, Feynman, and Dyson took quantum mechanics and special relativity for granted. Their attitude toward quantum mechanics—so concisely stated by Oppenheimer—is indeed a generational characteristic and contrasts sharply with that of Bohr and Dirac, the creators of modern quantum theory.

Dyson provided a major contribution to the understanding of the structure of field theories with his examination and classification of the higher-order contributions, his analysis of the structure of quantum field theories to all order of perturbation theory, and his formulation of the concept of renormalizability.[153] In this connection it should be remembered that by mid-1948 neither Feynman nor Schwinger had considered the problems connected with either the sufficiency or the consistency of the renormalization procedure in higher orders of perturbation theory. Their primary interest had been in obtaining predictions from QED which could be compared with experiments (e.g., the calculation of the magnetic moment of an electron, the radiative corrections to scattering processes, the level shifts in H and D, etc.). It was Dyson's contribution to indicate how Feynman's visual insights could be used to answer the question of whether charge and mass renormalization were sufficient to remove all the divergences in QED to all order of perturbation theory and to investigate what renormalizability implied for other field theories. He realized that Feynman diagrams (and their analogs in many-body theory) in addition to their intuitive appeal can also be viewed as a representation of the logical content of field theories (as stated in their perturbative expansions). Understanding this aspect of diagrammatics enabled Dyson to achieve deep insights into the structure of quantum field theories and thus to make his important contributions to the concepts of renormalization.

It is precisely the recovery of the possibility of "visualizing [atomic] physical phenomena in terms of the concepts of daily life"[154] which made Feynman's
contribution so telling. Starting with his Ph.D. dissertation in 1940, but particularly
after 1946 when he started writing up his results, Feynman's revealing reinterpretation of the formal and abstract rules of quantum mechanics—which allowed
transition amplitudes to be interpreted as being made up of contributions from alternative paths a particle could take in space-time—and his insights into the nature of
the probability calculus relevant to quantum mechanics, resulted in a totally novel,
visual way of describing atomic phenomena (Feynman 1948c, 1949a,b). Since the
ability to visualize is for most people the most important channel for understanding
the world, what Feynman made possible is for us to visualize and to conceptualize
the unseen.

It would be wrong to infer from the above that Schwinger, Feynman, and
Dyson were solely responsible for the impressive theoretical advances of the 1947–
1950 period. It is certainly the case that they were the principal contributors. But
we shall see in the subsequent chapters that others played influential and at times
crucial roles as well. Bethe is certainly one such person; Tomonaga another. As
Frank Manuel has observed, geniuses are those who, in their ability to synthesize,
overwhelm. Schwinger, Feynman, Dyson all have this overwhelming capability
to synthesize. And because we have come to learn QED and quantum field theory
from their syntheses, we associate the developments primarily with them.

Advances in quantum field theory (and in QED in particular) have always
been marked by an amalgamation of perturbation-theoretic with nonperturbation-
theoretic insights. The ideas of mass renormalization in QFT can be traced back
to Pauli and Fierz's seminal 1938 paper in which solutions of a Hamiltonian
describing a nonrelativistic extended charged particle interacting with the quantized electromagnetic field were obtained. Their work in turn was based on the
nonperturbative insights of Bloch and Nordsieck's analysis of the infrared divergences. The successes in the period from 1945 to about 1949 (the papers of Dyson
1949a,b,c; Salam 1950, 1951a,b; and Ward 1950a,b, 1951 mark the end of the
period) were based on the perturbative expansion of the solutions of quantum field
theories. Renormalizability was one such inference. By 1949 the emphasis had
turned to nonperturbative covariant formulations (the work of Salpeter and Bethe
1951; Nambu 1950; Källén 1952; Gell-Mann and Low 1951; and Schwinger
1951d marks the beginning of this period) partly because bound-state problems
were attracting attention (e.g., the relativistic calculations of the Lamb shift and of
the hyperfine structure of hydrogen, the structure of positronium), but also because
by then π-meson physics was becoming the center of interest and it was clear that
meson-nucleon interactions were much "stronger" than electrodynamic ones.

Since the 1950s many of the important advances in quantum field theory
have been the results of nonperturbative methods. Often, these were first developed
in many-body theory, for example, superconductivity and ferromagnetism.

In retrospect, perhaps the most important lesson that could have been learned from the period 1945 to 1952 was an appreciation of how rich and malleable a structure relativistic quantum field theory was and how readily it could accommodate the dramatic advances of that period (and subsequent ones!).

4.5 Conclusion

The Shelter Island conference marked a break in the history of theoretical physics. It placed experiments on the spectrum of hydrogen that indicated deviations from the predictions of the Dirac equation—the accepted dogma—at the center of theoretical interest. And by presenting *reliable* and *precise* values for these discrepancies, Lamb and Rabi posed a challenge to the theoretical community. Renormalization theory was the outgrowth of the discussions at Shelter Island. It allowed the difficulties that had plagued quantum field theories since their inception to be circumvented. The computational techniques devised by Schwinger, Feynman, and Dyson made possible unheard-of accuracy in the calculation of the electromagnetic properties of simple atomic systems. These developments completely dispelled whatever doubts there had been about the adequacy of quantum electrodynamics,[155] and gave renewed faith in the possibility of a field-theoretic explanation also for nuclear phenomena.

In the United States, these advances and the subsequent applications of the techniques were made by a new generation of theoretical physicists—many of whom had been deeply involved in the efforts to bring World War II to a victorious conclusion by the use of scientific devices and scientific methods. Shelter Island marked the coming into its own of the American theoretical physics community. This community, made up for the most part of young, energetic, American-trained scientists, flourished because of the huge infusion of government funds into physics during and after World War II and because the investigations and explanations of physical phenomena were becoming more complex and specialized, thus demanding a new division of labor.

It is a tribute to MacInnes's vision and sagacity that he had conceived the Shelter Island conference as a vehicle to "get out" these "younger men."

5. The Lamb Shift and the Magnetic Moment of the Electron

Self-confidence is an important ingredient that makes for a successful physicist.

—V. F. Weisskopf

5.1 Introduction

In a series of remarkable experiments performed in the late 1970s, Dehmelt and his associates at the University of Washington trapped single electrons for indefinite periods of time in what has become known as a Penning trap, an enclosure in which a homogeneous magnetic field and an electrostatic quadrupole field are superposed (Dehmelt 1981, 1983). Charged particles executing small orbits in such a Penning trap can be considered as an atomic system, one in which the atomic nucleus has been replaced by external trapping fields, that can be precisely and accurately adjusted. A single particle moving in a Penning trap rivals the hydrogen atom in its simplicity; and the properties of such single bound particles can be measured and calculated with extraordinary precision (Brown and Gabrielse 1986). For example, the magnetic moment of an electron—or rather its *g* factor—has been measured using the Penning trap and the value thus obtained,

$$\frac{g}{2} = 1.001\ 159\ 652\ 193\ (4), \tag{5.1.1}$$

is three orders of magnitude more accurate than the previous measurements using other techniques. The best result for the magnetic moment anomaly of the electron that had been reported by Rich and Wesley (1972) using earlier techniques was

$$\frac{g}{2} = 1.001\ 159\ 657\ 700\ (3500). \tag{5.1.2}$$

The measured value of Dehmelt and his associates is to be compared to a theoretical value that is also of exceptional precision,

$$\frac{g}{2} = 1.001\ 159\ 652\ 459\ (135). \tag{5.1.3}$$

The calculations which yielded this value included the quantum electrodynamic corrections to eighth order in the fine-structure constant $\alpha = e^2/\hbar c$ (Kinoshita 1984). The *g* factor of the positron has also been measured with an accuracy comparable to that attained with an electron. Its agreement with that of the electron provides a stringent test of CPT symmetry for leptons. Although single protons have been trapped and observed, at present only the properties of small clouds of protons have been measured. A comparison of the cyclotron frequency of such a cloud of protons with that of a small cloud of electrons yields the proton-electron mass ratio

$$\frac{m_p}{m_e} = 1836.\ 152\ 470\ (76).\qquad\qquad (5.1.4)$$

These are spectacular results. Furthermore, substantial increases in accuracy have been promised and are very likely to be effected in all the above measurements. A study of the physics involved indicates that these experiments rest on a fundamental assumption, namely, that nonrelativistic quantum mechanics is an accurate and reliable *phenomenological* theory which is used in the description and the analysis of the motion of nonrelativistic electrons and protons in a Penning trap in exactly the same way as Maxwell theory is used in the construction, description, and analysis of the electromagnetic fields in the trap and in the wave-guide cavities feeding into the trap. The same assumptions governed the Lamb shift experiment (Schweber 1989).

In this chapter I trace the historical evolution of this conceptualization of quantum mechanics, and make precise the sense in which quantum mechanics as used in the Lamb shift measurement and the earlier atomic beam experiments is a phenomenological theory. My intent is also to draw attention to the fact, previously emphasized by Cartwright (1983), Hacking (1983), Shapere (1984), and others, that experimental practice provides a continuity in techniques and assumptions that moderates some of Kuhn's claims about the discontinuous character of scientific revolutions, in particular the incommensurability of pre- and postrevolutionary discourse (Kuhn 1970). The chapter is organized as follows: Dirac's equation accounted for the essential features of the spectrum of hydrogen; it gave a theoretical underpinning for the Uhlenbeck-Goudsmit model of the electron's spin and also explained the value of the magnetic moment of the electron. This success polarized the interpretation of many of the experiments that were carried out during the 1930s to measure the spectrum of hydrogen more accurately. In section 5.2 we will look at some of these experiments. Sections 5.3 and 5.4 focus on the experiments carried out by Lamb and Retherford and by Nafe, Nelson, and Rabi at Columbia after World War II. Section 5.5 investigates the conceptualization of these measurements, and includes some philosophical remarks. Section 5.6 briefly presents Bethe's explanation of the Lamb shift. Section 5.7 reviews the relativistic calculations of the Lamb shift that were performed in 1947–1948, and section 5.8 covers Lewis's calculations of radiative corrections in the scattering problem (Lewis 1948).

5.2 The Experimental Situation during the 1930s

The energy levels of the hydrogen atom (and more generally of the hydrogenlike atoms) are essentially determined by the nuclear charge and its point-like character. The finite size of the nucleus, its mass, spin, and magnetic moment only give rise to small corrections to the formula first derived by Bohr in 1913:

$$E_n = -Z^2 hc\, R/n^2, \tag{5.2.1}$$

where R is the Rydberg wave number, h is Planck's constant, c the velocity of light, and n the principal quantum number whose possible values are 1, 2, 3,.... The Bohr theory had accounted for many of the observed lines in the hydrogenic spectrum. It however had failed to explain why the Balmer radiation emitted in a transition from a higher state to the $n = 2$ state consisted of a close doublet (fig. 5.2.1). This "fine structure" had first been observed in 1887 by Michelson and Morley. Sommerfeld in 1916 discovered that the application of the Bohr quantization rules to the *relativistic* hydrogen atom could explain the doublet structure: the relativistic variation of mass with velocity gave rise to a dependence of the energy on the eccentricity of the orbit. For the $n = 2$ level the circular ($\ell = 1$) and the elliptic orbit ($\ell = 0$) differed (for small Z) by

$$\Delta E_2 = \frac{1}{16}\alpha^2 Z^4\, hcR, \tag{5.2.2}$$

where $\alpha = e^2/hc \approx \frac{1}{137}$ is the fine structure constant. For hydrogen, $Z = 1$, the $n = 2$ separation, ΔE_2, amounts to 0.365 cm^{-1} and agreed qualitatively with the observed doublet separation (Kragh 1985). Schrödinger—who had been the first to write down the second-order Klein-Gordon equation, and had in fact postulated that equation to describe the wave properties of the electron—had discovered that the solution of this relativistic equation for the case of an electron moving in a

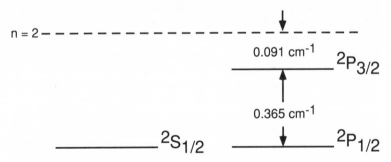

Figure 5.2.1 Fine structure of $n = 2$ levels of hydrogen according to the Dirac theory. The dotted line indicates the postion of the $n = 2$ level according to the Bohr theory.

Coulomb field yielded 8/3 the value obtained by Sommerfeld for the $n = 2$ level. Because of this discrepancy he abandoned the work. But some months later, he found that a non-relativistic version of the equation—the Schrödinger equation—yielded energy levels for the Coulomb field which agreed exactly with those of the Bohr theory (Mehra and Rechenberg 1987).[1] By including not only the relativistic mass variation, but also the effects of electron spin and magnetic moment as postulated by Goudsmit and Uhlenbeck and the effects of spin-orbit interaction as deduced by Thomas, it was found that wave mechanics predicted a fine structure separation equal to the Sommerfeld value. Dirac's equation for an electron moving in a Coulomb field automatically guaranteed the correct mass variation, spin, magnetic moment, and the corresponding spin-orbit interaction and yielded the following formula for the energy levels:

$$
E = m_0 c^2 \left[1 + \left(\frac{\alpha Z}{n - k + \sqrt{k^2 - \alpha^2 Z^2}} \right)^2 \right]^{-1/2} , \tag{5.2.3}
$$

which is of exactly the same form as the Sommerfeld result (see Biedenharn 1984). In eq. (5.2.3), n is the principal quantum number and $k = j + 1/2$ is a quantum number which has $1, 2, \ldots n$ as possible values. For each value of k or j (except $k = n$) there correspond two possible values of the orbital quantum number, $\ell = j + 1/2$ and $\ell = j - 1/2$ (for $k = n$, $\ell = j + 1/2 = n - 1$ only!). Note that the pairs of levels with $\ell = j \pm 1/2$ have *exactly* the same energy. The fine structure of the $n = 2$ levels according to Dirac theory are indicated in figure 5.2.2.

An enormous amount of effort was expended during the 1930s in spectroscopic work to establish the extent of the agreement of the observational data with the prediction of the Dirac theory (Lamb 1951; M. Morrison 1986). The spectral lines studied included the Balmer doublets of hydrogen, deuterium, and singly ionized helium. By 1938, deviations from the predictions of the Dirac theory for the H_α lines (the radiation emitted in the transition from $n = 3$ to $n = 2$; see fig. 5.2.2) had been observed (Houston 1937; R. C. Williams 1938), and Pasternack (1938) suggested that these results could be interpreted as indicating that the $2^2 S_{1/2}$ level was raised by about 0.03 cm^{-1} relative to the $2^2 P_{1/2}$ level. However, by 1940 the situation had become unclear again. Whereas W. W. Houston's and R. C. Williams's work had indicated probable discrepancies between theory and experiment, J. W. Drinkwater, O. Richardson, and W. E. Williams attributed these discrepancies to impurities in the source. Also, the theorists who attempted to give a theoretical explanation for Pasternack's shift in terms of a deviation of the electron-proton potential from the Coulomb law by virtue of mesonic effects were either frustrated by the smallness of their predictions or misled by the inadequacies in their theories due to divergence difficulties (Lamb 1983). Lamb later recalled that theorists "rather eagerly hailed the results of Drinkwater, Richardson and Williams which confirmed the Dirac theory" (Lamb 1951).

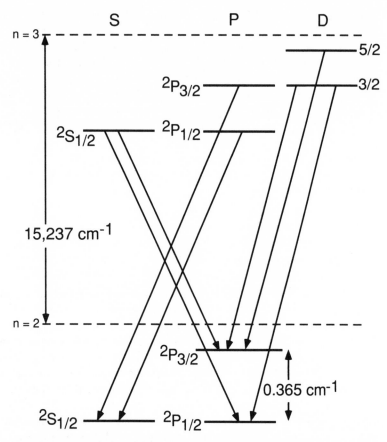

Figure 5.2.2 Energy levles of the hydrogen atom according to Dirac theory showing the component transitions of H_α line.

During the 1930s the situation with respect to the magnetic moment of the electron was the following. In 1925 Goudsmit and Uhlenbeck had postulated that an electron had an intrinsic angular momentum of $1/2\,\hbar$, and that associated with this spin angular momentum there is a magnetic dipole moment equal to $e\hbar/2mc$. This quantity is called the "Bohr magnetron" and has been given the symbol μ_0 in the standard notation of physics. If we call the spin magnetic dipole moment of the electron, μ_s, then Goudsmit and Uhlenbeck's postulate was the statement that $\mu_s = \mu_0$. The important physical question "is μ_s precisely equal to μ_0?" was addressed by Back and Landé in 1925. Their conclusion, based on numerous experimental investigations on the Zeeman effect in hydrogenlike and complex atoms, was that μ_s was equal to μ_0 within an experimental uncertainty of at least one part in a thousand (Bethe and Salpeter 1957). Kinster and Houston in 1934 made a study of the Zeeman effect in neon, and arrived at results which seemed to confirm the value 2 for the gyromagnetic moment, g, of the electron (g is the ratio of the electron's magnetic dipole moment in units of μ_0 to its spin angular momentum in units of h; $\mu_s = 1/2g\mu_0$) (Kinster 1934). It was not until 1947 that the discrepancy which is now described as the anomalous magnetic moment of the electron—the fact that g is not equal to 2—was demonstrated. Although the spectroscopic techniques of the thirties were primitive by the standards of the late forties, the result that μ_s was not equal to μ_0 could have been established much earlier if someone had asked the question. Kinsler and Houston's experiments were taken to corroborate Dirac's prediction and thereafter the equality $\mu_s = \mu_0$ was accepted as exhibiting a fundamental property of the electron. That this was indeed the case can be seen from the reports by Millman and Kusch of the precision measurements of nuclear magnetic moments using molecular beams (Millman and Kusch 1940). In the molecular beam experiments carried out until 1939, both the magnetic field, H, to which the molecule was subjected and the frequency, f_0, of the resonance were measured. From these data a nuclear moment could be calculated.[2] The measurement of the magnetic field, which had to be carried out with high precision and in vacuum, posed a major experimental problem. Millman and Kusch observed that one could avoid the need of measuring a magnetic field by measuring in the same magnetic field a nuclear resonance and the spin electron resonance. When measured in the same field, the ratio of the nuclear and electron spin resonance frequencies f_n/f_s would be equal to the ratio of nuclear and electron spin moments

$$\frac{f_n}{f_s} = \frac{g_n}{g_s}. \tag{5.2.4}$$

If g_s is assumed to be 2—which was then accepted as a fact since this was a prediction of the Dirac equation—g_n may be found.

Millman and Kusch reported the observed values of several nuclear magnetic moment computed on the basis of (5.2.4) with $g_s = 2$. Interestingly, the g_n values found by Millman and Kusch in this fashion were consistently 0.12% greater than those found in earlier work, where both f_n and H were measured and g_n was determined by

$$g_n = hf_n/\mu_0 H. \tag{5.2.5}$$

Kusch (1958) recalled that "at the time, this was a source of self-congratulation since it suggested that the field had in fact been measured to a considerably greater precision than we had any reasonable hope of achieving." The discovery by Nafe, Nelson, and Rabi in 1947 of a discrepancy in the hyperfine structure of hydrogen, and a suggestion by Breit that μ_s may not equal μ_0, made Kusch return to these old experiments.

The Williams and the Millman and Kusch experiments of 1940–1941 illustrate the weight of theoretical presuppositions. Willis Lamb, who as a theorist came to doubt the exactness of Coulomb's law and as an experimenter believed that he could devise an experiment that would test the accuracy of the prediction of the Dirac equation for the hydrogen atom, changed the situation dramatically.

The opening statement of the progress report Lamb submitted in 1946 for his "Experiment to Determine the Fine Structure of the Hydrogen Atom" outlined the motivation for undertaking his investigation:

> The hydrogen atom is the simplest one in existence, and the only one for which essentially exact theoretical calculations can be made on the basis of the fairly well confirmed Coulomb law of interaction and the Dirac equation for an electron. Such refinements as the motion of the proton and the magnetic interaction with the spin of the proton are taken into account in rather good approximate fashion. Nevertheless, the experimental situation at present is such that the observed spectrum of the hydrogen atom does not provide a very critical test either of the theory or of the Coulomb law of interaction between two point charges. A critical test would be obtained from a measurement of the fine structure of the $n = 2$ quantum state. (Lamb 1946)

It is to the Lamb-Retherford experiment that we turn to next.

5.3 Willis Lamb

Willis Lamb stands in the top rank of physicists both as a theorist and as experimenter. Together with Bloch, Fermi, and Rabi he is one of the last physicists who could master the whole of physics.

Lamb was born on July 12, 1913, in Los Angeles. His father was a telephone engineer, his mother a "deeply religious" woman. When he was young she was a "confirmed episcopalian."[3] Later on she became a congregationalist, and Lamb remembers attending Sunday school and Sunday evening services in which the sermons were based on "moving pictures." Still later, when the family moved to Oakland, she became a Christian Scientist. Lamb attended the University of California in Berkeley. He majored in chemistry and obtained his Bachelor of Science degree in 1934. However, his interests were much wider-ranging than his concentration indicated; while an undergraduate he took a fair number of physics and mathematics courses. As a junior he studied Ruark and Urey's book on quantum mechanics (Ruark and Urey 1930), and took a course from Williams in vector and tensor analysis that made a deep impression on him. In the course Williams presented a "coordinate-free formulation" of the subject and also talked about quantum mechanics based on Dirac's formulation. Lamb's conversion to theoretical physics dates from that period.

In the fall of 1934 Lamb enrolled as a graduate student in physics at Berkeley. He took Oppenheimer's course in quantum mechanics and by the end of the second semester he had been exposed to Dirac's equation and had received an introduction to quantum electrodynamics. He became a member of Oppeheimer's theory group, received a desk in LeConte Hall 219, and migrated with him each spring to Cal Tech. In 1935 on his first visit there, Lamb took P. S. Epstein's course on advanced quantum mechanics, which introduced him to Fermi's *Reviews of Modern Physics* article on quantum electrodynamics (Fermi 1932). That same summer he attended the Michigan Summer Symposium and heard lectures by Fermi (on neutron scattering), Bloch (on electron conduction in crystal lattices), Uhlenbeck (on β decay), and Goudsmit (on the theory of atomic spectra).

Berkeley was an exciting place in theoretical physics during Lamb's graduate studies there. Every year, two or three outstanding postdoctoral fellows would come to work with Oppenheimer. They and the graduate students working for Oppenheimer formed a close-knit intellectual community. The problems addressed were invariably at the frontiers of physics: nuclear physics, QED, general relativity, and after the publication of Yukawa's papers and the discovery of the "mesotron," meson theory. Weekly seminars alternated between Berkeley and Stanford, with Oppenheimer and Bloch not only providing the intellectual leadership but also the social management. After the weekly talk, everyone attending the seminar would go out to eat and the group would then convene for games and partying at Bloch's or Oppenheimer's house. Lamb got to be close to Bloch. He spent the summers from 1936 to 1939 and that of 1941 at Stanford, where Bloch had "kindly" provided him with a desk and library privileges. Each year there was a visiting summer school lecturer and Lamb got to know George Gamow, Edward Teller, Victor Weisskopf, John Van Vleck, Isidore Rabi, and many others who passed through Palo Alto on their summer travels. He also met most of the Stanford physicists: David Webster, Paul Kirkpatrick, Norris Bradbury, William Hansen, Russel Varian, Arnold Siegert, and Arnold Nordsieck (Lamb 1983).

The subject of Lamb's Ph.D. thesis was stimulated by one of the seminar talks at Stanford. He heard Siegert talk on his work on exchange currents (Siegert 1937) and realized that such currents would contribute to radiative interactions, and suggested that these contributions could be experimentally tested in the reaction $n + p \rightarrow D + \gamma$. This eventually became part of his thesis (Lamb 1983, p. 316).

In 1938 Lamb applied for an NRC fellowship with the intention of going to Wisconsin to work with Breit. Breit had signed the referee's report on Lamb's paper on exchange currents and had approved the paper for publication (Lamb and Schiff 1938). Oppenheimer had not been satisfied with Lamb's original version, and in fact had rewritten parts of it and had brought in Schiff to work on some further points. Relations between Oppenheimer and Lamb were never very smooth. As Lamb recalled, "To my regret, Oppenheimer did not care much for the kinds of problems I wanted to work on . . . . Oppenheimer did not suffer fools gladly. Sometimes I annoyed him because of my stupidity and ignorance. Mostly, he was as kind to me as I deserved. I wish I could have learned as much as he could have taught me" (Lamb 1983, pp. 315–316).

In retrospect, the problems that Lamb was interested in were important, but they were too difficult to work on given the state of physics at the time. Thus Lamb spent a great deal of effort calculating the nuclear forces predicted by various field theories. His first investigation consisted in exploring the nuclear forces generated by the Fermi interaction, that is, those produced by the exchange of pairs of electrons and neutrinos between nucleons. The resulting nucleon-nucleon potentials were highly singular "and then only if infinite integrals were interpreted with the help of convergence procedures" (Lamb 1983, p. 316). After the appearance of Yukawa's papers (Yukawa 1935; L. M. Brown 1981), Lamb worked on the nuclear forces generated by Yukawa's meson, and in particular on the deviations from Coulomb's law due to mesonic effects. The question of the validity of Coulomb's law became a recurrent theme in Lamb's researches (Lamb 1960).

In the summer of 1938 Lamb met Rabi for the first time during a visit by Rabi to Stanford. Lamb impressed Rabi and Rabi suggested to him that he apply for a job at Columbia. Lamb did so and was offered an instructorship. He resigned from his NRC fellowship and went to Columbia in the fall of 1938. Forty years later Rabi wrote Lamb: "I flatter myself as the first man to recognize your genius in the most practical way of giving you a job. It is one of my good deeds for which I have no regrets. Your coming to Columbia has turned out to be one of the great events in the history of physics. The Lamb shift and its theory were the great events that were decisive for the development of the Q.E.D. You also established yourself as one of the small group of physicists who could do both experiments and mathematical theory in the spirit of Enrico Fermi" (Rabi in ter Haar and Scully 1978).

In a talk at a Fermi Lab conference on "The Birth of Particle Physics" in 1980, Lamb outlined the steps which brought him to his famous experiment on the fine structure of hydrogen (Lamb 1983, pp. 317ff).

In 1939 Lamb became involved in a controversy with Herbert Fröhlich, Walter Heitler, and Boris Kahn (1939a,b) (see also Kahn 1941) about their meson field-theoretic calculation that indicated that the electric field around a proton should deviate from a pure $\frac{1}{r^2}$ Coulomb field. These theorists were actually trying to account for the deviations from the predictions of the Dirac equation for the spectrum of hydrogen that Houston (1937) and R. C. Williams (1938) had observed. They thought that in addition to the attractive Coulomb interaction between an electron and a proton, they could get a short-range repulsion by the mesotron charge distribution around the proton. Lamb (1939, 1940) argued that their result was possible only because they were working with a perturbative solution of a theory plagued with divergences. Although the differences were not resolved at the time, the discussion was useful because it kept Lamb "thinking about the hydrogen fine structure." It also made him aware of "the suggestion of Simon Pasternack (1938) that Houston's spectroscopic data could be interpreted in terms of a short range repulsion between electron and proton" (Lamb 1983). Being at Columbia, Lamb came into close contact with Rabi and members of the molecular beam laboratory. His attention was drawn briefly to the problem of the detection of metastable atoms in connection with a proposed atomic beam experiment (Cobas and Lamb 1944). His work with Cobas later "made an essential contribution to the hydrogen fine-structure work" (Lamb 1983, pp. 318–319).

During the war Lamb worked at the Columbia Radiation Laboratory (CRL) on microwave magnetron oscillators. Although Lamb's main job at CRL was as a theorist, his interest in different configurations for magnetron devices led him to "design, build, and test magnetrons," and he takes great pride in the fact that he was the first to construct "a continuous-wave magnetron oscillator to give a kilowatt of output at wave length shorter than 3cm." He learned about metal fabrication, high-vacuum technology, and other experimental techniques from Kusch. One of the main efforts at CRL was the design of a K-band magnetron, with a wavelength centered at 1.25 cm. After the project was well under way, Van Vleck pointed out that there was an H_2O absorption line at about 1.25 cm and that K-band radar would be seriously affected by high humidity. In order to assess the efficiency of the K-band radar, Lamb helped design an apparatus for measuring the absorption coefficient of water vapor as a function of the wavelength of the microwave radiation (Becker and Autler 1946; Lamb 1946). Lamb's involvement in the experimental determination of the absorption coefficient of centimeter waves in atmospheric water vapor was the point of departure for his experimental activities to measure the hydrogen fine structure.

"In teaching a summer session class in atomic physics in 1945 using a text book by Herzberg [Lamb] found references to some attempts made in 1932–35 to detect absorption of short wave length radio waves in a gas discharge of atomic hydrogen. At first it seemed . . . that these experiments had failed because of inadequate microwave techniques. [Lamb] thought of repeating them with the greatly improved facilities developed during the war" (Lamb 1983, p. 320). Lamb designed and constructed a continuous-wave magnetron with a wavelength centered at the "right" value of 2.74 cm, which corresponds to the wavelength of the transition from the $2^2S_{1/2}$ to $2^2P_{3/2}$ state.

By July 1946 Lamb had some definite notions on how to carry out an experiment to measure the hydrogen fine structure. He was fortunate to obtain Robert Retherford as a graduate student at the time. Before the war, Retherford had worked on atomic beams with Kellogg and was an expert on vacuum systems, dc amplifiers and experimental techniques in general. The history of their joint enterprise is chronicled in the Columbia Radiation Laboratory Reports from October 1946 to March 1948.

The references that Lamb had found in Herzberg were to two German papers dealing with attempts to detect the transition from the $2^2S_{1/2}$ or $2^2P_{1/2}$ level to the $2^2P_{3/2}$ level of hydrogen. In 1932, Betz (1932) had used a spark gap oscillator to generate the frequency of 2.74 cm that would induce a transition between these states. However, the power output of the spark gap oscillator was very low, and in addition it generated a continuous spectrum from which the desired frequency was selected by using interferometric methods. Betz's experiment consisted in passing the radiation through an absorption vessel containing a Wood's-type hydrogen discharge, and the attenuation was determined as a function of wavelength. Betz claimed to have observed an absorption around 2.74 cm. Three years later Haase (1935) repeated the experiments more carefully and failed to find any effect. Moreover, he pointed out that the rate of production of excited hydrogen atoms in the Wood's tube was so low that no detectable energy could have been absorbed by all of them. Furthermore, the experiment is "particularly difficult because in the Wood's discharge the absorption of the electron gas exceeds by far any expected absorption by the excited hydrogen atoms" (Lamb 1983).

The success of any absorption experiment depends on the possibility of obtaining a sufficient number of excited hydrogen atoms, and in particular of $2^2S_{1/2}$ atoms, to detect a measurable effect. Now the $2^2S_{1/2}$ state is metastable because single photon decay to the $1^2S_{1/2}$ level is forbidden by angular momentum conservation. Breit and Teller (1940) had calculated the lifetime of the state by double quantum emission to be about 1/7 second.[4] However, due to the degeneracy of the $2^2S_{1/2}$ and $2^2P_{1/2}$ levels, any perturbation will mix these states and allow a rapid decay to the ground state. Bethe (1933) in his *Handbuch der Physik* article had shown that due to the electric fields present in Wood's discharge tubes, the $2^2S_{1/2}$ state is effectively unstable. Due to the degeneracy of the two $j = 1/2$ states, one has a linear Stark effect instead of the usual quadratic one. Bethe computed that even a field as low as 10 volts per cm can reduce the lifetime from 1/10 second to about 10^{-8} second.

Lamb concluded that "in order to make precise measurements on the fine structure, it would appear necessary to increase the lifetime of the $2^2S_{1/2}$ state." Working in a field-free region was one possibility, but this was extremely difficult to achieve experimentally. Lamb explored the possibility of applying a strong magnetic field, so that the Zeeman splitting of the $2P_{1/2}$ and $2P_{3/2}$ states would be large enough so that for the electric field encountered in Wood's tubes the Stark effect would be quadratic, and hence not very destructive of the metastability. Lamb calculated "that electric fields as high as 10 volts per centimer would be permissible if a magnetic field of a few thousand gauss were present." However, he concluded that "even with the aid of a magnetic field it would be very difficult to detect the absorption of microwaves in a Wood's tube," because of the large absorption of microwaves due to the electron gas present in the tube, and the small absorption coefficient of the microwaves in the $2S_{1/2}$ state of hydrogen.

Lamb therefore considered an alternative method which would "attempt to form some kind of a stream of meta-stable hydrogen atoms which fall upon a detector sensitive to them. When R.F. (radio frequency) is absorbed by the atoms, they will go almost instantaneously to the ground state with the emission of a 1200 Å quantum. A decrease in the response of the detector will then be obtained. Calculations show that even intensities of the order of a micro-watt are sufficient to obtain an appreciable depletion of the stream of meta-stable atoms. On the other hand, an exceedingly large number of meta-stable H atoms would have to be present to give an absorption of a micro-watt of power. The proposed method is analogous to the molecular beam methods of Rabi."

The experimental problem thus became (1) the detection of metastable H atoms, and (2) the production of such atoms. Lamb's previous work with Cobas had introduced him to possible detection mechanisms. One way was to allow the atom to hit a metal surface and to observe the electron that is usually ejected in the process; or alternatively, to allow the metastable atom to hit a metallic surface whose work function is greater than the ionization energy of the $2S$ state, and to detect the H^+ ion that results from the collision with the surface. To obtain the $2S$ atoms, Lamb proposed to bombard atomic hydrogen obtained from a hot tungsten surface with electrons of the correct energy. Stark effects due to the electron collision were to be minimized by the application of a magnetic field.

The October 1946 report concluded with this statement:

> We have made quantitative estimates of the effects which should be observed, and the experiment seems feasible, if a detector with sufficient sensitivity can be found.
>
> An apparatus has been designed, and the parts are in process of construction in the shop. (Lamb 1946)

By January 1947 most of the apparatus had been built and a means of producing atomic hydrogen by thermally dissociating molecular hydrogen in a small tungsten furnace had been adopted. By March 1947 the apparatus had been rebuilt to incorporate several improvements in the source, the detector, the electron bombarder, and in the overall geometry and "tests [were] about to begin."

On April 26, 1947, the experiment was successful. Recognizing the historic importance of the accomplishment, Lamb and Rutherford in the June CRL Report gave a recapitulation of the general theory and background and included a succinct description of the experimental arrangement:

> Molecular hydrogen is thermally dissociated in a tungsten oven, and a jet of atoms emerges from a slit to be cross-bombarded by an electron stream. About one part in a hundred million of the atoms is thereby excited to the metastable $2^2S_{1/2}$ state. The metastable atoms (with small recoil deflection) move on out of the bombardment region and are detected by the process of electron ejection from a metal target. The electron current is measured with an FP-54 electrometer tube and a sensitive galvanometer. If the beam of metastable atoms is subjected to any perturbing fields which may cause a transition to any of the 2^2P states, the atoms will decay while moving through a very small distance. As a result, the beam current will decrease, since the detector does not respond to atoms in the ground

state.... Transitions may ... be induced by radiofrequency radiation for which $h\nu$ corresponds to the energy difference between one of the Zeeman components of $2^2S_{1/2}$ and any component of either $2^2P_{1/2}$ or $2^2P_{3/2}$. Such measurements provide a precise method for the location of the $2^2S_{1/2}$ state relative to the P states, as well as the distance between the later states.

We have observed an electrometer current of the order of 10^{-14} amperes which must be ascribed to metastable hydrogen atoms....

We have also observed the decrease in the beam of metastable atoms caused by microwaves in the wave length range 2.4 to 18.5 cm. in various magnetic fields.... The results indicate clearly that, contrary to theory but in essential agreement with Pasternack's hypothesis, the $2^2S_{1/2}$ state is higher than the $2^2P_{1/2}$ by about 1000 megacycles/second (0.033cm^{-1}).

When the experiment succeeded Lamb felt that it was of Nobel caliber. "I awakened the following morning with the realization that a very nice experiment had worked. It made for a very good feeling. I realized that the research was of the quality deserving a Nobel Prize" (Lamb 1983, p. 323). He was very pleased to hear that Kellogg, whom he admired and respected, had said "that was a hard experiment," which Lamb took to mean that Kellogg thought "it was too hard for him."[5] Indeed, very few experimentalists could have undertaken the experiment. Its success required the confluence of mastery of theory and brilliance in experimental techniques. When it was pointed out to Lamb that his achievement was not fortuitous but the convergence of a series of activities that he had initiated and carried through—his criticism of the Heitler-Fröhlich-Kahn paper, his study of the interaction of metastable atoms with metallic surfaces, his analysis and design of magnetrons, his extensive investigations of the Stark quenching of the metastable $2S$ hydrogen atoms, his analysis of the lifetime and Zeeman patterns of the $2S$ and $2P$ states, etc.—he commented, "Somebody was trying to tell me something."[6]

Waller, who made the presentation speech when Lamb was awarded the Nobel Prize in 1955, observed that "it does not often happen that experimental discoveries exert an influence on physics as strong and invigorating as did your work. Your work led to a re-evaluation and a reshaping of the theory of the interaction of electrons and electromagnetic radiation, thus initiating a development of utmost importance to many of the basic concepts of physics, a development the end of which is not yet in sight" (*Nobel Lectures in Physics*, 1942–1962).

On the occasion of Lamb's sixty-fifth birthday, Freeman Dyson wrote to congratulate him. He noted:

Your work on the hydrogen fine structure led directly to the wave of progress in quantum electrodynamics on which I took a ride to fame and fortune. You did the hard, tedious, exploratory work. Once you had started the wave rolling, the ride for us theorists was easy. And after we had zoomed ashore with our fine, fancy formalisms, you still stayed with your stubborn experiment. For many years thereafter you were at work, carefully coaxing the hydrogen atom to give us the accurate numbers which provided the solid foundations for all our speculations. . . .

Those years, when the Lamb shift was the central theme of physics, were golden years for all the physicists of my generation. You were the first to see that that tiny shift, so elusive and hard to measure, would clarify in a fundamental way our thinking about particles and fields. (In ter Haar and Franken 1978)

5.4 The Anomalous Magnetic Moment of the Electron

Norman Ramsey remembers visiting the Radiation Laboratory at MIT in the spring of 1945 and talking to Rabi about "major things" they could do after the war when they returned to Columbia. One of the ideas discussed turned out to be nuclear magnetic resonance. "The other thing we debated us doing—and decided it would be our highest priority experiment—was to do the hyperfine interaction in hydrogen and deuterium. And it was that out of which the Nafe-Nelson [experiment] started . . . . I remember hunting around gathering some of the old apparatus to do that. I . . . found the pieces of the apparatus [that Zacharias and Kellogg had built] in the attic."[7] In 1946 Ramsey became deeply involved with the founding of the Brookhaven National Laboratory and was appointed chairman of its physics department,[8] and Rabi took over the responsibility of overseeing the hyperfine experiment (Ramsey 1966). Thus at the same time as Lamb and Retherford were performing their landmark experiment on the fine structure of the hydrogen spectrum in the Pupin Laboratories at Columbia, another important experiment was being carried out there on the hyperfine structure of hydrogen and deuterium by Rabi and his students, Nafe and Nelson.

The magnetic interaction of the electron's magnetic dipole moment with the magnetic field of the nuclear dipole moment splits the $^2S_{1/2}$ ground state of hydrogen into two components—the state with the electronic and nuclear spins parallel, the $F = 1$ state, lying above the $F = 0$ state in which the spins are antiparallel. This splitting can easily be calculated from the properties of the proton and electron magnetic dipole moments. Using the measured value of g_n for the proton and an assumed $g_s = 2$ for the electron, the computed splitting between the $F = 0$ and $F = 1$ levels corresponds to a frequency of about 1416.90 ± 0.54 mc. In their experiment, Nafe, Nelson, and Rabi measured a splitting of 1421.3 mc. The uncertainties in both the measured and calculated values were such that the reality of the discrepancies could not be doubted.

A letter containing the results of their experiment was submitted to the *Physical Review* on May 19, 1947, and appeared in the June 15 issue. In it Nafe, Nelson, and Rabi noted the following:

> There is clearly an important difference between the measured and calculated values of ν_H and ν_D of about 0.26 percent compared with the probable error of the calculated value of 0.05 percent. The difference is five times greater than the claimed probable error in the natural constants. Whether the failure of theory and experiment to agree is because of some unknown factor in the theory of the hydrogen atom or simply an error in the estimate of one of the natural constants such as α^2, only further experiment can decide. (Nafe et al. 1947)

While carrying out his experiments on the hyperfine structure of H and D, Rabi enlisted the theoretical assistance of Gregory Breit, who had moved to Yale after the war. Breit computed the corrections to the Fermi formula that result from taking into account the nuclear motion. "I have worked out the corrections to be expected for the idealized case of a Dirac proton," Breit wrote Rabi at the end of May 1947. "The cube of the usual reduced mass ratio represents the corrections in the case. I am not through with examining the contributions due to the excess part of the proton's moment and the deuteron case."[9]

Breit reported the results of these calculations at Shelter Island. The importance of Rabi, Nelson, and Nafe's observations became more evident there, and Breit told Rabi after the conference that he "wanted to check [his calculations] again."[10] The further study of the discrepancy between observed and predicted magnitude of the hyperfine structure of hydrogen led Breit to suggest that the electron may possess an intrinsic moment different than μ_0 by the order of $\alpha \mu_0$, where α is the fine structure constant (Breit 1947b). Breit chronicled his researches in a letter to Rabi:

> After the work on the mass motion I have tried to account for the effect observed on the ground term from the point of view of the distortion of the wave function of the ground state. I have never quite finished the calculation but in the process of making it was struck by the fact that an anomaly of the moment would be a more natural explanation. After this I talked to you and Norman Ramsey as well as others at Brookhaven. Shortly afterwards I made estimates regarding the difference in electrodynamic shift for the hyperfine levels . . . The estimates looked hopeful but I did not trust them. Shortly afterwards I took a vacation. After the vacation I wrote the note ["Does the Electron Have and Intrinsic Moment" [Breit 1947b] and sent you a copy. You became interested and instigated the experiments [of Kusch and Foley].[11]

Indeed, upon receiving Breit's letter to the editor, Rabi wrote him back excitedly:

> It is extraordinarily interesting and pointed out to us a gross over-sight in our recent experiments. Kusch and Becker have been measuring the hyperfine structure of gallium in the $P_{3/2}$ and $P_{1/2}$ states of both isotopes. The results can be explained by the cosine x cosine coupling to an amazing accuracy to one part in 30,000. However, they neglected to measure the ratio of the atomic g-values in the same magnetic field so as to obtain the ratios of the atomic g's to high precision. Kusch will do this in the near future and I think this measuring will yield a very direct and accurate test of your suggestion of an intrinsic moment. We will do it any way since we are always looking for obvious things to do which will test fundamental principles, and after you made your suggestion we can't understand how we overlooked the op-portunity. I suppose it will always remain true that the obvious is hard to see.[12]

Until he had obtained Breit's letter to the editor, Rabi had "felt that the Nafe-Nelson discrepancy was due to an erroneous value of α." "On receiving your letter of September 24th and the copy of your manuscript of the Letter to the Editor," Rabi revealed to Breit, "I immediately got the idea that your hypothesis of a larger electron moment than μ_0 could be readily tested by comparing the g-values of the $P_{3/2}$ and $P_{1/2}$ states of gallium. I wrote you the same day telling you that fact. Your prediction of such a large change in the moment was a great help because I was worried about perturbations in the P states."[13]

Kusch and Foley undertook to investigate in detail the question of the existence of an anomalous magnetic moment of the electron (Kusch 1956). Kusch and Foley's initial experiment consisted in measuring the effect of a magnetic field on the two lowest states of gallium, thus obtaining the ratio of the Landé g-factors for these two levels (Kusch and Foley 1947). The absolute values of the g-factors could not be obtained with precision because of the difficulty in making precise absolute measurements of a magnetic field. In a second experiment, Foley and Kusch compared the g-factors of the two gallium states with the g-factor of the lowest state of Na (Kusch and Foley 1948).

The gallium atom consists of a number of closed electron shells with an additional electron in a $4p$ state. The lowest states thus form a P doublet. The separation between the $P_{1/2}$ and $P_{3/2}$ state is 174 cm^{-1}, and both states are present in an atomic Ga gas at room temperature. According to theory, one has $g = 2/3$ for the $P_{1/2}$ state and $g = 4/3$ for the $P_{3/2}$ state. In the same magnetic field, the Zeeman effect should therefore be twice as large for the $P_{3/2}$ state. Kusch and Foley found the ratio to be 2.00344 ± 0.00012.

This deviation might have been due to some configuration interaction of the two lowest states of Ga, so that they are not pure $P_{1/2}$ and $P_{3/2}$, respectively. However, these doubts were removed by the second experiment comparing the $P_{1/2}$ state to the ground state of sodium (Kusch and Foley 1948). The g factor of the latter should be 2. Furthermore, M. Phillips (1941) had shown that perturbation from other spectroscopic configurations is less than one part in 20,000. The ratio of the g-factors of Na and Ga $P_{1/2}$, which ought to be 3, was found to be 3.00732. This result can be explained by assuming that the electron spin has a magnetic moment of 1.00122 ± 0.00003 Bohr magnetrons, while the orbital motion of the electron has the normal moment of one Bohr magnetron per unit angular moment. Similarly, the original Kusch and Foley experiment comparing the two states of Ga can be explained by assuming a spin magnetic moment to be $1.00114_5 \pm 0.00004$.

On December 12, 1947, Rabi could write Breit: "I really believe that the electron moment is 12% greater than the Bohr magneton. This seems to be in almost exact agreement with Schwinger's calculation and accounts completely for the deviation of the hyperfine structure in hydrogen and deuterium....You will be interested to hear that Nafe and Nelson have measured the ratio of the electron moment in hydrogen and deuterium and find that it is unity to about one part in 50,000."[14] This comparison of the experimental *ratio* to the calculated *ratio* was particularly important, since most of the natural constants cancel out.

The experiments of Nafe, Nelson, and Rabi, and of Kusch and Foley, were of great importance in changing the status of the Dirac equation. Their significance was reflected in the award of a Nobel Prize to Kusch in 1955, a prize he shared with Lamb. But the intellectual triumphs were marred by some painful behind-the-scenes incidents. Breit was hurt by the way he was treated by Kusch and Foley and felt that he had not received adequate recognition for his contribution. Breit was probably oversensitive and perhaps a bit paranoid; nonetheless, the correspondence between Breit and Rabi in February 1948 indicates that the character, the pace, and the stakes in doing physics had changed from the prewar period. The reason for the exchange of letters was Breit's unhappiness "with the way my part in stimulating the work on the difference between g_s and g_L is appearing in publication. "The thing that worries me is largely not the personal situation concerned with myself but rather the general change from the practice of friendliness in dealing among scientists. I am writing to you about it because I have sent you the letter to the editor to which Foley and Kusch refer now as follows: 'These results are not in agreement with the recent suggestion made by Breit as to the magnitude of the intrinsic moment of the electron.'"

Breit went on to comment that he had heard Schwinger talk about his quantum electrodynamic calculations at the theoretical physics conference at Washington University in the fall of 1947, which "made it ethically difficult to publish additional material re Kusch and Foley's work and I was left with an incorrect 'prediction.'" Breit continued: "I believe that in this whole matter I have been acting in a decidedly social manner and have helped progress along. I do not understand why Foley and Kusch wish to make things difficult for me by mentioning the incorrectness of the quantitative side of my [letter to the editor] without mentioning the way it was useful. Will physicists have to expect in the future to be taken to task just because they are friendly and wish to get at the truth?...I am puzzled."[15]

Rabi promptly answered him stating that he agrees "and Kusch agrees that the reference to your work was somewhat ungraciously worded and could even give a wrong impression about the really friendly feeling which is current here about you and the important contributions which you made to our thinking."[16] Rabi even suggested writing a letter to the editor, outlining the "story of these events," signed by Breit, Foley, Kusch, and Rabi but nothing came of the matter.

5.5 The Magnetic Resonance Experiments

Lamb and Retherford's experiment used the resonance radio-frequency spectroscopic techniques that Rabi had developed at Columbia before World War II (Rigden 1985, 1986). In this method, atoms in a beam are irradiated and the radiation causes observable changes to take place in them. In Rabi's experiment, atoms that have absorbed radiation are deflected out of the beam by an inhomogeneous magnetic field. In Lamb's experiment essential use is made of the metastability of the $2^2S_{1/2}$ state of hydrogen. The $2^2S_{1/2}$ state cannot decay to the $1^2S_{1/2}$ ground state by single photon emission and has a lifetime of approximately 1/7 seconds (it can decay to the ground state by 2-photon decay). On the other hand, the $2p$ state decays to the $1s$ state with the emission of a photon of wavelength 1216 Å in 1.595×10^{-9} sec. Thus if a beam of atomic hydrogen in the $2^2S_{1/2}$ state could be prepared and a transition to a $2p$ state then induced by radio-frequency radiation, a decay to the $1s$ state takes place so quickly that the number of excited atoms in the beam is reduced. A detector which responds selectively to the excited hydrogen atoms in the $2^2S_{1/2}$ states then renders possible a measurement of the energy difference between the metastable $2^2S_{1/2}$ and the various $2p$ states. The present section investigates these types of experiments and comes to the important conclusion that the theory which is used to design and analyze these magnetic resonance experiments is different from that used to account for the value of the measured parameters.

A schematic diagram showing the principle of the first Rabi molecular beam magnetic resonance apparatus is sketched in figure 5.5.1. In these experiments the atoms or molecules were deflected by a first inhomogeneous field and refocused by a second one. When the resonance transition was induced in the C-field region, the occurrence could easily be recognized by the reduction of the intensity associated with the accompanying failure of refocusing (see Kusch and Hughes 1959).

Let me briefly outline the theoretical basis of the method. The foundations were laid in a classic paper by Rabi in 1937 in which he studied transitions induced when atoms or molecules in a molecular beam traversed a region of space

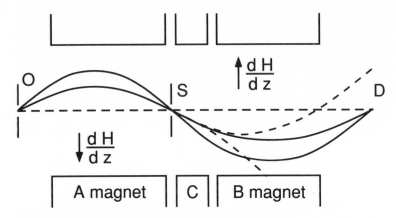

Figure 5.5.1 Schematic diagram showin the principle of the first molecular beam mangetic resonance apparatus. Two solid curves indicate two paths of molecules having different orientations that are not changed during passage through the apparatus. The two dashed curves in the region of the B magnet indicate two paths of molecules whose orientation has been changed in the C region so that the refocusing is lost.

in which the direction of the magnetic field present changed successively (Rabi 1937). For simplicity Rabi assumed that the field was oscillatory in time even though the problem to be solved concerned fields varying along the beam, rather than oscillating ones in time. As a consequence, Rabi's formulas are applicable to the resonance case with temporarily oscillating fields and the paper provided the fundamental theory for all molecular beam magnetic resonance experiments. The Hamiltonian for an atom (or molecule) in the C-region of the beam apparatus can be written as

$$H = H_0 + V,$$

in which H_0 includes all internal interactions of the atomic system and also the interaction of the atom with the externally applied steady magnetic (or electric) field, and V describes the interaction of the atom with the applied oscillating field. The problem to be solved is to find the solution of

$$H\psi = (H_0 + V)\psi = i\hbar \frac{\partial \psi}{\partial t}, \tag{5.5.1}$$

which describes the evolution of the system as it traverses the C region. If $\psi(t)$ is expanded in terms of the eigenstates ϕ_n of H_0, $H_0\phi_n = W_n\phi_n$,

$$\psi(t) = \sum_n c_n(t)\,\phi_n, \tag{5.5.2}$$

eq. (5.5.1) becomes

$$i\hbar \dot{c}_m(t) = \sum_n \langle \phi_m | H | \phi_n \rangle c_n(t). \tag{5.5.3}$$

The special case of two stable, long-lived eigenstates, ϕ_p and ϕ_q of H_0, between which a transition may occur is the canonical case for atomic beam experiments. The interaction term V arising from the oscillating field of angular frequency ω is assumed to have the matrix elements

$$V_{pq} = \langle \phi_p | V | \phi_q \rangle = \hbar b e^{i\omega t}$$
$$= \overline{V_{qp}} \tag{5.5.4}$$
$$V_{pp} = V_{qq} = 0. \tag{5.5.5}$$

The quantity b measures the "strength" of the transition matrix element and is independent of time. For this case eq. (5.5.3) becomes

$$i\hbar \, \dot{c}_p = W_p \, c_p + \hbar b \, c_q e^{i\omega t} \tag{5.5.6}$$
$$i\hbar \, \dot{c}_q = W_q \, c_q + \hbar \bar{b} \, c_p e^{-i\omega t}, \tag{5.5.7}$$

where W_p and W_q are the energy eigenvalues of the states ϕ_p and ϕ_q of the Hamiltonian H_0. If it is assumed that all atoms are in the state p when they enter the region C in which the oscillating field acts, that is,

$$|c_p| = 1 \qquad |c_q| = 0 \qquad \text{at} \qquad t = 0, \tag{5.5.8}$$

then the probability $|c_q(t)|^2$ that an atom will be in the state q after spending a time T in the oscillating field is

$$P_{qp} = \frac{|2b|^2}{|2b|^2 + (\omega_0 - \omega)^2} \sin^2 \left\{ \frac{1}{2} \left[|2b|^2 + (\omega_0 - \omega)^2 \right]^{1/2} T \right\}, \tag{5.5.9}$$

in which $\hbar \omega_0 = W_q - W_p$. The function P_{qp} is called the natural line shape. Its maximum value is 1, which is attained at resonance, $\omega = \omega_0$. Eq. (5.5.9) corresponds to a somewhat idealized case. In actual practice, the oscillating field in the C-region has a time dependence of $\cos \omega t$ rather than of $e^{i\omega t}$. Similarly, the existence of other states besides the states p, q for which V has nonvanishing matrix elements leads to an expression for P_{qp} somewhat different from (5.5.9). It has, however, been shown that the inclusion of all these effects leads to corrections to the formula (5.5.9) that can be neglected for the amplitudes of oscillatory field used in practice (Kusch and Hughes 1959, p. 56).

In the Lamb shift experiment the metastable atoms are subjected to a radio-frequency field between the source and the detector. When the frequency of that radiation is such that $\hbar\omega$ is equal or nearly equal to an energy difference between a Zeeman component of the $^2S_{1/2}$ level and a Zeeman component of one of the $^2P_{1/2}$ levels, the radiation may induce a transition to a nonmetastable level and the detected signal will decrease. According to quantum mechanics, the decay rate of the $2^2S_{1/2}$ state to one of the $2p$ states due to the radio frequency radiation is

$$\frac{1}{\tau_s} = \frac{2\pi e^2 \gamma S_0}{c\hbar^2} \frac{|<2p|V|2s>|}{(\omega - \omega_0)^2 + \gamma^2/2}, \tag{5.5.10}$$

where S_0 is the incident energy flux of the radiation of frequency ω, V is the operator representing the perturbation which in the present case is (to a very good approximation) equal to $\hat{\mathbf{e}} \cdot \mathbf{r}$, where $\hat{\mathbf{e}}$ is the direction of the electric polarization of the radio frequency field and $\mathbf{r}$ is the coordinate vector of the electron undergoing the transition; γ is the radiative damping constant for the $2p$ states, a **phenomenological** parameter introduced in the quantum-mechanical description to account for the natural decay of the $2p$ states, with τ the lifetime of the state. All these experiments make use of the fact that the dependence of the line shape on the frequency of the microwave field is of the form $\frac{1}{(\omega-\omega_0)^2+|b|^2}$.

Note that the resonance frequency, $\omega = \omega_0$—as determined by the maximum of the resonance curve—measures the energy difference of the *actual* states of the system. Conceptually the W_p, W_q that appear in eqs. (5.5.6) and (5.5.7) can therefore be thought of as referring to a phenomenologically described atomic system: the parameters entering that description are determined by the measurement.

The situation is clarified by a discussion of electron spin resonance. Here the spin moves under the joint action of a large uniform field $\mathbf{B}$ in the z-direction which is constant in time, $\mathbf{B} = B\hat{\mathbf{z}}$, and a small perpendicular, oscillatory field $\mathbf{b}(t)$ in the x-y plane:

$$\mathbf{b}(t) = |b|(\hat{\mathbf{x}}\cos\omega t + \hat{\mathbf{y}}\sin\omega t). \tag{5.5.11}$$

The spin motion is assumed to be governed by the Hamiltonian

$$H = -g_s \frac{e\hbar}{2mc} \frac{1}{2} \boldsymbol{\sigma} \cdot [\mathbf{B} + \mathbf{b}(t)], \tag{5.5.12}$$

where the gyromagnetic ratio is introduced as a *phenomenological* parameter. The uniform field causes the spin to process at frequency

$$\omega_s = g_s \frac{|eB|}{2mc}. \tag{5.5.13}$$

The line shape in this case is given by the Rabi formula

$$P_{-1/2} = \frac{\left(\frac{\omega_s b}{B}\right)^2}{(\omega - \omega_s)^2 + \left(\frac{\omega_s b}{B}\right)^2} \sin^2 \left\{ \left[(\omega - \omega_s)^2 + \left(\frac{\omega_s b}{B}\right)^2 \right]^{1/2} \frac{T}{2} \right\}. \quad (5.5.14)$$

If the field B is known, a measurement of the resonance frequency $\omega = \omega_s$ thus measures the phenomenological parameter g_s.

The situation is conceptually the same—though more complicated—in the case of the Lamb shift measurement. There, for technical reasons, transitions are induced between Zeeman components of the $2^2S_{1/2}$ and $2^2P_{1/2}$ levels of hydrogen. That level structure is assumed to be described by a Hamiltonian $H = H_0 + H'$, where H_0 is responsible for the structure of the unperturbed atom, and the perturbation due to the magnetic field is given by $H' = \mu_0 (\mathbf{L} \cdot \mathbf{B} + g_s \mathbf{S} \cdot \mathbf{B})$ with $\mu_0 = \frac{e\hbar}{2mc}$, and where $\mathbf{L} = \mathbf{r} \times \mathbf{p}/\hbar$ is the orbital angular momentum and $\mathbf{S}$ is the spin angular momentum of the electron measured in units of $\hbar$.

The behavior of the levels in the presence of the magnetic field $\mathbf{B}$ can be readily computed (Bethe and Salpeter 1957). For example, the energy of the $2^2S_{1/2}, m_s = \pm 1/2$ state in the presence of the field B is given by

$$E(2^2S_{1/2}, m_s; B) = E(2^2S_{1/2}) + 2\mu_0 g_s B m_s,$$

where the energy level in the absence of field, $E(2^2S_{1/2})$, is assumed known and can be treated as a phenomenological parameter. Similar but more complicated expressions can be derived for the energy levels $E(2^2P_{1/2}, m_j = \pm 3/2, B)$, $E(2^2P_{3/2}, m_j = \pm 1/2, B)$, and $E(2^2P_{1/2}, m_j = \pm 1/2, B)$ in terms of the unperturbed energies $E(2^2P_{3/2})$ and $E(2^2P_{1/2})$, that is, the levels in the absence of the field. Note no assumption is made about the location of the unperturbed levels, and in particular about the equality $E(2^2S_{1/2})$ and $E(2^2P_{1/2})$. The measurement of the resonance frequencies as functions of the magnetic field B yields the value of $E(2^2S_{1/2}) - E(2^2P_{1/2})$.

It should be stressed that *quantum electrodynamics is not invoked in obtaining this level shift*. What is assumed is that the hydrogen atom is composed of a proton and an electron, entities that can be characterized by their mass, charge, and magnetic moment, and that their electromagnetic interaction is essentially determined by classical electrodynamics. The structural problem is subsumed in the Hamiltonian H_0 (which is assumed to give a fairly accurate quantum-mechanical description of the system). After the energy parameters $E(1^2P_{1/2})$, $E(2^2S_{1/2})$, $E(2^2P_{1/2})$, etc., have been experimentally determined, the structural problem is analyzed in further detail.

It then becomes the task of a higher-level theory—quantum electrodynamics (QED)—to account for the observed level structure. Stated differently, it is QED that accounts for the value of some of the phenomenological parameters that were present in the theory that helped determine the experimental values. This strategy of phenomenological description is standard and has a long-standing history in physics. For example, Lorentz's microscopic theory was to account for the phenomenological parameters (such as the electric and magnetic susceptibilities that are present in the macroscopic Maxwell electromagnetic theory (see, for example, Van Vleck 1932 and Rosenfeld 1951).

We turn next to the quantum electrodynamical explanation of the Lamb shift given by Bethe.

5.6 Bethe's Calculation

Pasteur is said to have coined the aphorism that "chance favors the prepared mind." Bethe was certainly prepared to undertake the level shift calculation. In the spring of 1946, he had collaborated with Oppenheimer on a paper that investigated the infrared problem within Heitler's radiation damping theory (Bethe and Oppenheimer 1946). In the process he had again studied Pauli and Fierz's (1938) classic paper on the infrared problem. During the fall semester of the academic year 1946/47, Bethe taught a course on advanced quantum mechanics. The notes for the course are extant and indicate that its "program" included: Relativistic Theory (8 lectures); Radiation Theory, Self-Energy, and Heisenberg?[17] The latter presumably referred to Heisenberg's recent S-matrix work. Bethe's treatment of radiation theory followed that given in Heitler's (1945) *The Quantum Theory of Radiation*. His notes on the self-energy of an electron are extensive and confirm that he had looked at the problem carefully. The self-energy of a nonrelativistic particle to second-order in perturbation theory was shown to be equal to

$$\Delta W = \text{Self-energy} = -\sum_n \sum_\lambda \frac{|(n, 1_\lambda |H'|m)|^2}{E_n + k - E_m}$$

$$(n, 1_\lambda |H'|m) = e\hbar c \sqrt{\frac{2\pi}{k}} \int \psi_n^* \, \mathbf{p} \cdot \mathbf{e}_\lambda \, \psi_m / mc.$$

For a nonrelativistic free particle of momentum $p_o (m = p_o; n = p_o + \text{k})$, Bethe obtained the result that the self-energy is linearly divergent,

$$\Delta W = -\frac{e^2}{\hbar c} \left(\frac{p_o}{mc}\right)^2 \frac{2}{3\pi} \int_0^\infty dk, \qquad (5.6.1)$$

assuming that recoil could be neglected. The notes also contain the remark that for a Dirac particle—as a "one electron theory"—the contribution to the self-energy from the transverse waves diverges quadratically.[18]

Bethe had been asked to prepare a brief talk on the infrared problem for the Shelter Island conference. His presentation took place on the second day of the conference—and included a review of the papers by Bloch and Nordsieck (1937), Pauli and Fierz (1938), and Bethe and Oppenheimer (1946) (Breit 1947b). Bethe comments that after hearing Lamb's and Kramers's presentations and the discussions these generated he knew how to make a nonrelativistic calculation and was eager to do so throughout the conference.[19] He summarized these discussions in his paper:

> Schwinger and Weisskopf, and Oppenheimer have suggested that a possible explanation might be the shift of energy levels by the interaction of the electron with the radiation field. This shift comes out infinite in all existing theories, and has therefore always been ignored. However, it is possible to identify the most strongly (linearly) divergent term in the level shift with an electro-magnetic *mass* effect which must exist for a bound as well as for a free electron. This effect should properly be regarded as already included in the observed mass of the electron, and we must therefore subtract from the theoretical expression, the corresponding expression for a free electron of the same average kinetic energy. The result then diverges only logarithmically (instead of linearly) in non-relativistic theory: Accordingly, it may be expected that in the hole theory, in which the *main* term (self-energy of the electron) diverges only logarithmically, the result will be *convergent* after subtraction of the free electron expression.[6] This would set an effective upper limit of the order of mc^2 to the frequencies of light which effectively contribute to the shift of the level of a bound electron. I have not carried out the relativistic calculations, but I shall assume that such an effective relativistic limit exists. (Bethe 1947)

Bethe followed the general ideas of Kramers on mass renormalization. Explicitly this meant that in the quantum-mechanical treatment of the self-energy of a free particle one should interpret the second-order contribution ΔW, eq. (5.6.1), as a contribution to the mass of the charged particle. In other words, the zeroth and second order combine to give the energy, $\frac{p^2}{2m}$, of the particle with mass m

$$ W = W_0 + \Delta W = \frac{p^2}{2m} = \frac{p^2}{2(m_0 + \mu)} \tag{5.6.2}$$

$$ \Delta W = -\frac{p^2}{2m_0^2}\,\mu. \tag{5.6.3}$$

In eq. (5.6.2) $m = m_0 + \mu$ is the observed mass of the charged particle, m_0 is its "bare" mass, and

$$\mu = \frac{4}{3\pi} \frac{e^2}{\hbar c^3} \int_0^\infty dk \qquad (5.6.4)$$

is the linearly divergent contribution (5.6.1). In any experiment it is only the observed mass m that can be measured, and any distinction between m_o and μ is meaningless. All observable quantities must therefore involve only the observed mass. Hence any reference to a mass other than the total observed mass must be eliminated in all equations. This is what Kramers' mass renormalization meant.

The actual calculation of the nonrelativistic Lamb shift was made on a train ride from New York to Schenectady. Bethe had stayed in New York after the Shelter Island conference to visit his mother, and had gone on to Schenectady to consult for General Electric. The calculation is straightforward (Bethe 1947). The self-energy of an electron in a quantum state m, due to its interaction with the radiation field, is

$$W = -\frac{2e^2}{3\pi\hbar c^3} \int_0^K k\,dk \sum_n \frac{|\mathbf{v}_{mn}|^2}{E_n - E_m + k}, \qquad (5.6.5)$$

where $k = \hbar\omega$ is the energy of the photon that is emitted and reabsorbed by the electron, and

$$\mathbf{v} = \mathbf{p}/m = \frac{\hbar}{im}\nabla \qquad (5.6.6)$$

is the velocity operator of the electron. For a free electron, $\mathbf{v}$ only has diagonal elements and (5.6.3) is replaced by

$$W_0 = -\frac{2e^2}{3\pi\hbar c^3} \int k\,dk\,\mathbf{v}^2/k. \qquad (5.6.7)$$

Now W_o "represents the change of the kinetic energy of the electron for fixed momentum, due to the fact that electromagnetic mass is added to the mass of the electron. This electromagnetic mass is already contained in the experimental electron mass; the contribution (5.6.7) should therefore be disregarded. For a bound electron, $\mathbf{v}^2$ should be replaced by its expectation value $(\mathbf{v}^2)_{mm}$."[20] But since

$$\sum_n |\mathbf{v}_{mn}|^2 = (\mathbf{v}^2)_{mm}, \qquad (5.6.8)$$

the relevant part of the self-energy becomes

$$W' = W - W_0 = \frac{2e^2}{3\pi\hbar c^3} \int_0^K dk \sum_n \frac{|v_{mn}|^2(E_n - E_m)}{E_n - E_m + k}, \qquad (5.6.9)$$

which expression Bethe considered "the true shift of the levels due to interactions."
This expression diverges logarithmically as $K \to \infty$. Bethe, with characteristic
self-confidence, next assumed that a relativistic hole-theoretic calculation would
provide a natural cutoff for the frequency K at energies

$$K = mc^2. \qquad (5.6.10)$$

Upon performing the k integration,

$$W' = \frac{2e^2}{3\pi\hbar c^3} \sum_n |v_{mn}|^2(E_n - E_m) \, \ell n \, \frac{K}{|E_n - E_m|}. \qquad (5.6.11)$$

Since the argument in the logarithm is very large, Bethe assumed it to be constant
(independent of n) in first approximation. The sum over n can then be performed
and yields

$$\sum_n |p_{nm}|^2 \, (E_n - E_m) = 2\pi \, \hbar^2 e^2 Z \, \psi_m^2(0), \qquad (5.6.12)$$

where $\psi_m(0)$ is the wave function of the electron at the position of the nucleus. The
shift (to this approximation) is therefore non-zero only for S-states, and in this case

$$W'_n = \frac{8}{3\pi} \left(\frac{e^2}{\hbar c}\right)^3 \text{Ry} \, \frac{Z^4}{n^3} \, \ln \frac{K}{< E_n - E_m >_{Ave}}, \qquad (5.6.13)$$

where Ry is the Rydberg energy $\alpha^2 mc^2/2$, $\alpha = e^2/\hbar c$. This is the expression that
Bethe had obtained on his arrival at Schenectady. He was not quite confident of its
accuracy, because he was not quite sure of the correctness of a factor of $\sqrt{2}$ in his
expansion of the radiation operators in terms of creation and annihilation operators.
This he checked on Monday morning in Heitler's book. He also got Miss Steward
and Dr. Stehn from GE to evaluate numerically $< E_n - E_m >_{Ave}$ for the 2s state. It
was found to be 17.8 Ry, "an amazingly high value."[21] Inserting this into (5.6.13),
Bethe found

$$W_{2s} \approx 1040 \text{ megacycles}, \qquad (5.6.14)$$

"in excellent agreement with the observed value of 1000 megacycles" (Bethe
1947).

5.7 Relativistic Lamb Shift Calculations: 1947–1948

As soon as Bethe completed and circulated his calculation indicating that a major part of the Lamb-Retherford experimental result on the $2s$-$2p$ level shift in hydrogen could be explained as a nonrelativistic quantum-electrodynamical effect, the task at hand became to carry out a relativistic calculation using the full hole theoretic formalism to justify Bethe's introduction of the cutoff $K \approx mc^2$—a much more difficult undertaking.

French and Weisskopf at MIT continued their calculation, but now made use of Kramers' idea and thus simplified somewhat the subtraction of infinities. (Their approach will be detailed in section 5.8.) At Cornell, Bethe assigned the problem to one of his graduate students, Scalettar. Lamb at Columbia started on a hole-theoretical calculation during the early part of the summer of 1947, and was soon joined by Norman Kroll. Similarly, Fermi, who was spending the summer 1947 at Los Alamos, upon receiving a copy of Bethe's preprint explored a relativistic calculation. His first step was to understand the Bethe calculation—which he redid, in collaboration with Uehling, who was also visiting Los Alamos, but they obtained an expression for the Lamb shift which was 4/3 times larger than Bethe's formula. "The factor 4/3 [was] due ... to the inadequacy of our assumption that the [intermediate] states can be described by plane waves,"[22] Fermi wrote Uehling after speaking to Bethe. Furthermore:

> A point that is not explained in Bethe's paper but which he explained to us in Los Alamos is the procedure for justifying that the recoil of the light quantum can be disregarded.
>
> This can actually be done by using an only slightly more complicated sum rule and I do not understand why Bethe did not follow this more complete procedure[23] in writing his paper since it would have made the result more convincing.

Fermi continued:

> The point that still is quite unsatisfactory is of course the upper limit of the logarithm in Bethe's formula (11). Apparently several people (Bethe, Weisskopf and Schwinger) have tried unsuccessfully to carry out a relativistic calculation of this upper limit. Also Teller and I tried the same and we believe that we have a method that seems to be practical though probably far from simple.
>
> This method consists in describing the [intermediate] ... state n as plane waves plus a first approximation [Coulomb] correction which is necessary and sufficient to correct for the factor 4/3 discussed above.[24]

1. Dirac, by Feynman. (Courtesy Millikan Library Archives, California Institute of Technology)

2. Dirac, Sugiura, and Oppenheimer. Göttingen, 1928. (Courtesy AIP Niels Bohr Library, Uhlenbeck Collection)

3. University of Michigan summer session, 1929. (Courtesy AIP Niels Bohr Library)

Theory of the Positron

by P. A. M. Dirac.

The recent discovery of the positively charged electron or _positron_ has revived interest in an old theory about the states of negative kinetic energy of an electron, as the experimental results that have been obtained so far are in agreement with the predictions of the theory.

The question of negative kinetic energies arises as soon as one considers the motion of a particle according to the principle of restricted relativity. In non-relativistic theory the energy W of a particle is given in terms of its velocity v or its momentum p by

$$W = \tfrac{1}{2}mv^2 = \frac{1}{2m}p^2,$$

which makes W always positive, but in relativistic theory this formula must be replaced by

$$W^2 = m^2c^4 + c^2p^2$$

$$W = c\sqrt{m^2c^2 + p^2},$$

or

which allows W to be either positive or negative.

One usually makes the extra assumption that W must always be positive. This assumption is permissible in the classical theory, where variables always vary continuously, since W can then never change from one of its positive values, which must be $\geq mc^2$, to one of its negative values, which must be $\leq -mc^2$. In the quantum theory, however, discontinuous changes of a variable may take place, so that W may change from a positive to a negative value. It has not been found to be possible to set up a

4. Dirac's ms for the 1933 Solvay Conference

5. Participants at the Shelter Island Conference. *Left to right*: I. I. Rabi, L. Pauling, J. Van Vleck, W. E. Lamb, Jr., G. Breit, D. MacInnes, K. K. Darrow, G. E. Uhlenbeck, J. Schwinger, E. Teller, B. Rossi, A. Nordsieck, J. Von Neumann, J. A. Wheeler, H. A. Bethe, R. Serber, R. E. Marshak, A. Pais, J. R. Oppenheimer, D. Bohm, R. P. Feynman, V. F. Weisskopf, H. Feshbach. (Not in photo: H. A. Kramers.) (Courtesy of Archives, National Academy of Sciences)

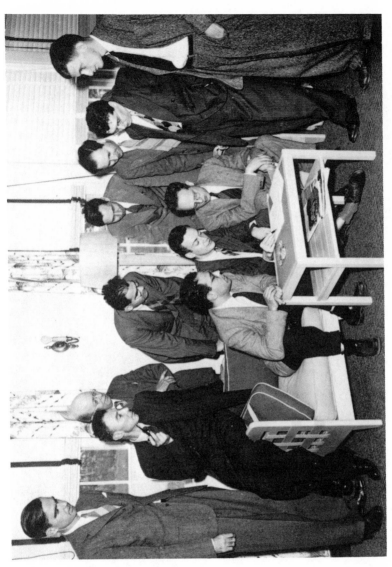

6. Feynman explaining a point at Shelter Island. *Left to right:* Standing—W. Lamb, Jr., K. K. Darrow, V. Weisskopf, G. E. Uhlenbeck, R. E. Marshak, J. Schwinger, and D. Bohm. Seated—J. R. Oppenheimer, A. Pais, R. P. Feynman, and H. Feshbach. (Courtesy of Archives, National Academy of Sciences)

7. Tomonaga at Riken. (Courtesy of Tomonaga Memorial, University of Tsukuba)

8. Tomonaga's self-portrait. (Courtesy of Tomonaga Memorial, University of Tsukuba)

9. Tomonaga, 1960. (Courtesy of Tomonaga Memorial, University of Tsukuba)

10. Tomonaga, 1970. (Courtesy of Tomonaga Memorial, University of Tsukuba)

11. Schwinger at the MIT Radiation Laboratory, ca. 1943. (Courtesy of the MIT Museum)

12. Schwinger, Inglis, and Teller (with son Paul), at Los Alamos, 1945. (Courtesy AIP Niels Bohr Library, Fermi Film Collection)

13. The participants at the tenth Washington Conference, November 1947. Oppenheimer is seated seventh from the left; Wheeler and Feynman are on Oppenheimer's right, Schwinger and Breit on his left. (Courtesy of the Archives, California Institute of Technology)

14. Schwinger at Harvard University. (Courtesy AIP Niels Bohr Library)

TYPEWRITER SYMBOLS

*	plus	Can't	write	the	other
-	minus	"	"	"	"
;	times	"	"	"	"
/	divide by	"	"	"	"
#	equals	"	"	"	"
$	integral	"	"	"	"
	differancal	"	"	"	""
?a$b?	definite integral	"	"	"	"
¢	pi # 3.1416		"		"
?	infinity	"	"	"	"
$x^{"}$	x square				
x^2	x "				

EXAMPLE(SEE BELOW)

$$\$(x"*a")^{\frac{1}{2}} \# \tfrac{1}{2}x(x"*a")*a":\tfrac{1}{2}:\log_e(x*(x"*a"))$$

15. Typewriter symbols for Feynman's table of integrals, 1933. (Courtesy of the Archives, California Institute of Technology)

COMPLEX NUMBERS

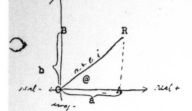

Since any numbr, real, or pure or complex imaginary, is of the form a*bi and can shown on the graph adoining.

On the figure:

tan@ # b/a
nat(b/a) # @

Letting OR # r: sin@ # b/r
cos@ # a/r

a/r * ib/r # cos@ * i:sin@

a * bi # r(a/r * ib/r) # r(cos@ * i:sin@)

But: (see page 54 of M):   e$^{i@}$ # cos@ * i:sin@

So: a*bi # re$^{i@}$ But; @ # nat b/a

Or; a*bi # re$^{i:nat(b/a)}$

Now will this work for real numbers and in what way?

in this case b # 0 : Therefore nat0 # 0 or ¢ or 2¢ et
     a # re^{0} # re$^{2¢i}$ # re$^{¢i}$

Since e^{0} # 1 ; and ; e$^{¢i}$ # -1 what is (r)

r # (a"*b")$^{\frac{1}{2}}$ from the diagram; rt. tri. OAR

So the complete answere is a*bi # (a"*b")$^{\frac{1}{2}}$e$^{i:nat(b/a)}$

If (a"*b")$^{\frac{1}{2}}$ # m ; and nat(b/a) # n
all numbers are of the form;
                          me^{in}

Or in other words all numbers can be shown to be a certain constant times that remarkable numbr (e) to another constant times that other remarkable numbr i, power. which is truly a remarkable thing.

Thus what is i in this formula
     firstly a # 0   and b # 1

Or m # (a"*b")$^{\frac{1}{2}}$ # *1

and n # nat(1/0) # nat? # ¢/2 or 3¢/2 or 5¢/2 etc.
So i . e$^{i/2}$

16. From Feynman's "Calculus for the Practical Man," 1933. (Courtesy of the Archives, California Institute of Technology)

17. A Feynman entry in the Feynman-Welton notebook, 1936. (Courtesy of the Archives, California Institute of Technology)

18. Los Alamos: the weekly gatherings. Feynman is sitting to the right of Oppenheimer.
(Courtesy AIP Niels Bohr Library)

19. W. Lamb, A. Pais, R. Feynman, H. Feshbach, and J. Schwinger conferring at
Shelter Island, June 1947. J. A. Wheeler is standing against the wall. (Courtesy AIP
Niels Bohr Library)

Geometry of Dirac Equ. 1 dimension

Prob = Squ. of Sum of contrib each path
Paths zig zag at light velocity.
Contributes $i\epsilon$ factor for each reversal
and factor $e^{\frac{i}{\hbar}\int_{Path} A\,dx + \phi\,dt}$ if fields are present.

$v = x+t$
$u = x-t$.

cont. Paths	$i\epsilon$	Relation 1
3	$(i\epsilon)^3 \cdot NM$	$-ab$
5	$(i\epsilon)^5 \cdot \frac{M(N-1)}{2} \cdot \frac{M(M-1)}{2}$	$+\frac{a^2 b^2}{2!^2}$
⋮		$-\frac{a^3 b^3}{3!^2}$ ..

$\therefore \; J_0(\sqrt{ab}) = J_0(\sqrt{t^2 - x^2})$

$+ \delta(t^2 - x^2) + \delta(t+x)$

Green Solution of $\dfrac{\partial^2}{\partial t^2}\phi - \dfrac{\partial^2 \phi}{\partial x^2} = \rho.$

$\left(ab = m^2 \atop \text{for mass} \right)$

interaction

$\psi_L(x,t) = \psi_L(x-\epsilon, t-\epsilon) + \psi_R(x-\epsilon, t-\epsilon)\cdot i\epsilon$

$\psi_R(x,t) = \psi_R(x+\epsilon, t-\epsilon) + \psi_L(x+\epsilon, t-\epsilon)\cdot i\epsilon.$

Most important paths
have equal left turns
& right turns.
n turns occur about
every unit (compton mass
length) length
$\therefore$ oscillations at
frequency about unit
$\therefore e^{is}$ $s = \frac{mc}{\hbar} \cdot \frac{Path}{time}$

20. Paths for spin ½ particle in 1 space-time. (Courtesy Millikan Library Archives, California Institute of Technology)

Theory of Positrons

(1)

We shall consider that when an electron travels along as proper time increases so does true time. For a positron proper time increases as true time decreases. This is classically. Quantum mechanically the situation is that the wave function has a phase (in $e^{-i\varphi}$ define φ as phase) which increases as you move in positive true time for an electron and increases in negative true time for a positron. First then suppose a potential at $\boxed{X}$ (at very small element of Vol) for space time. to scatter $\otimes$ an electron (at A) exists and is present as time goes on, the phase increasing as indicated by the arrows. Now Dirac says $\boxed{X}$ sends $\longleftrightarrow$ A into states of positive and negative energy both of which spread upward in time. ($\mathcal{I}e \pm$ phase change with t). We say instead that $\boxed{X}$ scatters a wave B toward future representing scattered electron, and a wave C toward past representing (in quantitative sense made more clear later) a positron with which the electron may have annihilated by action of the potential at $\boxed{X}$.

arrows on waves indicate direction of increasing phase
∴ A, B electron states
C "positron"

THIS THEORY

DIRAC THEORY.

Now suppose the electron wave function arriving at x by $\psi(x)$. (Thus $\psi(x)$ contains only plus energy components) (If a positron were to arrive at x, it would come from a wave from the future of x or would give a $\psi(x)$ with only — energy components — This general case considered later). The source of the scattered wave is then $\frac{i}{e} \gamma_\mu A_\mu(x)\psi(x)$ or as we will write $\frac{i}{e} A(x)\psi(x)$ (the / meaning an operator).

The amplitude (scattered) arriving at 2 (electron) is say $\int \phi(2,x) \frac{i}{e} A(x)\psi(x) d\tau_x$ (1st order in A)

" " " 3 (positron) " $-\int \phi(3,x) \frac{i}{e} A(x)\psi(x) d\tau_x$ (...)

Here $\phi(A,B)$ is a function (operator) of A, B which we have therefore defined when $t_A > t_B$ and $t_A < t_B$. It is a function only of R_{AB} and t_{AB}. say $\phi(R_{AB}, t_{AB})$ We show later it is complex, and so time symmetrical $\phi(R_{AB}, t_{AB}) = \phi(R_{AB}, -t_{AB})$.

(It is $\phi(R_{AB}, t_{AB}) = (\gamma_\mu \frac{\partial}{\partial x_\mu} + \mu) I (R_{AB}, t_{AB})$ where I is a number (no operator) time symmetrical outgoing wave solution of $(\square^2 - \mu^2)I = 0$ except at B where there is unit source $\delta(R_A - R_B)\delta(t_A - t_B)$ and therefore involves .... function of $R_{AB}^2 - t_{AB}^2$).

21. From Feynman's "Theory of Positrons" notes, fall 1947. (Courtesy Millikan Library Archives, California Institute of Technology)

22. Feynman in Kyoto, 1955. Left to right: Mrs. Yukawa, Hayakawa, Feynman, Yukawa, Mano, and Kobayasi. (Courtesy AIP Niels Bohr Library, *Physics Today* Collection)

23. Richard Feynman, 1959. (Courtesy AIP Niels Bohr Library, *Physics Today* Collection)

24. Freeman Dyson with his parents, 1928.
(Courtesy Alice Dyson)

25. Alice and Freeman Dyson with their father, George, 1929.
(Courtesy Alice Dyson)

26. Freeman Dyson at Winchester, 1937. (Courtesy Alice Dyson)

THE INSTITUTE FOR ADVANCED STUDY
OFFICE OF THE DIRECTOR
PRINCETON, NEW JERSEY

Nolo contendere

R. O.

27. Oppenheimer's capitulation. (Courtesy Freeman Dyson)

28. Weisskopf and Dyson, 1952, on boat to Copenhagen. (Courtesy AIP Niels Bohr Libary)

Fermi gave the problem "of the electromagnetic energy level shift in the relativistic case" to Marvin L. Goldberger, who was a graduate student at Chicago at the time. Goldberger wrote Bethe in early October 1947 to ask him whether "our work is sufficiently different to warrant both Mr. Scalettar and me to work on the problem. Clearly, if our work is merely repetition of his, we will drop our program." The approach Goldberger was to employ was the Fermi-Teller proposal to use "for the intermediate state [in the hole theoretic generalization of the $\Sigma |p_{mn}|^2 (E_n - E_m)$ term in the Bethe formula] the first order Coulomb perturbation of the plane waves . . . [since] with this device the problem appears to be not too difficult."[25]

Bethe promptly answered him and informed him of the following:

> We are using a very similar method to yours which effectively amounts to a Born approximation on the intermediate state. However, the calculation is by no means simple even with this method. . . . In some calculations which I did in August, I was able to . . . demonstrate the convergence of the result. Moreover, I found that the result is similar to the nonrelativistic case. Scalettar is now checking my arguments and especially calculating explicitly the result in order to obtain the numerical value. There are approximately twenty different terms which have to be integrated. . . .
>
> Because of the considerable complication of the calculation I should find it desirable that the calculation be done at several places independently. You may be able to find a simple method. The main reason against further duplication is that in addition Scalettar, also Weisskopf and Lamb are engaged in similar calculations.[26]

Evidently Goldberger dropped the problem. In addition to Weisskopf and French, Kroll and Lamb, and Bethe and Scalettar, others also began work on the problem. Schwinger, who had gotten married right after Shelter Island and for nearly two months thereafter traveled throughout the United States on his honeymoon, started on such a calculation in late July; it will be presented in chapter 7. In Switzerland, Jost and Luttinger "calculated the [relativistic] line shift [for a spin 0 particle] and also found that it is finite."[27] And in Japan, Tomonaga and his collaborators, and independently Nambu, started on such a calculation early in 1948 (Tomonaga 1948; Nambu 1949; Hayakawa 1988). All these workers proved that a hole-theoretical calculation of the $2^2 S_{1/2} - 2^2 P_{1/2}$ displacement gave a convergent answer. However they all also concluded that the formal relativistic invariance of the Dirac, Heisenberg, Pauli, and Weisskopf formulation of quantum electrodynamics "is to some degree illusory in that all self-energies diverge logarithmically, so that the difference of two energies such as $W(2^2 S_{1/2})$ and $W(2^2 P_{1/2})$, although finite, is not necessarily unique" (Kroll and Lamb 1949, p. 388).

Two new problems arise in a relativistic theory. The first is to define precisely the "free electron self-energy" to be subtracted. The second is that unless care is taken, the level shift seems to come out larger than the nonrelativistic (and observed) effect by a factor of the order $1/\alpha^2$.

The most natural "free electron self-energy" expression to subtract would be $\delta mc^2 \beta_{ave}$, where δmc^2 is the self-energy of an electron at rest, and β_{ave} is the expectation value of the Dirac operator β for the wave function of the bound state whose energy is to be calculated:

$$\beta_{ave} = \int \psi_0^* \, \beta \psi_0 \, d\tau. \tag{5.7.1}$$

The justification for this procedure is that in any covariant theory, the self-energy of a free electron must be equivalent to a change of its rest mass, and therefore to an extra term in the Dirac Hamiltonian of the form

$$\delta H = \beta \, \delta mc^2. \tag{5.7.2}$$

Or conversely, in the description of an electron interacting with the quantized electromagnetic field, one should write for the mass of the electron entering in the Hamiltonian

$$m_0 = m - \delta m, \tag{5.7.3}$$

where m_o is the bare mechanical mass, and m is the observed, experimental mass. The Hamiltonian is therefore

$$H = c\alpha \cdot \mathbf{p} + \beta mc^2 + H_0^{em} + H_{int} - \delta mc^2 \beta, \tag{5.7.4}$$

with H_0 the Hamiltonian for the free radiation field and H_{int} the interaction Hamiltonian of the electron with the radiation field.

Now, for a free electron of energy E, the expectation value of the Dirac operator β is given by

$$u^* \beta u / u^* u = m/E. \tag{5.7.5}$$

Hence for the procedure to be satisfactory, the self-energy of the free electron must be of the form $\delta mc^2 m/E$.

Weisskopf in his 1934 paper on the electron's self-energy calculated the self-energy for an electron at rest to be equal to

$$W = \frac{3}{2\pi} \frac{e^2}{hc} mc^2 \lim_{K \to \infty} \ln \frac{2K}{mc^2} + \text{finite terms} \tag{5.7.6}$$

and indicated that "the self-energy of free electron in motion can be obtained by a Lorentz transformation from [this expression]. The direct calculation from the above methods is ambiguous because it leads to a difference of terms, each of which diverges quadratically. The factor of the logarithmically divergent difference of these terms depends essentially on the way in which the infinite terms are subtracted." The expression for the self-energy of a free electron of energy E which Weisskopf calculated in 1934 was in fact not relativistically covariant.

Upon starting his relativistic calculation of the level shift in hydrogen in the summer of 1947, Lamb recalculated the self-energy of a moving free electron and reported his results at a cosmic ray conference held at Brookhaven National Laboratory in September 1947.[28] He found that the divergent part of the self-energy of a free electron of energy

$$E = c \sqrt{\mathbf{p}^2 + m^2 c^2} \tag{5.7.7}$$

is given by

$$\frac{3e^2}{2\pi hc} mc^2 \frac{mc^2}{E} \ln \frac{K}{mc^2} + \text{finite terms},$$

where K is the upper limit at which the integral over virtual photons (including Coulomb) is cut off. However, the finite terms did not have the covariant $\frac{mc^2}{E}$ dependence. The finite part of a divergent integral depends on the exact way the integral is cut off, that is, the manner in which the upper limit is chosen. It is possible to choose the upper limit to be a suitable function of p, the electron momentum, so that the finite terms are also proportional $\frac{mc^2}{E_p}$, but such a choice would be very inconvenient for the calculation of the self-energy of a bound electron. In such calculations one "is virtually forced to take the integrals over a sphere in k-space of radius K" (Weisskopf 1939). It would be an accident if such a symmetric integration led to a covariant result for the finite terms in the self-energy. Bethe asked Lennox, then a graduate student at Cornell, to do this calculation for the free electron.[29] The result he obtained was

$$W = \frac{e^2}{2\pi hc} \frac{mc^2}{E} mc^2 \left[3 \ln \frac{2K}{2mc^2} - \frac{1}{2} + \frac{4}{9} \left(\frac{p}{mc} \right)^2 \right], \tag{5.7.8}$$

the last term indicating, as expected, that the finite terms do not have the proper covariance if the integration is performed symmetrically in k space. A straightforward application of the above subtraction procedures would then lead to the (incorrect) result that there are radiative corrections to the motion of a free electron of order $\alpha (\frac{p}{mc})^2 mc^2 \sim \alpha E_{\text{kinetic}}$. For the bound state problem it would give a spurious term of order $\alpha < (\frac{p^2}{2m}) >$, that is, of order α Ry, which is $1/\alpha^2$ larger than the term Bethe calculated for the Lamb shift (Bethe 1948).

The hole-theoretic calculations of the Lamb shift were intricate and involved, and the lack of covariance made them suspect. "Most of our attempts at relativistic cut-off have run into trouble. How is the calculation of the line shift?"[30] Bethe inquired of Weisskopf in early November 1947. "Our line shift calculations have also run into some slight trouble," Weisskopf answered, "but they seem to be overcome and they change the upper limit a little compared to the value which I gave you last time. I will not publish any value at this time, because of the high fluctuations."[31]

A big breakthrough occurred with Schwinger's calculations that connected the line shift calculation with the radiative correction to the electron's magnetic moment. Rumors of Schwinger's results began circulating in early November. [32] By mid-November Weisskopf wrote Bethe: "Yes, Schwinger has a theory of corrections to the g-factor of the electron on the basis of the same idea as the line shift. It seems to work out all right and gives rise to a small positive addition to the g-value."[33]

By early December, Bethe talked about "Schwinger's new theory of Quantum Electrodynamics" at a joint theoretical seminar of Rochester and Cornell.[34] Bethe had learned of Schwinger's work from Weisskopf, who had given a seminar on it at Cornell on December 1.[35] Weisskopf in turn had been exposed to Schwinger's research at the weekly Harvard-MIT theoretical seminar at which "Julian Schwinger dominated the proceedings with his elegant and well organized lectures on his rapidly evolving 'manifestly covariant' version of quantum electrodynamics."[36] But Schwinger's insight into the connection between the anomalous moment and the Lamb shift created new problems. At the end of December 1947 Weisskopf wrote Oppenheimer: "Julian [Schwinger] and I are rather disturbed over the lack of relativistic invariance in our results. The additional magnetic moment of the electron is different when you calculate it the way Schwinger did, namely in a magnetic field, and the way we did it, namely as a split of P-states in the hydrogen atom. We did not make a mistake, since Julian has calculated both things himself, and I rely on his arithmetic more than on my own."[37] And he added in a Bohrian vein: "I am somewhat doubtful that this problem can be solved, and I think it is a limit in principle of the theory. Julian thinks that he may find a way out, but he hasn't yet."[38] Schwinger did find a way out, by formulating a fully covariant version of the theory, which will be presented in section 5 of chapter 7.

By the January 1948 meeting of the American Physical Society, much progress had been made. Feynman had developed his invariant cutoff methods to the point that the answers obtained for the anomalous magnetic moment from a calculation of radiative corrections in a Coulomb field (Lamb shift) and in an external magnetic field agreed with each other. Schwinger and Weisskopf had fully elucidated the hole-theoretic relativistic calculation of the Lamb shift and were in the process of obtaining a numerical value for it. Schwinger's method consisted "in modifying the Hamiltonian to unite the electromagnetic mass with the mechanical mass of the electron," whereas in Weisskopf's method "the elimination of the infinite self-energy" was obtained "by a suitable subtraction of the free electron mass" (Schwinger and Weisskopf 1948). This latter method was used by French and Weisskopf in their calculation of the Lamb shift, to which we now turn.

5.8 The French and Weisskopf Calculation

Spurred by Lamb's experimental finding and by Bethe's paper, French and Weisskopf intensified their efforts to calculate the $2s$-$2p$ level shift hole theoretically, a problem they had begun working on in the winter of 1946. The calculation was completed in the spring of 1948 and is contained in the thesis French submitted in May of that year to MIT. The problem addressed was described as follows: "We consider an electron in a stationary state of an externally applied time-independent electrostatic or magnetic field. The state of the vacuum in this system will be that where all the negative energy states are filled. Thus the physical situation in which we shall be interested will be that where all the negative energy states and one positive energy state are filled" (French 1948, p. 8).

The Hamiltonian for the system was taken to be

$$H = H_{electron} + H_{photon} + H_{int} + H_{coul} \tag{5.8.1}$$

$$H_{(elect)} = c\boldsymbol{\alpha} \cdot (\mathbf{p} - \frac{e_0}{c}\mathbf{A}_0) + \beta m_0 c^2 + e_0\phi_0 = H_0, \tag{5.8.2}$$

where $\mathbf{A}_0$, ϕ_0 correspond to the static external field in which the electron finds itself:

$$H_{(photons)} = c \sum_{\substack{\lambda=1,2 \\ \kappa}} \kappa \, b_\lambda^*(\boldsymbol{\kappa}) b_\lambda(\boldsymbol{\kappa}) \tag{5.8.3}$$

$$H_{(interaction)} = -e \sum_{\kappa,\lambda} B_\kappa \, \boldsymbol{\epsilon}_\lambda \cdot \boldsymbol{\alpha} \left\{ e^{i\boldsymbol{\kappa}\cdot\mathbf{r}/\hbar} \, b_\lambda(\boldsymbol{\kappa}) + e^{-i\boldsymbol{\kappa}\cdot\mathbf{r}/\hbar} \, b_\lambda^*(\boldsymbol{\kappa}) \right\}, \tag{5.8.4}$$

where $b_\lambda(\boldsymbol{\kappa})$, $b_\lambda^*(\boldsymbol{\kappa})$ are the usual annihilation and creation operators for a transverse photon of momentum $\boldsymbol{\kappa}$ and polarization λ, and B_κ

$$B_\kappa = \left[\frac{2\pi\hbar^2 c}{\kappa}\right]^{1/2} \tag{5.8.5}$$

is the normalization factor for the expansion of the radiation field operator $A(\mathbf{r})$,

$$A(\mathbf{r}) = \sum_{\substack{\lambda=1,2 \\ \kappa}} B_\kappa \boldsymbol{\epsilon}_\lambda \left\{ e^{i\boldsymbol{\kappa}\cdot\mathbf{r}/\hbar} \, b_\lambda(\boldsymbol{\kappa}) + e^{i\boldsymbol{\kappa}\cdot\mathbf{r}/\hbar} \, b_\lambda^* \boldsymbol{\kappa} \right\}. \tag{5.8.6}$$

In eq. (5.8.1),

$$H_{coul} = \frac{1}{2} e^2 \int \frac{\rho(\mathbf{r})\rho(\mathbf{r}')}{|\mathbf{r} - \mathbf{r}'|} d^3 r d^3 r' \tag{5.8.7}$$

is the electrostatic Coulomb interaction between the electrons. The significant energy of any state was defined as the difference of the eigenvalue of H in that state and the vacuum energy. Thus French and Weisskopf calculated

$$W = W_{vacuum+1} - W_{vacuum}. \tag{5.8.8}$$

French and Weisskopf (FW) next derived an elegant formula for W valid to order e^2

$$W = W^X + W^N \tag{5.8.9}$$

$$W^X = \frac{\alpha c^2}{4\pi^2} \int \frac{d^3 k}{k^2} \sum_{J\lambda}{}' \frac{A_{0JJ0}^\lambda}{(E_o - E_J - c\kappa\delta_J)} \tag{5.8.10}$$

$$W^N = -\frac{\alpha c}{2\pi^2} \int \frac{d^3 \kappa}{\kappa^2} \sum_{J^-\lambda}{}' A_{00JJ}^\lambda, \tag{5.8.11}$$

where the primed sum $\sum_\lambda'$ means

$$\sum {}' F(\lambda) = F(1) + F(2) + F(3) - F(4) \tag{5.8.12}$$

and where

$$A_{klmn}^\lambda = \left\langle \Psi_k^* \, \alpha_\lambda \, e^{-i\boldsymbol{\kappa}\cdot\mathbf{r}/\hbar} \, \Psi_\ell \right\rangle \left\langle \Psi_m^* \, \alpha_\lambda \, e^{i\boldsymbol{\kappa}\cdot\mathbf{r}/\hbar} \, \Psi_n \right\rangle, \tag{5.8.13}$$

with $\alpha_1, \alpha_2, \alpha_3$ the usual Dirac matrices and $\alpha_4 = 1$. $\sum_{J^-}$ means a sum over negative energy states and $\delta_J = E_J/|E_J| = \pm 1$; k,l,m,n denote one electron states in the external field A_0, ϕ_0. FW call W^X the exchange part, and W^N the nonexchange part of W. In terms of Feynman diagrams, W^X corresponds to the contribution of the "vertex" diagrams (fig. 5.8.1) and W^N that of the vacuum polarization diagram (fig. 5.8.2).

The terms in W for $\lambda = 1, 2$ are the contributions from H_{int}, that is, from the transverse photons; those for $\lambda = 3, 4$ are those from H_{coul}.

Both W^X and W^N are divergent. The difficulty encountered by FW in their calculations before Shelter Island was that not only was W^X itself divergent but "that the difference between the values of W^X evaluated for two different states is

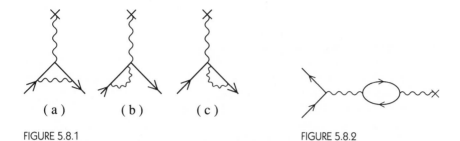

FIGURE 5.8.1   FIGURE 5.8.2

in general divergent." Thus, since the divergence depends in general upon the state, "a new physical idea" is needed to give finite values for the difference in energy of two levels (French 1948, p. 27). The new idea was provided by Kramers, which FW incorporated into their calculation in the following manner: In general, part of the effect of the radiative corrections is to generate in the perturbation energy terms of the form

$$\delta_1 \, e_0 \left\langle [-\boldsymbol{\alpha} \cdot \mathbf{A}_0 + \phi_0] \right\rangle_{av}$$

and

$$\delta_2 \left\langle \beta m_0 c^2 \right\rangle_{av},$$

where δ_1 and δ_2 are of order $\alpha = \frac{e^2}{\hbar c}$, and $< O >_{av}$ means the expectation value of the operator O in the one-electron state under consideration. These radiative corrections can therefore be considered as changing the (unperturbed) Hamiltonian H_0,

$$H_0 = c \, \boldsymbol{\alpha} \cdot \mathbf{p} + \rho \, e_0[-\boldsymbol{\alpha} \cdot \mathbf{A}_0 + \phi_0] + \beta m_0 c^2, \qquad \textbf{(5.8.14)}$$

to

$$H'_0 = c \, \boldsymbol{\alpha} \cdot \mathbf{p} + (1 + \delta_1)e_0[-\boldsymbol{\alpha} \cdot \mathbf{A}_0 + \phi_0] + (1 + \delta_2)m_0 c^2 \beta. \qquad \textbf{(5.8.15)}$$

The effect of these additional terms is to "renormalize" the parameters e_0 and m_0 appearing in H_0 to the value

$$m = m_0(1 + \delta_1) \qquad \textbf{(5.8.16)}$$

$$e = e_0(1 + \delta_2), \qquad \textbf{(5.8.17)}$$

which are then identified with the "observable" mass and charge. Alternatively, the Hamiltonian (5.8.14) can be reexpressed in terms of the observed mass $m = m_0 + \delta m$,

$$H + c\boldsymbol{\alpha}\cdot\mathbf{p} + \beta mc^2 + H_{em} + H_{int} - \delta mc^2\beta, \tag{5.8.18}$$

and δm is determined so that a free electron has the correct observable mass m.

The charge renormalization term that comes from W^N had previously been discussed by Weisskopf in 1936 and can readily be evaluated (Uehling 1935).[39] After an infinite charge renormalization, the contribution of W^N, to first order in the external field, was computed by FW to be

$$\Delta W^N = -\frac{\alpha}{15\pi}\left[\frac{\hbar}{mc}\right]^2 e\left\langle\nabla^2\phi_0 - \boldsymbol{\alpha}\cdot\nabla^2\mathbf{A}_0\right\rangle_{Av}. \tag{5.8.19}$$

The mass renormalization terms were, however, more troublesome. For a free electron, that is, when $A_0 = \phi_0 = 0$, the perturbation ought to give exclusively a masslike term W_0^X, that is, a term of the form $\delta_2\beta m_0c^2$. As noted earlier, the divergent part of the self-energy,

$$\frac{3\alpha}{2\pi}\left[\beta m_0c^2\right]_{av}\int_0^\infty\frac{d\kappa}{\kappa},$$

indeed has this form. The finite terms, however, do not have this form (Weisskopf 1939)—a consequence of the divergent nature of the theory and the lack of manifest relativistic invariance in making the computations.

Two ways are open to rectify the situation: (1) devise an invariant calculational scheme that guarantees that the self-mass δm of a free electron is invariant, or (2) devise a (possibly noncovariant) method which guarantees to give zero for the self-energy of a free electron in motion calculated with the Hamiltonian (eq. 5.8.18).

French and Weisskopf chose the second method and specified that one is to subtract from the self-energy of the bound electron simply the self-energy of a free electron wave packet identical to the bound-state wave function.[40] Thus, if one expands the bound-state wave function in plane waves

$$\psi_0 = \sum_{\mathbf{p},s} a(\mathbf{p},s)\,u_s(\mathbf{p})\,e^{i\mathbf{p}\cdot\mathbf{r}/\hbar}, \tag{5.8.20}$$

where s distinguishes the states of different spin and sign of the energy ($s = 1,2,3,4$), the free electron self-energy subtracted by French and Weisskopf is given by

$$W_{free} = \sum_{\mathbf{p}}\sum_{ss'} a^*(\mathbf{p},s)a(\mathbf{p},s')(\mathbf{p}s|W|\mathbf{p}s'), \tag{5.8.21}$$

where the last factor represents the matrix element of the free electron self-energy. Besides the usual diagonal elements $s = s'$, there are also "off-diagonal elements corresponding to transitions from positive to negative energy states since the expansion (5.8.20) of the Coulomb wave functions has negative energy plane wave components. The method was further refined, and FW proceeded on the assumption that a charge symmetrized[41] W_0^X evaluated for the "bound" electron state is the mass term to be subtracted from W^X (see sect. 3 of FW 1949).

In a Lamb shift calculation, the states for which the radiative corrections are calculated are nonrelativistic, that is, states whose energy, exclusive of the rest mass energy, is small compared to mc^2, and whose momenta $\bar{p}$ are small compared to $\mu = mc$. In evaluating eq. (5.8.9), Ψ_0 is treated exactly (i.e., as a state of the electron in the field A_0, ϕ_0), but the intermediate states are expanded in powers of A_0, ϕ_0, regarding A_0, ϕ_0 for these states as a perturbation on the free-particle Hamiltonian, $c\boldsymbol{\alpha} \cdot \mathbf{p} + mc^2$. This procedure is not accurate for low-lying intermediate states but is satisfactory for higher states in the continuum. To circumvent this difficulty, FW split the range of κ values at some convenient intermediate point $\delta\mu$. For $\kappa < \delta\mu$, the dipole approximation can be made in computing W^X, for $\lambda = 1, 2$, and this term then yields the Bethe contribution. For $\kappa > \delta\mu$ the intermediate states can be approximated by plane wave states, and the contribution evaluated. After lengthy calculations, done in two different ways, French in his dissertation obtained for the case of a Coulomb field, $e\phi = V$, $\mathbf{A} = 0$, the following result for the level shift:

$$\Delta W = \frac{\alpha}{3\pi} \left(\frac{\hbar}{mc}\right)^2 \left[\langle \nabla^2 V \rangle_{Av} \left\{ \int_{\kappa_0}^{\mu} \frac{d\kappa}{\kappa} - \ln 2 + \frac{5}{6} - \frac{1}{5} \right\} \right.$$
$$\left. + \frac{1}{\hbar} [\nabla V \cdot \boldsymbol{\sigma} \times \mathbf{p}]_{Av} \left\{ -\frac{3}{4} \right\} \right].$$

(5.8.22)

Here $c\kappa_0$ is the Bethe lower limit. The second term has the characteristic form of a spin-orbit coupling and corresponds to the Schwinger result that the electron has an anomalous magnetic moment. If the electron is assumed to possess an additional spin magnetic moment $\delta \frac{e\hbar}{2mc}$, then the term

$$\delta \frac{e\hbar}{2mc} \boldsymbol{\sigma} \cdot (\mathcal{H} + \frac{1}{c}\mathcal{E} \times \mathbf{v}) = \delta \left\{ \frac{e}{mc} \mathbf{s} \cdot \mathcal{H} - \frac{e\hbar}{2m^2c^2} \mathcal{E} \cdot \mathbf{r} \times \mathbf{p} \right\}$$

(5.8.23)

must be added to the Hamiltonian. When there is no external magnetic field present, $\mathcal{H} = 0$. Hence the second term in eq. (5.8.22) indicates that the electron behaves as if it had an additional contribution to $g/2$ of magnitude

$$\delta = \frac{\alpha}{2\pi}.$$

(5.8.24)

By the spring of 1948 four Lamb shift calculations had been completed: by French and Weisskopf, by Kroll and Lamb, by Schwinger, and by Feynman. The first sets of authors had used a hole-theoretic approach. Schwinger and Feynman, on the other hand, had used four-dimensional covariant methods, and had found that the radiative corrections to the motion of an electron in an external electromagnetic field $(\phi, \mathbf{A})$ due to quanta of energy greater than k_1, where

$$\text{Ry} \ll k_1 \ll mc^2,$$

added to the one-particle Hamiltonian a term

$$
H_1 = \frac{e^2}{3\pi\hbar c}\left(\frac{\hbar}{mc}\right)^2 \Box^2 e(\phi - \boldsymbol{\alpha} \cdot \mathbf{A})\left(\ln\frac{mc^2}{k_1} + a - \ln 2\right)
$$
$$
- \frac{e^2}{2\pi\hbar c}\frac{e\hbar}{2mc}\beta\,(\boldsymbol{\sigma} \cdot \mathcal{H} - i\boldsymbol{\alpha} \cdot \mathcal{E}),
$$
(5.8.25)

where $\Box^2$ is the d'Alembertian

$$
\Box^2 = \nabla^2 - \frac{1}{c^2}\frac{\partial^2}{\partial t^2},
$$
(5.8.26)

$\mathcal{H}$ is the external magnetic field, $\mathcal{H} = \nabla \times \mathbf{A}$, and $\mathcal{E}$ is the external electric field. The constant a was 5/8 according to Feynman, 3/8 according to Schwinger.

If eq. (5.8.25) is combined with the effect of the quanta of energy less than k_1, the level shift for the case of a Coulomb field $(\mathbf{A} = 0, -e\phi = Ze^2/r)$ is given by

$$
W = \frac{8}{3\pi}\left(\frac{e^2}{\hbar c}\right)^3 \text{Ry}\frac{Z^4}{n^3}\left(\ln\frac{mc^2}{16.721\text{Ry}\,Z^2} + a - \ln 2\right)
$$

$$\text{for}\,S\text{-states} \qquad (5.8.27)$$

$$
W = \frac{8}{3\pi}\left(\frac{e^2}{\hbar c}\right)\text{Ry}\frac{Z^4}{n^3}\left(\frac{3}{8\ell(\ell + 1)(2\ell + 1)}\ell \cdot \boldsymbol{\sigma} + N\right)
$$

$$\text{for}\,\ell \neq 0 \qquad (5.8.28)$$

(N is the small contribution from eq. 5.6.9 for $\ell = 0$). The value of a found by these various authors is given in table 5.8.1 taken from Bethe's 1948 Solvay report.[42] The contribution $-\frac{1}{5}$ to a in the table is that due to the Uehling term.

Table 5.8.1
Results for 2s state of hydrogen, in megacycles.

Author	Value of a	Shift of 2s	$2s = 2p$
Schwinger	$3/8 = 0.375$	1003.13	1016.11
Feynman with polarization	$5/8 - 1/5 = 0.425$	1009.91	1022.89
Feynman without polarization	$5/8 = 0.625$	1037.03	1050.01
Lamb	$3/4 - 1/5 = 0.550$	1026.86	1039.84
Weisskopf and French	$5/6 - 1/5 = 0.633$	1038.15	1051.13
Lamb without polarization	$3/4 = 0.750$	1053.98	1066.96

French and Weisskopf were the first to arrive at the correct result. There then followed what Weisskopf has called a "tragicomical" episode (Weisskopf 1983, p. 75). FW showed their result to Schwinger and Feynman, who also had calculated the Lamb shift but, as indicated in table 5.8.1, had found a result differing from that of FW by a small additive numerical constant. Although in April, Feynman's and Schwinger's answer also differed from each other, by the end of the summer they had obtained the same answer. Weisskopf lost faith in the accuracy of the FW result. "The trouble was that both of them got the same result. Having both Feynman and Schwinger against us shook our confidence, and we tried to find a mistake in our calculation, without success" (Weisskopf 1983, p. 75).

In December 1948, Weisskopf wrote Oppenheimer:

> I have not too much to report about electrodynamics. The present state is as follows: Schwinger has essentially conformed his calculations to the ones of Feynman and given up any fancier requirements than the ones used in Feynman's calculation and therefore gets exactly the same result as Feynman, which as you know differs from our own. We could locate the trouble to that extent that the difference between the two methods is an integral which, with all reasonable methods of evaluating (like the Feynman method of introducing heavy light quanta) is identically zero. We do not know the rather subtle reason why we find some difference at the end. It seems to me, however, that the Schwinger-Feynman result will be the right one, since it retains relativistically invariant forms during a larger part of the calculation.[43]

Feynman, who had likewise been puzzled by the discrepancy, wrote a letter to Weisskopf shortly before Christmas 1948 indicating that he felt that the FW results were incorrect:

> Maybe this summer... I will try to figure out why we get a different answer for self-energy. The places where there may be trouble are:
>
> (1) to use the correction for the radiationless scattering in the self-energy problem may not be precise, (2) as you suggest, there may be some error in using a small mass for the quanta to avoid the lower limit in the integrals. I think it could all be straightened out very easily if you would calculate the radiationless scattering problem by your methods. If you still get 5/6, then I think you are wrong; but if you get 5/8 for this problem and 5/6 for the self-energy problem, I resign. In addition, if the latter turns out to be the case you will be able to see what the difference is in the two problems which leads to the different numerical value. Alternatively I could calculate this self-energy problem with much greater precision and verify that my methods of calculating the radiationless scattering first and simply transferring the results is O.K.[44]

The puzzle was finally resolved by French: "The source of discrepancy is the way in which the joining to the Bethe non-relativistic result is done." Feynman in his calculation had given the photons a small mass λ and integrated down to $k = 0$ and had obtained a level shift proportional to $(\ln m/\lambda - 3/8) (0|\nabla^2 V|0)$. However, care must then be exercised in calculating and joining the contributions of the longitudinal photons—for which retardation effects cannot be neglected—with the nonrelativistic result of Bethe. Feynman had not taken into account this nonrelativistic longitudinal contribution, which does not vanish when the joining is made with a nonvanishing photon mass λ.[45]

In a footnote to his positron theory paper, Feynman (1949b) noted that FW repeatedly pointed out to him that his published result for the Lamb shift (Feynman 1948b) "was in error." Feynman went on to say that: "The author feels unhappily responsible for the very considerable delay in the publication of French's result occasioned by the error." The delay in FW publishing their result resulted in Kroll and Lamb's calculation appearing in print a few months earlier than FW's[46] (Kroll and Lamb 1949).

Interestingly, the same joining mistake as Feynman's had been made by Schwinger. The delay in the publication of French and Weisskopf's result was also due to a disagreement with Schwinger's result. John Blatt who was at MIT at the time recalls asking Julian Schwinger for the reason of the disparity and he "distinctly remembers Schwinger's shrug of the shoulders, along with words somewhat like: "Well, if you do not keep the calculation explicitly covariant, anything can happen."[47] Later on, after Schwinger's mistake had been discovered Weisskopf told Blatt, "Er ist zum Kreuze gekrochen," referring to the "crawling to the cross" at Canossa in the investiture contest with the pope.[48] Schwinger also acknowledged his responsibility for the delay in the publication of FW's calculation—but much less forthrightly than Feynman had—in a letter to the editor reporting on his results on the radiative corrections for the Coulomb case (Schwinger and Feldman 1949).[49]

As Weisskopf was to remark, "Self-confidence is an important ingredient that makes for a successful physicist" (Weisskopf 1983, p. 75).

This failure of nerve robbed Weisskopf of the credit he so richly deserved in being the first, with French, to calculate relativistically the value of the Lamb shift. Weisskopf had taken seriously the discrepancy between the Dirac theory and the experimental data of Houston and R. C. Williams on the spectrum of hydrogen, and he was the first to recognize that a hole-theoretic calculation of the $2s - 2p$ level shift could be carried out. In fact, he had put French to work on the problem in October 1946, well before he knew of Lamb's experiment.

5.9 Radiative Correction to Scattering

Bethe's calculations had indicated that in hole theory, after mass renormalization, a finite value would be obtained for the radiative corrections (to order e^2) to the energy levels of an electron in a bound state in a Coulomb field.

The question immediately arose whether the radiative corrections to the cross section for the scattering of electron in a Coulomb field would also be finite. Since the cross section is an observable quantity, a satisfactory theory must give a finite value for it. That a finite result would be obtained was almost guaranteed, since a close relation exists between shifts in energy levels and radiative corrections to scattering. If $(\mathbf{p}'|V|\mathbf{p})$ is the matrix element of the unperturbed potential for scattering from $\mathbf{p}$ to $\mathbf{p}'$ and $(\mathbf{p}'|\delta V|\mathbf{p})$ the radiative corrections to it, then the level shift for the state $\psi(\mathbf{p})$ is given by

$$\Delta E = \int d^3 p' \int d^3 p \psi^*(\mathbf{p}')(\mathbf{p}'|\delta V|\mathbf{p})\psi(\mathbf{p}) \qquad (5.9.1)$$

according to perturbation theory.

Since a finite result had been obtained for ΔE it was likely that $(\mathbf{p}'|\delta V|\mathbf{p})$ would also be finite. On the other hand, Dancoff had calculated $(\mathbf{p}'|\delta V|\mathbf{p})$ in 1938/39 and had obtained a divergent expression for it (Dancoff 1939).

Oppenheimer gave the problem of calculating the radiative corrections to the radiationless scattering cross section of an electron by a Coulomb field to Hal Lewis at the end of the summer of 1947. Lewis, together with Robert Finkelstein,

Saul Epstein, Leslie Foldy, and Sig Wouthuysen, had worked with Oppenheimer in Berkeley and had come to Princeton with him when he became the director of the Institute for Advanced Study.[50] The five of them shared an office that adjoined Oppenheimer's, and so "were pretty much in each other's pockets."[51] Epstein tells of Wouthuysen deciding after a while to work at home: "He got tired of coming in every day and being told by someone how to overcome his difficulties of the day before."[52]

When they came to Princeton, Oppenheimer lectured to them on the presentations at the Shelter Island conference, and on Bethe's calculations (nonrelativistic and relativistic) of the Lamb shift. Interestingly, all of them (including Oppenheimer) had heard Kramers' lecture on mass renormalization in classical electrodynamics during his visit to Berkeley in the spring of 1947,[53] but evidently without any impact on them. Lewis established that $(\mathbf{p}'|\delta V|\mathbf{p})$ was indeed finite. He discovered that Dancoff had neglected to take into account two terms required by second-order perturbation theory. He showed that the off-diagonal matrix elements of the mass correction term $\beta \delta mc^2$ in eq. (5.8.17), corresponding to transitions between two states of the same momentum but opposite sign of the energy, exactly compensate the divergent terms in Dancoff's calculation: "Dancoff's divergent terms arise entirely from the fact that the mass term is not diagonal in the Dirac Hamiltonian, so that a small (but divergent) increment in mass mixes positive and negative energy states, and makes new transition schemes possible" (Lewis 1948, p. 175).

Lewis was aided in his calculations by Epstein's insight "that it would be easy to identify the mass renormalization terms in the 2d order radiative corrections since it should just be the 2d order terms one gets by replacing m by $m + \delta m$ in the lowest order result."[54] Epstein had related his observation to Hal Lewis "with the result that shortly thereafter there was Oppenheimer looming over me and saying something like 'So you're trying to horn in, are you?'"[55] Epstein doesn't remember his reaction to this, "but whatever it was it was soon turned to pleasure" when he heard Oppenheimer on the phone telling Bethe that the problems were all solved and explaining to him Epstein's observation. Oppenheimer was very helpful thereafter and suggested to Epstein that he prepare a separate note on the subject (Epstein 1948).

Lewis's work was important. It showed the effectiveness and consistency of mass renormalization and indicated that hole-theoretic QED was finite to order e^2 after mass and charge renormalization. The same result was obtained by Schwinger (1948b) and by Koba and Tomonaga (1948) in Japan. They also discovered the error in Dancoff's calculation, and proved that the amplitude for the scattering of an electron in a Coulomb field does not contain any ultraviolet divergencies (Tomonaga 1948; Hayakawa 1988b).

Schwinger (1948b) in his letter to the editor of the *Physical Review* announcing the $\alpha/2\pi$ correction to the electron's magnetic moment indicated that his calculation of "the finite radiative correction to the elastic scattering of electrons by a Coulomb fields provides a satisfactory termination to a subject that has been beset with much confusion."

It should be recalled that it had been Oppenheimer who had suggested to Sidney Dancoff in 1938 that he look at the problem of radiative corrections to scattering after Bloch and Nordsieck and Pauli and Fierz had shown how to overcome the infrared divergencies in quantum electrodynamics. After Lewis had finished his calculation, Oppenheimer had occasion to write to Arnold Nordsieck, who was a colleague of Sidney Dancoff in the Department of Physics at the University of Illinois. Oppenheimer inquired of him: "Does Sid know that Lewis, with a great deal of exhortation from one, finally cleared up his radiative corrections problem and showed how to get the finite answer? That should be out in one the January *Phys. Rev.* and really reveals a comedy of errors in the old work."[56] He continued: "Our present agonies are over covariance. In spite of favorable earlier indications, the self-energy in not ... covariant in the present theory, and the remaining terms ... do not look and cannot be proven covariant, and when tested seem not always to give covariant results."[57]

Tomonaga, Schwinger, and Feynman devised new ways to obtain covariant results in perturbation theory. The next three chapters explore their contribution.

6. Tomonaga and the Rebuilding of Japanese Physics

6.1 Introduction

In the previous chapter an account was given of the developments in quantum electrodynamics in the aftermath of the Shelter Island conference. The focus was on contributions made in the United States: the experiments carried out by Lamb, Rabi, Kusch, and their coworkers, and the theoretical advances made by Bethe, Weisskopf, Schwinger, Feynman, and their students. That the advances in QED were made principally in the United States is not surprising. The United States was not ravaged by the war the way Great Britain, the USSR, Holland, France, Japan, and the other combatants had been. But other factors were also important. The crucial experiments by Lamb and Kusch depended on an instrumental tradition that had been developed at Columbia before the war—namely, Rabi's molecular beams—and on apparatus and techniques to generate and detect microwave radiation that had been perfected in the United States during the war. In addition, graduate education in physics in the United States provided the level of theoretical expertise that was required of experimenters in order to design and carry out these difficult experiments. Given the precise and reliable data generated by the Lamb and Rabi experiments—data that undermined the predictions of the Dirac equation—and given the prevailing philosophical outlook that theories must explain numbers, the leading theorists there had to meet the challenge posed by these numbers.

Of course, this view of theoretical physics was not unique to the United States. In the present chapter I explore the response by theorists in Japan to the data generated by the Lamb shift and the hydrogen hyperfine structure experiments. There, Tomonaga and his school did achieve important progress. However, these advances were also made independently by Schwinger. They were not known to Feynman, they did not influence Schwinger, and Dyson has remarked that their effect on him was psychological rather than technical. Thus the developments in the United States and in Europe proceeded largely independently of the Japanese work. The impact of contributions from Japanese theorists only began to be felt in the West during the fifties.

The story of the postwar Japanese contributions to quantum field theory is perhaps the most interesting part of the developments. Nowhere else, I believe, are the style and the approach to doing physics so influenced by cultural and national factors and by philosophical and ideological commitments as in the case of the Japanese theorists. The major events in the history of modern Japan—the Meiji Restoration, World War I, the post-World War I depression—deeply affected the practice of Japanese physics. World War II was another watershed. The hardships the Japanese bore as a result of the protracted conflict would have been difficult to bear under most circumstances. But they were aggravated by the tight blockade the U.S. Navy was able to impose toward the end of the war and by the relentless bombings the U.S. Air Force inflicted on Japanese population centers. The fire raids that incinerated Tokyo were merely the most "successful" of such attacks. Only the five or so cities that had been set aside as possible targets for atomic bombs escaped saturation aerial bombings. These culminated with the leveling of Hiroshima and Nagasaki. Japan's unconditional surrender—the price of total defeat—stopped the carnage. The conditions under which the stunned nation reconstituted itself were dismal. For several years after the armistice was signed, Japan was in utter despair, a mood nurtured by the widespread devastation and the severe food shortages. The recovery from the wounds of this tragic war became a communal enterprise. The Japanese saw themselves as a large family whose members were to help one another in the national reconstruction.

This sense of familial responsibility permeated the group of young scholars who gathered around Tomonaga in Tokyo in 1946. Meeting regularly in the "half burned" quonset hut that constituted Tomonaga's Tokyo lodgings,[1] they saw themselves as rebuilding Japanese theoretical physics. In contrast to the American case, where the advances were made principally by individuals working by themselves, the success of Tomonaga's efforts after the war was due in part to the contributions of the young Japanese physicists working with him. In an important sense it was successful because it was a group effort.

The brevity of my account of Tomonaga's work should not be interpreted as a judgment of the importance of it. Tomonaga's stature as a physicist is immediately apparent from a perusal of his *Scientific Papers* (Tomonaga 1973, 1976). My limited presentation only reflects the fact that I am not able to do justice to his story. Tomonaga's intellectual biography is intimately connected with the history of the Japanese contributions to quantum field theory during the 1930s and 1940s. These demand a detailed exposition that I am unable to give.[2]

6.2 Theoretical Physics in Japan

Japan emerged from the rigidly enforced isolation that had characterized the Tokugawa period as a result of the Meiji Restoration of 1868. The Restoration was a response to the inability of the feudal central government—the *bakufu*—and of the feudal lords that ruled the many fiefdoms that made up Japan—the *daimyō*—to meet the internal and external threats Japan was facing at the time. The last phase of the Tokugawa shogunate was marked by a general deterioration of the bakufu's control over domestic affairs, and it found itself unable to arrest the many agrarian uprisings that had erupted. The arrival of Western powers on Japanese soil and the enormous pressures they exerted for open trade further weakened the bakufu. The Restoration was instigated by the lower samurai and merchant classes and it altered the political structure of Japan. The power of the daimyō was given over to a strong, new, centralized national government run by a group of ministers, and the vertical class system of samurai, peasant, artisan, and merchant was abolished (Fairbank 1974; Faribank et al. 1973; Reischauer and Craig 1977; Reischauer and Fairbank 1973; Beasley 1974; Koizumi 1975). The leaders of the Restoration believed they had a "mandate to create, through bureaucratic means, a powerful, wealthy and autonomous Japanese nation" (Najita 1974, p. 70), and they promulgated a program of rapid industrialization and modernization under the slogan "*Fukoku kyo-hei*" ("wealth and military strength for Japan"). The goals were to make Japan economically independent and capable of defending itself militarily. Although the reforms of the Meiji Restoration were "to a large extent based upon Western ideas, [Japan] did not seek to become a Western nation. Rather, Japan sought to become a nation capable of coping with external encroachments and of competing with the West, a nation of people who could be proud of themselves before Western people" (Koizumi 1975, p. 6).

The reforms were instituted by imperial fiat, since all power still flowed directly from the emperor. A new taxation system was established, light and heavy machinery were imported, patent rights purchased, a national railway system built, and extensive capital investments made in steel mills, shipyards, mines, and munitions factories. Japan's educational system was revamped, new universities established, and Western-style learned societies formed. Foreign teachers and technical experts were brought in, and large numbers of students were sent to Europe to study Western science and technology (Koizumi 1975). The value of science as a source of power for the nation had been recognized early. This initial phase of Japan's modernization culminated with the Sino-Japanese War of 1894–1895 and the Russo-Japanese War of 1904–1905. Its victory over China was the signal that Japan had emerged from feudal obscurity; Japan now spoke "less of survival and more of pulling abreast with and surpassing the West" (Koizumi 1975, p. 25). With its stunning victory over tsarist Russia, Japan became one of the most powerful countries in the world. The Japanese believed that through science "Japan had already pulled itself up by its bootstraps. It had survived and now science was to lead it to the heights of civilization" (Koizumi 1975, p. 29).

By the end of the first decade of the twentieth century, the educational reforms had produced a cadre of highly trained mathematics and science teachers, who were educating elementary and high school students using modern texts written in Japanese. Similarly, Japanese scholars were replacing foreigners at the universities (Koizumi 1975). World War I cut off Japan from its traditional sources of materials and equipment and led it to embark on a course designed to achieve self-sufficiency. However, the earthquake of 1923 devastated many urban centers, destroyed a considerable number of factories, and contributed additional economic problems to a nation already in the midst of a postwar recession. The global economic depression that followed the collapse of the financial markets in 1929 hit Japan very hard, and its ambitious plans had to be scrapped. The consequences were a takeover by militarists and the establishment of a militantly nationalist, totalitarian government that attacked rationalism, suppressed progressive political movements,[3] and demanded total loyalty from the citizenry. The new government decided to extend the boundaries of Japan's colonies on the Asian mainland,[4] thus propelling it toward the tragedy of World War II. Many physicists responded to this ultranationalism by adopting Marxism, whose universalistic features they found attractive.[5] Marxism, in turn, deeply affected the approach taken by some of the leading Japanese physicists[6] to the problems of "elementary particle" physics.

By 1930 Japan could claim to have a very competent physics establishment. Even though Nagaoka, Hantaro, and Ishiwara Jun[7] had not been able to create a Japanese school of theoretical physics, by the late twenties in Japan as elsewhere the importance of theory, and particularly of quantum mechanics, had been recognized; theoretical physics was therefore given a higher status. Greater emphasis was also placed on the study of cosmic rays—partly because of their bearing on meteorology—and on research in nuclear physics.

The history of fundamental physics in Japan is intimately related to the career of Nishina Yoshio, one of Japan's leading physicists, and certainly its most influential scientist during the second quarter of the twentieth century. Nishina graduated from Tokyo Imperial University in 1916 with an engineering degree, and remained there for two more years of graduate studies. He then accepted a position at the recently founded Institute for Physical and Chemical Research (Rikagaku Kenkyusho, usually abbreviated as "Riken").[8] In 1921 Nishina was sent to Europe to study physics. He first visited Cambridge and worked with Rutherford. The following year he spent a semester in Göttingen; while there his interests shifted from experimental to theoretical physics. From Göttingen he went to Bohr's institute. He stayed there for six years and became an active participant in the quantum-mechanical revolution. A visit to Hamburg in the fall of 1927 led to a collaboration with Rabi. Upon his return to Copenhagen in the spring of 1928, he worked with Oskar Klein on the description of the scattering of photons off free electrons using the newly formulated Dirac equation to describe the electron. Together they derived an expression for the scattering cross section that became known as the Klein-Nishina formula (Heitler 1936).

Research in modern quantum mechanics in Japan was initiated by Nishina when he returned to Riken in December 1928. He also accepted a position at the Tokyo College of Literature and Science and started lecturing on recent developments in physics. As a physicist with an international reputation, Nishina was invited to lecture at the leading Japanese universities. In May 1931 he delivered a set of lectures on quantum mechanics at Kyoto University that were attended by Yukawa, Tomonaga, and Sakata. In the spring of 1931 Nishina was appointed chief researcher at Riken, and that summer he established a modern physics laboratory there. Nishina went against the grain of accepted Japanese practice in the running of his laboratory. In a society deeply conscious of tradition, status, and origins, he disregarded the university background of his assistants, caring only about their ability, accomplishments, and intellectual powers. His keen intellect and supportive attitude helped create a stimulating environment at Riken in which the most recent advances in physics were discussed and thoroughly analyzed. His laboratory became the locus of an active intellectual community.[9] He gathered around him the most promising young physicists in Japan—among them Tomonaga, Sakata, Kobayashi, and Umeda—and trained many of the most gifted and imaginative Japanese physicists of that generation.

The first investigations in nuclear physics in Japan were performed in Nishina's laboratory. Work in cosmic ray physics was begun there in 1932 after a cloud chamber and counters had been built. The following year a (somewhat unstable) Van de Graaff electrostatic generator yielding 0.7 MeV protons was assembled, and in the late thirties a 60-inch cyclotron was constructed. As was the case at the leading American universities and in Bohr's institute, building cyclotrons and designing and analyzing nuclear physics and cosmic ray experiments brought the theorists and experimenters at Riken in close contact.[10] Thus, largely as a result of Nishina's influence, there developed at Riken a tradition for theorists to get the numbers out. This was not the accepted norm elsewhere in Japan until the 1950s. In most government universities, research was supported through the Koza-hi budgets. Each professorial "chair," or Koza, had a separate budget attached to it, and usually consisted of a professor, an assistant professor, and one or more assistants, all of whom were tenured. Control over this budget was in the hands of the professor, and each chair formed "a little kingdom in which the professor exercised jurisdiction over his staff" (Hirosige 1963).

6.3 Tomonaga

Tomonaga Sin-itiro was a member of the generation of physicists that grew up under Nishina's influence. He was born in Tokyo on March 31, 1906, the oldest son of a well-known philosophy professor. As a child he was sickly, and was frequently absent from school. Very sensitive, and not very good at gymnastics, he was often bullied by his classmates, who called him a crybaby. In an autobiographical fragment written during the 1960s,[11] Tomonaga recorded some of his recollections of the science experiments he performed as a young boy. When he was in third grade he devised a pin-hole camera using a small opening in one of the wooden storm windows of his house. He recalled adding a convex lens to the camera and being very surprised when he obtained a smaller rather than a larger picture, contrary to his expectations. In his reminiscences he also described his fourth-grade chemistry demonstrations, and recounted filling balloons with the hydrogen gas he had obtained by electrolytically decomposing water, and releasing them and watching them ascend. When Tomonaga was eleven he had a good friend who was also interested in science, and the two of them would build simple electric circuits with parts scavenged from junkyards. Only the batteries would be bought. They designed and put together a telegraph set and an electric doorbell. Once, out of curiosity to see what would happen, they plugged the doorbell directly into the wall outlet: the resulting meltdown remained a well-kept secret between them. Optics held a special fascination for Tomonaga. He remembered being given some footage of a film of the Russo-Japanese War as a teenager. Since he didn't have a projector, he decided to build one. When he discovered that a large condensing lens was required that was quite expensive, he improvised using a specially shaped translucent container filled with water, "and it worked." After he had taught himself some photochemistry, he devised a scheme for making slides from regular photographic films using gels that he had prepared from seaweeds. When he was twelve he was given a small microscope with a 20-power magnification, which he transformed into one with a magnification of 200 by adding a small spherical lens that he made by melting some glass tubing. The challenge was always to construct the apparatus from scratch, usually using bits and pieces found in the local junkyard, and making everything with his own hands. Finished, manufactured apparatus had no great appeal. He enjoyed solving the problems connected with the design and the construction of a piece of equipment, and he derived enormous satisfaction from its successful operation. Making gadgets remained a hobby throughout his life.

Tomonaga's family moved to Kyoto in 1913 when his father was appointed professor of philosophy at the Imperial University there. He attended the Third Higher School, "a renowned senior high school which had educated a number of personalities in prewar Japan."[12] He was an outstanding student, but was greatly bothered by his somewhat poorer performance in English and calligraphy.[13] Yukawa Hideki was a student in the same class as Tomonaga and he became one of Tomonaga's "intimate friends."[14] The newspapers' coverage of Einstein's visit to Japan in 1922 kindled Tomonaga's interest in the theory of relativity, and he attempted to read Ishiwara's book on the subject at the time, but it proved to be too difficult.

Yukawa and Tomonaga both entered Kyoto University in 1923 and majored in physics. The Kyoto chair in theoretical physics was then held by Professor Tamaki, who worked on problems in relativity and hydrodynamics, but who had no interest in the old quantum theory. Although Kyoto was known as a conservative university, it was open to new ideas. In their senior year, with Tamaki's encouragement, Tomonaga and Yukawa embarked on a project to learn quantum mechanics. They read the papers of Heisenberg, Dirac, Jordan, Schrödinger, and Pauli that had laid the foundations of the field, and explained them to each other. Shortly after their graduation from Kyoto University, the two traveled to Tokyo to attend the lectures that Dirac and Heisenberg gave at Riken during their visit to Japan in 1929. "These lectures . . . made a strong impression on me. Heisenberg and P.A.M. Dirac were scholars who greatly influenced me," Tomonaga later asserted.[15] Although the formal parts of these lectures presented no problems, some of the subtler aspects of the theory—such as the interpretation of the probabilities in quantum mechanics—proved refractory. Nishina's lectures on quantum mechanics in Kyoto in 1931 helped clarify some of these.

Tomonaga completed the work for his *Rigakushi* (bachelor's degree) in physics in 1929. With Japan in the throes of an economic depression and no prospect for a job, Tomonaga decided to stay at Kyoto Imperial University for graduate work. He became an unpaid assistant in Tamaki's group and remained in Kyoto for three more years. When Nishina lectured in Kyoto in 1931, Tomonaga came to his attention. Sakata Shoishi, who was then also a student at Kyoto, recalled that Tomonaga and Yukawa asked the most questions after Nishina's lectures. Nishina invited Tomonaga to work at Riken; he accepted and was appointed a (tenured) paid assistant in Nishina's laboratory. Nishina became a father figure and role model for Tomonaga, and Tomonaga flourished under Nishina's tutelage. He quickly assumed the position of house theorist in Nishina's laboratory and acquired the reputation of being a "virtuoso" in making calculations (Darrigol 1988a, p. 7). "I was very intimate with the experimentalists," is the way Tomonaga phrased it many years later (Brown et al. 1980, p. 8). He became a highly valued and respected colleague, and in the process grew more self-confident. In one of his *Reminiscences*, on the occasion of Minoru Kobayasi's sixtieth birthday in 1968, Tomonaga described the atmosphere in Nishina's group:

> The Nishina Laboratory in those days was full of freshness. All the members were young; even our great chief Nishina was still in his early forties. We all got together after lunch every day, an eager group of people discussing various matters, not only physics but also such things as plans for beer parties, excursions and so on. . . .
>
> When Kobayasi came to Tokyo [in 1935] he had to find lodgings somewhere not far from the Institute. Fortunately, there was a vacant room in the lodging house where I lived, so I recommended this room to him. Living in the same house, which was only a five-minute walk to the Institute, Kobayasi and I used to go once more after the evening meal to the laboratory to continue the work left unfinished during the day, or to read articles together and discuss the content from every angle. But I must confess that our course was not infrequently shifted towards a movie theater or to a café or beer house. (Tomonaga 1976, p. 465)

Tomonaga learned "to get the numbers out" at Riken. His first initiation into phenomenology involved calculations to understand better the properties of the recently discovered neutron. He investigated the properties of the deuteron that resulted from various phenomenological models of the neutron-proton interaction, and also computed the neutron-proton scattering cross section for the potentials representing these models. This work was reported at the fall 1932 and spring 1933 meetings of the Riken staff. After the discovery of the positron in 1932, Nishina, who was an expert in Dirac theory, suggested that pair production by γ-rays in the Coulomb field of a nucleus was the process responsible for the production of the positrons in Anderson's cloud chamber. Nishina asked Tomonaga to calculate the cross section for this process in positron theory. The results he obtained (Nishina and Tomonaga 1933) were equivalent to those calculated by Bethe, Heitler, and others. This was the first of Tomonaga's many contributions to quantum electrodynamics.

Tomonaga kept abreast of developments in QED by studying all the papers that were being published. In his Nobel Prize lecture, Tomonaga told of the impact that Dirac's paper on "Relativistic Quantum Mechanics" had on him. Tomonaga read it shortly after its publication in the *Proceedings of the Royal Society* in the spring of 1932. In that paper, Dirac tried to assign different roles to the charged particles and to the electromagnetic field in QED. Dirac expressed the view that "the role of the field is to provide a means for making observations of a system of particles" and that therefore "we cannot suppose the field to be a dynamical system on the same footing as the particles" (Dirac 1932).

> This paper of Dirac's attracted my interest because of the novelty of the philosophy and the beauty of its form. Nishina also showed a great interest in this paper and suggested that I investigate the possibility of predicting some new phenomena by this theory.... I started computations to see whether the Klein-Nishina formula could be derived from this theory or whether any modification of the formula might result. I found out immediately, however, without performing the calculations through to the end, that it would yield the same answer as the previous theory. This new theory of Dirac's was in fact mathematically equivalent to the Heisenberg-Pauli theory and I realized during the calculation that one could pass from one to the other by a unitary transformation. The equivalence of these two theories was also discovered by Rosenfeld [1932] and Dirac-Fock-Podolsky [1932b] and was soon published in their papers. (Tomonaga 1971, p. 713)

Dirac deeply influenced Tomonaga, and through Tomonaga he had a pervasive influence on an entire generation of Japanese physicists. During the summer of 1935, Nishina, Kobayasi, Tamaki, and Tomonaga—the theoretical group of the Nishina Laboratory—"devoted" themselves to translating Dirac's "famous textbook of the quantum theory" into Japanese.

> We three rented a small villa at Karuizawa, a famous summer resort of Japan where Nishina stayed with his family. As soon as we started work, we found how difficult it was to translate English into Japanese which has a completely different sentence structure. The work of translation was really heavy labour, and sometimes we became so tired that we all became

bad humoured and disputes often arose over trifling matters. But we made it a rule to take a rest on Sundays and on finishing every chapter, to make excursions to neighbouring hills and meadows. The beautiful landscapes and refreshing air were so effective that we all recovered our good humour and we were able to continue our hard work. We believe that the Japanese edition of Dirac's book has been and will continue to be appreciated by many physics students of our country. (Tomonaga 1976, pp. 466–467)

In the mid-thirties, Tomonaga's interests shifted once again to nuclear physics. However, the results of his researches proved disappointing and he decided to go to Europe to finish his training. In May 1937 he went to Leipzig to work with Heisenberg. At the time, Heisenberg was the only one among the founders of quantum mechanics who was concerned with problems in nuclear physics and cosmic rays. When Tomonaga met Heisenberg he asked him to advise him on a topic for research. Heisenberg's reply was: "You plan to stay here for two years. I think it would be good for you to treat something having a concrete basis, at least for the first year. I am interested in Yukawa's theory, but its foundation does not seem to be very clear. I suggest that you do not work on it, but on some less speculative topic" (Tomonaga 1978 in Brown 1980, p. 3).

Actually, before leaving Japan, Tomonaga had chosen a problem to work on in Leipzig: the description in the Bohr liquid drop model of the heating of a nucleus when it absorbs a neutron. He was planning to extend this model by including in it heat conduction and viscosity. Although the results Tomonaga obtained were not particularly encouraging—the calculated viscosity of nuclear matter using a crude Fermi-Thomas model turned out to be too large to accommodate the oscillations anticipated by Bohr—Heisenberg nonetheless urged him to publish (Tomonaga 1938). He told him that "it is very common for a physicist to start from a very crude model, without knowing whether his assumptions are valid. You have no need to worry about your result" (Tomonaga 1978 in Brown 1980, p. 3). Heisenberg felt strongly that crude models could give important insights, and he reminded Tomonaga that such models had often played an important role in the history of physics. This work became a major portion of the thesis Tomonaga submitted in 1939 to Tokyo University for the degree of *Rigakuhakushi* (Doctor of Science).

After the discovery of the mesotron in 1937, and its identification by Yukawa and Stueckelberg with Yukawa's U-particle, Tomonaga began working on "meson" theory. The first problem he attacked was the discrepancy between the measured lifetime of the mesotron and its calculated value from Yukawa's theory. Tomonaga tried to modify the theory by assuming that the mechanism for mesotron decay was through a Fermi β-decay interaction, rather than through a direct coupling of the mesotron with an electron-neutrino pair.[16] The result of the calculation was entered in the diary Tomonaga kept while in Leipzig: "As I went on with the calculation, I found the integral diverged—was infinite. After lunch I went for a walk. The air was astringently cold and the pond in Johanna park was half frozen, with ducks swimming where there was no ice. I could see a flock of other birds. The flower beds were covered with chestnut leaves against the frost. ... Walking in the park I was no longer interested in the existense of the neutron, neutrino" (quoted in Schwinger 1983, p. 358).

Tomonaga's poetic evocations reflect his keen aesthetic sensibilities, and reveal the emotional satisfaction he obtained from his remarkable communion with nature.[17] The diary also records Tomonaga's fits of depression:

> It has been cold and drizzling since morning and I have de-
> voted the whole day to physics in vain. As it got dark I went
> to the park. The sky was grey with a bit of the yellow of
> twilight in it. I could see the silhouetted white birch grove
> glowing vaguely in the dark. My view was partly obscured by
> my tired eyes; my nose prickled from the cold and upon re-
> turning home I had a nosebleed. After supper I took up my
> physics again, but at last I gave up. Ill-starred work indeed!
> Recently I have felt very sad without any reason, so I went to
> a film. . . . Returning home I read a book on physics. I don't un-
> derstand it very well. Meanwhile I suffer. . . . Why isn't nature
> clearer and more directly comprehensible? (Quoted in Schwinger
> 1983b, p. 358)

Deeply troubled Tomonaga had written to Nishina:

> I complained in emotional words to Professor Nishina about
> the slump in my work, whereupon I got his letter in
> reply this morning. After reading it my eyes were filled with
> tears. . . . He says: only fortune decides your progress in achieve-
> ments. All of us stand on the dividing line from which the future
> is invisible. We need not be too anxious about the results, even
> though they may turn out quite different from what you expect.
> Bye-and-bye you may meet a new chance for success." (Quoted
> in Schwinger 1983b, p. 358)

While Tomonaga was in Leipzig, Yukawa kept him posted of his researches in "meson" theory by sending him the papers he and his coHabo-
rators were writing. Tomonaga studied them and was responsible for getting
Heisenberg—who at the time was working on cosmic ray problems—interested
in meson theory. Heisenberg began making some calculations of the properties of
mesotrons. The great puzzle then was how to reconcile a strong nucleon-nucleon
interaction that was to be mediated by the exchange of mesons between the

nucleons, with the small observed meson-nuclear cross sections, and hence the presumably weak meson-nucleon interactions.

Heisenberg was able to show that in a nonperturbative treatment of a classical model of the interaction of a meson with the nucleon's spin, self-field effects—arising from the cloud of mesons that "dressed" the nucleon—could produce a small effective meson-nucleon interaction, even though the "bare" meson-nucleon interaction was quite large. In his Nobel speech Tomonaga recalled: "Heisenberg, in this paper published in 1939, emphasized that the field reaction would be crucial in meson-nucleon scattering. Just at that time I was studying at Leipzig, and I still remember vividly how Heisenberg enthusiastically explained this idea to me and handed me the galley proofs of his forthcoming paper. Influenced by Heisenberg, I came to believe that the problem of field reactions far from being meaningless was one which required a frontal attack" (Tomonaga 1971, p. 717).

Tomonaga thought about how to incorporate Heisenberg's classical insight into a quantum-mechanical treatment. Although Tomonaga had hoped to be able to carry out this work in Leipzig in what was to be the final year of his stay there, World War II broke out on September 1, 1939. A few days later, Tomonaga boarded a Japanese ship in Hamburg to return home.[18]

Back in Japan, Tomonaga began working on his ideas regarding a quantum field theoretical treatment of the meson-nucleon interaction that would take into account Heisenberg's observation that field reaction effects played an important role. Reading the *Physics Abstracts* led him to Wentzel's investigation of a simple model of the meson-nucleon system using strong coupling methods (Wentzel 1940). In his paper Wentzel had carried out a canonical transformation to "dress" the nucleon with a virtual cloud of mesons. This canonical transformation transformed the original Hamiltonian that described a nucleon interacting with charged mesons:

$$H = H_0 + gH' \tag{6.3.1}$$

into the form

$$UHU^{-1} = \overline{H} + g^{-2}H''. \tag{6.3.2}$$

Wentzel discovered that $\overline{H}$, the new unperturbed Hamiltonian that described the "physical" nucleon, predicted stable excited states with higher spin, charge, and energy than the usually observed nucleon states. Even though Wentzel's results indicated that meson-nucleon scattering would be suppressed, the derived cross section did not fit the extant data. In his concluding remarks, Wentzel suggested that perhaps an intermediate coupling approximation might still suppress meson-nucleon scattering but yield cross sections in better agreement with experiments.

The canonical transformation Wentzel had used was similar to the one that Pauli and Fierz (1938) had introduced in their treatment of the infrared divergences in the quantum field theoretic description of the interaction of a charged particle with the radiation field. In that case the dressing transformation "attached" to a moving charge particle a cloud of photons that propagated with it. These "bound" photons were responsible for the Biot-Savart field, that is, the magnetic field, produced by the particle. In both the electromagnetic and the nuclear problem the canonical transformation introduced states with an arbitrary number of quanta into the description of the "physical" particle. Tomonaga devised an alternative way of incorporating states with arbitrary numbers of quanta.[19] The state vector describing a single nucleon in charged scalar meson theory can be represented in Fock space by amplitudes

$$\Phi_{m,n}(k_1, k_2, k_3, \cdots k_m; l_1, \cdots l_n)$$
$$m, n = 0, 1, 2, 3, 4 \ldots,$$

with $k_1, \cdots k_m$, and $l_1, \cdots l_n$ the momenta of the positively and negatively charged mesons, respectively. Tomonaga approximated these amplitudes as a product of functions

$$g(k_1)g(k_2) \cdots g(k_m)f(l_1)f(l_n)$$

and then determined the best one-particle functions g and f variationally; Tomonaga's method was in effect a field-theoretic Hartree approximation. The treatment was later generalized to the case of scattering of mesons by nucleons (Miyamiza and Tomonaga 1942; Tomonaga 1943a). The techniques that Tomonaga learned, particularly the use of canonical transformations, and the insights he obtained while carrying out these meson theoretic calculations proved to be very valuable in his researches on QED in 1946.

Fukuda Hiroshi, one of Tomonaga's oldest students, first met Tomonaga in the fall of 1939 when he attended a seminar given by him at Riken. He remembers Tomonaga as being extremely skinny at the time, probably weighing less than a hundred pounds. He was deeply impressed by the fact that Tomonaga gave this seminar—which lasted several hours and during which he wrote down many "very long equations" on the blackboard—without ever referring to any notes. Fukuda also noticed that Nishina, while clearly in charge, had full confidence in Tomonaga and believed whatever Tomonaga said. But what touched Fukuda most deeply was the fact that after he had been introduced to Tomonaga as a young student from Kyoto University, Tomonaga took him aside and explained to him not only the mathematical formalism he had used during his presentation, but also the physics behind the formalism.[20]

In 1940 Tomonaga was appointed professor of physics at the Tokyo College of Science and Literature.[21] His gifts as a teacher became legendary. His approach was socratic, and students said of him that "he was like a magician." Tomonaga himself was fond of saying that "if you formulate the problem correctly, that is, if you ask the right question, the answer emerges spontaneously."[22]

6.4 The War Years

Japanese theorists worked on foundational problems even in the midst of the war. In 1942 Taketani published his Hegelian "three stage" theory of the evolution of scientific knowledge. In this scheme, the first, "phenomenological," stage was concerned with experimental facts. In the second stage, which Taketani called "substantialistic," the identification of the ontology—the objects or particles—to which natural laws apply was the focus of interest. In the third, the "essentialistic" stage, the general relation between the structure of the objects and the derived laws was the primary concern. He illustrated these notions with examples from the history of physics. Taketani's work, even though philosophical in character, influenced many of the theorists working in elementary particle physics, in particular Sakata and Yukawa. The Marxist methodological inclination of Taketani and Sakata "helped diversify the directions of research in Japanese war-time quantum field theory" (Darrigol 1988a, p. 15). Influenced by Taketani and Sakata, Yukawa in 1942 published a paper in which he gave a "comprehensive consideration about the quantum theory of wave fields" and stressed the unsatisfactory nature of the theory with respect to relativistic covariance. To remedy the difficulties encountered in the Pauli-Heisenberg version of quantum field theory, Yukawa proposed a revolutionary reformulation of the subject in which the usual linkage between cause and effect was relaxed.[23] In his presentation Yukawa made use of closed four-dimensional surfaces in space-time which he drew as circles (*maru*). The maru came to symbolize the nonseparability of cause and effect in Yukawa's theory.

Stimulated by the article of Yukawa (1942), Tomonaga also undertook to analyze the foundations of quantum field theory. However, Tomonaga thought "that it might be possible to remedy the unsatisfactory, unpleasant aspect of the Heisenberg-Pauli theory (without introducing such drastic changes as Yukawa and Dirac had done).... In other words, ... [that] it should be possible to define a relativistically meaningful probability amplitude, which would be manifestly relativistically covariant, without being forced to give up the causal way of thinking" (Tomonaga 1971, p. 714).

Whereas Yukawa had suggested revolutionary measures—for example, giving up causality—to overcome the problems encountered in the unification of quantum mechanics and special relativity, Tomonaga was guided by his conservative stance and believed that advances were made by limited, incremental, evolutionary steps. The concluding remarks of his 1942 paper on meson theory were illustrative of that outlook: "The field equations of the present theory must ultimately be regarded, strictly speaking, as having no finite solutions at all. It may be, nevertheless, expected that if we interpret the present theory correctly, it will be a good approximation of the forthcoming theory, and for each solution of the fundamental equation of the latter theory the corresponding solution in the present theory will exist in some way" (Miyamiza and Tomonaga 1943).

Tomonaga thought the reason that quantum field theory was so unsatisfactory was "that it had been built up in a way much too analogous to the ordinary non-relativistic mechanics." As usually formulated, the theory was divided into two parts, one that gave kinematical relations between various quantities at the same instant of time—for example, the equal time commutation rules—and another that determined the causal relations between quantities at different instants of time. The Schrödinger equation belonged to the latter category. In this canonical approach to quantum field theory the concept of "the same instant of time" played a central role and was responsible for its lack of covariance. Tomonaga took it as his problem "to build up the theory on the basis of concepts having relativistic space-time meaning" (Tomonaga 1943b, 1946).[24] The clue for how this was to be done had been given by Dirac in his 1932 paper, "The Lagrangian in Quantum Mechanics." In the last section of that paper, entitled "Applications to Field Dynamics," Dirac noted that in the field-theoretic case the relativistic invariant generalization of the transformation function $(q_t | q_T)$ of particle theory is a "generalized transformation function"

$$\left(v(xyz), \sigma | v'(x'y'z'), \sigma' \right)$$

that depended on "two separate (three-dimensional) surfaces [σ and σ'], each extending to infinity in the space directions and lying entirely outside any light-cone having its vertex on the surface" (Dirac 1932). Here $v(xyz)$ is the quantity specifying the field.[25] Dirac had come to this insight from his relativistically invariant many-time formulation of quantum electrodynamics (Dirac et al. 1932a), which was outlined in section 5 of chapter 1.

In order to give a relativistically invariant description of two interacting fields described by the Hamiltonian

$$H = H_0 + H_{12} \tag{6.4.1}$$

$$H_0 = H_{01}(v_1, \lambda_1) + H_{02}(v_2, \lambda_2), \tag{6.4.2}$$

Tomonaga proceeded in complete analogy with the treatment of QED given by Dirac, Fock, and Podolsky (1932) and by Rosenfeld (1932a).[26] He first performed a unitary transformation[27]

$$U = \exp(iH_0 t) \tag{6.4.3}$$

of the Schrödinger representation operators v_1 and v_2 and their canonically conjugate momenta λ_1 and $\lambda_2^\dagger$ that described the fields

$$V_r = U v_r U^{-1} \quad \Lambda_r = U \lambda_r U^{-1} \tag{6.4.4a}$$

and of the state vector

$$\Psi = U\psi, \tag{6.4.4b}$$

where ψ satisfies the Schrödinger equation

$$i\hbar \partial_t \psi = H\psi. \tag{6.4.5}$$

The transformed operators obey free field equations, "the equations which the fields must satisfy when they are left alone." The invariant commutator of $V(xyzt)$ with $V(x'y'z't')$, which now refers to two times, t and t', can readily be computed since the V's are free fields. The transformed state vector Ψ obeys the equation

$$H_{12}(V_1, V_2, \Lambda_1, \Lambda_2)\Psi = \int dx\,dy\,dz\,\mathcal{H}_{12}(V_1, V_2, \Lambda_1, \Lambda_2)\Psi$$
$$= i\hbar\partial_t \Psi. \tag{6.4.6}$$

This is the analog of the equation

$$\sum_n H_{12n}(q_n, p_n, \mathcal{A}(q_n, t))\Phi = i\hbar\partial_t\Phi \tag{6.4.7}$$

that Dirac, Fock, and Podolsky had obtained in their description of the interaction of charged particles with the quantized radiation field. Both eqs. (6.4.6) and (6.4.7) still make reference to a single time variable, and hence single out a particular Lorentz frame. Dirac had generalized the theory by assigning to each particle its own time variable and introducing a state vector $\Phi(q_1t_1, q_2t_2, q_3t_3, \ldots, q_Nt_N)$ that depended on the times of all the different particles and was assumed to satisfy

$$H_{12n}(q_n, p_n, \mathcal{A}(q_n, t))\Phi = i\hbar\partial_t\Phi$$
$$n = 1, 2, 3, 4, \ldots N. \tag{6.4.8}$$

Eq. (6.4.7) is then the equation satisfied by $\Phi(t_1, t_2, t_3, \ldots t_n)$ when $t_1 = t_2 = t_3 = \ldots t_n = t$.

In the field-theoretic context the role that a charged particle played is taken over by a small volume element of space. Tomonaga assigned to each spatial volume element a "local time" t_{xyz} and supposed that the state vector $\Psi[\sigma]$, which is now a functional of the 3-dimensional surface σ in space time defined by $t = t_{xyz}$, satisfies the infinitely many simultaneous equations

$$\mathcal{H}_{12}(x, y, z, t)\Psi = i\hbar \frac{\delta}{\delta t_{xyz}} \Psi \qquad (6.4.9)$$

that are the analogue of the N equations (6.4.8). The system of eq. (6.4.9) is integrable when the surface is spacelike. Since $\mathcal{H}_{12}$ is a scalar and $\delta/\delta t_{xyz}$ is independent of any frame of reference, Eq. (6.4.9) "has now a perfect space-time form": the theory thus formulated is completely covariant. Tomonaga proceeded to make some observations concerning the integration of eq. (6.4.9) and exhibited explicitly the generalized transformation function that Dirac had intuited in his 1932 paper.

In his concluding remarks, Tomonaga commented that in this "new super-many-time" formalism the theory is still divided into two parts. One part

> gives the laws of behavior of the fields when they are left alone, and the other gives the laws determining the deviation from this behavior due to interactions. This way of separating the theory can be carried out relativistically. Although in this way the theory can be brought into a more satisfactory form, no new contents are added thereby. So the well-known divergence difficulties of the theory are inherited also by our theory.... Thus a more profound modification of the theory is required in order to remove this fundamental difficulty.

But Tomonaga could not investigate possible modifications because by 1943 the war was coming closer to Japan's shores and research related to military needs was being thrust onto scientists. Tomonaga stopped working on particle physics after 1943 and became involved in research on the properties of magnetrons and the behavior of microwaves in waveguides and cavity resonators. Miyamiza Tatsuoki, one of Tomonaga's colleagues, who had been asked by Nishina to join a Riken-Army project to develop efficient power sources for microwave radiation, recalled:

> One day our boss, Dr. Nishina, took me to see several engineers at the Naval Technical Research Institute. They had been engaged in the research and development of powerful split anode magnetrons, and they seemed to have come to a concrete conclusion about the phenomena taking place in the electron cloud.... Since they were engineers their way of thinking was characteristic of engineers and it was quite natural that they spoke in an engineer's way, but unfortunately it was... completely foreign to me at the beginning.... Every time I met

> them, I used to report to Tomonaga how I could not understand
> them, but he must have understood something...because after
> a month or so, he showed me his idea...[of] applying the idea
> of secular perturbation theory, well known in celestial mechan-
> ics and quantum theory, to the motion of the electrons in the
> cloud.... I remember that the moment he told me I said, "This is
> it." Further investigation actually showed that the generation of
> electromagnetic oscillations in split anode magnetrons can be es-
> sentially understood by applying his idea. (Quoted in Schwinger
> 1983b, p. 365)

In a brief reminiscence of his wartime association with Tomonaga,
Miyamiza indicated that "Tomonaga had a very deep insight into the problem"
and that he could combine powerful mathematical techniques with elegant physical reasoning. Miyamiza also recalled that Tomonaga "would always talk about
his findings, but only after he had obtained his solution."[28] The way Tomonaga
worked in those days implied that there was nothing left for students to do after
he was finished. Tomonaga also worked on the analysis of microwave junctions
and the description of microwave circuits, and "developed a unified theory of
the systems consisting of wave guides and cavity resonators."[29] He applied the
scattering matrix concepts that Wheeler and Breit had developed for describing
nuclear reactions to the solutions of Maxwell's equations in the waveguide context. The main task of any theory of waveguide junctions is to find the amplitudes
and phases of the waves propagating in and out of the openings of a cavity res-
onator or waveguide junction. Tomonaga solved this problem in an important and
impressive paper, "A General Theory of Ultra-short Wave Circuits," in which he
showed that it was possible to obtain

> relations [between the amplitudes and phases of the various
> waves at the opening of a wave guide junction]...without solv-
> ing the Maxwell's equations explicitly, [and that these] give so
> many informations about the circuit that the general considera-
> tion alone often suffices to find the properties of the circuit.
> Even when this is not the case, further experimental de-
> termination of a few characteristic constants is usually sufficient
> for many purposes. (Tomonaga 1976, p. 2)

This, incidentally, was a viewpoint very similar to the philosophy expressed by the renormalization program.

Tomonaga introduced into radar circuit analysis the concept of the "characteristic matrix," which was to play a role analogous to "impedance" in ordinary circuits. Tomonaga derived the properties of the characteristic matrix that
followed from conservation of energy, and showed how to describe a given

junction in terms of its characteristic matrix. But he cautioned that "the final decision, however, whether or not the new concept is here preferable to impedance should of course be given not only by a theoretical physicist but also by general electroengineers" (Tomonaga 1976, p. 2).

Besides carrying on wartime research that took up a great deal of his time, Tomonaga was also teaching at Tokyo University of Education (Bunrika), whose faculty he had joined as a professor of physics. Occasionally he also lectured at Tokyo University. During World War II, undergraduates majoring in physics were required to assist in military research at one of the research institutes after they had completed a year and half of studies. Tokyo University, however, provided courses that allowed students to continue their studies while engaged in war work. Each course consisted of five lectures given on alternate Sundays, "which were the only official holidays during wartime" (Hayakawa 1988b, p. 44). Tomonaga gave such a Sunday course on quantum mechanics in 1944; its content was essentially the material to be found in the first volume of his treatise on quantum mechanics (Tomonaga 1962). Hayakawa Satio, who was a student in the course, remembers that during the last lecture Tomonaga wrote down the names of some prominent physicists—Bohr among them—and their age when they made their important contributions. All of them were young, and Tomonaga encouraged his audience to follow in their footsteps.[30] He then wrote down Planck's name and that of some other physicists who had made their important contributions when they were older and he encouraged himself to emulate them.

6.5 The Postwar Years

Japan's surrender was accepted by the Allies on August 14, 1945. It came in the wake of the leveling of Nagasaki and a few days after Russia's entry into the Far Eastern war. The end of hostilities brought about not only the collapse of Japan's military establishment: essentially all governmental institutions ceased functioning. Serious food shortages developed that were aggravated by the fact that the rationing system that had operated during war broke down, leaving the population near starvation. Being hungry, searching for food, and trying to find shelter became the constants of life in Japanese cities after the armistice. Nambu, who was a young assistant at Tokyo University after the war, tells of living in the office he shared with Koba, eating and working there during the day and sleeping on his desk at night (Brown 1986). He distinctly remembers that on one occasion he was too weak to walk to attend Tomonaga's seminar. Yet despite these abysmal conditions, the Tokyo theoretical physics community reconstituted itself amid the devastation. And at the center of the effort stood Tomonaga.

In a *Monologue* written in 1955 in which he looked back over the previous decade, Tomonaga described the situation after the war as chaotic, and revealed that he and his family suffered greatly from the lack of food. "Although the war had ended we had no food and no house." Given these circumstances, he "wanted to be engaged in something that required no thinking."[31] He therefore set himself the task of translating into English all his wartime researches—the super-many-time theory, his papers on intermediate coupling meson theory, and his reports on magnetrons and microwave circuits. He himself typed the translations on the backside of previously used papers. The general situation had not improved when the job was completed a few months later. As he did not believe that quantum field theory was relevant to the solution of the food shortage problem, he began studying photosynthesis. For a while all his energies were devoted to activities in that field. He organized a research group at Riken to investigate ways to overcome quickly the food shortages by using photosynthetic processes. They were not successful and in the spring of 1946 he returned to the problems of quantum field theory. He used the money he received for being awarded the Asahi ("Morning Sun") Prize to rebuild and refurbish a little quonset hut he had been given as his office on the campus of the Tokyo University of Education. He also bought ten tatami mats, and the quonset hut became not only his residence but also the site of the seminar he organized. Before then, it had taken him over four hours to commute from his living quarters to the university. However, the time had been well spent: the solution to the problem of how to formulate the subsidiary condition in the super-many-time formulation of quantum electrodynamics was worked out during one of these trips.

Hayakawa recalls that one day in April 1946 Tomonaga called all the young people who had been working with him to meet at his office at the Otsuka campus. That was the first time that many of the researchers from Tokyo Imperial University[32] met Tati, Kanesawa, and the other physicists of the Tokyo University of Education (Bunrika Daigaku). Tomonaga told them of his intention to implement "by the efforts of this group" the super-many-time theory that he had developed in 1943 and "proposed having meetings once a week to discuss developments achieved by the participants" (Hayakawa 1988b).

Hayakawa kept a record of the presentations made at these meetings and of his discussions with Koba and Miyamoto in notebooks that are still extant. During the spring of 1946 Tomonaga and his associates worked out the formulation of quantum electrodynamics for spin 0, spin 1/2, and spin 1 charged fields in the "super-many-time" formalism. For the case of spin 0 and spin 1, this required introducing terms into the interaction Hamiltonian that depended on timelike vectors normal to the spacelike surface, and proving that the Tomonaga equation was still integrable. Tomonaga and his associates also worked out how the gauge fixing auxiliary condition had to be generalized, and verified the compatibility of this new auxiliary equation with the Tomonaga equation. By early August everything was in place and

> Tomonaga decided to publish a series of several papers. Although these works were accomplished through the collaborative efforts of those who participated in the discussions, he suggested that not too many authors should appear on a single

paper.... A number of papers were written and were submitted to the *Journal of the Physical Society of Japan* and *Progress of Theoretical Physics* taking into account the low printing capacity that would prevent the editors from publishing a flow of papers. (Hayakawa 1988b)

The disposition of the authorships of these papers was as follows:

Tomonaga (1946)[33]
Hayakawa, Miyamoto,
and Tomonaga   (see Hayakawa et al. 1947a,b)
Koba, Tati, and Tomonaga   (see Koba et al. 1947a,b)
Kanezawa and Tomonaga   (Kanezawa 1948)
Koba, Oishi, and Sasaki   (Koba 1948)
Miyamoto   (Miyamoto 1948)

The contents of these papers were reported at the meeting of the Physical Society of Japan that was held on November 21 and 22, 1946, at Kyoto University. In his summary talk Tomonaga stated that he had tried "to formulate the existing theory as transparently as possible without touching the difficulties of the theory" (Hayakawa 1988b). At that same Kyoto meeting, Sakata presented a paper in which he outlined his views concerning the divergence difficulties. He advanced the view that these stemmed from the fact that "mutually related interactions were treated as if they were separable." He noted that ways of taking the interrelationship into account had previously been indicated by Møller and Rosenfeld, who had shown that by an appropriate choice of the meson-nucleon coupling constants, a "mixture" of a pseudoscalar and a vector meson theory (with quanta of the same mass) could be made to yield nucleon-nucleon interaction potentials that were less singular than those generated by the "separate" theories. Sakata and his students had also quantized a nonlinear modification of the electromagnetic field that had been advocated by Bopp, and had shown that it corresponded to a "mixture" of a neutral massless vector field and a neutral massive vector field that made a negative contribution to the total energy of the quantized field system. Sakata also reported that his associate, Hara Osamu, had found that the second-order self-energy of an electron that interacted with both the electromagnetic field and a scalar field would be finite if the coupling to the electromagnetic field, e, and that to the scalar field, f, were related by the equation

$$e^2 = 2f^2. \tag{6.5.1}$$

Although Tomonaga doubted that the cancelation of the logarithmic divergences would continue to hold in higher orders, Sakata's and Hara's insights played an important role in the further developments of Tomonaga's program.

These and other methods of handling the divergences were discussed at the meetings of the Tomonaga seminars during the fall of 1946. Hayakawa's notebooks record a series of talks by Miyazawa, the most senior of Tomonaga's associates, that reviewed Wentzel's limiting process (Wentzel 1933), Heitler's damping theory (Heitler 1941; Wilson 1941), and other procedures to overcome the divergences.[34] In the spring of 1947, stimulated by the recently published paper of Bethe and Oppenheimer (1946) on the inadequacies of the Heitler damping theory in dealing with the infrared divergences of QED, Tomonaga's seminar carefully reviewed the work of Bloch and Nordsieck, Pauli and Fierz, and Dancoff's paper on the treatment of radiative corrections to the scattering of an electron by a Coulomb field. Tomonaga posed the question whether the Sakata C-meson approach[35] would eliminate the divergence that Dancoff had encountered in the problem of the radiative corrections to Rutherford scattering. Koba, Ito, and Tomonaga proceeded to calculate the radiative corrections and arrived at the conclusion that the contributions would be finite provided

$$f^2 = (7/9)2e^2. \tag{6.5.2}$$

But this relationship would imply that the self-energy of a free electron would again diverge to lowest order. They reported this conclusion in a letter submitted to the *Progress in Theoretical Physics* (Koba and Tomonaga 1948) dated November 1, 1947.

While carrying out these calculations, Tomonaga obtained his "first information concerning the Lamb shift ... through the popular science column of a weekly U.S. magazine."[36] Shortly thereafter, he read the issue of the *Physical Review* that contained Bethe's nonrelativistic Lamb shift calculation and immediately appreciated its importance. "This information about the Lamb shift prompted us to begin a calculation more exactly than Bethe's tentative one" (Tomonaga 1971, p. 723). On October 11, 1947, Tomonaga gave a report on Bethe's calculation in his seminar. At that same seminar Tomonaga presented "a 'self-consistent' subtraction method which aims at disposing of the infinities in a self-consistent manner and obtaining finite results for various processes" (Hayakawa 1988b). In this he was "following Bethe's idea" (Nambu 1949). Ivar Waller, in his speech introducing the Nobel laureates for the award of the prize, put it thus: "As soon as Tomonaga knew about the Lamb shift experiment and Bethe's paper he realized that an essential step to be taken was to substitute the experimental mass for the fictive mechanical mass which appeared in the equations of quantum electrodynamics and to perform a similar renormalization of the charge. The compensating terms which had then to be introduced in the equations should cancel the infinities."

Tomonaga's subtraction method—the mass renormalization procedure—was implemented within his super-many-time theory by using a covariant canonical transformation suggested by the work of Pauli and Fierz (1938) and Wentzel (1940). He also remarked at the time that "the Lamb shift and radiative corrections could be more unambiguously worked out in the covariant formalism" (Hayakawa 1988b), because the electromagnetic mass could be exhibited and identified explicitly. Furthermore, because the nature of the various terms could be readily identified (e.g., as a self-energy contribution or as a vacuum polarization contribution), "what took a few months in a Dancoff type calculation could [now] be done in a few weeks" (Tomonaga 1971, p. 718). Tomonaga initially called renormalization theory "readjustment theory"; later he characterized the method as an "amalgamation theory" (Tomonaga 1949). He viewed the procedure as provisional: the self-consistent subtraction method made it possible to avoid the divergence difficulties; it did not solve them.

The first problem that was addressed within the new formalism was Dancoff's problem of the radiative corrections to the Rutherford formula: Owing to this new, more lucid method, we noticed that among the various terms appearing in both Dancoff's and our previous calculation, one term had been overlooked. There was only one missing term, but it was crucial to the final conclusion. Indeed, if we corrected this error, the infinities appearing in the scattering process of an electron due to the electromagnetic field and cohesive force field cancelled completely, except for the divergences of the vacuum polarization type" (Tomonaga 1971, p. 723). The calculation was completed in late December 1947 and a letter sent to *Progress in Theoretical Physics* (Koba and Tomonaga 1948).

Tomonaga also established that this same contact transformation could be applied to the computation of the level shift, thereby extending Bethe's calculation and justifying his introduction of a cutoff.[37] In early January 1948, Fukuda and Miyamoto undertook a relativistic calculation of the Lamb shift "under the guidance of Prof. Tomonaga" (Miyamoto 1988). Starting from the Tomonaga equation

$$\left\{ \mathcal{H}_I + V - i\hbar \frac{\delta}{\delta t_{xyz}} \right\} \Psi = 0, \qquad (6.5.3)$$

with $\mathcal{H}_I$ the interaction Hamiltonian of the electron-positron field with the radiation field

$$\mathcal{H}_I = -ie\, \bar{\psi} \gamma_\mu \psi\, A_\mu \qquad (6.5.4)$$

and V that with the external (Coulomb) field

$$V = -ie\, \bar{\psi} \gamma_\mu \psi\, A_\mu^e, \qquad (6.5.5)$$

they performed the canonical transformation

$$\Psi = \left\{ \exp\left(-\frac{i}{\hbar} \int d^4 x' \mathcal{H}_I \right) \right\} \Psi' \qquad (6.5.6)$$

that "dresses" the electrons. They eliminated the electron self-energy by a mass renormalization, or in their terminology by a self-consistent subtraction. As Miyamoto later recalled: "This method of calculation was quite new. We were sometimes at a loss how to calculate, since there were many new things to be solved. At first we attempted to use a relativistic covariant way of calculating in various ways, but we were always unsuccessful. Then we decided to evaluate in a way which is similar to convential perturbation theory. Prof. Tomonaga also calculated some fundamental things, and gave us many valuable suggestions. The calculation was very tedious" (Miyamoto 1988). They also were plagued by the problem of gauge invariance: "We discussed many times this problem. We tentatively dropped the quadratic divergence corresponding to [the] photon self energy, and preserved the remaining finite term. Prof. Tomonaga considered that even this finite term was dubious" (Miyamoto 1988).

In early April 1948 Tomonaga sent Oppenheimer a letter summarizing both the accomplishments of his group and the difficulties they had encountered. Tomonaga also reported the status of various calculations at that point in time. Oppenheimer answered with a telegram urging him to "write a summary account of present state and views for prompt publication *Physical Review*." In addition, Tomonaga received a letter from Pais informing him of his recent work on the self-stress of the electron and of further developments in the theory of the compensating scalar field, what Sakata had called the C-field and Pais the f-field (Pais 1945, 1946). Pais referred him to the recent papers by Wightman (1947) and Zirzel (1947) "in which it is pointed out that a range of the f-field which is large enough to give the right order of magnitude for the proton-neutron mass differences is in disagreement with the symmetry of the atomic nuclei." He added that he had "found no way out of the difficulty."[38]

In September 1948 Fukuda, Miyamoto, and Tomonaga obtained an expression for the Lamb shift (Fukuda et al. 1949) that agreed with the result of French and Weisskopf and Kroll and Lamb. Although the calculation had started out using covariant methods to identify and eliminate the divergences, the subsequent steps reverted to noncovariant techniques.[39]

Tomonaga is a product of Kyoto University. Kyoto University prides itself for nurturing liberalism and antiauthoritarianism and also for encouraging individualism in intellectual activities. Its ethos is to be contrasted with the values and attitudes fostered at Tokyo University—the nursery of the high-level government bureaucrats of Japan. The Kyoto spirit dictates that if everybody is doing the same thing, you do something else. By the fall of 1948 it seemed to Tomonaga that everybody was doing quantum electrodynamics and that it was time therefore to leave this fashionable field. Prior to doing so, he gave a series of lectures on "The Developments of Elementary Particle Physics: Discussions Centering on the Divergence Difficulties" (Tomonaga 1971). His purpose was to present the recent advances made in QED "with a view to searching for the whereabouts of the path overcoming the [divergence] difficulties and leading to a truly flawless elementary particle theory" (Tomonaga 1971, p. 512).

His point of departure was the observation that it had "been experimentally ascertained that the effects of the interaction between electrons and photons is actually small." He reviewed the predictions made by QED and analyzed the divergences that were encountered in higher-order calculations, in particular in those of the "apparent" energy of an electron and of the apparent charge of a test body. He showed that the effects of the interaction between electrons and photons to lowest order gave rise to an apparent mass modification, δm, of the electron, and an apparent charge modification, δe, of the test body creating an external field. He then demonstrated that the "ultraviolet divergence appearing in scattering problems is of the same source as the infinities appearing in the apparent mass and apparent charge of an electron." He next turned to a discussion of "amalgamation," that is, renormalization, theory.

> The mass and charge of the electrons which we could actually observe [are] the corrected quantities $m + \delta m$ and $e + \delta e$. Consequently, [even] though the theoretical values of $m + \delta m$ and $e + \delta e$ may be infinitely large, their actual values are finite. And consequently again, the infinity-difficulties in . . . [QED] calculations can be evaded by substituting the finite experimental values instead of the theoretical ones $m + \delta m$ and $e + \delta e$. . . . In a word, we may say that we have lumped all the infinity-difficulties into the self-energy of a free electron, and the problem of vacuum polarization. (Tomonaga 1971, pp. 541–542)

Moreover, the results thus computed agree well with experiments. However, since only the Lamb shift and the corrections to scattering had been computed, and these only to lowest order in perturbation theory, "it is not yet clear how things would stand in [a] wider class of problems and if further higher-order corrections are considered" (Tomonaga 1971, p. 544).

Tomonaga then asked: What about the infinities in the expressions for the self-energy of an electron and for the vacuum polarization? He observed that with the introduction of the Sakata-Pais cohesive field, "the self-energy of a free electron can be kept from being infinite" by making a suitable choice of the interaction constant; and that in the calculation of the Lamb shift and the radiative corrections to scattering, all divergences not related to vacuum polarization were removed by the c-field: "In this sense the c-meson theory is the only theory which has succeeded in removing divergences. Even here, however, the infinity of vacuum polarization remains inexorable" (Tomonaga 1971, p. 543).

In his concluding remarks, Tomonaga suggested that the theory was inconsistent. Quantum field theory arrived at the concept of an elementary particle "only when no interaction exists at all," or at best when the interaction between elementary particles can be made arbitrarily small. But the theory, "far from justifying this . . . has it that the effects of the interaction are always infinitely large." Tomonaga attributed the source of this inconsistency to the uncritical borrowing of the concept of fields and that of interactions between fields from the classical (unquantized) field theory. It seemed to Tomonaga that "in the correct theory, such concepts as a field independent of others on the one hand, and the interaction on the other require fundamental modification or restriction."

How is this to be done? This is a question that could not be answered "now." The "real" solution probably still lies far ahead, and might involve a fundamental change in our conceptualization of nature. It might require the ascription of a discontinuous structure to time and space, or it might require a radical modification of our concepts of elementary particles and their mutual interactions. Furthermore, Tomonaga believed that the right approach could not be ascertained on the basis of theory alone: "We are too powerless to make assumptions based only on reasoning. We must beg instruction from Nature herself. The recent development of experimental technique has not only given us precise knowledge as to the energy level shifts, but has also enabled us to produce mesons in great quantity within the laboratory.... Great indeed, will be the contributions of these experiments to the construction of a correct theory of elementary particles" (Tomonaga 1971, p. 545).

7. Julian Schwinger and the Formalization of Quantum Field Theory

7.1 Introduction

In 1978, Columbia University celebrated the fiftieth anniversary of the Pupin Laboratories,[1] the building that houses its physics department, and used the occasion to honor the many contributions of I. I. Rabi. The celebration was marked by a convocation and by three symposia whose themes were "Science, Technology and Education," "Science and Government," and "Physics in the Future."[2] The speakers at the third symposium were Freeman Dyson, Edward Purcell, Leon Lederman, and Julian Schwinger. The first three took their assignment at face value and spoke of their vision of physics in the future. Schwinger, on the other hand, told the distinguished audience:

> When I first saw the preliminary title of this session, Physics in the 21st Century, my reaction was one of total horror. Admittedly that was somewhat tempered by the ultimate title, Physics in the Future, since I had no longer to traverse a gap of 23 years. Nevertheless the essential difficulty remained. Reliable predictions of the future are possible only to the extent that they constitute modest extrapolations of present knowledge. But every scientist knows that the real fascination of research lies in the totally unexpected development, the revelation that tears the fabric of supposedly sacred theory. Accordingly I decided instead to look *backward* and recount an episode in the history of physics, both for its intrinsic interest, and for the intriguing ["fascinating" crossed out] parallels that it presents with the present day situation in high energy physics, in the expectation that therein resides some useful lessons for physics in the future.[3]

Schwinger proceeded to talk about the rise of atomism in the nineteenth century and focused on some episodes in the history of the kinetic theory of matter, "the understanding of heat as a form of motion." His history lesson contained illustrations from the lives of Herapath and Waterston, summaries of the contributions of Maxwell, Boltzmann, Gibbs, and Einstein, and observations concerning the reception of their ideas by the scientific community. Schwinger had included his remarks concerning Herapath and Waterston to illustrate "the pettiness of individual men and the arrogance of institutions," and he intended to make clear by his presentation of the later developments that the proper arena of physics was "the competition of ideas, in which the ultimate referee is the response of Nature to the conflicting assertions made by rival theories." Drawing parallels between the history of atomism at the turn of the nineteenth century and the "present day situation in high energy physics" Schwinger indicated that "for what it's worth, my own view has been that of J. Willard Gibbs: While not questioning the ultimate emergence of inner structure, refrain as far as possible from making specific, speculative hypothesis about that structure." He concluded his disquisition with the exhortation: "We are again in a situation that cries out for the wisdom of Boltzmann; who sees the future? Let us have free scope in all directions of research; away with all dogmatism."[4]

Schwinger's lecture on that occasion was revealing of the man: he has always marched to the beat of his own drum, often charting the direction others are to take. His inclinations are conservative, and he has never abandoned lightly what has been hard won. Yet indeed for him "the real fascination of research lies in the totally unexpected development, the revelation that tears the fabric of supposedly sacred theory." The Lamb shift experiment had given him an instance of the "totally unexpected," namely the breakdown of the Dirac theory. Schwinger's researches in the period from 1947 to 1951 were hugely successful and were guided by his conservative outlook. His aim was to determine to what extent quantum electrodynamics could account for the observed deviations from the Dirac theory when the requirements of relativistic invariance and gauge invariance were rigidly enforced and the ideas of mass and charge renormalization were incorporated into the existing formalism. His efforts culminated in the acceptance of quantum field theory as the proper representation of microscopic phenomena, and he was awarded the Nobel Prize for this achievement. In the opening remarks of the lecture he delivered in Stockholm on December 11, 1965, upon receiving the Nobel Prize, Schwinger noted that "the relativistic quantum theory of fields was born some thirty-five years ago through the paternal efforts of Dirac, Heisenberg, Pauli and others. It was a somewhat retarded youngster, however, and first reached adolescence seventeen years later, an event which we are gathered here to celebrate" (Schwinger 1966, p. 949).

The theory achieved adolescence because in Schwinger's hands it gave a satisfactory account of several experimental phenomena that earlier formulations had been unable to explain. Most earlier attempts to improve upon Heisenberg and Pauli's formulation of electrodynamics had concentrated upon the elimination of the divergence difficulties from the theory and had effected this elimination by arbitrary subtraction prescriptions that lacked both experimental confirmation and theoretical plausibility. Schwinger avoided such mutilation of the theory. "He merely reformulated it without the addition of fundamentally new concepts, in a way that was manifestly covariant with respect to Lorentz and gauge transformations. The divergences were not eliminated, but were isolated in expressions which are unobservable and cleanly separated from finite observable effects" (Dyson 1949a).

7.2 The Young Schwinger

Julian Schwinger's family was a "typical middle class Jewish family."[5] Benjamin Schwinger, Julian's father, was born in Newsandez, a small village in the foothills of the Carpathian Mountains in one of the provinces of the Austro-Hungarian empire. He came to the United States by himself when he was in his early teens around 1880. Having to learn a trade while supporting himself prevented him from obtaining more than a rudimentary education. But he was bright, able, and enterprising and he became a successful designer of women's clothing.

The family of Schwinger's mother came from Lodz, a large manufacturing town in eastern Poland, which was then a part of Russia. At the end of the last century, Lodz was one of the foremost cultural and intellectual centers of Eastern European Jewry. The family emigrated to the United States when Belle, Schwinger's mother, was a very young child. Schwinger's grandfather, although raised as an orthodox Jew—a tradition he maintained—had also been steeped in Lodz's cosmopolitan outlook. He was a clothing manufacturer who, after coming to the United States, often traveled to Europe to keep himself *au courant* of the latest in fashions. Belle started working in her father's clothing manufacturing business upon graduating from high school.

After they were married, Schwinger's parents lived near Morris Park in Harlem, then a well-to-do Jewish neighborhood. Their first son, Harold, was born in 1911. Some six years later they moved to a large apartment at 640 Riverside Drive near 141st Street. Belle's parents lived in an apartment next to them. Julian Seymor was born on February 12, 1918, which is, as he later noted in the biographical note attached to his Nobel Prize speech, "just five score and nine years after the birthday of Abraham Lincoln" (Schwinger 1966, p. 113). The family was well off, for Benjamin was a talented couturier and his business had prospered. Harold remembers that the family had in its employ a German nursemaid and a Hungarian maid. Life became more difficult after the crash of 1929, for Benjamin lost his business. But his talents and reputation were such that he readily found employment. "My father was a very wonderful designer of women's clothing, coats and suits. . . . He had an eye for lines, design,"[6] Harold asserts. Even though the family was not as affluent in the 1930s as before, it was still relatively well off.

Benjamin was hard-working—and the demands of his business were always great. He thus spent less time with his children than did his wife. It was Belle who was the disciplinarian in the family. She was the one who nurtured her children's artistic talents, exposed them to the vast cultural resources of New York City, and spurred them on. Harold remembers her as a "very fastidious lady" and "very protective of her children."

While growing up, Harold and Julian shared a large room. Both of them attended PS 186 on 145th Street between Broadway and Amsterdam Avenue, an easy five-block walk from their home. Harold recalls his teachers complaining about his "not living up to his potentials"—and that they did the same with

Julian. Although Julian was exceedingly precocious and clearly very talented, he was not considered the bright one in the family: "The one who got all the recognition and all the prizes was his older brother."[7] Harold was an outstanding student: he was the valedictorian of his junior high school class, and he finished in two years the graduation requirements at Townsend Harris, New York's premier high school for gifted students. Julian followed in Harold's footsteps. But it was only after Schwinger came under Rabi's tutelage at Columbia that he too became an outstanding student in *all* his studies.

As a teenager, Schwinger seemed in many ways to be a caricature of Wilhelm Ostwaldt's *Grosser Männer*: precocious, impatient with literary studies, highly imaginative, and exceedingly quick. Harold remembers the young Julian as "shy and retiring" and that at a very early age he "had his nose in books." When Julian was eight or nine he attended the camp in which Harold was a swimming counselor. Harold relates that Julian spent most of his time in his bunk reading his books, and that he had to "take" Julian to the lake to teach him how to swim.

As Harold had done before him, Julian attended Townsend Harris, which was then probably the outstanding high school for academic studies in the United States. He was fourteen when he entered Townsend Harris in 1932. The high school was located on Amsterdam Avenue and 136th Street on the campus of the City College of New York (CCNY). Students normally completed the degree requirements in three years, and graduation from Townsend Harris automatically guaranteed entrance to CCNY.[8] The teachers at Townsend Harris were unusually able, and many of them were eminently suited by training and competence for college teaching positions. They taught at Townsend Harris because college instructorships were very scarce during the Depression. When Schwinger attended Townsend Harris one of his physics teachers was Irving Lowen. At the time, Lowen was working for his Ph.D. in theoretical physics at New York Univeristy (NYU), and he became an instructor there a few years later.

In the autobiographical statement attached to his Nobel Prize lecture, Schwinger indicates that the principal direction of his life "was fixed at an early age by an intense awareness of physics and its study became an all-engrossing activity."[9] His precocity and genius in physics made him a living legend even while in high school. William Rarita recalls that "one day my friend Irving Lowen, then a physics teacher at Townsend Harris, . . . told me he had a genius in his class [whose] name was Julian Schwinger."[10] Bernard Feld, who entered City College in 1935, remembers that already then there was a story circulating that Lowen had "come across this kid sitting in the library [of CCNY] reading the *Physical Review*. He looked over his shoulder [and saw that he] . . . was reading Dirac. So Lowen thought, well, here's another one of these smart aleck kids that we get every once in a while. So he quizzed him about what he had been reading and Schwinger allegedly was not only capable of telling him what he was reading but also told him what needed to be done to complete what Dirac hadn't completed in this particular paper."[11] Lowen introduced Schwinger to his teacher at NYU, Otto Halpern, and a year and a half later Halpern and Schwinger collaborated on a paper together (Halpern and Schwinger 1935).

In the fall of 1934, after graduating from Townsend Harris, Schwinger entered CCNY. During the 1930s the colleges of the City of New York—CCNY, Brooklyn College, Hunter College, and Queens College—were remarkable institutions. Only high school graduates who had earned a grade point average above a high minimum were admitted. For most of the students, enrollment in these tuition-free institutions was their only means to obtain a college education. The students were bright, ambitious, and hardworking. Many of them were Jewish, children of immigrant parents, from families in which scholarly activities and respect for learning were two of the highest values. A college education was the means for upward mobility; and what a better investment than free tuition! Among the graduates of these institutions during the 1930s are to be found some of the outstanding scholars and scientists of their generation.[12]

The city colleges could boast of their impressive faculties. At CCNY, Morris R. Cohen taught in the department of philosophy, and Emile Post was a professor of mathematics. The physics department was staffed by outstanding teachers, among them Mark Zemansky, Henry Semat, Simon Sonkin, and Robert Wolf. The senior faculty usually taught the advanced physics courses and gave the lectures in the introductory courses. The recitation sections of the introductory physics courses were taught by graduate students from Columbia and NYU. Lloyd Motz, who was studying for his Ph.D. at Columbia, was also an instructor in the physics department at CCNY. Harold Schwinger and Lloyd Motz had been classmates at CCNY in the late twenties. Harold had told Motz: "I have a younger brother [who] people say is a genius—would you get to know him."[13] Motz took Schwinger under his wing. He immediately realized that Schwinger as a freshman knew too much physics to have to take the introductory courses. He went to see Corcoran, the chairman of the physics department, and asked him to let Schwinger take some of the advanced courses. "Over my dead body," is what Corcoran is alleged to have said. "As long as I'm chairman of this department no smart ass kid is going to be allowed to skip taking my course in elementary physics."[14]

The extent of Corcoran's poor judgment can be gauged from the fact that Schwinger wrote his first paper while a freshman at City College. Its title was "On the Interaction of Several Electrons." In it, Schwinger generalized the Dirac-Fock-Podolsky (Dirac et al. 1932b) many-time formulation of quantum electrodynamics by describing the electrons *field theoretically* by second-quantized *field operators*. What Schwinger would later call the interaction representation (Schwinger 1948b) was introduced for *all* the field operators (electromagnetic as well as the matter field operators), and the interaction between charged particles was computed to order e^2. Although he never published it, the manuscript was typed and it is a testament to his impressive maturity in physics at age 16.[15]

Morton Hamermesh,[16] who was also a student at City College at that time, recalls that Schwinger and he were enrolled in a course on "Modern Geometry" taught by an "old dodderer" named Frederick B. Reynolds, who was the head of the mathematics department:

> It was a course in projective geometry from a miserable book by a man named Graustein from Harvard, and Julian was in the class, but it was very strange because he obviously never could get to class, at least not very often, and he didn't own the book. That was clear.[17] And every once in a while he'd grab me before class and ask me to show him a copy of this book and he would skim through it fast and see what was going on. And this fellow Reynolds, although he was a dodderer was a very mean character. He used to send people up to the board to do a problem and he was always sending Julian to the board to do problems because he knew he'd never seen the course. And Julian would get up at the board, and of course, projective geometry is a very strange subject. The problems are trivial if you think about them pictorially, but Julian never would do them this way. He would insist on doing them algebraically and he'd get up at the board at the beginning of the hour and he'd work through the whole hour and he'd finish the thing and by that time the period was over and anyway, Reynolds didn't understand the proof, and that would end it for the day ... that was my introduction to Julian.[18]

Schwinger did not do well at CCNY. He spent most of his time in the library reading advanced physics and mathematical texts, and rarely went to his classes. Herman Feshbach, who was one year ahead of Schwinger at CCNY, remembers him as a shy, yet friendly and social teenager, endowed with amazing powers. He has a vivid recollection of Schwinger as a sophomore in 1935 giving a clear and illuminating lecture on general relativity to the mathematics club.[19] Schwinger "read everything there was."[20] He had studied Dirac's *Quantum Mechanics* by himself before coming to CCNY, and as a freshman he was reading the recently published papers in quantum field theory by Dirac, Heisenberg, Møller, Oppenheimer, Pauli, Fermi, and others. He did interact with some of the tutors who were studying for their Ph.D. at Columbia or NYU—Lloyd Motz, Hy Goldsmith, Irving Lowen— but "Dirac and all his papers were by far the overwhelming influence." In fact, Schwinger describes himself as "unknown to him, a student of Dirac."[21] He often went browsing in the secondhand bookshops on 4th Avenue near the old Wanamaker department store and in the huge Barnes and Noble bookstore on 5th Avenue and 15th Street, and frequently bought texts he thought looked interesting. The bookshelves in his office at UCLA still contain some of these books: E. B. Wilson's *Advanced Calculus*, H. Weyl's *The Theory of Groups and Quantum Mechanics*, V. Volterra's *Theory of Functionals*, and tucked away in a bottom corner, *Studies in the Psychology of Sex, Studies in the Sexual Impulse*, and *Love and Pain*, all by Havelock Ellis.

Lloyd Motz, the person who probably knew Schwinger best when he was at CCNY, remembers him as "very, very shy, introverted, gentle, kind and musical."[22] Schwinger still plays the piano rather well. Motz's evaluation of him at the time was that "he was so far above anybody else—there was no way to compare him to anyone. It appeared to me that there was nothing he didn't know—in mathematics or physics."[23] Motz recalls that during the fall of 1934, Schwinger was working on the quantum mechanical description of the behavior of spins in magnetic fields. At the time Schwinger was also helping Wills, one of the faculty members at CCNY, and was calculating wave functions for him.

Schwinger's grades at City College reflected his erratic class attendance. Although he had no difficulty getting A's in mathematics and physics—even though he rarely went to class—the same was not the case in his other courses. The matter got serious enough for Motz to bring Schwinger's problems to Rabi's attention. Rabi vividly remembered the episode and told it many times:

> There was this paper of Einstein, Podolsky and Rosen, and I was reading it. And one of my ways of trying to understand something was to call in a student, explain it to him and then argue about it. And [in this case] this was Lloyd Motz who was also at that time, I think an instructor or something at City College . . . we talked about it. And then he said there was somebody waiting for him outside, so I said call him in. So he called him in and there was a young boy there. [I] asked him to sit down and we continued. And then at one point there was a bit of an impasse and this kid spoke up and used the completeness theorem [of orthogonal functions] to settle an argument. . . . I was startled. What's this, what's this? So then I wanted to talk to him which I did. . . . It turned out I was told he was having difficulty at City College. I knew the people he was having difficulty with and it's an honorable thing to have difficulty with those people. One of his professors, in English, [was] an old friend of mine, . . . and he wasn't doing well in English.[24] So I asked him why is this? He said, I don't have the time to do the themes. Well, anyway, I suggested transferring to Columbia and then got a transcript from City College, and took it to one of the officials—I forget who it was. Now what about a fellowship, scholarship for this guy? He looked at the transcript and said, on the basis of this we wouldn't admit him. I [then] said something very tactless—I said, suppose he were a football player. I suspect it was the wrong thing to say, but I was never very tactful. Still the problem remained. [Since] he had written something on quantum electrodynamics . . . [and] Hans Bethe happened to pass through[25] I gave it to him and said what do you think of this? He thought very well of it and on this basis I just

simply overrode them and he was admitted. Everything changed for Julian after that. He actually became a member of Phi Beta Kappa. A reformed character.[26]

Norman Ramsey, who was a graduate student at Columbia University at the time and was a member of the Phi Beta Kappa election committee, remembers the difficulty Schwinger's election presented: he had obtained an F in Victor LaMer's chemistry course. Ramsey had taken this same LaMer course and recollects that

LaMer was a nice guy but it was really quite a dullish course and his examination was equally dull. It was always based on the notation—that was the course. Julian . . . took the course, stayed through one lecture and decided it wasn't worthwhile. He could learn more in the library. And in fact that was his usual procedure. . . . He would invent answers never [having] seen the original course, but he would just improvise. [It] made the problem of people who graded his exams usually fairly difficult. But he got LaMer's examination and the typical question was: Prove $d\varepsilon$ is equal to $d\xi + d\eta$. But he didn't know what $d\varepsilon$ or $d\eta$ were. That wasn't defined. You had to take the lecture[s] . . . to do that. So he got an F. My understanding of that story . . . was that LaMer went to [Rabi] and said what shall I do. This genius you've been boasting about just failed my course. And [Rabi] said to LaMer, did he really fail? And LaMer said, failed cold. And [Rabi] said to LaMer, are you a mouse or a man? "Give him an F."[27]

Rabi claims he didn't tell LaMer the grade, but that otherwise the story is accurate: "I didn't like LaMer. He was a great reactionary, a great opponent of Harold Urey and so on. So if he was going to flunk Julian I think that's something he, LaMer, deserved."[28]

Schwinger was elected to Phi Beta Kappa, but only after a "big argument."[29] There is a fitting conclusion to the LaMer story. Some years later in 1948, Schwinger was asked to give an invited paper to the APS meeting in New York, which was held at Columbia University. As Ramsey remembers it:

This was the only time in the history of the Physical Society [that] the following occurred: Darrow always scheduled theoretical talks in small rooms. It was scheduled in the smallest lecture room [in Pupin]. It was crowded two hours before [Schwinger] was to give [his] talk, so there was real objection. It was then put in the biggest lecture in the physics department and that was crowded an hour before—the word had then spread of this exciting development that [Schwinger] had done in quantum electrodynamics. So it was rescheduled for a third time in MacMillan Theatre, the biggest place that Columbia has, and indeed then the people could all manage to fit... practically the whole Physical Society... into that room.... I got to two of them—I missed the first one.

... It was a superb lecture. We were impressed. And as we walked back together—Rabi and I were sitting together during the lecture—Rabi invited me to the Columbia Faculty Club for lunch. We were in the elevator coming back from this, who should happen to walk into the elevator with us but LaMer. And as soon as Rabi saw that a mischievous gleam came in his eye and he began by saying that was the most sensational thing that's ever happened in the American Physical Society. The first time there's been this three repeats—it's a marvelous revolution that's been done—LaMer got more and more interested and finally said, "Who did this marvelous thing?" And Rabi said, "Oh you know him, you gave him an F. Julian Schwinger."[30]

Schwinger's talents as a lucid and polished lecturer were already evident when he was an undergraduate. Rabi would ask him to lecture in his graduate course on quantum mechanics whenever he had to go away. "I can assure you that it was a great improvement. He is a much better teacher than I ever was," Rabi declared. "One of the people in that same class was Bob Marshak and it was a bad time for being in the same class with Julian, no matter how clever you were."[31]

Already as a freshman at CCNY, Schwinger regularly attended the weekly theoretical seminar that Rabi and Breit ran at Columbia on Wednesday evenings. Lloyd Motz had taken him there when he first had met him. Rarita remembers the young Julian presenting to the seminar the content of the recently published papers by Born and Infeld (1934a, b). It was a "well delivered talk in a clear resonant voice."[32] After Breit left for Wisconsin in 1935, Halpern took his place. Otto Halpern who had come from Germany in 1933 upon Hitler's dismissal of all Jews in university positions, had obtained a position at NYU as a professor of theoretical physics. Hamermesh, who studied under Halpern, remembers him as "a rather testy fellow" who "couldn't be bothered with his [Ph.D.] students until they'd ripened." The joint seminar "was a sort of battlefield... Halpern would take on anybody and... loved getting into arguments. He would just take the greatest pleasure in taking on Gene Feenberg, or Fermi, when Fermi came, or anybody else. It just didn't matter. They were just violent fights."[33]

Schwinger, although a 16-year-old undergraduate, clearly must have impressed Halpern. They collaborated on the problem of the polarization of electrons in double scattering experiments—the young Julian doing extensive and "difficult" calculations—and published their conclusions as a letter to the *Physical*

Review (Halpern 1935). This was Schwinger's first publication. It was an important initiation. Scattering theory became the focus of many of Schwinger's subsequent researches. He would in fact soon apply what he had learned about double scattering in atomic scattering in his investigations of nuclear scattering. That same year, after having read Fermi's article on β-decay, Schwinger collaborated with Motz and calculated the lifetime of a neutron in the Konopinski-Uhlenbeck theory of β-decay (Motz and Schwinger 1935).

During his senior year at Columbia, Schwinger worked on the problem of the magnetic scattering of slow neutrons by atoms. At the time, the magnetic moment of the neutron had not been measured directly. Its value had been inferred from the magnetic moment of the deuteron on the assumption of simple additivity of the magnetic moments of neutron and proton. Bloch, in 1936, had suggested a direct method of measuring the neutron's magnetic moment that depended on the fact that an atom will scatter a neutron by virtue of the nuclear interaction with the neutron *and* the magnetic coupling between the atomic electrons and the neutron spin. Schwinger had read Bloch's paper (1936) and had come to the conclusion that Bloch's treatment of the magnetic interaction between the neutron and the electrons was inadequate. One could not describe the magnetic interaction as a "classical interaction" between magnetic dipoles. Schwinger proceeded to formulate a fully quantum-mechanical treatment of the problem using the "correct Dirac value [for] the current density of the electrons." He found that neutrons scattered from an unpolarized beam would be partially polarized by virtue of the magnetic interaction—and that the thus produced polarization could be detected by a second scattering. He worked out a formula for the neutron intensity after double scattering from magnetized iron plates, and suggested various experiments for producing and detecting a polarized beam of neutrons. He wrote up his researches, and in early January 1937 he sent a manuscript entitled "The Magnetic Scattering of Neutrons" to the *Physical Review*. In the concluding paragraph of the paper, Schwinger expressed "his indebtedness to Professor I. I. Rabi and Professor E. Fermi for helpful discussions and suggestions, and to Professor F. Bloch for an interesting conversation on this subject" (Schwinger 1937a).

Bethe was the referee and submitted the following comments on the paper by the 19-year-old Schwinger:

> This paper is no doubt a very important paper—but from the way it is written up, it is not apparent which are the really important parts. The main point seems to me that Schwinger shows that Bloch's formula for the magnetic scattering of neutrons is incorrect. *However he does not draw any attention to this point which, I think, is too great modesty on his part* [emphasis added]. It would be much easier for the reader if the correct formula and Bloch's formula were contrasted in the paper.
>
> Quite generally, the paper suffers somewhat from too great complication at places where it is not necessary. E.g., the

calculation of the scattering in the first section starting from the very foundations of scattering theory does not seem necessary after ten years of applications of the Born method. It would, in my opinion, be entirely sufficient to say that the amplitude of the scattered wave is equal to the contribution of the nuclear scattering plus that of the magnetic scattering, the latter being calculable from the Born method, using plane waves for the neutrons. Instead, I should like to see the fact emphasized that a different result is obtained by using the correct Dirac value of the current density and the corresponding magnetic field, than by using the "classical interaction" between two magnetic dipoles which apparently was used by Bloch.

Bethe concluded his report by stating that it seemed to him "that the paper would gain considerably in readability and applicability to experimental data if it were rewritten."[34] The editor thought otherwise, since evidently Schwinger's manuscript was printed in its original form.

Most of the characteristics that the mature Schwinger exhibits are already present in this paper: it is an important *physical* problem that drives the inquiry; the tendency is "to start from the very foundations"; the solution is elegant, the methods used are powerful; contact is made with experimental data, and suggestions for empirical tests are given.

Edward Teller, who was visiting Columbia in the spring of 1937, suggested that Schwinger's researches on the scattering of neutrons could be submitted as a Ph.D. thesis if developed further. Schwinger worked with Teller on the theory of the scattering of neutrons by molecular ortho- and parahydrogen (Schwinger and Teller 1937a)[35] and showed that the scattering of neutrons by ortho- and parahydrogen could yield information about the spin dependence and the range of the neutron-proton interaction. In the ortho-hydrogen molecule, the spins of the two protons are parallel; in parahydrogen, they are antiparallel. The scattering of a neutron by the two types of molecules will differ—particularly when the neutron's wavelength is of the same order of magnitude as the separation between the protons—because the amplitude for ortho- and for parahydrogen involve differing and interfering combinations of the triplet and singlet neutron-proton amplitudes. Schwinger and Teller pointed out that this interference phenomenon could be exploited to disentangle the zero-energy triplet and singlet neutron-proton scattering amplitudes.

The fact that Schwinger had written his Ph.D. dissertation before receiving his bachelor's degree is indicative of his remarkable talents. Incidentally, neither his arduous course work nor his extensive research activities nor his shyness prevented the young Julian from being interested in, or having time for, members of the opposite sex. Lloyd Motz tells of Julian as an undergraduate being courted by and dating a girl he later realized was the wrong person for him.[36]

After receiving his B.S. degree from Columbia in 1936, Schwinger continued his graduate studies there. The department of physics at Columbia

University at that time was a remarkable place, with Isidore Rabi its moving force. "Under Rabi's leadership, the Pupin Laboratories became an institution, and the institution a tradition—locally, nationally, and globally."[37] The people working with Rabi on his own experiments and on the other projects that he initiated, stimulated, and encouraged were a motley crew. Some were on the faculty of the various city colleges—J. Zacharias at Hunter College, S. Millman (who had been Rabi's student earlier in the decade) at Brooklyn College, M. Zemansky at City College; many were Ph.D. students at Columbia—J. Manley, Don Hamilton, J. Kellogg, N. Ramsey, P. Rosenberg, J. Kelley, V. Cohen, M. Fox; others were Ph.D. candidates at NYU—M. Hamermesh; and some were undergraduates at Columbia.

All were deeply committed to physics; all were spurred on by the brilliance, wit, charisma—and at times the wrath—of Rabi (Millman 1977; Rigden 1987). Morton Hamermesh reminisces that: "Columbia was an amazing place in those days...a golden age for Physics. They didn't care whether you had a job there, whether you were a student there. You simply walked in and worked. No one paid me...I didn't need them to pay me. I had an assistantship at NYU,...I knew lots of people who worked at Columbia who had absolutely nothing to do with the place. You just knew somebody and so you came to the party."[38] Schwinger became actively involved in the experimental activities at Columbia. He worked with John Manley and Hy Goldsmith on neutron energy levels (Manley 1937). Manley writes:

> The experimental work was done by me. Goldsmith and Schwinger would usually come to the lab about dinner time and the three of us would discuss the data, often until late. The division of labor was quite clear. I had considerable experimental experience but was just shifting from molecular beams to neutron physics. Goldsmith read avidly and knew the literature.
>
> Schwinger was very able in math and could analyze experimental procedure expertly.... No new theory was involved, just rigorous understanding of all relevant processes. It was a sympathetic, understanding relation in which all points, experimental and theoretical, were discussed until each felt completely familiar with the meaning of our procedures and of the results obtained. We educated each other.[39]

Schwinger also became involved with Rabi's atomic beam experiments and, stimulated by the experimental arrangement used in these experiments, wrote a paper that has become a classic on the behavior of an arbitrary magnetic moment in harmonically time-varying magnetic fields (Schwinger 1937b). This early association with Rabi and his atomic beam school resulted in "a protracted fascination with atomic and nuclear moments, and more generally, the quantum theory of angular momentum" (Schwinger 1978).

Shortly after Schwinger had begun his graduate studies, it became evident to Rabi—who was both the senior experimentalist and theoretician at Columbia—that he "had about had everything at Columbia that we could offer—by we, as theoretical physics is concerned, is me."[40] So Rabi got Schwinger a traveling fellowship from Columbia for the academic year 1937/38. The plans were for Schwinger to spend six months at Wisconsin to study with Breit and Wigner, and then to go on to Berkeley for another six months to work with Oppenheimer. Schwinger found the atmosphere at Wisconsin so pleasant that he did not go to California. "I kind of liked it. So I just stayed. I didn't know what I was missing."[41] It was at Wisconsin that Schwinger developed his characteristic working style: staying up at night and sleeping during the day. Although he had always been a night owl and liked sleeping late, the reversal became total at Wisconsin. Rabi and Schwinger have elucidated how this came about. Rabi recalled visiting Wisconsin after Schwinger was there for a while, and asking him, "How are you doing?" and Schwinger replying, "Oh everything is fine."

"[Are you] getting anything out of Breit and Wigner?"

"Oh yes, they're very good, very good."

Rabi then asked Breit and Wigner: "How is Julian doing?" "We never see him," was their reply.[42] Rabi's theory was that "they [Breit and Wigner] are great physicists, but . . . they wanted [Schwinger] to work along their lines. Being polite, not a man to argue, he just didn't show up when they were there. He is such a gentle soul, he avoided the battle by working at night." Schwinger concurred: "I think that goes to the heart of the matter. . . . That's my principle. Don't fight. Just don't be around."[43] In private, Schwinger indicates that he started staying up at night because of a strong feeling of not wanting to be "dominated" by Wigner and Breit. Incidentally, he had felt that Rabi also had wanted to do so, and that was one of the reasons he left Columbia.[44]

Although Breit and Wigner did not see much of Schwinger, he did participate in the seminar they ran. On October 6, 1937, Schwinger gave the first lecture in the seminar and reported on his work with Teller on the scattering of neutrons by ortho- and parahydrogen. He had in the meantime extended this work to show that neutron-hydrogen scattering experiments could determine the spin of the neutron (Schwinger and Teller 1937b). On December 12, 1937, and January 12, 1938, Schwinger gave two more seminars on the magnetic scattering of neutrons, and on May 11, 1938, he presented "Flugge's Calculations on Deuteron Reactions." The seminar was clearly very stimulating due to the presence of Breit and Wigner. It included talks on theory by J. K. Knipp ("Phenomena Associated with β Decay" and "Quadrupole-Quadrupole Interacting Forces"); S. Share ("Wick's Theory of Scattering of Neutrons in Crystals"); E. Wigner ("Physical Interpretation of Relativistic Equations with Several Times" and "Majorana's Theory of Negative Energy States [Papers by Majorana, Dirac, Furry, and Wigner's improvement]"); G. Breit ("Approximately Relativistic Equations," "Relativistic Corrections for the Deuteron," and "Astrophysical Speculations of Strömgren, Garnor and Others"), as well as reports on recent experimental developments by W. Kanne ("Gamma and Beta Rays from the Isotopes of Li, B. F"); E. R. Wicher ("The Fine Structure of Hydrogen"); L. Ragan ("Induced Radioactivity").[45]

Rabi characterized Schwinger "a changed man" when he returned to Columbia after his year at Wisconsin. He became deeply involved in the experimental activities in Rabi's laboratory and became the "house theorist" for the experiments in nuclear physics being carried out at Columbia at the time. Schwinger himself suggested, designed, and helped run an experiment on the neutron-proton scattering cross section (V. Cohen et al. 1939). He is particularly proud of the result obtained by the collaboration since the number is "better than Fermi's by 50% and is still accurate."[46]

Rabi stressed that Schwinger was of great importance to his group. He was "very good, in the sense [that when] you ask[ed] him to do something,... you got a number. Feenberg [a theoretician at NYU during this same period] would give you a formula, but Julian was very close to experiments, especially through Goldsmith."[47] Schwinger's papers from that period reflect not only his wide interests—the scattering of neutrons by hydrogen molecules, by deuterons, the quadrupole moment of the deuteron, the effect of tensor forces on the scattering of neutrons by protons, the widths of nuclear energy levels—but also attest to his talents as a superb phenomenologist. Nor do the papers give a full view of the range of his interests and involvements. For that one must study the abstracts of the papers he presented at meetings of the American Physical Society, and the many calculations that never saw the light of day. Morton Hamermesh, who was then just starting to do research, recollects:

> I remember very well that the first piece of research that I worked on was something having to do with...a paper on the widths of nuclear energy levels by Manley, Goldsmith and Schwinger (Manley et al. 1939).... I was looking for something to do and I came there and Julian said, "Well, there is this thing that should be worked out and I haven't done it because it takes a little more work. Here's what you do." It was some calculation about albedo of neutrons and I didn't know what he was talking about. I mean, I really was miserably educated. And then a process began which recurred very often after that. The calculation involved a great deal of information about Bessel functions and then here's how you get the solution. And then he went on this way for about, I think, it must have been five days. He put in an enormous number of hours, by which time he had reproduced at least three-quarters of Watson's *Bessel Functions*, but this apparently was a technique he liked to use. It was very helpful to me because I began to see how you did physics. At least math-

ematical physics. And I remember the great pride I had when my name appeared in an abstract of a paper in the *Physical Review*...[Hamermesh and Schwinger 1939]. It was along about this time that I started to help Julian and Goldsmith and Cohen, Bill Cohen, in some further experiments in nuclear physics, including of course, the neutron-proton interaction...Life was very strange. I would work up at NYU or City College, come to Columbia around three o'clock, start doing calculations. Julian would appear sometime between four and six and we would have a meal which was my dinner and his breakfast, and then we would begin the evening's work. We were experimenters, if you can call us that. That is, we were capable of putting foils in front of a radon beryllium source and measuring transmissions through them and activations,...grabbing the foils, running down the hall of Pupin—it was on the top floor—running like crazy, putting the foil on a counter, and taking a reading. And then we would run back, put them up again, and start doing theoretical work. And we would work rather strange hours. It seemed to me we would work usually to something like midnight or one A.M. and then go out and have a bite to eat. This would mean two or three hours during which I would get educated on some new subject. I learned group theory from Julian,...as I recall, I had all of Wigner's book given to me, plus a lot more at the time. This was a regular process we went through and I think this must have gone on for a year or so and we started doing calculations on ortho-para deuterium and on ortho-para hydrogen scattering or neutrons, and this involved just an unbelievable amount of computation. And we would work on this nights.[48]

Even though Schwinger was a "changed man" upon his return from Wisconsin, some of his former traits lingered on. Rabi remembers Schwinger taking a course in statistical mechanics with Uhlenbeck, who was visiting Columbia during the academic year 1938/39. Uhlenbeck, who was steeped in the Dutch tradition, gave an oral final examination. During the final week of his visit Uhlenbeck came to Rabi and asked him what to do about Schwinger. "What do you mean?" asked Rabi. "Well," replied Uhlenbeck, "Schwinger is taking the course, but he has never attended any of the lectures and he hasn't even shown up to make arrangements for the examination!" "Well, that's bad," thought Rabi. So he got a hold of Schwinger and said, "Now look, Julian, that's not polite," and Schwinger, "not being a man to give offense," went to Uhlenbeck to arrange for an examination. He evidently tried to obtain one with him for ten o'clock at night. But Uhlenbeck "is no weak character" and the time was set for ten o'clock in the morning," and to this sacrifice

Julian appeared."[49] Uhlenbeck later told Rabi that what surprised him was that not only did Schwinger answer all his questions but that "he had done a little research in the field." And what astonished Uhlenbeck even more was that he presented these results in Uhlenbeck's notation even though "he hadn't appeared in class."

Schwinger thinks back fondly to his days at Columbia. Many "people came through"—Bloch, Dirac, Fermi, Heisenberg, Pauli, Teller, Uhlenbeck— exposing him to what was going on at the frontiers of physics. And as the acknowledgments in his papers indicate, visits were opportunities for extended discussions. "For a guy who is fundamentally shy I did get around." By "getting around" Schwinger means "talking."[50]

Schwinger's outstanding abilities were rapidly recognized by the theoretical physics community. At Columbia he became known as the person "who would help you if you had any problems."[51] Pauli wrote him a very complimentary letter upon receiving his paper on the motion of a spin J magnetic moment in a time-dependent magnetic field.[52] In the summer of 1938, Kramers invited him to come to Leyden for a year. Van Vleck thought that Schwinger was "the outstanding candidate" for a three-year appointment to Harvard's Society of Fellows, and in the fall of 1938 wrote Rabi to "get [his] ideas concerning putting up Schwinger."[53] However, "on thinking over this matter of Schwinger rather more thoroughly and talking with Schwinger," Van Vleck concluded "that it seemed to [him] best that Schwinger continue at Columbia," since Rabi could look after him there. Moreover, "there are one or two other excellent candidates in theoretical physics to whom the award of the fellowship will mean more than to Schwinger in as much as he has every advantage at Columbia. . . . [W]e can do little, if anything, for him that you [Rabi] cannot also do at New York."[54] Actually Rabi did not keep Schwinger at Columbia, but suggested that he apply for an NRC fellowship and go to work with Pauli. But Schwinger thought Oppenheimer "a more interesting physicist" and so, in the fall of 1939, he went to Berkeley.[55] Relations between Oppenheimer and Schwinger were strained at first. According to Rabi, Oppenheimer "was terribly disappointed" initially, and almost came to the point of writing a letter to the National Research Council suggesting that Julian go elsewhere. Rabi explained: "It took a man like Oppenheimer quite a bit to get used to Julian. Pauli once referred to Oppenheimer's students as being Zunicker. Somebody who knows enough German knows what he means—people who nod their heads. Julian wasn't that way—that and his hours. However, Oppenheimer thought better of it and he soon learned to not only respect but to love him."[56] Schwinger told of the relationship as follows: "Oppenheimer was a dominating personality, but I didn't want to be dominated. So to some extent I would back away and say: I have to think about this my own way."[57] Schwinger also remembers startling Oppenheimer with his "strange choice of physics problems." When he came to Berkeley he was asked to give a colloquium and he talked about the absorption and dispersion of sound waves, a subject that Morton Hamermesh had gotten him interested in.[58]

After the lecture Oppenheimer told him, "You do some very strange physics."[59]

But Schwinger readily admits that Oppenheimer's influence on him "was enormous.... He steered me into areas of physics that I hadn't thought of: cosmic rays, meson spin assignments and so on."[60]

Schwinger stayed in Berkeley for two years, the first as an NRC fellow, the second as a research associate to Oppenheimer. His stay was enormously productive. Oppenheimer, and the large number of doctoral students and postdoctoral fellows working under his tutelage, made for a very stimulating atmosphere. Schwinger characterized the group of young theoretical physicists at Berkeley in 1939—Dancoff, Snyder, Volkoff, Schiff, Bohm, Morrison, Serber—as "remarkable." Moreover, it was a congenial intellectual community. At Berkeley, perhaps more than anywhere else, Schwinger "really got around." He collaborated extensively—with Corben, Gerjouy, Oppenheimer, Rarita—and worked on a wide range of subjects. The underlying quest was always to understand the nuclear forces better. The investigations ranged from phenomenological analyses of empirical data on the deuteron and light nuclei to extensive field-theoretic calculations using differing spin and charge assignments for the "mesotrons" involved.

Before Schwinger came to Berkeley, a paper by Kemmer (1938) on a meson field theoretical model of the nuclear forces had made him aware of the possibility of the presence of tensor forces in the neutron-proton interaction. He remembers saying to himself after reading the paper, "Why doesn't somebody do something about this."[61] An analysis of the electromagnetic properties of the deuteron when tensor forces are present led him to predict the existence of the deuteron's quadrupole moment—before it had been measured by Kellogg, Rabi, Ramsey, and Zacharias (Kellogg et al. 1939, 1940). Tensor forces, and the kind of "mesotrons" and meson-nuclear couplings that could give rise to them, became a central focus of the investigations Schwinger carried out during the Berkeley period. These researches have one other feature in common: no matter how theoretical or abstract the starting point—whether formulating the quantum field theory of spin 3/2 particles or the solutions of the Proca equations for a charge spin 1 particle moving in a Coulomb field—contact is always made with "numbers" and experiments. The "numbers" came from a variety of experiments: cosmic ray shower phenomena, neutron-proton and neutron-deuteron scattering at low energies, molecular beam experiments, and, from the data on the binding energy, spin and magnetic moment of the light nuclei. And in the process of absorbing the "numbers," Schwinger mastered the intricate details of the experiments—the apparatus involved as well as the analysis and reduction of the data—from which the "numbers" had been obtained! H. C. Corben, with whom Schwinger collaborated on two papers dealing with the electromagnetic properties of spin 1 "mesotrons" (Corben and Schwinger 1940a,b), writes: "Oppenheimer suggested that I team up with Julian Schwinger.... Julian was usually dressed by 4 P.M. but did not want to start working until evening, causing Oppe to remark that our wave-functions did not overlap. Gradually I went to bed later and I think he arose a little earlier and we spent many hours together. He was only 20 at the time and his ability and knowledge astounded me."[62]

William Rarita, another of Schwinger's collaborators, was at Berkeley on a sabbatical leave from Brooklyn College during Schwinger's second year there. Oppenheimer had proposed to Rarita that he work on the problem of photodisintegration of the deuteron when the neutron-proton interaction includes tensor forces.[63] "Before long Schwinger worked out a special case of the problem and asked me to compare my results with him. In this way we started to work together. Schwinger worked the night shift. He got up about 2 in the afternoon and went to the seminar at 4 P.M. After dinner we talked and worked until 10 P.M., when I went to bed. He continued to work until 5 in the morning." That collaboration resulted in two papers (Rarita and Schwinger 1941a,b). Later that year Oppenheimer suggested to Rarita that he help Schwinger update his paper on the application of tensor forces to the binding energy of the deuteron and the scattering of neutrons on protons. "This collaboration led to two long papers on tensor forces" (Rarita and Schwinger 1941b,c). When Rarita began looking into the problem of the relativistic quantum-mechanical description of spin 3/2 particles, he discussed matters with Schwinger. "It was not long before Julian had an idea and again we worked together"[64] (Rarita and Schwinger 1941d).

Forty years after leaving Berkeley, Schwinger reminisced about his stay there. He indicated that two pieces of research carried out there were of great importance later when he worked on quantum electrodynamics in the post-World War II period. The first was a collaboration with Oppenheimer (Oppenheimer and Schwinger 1939) "that used quantum electrodynamics to describe the electron-positron emission from an excited oxygen nucleus, which emphasized [for Schwinger] the physical reality of . . . virtual photon processes" (Schwinger 1983a, p. 330). The second was his work on strong-coupling mesotron theory (Oppenheimer and Schwinger 1941; Schwinger 1970), through which he gained experience in using canonical transformations to extract the physical consequences of a field theory.

Yukawa had proposed his meson theory of nuclear forces in 1935, and within a year the "mesotron" was found. However, as we now know, it was the "wrong" meson (the muon rather than the π meson). This "mesotron" exhibited weak nuclear interactions in its passage through the atmosphere (Cassidy 1981; Galison 1983). Field theorists were thus trying to find a meson theory that yielded strong nucleon-nucleon forces yet a small meson-nucleon cross section. Heitler (1940) and Bhabha (1940) independently indicated how this could come about. They observed that the cross-section for the scattering of a neutral scalar meson on a nucleon would be quite small, because the direct and crossed Born terms very nearly cancel. For a charged meson the cross section is large, because only one of the Born terms exists. They pointed out that, if in addition to the nucleons there existed low-lying excited states of the nucleon—"isobars"—with charge $+2e$ and

$-1e$, then there would be two Born terms which could nearly cancel, and thus produce a small cross section as in the neutral case. Somewhat earlier, Heisenberg (1939) had noted a different mechanism. He had calculated the scattering of a neutral vector meson coupled to the spin of a nucleon and had found that the large Born approximation result was greatly suppressed by "reaction effects." He observed that the self-field of the nucleon, that is, that part of the meson field which is attached to the nucleon and carried with it, was responsible for an increased inertia of the spin motion, and resulted in a cross section of the order of a^2 where a is the assumed radius of the nucleon.

The next important step was taken by Wentzel, who in 1940 showed that the simplest nontrivial static model, that of a charged scalar meson field coupled to a fixed scatterer, could be solved quantum mechanically in the limit of a large coupling constant.

He found that the solution exhibited isobars; however, these isobars did not result in a particularly small cross section because their excitation energy was not small enough. Oppenheimer and Schwinger (1941) extended Wentzel's strong coupling calculations to the case of neutral pseudoscalar mesons. They found that Heisenberg's (1939) classical result for the scattering was reproduced, and noted that the solution gave rise to isobars that had been overlooked by Heisenberg and reduced the cross section for meson-nucleon scattering by the Heitler-Bhabha mechanism. "Stimulated by Wentzel's pioneering work on strong coupling theory," Schwinger gave a more extensive treatment of the strong coupling limit of the charged scalar mesotron field. He observed that the large coupling "should serve to bind mesotrons in stationary states around the nucleon" and give rise to isobaric states, which he exhibited by using a series of canonical transformations (Schwinger 1970).

Schwinger left Berkeley in the summer of 1941 to accept an instructorship at Purdue. That summer, he was one of the lecturers at the University of Michigan Summer Symposium. The other lecturers were Wolfgang Pauli, Frederick Seitz, and Victor Weisskopf. Schwinger's friendship with Weisskopf dates from that summer. In September 1946, when Rabi was trying to have Schwinger appointed as a full professor in his department, he made the following point in his letter to the president of Columbia:

> As evidence of the high position which he [Schwinger] has held among physicists, he was appointed to lecture at the University of Michigan for the Summer Symposium of 1941. This symposium, which is designed for a distinguished clientele, lasts approximately eight weeks and furnishes a congregating place for advanced theoretical physicists from all over America. It is thus a distinct honor to be invited to one of the lecturers and many of the outstanding physicists of pre-war days made many special trips to America for this occasion.[65]

At Purdue, as elsewhere, the teaching situation in physics had become desperate because many of the faculty members had left for war work during the 1940/41 academic year. Schwinger's teaching responsibilities included taking at least one section of the introductory physics course. While it was easy to schedule graduate courses for the end of the day to accommodate his habits, that was not possible for the undergraduate classes he had to teach. Some concessions were made by having his undergraduate classes meet in the early afternoon. But to insure his presence, the department secretary had orders to telephone him and wake him at noon on the days that he had to meet his classes.

In the spring of 1942, Robert G. Sachs, a young theorist who had just finished a postdoctoral fellowship with Edward Teller, was appointed to an instructorship at Purdue to help with the teaching. He came to know Schwinger fairly well while the two of them were at Purdue. "Our personal association during that brief period was quite close [and] . . . his visits to our house constituted a substantial fraction of his social life. His work habits did not allow for much social life since he slept all day, had a breakfast of steak, French fries and chocolate ice cream and started work at 7 or 8 P.M. He would be leaving for home in the morning when other members of the department were arriving."[66]

Sachs was struck by Schwinger's predilection for large cars. "Schwinger owned a LaSalle when he came to Purdue, but later acquired a large Cadillac in addition." Sachs also remembers celebrating Schwinger's birthday with him in February 1942. "We had to spend the whole time trying to cheer him up because he had already reached the grand old age of 24 and not yet made the required great discovery expected of him."[67]

As had been the case at Berkeley, Schwinger collaborated with the other theorists at Purdue. During his stay in Berkeley, Schwinger had investigated with Gerjuoy the effect of tensor forces on the three body nuclear system. They had tried to determine the amount of D state present by calculating the binding energy of the triton, but the results were ambiguous. In their paper they had also presented a method for characterizing the most general form of the three-body wave function, following a technique established earlier by Schwinger in a paper with Rarita (Gerjuoy and Schwinger 1941, 1942; Rarita and Schwinger 1941b). After reading the Gerjuoy-Schwinger papers, it occurred to Sachs that the magnetic moment of a nucleus might be a much more sensitive gauge of the admixture of states and convinced himself that calculating the magnetic moment of the triton using the Gerjuoy-Schwinger general form of the wave function would be a relatively easy matter. When Sachs asked Schwinger whether he had done the calculation, Schwinger said, "No, but let's do it."[68] The work was done quickly and presented at the Baltimore meeting of the American Physical Society in May 1942 (Sachs and Schwinger 1942). However, both of them became too involved with war work to write up their results in a paper. It was only after the war that the work was prepared for publication (Sachs and Schwinger 1946).

7.3 The War Years

Schwinger's contributions to the war effort were determined by his being at Purdue. An active program in semiconductor research was being carried out there by Lark-Horowitz for the Radiation Laboratory in order to develop better rectifiers for the detection of radar. In 1942 Schwinger and several other theorists at Purdue were asked to join a Rad Lab project on the propagation of microwave radiation under Hans Bethe's direction.

When World War II broke out in September 1939, Hans Bethe was not yet a citizen of the United States. After the fall of France in the spring of 1940, he was desperate "to make some contributions to the war effort." Unable to get clearance to work on classified projects, he devised a project of his own: "I chose the subject of the penetration of armor by projectiles" (Bernstein 1979, p. 61). He went to the *Encyclopaedia Britannica* and read the articles on armor penetration and armor manufacture. Two of his engineering colleagues at Cornell, Professors Goodyear and Winter, suggested to him the books to read on the theory of elasticity and of large buckling of plates—and he quickly became "something of an expert on these matters" (Bernstein 1979, p. 61). He worked out a theory of armor penetration that became the point of departure of much work on the subject during the war and proved helpful to tank manufacturers.[69] While completing this project, Bethe also worked on the theory of shock waves in gases. He obtained a solution to the problem "of the shock wave detached from the nose of the projectile which occurs at velocities slightly above the velocity of sound"[70] and collaborated with Teller on the problem of the approach to equilibrium of the gas behind the shock.[71] Bethe and Teller produced a theory of the process that later became the basis for the use of shock waves to investigate the properties of gases (Bernstein 1979, pp. 64–65). In March 1941 Bethe became a citizen of the United States, and on the day following Pearl Harbor he received his clearance to work on classified military projects. The first such project he became involved with dealt with microwaves. The extensive use of microwaves in offensive and defensive weapons made it imperative to study in great detail the theory of these waves and of the apparatus used to generate and detect them. The directorate of the Radiation Laboratory at MIT had recognized that a well-developed theory would save months of experimental labor that would otherwise have to be spent in searching, by trial and error, for the best size, shape, and arrangement of apparatus for a given purpose. They therefore asked Bethe to work on the theory of microwave propagation.

Bethe's initial work was confined to problems that were integrable in closed form. However, most of the experimental problems involved cavities, waveguides and similar devices of shapes that did not permit exact integration. Since the actual geometries were often close to simple shapes that had been treated exactly, the use of some kind of "perturbation theory" was indicated. Bethe developed the foundations for such a treatment in a paper entitled "Theory of Diffraction by a Small Hole," which he submitted to DuBridge in January 1942.[72] In that paper he solved "the problem of the effect of a small hole in a cavity upon the oscillation of that cavity." He proved that the usual Kirchhoff theory of diffraction, based on Huygens' principle, failed "completely" when the size of the hole is small compared with the wavelength, that is, in the case of relevance for microwave work. His solution also indicated that much less radiation was emitted through a small hole than would be expected from Kirchhoff's formula, the power being reduced approximately in the ratio (radius of hole/wavelength).[73]

The theory developed by Bethe made it possible to study theoretically the coupling of any number of cavities joined by small circular holes in their common plane boundaries. To make contact with experimental practice and with actual devices, the theory needed to be extended to encompass noncircular holes, holes in curved surfaces, and holes of sizes comparable to the wavelength of the radiation or even larger. This last problem was of importance for the coupling of waveguides to cavities, and in the computation of the angular distribution of the emission radiation by horns.

In February 1942 Bethe submitted a proposal to extend his earlier work for support by the Radiation Laboratory. He characterized the work as "not direct war research but rather [as] fundamental research which may serve as a basis for directly useful investigations," such as those currently undertaken at MIT. It therefore seemed to him "entirely feasible and, in fact, desirable to have the work done by various theoretical physicists at their home universities" in view of the urgent necessity for training more graduate students in physics. Bethe gave as an additional argument the fact that if the work were carried out in this way its cost would be reduced by "a factor of five or ten." He proposed that the following theoretical physicists be asked to cooperate in the project since all of them were acquainted with the problem "and have already started work on some of the questions" mentioned in the proposal:

Dr. Julian Schwinger, Purdue University
Dr. E. S. Akeley, Purdue University
Dr. J. F. Carlson, Purdue University
Dr. J. K. Knipp, Purdue University
Dr. R. E. Marshak, University of Rochester

The proposal asked for support of the investigators during the summer, for travel funds so that Marshak could visit Ithaca every month and that one of the Purdue physicists could travel to Ithaca every two months, and finally for funds so that a conference of all workers on the project could be held two or three times a year.[74]

This was the beginning of Schwinger's association with the Rad Lab. He came to Cornell several times to consult with Bethe. The work, carried out during 1942, resulted in a report entitled "Transmission of Irises in Wave Guides" whose authors were H. A. Bethe (Cornell), J. Schwinger and J. F. Carlson (Purdue), and L. J. Chu (MIT).[75]

When Los Alamos was organized in 1943 to build an atomic bomb, Oppenheimer asked Schwinger to join the laboratory there but he declined: "I would like to think that I had a gut reaction against [going]. I was probably the only active theoretical nuclear physicist who wasn't there. There must have been some deep instinct to stay."[76]

Perhaps his fear of being "dominated" was part of his gut reaction! However, as many of the leading theorists at the Rad Lab were leaving to go to Los Alamos, Schwinger was asked to come to MIT and work there full time. This he did in the fall of 1943. But before going to Cambridge, Schwinger spent the summer in Chicago at the Met Lab. Bernard Feld, who knew Schwinger from Columbia, recalls:

> I was in Chicago and Julian arrived at Metallurgical Laboratory to spend ... two months in the summer of 1943. I'd been there a year and I was pretty sophisticated by then. In fact, I was doing experiments. That was the period in Chicago after the chain reaction had been proved and what the people in Chicago were doing was designing the Hanford Reactor. Now that was not easy because nuclear engineering didn't exist. They were really discovering and inventing nuclear engineering. This was Wigner and Szilard and Fermi and Wheeler and a crowd of young people who were working with them. They were all physicists but in a pinch a physicist, a theoretical physicist makes a pretty good engineer. In fact, they were all first-class engineers. No question ... I didn't realize that he'd been doing engineering already at the Radiation Lab, but, Julian is also a pretty good engineer and he demonstrated it that summer. I guess there were a number of problems that remained to be solved, mainly to help in the design of Hanford, and these were in transport theory and some of them were more or less difficult and Julian was working on some of them. The things I was doing were not really directly connected with the Hanford thing and the design of Hanford was the most important. Anyhow, there was the need of some kind of a match between Julian and the Metallurgical Laboratory, because Julian didn't come in until sometime in the evening and most people were already gone, so the contact wasn't well established, so I sort of elected myself in that period to be the matchmaker. I had known Julian from Columbia, slightly, but at least well enough that I figured that I might be able to help in making the match between Julian and

the project. It was really a very interesting process because what would happen was that I would go around in the afternoon—not every afternoon, but occasionally—go around to my friends in Wigner's group and sort of try to smell out what were the problems with which they were having trouble. Things that were giving them some difficulty. And then sometime in the late evening, maybe 10 o'clock or 11 o'clock or something, I would wander off to Julian's office and wander in and Julian would be sitting there at his desk, typically. This was a very hot summer in Chicago, and Julian was a very fastidious dresser in those days. He never took off his coat. In Chicago I never saw Julian without his jacket. He'd be sitting there with his white shirt and tie, tie never even loosened, jacket on, with a pad and a paper, he'd be scribbling furiously, working on some problem on the pad, with his handkerchief, supersaturated handkerchief in the left hand, mopping the sweat off his brow as he worked, and I used to wander in and sit down, and wait and at some point Julian would pause to catch his breath and I would kind of interrupt him and try to get his attention away from whatever he was doing, and I usually succeeded and not only because I was a pretty persistent guy, but because Julian is a nice guy and if you sort of bother him, he'll pay some attention to you, and after a while I would get him interested in the particular problem I had in mind. I'd start talking about it and Julian would get interested and then he would go to work on it. He'd get up to the blackboard and I would start taking notes. As Julian worked on the problem, I would be taking notes and sometimes, you know, that could be pretty hectic. I don't know any of you who saw Julian work in those days. Julian is ambidextrous. He has a blackboard technique that uses two hands, and frequently, when he really got carried away, he would be solving two equations, one with each hand, and trying to take notes could be a hectic job. Well, at some point, either we would finish the problem or the dawn would start to break in the eastern horizon, and we would decide it was time to quit and then often, we would go to have breakfast together. We would get into Julian's sleek black Cadillac and go to the nearest all-night eatery, where we would both have breakfast of, I think, it was a steak . . . and then go off to our respective beds.[77]

When Schwinger arrived in Cambridge in the fall of 1943 to work at the MIT Rad Lab, he rented a room in a house that Robert ("Bob") Marshak and Nathan Marcuvitz had sublet from a Harvard biology professor. Schwinger had known Marshak from his undergraduate days at Columbia and from his previous collaboration with Bethe. Schwinger and Marcuvitz "got to chatting," and became good friends.[78]

Schwinger joined the theoretical group under George Uhlenbeck's leadership and became "the stimulus"[79] for the waveguide work that was carried out by several members of the group, in particular J. Frank Carlson, A. E. Heins, Harold Levine, P. M. Marcus, and David Saxon. Harold Levine had come to the Rad Lab early in 1944 after completing his Ph.D. at Cornell. He recalls that "Schwinger would arrive late afternoon—have dinner—come back, and be available for help." Although Levine found Schwinger "quite formal," he adds that he was also "approachable and available, and interactions were a little informal."[80] David Saxon, who was a graduate student at MIT before joining the theoretical group, indicates that Schwinger was held in "great awe" by the members of the theory group, but that those who overcame the barriers of Schwinger's shyness and their sense of inferiority found that a bond of "great camaraderie and great friendship" developed.[81]

Schwinger's working habits at the Rad Lab were the same as they had been elsewhere: he became the "night research staff." He would arrive late in the afternoon and work through the night.

While at the Radiation Laboratory, Schwinger worked on a number of problems, all connected with the theory of waveguides. (See Oliner 1984.)

In the biographical note submitted on the occasion of receiving the Nobel Prize, Schwinger stated: "I first approached radar problems as a nuclear physicist; soon I began to think of nuclear physics in the language of electrical engineering."[82] The person who influenced him to think in the language of electrical engineering was Nathan Marcuvitz. Five years older than Schwinger, Marcuvitz had been trained as an electrical engineer at the Brooklyn Polytechnical Institute and was a member of division 4.3, the advanced development group headed by Ed Purcell. He and Schwinger would often discuss waveguide problems, with Marcuvitz expounding the engineer's viewpoint. As a result of these discussions, they reformulated waveguide theory—which was usually formulated in terms of partial differential equations and boundary conditions— in the language of transmission lines and networks that used ordinary differential equations. The mathematical techniques employed involved expansions in appropriate complete sets of functions. This nodal point of view was then proved to be equivalent to the network point of view. Microwave theory was thus translated into engineering terminology. Their method of collaboration consisted in eating together, around 7 P.M.— dinner for Marcuvitz, breakfast for Schwinger—and chatting for a while afterwards at the lab. Thereafter Schwinger would work through the night and leave some notes on his desk for Marcuvitz to read in the morning, and the discussion would resume in the evening. Schwinger would also talk about some of the other things he was working on, for example, variational methods and scattering theory for microwaves. Marcuvitz remembers finding a set of notes marked "A new era dawns" the morning after Schwinger had rediscovered Weiner-Hopf techniques for solving some problems in scattering theory.[83]

Marcuvitz, who later edited the *Waveguide Handbook* in the *Radiation Laboratory Series,* noted in its preface:

> In the years 1942 to 1946 a rather intensive and systematic exploitation of both the field and network aspects of microwave problems was carried out at the Radiation Laboratory at MIT by a group of workers among whom J. Schwinger played a dominant role. By means of an integral-equation formulation of field problems, Schwinger pointed the way both in the setting up and solving of a wide variety of microwave problems. These developments resulted in a rigorous and general theory of microwave structures in which conventional low-frequency electrical theory appeared as a special case. (Marcuvitz 1950, p. vii)

Schwinger gave lectures on the theory of waveguides that were attended by a small group of colleagues. Notes on these lectures were taken by David Saxon, and they were duplicated and circulated among the participants. An abbreviated set of notes—the war ended before the series had finished, and some of the lectures notes were never written up—were eventually published in 1965 by Schwinger and Saxon under the title *Discontinuities in Wave Guides.* Since the notes were reprinted as they were originally issued with inaccuracies not corrected, the book conveys "the flavour of the state of wartime study" (Saxon and Schwinger 1965, p. ix). The lectures exhibit Schwinger's formidable analytical powers. In them, he also demonstrates his ability to be equally at home in the world of physics, of applied science, of engineering, and of applied mathematics.

For a particular waveguide junction, one usually only wants to know the amplitudes of the waves coming out of all the arms, given the amplitude of the wave entering a particular arm. Schwinger's initial set of lectures indicates how such a "far field description of obstacles" is given and examines "certain characteristics of the scattered field which arise in consequence of the law of conservation of energy and the form of the differential equations satisfied by the field and which are independent of the detailed nature of the obstacle" (Saxon and Schwinger 1965, pp. 7–16). These "characteristics" are the unitarity and symmetry of the "scattering matrix" (as the S-matrix was then known) that relates the scattered and incident waves. Schwinger had read Wheeler's and Breit's work on the scattering matrix (Wheeler 1937; Breit 1941) and had appreciated its importance and usefulness. He had also attended the lectures on meson theory that Pauli (1946b) gave at the Rad Lab in the fall of 1944 in which Heisenberg's detailed investigation of the S-matrix (Heisenberg 1943a,b) had been presented. But he quickly proceeded from these formal considerations to the practical problems at hand: the computation of the elements of the scattering matrix—in that context, the "reactance" and "susceptance" matrix—for various configurations of immediate relevance in the design of radar sets. To this end he introduced and deployed powerful mathematical techniques that became characteristic features of many of his subsequent inquiries: Green's functions and variational methods. In a memorial lecture for Sin-itiro Tomonaga delivered in 1980, Schwinger mused "that those years of distraction were not without their useful lessons": "The wave guide investigations showed the utility of organizing a theory to isolate those inner structural aspects that are not probed under the given experimental circumstances. That lesson was soon applied to the effective-range description of nuclear forces. And it was this viewpoint that would lead to the quantum electrodynamic concept of self-consistent subtraction or renormalization" (Schwinger 1983b, pp. 365–366).

But war work and waveguides were not an all-consuming activity. Schwinger also worked on the problem of synchrotron radiation, which taught him "the importance of describing relativistic situations covariantly—without specialization to any particular coordinate system" (Schwinger 1983b, p. 366). Some forty years later Schwinger recalled:

> As the war in Europe approached its end, the American physicists responsible for creating a massive microwave technology began to dream of high-energy electron accelerators. One of the practical questions involved was posed by the strong radiation emitted by relativistic electrons swinging in circular orbits. In studying what is now called synchrotron radiation, I used the reaction of the field created by the electron's motion. One part of that reaction describes the energy and momentum lost by the electron to the radiation. The other part is an added inertial effect characterized by an electromagnetic mass. I have mentioned the relativistic difficulty that electromagnetic mass usually creates. But, in the covariant method I was using, based on action and proper time, a perfectly invariant form emerged. Moral: To end with an invariant result, use a covariant method and maintain covariance to the end of the calculation. And, in the appearance of an invariant electromagnetic mass that simply added to the mechanical mass to form the physical mass of the electron, neither piece being separately distinguishable under ordinary physical circumstances, I was seeing again the advantage of isolating unobservable structural aspects of the theory. Looking back at it, the basic ingredients of the coming quantum electrodynamic revolution were then in place. Lacking was an experimental impetus to combine them and take them seriously. (Schwinger 1983b, p. 366)

Schwinger also found time to work with Hamermesh to write up their researches on the scattering of neutrons off molecular ortho- and parahydrogen

(Hamermesh and Schwinger 1947). Hamermesh recalls Schwinger showing him some of the work he had accumulated during that period: "He apparently worked on a lot of problems, solved them and then squirreled them away for possible future publication. One of these was the scattering of neutrons in the coulomb field of the nucleus, another was the essence of his long paper on angular momentum, using creation and annihilation operators."[84]

Schwinger was then also courting his wife-to-be, Clarice Carrol, and the two of them would often visit friends, in particular the Hamermeshes. Interestingly, Nathan Marcuvitz remembers Schwinger telling him at that time that he would quit physics at forty and become a composer.

Just after the Trinity test, Schwinger spent a week at Los Alamos lecturing on nuclear physics. "It was a good time to get here. Every one was euphoric."[85] Feynman, who had been at Los Alamos since early 1943, gave the following account of Schwinger's visit:

> It was not until I went to Los Alamos that I got a chance to meet Schwinger. He had already a great reputation because he had done so much work . . . and I was very anxious to see what this man was like. I'd always thought he was older than I was because he had done so much more. At the time I hadn't done anything. And he came and he gave us lectures. I believe they were on nuclear physics. I'm not sure exactly the subject, but it was a scene that you probably have all seen once. The beauty of one of his lectures. He comes in, with his head a little bit to one side. He comes in like a bull into a ring and puts his notebook down and then begins. And the beautiful, organized way of putting one idea after the other. Everything very clear from the beginning to the end. You can imagine for a lecturer like me, what a sensation it was to see such a thing. I was supposed to be a good lecturer according to some people, but this was really a masterpiece. Each one of the lectures was a great discourse while what I did was a talk on something. So I was very impressed and the times I got then to talk to him, I learned more.[86]

Schwinger had gone to Purdue University in 1941 as an instructor and was promoted to an assistant professorship in 1942. He was on leave from Purdue during his stay at the Rad Lab. When in 1944 universities began competing with one another for the outstanding talents in physics, Schwinger was courted by a number of them, in particular by Harvard. In the summer of 1944 the Department of Physics there had obtained permission to make a "new major appointment." It opened a quest for "a man of the highest distinction" which "[led] automatically to candidates who are either in theoretical or experimental nuclear physics."

Since the department was not in a position to defray the cost of new apparatus nor to pay for the supporting personnel necessary to attract an experimenter, it decided to look for a theorist who would also serve as a "theoretical advisor" to the experimental program that Bainbridge was overseeing using the Harvard cyclotron.[87]

On December 28, 1944, Van Vleck, the chair of the department, wrote E. P. Wigner, G. Breit, F. Bloch, Lee DuBridge, and E. Fermi to tell them that Harvard "is considering the possibility of offering Dr. Julian Schwinger an appointment in the field of theoretical physics and is presently engaged in collecting as much information as it can about Dr. Schwinger." In mid-January, Van Vleck also wrote Pauli asking him "to be so kind as to favor us with your candid opinion of the merits of Dr. Schwinger's work in this field [theoretical physics], of his abilities and his potentialities, and of his general adaptability to a place on the University faculty. We should also be glad of suggestions of other men in this field, of equal or greater ability." Some concern had been expressed about Schwinger's "reputation as a night person" and Van Vleck asked Morton Hamermesh, who was working at the Harvard Radio Research Laboratory under Van Vleck, whether he thought Schwinger could really meet classes. Hamermesh assured him that "Julian could get to classes by 11 A.M. or noon."[88]

By early spring the search had narrowed, and Harvard was only considering Bethe and Schwinger. The choice between these two men turned out to be a difficult one. Many of the people who were asked to compare the two could not decide whom they would prefer. Pauli opted for Schwinger. Wigner could not make a choice. Fermi thought very highly of both, but since he knew Bethe better than Schwinger it was "perhaps for this reason that if it were my choice, I would rather have Bethe." Oppenheimer thought that "Bethe is probably the more productive ... and by far the more mature. Schwinger has perhaps the finer mind, and my own bet would be ... that he is the more likely of the two to make contributions of the most fundamental sort. I think you should settle the problem essentially on the basis of whether you want a young man or a mature one, since each in his class is the best you can get."[89] Given Oppenheimer's reply, Van Vleck wrote to Pauli again for wisdom. Pauli answered him that "the question, whether it is preferable to appoint a younger or an older man, has no unique answer," but the tenor of the letter made it clear that he thought Schwinger the best candidate for the job. The choice between the two men proved so difficult that the Department of Physics took the unusual step of recommending "Dr. Hans Albrecht Bethe or Dr. Julian Schwinger." Van Vleck, in his letter to the dean, indicated that the department "realizes that normally it would recommend but a single candidate for a given post. In the present case, however, the decision between the two men proves so exceedingly difficult that the department would welcome the opinion of the ad hoc committee as to the better candidate and would be glad to abide by its decision." Evidently the ad hoc did decide, for in late June, Van Vleck wrote Pauli to thank him for his letter "comparing Schwinger and Bethe" and to inform him that although "still confidential, but it is probable that Schwinger will ultimately be a member of our staff."[90]

In the fall of 1945 Schwinger accepted an appointment as an associate professor at Harvard University, to commence in February 1946. In the fall of 1946, Rabi tried to entice him to come to Columbia with an offer of a full professorship at an annual salary of $7,500. In his letter to the acting president of Columbia, Rabi commented that "Schwinger is a very young man and his appointment is consequently somewhat unusual. However, we all feel very strongly and the enclosed documents give some of the basis for the view that Dr. Schwinger is far and away the leading young theoretical physicist in the country, if not in the world."[91]

A year later he was offered a full professorship at Berkeley, and Harvard promptly promoted him. He refused the Berkeley offer because he thought Oppenheimer would be staying and Oppy would certainly have "dominated" him. Oppenheimer had not told him that he had accepted the directorship of the Institute of Advanced Study in Princeton. Schwinger believes that he would probably have gone to Berkeley had he known that Oppy was leaving.[92]

Cambridge turned out to be a congenial place for Schwinger. During the late forties, helped by the friendship of Weisskopf and Schwinger, the theoretical physicists at Harvard and MIT formed a close-knit community. The weekly joint Harvard-MIT theoretical physics seminar brought the community together for stimulating discussions of the latest advances. Many of MIT's graduate students and postdoctoral fellows "and a fair fraction of its faculty" would regularly attend Schwinger's courses at Harvard "what ever the subject may have been" (Feshbach 1979).

Harvard provided Schwinger with outstanding graduate students. They formed the audience for his brilliant lectures, and he became the thesis adviser for many of them. Harvard became one of the most important training centers for theoretical physics during Schwinger's stay there. When he first started lecturing in 1946/47, Schwinger gave a course in waveguides as well as a two-semester course in special topics in theoretical physics, which dealt primarily with nuclear physics. The following year, in 1947/48, Schwinger lectured on quantum mechanics, in addition to giving a course on "Advanced Theoretical Nuclear Physics," the completion of the course he had begun the previous year. The nuclear physics lectures were written up by John Blatt (1946/47) and made available to a wide audience. These notes "form an excellent introduction to the applications of quantum mechanics, developing a number of elegant methods of wide applicability. They contain many results specifically important for nuclear physics, many of which were never published or were later rediscovered. . . . It is difficult to exaggerate the impact of these lectures on the generation of physics graduate students in the late forties and fifties by which time a substantial fraction of the notes had been incorporated into the general background material all practicing theorists were expected to know" (Feshbach 1979).

7.4 Shelter Island and Its Aftermath

Schwinger remembers that in the late thirties he had been aware of the discrepancies between the Dirac theory and the experimental data on the fine structure of hydrogen, and that he had studied Pasternack's 1938 paper—"at the time I read everything . . . I lived in the library"[93]—but that he had not worked on the problem. When Schwinger was in Berkeley, the prevalent attitude toward field theory, and toward QED in particular, was not conducive to working on these subjects. "Oppy felt that QED had to break down at higher energies. [He believed that] things can't be infinite, QED must break down some place. The development of the renormalization theory—which seems so trivial now—was flying in the face of the almost universal belief that the theory was going to break down and that we need a new theory. We overdo the bit, 'They should have done it'. Everybody says if Dancoff had gotten it right this all would have happened ten years earlier— but this underestimates the difficulties involved."[94]

Shelter Island brought Schwinger back to quantum electrodynamics. Before the conference, Weisskopf had spoken to him "a little bit" about the Lamb-Retherford experiment. "The Lamb bit had been sort of brooded about. Vicki Weisskopf and I were talking about it on the way to Shelter Island. We came on the same train from Boston. Both of us were certainly up enough on field theory and the rest, to appreciate that self-energy effects were logarithmically divergent and that if you took the difference of two fine structure levels—that in itself would make it converge—never mind anything else. So we knew it was going to work. All this had been talked about before Shelter Island."[95]

Schwinger has vivid recollections of the conference: "Everybody was highly euphoric. Everybody was talking physics after five years. The facts were incredible: to be told that the sacred Dirac theory was breaking down all over the place!"[96] The reports of the Columbia experiments—by Lamb and Retherford, and by Nafe, Nelson, and Rabi—on the first day of the conference made a deep impression on Schwinger. He recalls listening to Kramers on the next day and "not being impressed. What Kramers was saying then I knew already because it was the classical stuff. It was a repeat." Schwinger had studied Kramers' paper at the time of its publication and had been "repelled by it." "I didn't think that refining things was to be done at the classical level." He had also read the work of Kramers' students, Serpe and Opechowski, who had attempted in 1940 and 1941 to extend Kramers' classical insights to simple quantum-mechanical systems. But again he was not impressed. He was convinced that relativistic quantum electrodynamics provided the right approach. He also remembers reading Dirac (1939a) on negative probabilities and asking what does it mean and "consciously moving away of it. I was the one guy who was not trying desperately to change field theory, but asking what was it really saying. . . . I am a field theorist and a conservative one in the creative sense of not willing to abandon what has been hard won."[97]

Schwinger does not recall "actually saying anything at Shelter Island" (Schwinger 1983a, p. 331) but others do. Bethe in particular in his paper on the nonrelativistic Lamb shift (Bethe 1947) acknowledges remarks made by Schwinger during the conference to the effect that a finite result for the level shift would emerge from a relativistic calculation.

A few days after the Shelter Island conference, on June 8, 1947, Schwinger abandoned his bachelor quarters, gave up smoking, and married Clarice Carrol of Boston. After the wedding, which was a "quite posh affair,"[98] they embarked on an extended honeymoon, a trip around the country that occupied most of the summer. "Not until September did [Schwinger] set out on the trail of relativistic quantum electrodynamics. But [he] knew what to do" (Schwinger 1983a, p. 332). When he started upon his calculations he was aware of Bethe's nonrelativistic calculation of the Lamb shift. Preliminary to making a relativistic calculation of the Lamb shift, he tried to understand the nonrelativistic case better. He recast ordinary perturbation theory so it would allow him to make use of Kramers' suggestion. In essence his method was a generalization of Pauli and Fierz's (1938) treatment of the infrared problem. Similar results were obtained independently by Welton (1948). Schwinger's point of departure was the quantum mechanical description of a charged particle interacting with the radiation field and with an external potential $V(\mathbf{r})$.[99] The Schrödinger equation for the system is

$$ih\ \partial_t\ \Psi(t) = \left[\frac{1}{2m_0} \left(p + \frac{e}{c}A \right)^2 + V(r) + H_{rad} \right] \Psi(t). \qquad (7.4.1)$$

Upon making the contact transformation

$$\Psi'(t) = e^{iH_{rad}t/\hbar}\Psi(t) \qquad (7.4.2)$$

$$A(t) = e^{iH_{rad}t/\hbar}Ae^{-iH_{rad}t/\hbar}, \qquad (7.4.3)$$

the new state vector satisfies the equation

$$i\hbar \frac{\partial \Psi'}{\partial t} = H'\Psi'(t) \qquad (7.4.4)$$

with

$$H' = \frac{\mathbf{p}^2}{2m_0} + V(\mathbf{r}) + \frac{e}{m_0 c}(\mathbf{p} \cdot \mathbf{A}(t)) + \frac{e^2}{2m_0 c^2}\mathbf{A}^2(t). \qquad (7.4.5)$$

Schwinger next performed a further contact transformation to remove the virtual effects induced by the $\mathbf{p} \cdot \mathbf{A}(t)$ term:

$$\Psi(t) \rightarrow e^{-iS(t)}\Psi(t) \qquad \text{(the primes have been dropped)} \qquad (7.4.6)$$

with

$$\hbar\frac{\partial S}{\partial t} = +\frac{e}{m_0 c}\mathbf{p}\cdot\mathbf{A}(t). \qquad (7.4.7)$$

If $\mathbf{A}(t)$ is derived from the Hertz potential $\mathbf{Z}(t)$

$$\mathbf{A}(t) = \frac{\partial\mathbf{Z}(t)}{\partial t}, \qquad (7.4.8)$$

then (since $\mathbf{p}$ is a time-independent operator)

$$S(t) = +\frac{e}{\hbar m_0 c}(\mathbf{p}\cdot\mathbf{Z}(t)). \qquad (7.4.9)$$

Now

$$e^{iS}\frac{\partial e^{-iS}}{\partial t} = -i\dot{S} + \frac{1}{2}\left[S,\dot{S}\right] \qquad (7.4.10)$$

and

$$e^{iS}\dot{S}e^{-iS} = \dot{S} + i\left[S,\dot{S}\right], \qquad (7.4.11)$$

with the series terminating because $[A, A] = c$ number. Therefore

$$i\hbar\frac{\partial\Psi(t)}{\partial t} = \left[\frac{p^2}{2m_0} + \frac{i\hbar}{2}[S,\dot{S}] + e^{iS}V(\mathbf{r})e^{-iS} + \frac{e^2}{2m_0 c^2}e^{iS}\mathbf{A}^2 e^{-iS}\right]\Psi(t). \qquad (7.4.12)$$

Upon substituting (7.4.9), (7.4.10) into (7.4.11), Schwinger found

$$i\hbar\frac{\partial\Psi(t)}{\partial t} = \left[\frac{p^2}{2m_0} + \frac{1}{2}i\hbar\left(\frac{e}{\hbar m_0 c}\right)^2 [\mathbf{p}\cdot\mathbf{Z}, \mathbf{p}\cdot\mathbf{A}] + e^{iS}V(\mathbf{r})e^{-iS}\right]\Psi(t)$$

$$= \left[\frac{p^2}{2m_0} + \frac{1}{2}i\hbar\left(\frac{e}{\hbar m_0 c}\right)^2 [\mathbf{p}\cdot\mathbf{Z}, \mathbf{p}\cdot\mathbf{A}] + V\left(\mathbf{r} + \frac{e}{mc}\mathbf{Z}\right)\right]\Psi(t),$$

$$(7.4.13)$$

where the A^2 term has been omitted. In dipole approximation, with

$$\mathbf{A}(t) = \sum_{k\mu} \varepsilon_{k\mu} c \sqrt{\frac{\hbar}{2\omega_k}} \left[a_{k\mu} e^{-i\omega_k t} + a^*_{k\mu} e^{i\omega_k t} \right]$$

$$\mathbf{Z}(t) = i \sum_{k\mu} \varepsilon_{k\mu} c \sqrt{\frac{\hbar}{2\omega_k^3}} \left[a_{k\mu} e^{-i\omega_k t} - a^*_{k\mu} e^{i\omega_k t} \right], \qquad (7.4.14)$$

the commutator of A and Z is

$$[A_\ell, Z_m] = -\frac{i}{3}\delta_{\ell m} \int \frac{\hbar c^2}{\omega^2} \frac{8\pi\omega^2 d\omega}{8\pi^3 c^3}. \qquad (7.4.15)$$

Upon introducing limits ω_0 and ω_1 for the integration whenever they were needed for convergence, Schwinger deduced that

$$[A_\ell, Z_m] = -\frac{i\hbar\omega_1}{3\pi^2 c}\delta_{\ell m}. \qquad (7.4.16)$$

In terms of the "minimum length" $a = \frac{2\pi c}{\omega_1}$, the commutator is equal to

$$\frac{1}{2} i\hbar \left(\frac{e}{\hbar m_0 c} \right)^2 [\mathbf{p} \cdot \mathbf{Z}, \mathbf{p} \cdot \mathbf{A}] = -\frac{\mathbf{p}^2}{2m_0} \frac{\delta m}{m_0} \qquad (7.4.17)$$

with

$$\delta m = \frac{2e^2}{3\pi c^2 a} = \frac{8}{3} \frac{e^2}{ac^2}. \qquad (7.4.18)$$

The electron thus has a kinetic energy

$$\frac{\mathbf{p}^2}{2m_0} \left(1 - \frac{\delta m}{m_0} \right) \equiv \frac{\mathbf{p}^2}{2m}. \qquad (7.4.19)$$

Equation (7.4.19) defines the "mass renormalization," with $m = m_0 + \delta m$ the "observed" mass. This, incidentally, is the Pauli-Fierz result of 1938. The effects of the radiative corrections are included in the term

$$e^{iS} V(\mathbf{r}) e^{-iS} = V\left(\mathbf{r} + \frac{e}{mc} \mathbf{Z} \right). \qquad (7.4.20)$$

Schwinger obtained the following Schrödinger equation for the charged particle upon expanding $V(\mathbf{r} + \frac{e}{mc}\mathbf{Z})$

$$i\hbar\frac{\partial\Psi}{\partial t} = \left[\frac{\mathbf{p}^2}{2m} + V(\mathbf{r}) + \frac{e}{mc}\mathbf{Z}\cdot\nabla V(r) + \frac{1}{2}\left(\frac{e}{mc}\right)^2 (\mathbf{Z}\cdot\nabla)^2 V(\mathbf{r}) + \cdots\right]\Psi.$$

$$(7.4.21)$$

Note that $\mathbf{Z}\cdot\nabla V$ is proportional to the force acting on the electron, that is, to its *acceleration*; and that therefore there are no virtual interactions for a free *electron*. For a free electron, all radiative effects have been incorporated into the mass renormalization. When no photons are present initially or finally, the Schrödinger equation that governs the system is

$$i\hbar\frac{\partial\Psi}{\partial t} = \left[\frac{\mathbf{p}^2}{2m} + V(\mathbf{r}) + \frac{1}{2}\left(\frac{e}{mc}\right)^2 \langle(\mathbf{Z}\cdot\nabla)^2\rangle_{vac} V(\mathbf{r}) + \cdots\right]\Psi(t), \qquad (7.4.22)$$

since $< \mathbf{Z} >_{vac} = 0$. Now

$$\left\langle\left[\frac{e}{mc}\mathbf{Z}\right]^2\right\rangle_{vac} = \frac{e^2}{m^2c^2}\sum_h \frac{\hbar c^2}{2\omega_k^3}$$

$$= \frac{2\alpha}{\pi}\left(\frac{\hbar}{mc}\right)^2 \ell n\frac{\omega_1}{\omega_0}, \qquad (7.4.23)$$

so that (7.4.20) becomes

$$i\hbar\frac{\partial\Psi}{\partial t} = \left[\frac{\mathbf{p}^2}{2m} + V(\mathbf{r}) + \frac{\alpha}{3\pi}\left(\frac{\hbar}{mc}\right)^2 \ln\frac{\omega_1}{\omega_0}\nabla^2 V(\mathbf{r})\right]\Psi(t). \qquad (7.4.24)$$

For the case $V(\mathbf{r}) = -Ze^2/r$, $\nabla^2 V = 4\pi Ze^2\delta(\mathbf{r})$, and the third term on the right-hand side of (7.4.24) gives the Bethe-Lamb shift. The treatment fails as $\omega_1 \rightarrow \infty$ because the theory is nonrelativistic and because a dipole approximation has been made and recoil has been neglected. However, the low-frequency divergence should not occur. It is the analog of the divergence encountered by Bloch and Nordsieck (1937). It occurs because the second-order effect of the $\mathbf{Z}\cdot\mathbf{A}(t)$ term in eq. (7.4.21) has not been taken into account. Schwinger computed this contribution and found it to be equal to

$$\langle H_A\rangle = -\left(\frac{e}{mc}\right)^2 \int_{\omega_0}^{\omega_1=\infty}\sum_n \frac{c^3\hbar}{2\omega^3}\frac{(0|\nabla V|n)(n|\nabla V|0)}{E_n - E_0 + \hbar\omega}\frac{8\pi\omega^2 d\omega}{8\pi^3 c^3}, \qquad (7.4.25)$$

where the sum $\sum_n$ is over all atomic states. The integral can readily be performed. It converges at the upper limit and yields

$$-\frac{2\alpha}{3\pi}\left(\frac{\hbar}{mc}\right)^2 \sum_n \frac{|<0|\nabla V|n>|^2}{E_n - E_0} \ln \frac{|E_n - E_0|}{\hbar\omega_0}. \tag{7.4.26}$$

Now

$$<|\nabla V|n> \; = \; \frac{i}{\hbar}<0|[\mathbf{p}, H]|n>$$

$$= \; \frac{i}{\hbar}(E_n - E_0)<0|\mathbf{p}|n>$$

$$H \; = \; \frac{\mathbf{p}^2}{2m} + V. \tag{7.4.27}$$

The sum is thus equal to

$$\sum_n [(0|\mathbf{p}|n)(n|\nabla V|0) - (0|\nabla V|n)(n|\mathbf{p}|0)] \cdot \ln \frac{|E_n - E_0|}{\hbar\omega_0}, \tag{7.4.28}$$

which without the ℓn term would simply be $<\nabla^2 V>$. Since the logarithmic factor varies slowly, it is replaced by a constant, in which case the sum (7.4.28) is equal to

$$\ln \frac{\Delta E}{\hbar\omega_0} \langle \nabla^2 V \rangle, \tag{7.4.29}$$

where ΔE is a suitably weighted average of the excitation energies. Bethe had found $\Delta E \sim 16.72$ Ry for the $2S$ state of H. Combining both terms now gives exactly Bethe's formula,

$$<H> = \frac{8}{3\pi}\alpha^3 \frac{Z^4}{n^3} \ln \frac{\hbar\omega_1}{\Delta E}, \tag{7.4.30}$$

and a level shift of ~ 1040 mc in $2S$ level if $\hbar\omega_1 = mc^2$.

The advantages of Schwinger's approach are that

1. The mass renormalization and the reactive effects of radiation are clearly separated from the beginning.

2. The theory works with the "observed" mass rather than the mechanical mass.

3. If recoil is included (i.e., if the dipole approximation is not made) it gives convergent integrals without further manipulations.

4. It shows automatically that a constant potential will give no effect and therefore eliminates any term of order αRy and leaves only terms of order α^3Ry.

Having convinced himself that his method of canonical transformations for eliminating virtual effects when applied to the case of a nonrelativistic charged particle interacting with the radiation field yielded Bethe's result for the Lamb shift, Schwinger proceeded to the much more difficult hole-theoretic description. In parallel with his calculation of the radiative corrections for a spin 1/2 Dirac field interacting with the electromagnetic field and an external Coulomb field— the "Coulomb case"—Schwinger considered the simpler case of a spin 1/2 field interacting with a finite mass scalar field, with an interaction term

$$H_{rad} = g \int \left[\psi^+ \beta \psi - (\psi^+ \beta \psi)_{vac} \right] \phi d\tau. \qquad (7.4.31)$$

This "toy" model allowed Schwinger to become familiar with the intricacies involved in identifying self-energy terms, and also with the integrations and manipulations needed to obtain the effective one-particle Hamiltonian. It also permitted him to make use of Pais' method of compensating fields (Pais 1946). Pais had found that one could obtain a *finite* self-energy for the electron to order α if one assumed that the electron couples not only to photons but also to neutral scalar mesons. He had reported on his theory at Shelter Island. (Essentially the same method had been developed by Sakata during the war [Sakata 1947].)

The notes of Schwinger's calculations are extant. They give proof of his awesome computational powers. Starting from the Hamiltonian for the coupled field system, Schwinger proceeded to make a canonical transformation to eliminate virtual effects to lowest order. The self-energy terms were identified and a mass renormalization was then performed. The remaining interaction Hamiltonian was analyzed to extract the pieces that were responsible for the dynamics of a system consisting of one electron and no photons. From this one-particle Hamiltonian, the divergent contribution due to the vacuum polarization by the external Coulomb was isolated and removed by a charge renormalization. His method was general in the sense that it yielded a *consistent* quantum electrodynamics to order $e^2/\hbar c$, and exhibited a divergence-free Hamiltonian to order $e^2/\hbar c$ that could be taken as the starting point to describe quantum mechanically any system consisting of electrons, positrons, and photons in the presence of an external Coulomb field. What previously had been pieces of a theory became welded and unified into a consistent and coherent quantum electrodynamics to order α.

To obtain the expression from which the Lamb shift can be calculated, lengthy computations involving properties of solutions of the Dirac equation, traces over photon polarizations, and integrations over photon energies had to be performed. These were carried out fearlessly and seemingly effortlessly. Schwinger just plowed ahead. Often, involved steps were carried out mentally and the answer was written down. And most important, the lengthy calculations are error free!

Schwinger used the radiation gauge for the description of the electromagnetic field (Wentzel 1943, pp. 120–130). The Hamiltonian for the "Coulomb case"—the terminology is that of Schwinger— that is, for a Dirac field interacting with the radiation field and an external Coulomb field, was taken to be

$$H = H_0 + H_{rad} + H_{coul} + H_{stat}$$
$$H_0 = H_{mat} + H_{field} \tag{7.4.32}$$

$$H_{mat} = \sum_m E_m(a_m^+ a_m - \epsilon_m) \qquad \epsilon_m = \begin{cases} 1 & E_m < 0 \\ 0 & E_m > 0 \end{cases} \tag{7.4.33}$$

$$H_{field} = \sum_k \hbar \omega_k A_{k\lambda}^+ A_{k\lambda}, \tag{7.4.34}$$

with $a_r, a_r^+, A_{k\lambda}, A_{k\lambda}^+$ the electron-positron and photon creation and annihilation operators that satisfy the commutations rules

$$[a_r, a_s^+]_+ = \delta_{rs} \tag{7.4.35}$$

$$\left[A_{k\lambda}, A_{k'\lambda'}^+ \right] = \delta_{kk'} \delta_{\lambda\lambda'}. \tag{7.4.36}$$

The interaction Hamiltonian H_{rad} is given by

$$H_{rad} = -\frac{1}{c} \int \mathbf{j} \cdot \mathbf{A} d\tau, \tag{7.4.37}$$

with

$$\mathbf{A}(\mathbf{r}) = c \sum_{k\lambda} \sqrt{\frac{2\pi\hbar}{\omega_k}} \mathbf{e}_{k\lambda} \left(A_{k\lambda} e^{i k \cdot r} + A_{k\lambda}^+ e^{-i k \cdot r} \right) \tag{7.4.38}$$

and

$$\mathbf{j}(\mathbf{r}) = -ec[\psi^+ \boldsymbol{\alpha} \psi(\mathbf{r}) - < \psi^+ \boldsymbol{\alpha} \psi(\mathbf{r}) >_{vac}]$$
$$= -ec \sum_{mn} (a_m^+ a_n - \delta_{mn} \epsilon_m)(u_m^+ \boldsymbol{\alpha} u_n) e^{\frac{i}{\hbar}(\mathbf{p}_n - \mathbf{p}_m) \cdot \mathbf{r}}, \tag{7.4.39}$$

where u_m are Dirac spinors, solutions of $(c\alpha \cdot \mathbf{p}_n + \beta mc^2)u_n = E_n u_n, E_n = \pm c \sqrt{\mathbf{p}_n^2 + m^2c^2}$, the $+$ or $-$ sign depending whether n is a positive or a negative energy solution. Schwinger next introduced the notation

$$(m|O|n) = \int d\tau \, u_m^+ e^{-i\mathbf{p}_m \cdot \mathbf{r}} O u_n e^{i\mathbf{p}_n \cdot \mathbf{r}}, \tag{7.4.40}$$

which allowed him to write H_{rad} as

$$H_{rad} = ec \sum_{mnk\lambda} \sqrt{\frac{2\pi\hbar}{\omega_k}} (a_m^+ a_n - \delta_{mn}\epsilon_m)$$
$$\left[A_{k\lambda}(m|\mathbf{e}_{k\lambda} \cdot \alpha e^{i\mathbf{k}\cdot\mathbf{r}}|n) + A_{k\lambda}^+(m|\mathbf{e}_{k\lambda} \cdot \alpha e^{-i\mathbf{k}\cdot\mathbf{r}}|n) \right]. \tag{7.4.41}$$

The interaction of the electron-positron field with the Coulomb field of a nucleus of charge Ze is described by H_{coul} with

$$H_{coul} = \int \frac{Ze}{r} \rho \, d\tau, \tag{7.4.42}$$

where the charge density operator is

$$\rho(\mathbf{r}) = -e \left[\psi^+ \psi(\mathbf{r}) - < \psi^+ \psi(\mathbf{r}) >_{vac} \right]$$
$$= -e \sum_{mn} (a_m^+ a_n - \delta_{mn}\epsilon_m)(u_m^+ u_n) e^{\frac{i}{\hbar}(\mathbf{p}_n - \mathbf{p}_m)\cdot\mathbf{r}}. \tag{7.4.43}$$

In terms of the notation of eq. (7.4.40), H_{coul} can be written as

$$H_{coul} = -\sum_{mn} (a_m^+ a_n - \delta_{mn}\epsilon_m)(m|\frac{Ze^2}{r}|n). \tag{7.4.44}$$

Similarly,

$$H_{stat} = \frac{1}{2} \int \frac{\rho(\mathbf{r})\rho(\mathbf{r}')}{|\mathbf{r} - \mathbf{r}'|} d\tau d\tau'$$
$$= \frac{2\pi c^2}{2} \sum_{k,\lambda} \frac{1}{\omega_k^2} \int \rho(\mathbf{r}) e^{i\mathbf{k}\cdot\mathbf{r}} \rho(\mathbf{r}') e^{i\mathbf{k}\cdot\mathbf{r}'} d\tau \, d\tau'$$
$$= \frac{e^2 c^2}{2} \sum_{k\lambda mn} \frac{2\pi}{\omega_k^2} \left(a_m^+ a_n - \delta_{mn}\epsilon_m\right)\left(a_r^+ a_s - \delta_{rs}\epsilon_r\right)\left(m|e^{i\mathbf{k}\cdot\mathbf{r}}|n\right)\left(r|e^{-i\mathbf{k}\cdot\mathbf{r}'}|s\right).$$
$$\tag{7.4.45}$$

In analogy with the nonrelativistic case, Schwinger performed a canonical transformation,

$$
\begin{aligned}
H' &= e^{iS} H e^{-iS} \\
&= H + i[S,H] + \frac{i^2}{2}[S,[S,H]] + \frac{i^3}{3!}[S,[S,[S,H]]] + \cdots \\
&= H_0 + i[S,H_0] - \frac{1}{2}[S,[S,H_0]] + \cdots \\
&\quad + H_{rad} + i[S,H_{rad}] + \\
&\quad + H_{stat} + \\
&\quad + H_{coul} + i[S,H_{coul}] + \cdots ,
\end{aligned}
\tag{7.4.46}
$$

with S so chosen that H_{rad} is eliminated. In other words, S is determined such that

$$
i[S,H_0] + H_{rad} = 0.
\tag{7.4.47}
$$

The transformed Hamiltonian (7.4.46) therefore becomes

$$
\begin{aligned}
H' &= H_0 + H_{coul} + H_{stat} + \frac{i}{2}[S,H_{rad}] \\
&\quad + i[S,H_{coul}] - \frac{1}{2}[S,[S,H_{coul}]] + \cdots .
\end{aligned}
\tag{7.4.48}
$$

The solution of (7.4.47) is readily found to be [100]

$$
\begin{aligned}
S = -iec \sum_{mnk\lambda} \sqrt{\frac{2\pi\hbar}{\omega_k}} (a_m^+ a_n - \delta_{mn}\epsilon_n) \\
\left[A_{k\lambda} \frac{(m|e_{k\lambda}\cdot\alpha e^{ik\cdot r}|n)}{E_m - E_n - \hbar\omega_k} + A_{k\lambda}^+ \frac{(m|e_{k\lambda}\cdot\alpha e^{-ik\cdot r}|n)}{E_m - E_n + \hbar\omega_k} \right].
\end{aligned}
\tag{7.4.49}
$$

The stage is now set for carrying out the computations that are required to obtain H'. In particular, Schwinger calculated that

$$
\begin{aligned}
&H_{stat} + \frac{i}{2}[S,H_{rad}] \\
&= \frac{e^2 c^2}{2} \sum_{k\lambda m,n,s,r} (a_m^+ a_n - \epsilon_m \delta_{mn})(a_r^+ a_s - \delta_{rs}\epsilon_r) \left[\frac{1}{\hbar\omega_k} (m|e^{ik\cdot r}|n)(r|e^{-ik\cdot r}|s) \right. \\
&\quad \left. - \frac{(m|e_{k\lambda}\cdot\alpha e^{ik\cdot r}|n)(r|e_{k\lambda}\cdot\alpha e^{-ik\cdot r}|s)}{\hbar\omega_k - E_m + E_n} - \frac{(m|e_{k\lambda}\cdot\alpha e^{ik\cdot r}|n)(r|e_{k\lambda}\cdot\alpha e^{-ik\cdot r}|s)}{\hbar\omega_k + E_r - E_s} \right]
\end{aligned}
\tag{7.4.50}
$$

$$= \sum (a_m^+ a_n - \delta_{mn} \epsilon_m)(a_r^+ a_s - \delta_{rs} \epsilon_r)(mr|U|ns), \qquad (7.4.51)$$

with $(mr|U|ns)$ defined by the terms within the brackets on the right-hand side of eq. (7.4.50). $H_{stat} + \frac{i}{2}[S, H_{rad}]$ gives rise to an interaction between two charged particles, and to a self-energy of a single particle. The self-energy arises from the one-particle part of $(a_m^+ a_n - \delta_{mn} \epsilon_m)(a_r^+ a_s - \delta_{rs} \epsilon_r)$, and so Schwinger decomposed the term into two pieces,

$$H_{stat} + \frac{i}{2}[S, H_{rad}] = H_{self\text{-}energy} + H_{int} \qquad (7.4.52)$$

with

$$H_{self\text{-}energy} = \sum (a_m^+ a_n - \delta_{mn} \epsilon_n)(m|H_{s.e.}|n). \qquad (7.4.53)$$

The decomposition amounts to what later would be called normal ordering by Wick (1949). $(m|H_{s.e.}|n)$ is readily verified to be equal to

$$(m|H_{s.e.}|n) = \sum_+ (mr_+|U|r_+n) - \sum_- (r_-m|U|nr_-), \qquad (7.4.54)$$

with U defined by eq. (7.4.51). The subscripts $+$ and $-$ indicate the sum is to be carried out over positive energy and negative solutions of the Dirac equation, respectively.

Explicitly,

$$(m|H_{self\text{-}energy}|n) = \frac{e^2 c^2}{2} \sum \frac{2\pi\hbar}{\omega_k}$$

$$\left[\frac{1}{\hbar\omega_k}(m|e^{i\mathbf{k}\cdot\mathbf{r}}|r_+)(r_+|e^{-i\mathbf{k}\cdot\mathbf{r}}|n) - (m|e^{-i\mathbf{k}\cdot\mathbf{r}}|r_-)(r_-|e^{+i\mathbf{k}\cdot\mathbf{r}}|n)\frac{1}{\hbar\omega_k} \right.$$

$$- \frac{(m|\mathbf{e}_{k\lambda}\cdot\boldsymbol{\alpha}\,e^{i\mathbf{k}\cdot\mathbf{r}}|r_+)(r_+|\mathbf{e}_{k\lambda}\cdot\boldsymbol{\alpha}\,e^{-i\mathbf{k}\cdot\mathbf{r}}|n)}{\hbar\omega_k - E_m + E_{r+}}$$

$$+ \frac{(m|\mathbf{e}_{k\lambda}\cdot\boldsymbol{\alpha}\,e^{-i\mathbf{k}\cdot\mathbf{r}}|r_-)(r_-|\mathbf{e}_{k\lambda}\cdot\boldsymbol{\alpha}\,e^{+i\mathbf{k}\cdot\mathbf{r}}|n)}{\hbar\omega_k - E_{r-} + E_n}$$

$$- \frac{(m|\mathbf{e}_{k\lambda}\cdot\boldsymbol{\alpha}\,e^{i\mathbf{k}\cdot\mathbf{r}}|r_+)(r_+|\mathbf{e}_{k\lambda}\cdot\boldsymbol{\alpha}\,e^{-i\mathbf{k}\cdot\mathbf{r}}|n)}{\hbar\omega_k + E_{r_+} - E_n}$$

$$\left. + \frac{(m|\mathbf{e}_{k\lambda}\cdot\boldsymbol{\alpha}\,e^{-i\mathbf{k}\cdot\mathbf{r}}|r_-)(r_-|\mathbf{e}_{k\lambda}\cdot\boldsymbol{\alpha}\,e^{i\mathbf{k}\cdot\mathbf{r}}|n)}{\hbar\omega_k + E_m - E_{r-}} \right]. \qquad (7.4.55)$$

Only the diagonal term $m = n$ contributes, and Schwinger evaluates the self-energy term for

$$\mathbf{p}_m = \mathbf{p}_n = \mathbf{p}. \qquad (7.4.56)$$

By introducing the usual projection operators for positive energy and negative energy solutions of the Dirac equations, and with $\mathbf{p}' = \mathbf{p} + \hbar\mathbf{k}$, he found that $H_{self\text{-}energy}$ could be reduced to the expression

$$
\begin{aligned}
H_{self\text{-}energy} = \frac{e^2 c^2}{2} \sum_{\mathbf{k}\lambda} \frac{2\pi\hbar}{\omega_k} \Bigg[& \frac{1}{\hbar\omega_k} \frac{E_{p'} + H_{p'}}{2E_{p'}} \\
& - \frac{1}{\hbar\omega_k} \frac{E_{p'} - H_{p'}}{2E_{p'}} - \frac{1}{\hbar\omega_k + E_{p'} - H_p} \mathbf{e}_{\mathbf{k}\lambda} \cdot \boldsymbol{\alpha} \frac{E_{p'} + H_{p'}}{2E_{p'}} \mathbf{e}_{\mathbf{k}\lambda} \cdot \boldsymbol{\alpha} \\
& + \frac{1}{\hbar\omega_k + E_{p'} + H_p} \mathbf{e}_{\mathbf{k}\lambda} \cdot \boldsymbol{\alpha} \frac{E_{p'} - H_{p'}}{2E_{p'}} \mathbf{e}_{\mathbf{k}\lambda} \cdot \boldsymbol{\alpha} \\
& + \mathbf{e}_{\mathbf{k}\lambda} \cdot \boldsymbol{\alpha} \frac{E_{p'} - H_{p'}}{2E_{p'}} \mathbf{e}_{\mathbf{k}\lambda} \cdot \boldsymbol{\alpha} \frac{1}{\hbar\omega_k + E_{p'} + H_p} \\
& - \mathbf{e}_{\mathbf{k}\lambda} \cdot \boldsymbol{\alpha} \frac{E_{p'} + H_{p'}}{2E_{p'}} \mathbf{e}_{\mathbf{k}\lambda} \cdot \boldsymbol{\alpha} \frac{1}{\hbar\omega_k + E_{p'} - H_p} \Bigg]. \qquad (7.4.57)
\end{aligned}
$$

Upon summing over the polarization states eq. (7.4.53) was equal to

$$
\begin{aligned}
H_{self\text{-}energy} = \frac{e^2 c^2}{2} \sum_{\mathbf{k}} \frac{4\pi\hbar}{\omega_k} & \frac{1}{\hbar\omega_k E_{p'} + c^2 \mathbf{p}' \cdot \hbar\mathbf{k}} \left[H_{p'} \left(1 + \frac{c^2 \hbar\mathbf{k} \cdot \mathbf{p}'}{\hbar\omega_k E_{p'}} \right) - H_p \right. \\
& \left. + \left(\hbar\omega_k + E_{p'} \right) \frac{\beta mc^2 + (c^2/\omega_k^2)c\mathbf{k}' \cdot \mathbf{p}'\boldsymbol{\alpha} \cdot \mathbf{k}}{E_{p'}} \right], \qquad (7.4.58)
\end{aligned}
$$

which could be further reduced to give the following expression for $H_{self\text{-}energy}$:

$$H_{self\text{-}energy} = \sum_n (a_n^+ a_n - \epsilon_n)\delta mc^2 \frac{m_0 c}{\sqrt{m_0^2 c^2 + \mathbf{p}^2}}. \qquad (7.4.59)$$

δm is logarithmically divergent, and apart from some numeral factors Schwinger obtained the same expression for it as the one previously derived by Weisskopf (1934, 1939). Schwinger then removed $H_{self\text{-}energy}$ by a mass renormalization. What is meant by this is that $H_{self\text{-}energy}$ is combined with H_{matter}, with the result that to order $e^2/\hbar c$, the energy of an electron of momentum $\mathbf{p}$ is

$$c\sqrt{m_0^2 c^2 + \mathbf{p}^2} + \delta mc^2 \frac{m_0 c}{\sqrt{m_0^2 c^2 + \mathbf{p}^2}} \simeq c\sqrt{(m_0 + \delta m)^2 c^2 + \mathbf{p}^2}, \qquad (7.4.60)$$

and $m_0 + \delta m$ is identified with the observed mass of the electron.

To obtain the radiative corrections to the Dirac equation description of an electron in a Coulomb field—and in particular to compute the energy shift induced by the radiative corrections for an electron in a bound state around a nucleus of charge Ze—Schwinger proceeded to express

$$H_{int} = \sum_{mnrs}(a_m^+ a_n - \delta_{mn}\epsilon_m)(a_r^+ a_s - \delta_{rs}\epsilon_r)(mr|U|ns)$$
$$- \sum_{mn}(a_m^+ a_n - \delta_{mn}\epsilon_m)(m|H_{self}|n) \tag{7.4.61}$$

in terms of creation and annihilation operators that create and annihilate electrons and positrons in one-electron states that are eigenstates of $c\boldsymbol{\alpha}\cdot\mathbf{p} + \beta mc^2 - \frac{Ze^2}{r}$, that is, for electrons or positrons in a Coulomb field. Schwinger did this in a very elegant fashion. He introduced the creation and annihilation operators, b and b^+, that create and annihilate electrons in one-particle Coulomb states, that is, of solutions of the Hamiltonian $c(\boldsymbol{\alpha} \cdot \mathbf{p} + \beta mc) - Ze^2/r$, through the equations

$$a_n = \sum_{\alpha}(n|\alpha)b_\alpha$$
$$a_n^+ = \sum_{\alpha}b_\alpha^+(\alpha|n). \tag{7.4.62}$$

In terms of the b operators,

$$H_{int} = \sum(b_\alpha^+ b_\beta - d_{\alpha\beta})(b_\rho^+ b_\sigma - d_{\rho\sigma})(\alpha|m)(\rho|r)(mr|U|ns)(n|\beta)(s|\sigma)$$
$$- \sum(b_\alpha^+ b_\beta - d_{\alpha\beta})(\alpha|m)(m|H_{self}|n)(n|\beta), \tag{7.4.63}$$

with

$$d_{\alpha\beta} = \sum_{n_-}(\beta|n_-)(n_-|\alpha). \tag{7.4.64}$$

The expectation value of H_{int} in the state $b_\xi^+|0>$, where $|0>$ is the vacuum state in the presence of the Coulomb field (i.e., $|0>$ is ground state of the Hamiltonian $H_0 + H_{coulomb}$) is readily evaluated. Actually, Schwinger computed $< 0|b_\xi H_{int}b_\xi^+|0> - < 0|H_{int}|0>$. Furthermore, since he was interested only in the lowest-order corrections, he approximated $(s|\beta)$ by

$$(s|\beta) \simeq \delta_{s\beta} + \frac{1}{E_s - E_b}\left(s|\frac{Ze^2}{r}|b\right). \tag{7.4.65}$$

After a long calculation Schwinger obtained the result

$$
\begin{aligned}
(H_{int})_{\text{vacuum}+\xi} - (H_{int})_{\text{vacuum}} &= \frac{58}{45}\alpha Z e^2 \left(\frac{\hbar}{mc}\right)^2 |\psi_\xi(0)|^2 \\
&- \frac{1}{2\pi}\alpha Z e^2 \left(\frac{\hbar}{mc}\right)^2 \left(\xi\left|\frac{\boldsymbol{\sigma}\cdot\mathbf{L}}{r^3}\right|\xi\right).
\end{aligned}
\tag{7.4.66}
$$

Similarly, Schwinger proceeded to calculate the one-particle part of $-\frac{1}{2}[S,[S,H_{coul}]]$. This term contains a divergent contribution, due to the polarization of the vacuum by the external Coulomb held, which Schwinger eliminated by a charge renormalization as Uehling (1935) and Serber (1936) had done.

Sometime in mid-September 1947, in discussions with Ramsey, Schwinger became aware of Breit's suggestion that perhaps the electron had an intrinsic magnetic moment that was different from the value predicted by the Dirac equation (Breit 1947a). Norman Ramsey, who had come to Harvard in early September after having accepted an associate professorship there, remembers Schwinger telling him that he thought he could calculate the anomalous magnetic moment. He indicated that "it would be very, very much more difficult than the [Bethe nonrelativistic] Lamb shift calculation, because it was intrinsically going to be a relativistic calculation, and it was going to be a lot of work." Before undertaking it he wanted to be convinced that the effect was real, so he "cross-examined" Ramsey quite a bit about the Nafe-Nelson experiment. Ramsey, who had discussed the experiments with Rabi and Breit at Brookhaven, assured him that it was worth working on. "I believe the experiment fully. It is a big effect," Ramsey told Schwinger.[101] Thereupon Schwinger—while still working on the "Coulomb case"—embarked on a calculation of the radiative correction to the motion of an electron in an external magnetic field. The Hamiltonian for that system is the same as (7.4.39), but with H_{coul} replaced by H_{ext}, with

$$
H_{ext} = \frac{1}{c}\int \mathbf{j}\cdot\mathbf{A}^e d\tau,
\tag{7.4.67}
$$

$\mathbf{A}^e$ being the vector potential of the classical, prescribed magnetic field, and $\mathbf{j}$ the current operator for the electron-positron field, eq. (7.4.36). Schwinger once again performed the contact transformation (7.4.46) to eliminate the virtual transitions induced by H_{rad}. The transformed Hamiltonian becomes

$$
H' = H_0 + H_{ext} + H_{stat} + \frac{i}{2}[S,H_{rad}] + i[S,H_{ext}] + \frac{i^2}{2}[S,[S,H_{ext}]] + \cdots
\tag{7.4.68}
$$

The self-energy subtraction is the same as in the case of the Coulomb field, and the previous calculation could therefore be taken over unchanged. The $-\frac{1}{2}[S,[S,H_{ext}]]$

term in (7.4.64) gives rise to a one-particle contribution of the form

$$-\frac{1}{2}[S,[S,H_{ext}]]_{1\,particle} = \sum (a_m^+ a_n - \epsilon_n \delta_{mn})(m|H_e'|n). \qquad (7.4.69)$$

A divergent vacuum polarization contribution is again removed by a charge renormalization. The finite part of H_e' was found to be equal to

$$H_e' = -\frac{2}{3\pi}\frac{e^2}{\hbar c}(\mu_0 \boldsymbol{\sigma} \cdot \mathbf{H}^e + e\boldsymbol{\alpha} \cdot \mathbf{A}^e) \cdot \frac{1}{4} \cdot \int_0^1 x\,dx \left[1 - \frac{x}{\sqrt{x^2+1}}\right]. \qquad (7.4.70)$$

Here was the proof that radiative corrections could give rise to an anomalous magnetic moment. The electron behaved as if it had a moment of magnitude $\mu_o + \delta\mu$ with $\delta\mu$ the coefficient of the $\boldsymbol{\sigma} \cdot \mathbf{H}^e$ term in (7.4.66). As in the Coulomb case calculation, to this contribution must be added that of the one-particle contribution of H_{int}, where the one-particle state to be considered is that of the charge in the magnetic field $\mathbf{A}^e$. Again, only contributions to first order in $\mathbf{A}^e$ were retained. But before the calculation could be completed, Schwinger went to Washington in mid-November to attend the tenth Washington Conference on Theoretical Physics.[102] Schwinger there gave a report on the status of his calculations that got "Oppy very excited." Feynman, who attended the conference, wrote his friend Corben that, although he didn't understand most of Schwinger's lecture, "One thing Schwinger pointed out was that the discrepancy in the hyperfine structure of the hydrogen noted by Rabi, can be explained on the same basis as that of the electromagnetic-self energy, as can the line shift of Lamb."[103] Upon his return from Washington, Schwinger completed his calculation and found that H_{int}'s contribution was equal to

$$\begin{aligned} H_e'' = \frac{2}{3\pi}\frac{e^2}{\hbar c}\int dx & \left\{ \frac{1}{\sqrt{x^2+1}} - \frac{1}{2}\frac{x^2}{(x^2+1)^{3/2}} \right. \\ & \left. -\frac{2}{(x^2+1)^{3/2}} + \frac{3x^2}{(x^2+1)^{5/2}} \right\} e\boldsymbol{\alpha} \cdot \mathbf{A}^e \\ & + \left\{ \frac{1}{2}\frac{x^2}{(x^2+1)^{3/2}} + \frac{2}{(x^2+1)^{3/2}} - \frac{3x^2}{(x^2+1)^{5/2}} \right\} \mu_0 \boldsymbol{\sigma} \cdot \mathbf{H}^e. \end{aligned} \qquad (7.4.71)$$

When the two terms (7.4.66) and (7.4.67) are combined, the result is

$$H_e' + H_e'' = \frac{1}{2\pi}\frac{e^2}{\hbar c}\mu_o \boldsymbol{\sigma} \cdot \mathbf{H}^e - \frac{1}{6\pi}\frac{e^2}{\hbar c}e\boldsymbol{\alpha} \cdot \mathbf{A}^e \qquad (7.4.72)$$

The coefficient of the $\boldsymbol{\sigma} \cdot \mathbf{H}^e$ term indicates that by virtue of radiative corrections the electron acquires a contribution to its gyromagnetic ratio of magnitude $2 \cdot \frac{\alpha}{2\pi}$.

The predicted additional magnetic moment accounted for the hyperfine structure measurements of Nafe, Nelson, and Rabi (1947) and of Nagel, Julian, and Zacharias (1947).

The importance of Schwinger's calculation cannot be underestimated. In the course of theoretical developments there sometimes occur important calculations that alter the way the community thinks about particular approaches. Schwinger's calculation is one such instance. By indicating, as Feynman had noted, that "the discrepancy in the hyperfine structure of the hydrogen atom ... could be explained *on the same basis* as that of the electromagnetic self-energy, as can the line shift of Lamb" [emphasis added] Schwinger had transformed the perception of quantum electrodynamics. He had made it into an effective, coherent, and consistent computational scheme to order e^2.

7.5. The APS Meeting and the Pocono Conference

The news of the magnetic moment calculation spread rapidly. Schwinger talked about it at Columbia on his way back to Boston from the Washington conference. In a letter to Schwinger in which he indicated how much he enjoyed his visit, Rabi commented that he found it "very regretful and melancholy" that he [Schwinger] should spend his days "in self-imposed exile, in a barren land where fish is consumed as a brain food, in large quantities, with results that fall short of highest expectations."[104] Rabi still hoped to lure Schwinger back to Columbia. On December 2, 1947, Rabi wrote Bethe: "It certainly seems very likely that the g value of the electron is greater than 2 by slightly over 1/10 of 1% and that the Schwinger theory of our hyperfine structure anomaly is as correct as your theory of the Lamb-Retherford effect—God is great!" Two days later Bethe wrote Rabi back: "I have heard about Schwinger's theory and find it very wonderful. Nobody so far, has been able to give me a complete account of his theory of the hyperfine structure or of the g-factor. But I am sure it is alright. It is certainly wonderful how these experiments of yours have given a completely new slant to a theory and how the theory has blossomed out in a relatively short time. It is as exciting as in the early days of quantum mechanics."[105]

Responding to Oppenheimer's inquiry about possible sites and dates for the followup NAS conference to Shelter Island, Rabi suggested that since "so much has happened since the Shelter Island meeting that it might be advisable to set the date for this [second] meeting somewhat ahead, such as [late] February. Schwinger's theory seems to be very far advanced. He came down on Tuesday and told us about it. The applications of these ideas to meson theory will be well under way by then and the discussion should be very exciting and fruitful."[106]

Schwinger's results on the energy shift in a homogeneous magnetic field—which is the prediction of an additional magnetic moment of the electron—and on the radiative corrections to scattering and bound states in the Coulomb field of a nucleus were communicated to the *Physical Review* in late December 1947.

In his letter to the editor, Schwinger indicated that finite results had been obtained for the radiative corrections to scattering in a Coulomb field, thereby ending the "confusion" that had been generated by Dancoff's (1939) paper. Although he had obtained a finite result for the Lamb shift, Schwinger did not give a value for it: "The values yielded by our theory differ only slightly from those conjectured by Bethe on the basis of a non-relativistic calculation, and are thus in good accord with experiment" (Schwinger 1948b). The reason for Schwinger's not quoting a precise number for the Lamb shift was that the Coulomb and the magnetic field calculations did not agree with one another. Schwinger had obtained an additional moment of $\alpha/2\pi$ in the externally applied homogeneous magnetic field, but in the Coulomb case the spin-orbit coupling had a wrong factor of 1/3 to be identified with this particular value for the additional magnetic moment; relativistic invariance was violated by the noncovariant hole-theoretic methods that were being used. Schwinger therefore embarked on a formulation that was manifestly covariant and kept the relativistic covariance manifest throughout the calculation.

> I have a distinct memory of sitting on the porch of my new residence during what must have been a very late Indian summer in the fall of 1947 and with great ease and great delight arriving at invariant results in the electromagnetic-mass calculation for a free electron. I suspect this was done with equal-time interaction.
>
> The spacelike generalization, to a plane, and then to a curved surface, took time, but all that was in place at the New York meeting. (Schwinger 1983a, p. 336)

The mathematical basis of the covariant formulation was only slightly different from that of the quantum electrodynamics as presented in Wentzel's *Einführung* (1943). The chief innovation was the introduction of a state vector $\Psi[\sigma]$, a functional of a general 3-dimensional surface σ in space-time, which represented the state of the system as determined by the specification of a set of commuting field variables on the surface σ. In Wentzel's book and in earlier formulations Ψ had been defined only for a surface whose equation in some frame was $t = $ constant. In Schwinger's formulation Ψ was defined for every surface which is spacelike in the sense that every two points on it are separated by a spacelike interval. The Schrödinger equation was generalized and became a functional derivative equation describing the variation of $\Psi[\sigma]$ with σ:

$$\hbar c \frac{\delta \Psi[\sigma]}{\delta \sigma(x)} = \mathcal{H}(x)\Psi[\sigma], \qquad (7.5.1)$$

where $\mathcal{H}(x)$ is the interaction energy density of the fields at a point x of spacetime, and $\delta\Psi/\delta\sigma(x)$ is the functional derivative of Ψ with respect to a small variation of σ in the neighborhood of x. The important fact is that unlike the old Schrödinger equation, eq. (7.5.1) is manifestly Lorentz covariant.

The New York APS meeting took place at the end of January 1948. Schwinger's invited lecture on "Recent Advances in Quantum Electrodynamics" was given on Saturday, January 31. K. K. Darrow made the following entry in his diary for that day: "Third and last day of the great meeting > 1600 registrants. I heard no paper but Schwinger's, given too rapidly for my apprehension but given with great gusto which implied a great advance."[107] Schwinger's lecture created quite a stir. The dean of the faculty at Harvard informed the Board of Overseers that "Professor Schwinger's lecture, reporting on some small but basically very interesting corrections to the magnetic moment of the electron which arise from self-energy, attracted such a large gathering that it was necessary for him to repeat it a second time. It also received considerable attention by the press."[108] Freeman Dyson was in the audience. He had come to the United States in September 1947 and was then a graduate student at Cornell working with Bethe and had just finished a Lamb shift calculation for a spin zero charged particle. He wrote his parents the following account of Schwinger's lecture:

> The great event came on Saturday morning, and was an hour's talk by Schwinger, in which he gave a masterly survey of the new theory which he has the greatest share in constructing and at the end made a dramatic announcement of a still newer and more powerful theory, which is still in embryo. This talk was so brilliant that he was asked to repeat it the afternoon session, various unfortunate lesser lights being displaced in his favour. There were tremendous cheers when he announced that the crucial experiment had supported his theory; the magnetic splitting of two of the spectral lines of gallium (an obscure element hitherto remarkable only for being a liquid metal like mercury) were found to be in the ratio 2.00114 to 1; the old theory gave for this ratio exactly 2 to 1, while the Schwinger theory gave 2.0016 to 1.[109]

Schwinger gave the first presentation of his covariant formulation at the Pocono conference. Schwinger spoke on Wednesday, March 31. His talk turned out to be an all-day affair, except for a brief presentation by Fermi on the electron-neutron interaction.[110] As recorded by Wheeler, Schwinger's opening remarks stressed that

> in order to get reasonable answers in the present form of electrodynamics one must use a subtraction procedure which is both gauge and relativistically invariant. Only then can one identify infinite terms in a sensible way. The following treatment of the

quantized electromagnetic and electron pair field satisfies this criterion. Guided by Dirac's theory of the interaction between a finite number of electrons and the electromagnetic field, it treats the case of an indefinite number of particles. Each small volume of space is now to be handled as a particle.[111]

Before presenting the details of his formulation of quantum electrodynamics, Schwinger outlined the quantization of the free electromagnetic field and of the free Dirac field, which was done in order to introduce his notation. (That notation is essentially the same as in QED I [Schwinger 1948d] and is summarized in table 7.1.)

Table 7.1
Electromagnetic Field in Vacuum

Space-time point $x_\mu = (\mathbf{r}, \, ict)$; $x_0 = (1/i)x_4 = ct$; $d\omega = dx_0 \, dx_1 \, dx_2 \, dx_3$. The four-vector potentials of the electromagnetic field are $A_\mu = (\mathbf{A}, i\phi)$ and their equations of motion are $\Box^2 A_\mu = 0$. State vectors satisfy the subsidiary condition:

$$\frac{\partial A_\mu}{\partial x_\mu} \Psi = 0$$

Commutation rules:

$$[A_\mu(x), A_\nu(x')] = \frac{\hbar c}{i} \delta_{\mu\nu} D(x - x').$$

D satisfies $\Box^2 D = 0$ $\qquad$ $D(\mathbf{x}, 0) = 0$ $\qquad$ $\frac{1}{c}\frac{\delta D}{\delta t}(\mathbf{x}, 0) = \delta(\mathbf{x})$

$$D(\mathbf{x}, t) = \frac{1}{(2\pi)^3} \int d^4 k \, e^{ik_\mu x_\mu} \frac{k_0}{|k_0|} \delta(k_\mu k_\mu)$$

$$= \frac{1}{2\pi} \frac{x_0}{|x_0|} \delta(x_\mu^2).$$

The singular function $D^{(1)}(\mathbf{x},t)$ also satisfies $\Box^2 D^{(1)} = 0$, but is an even fuction:

$$D^{(1)}(\mathbf{x}, t) = \frac{1}{(2\pi)^3} \int d^4 k \, e^{ik_\mu x_\mu} \delta(k_\mu k_\mu).$$

Decomposition into positive and negative frequency parts:

$$A_\mu(x) = A_\mu^{(+)}(x) + A_\mu^{(-)}(x).$$

$A_\mu^{(+)}$ contains only $e^{-ik_o t}$; $A_\mu^{(-)}$ only $e^{+ik_o t}$.

For $\mu = 1, 2$ $A_\mu^{(+)}(x)$ destroys a photon: $A_\mu^{(-)}(x)$ creates a photon. The vacuum expectation value of the anticommutator* $\{A_\mu(x), A_\nu(x')\} = A_\mu(x)A_\nu(x') + A_\nu(x')A_\mu(x)$ is given by

$$\langle\{A_\mu(x), A_\nu(x')\}\rangle_o = \hbar c \, D^{(1)}(x - x')\delta_{\mu\nu}$$

Matter Field. No Radiation. The electron field is described by the 4-component spinor ψ that satisfies $(\gamma_\mu \frac{\partial}{\partial x_\mu} + \kappa)\psi = 0$. The γ_μ are hermitian with $\gamma_\mu\gamma_\nu + \gamma_\nu\gamma_\mu = 2\delta\mu\nu$. The adjoint spinor is $\bar{\psi} = \psi^+\gamma_4$ with ψ^+ the hermitian conjugate of ψ. The charge conjugate operator is $\psi' = C\bar{\psi}$, where $C^{-1}\gamma_\mu C = \gamma_\mu^T, C^T = -C, C^+ C = 1$.

Anticommutations relations:

$$\{\psi_\alpha(x), \bar{\psi}_\beta(x')\} = i\left(\gamma_\mu \frac{\partial}{\partial x_\mu} - \kappa\right)_{\alpha\beta} \Delta(x - x')$$

$$\equiv iS_{\alpha\beta}(x - x'),$$

with

$$(\Box^2 - \kappa^2)\Delta = 0 \qquad \Delta(\mathbf{x}, 0) = 0 \qquad \frac{1}{c}\frac{\partial\Delta}{\partial t}(\mathbf{x}, 0) = \delta(\mathbf{x})$$

$$\Delta(\mathbf{x}, t) = \frac{i}{(2\pi)^3}\int e^{ik_\mu x_\mu}\frac{k_0}{|k_0|}\delta(k_\mu^2 + \kappa^2)d^4 k.$$

Decomposition into positive and negative frequency parts:

$$\psi(x) = \psi^{(+)}(x) + \psi^{(-)}(x).$$

$\psi^{(+)}$ destroys an electron, $\psi'^{(+)}$ destroys a positron, $\overline{\psi'^{(+)}}$ creates a positron, $\overline{\psi^{(+)}}$ creates an electron, etc. The vacuum is defined by $\psi^{(+)}(x)\Psi_0 = \psi'^{(+)}(x)\Psi_0 = 0$. The symmetrized cuurent is given by

$$j_\mu(x) = +\frac{iec}{2}[\bar{\psi}(x)\gamma_\mu, \psi(x)].$$

---

* In the Pocono notes, the anticommutator is denoted by $[A, B]_+ = AB + BA$. I have used Schwinger's notation of QED I, and have denoted $[A, B]_+$ by $\{A, B\}$.

Schwinger next turned his attention to the case of the interacting fields. Their dynamics is governed by the Schrödinger equation,

$$i\hbar\frac{\partial \Psi}{\partial t} = (H_1 + H_2 + H_{12})\Psi$$

$$= (H_0 + H_{12})\Psi. \tag{7.5.2}$$

In this representation the operators are time independent; H_1 is the Hamiltonian for the free radiation field, H_2 that of the free matter field, and H_{12} their interaction energy:

$$H_{12} = -\frac{1}{c}\int j_\mu A_\mu dV \tag{7.5.3}$$

To bring this equation into a "conspicuously invariant form" Schwinger, "following Dirac and Tomonaga," made the contact transformation

$$\Psi_{\text{new}} = e^{\frac{i}{\hbar}(H_1 + H_2)t}\Psi_{\text{old}}. \tag{7.5.4}$$

Ψ_{new} thus satisfies

$$i\hbar\frac{\partial \Psi_{\text{new}}}{\partial t} = e^{\frac{i}{\hbar}H_0 t}H_{12}e^{-\frac{i}{\hbar}H_0 t}\Psi_{\text{new}}$$

$$= H_{12}(t)\Psi_{\text{new}}. \tag{7.5.5}$$

The operators are now time dependent, with

$$\psi(\mathbf{r}, t) = e^{\frac{i}{\hbar}H_1 t}\psi(\mathbf{r})e^{-\frac{i}{\hbar}H_1 t} \tag{7.5.6}$$

and

$$H_{12} = \int \mathcal{H}(\psi(\mathbf{r}, t), A(\mathbf{r}, t)) dV. \tag{7.5.7}$$

Schwinger then regarded eq. (7.5.4) the result of setting "times equal in an ∞ set of equations of the [Dirac] many time formalism." He introduced a time for each point of a spacelike surface in space-time, and considered the original $\Psi(t)$ as being generalized by the sequence of steps

$$\Psi(t) \longrightarrow \Psi(t_1, t_2 \cdots) \longrightarrow \Psi(t(x))$$
$$\text{Dirac} \qquad \text{Dirac-Tomonaga.}$$

Since $H_{12}(t)$ is the sum of the Hamiltonian densities at all points of the surface

$$H_{12}(t) = \sum_{\mathbf{x}} \mathcal{H}(\mathbf{x}, t)\Delta V, \tag{7.5.8}$$

the basic equation is replaced by

$$i\hbar\frac{\partial \Psi}{\partial t(\mathbf{x})} = \mathcal{H}(\mathbf{x}, t)\Delta V\Psi. \tag{7.5.9}$$

Schwinger next defined the functional derivative of a functional $F[\sigma]$ of the space-like surface σ with respect to the surface σ at the point xt, as

$$\frac{\delta F[\sigma]}{\delta\sigma(\mathbf{x})} = \lim_{\delta\omega\to 0} \frac{F[\sigma'] - F[\sigma]}{\delta\omega}, \tag{7.5.10}$$

where the surface σ' deviates from σ only in a neighborhood of the point $\mathbf{x}$, and $\delta\omega$ is the volume enclosed between the two surfaces. Eq. (7.5.8) can then be written in the covariant form

$$i\hbar c\frac{\delta\Psi[\sigma]}{\delta\sigma(x)} = \mathcal{H}(\mathbf{x}, t)\Psi[\sigma]. \tag{7.5.11}$$

Since the equations for different points x on the spacelike surface must be compatible, integrability requires that

$$[\mathcal{H}(x), \mathcal{H}(x')] = 0 \tag{7.5.12}$$

for any two space-time points x, x' on σ.

Eq. (7.5.11) has become known as the Tomonaga-Schwinger equation. For quantum electrodynamics it reads

$$i\hbar c\frac{\delta\Psi[\sigma]}{\delta\sigma(x)} = -\frac{1}{c}j_\mu(x)A_\mu(x)\Psi[\sigma], \tag{7.5.13}$$

to which must be added the supplementary condition

$$\Omega\Psi(\sigma) = 0 = \left(\frac{\partial A_\mu}{\partial x_\mu} + \frac{1}{ic}\int_\sigma D(x - x')j_\mu(x')d\sigma_\mu\right)\Psi(\sigma). \tag{7.5.14}$$

Wheeler's notes indicate that Schwinger emphasized the fact "these equations contain nothing more than the Heisenberg-Pauli formalism and would not be required if one knew how to carry out Heisenberg-Pauli calculations consistently."

After a brief discussion of the gauge invariance of the equations, Schwinger turned to applications of the formalism. The "properties of an electron as modified by its own radiation field" was the first problem addressed. To zeroth order in the coupling constant, e, there is no interaction between the charged-matter field and the radiation field, and the state vector $\Psi[\sigma]$ is constant. The interaction $\mathcal{H}(x)$ corresponds to the emission or absorption of a light quantum by an electron or the emission or absorption of a light quantum in the course of creation or annihilation of an electron-positron pair. Now a free electron cannot emit or absorb a light quantum, nor can a free electron-positron pair annihilate with the emission of a single quantum, since it is impossible to simultaneously satisfy the energy and momentum conservation laws under these circumstances. For a free electron, the effects produced by $\mathcal{H}(x)$ are virtual. The physical effects of the $j_\mu A_\mu$ interaction only exhibit themselves to second-order in such processes as the virtual emission and subsequent absorption of a light quantum by the matter field. These second-order terms produce the interaction between different particles and the self-energy of a single particle. Schwinger therefore performed a canonical transformation to obtain an equation from which the first-order interaction terms have been eliminated and replaced by the second-order couplings they generate. He defined Ψ' by

$$\Psi'[\sigma] \rightarrow e^{-iS[\sigma]}\Psi[\sigma]$$

and noted that $\Psi'[\sigma]$ satisfies (dropping the prime)

$$\hbar c \frac{\delta \Psi}{\delta \sigma(x)} = \hbar c \; e^{iS}\frac{\delta e^{-iS}}{\delta \sigma(x)}\Psi = e^{iS}\mathcal{H}(x)e^{-iS}\Psi. \qquad (7.5.15)$$

$S[\sigma]$ is determined by the requirement that

$$\hbar c\frac{\delta S[\sigma]}{\delta \sigma(x)} = \mathcal{H}(x). \qquad (7.5.16)$$

To order e^2, eq. (7.5.15) therefore reduces to

$$\hbar c \frac{\delta \Psi[\sigma]}{\delta \sigma(x)} \cong \frac{i}{2}[S[\sigma], \mathcal{H}(x)]\,\Psi[\sigma]$$

$$= \mathcal{H}'(\sigma, x)\Psi[\sigma]. \qquad (7.5.17)$$

Schwinger chose the particular solution of (7.5.16) of interest to be the one for which $e^{iS} \rightarrow 1$ as $\sigma \rightarrow \infty$ or $\sigma \rightarrow -\infty$, which solution guarantees that there are no real first-order effects. Explicitly,

$$S[\sigma] = \frac{1}{2\hbar c}\int_{-\infty}^{+\infty} \epsilon(\sigma, x')\mathcal{H}(x')dw', \qquad (7.5.18)$$

where

$$\epsilon(\sigma, x') = \begin{cases} +1 \text{ if } x' \text{ is earlier than } \sigma \\ -1 \text{ if } x' \text{ is later than } \sigma. \end{cases} \tag{7.5.19}$$

With this $S[(\sigma)]$, the Hamiltonian $\mathcal{H}'(x, \sigma)$ is given by

$$\mathcal{H}'(\sigma, x) = -\frac{i}{4\hbar c^3} \int_{-\infty}^{+\infty} \left[j_\mu(x) A_\mu(x), j_\nu(x') A_\nu(x') \right] \epsilon(\sigma, x') dw'. \tag{7.5.20}$$

Actually $\mathcal{H}'(x, \sigma)$ is independent of σ, since $[\mathcal{H}(x), \mathcal{H}(x')]$ vanishes everywhere on σ except at the point $x' = x$. The commutator in (7.5.19) can be simplified by noting that

$$[j_\mu(x) A_\mu(x), j_\nu(x') A_\nu(x')]$$

$$= \frac{1}{2}[j_\mu(x), j_\nu(x')]\{A_\mu(x), A_\nu(x')\} + \frac{1}{2} [A_\mu(x), A_\nu(x')]\{j_\mu(x), j_\nu(x')\}, \tag{7.5.21}$$

which in turn can be written as

$$= \frac{1}{2}[j_\mu(x), j_\nu(x')]\left(\{A_\mu(x), A_\nu(x')\} - <\{A_\mu(x), A_\nu(x')\} >_0\right)$$

$$+ \frac{\hbar c}{2}[j_\mu(x), j_\mu(x')]D^{(1)}(x - x') + \frac{i\hbar c}{2}\{j_\mu(x), j_\mu(x')\} D(x - x'). \tag{7.5.22}$$

The Hamiltonian density (4.19) can therefore be reexpressed as follows:

$$\mathcal{H}(\sigma, x) = -\frac{i}{8\hbar c^3} \int ([j_\mu(x), j_\nu(x')] - <[j_\mu(x), j_\nu(x')] >_0)$$

$$\cdot (\{A_\mu(x), A_\nu(x')\} - <\{A_\mu(x), A_\nu(x')\} >_0) \cdot \epsilon(\sigma, x') dw'$$

$$- \frac{i}{8\hbar c^3}\hbar c \int ([j_\mu(x), j_\mu(x')]D^{(1)}(x - x')$$

$$+ \{j_\mu(x), j_\mu(x')\}D(x - x'))\epsilon(\sigma, x') dw'$$

$$- \frac{i}{8\hbar c^3} \int < [j_\mu(x), j_\nu(x')] >_0 \cdot (\{A_\mu(x), A_\nu(x')\}$$

$$- <\{A_\mu(x), A_\nu(x')\} >_0)\epsilon(\sigma, x') dw'$$

$$= \mathcal{H}'^{(0)}(x, \sigma) + \mathcal{H}'^{(1)}(x, \sigma) + \mathcal{H}'^{(2)}(x, \sigma). \tag{7.5.23}$$

The first term, $\mathcal{H}^{(0)}$, describes processes involving real photons. The second term, $\mathcal{H}^{(1)}$, only involves virtual photons, with the part referring to two charged particles ($\{j_\mu, j_\mu\}$) giving exactly the Møller interaction. The last term, $\mathcal{H}^{(2)}$, describes the photon self-energy.

Schwinger proceeded to consider the terms that gave rise to the photon and electron self-energy. He remarked that the procedures he was going to apply to eliminate the divergences when dealing with one-electron and one-photon problems could be extended to "arbitrarily many" particles. At that point Bohr interjected that "one may not be able to treat all physical problems without a fundamental new idea" (Wheeler 1948, p. 28).

Schwinger then turned to the computation of the photon self-energy, which is determined by the last term

$$H^{(2)}_{PSE} = -\frac{i}{2\hbar c^3} \int^\sigma dw' < [j_\mu(x), j_\nu(x')] >_0 \left(A_\mu(x)A_\nu(x') \right.$$
$$\left. - < A_\mu(x)A_\nu(x') >_0 \right). \tag{7.5.24}$$

He stated that he "will show [that the photon self-energy] = 0 [even though] actually it is ∞ in any particular Lorentz frame but transforms as if zero."[112] His argument was as follows: "Consider a typical term in $A_\mu(x)$, say $e^{ik_\mu x_\mu} a_\mu$, where by the Lorentz condition $k_\mu a_\mu \Psi = 0$, and show that its contribution to

$$\int < [j_\mu(x), j_\nu(x')] >_0 A_\nu(x')dw'$$

vanishes."

The vacuum expection value of the commutator of two current operators is given by

$$\langle [j_\mu(x), j_\nu(x')]\rangle_0 = \frac{ie^2c^2}{2} Sp \{ S^{(1)}(x' - x)\gamma_\mu S(x - x')\gamma_\nu$$
$$- S^{(1)}(x - x')\gamma_\mu S(x' - x)\gamma_\nu \} \tag{7.5.25}$$
$$= -4ie^2c^2 \left\{ \frac{\partial \Delta}{\partial x_\mu}\frac{\partial \Delta^{(1)}}{\partial x_\nu} + \frac{\partial \Delta}{\partial x_\nu}\frac{\partial \Delta^{(1)}}{\partial x_\mu} \right.$$
$$\left. - \delta_{\mu\nu} \left[\frac{\partial \Delta}{\partial x_\lambda}\frac{\partial \Delta^{(1)}}{\partial x_\lambda} - \kappa_0^2 \Delta\Delta^{(1)} \right] \right\}$$

$$\Delta = \Delta(x - x'), \Delta^{(1)} = \Delta^{(1)}(x - x'). \tag{7.5.26}$$

Schwinger proceeded to show that the integral (7.5.23) has the structure

$$C_1 k_\mu k_\nu a_\nu + C_2 k_\lambda k_\lambda a_\mu,$$

where C_1 and C_2 are "divergent multiplicative factors."

But since $k_\nu a_\nu \Psi = 0$ and $k_\lambda k_\lambda = 0$, the terms giving rise to the photon self-energy vanish. Schwinger further noted that "there are several ways to make the integrals give finite multiplicative factors but there is no point to examine them in detail because the above arguments show that if $H_{PSE}^{(2)}$ is finite it is zero." Oppenheimer then remarked "that in a covariant gauge invariant theory, there is no way to put in a finite value for the photon mass, hence there is no need to calculate $H_{PSE}^{(2)}$ except as a test of consistency" (Wheeler 1948, p. 30).

Schwinger next addressed the computation of the self-energy of an electron that arose from the middle term in eq. (7.5.21). The term involving the commutator of two current operators,

$$[j_\mu(x), j_\mu(x')] = ie^2c^2[\overline{\psi}(x)\gamma_\mu S(x - x')\gamma_\mu\psi(x') - \overline{\psi}(x')\gamma_\mu S(x' - x)\gamma_\mu\psi(x)],$$

$$(7.5.27)$$

produces only one-electron transitions; the second, involving the anticommutator, gives rise to both one-electron and two-electron interactions. To separate out the one-electron contribution from this quadrilinear expression in $\overline{\psi}$ and ψ, Schwinger argued that one must choose one factor $\overline{\psi}$ and one factor ψ as actually effecting the transition of the electron from the initial to the final state, and then allow "the other two factors to represent successive emission and absorption of a particle in one of the vacuum states." Equivalently, Schwinger allowed one factor $\overline{\psi}$ and one factor ψ to remain as operators, while the product of the other two factors was replaced by its vacuum expectation value. Since the two factors that remain as operators in an expression of the form $\overline{\psi}\psi\overline{\psi}\psi$ can be chosen in four ways according to the schemes

$$\underline{\overline{\psi}\psi}\,\overline{\psi}\psi \qquad \overline{\psi}\psi\,\underline{\overline{\psi}\psi} \qquad \overline{\psi}\,\underline{\psi\overline{\psi}}\,\psi \qquad \underline{\overline{\psi}\psi\overline{\psi}\psi}$$

the four expressions thus obtained must be added together to give the one-particle part of the quadrilinear expression. Schwinger then showed that the one-particle contribution reduced to the expression

$$H_{ESE} = \frac{1}{2}\left[\overline{\psi}(x)\mathcal{X}(x) + \text{charge conjugate}\right],$$

where

$$\mathcal{X}(x) = \frac{ie^2}{2}\int^\sigma dx'[D(x-x')\gamma_\mu S^{(1)}(x-x')\gamma_\mu + D^{(1)}(x-x')\gamma_\mu S(x-x')\gamma_\mu]\psi(x').$$

$$(7.5.28)$$

Evaluating this expression "for a typical plane wave component of $\psi(x)$ say $ae^{ip_\mu x_\mu}$ with $(i\gamma_\mu p_\mu + \kappa)a = 0$ and $p_\mu^2 = -\kappa^2$," Schwinger obtained the result

$$\mathcal{H}(x) = \frac{e^2}{32\pi^2} \int_{-\infty}^{+\infty} \frac{d\xi}{|\xi|} \int_{-1}^{+1} d\eta 2\left(1 - \frac{1-\eta}{4}\right) \cdot exp\left[\frac{i\kappa^2}{4} \frac{1+\eta}{1-\eta}\xi\right] \cdot \psi(x),$$
(7.5.29)

$$= \delta mc^2 \psi(x),$$
(7.5.30)

where δ is logarithmically divergent but "the result *is independent of state of motion of electron.*"

Explicitly,

$$\delta mc^2 = \frac{3}{2\pi}\alpha \; mc^2\left[\log\frac{2}{\kappa\sqrt{\epsilon}} - \frac{1}{2}C - \frac{1}{6}\right],$$
(7.5.31)

where $\alpha = \frac{1}{4\pi}\frac{e^2}{\hbar c}$, and ϵ is the lower limit of ξ integration and $C = .577$.

This is the "same result as Weisskopf's calculation of the self-energy of an electron apart from some numerical errors in his paper" (Wheeler 1948, p. 35). Schwinger also inquired "how cancellation comes about when one uses the f-field of Pais." To answer the question, he computed the self-energy of an electron when the interaction Hamiltonian (7.5.12) is replaced by $j_\mu A_\mu + j_0 A_0$, with A_0 a scalar f-field

$$j_0 = \frac{fc}{2}(\overline{\psi}\psi + \overline{\psi}'\psi')$$
(7.5.32)

and

$$[A_0(x), A_0(x')] = \frac{\hbar c}{i}\Delta(x - x'; \mu),$$
(7.5.33)

where $\Delta(x; \mu)$ is a Δ function with the mass μ of the scalar meson. The result he found was that

$$\delta m = \delta_e m + \delta_f m$$

$$= \frac{\alpha}{2\pi}m\left\{\frac{1}{4}\int_{-1}^{+1}(-\eta)\ln\left(1 + \frac{2\mu^2}{\kappa^2}\frac{1-\eta}{(1+\eta^2)}\right) - 1\right\},$$
(7.5.34)

which is finite. If the f-field has a zero-rest mass quanta, $\mu = 0$, then

$$\delta m = -\frac{\alpha}{2\pi}m.$$
(7.5.35)

On the other hand, if μ is large, with $\mu \gg \kappa$,

$$\delta m = \frac{3}{2\pi} \alpha m \left[\ln \frac{\mu}{\kappa} + \frac{5}{4} \right]. \tag{7.5.36}$$

This concluded the first part of Schwinger's lecture. Fermi then made a brief presentation of his interpretation of the recent electron-neutron interaction experiments, and the conference adjourned for lunch. When the conference reconvened, Schwinger "resumed" his lecture and turned to an analysis of the radiative corrections to the motion of "electrons in a given external field," the computation of which "will yield the magnetic moment of the electron and the Lamb-Retherford shift of the energy levels of an atom" (Wheeler 1948, p. 37). The basic equation is now

$$\hbar c \frac{\delta \Psi}{\delta \sigma} = \{\mathcal{H}(x) + \mathcal{H}^{ext}(x)\}\Psi \tag{7.5.37}$$

$$\mathcal{H}(x) = -\frac{1}{c} j_\mu(x) A_\mu(x) \tag{7.5.38}$$

$$\mathcal{H}^{ext}(x) = -\frac{1}{c} j_\mu(x) A_\mu^{ext}(x). \tag{7.5.39}$$

The canonical transformation

$$\Psi \rightarrow e^{-iS} \Psi \tag{7.5.40}$$

with

$$S(\sigma) = \frac{i}{\hbar c} \int_{-\infty}^{\sigma} \mathcal{H}(x') dw' \tag{7.5.41}$$

to second order results in the equation

$$\hbar c \frac{\delta \Psi}{\delta \sigma(x)} = \left\{ \mathcal{H}^{(0)}(x, \sigma) + \mathcal{H}''^{(1)}(x, \sigma) \right. \\ \left. + \mathcal{H}''^{(2)}(x, \sigma) + e^{iS} \mathcal{H}^{ext}(x) e^{-iS} \right\} \Psi(x, \sigma). \tag{7.5.42}$$

As before, the first three terms account for the electron self-energy, the interaction of two charged particles, the photon self-energy, and two-photon processes. In the expansion

$$e^{iS} \mathcal{H}^{ext} e^{-iS} = \mathcal{H}^{ext}(x) + i \left[S, \mathcal{H}^{ext} \right] + \frac{i^2}{2} \left[S_1 S, \mathcal{H}^{ext} \right] + \cdots, \tag{7.5.43}$$

the various terms reflect the modification of the coupling between the matter field and the external radiation by radiative effects.

Schwinger then noted that by "modifying the original canonical transformation which was made in passing from the time independent ψ and A_μ operators to the Dirac-Tomonaga form, one can eliminate all the self-energy effects discussed above. This is equivalent (to this order in the perturbation theory) to putting a new mass in the Dirac equation and dropping the operator which accounts for the self-energy term from the interaction Hamiltonian density [eq. (7.5.34)]." Breit in his notes recorded: "Makes unitarity transformation such that $(\gamma_\mu \frac{\partial}{\partial x_\mu} + \kappa + \delta\kappa)\psi = 0$ occur."[113]

Explicitly, to second order in S,

$$\hbar c \frac{\delta\Psi}{\delta\sigma} = \left\{ \mathcal{H}_{int}(x) + \mathcal{H}^{ext}(x) + i[S, \mathcal{H}^{ext}] - \frac{1}{2}[S, [S, \mathcal{H}^{ext}]] + \cdots \right\} \Psi,$$

(7.5.44)

where $\mathcal{H}_{int}$ are the terms that remain after the $\delta mc^2 \overline{\psi}\psi$ term is removed:

$$\mathcal{H}_{int} = \frac{i}{2}[S(\sigma), \mathcal{H}(x)] - \mathcal{H}_{self-energy}(x).$$

(7.5.45)

Schwinger removed the virtual effects due to $\mathcal{H}_{int}$ by performing a second contact transformation

$$\Psi \to e^{-iT(\sigma)}\Psi,$$

(7.5.46)

with T so chosen that

$$\hbar c \frac{\delta T}{\delta\sigma} = \mathcal{H}_{int}(x).$$

(7.5.47)

The result is

$$\hbar c \frac{\delta\Psi}{\delta\sigma(x)} = \left\{ \mathcal{H}^{ext}(x) + i[T, \mathcal{H}^{ext}] - \frac{1}{2}[S, [S, \mathcal{H}^{ext}]] + i[S, \mathcal{H}^{ext}] \right\} \Psi.$$

(7.5.48)

The terms

$$R = i[T, \mathcal{H}^{ext}] - \frac{1}{2}[S, [S, \mathcal{H}^{ext}]]$$

(7.5.49)

must be treated together, because only this combination is a function of the point x that does not depend on any other point of the surface σ. Schwinger gave a lengthy and involved proof of this fact. Being interested in the properties of an electron in the presence of the external field, Schwinger picked out the terms that corresponded to "one electron, no photons" contributions in $[S, (\sigma'), [\mathcal{H}(x'), \mathcal{H}^{ext}(x)]]$. Essentially this means obtaining the one-electron, no-photon terms in

$$\left[[j_\lambda(x'')A_\lambda(x''), j_\nu(x'), A_\nu(x')], j_\mu(x)A_\mu^{ext}(x) \right].$$

The answer he obtained is

$$R = -\frac{i}{16}\frac{e^2}{\hbar c} e A_\nu^{ext}(x)\bigg|_{\text{all space-time}} \tag{7.5.50}$$
$$dw_\xi \, dw_{\xi'} \left\{ \overline{\psi}(x + \xi')\gamma_\mu S^{(1)}(\xi')\gamma_\nu S(\xi)\gamma_\mu \psi(x - \xi)D(\xi + \xi') \right\},$$

with

$$\xi' = x'' - x'; \qquad \xi = x' - x$$

plus two other terms obtained by permutation of labels on the S's and D, minus three more terms charge conjugate to the first three terms. The calculation of the right-hand side of (7.5.41) is not given in either Wheeler's or Breit's notes, which only say that "the result is divergent a) at low frequencies and b) high frequencies."

Inserting the Fourier representation for $\psi(x)$ and $\overline{\psi}(x')$, $\psi(x) \sim e^{ip_\mu x_\mu}$ and $\overline{\psi}(x) \sim e^{-ip_\mu x_\mu}$, Schwinger found

$$R = -\frac{i}{2^4 \pi}\frac{e^2}{4\pi\hbar c}\frac{e}{\kappa^2} A_\nu^{ext}\overline{\psi}(x)$$

$$\times \left\{ 2\gamma_\nu(2\kappa^2 + (\Delta p)^2)\int_{-1}^{+1} \frac{dy}{1 + \frac{(\Delta p)^2}{4\kappa^2}(1 - y^2)}(\ln z_0 - 1) \right.$$

$$+ \gamma_\nu\frac{(\Delta p)^2}{2}\int_{-1}^{+1} \frac{dy}{1 + \frac{(\Delta p)^2}{4\kappa^2}(1 - y^2)} - 2\kappa^2\gamma_\nu\left(\ln\frac{1}{\epsilon} + C + 1\right)$$

$$\left. + \kappa\Delta p_\lambda\sigma_{\lambda\nu}\int_{-1}^{+1} \frac{dy}{1 + \frac{(\Delta p)^2}{4\kappa^2}(1 - y^2)} \right\} \psi(x). \tag{7.5.51}$$

In (7.5.42),[114] z_0 is the lower limit of the divergent integral (the infrared divergence), ϵ is the upper limit (the ultraviolet divergence), and

$$\sigma_{\lambda\nu} = \frac{1}{2}(\gamma_\lambda\gamma_\nu - \gamma_\nu\gamma_\lambda) \tag{7.5.52}$$

$$\Delta p_\mu = p_\mu - p'_\mu. \tag{7.5.53}$$

Schwinger removed the ultraviolet divergent term—a contribution to R of the form $-\frac{1}{c}\delta e \, j_\mu A^e_\mu(x)$—by a charge renormalization and noted that the charge renormalization factor is also infra-red divergent "but probably will not be when one adds the second order effects of the term $[S, \mathcal{H}^{ext}]$. Such a cancellation has been shown ... for the Δp^2 term in R which gives the Lamb-Retherford Shift."

Schwinger then proceeded to evaluate the expression in the limit $(\Delta p)^2 \ll \kappa^2$—the condition valid in the low Z atomic situation—and found that to order $(\Delta p)^2$ the expression for R reduced to

$$-\frac{1}{3\pi}\alpha\frac{8}{3}\left(\ln z_0 - 1 + \frac{3}{8}\right)\frac{1}{\kappa^2}\frac{1}{c}j_\mu\square^2 A^{ext}_\mu$$

$$+\frac{1}{2\pi}\alpha\frac{e\hbar}{2mc}\sum_{\mu<\nu}F_{\mu\nu}\frac{1}{2}(\bar{\psi}\sigma_{\mu\nu}\psi - \text{charge conjugate}). \tag{7.5.54}$$

"The first line is the analogue of Bethe's result for the Lamb shift except that there is no divergence at high frequencies." There is, however, one at low frequencies. But, if the second-order effects of $[S, \mathcal{H}^{ext}]$ are added they give a compensating infra-red divergence. The effect is to replace $\ln z_0 - 1$ by $\ln \frac{mc^2}{2\Delta W}$ where ΔW is the Bethe mean excitation energy. Schwinger thus obtained the following result for the $^2S_{1/2} - ^2P_{1/2}$ splitting:

$$\frac{8\alpha}{9\pi}\left[\ln\frac{mc^2}{\Delta W} - \underline{\ln 2 + \frac{3}{8}} + \frac{1}{8}\right], \tag{7.5.55}$$

where the underlined terms were not present in Bethe's earlier treatment. Schwinger commented that "if the second order term were calculated a little better this result should be good for the Lamb shift for the K electron even in uranium" (Wheeler 1948, p. 43).

The expression (7.5.45) did not include the vacuum polarization contribution, the Uehling term. The result obtained by Schwinger disagreed with the expression reported by Feynman at the conference for the Lamb shift. Feynman's result had the 3/8 replaced by 5/8.[115]

What is clear from Wheeler's and Breit's notes on Schwinger's talk at the Pocono conference is that the elimination of the divergences in Schwinger's scheme was not a straightforward matter. The explicit covariance of the formalism and the invariant integration methods were of crucial importance in identifying various terms—particularly the divergent contributions—but there were a great many ambiguities in the procedure. Schwinger and Feynman had discussions at Pocono to clarify the nature of the divergent contributions. Feynman had *consciously* omitted closed loop vacuum polarization contributions. His diagrammatic approach, however, allowed the identification of divergences more straightforwardly than in Schwinger's schemes. Schwinger recalls that during one of their discussions "while they were horsing around about vacuum polarization," Oppenheimer came over to them and sadly said that "he never had deep thoughts about physics."[116] Schwinger's formalism could not have been extended easily to higher orders—and it is not clear that anyone besides Schwinger could have done what he did. Schwinger himself reports that an irritated critic once said: "Other people publish to show you how to do it, but Julian Schwinger publishes to show you that only he can do it."[117] But it should be noted that Schwinger emerged from the mathematical jungle with an experimentally confirmed numerical value for the magnetic moment of the electron and a number for the Lamb shift. The impact of Schwinger's theory can be gauged from the coverage Schwinger received in the public press. In late June 1948 Schwinger went to Pasadena to give an invited lecture at an APS meeting there. The *New York Times* reported his talk under the headline: "SCHWINGER STATES HIS COSMIC THEORY. Physicists Awed as Harvard man of 30 Tells Version of Electrodynamic Forces."

> Leading physicists from many parts of this country and abroad listened with fascination today to a 30-year old Harvard Professor of Physics expound a new theory through the symbols of higher mathematics about the interaction of energy and matter, the two basic phenomena in which the physical universe is manifested to man.
>
> The new theory was presented by Dr. Julian Schwinger, whom American physicists regard as the heir-apparent to the mantle of Einstein. While not yet complete the Schwinger theory outlined today at the opening of sessions of the meeting of the American Physical Society is looked upon by top flight theoretical physicists as the most important development in the last twenty years in our basic understanding of the cosmic forces holding the material universe together....
>
> Dr. Schwinger's theory has already found its first support in experiments carried out at Columbia University. Its further elaboration to include all the present known particles of the material universe awaits further experimental observations....
>
> Thus far the theory can be applied only to energies up to 100,000,000 volts. It is expected, however, that it will be applicable to higher energies, as more knowledge is obtained through observations of the cosmic rays and production of high-energy particles with the super-cyclotrons now being built.[118]

7.6. The Michigan Summer School

After the Pocono conference, Schwinger started to write up his results. "Quantum Electrodynamics I" (Schwinger 1948d) was "written up quickly"[119] and a first draft of "Quantum Electrodynamics II" (Schwinger 1949a) was completed in May 1948.[120] These served as the notes for the first part of the Michigan Summer School lectures that Schwinger delivered in Ann Arbor from July 19 to August 7.

Freeman Dyson, who had just completed a year of graduate studies at Cornell, was one of the students in Schwinger's course. The day after Schwinger's arrival Dyson wrote his parents: "Yesterday the great Schwinger arrived, and for the first time I spoke to him; with him arrived a lot of new people who came to hear him especially. His talks have been from the first minute excellent; there is no doubt he has taken a lot of trouble to polish up his theory for presentation at this meeting. I think in a few months we shall have forgotten what pre-Schwinger physics was like."[121]

The title of Schwinger's course was "Recent Developments in Quantum Electrodynamics" (Schwinger 1948f) and the notes for it are extant. Except for minor variations, chapter 1 of the notes is identical with "QED I" (Schwinger 1948d). Chapter 2, entitled "Applications of the Theory" contains most of "QED II" (Schwinger 1949a) and some additional materials relating to the radiative corrections for an electron in a Coulomb field (Lamb shift, scattering). The notes for this part of the course were written up by David Park. Schwinger had outlined his views about quantum electrodynamics in the introduction to QED I, and he reviewed them in his opening lecture. He pointed out that the objectionable aspects of QED were encountered in virtual processes involving particles with ultra-relativistic energies. The two basic phenomena of this type, which both gave rise to divergences, were the polarization of the vacuum and the self-energy of the electron.

> The two phenomena are quite analogous and essentially describe the interaction of each field [i.e., of the electromagnetic field and the electron-positron field] with the vacuum fluctuations of the other field. The effect of these fluctuation interactions is merely to alter the fundamental constants e and m, although by logarithmically divergent factors. However, it may be argued that a future modification of the theory, inhibiting the virtual creation of particles that possess energies many orders of magnitude in excess of mc^2, will ascribe a value to these logarithmic factors not vastly different from unity. The charge and mass renormalization factors will then differ only slightly from unity, as befits a perturbation theory, in consequence of the small value of the coupling constant $e^2/4\pi hc = 1/137$.

Schwinger then continued:

> We may now ask the fundamental question: Are all the physically significant divergences of the present theory contained in the charge and mass renormalization factors? Will the consideration of interactions more complicated than these simple vacuum fluctuation effects introduce new divergences; or will all further phenomena involve only moderate relativistic energies, and thus be comparatively insensitive to the high energy modifications that are presumably introduced in a more satisfactory theory? (Schwinger 1948d, p. 1)

Schwinger's lectures represented an attempt to supply at least partial answers to these questions, which had "acquired an immediate importance in view of recent conclusive evidence that the electromagnetic properties of the electron are not fully described by the Dirac wave equation." Schwinger was not to answer the question that he had posed: whether mass and charge renormalizations were sufficient to remove the divergences of QED to all order of perturbation theory. But a member of the audience, Freeman Dyson, made it *his* problem, and solved it.

The physical content of the formalism Schwinger had developed had been considerably clarified and refined as a result of discussions with Feynman at Pocono and with Weisskopf in Cambridge. Welton, a postdoctoral fellow working with Weisskopf, in the spring and summer of 1947 had elaborated a simple physical picture for the self-energy of a (nonrelativistic) charged particle which, in his model, arose from the interaction of the charge with the vacuum fluctuations of the electromagnetic field. Welton had also given a simple and intuitive explanation of the Bethe Lamb-shift calculation on this basis (Welton 1948; see also Feshbach 1950). The viewpoint was generalized by Weisskopf and made into an *anschaulisch* exposition of the physics of the divergences encountered in quantum electrodynamics (Weisskopf 1949). Schwinger outlined this viewpoint in the introduction to his notes.

Discussing first "vacuum polarization," he noted that in the language of perturbation theory, the phenomenon considered

> is the generation of charge and current in the vacuum through the virtual creation and annihilation of electron-positron pairs by the electromagnetic field.... When the electromagnetic field is that of a given current distribution, one obtains a logarithmically divergent contribution to the vacuum polarization which is everywhere proportional to the given distribution.... Thus the physically significant divergence arising from the vacuum polarization phenomenon occurs in a factor that alters the strength

of all charges, a uniform renormalization that has no observable consequence other than the conflict with the empirical finiteness of charge. (Schwinger 1948f, p. 2)

If the electromagnetic field that generates the virtual electron-positron pairs in the vacuum is that of a light quantum, the vacuum polarization effects are equivalent to ascribing a proper mass to the photon. This proper mass, however, must be zero in a properly gauge-invariant theory. "The failure to obtain this result from a gauge-invariant formulation can only be ascribed to a faulty application of the theory." Similarly the interaction of the charged-matter field with the electromagnetic field vacuum fluctuations produces the self-energy of an electron. "The mechanism here under discussion is commonly described as the virtual emission and absorption of a light quantum by an otherwise free electron, although an equally important effect is the partial suppression, via the exclusion principle, of the coupled vacuum fluctuations of the electromagnetic and matter fields." Schwinger stressed that in a Lorentz-invariant theory, self-energy effects for a free electron "can only result in the addition of an electromagnetic proper mass to the electron's mechanical proper mass" (Schwinger 1948f, p. 3).

The first part of Schwinger's course consisted in a detailed exposition of QED I. In the rest of the course Schwinger presented materials taken from a first draft of QED II that included covariant calculations of the self-energy of an electron and of the polarization of the vacuum by an external current. Freeman Dyson, who had studied Wheeler's notes of Schwinger's presentation at the Pocono conference, wrote Hans Bethe after the summer school ended that apart from some improvement in the handling of the gauge invariance in the calculation of the photon self-energy and vacuum polarization, "there was not much new in Schwinger's presentation of his theory."[122]

The discussion of the phenomenon of vacuum polarization had indeed been considerably improved since Pocono. Schwinger had worked some more on the problem for its presentation in QED II (Schwinger 1949a), which was concerned with the computation of the modification of the vacuum in the presence of an external current $J_\mu(x)$. The effect of the current is to change the unperturbed vacuum Ψ_0 to $\Psi[\sigma]$, which vector obeys the Tomonaga-Schwinger equation

$$i\hbar c \frac{\delta \Psi[\sigma]}{\delta \sigma(x)} = -\frac{1}{c} j_\mu(x) A_\mu(x) \Psi[\sigma]. \tag{7.6.1}$$

In (7.6.1), A_μ is the external electromagnetic field satisfying

$$\Box^2 A_\mu(x) - \frac{\partial}{\partial x_\mu} \left(\frac{\partial A_\nu(x)}{\partial x_\nu} \right) = -\frac{1}{c} J_\mu(x). \tag{7.6.2}$$

If $J_\mu(x)$ vanishes in the remote past, then

$$\lim_{\sigma \to -\infty} \Psi[\sigma] = \Psi_0. \tag{7.6.3}$$

To first order, the solution of (7.6.1) is given by

$$\Psi[\sigma] = \left\{ 1 + \frac{i}{\hbar c} \int_{-\infty}^{\sigma} j_\mu(x')A_\mu(x')d\omega' \right\} \Psi_0, \tag{7.6.4}$$

which Schwinger rewrote as

$$\begin{aligned}
= \Bigg\{ 1 &+ \frac{i}{2\hbar c} \int_{-\infty}^{+\infty} j_\mu(x')A_\mu(x')\epsilon[\sigma,\sigma']d\omega' \\
&+ \frac{i}{2\hbar c} \int_{-\infty}^{+\infty} j_\mu(x')A_\mu(x')d\omega' \Bigg\} \Psi_0.
\end{aligned} \tag{7.6.5}$$

The first integral on the right-hand side of (7.6.4) is symmetrical, the second anti-symmetrical in past and future. Schwinger noted that "physically, this distinction reflects the fact that in real processes, the final state is different from the initial state, whereas virtual processes produce no net change in the system, and are therefore symmetrical with respect to past and future" (Schwinger 1948f, p. 63).

Since he was interested in the situation in which only virtual pairs are produced, Schwinger set equal to zero the second integral in (7.6.4) so that the operator changing Ψ_0 to $\Psi[\sigma]$ becomes

$$S[\sigma] = 1 + \frac{i}{2\hbar c} \int_{-\infty}^{+\infty} j_\mu(x')A_\mu(x')\epsilon[\sigma,\sigma']d\omega'. \tag{7.6.6}$$

Since this operator should be unitary, Schwinger replaced it with

$$e^{-iS[\sigma]} \equiv e^{\frac{i}{2\hbar c} \int_{-\infty}^{+\infty} j_\mu(x')\epsilon[\sigma,\sigma']d\omega'}, \tag{7.6.7}$$

which to first order reduces to (7.6.6). The expectation value in the vacuum of the perturbed current vector is then given by

$$\begin{aligned}
\langle j_\mu(x) \rangle &= (\Psi[\sigma], j_\mu(x)\Psi[\sigma]) \\
&= (\Psi_0, e^{iS} j_\mu(x)e^{-iS}\Psi_0) \\
&= (\Psi_0, (j_\mu(x) + i[S, j_\mu(x)] + \cdots)\Psi_0) \\
&\approx (\Psi_0, j_\mu(x)\Psi_0) + \frac{i}{\hbar c^2} \int_{-\infty}^{+\infty} < [j_\mu(x), j_\nu(x')] >_0 A_\nu(x')\epsilon[\sigma,\sigma']d\omega'.
\end{aligned} \tag{7.6.8}$$

The expression on the right-hand side of (7.6.7) is essentially the same as the one Schwinger had encountered in the photon self-energy problem and had discussed at the Pocono conference. Wentzel (1948), who had redone the calculation, had pointed out to Schwinger that when the vacuum expectation value of $\epsilon(x - x')[j_\mu(x), j_\nu(x')]$ is computed explicitly according to Schwinger's prescription, some non-gauge-invariant terms are obtained. To meet this objection Schwinger had devised a new way to do the calculation, which he presented at Ann Arbor. Dyson outlined Schwinger's new approach in a letter to Bethe, who was going to Europe:

> For the benefit of Europeans whom you may meet, and in case you have not seen this previously, I will summarise the argument.
>
> The operator defining the total charge in a system is
>
> $$Q = \int j_\mu(x')d\sigma'_\mu$$
>
> and is a constant of the motion independent of the surface σ over which it is measured. In particular, it must commute with all observable quantities, and so for every point x
>
> $$\int [j_\mu(x'), j_\nu(x)]d\sigma'_\mu = 0$$
>
> Now if x happens to lie on the surface σ, then $[j_\mu(x'), j_\lambda(x)]$ is zero for $x' \neq x$ and has only a δ-function singularity at $x' = x$. Hence in this case, for every continuous function $f(x')$
>
> $$\int f(x')[j_\mu(x'), j_\lambda(x)]d\sigma'_\mu = f(x)\int [j_\mu(x'), j_\lambda(x)]d\sigma'_\mu = 0 \qquad (I)$$
>
> The surface integral can be rewritten as a 4-volume integral
>
> $$\int f(x')\frac{\partial}{\partial x'_\mu}\{\epsilon(x, x')[j_\mu(x'), j_\lambda(x)]\}\, dx' = 0 \qquad (7.6.9)$$
>
> and since this is true for every function $f(x')$,
>
> $$\frac{\partial}{\partial x'_\mu}\{\epsilon(x, x')[j_\mu(x'), j_\lambda(x)]\} = 0. \qquad (II)$$

Dyson claimed that this equation can be verified by direct expansion in momentum space. Now "the unpleasant term in the polarization of the vacuum turns out to be just the vacuum expectation value of II. So it can be eliminated by using the operator equation first and taking the vacuum expectation value only at the end of the argument."[123]

In the rest of his summer school lectures, Schwinger outlined the calculations of the radiative corrections for an electron in a Coulomb field. The value for the $S_{1/2} - P_{1/2}$ interval given in the notes is

$$\frac{8\alpha^3}{3\pi} \frac{Z^4}{n^3} Ry \left(\ln \frac{mc^2}{\Delta E} - \ln 2 + \frac{3}{8} - \frac{1}{5} + \frac{1}{2} \right) = 1040 Mc,$$

in which the 1/5 comes from vacuum polarization and the 1/2 is the magnetic moment effect.

Schwinger's field theoretical activities during the rest of the summer of 1948 consisted in writing a draft of QED III. The state of affairs at the end of the summer can be inferred from the manuscript he submitted for the "Mayer Nature of Light" competition.

7.7 The Charles L. Mayer Nature of Light Award

In 1943 Charles L. Mayer presented the National Academy of Science (NAS) money for "a prize of $4,000 to be awarded in the period 1943–1945 for an outstanding contribution to our basic understanding of the nature of light and other electromagnetic phenomena which provides a unified understanding of the two aspects of these phenomena which are at present separately described by wave and corpuscular theories."[124] The Council of the NAS charged the National Science Fund, which had been established by the NAS in 1941 "for the Promotion of Human Welfare through the Advancement of Science," to administer the award. The executive committee of the National Science Fund in turn created an advisory committee to help select the recipient of the award. In May 1946 William J. Robbins, the chairman of the executive committee of the National Science Fund, wrote E. U. Condon, I. I. Rabi, and R. A. Millikan asking them to serve on the advisory committee. K. K. Darrow was subsequently added as the fourth member of the committee.[125]

Evidently, during the deliberations of the advisory committee the conclusion was reached to award two prizes of $2,000 each, rather than a single prize of $4,000. The announcement for "The Charles L. Mayer Nature of Light Awards" stated that two prizes would be awarded in 1946 for contributions published in the calendar years 1943, 1944, or 1945 or submitted in manuscript to the National Science Fund before January 1, 1946. The conditions for the awards were the following:

1. One prize of $2,000 for a contribution that provides "in terms *intelligible to the community of scientists at large* a unified understanding of the two aspects of these phenomena which are at present jointly described by wave and by corpuscular theories."

2. One prize of $2,000 for a "comprehensive contribution to a logical, consistent theory of the interaction of charged particles with an electromagnetic field including the interaction of particles moving with high relative speeds."

3. All contributions must be in the hands of the Advisory Committee no later than January 1, 1946.[126]

Twenty-eight entries were received. A cursory review of all the contributions indicated that "6 should be considered for Award No 1, 13 for Award No 2, and 6 appear to merit consideration under the terms of both Awards." Three of the entries appeared unworthy of further consideration. The submittals included contributions by John Griffith on "The Creation of the Solar System" and by Immanuel Velikovsky on the "Velocity of Light in Relation to Moving Bodies." At a meeting attended by Condon, Millikan, and Rabi on April 23, 1946, the committee, having read all the entries, concluded that four papers deserved consideration for a possible award. These were

1. C. J. Eliezer, "The Interaction of Particles and an Electromagnetic Field"

2. Peter Havas, "On the Interaction of Radiation and Matter"

3. Giulio Racah, "On the Self-Energy of the Electron"

4. J. A. Wheeler and R. P. Feynman, "Classical Electrodynamics in Terms of Action at a Distance"

All these papers related to the No 2 award. The members of the committee felt "that none of the papers met the criteria for award No 1 that they be 'intelligible to the Community of Scientists at large.' "[127] A subsequent careful reading of the four papers led to the conclusion that "the number 2 prize be awarded to Dr. Eliezer."[128]

The committee also decided to propose to the donor that the "prize be opened for another two or three year period and be awarded for some contribution to original knowledge in this general field." In his reply to the committee's recommendation, Charles Mayer suggested that they "simply extend the competition for such a period of time," and "award the remaining $2,000 to the scientist who . . . makes the best contribution either for the first or second subject." The committee accepted the suggestion.[129]

In 1947 K. K. Darrow resigned from the advisory committee and suggested that in his place "some young theoretical physicist be chosen," for example, "R. P. Feynman, R. Serber, or Arnold Nordsieck," whom he had met and seen in action at the Shelter Island conference. He also recommended that the future committee "be relieved of the burden of examining unpublished papers."[130]

In August 1947 Feynman was invited to become a member of the advisory committee, and he accepted the assignment.[131] An announcement for the extension of the "Charles L. Mayer Nature of Light Award" was issued, indicating that a prize of $2,000 would be presented in 1948 for a contribution satisfying the previously announced criterion 1 or criterion 2. The closing date for submission of materials was initially set for July 1, 1948, but was extended to October 1, 1948.

In April 1948 Millikan resigned from the committee "no longer feeling equal to the task," and recommended that Paul S. Epstein replace him. The recommendation was accepted by all the parties involved.[132]

Sometime during the summer of 1948, Schwinger became aware of the Mayer prize. When he noticed that Feynman was on the committee for the award of the prize "and therefore presumably ineligible to receive it," he decided "that someone out there had [him] in mind." According to Schwinger it was a case of "it ain't the money, it's the principle of the thing" (Schwinger 1983a, p. 342). On October 1 he submitted his contribution.

Soon after the closing date, the members of the advisory committee (Condon, Epstein, Feynman, and Rabi) received copies of the six manuscripts that had been submitted as contributions for the Charles L. Mayer Nature of Light Award–1948. Three were "crank" contributions that had been submitted to the earlier Mayer competition of 1946. The other three were J. M. Jauch and K. M. Watson, "Phenomenological Quantum-Electrodynamics, Parts I, II, III"; Norman Kroll and Willis E. Lamb, "On the Self-Energy of a Bound Electron"; Julian Schwinger, "On Quantum Electrodynamics and the Magnetic Moment of the Electron, Parts I, II and III."

In March 1949 the members of the advisory committee received a letter informing them that all of them had selected Dr. Julian Schwinger as their choice for the Charles L. Mayer Nature of Light Award–1948.[133] The presentation of the award to Schwinger was made at a special meeting of the Physics and Applied Science Colloquium at Jefferson Physics Laboratory at Harvard on Monday, May 9, 1949. The citation of the award, which had been prepared by Rabi, stated that

> Dr. Schwinger has extended the scope and power of quantum electrodynamics of Heisenberg and Pauli by reformulating it in a relativistically covariant form.
>
> In this way he was able to eliminate in a systematic manner the infinities which made the statements of this theory ambiguous in application. The application of this theory to the Lamb shift of the $2S$ level in hydrogen gives a theoretical basis to the explanation of this effect by H. A. Bethe. The further application of the theory to the intrinsic magnetic moment of

the electron, which was discovered by Kusch and Foley, gave results in agreement with experiment.

The methods Schwinger has introduced have found further application in radiation theory, collision theory, and in the theory of mesons and nucleons.[134]

The contribution Schwinger submitted to the National Science Fund consisted of three manuscripts. The first two were those of Quantum Electrodynamics I and II (Schwinger 1948d, 1949a). The third was a 25-page typewritten manuscript entitled, "Quantum Electrodynamics III: Modification of Particle Electromagnetic Properties." It was a draft of Schwinger's Quantum Electrodynamics III, the final manuscript of which was submitted to the *Physical Review* on May 26, 1949.[135] The manuscript was 25 pages long because this was how far Schwinger had gotten in his writing by October 1, 1948, the closing date for the submission of materials.

The draft of QED III that Schwinger submitted for the Mayer prize gives a clear indication of how far he had developed his formalism by the fall of 1948. QED II was concerned with a discussion of the polarization of the vacuum by a prescribed external current. QED III considered the considerably more complicated situation "in which the original current is that ascribed to an electron or positron, a dynamical system, and an entity undistinguishable from the particles associated with the matter field vacuum fluctuations." Furthermore, the altered properties of an electron or a positron were also computed in the presence of an external field, so as to compare them "with the experimental indications of deviations from the Dirac theory."[136]

The first part of the manuscript is an elaboration of the formulas (7.6.4)–(7.6.7). It consists of a computation of the one-particle contribution to the current operator in the case $\mathcal{H}(x)$, the Hamiltonian density in the Tomonaga-Schwinger equation, describes the interaction of the electron-positron field with the quantized electromagnetic field. A slightly revised version of this material can be found in the first part of the published version of QED III (Schwinger 1949c). It culminated with an expression for the second-order correction to the one-particle current operator

$$\left(\delta j_\mu^{(2)}(x)\right)_1 = \frac{\alpha}{3\pi}\left(\log\frac{\kappa}{2k_{\min}} + \frac{17}{40}\right)\frac{1}{\kappa^2}\Box^2 j_\mu(x) + \frac{\alpha}{2\pi}c\frac{\partial}{\partial x_\nu}m_{\mu\nu}(x) \qquad \textbf{(7.7.1)}$$

valid under the conditions $\frac{1}{\kappa^2}\Box^2 j_\mu \ll j_\mu$, $\frac{1}{\kappa^2}\Box^2 m_{\mu\nu} \ll m_{\mu\nu}$. Eq. (7.7.1) is identical to the expression (1.124) in QED III.[137] The second term corresponds to the anomalous magnetic moment contribution. As in QED III, Schwinger proceeded to discuss the infrared divergence that is present in eq. (7.7.1)—and its removal by the consideration of processes involving the emission of soft quanta. The second part of the QED III manuscript submitted for the Mayer prize considered the interaction of the electron-positron field with both an external electromagnetic field and the quantized radiation field. The external field could be either the Coulomb field of a nucleus or that associated with a macroscopic apparatus. The description of this situation in the interaction representation is given by

$$\hbar c \frac{\delta \Psi[\sigma]}{\delta \sigma(x)} = \left[-\frac{1}{c} j_\mu(x) A_\mu(x) - \frac{1}{c} j_\mu(x) A_\mu^{(e)}(x) \right] \Psi[\sigma], \qquad (7.7.2)$$

where $A_\mu^{(e)}(x)$ is the external electromagnetic field acting on the matter field current distribution:

$$\Box^2 A_\mu^{(e)}(x) = -\frac{1}{c} J_\mu(x) \qquad (7.7.3)$$

$$\frac{\partial A_\mu^{(e)}(x)}{\partial x_\mu} = 0. \qquad (7.7.4)$$

Schwinger then presented "a treatment that does not regard the external potential as a small perturbation," by transforming to a Heisenberg representation with respect to the external potential:

$$\Psi[\sigma] = U[\sigma] \mathbf{\Psi}[\sigma], \qquad (7.7.5)$$

where $U[\sigma]$ is chosen to satisfy

$$\hbar c \frac{\delta U[\sigma]}{\delta \sigma(x)} = -\frac{1}{c} j_\mu(x) A_\mu^{(e)}(x) U[\sigma]. \qquad (7.7.6)$$

The new matter field operator

$$\mathbf{\psi}^{(e)}(x) = U^{-1}[\sigma] \psi(x) U[\sigma] \qquad (7.7.7)$$

then obeys the equation of motion

$$\left[\gamma_\mu \left(\frac{\partial}{\partial x_\mu} - \frac{ie}{\hbar c} A_\mu^{(e)}(x) \right) + \kappa_0 \right] \mathbf{\psi}^{(e)}(x) = 0, \qquad (7.7.8)$$

and the equation of motion for $\mathbf{\Psi}[\sigma]$ reads

$$\hbar c \frac{\delta \mathbf{\Psi}[\sigma]}{\delta \sigma(x)} = -\frac{1}{c} \mathbf{j}_\mu(x) A_\mu(x) \mathbf{\Psi}[\sigma], \qquad (7.7.9)$$

which is formally analogous to the equation of motion for the state vector in the absence of an external potential, except that the operators $\psi^{(e)}(x)$ and $\overline{\psi}^{(e)}(x)$ that enter in the expression for the current operator $j_\mu(x)$ obey eq. (7.7.7) rather than the Dirac equation for a free particle and satisfy the commutation rules

$$\left\{ \psi_\alpha^{(e)}(x), \overline{\psi}_\beta^{(e)}(x') \right\} = \frac{1}{i} S_{\alpha\beta}^{(e)}(x, x') \qquad (7.7.10)$$

with

$$\left[\gamma_\mu \left(\frac{\partial}{\partial x_\mu} - \frac{ie}{\hbar c} A_\mu^{(e)}(x) \right) + \kappa_0 \right] S^{(e)}(x, x') = 0. \qquad (7.7.11)$$

Schwinger also derived an integral equation for $S^{(e)}(x, x')$ that embodies not only eq. (7.7.10) but also the boundary condition that $S^{(e)}$ must satisfy

$$\begin{aligned} S^{(e)}(x, x') &= S(x - x') \\ &+ \frac{ie}{\hbar c} \int_{-\infty}^{+\infty} d\omega' S(x - x'') \gamma_\mu A_\mu^{(e)}(x'') S^{(e)}(x'', x'), \end{aligned} \qquad (7.7.12)$$

from which it can be inferred that

$$S^{(e)}(x, x') = 0 \quad \text{for} \quad (x_\mu - x'_\mu)^2 > 0, \qquad (7.7.13)$$

"expressing the fact that the velocity of light is the limiting particle speed even in the presence of an external field." The representation in which the matter field spinors obey equations that correspond to a particle moving under the influence of the external potential, eq. (7.7.7) was later rediscovered by Furry (1951) and has become known as the Furry representation.

This initial version of QED III was of importance in Schwinger's later work. The integral equation (7.7.11), an equation related to the propagation of a "particle" in space-time, was probably the first example of an equation obeyed by quantities that would become central to Schwinger's formulation of quantum field theory: Green's functions.

7.8 Wentzel's and Pauli's Criticism

While Schwinger was busy writing up some of his researches, extending his formalism and sharpening his computational methods, the world at large was critically examining his results.

Some of the most trenchant criticisms came from Gregor Wentzel and Wolfgang Pauli. Wentzel was an outstanding theoretical physicist who had made important contributions to quantum mechanics and quantum field theory.[138] During the thirties he had written an important series of papers on methods to overcome the divergence difficulties in quantum electrodynamics, and had made a penetrating analysis of the use of strong coupling methods in meson theory (Wentzel 1940, 1941). During the war he had written a book on the quantum theory of wave fields that became a classic as soon as it was published (Wentzel 1943). In 1947 he accepted a professorship at the University of Chicago, and in the fall of that year he was invited to attend the Pocono conference.

Wentzel had studied the developments following the Shelter Island conference carefully and had arrived at a somewhat pessimistic assessment of the successes. Commenting on Schwinger's calculation of the Lamb shift (Schwinger 1948c) and of the anomalous magnetic moment of the electron (Schwinger 1948c), calculations that explained these effects in terms of field-dependent parts of the self-energy of the electron, Wentzel noted that

> it seems that the concept of electromagnetic self-energy now acquires a more than merely mathematical significance. However, the field-dependent terms can, at best, be defined as finite parts of the still diverging, total self-energy of the electron which has to be eliminated from the Hamiltonian by a formal readjustment. Therefore, there is still but little hope for a final and satisfactory solution of the self-energy problems within the framework of the conventional quantum theory of fields. (Wentzel 1948, p. 1070)

Upon his return from the Pocono conference, Wentzel went through the content of Schwinger's lecture and came to the conclusion that Schwinger's proof of his claim "that, in his new formulation of quantum electrodynamics, the self-energy of the photon vanishes identically, . . . is highly objectionable." He had calculated the photon self-energy using the same methods as used by Schwinger in computing the electron self-energy, and had found that "the self-energy of the photon turns out to be finite but not zero, whereas other methods of calculation yield infinite values" (Wentzel 1948, p. 1071). A finite rest mass of the photon, however, destroys the gauge invariance of the theory, so that the electromagnetic potentials will appear as being observable quantities. Wentzel therefore concluded that "it seems questionable to what extent the predictions of such a theory in higher order effects are trustworthy." He submitted his findings to the *Physical Review* on June 29, 1948, and sent a preprint of his article to Schwinger. Schwinger by that time had already worked out an improved computational scheme for the photon self-energy that made explicit use of the gauge invariance of the theory. He had presented his new way of calculating at the Michigan Summer School and had incorporated it in his manuscript for QED II. Schwinger sent Wentzel his new derivation, but Wentzel was not convinced. In a note added in proof to his *Physical Review* paper, Wentzel commented that Schwinger's new argument "again involves an integration by parts which, in my opinion, is not legitimate mathematically" (Wentzel 1948, p. 1071; note after footnote 3).

Schwinger's manuscript for QED II was finished in late October and was received by the *Physical Review* on November 1, 1948. Schwinger sent a preprint of the paper to Pauli. It was carefully analyzed by Pauli and his associates at the ETH in Zurich—Jost, Luttinger, Villars—and resulted in an incisive criticism of Schwinger's approach and an important clarification of the "philosophy" of the new approach. Pauli's criticism was contained in a 16-page letter he sent Schwinger at the end of January 1949.[139] Pauli also sent a copy of his letter to Schwinger to Bethe, who had asked him to comment on a paper by Epstein, who had found a result similar to Wentzel's for the electron's self-stress.[140] Using Schwinger's techniques, Epstein had computed the self-stress of a free electron of zero momentum and had found that to order $\alpha = e^2/4\pi hc$ it is equal to

$$S(0) = \int T_{11}(x)d^3x$$

$$= -\left(\frac{\alpha}{2\pi}\right)m, \tag{7.8.1}$$

where T_{11} is xx component of the energy-momentum tensor for electrons interacting with the electromagnetic field. Now, an observer moving in the x-direction with constant velocity relative to another considered at rest finds the energy of the electron to be

$$E = \frac{E(0) - \frac{v^2}{c^2}S(0)}{\sqrt{1 - \frac{v^2}{c^2}}}, \tag{7.8.2}$$

so that a non-zero self-stress would contradict the relation between the energy and momentum of the particle in the moving frame and the energy ($E(0) = mc^2$) and momentum ($\mathbf{p} = 0$) in the rest frame:

$$E = E(0)/\sqrt{1 - \frac{v^2}{c^2}}. \tag{7.8.3}$$

In his letter to Bethe, Pauli pointed out that a confusion had arisen because it was believed that "the actual results of the Schwinger-Tomonaga formalism are derived by the use of nothing else "than conventional perturbation theory." "But," Pauli stressed, "this is not at all the case."

The way in which the infinities are treated in order to bring forth
something finite is so essential, that theories in which the handling of infinities is made in different ways cannot be considered "equivalent." Particularly the relativistic invariant form of
perturbation theory, used by Schwinger, Tomonaga and others,
leads first to formulas which are indetermined, for the reason that
they contain integrals over products of functions with two different types of singularities on the light cone, namely the pole-type
and the δ-type. Without use of new assumptions this formalism
is only an empty scheme.[141]

In a footnote, Pauli added, "This point of view, which seems to me of great help
in clearing the logical situation, was stressed very much by Stückelberg." In his
letter to Schwinger, Pauli was more explicit:

In the following I would like to tell you of some considerations
stimulated by a careful study of your "Part II" which you were
so kind as to send me. Although I agree with the final results
of your paper, it seems to me necessary to reformulate some
of the arguments in a different way in order to avoid confusion
and critical controversies already coming up in the literature.
I wish to stress the circumstances that products of fonctions
[*sic*] (or their derivatives) of which one has a singularity of
the type $\delta(x_\mu x_\mu)$ the other of the type $(x_\mu x_\mu)^{-1}$ at the light
cone are not well defined mathematical symbols and therefore
integrals over the four dimensional x'-space of such products
at the point $x - x'$ and external fields at the point x' (see (2)
below) have not an *a priori* meaning. The same holds for the
corresponding integrals over the four dimensional momentum
space, if its different invariant or covariant summands do not
converge separately. Every "evaluation" of expressions of this
type is not a "computation" in the ordinary sense, but rather a
new *definition*, which can only be made precise by referring to
a certain limiting process, in the course of which the singular
functions are first replaced by regular functions (analogous to
the well known defintion of Dirac's δ-function which, however,
turns out as a too particular case for the problems investigated
by you). The uniqueness of sensible rules for these definitions
has to be particularly investigated. A particular way to handle
an a priori indetermined (non defined) mathematical expression
is according to my opinion an essential encroachment upon the
fundaments of quantum electrodynamics and its consequences

should not be considered as simple deductions based on known principles. On the other hand this is just the reason that I consider your invariant handling of infinities as a progress in comparison with the non-invariant handling of the infinities in the older conventional perturbation theory.

A typical example of the occurrence of yet undefined mathematical symbols in your results we meet in the problem of vacuum-polarization by external fields in which the necessary requirement of gauge invariance of the final results for all actually observable physical effects plays such an important role. According to your equation (2.19) the additional current is determined by the tensor[1]

$$K_{\mu\nu}(x) = \frac{\partial\overline{\Delta}}{\partial x_\mu}\frac{\partial\Delta^{(1)}}{\partial x_\nu} + \frac{\partial\overline{\Delta}}{\partial x_\nu}\frac{\partial\Delta^{(1)}}{\partial x_\mu} - \delta_{\mu\nu}\left(\frac{\partial\overline{\Delta}}{\partial x_\lambda}\frac{\partial\Delta^{(1)}}{\partial x_\lambda} + m^2\overline{\Delta}\Delta^{(1)}\right) \qquad (1)$$

and given by

$$< j_u(x) > = -4e^2\int K_{\mu\nu}(x - x')A_\nu(x')d^4x' \qquad (2)$$

One sees immediately that the condition of the invariance of the latter expression with respect to the gauge transformation

$$A_\nu \rightarrow A_\nu + \frac{\partial f}{\partial x^\nu}$$

has the form

$$\frac{\partial K_{\mu\nu}}{\partial x^\nu} = 0 \qquad (3)$$

which is equivalent to your eq. (2.29).

In your attempt to "prove" this equation (p. 20) you did not mention the important circumstance that the space-like surface, where your $\frac{\partial\epsilon(x-x')}{\partial x_\mu}$ does not vanish, contains also the point $x - x' = 0$, which can give a contribution to the left side of (3). For the latter one gets indeed

$$\frac{\partial K_{\mu\nu}}{\partial x^\nu} = \frac{\partial\overline{\Delta}}{\partial x_\nu}\Box^2\Delta^{(1)} + \Box^2\overline{\Delta}\frac{\partial\Delta^{(1)}}{\partial x_\mu} - \frac{\partial}{\partial x_\mu}\left(m^2\overline{\Delta}\Delta^{(1)}\right) \qquad (4)$$

Using now the wave equations (1.58) and (A.6) for $\Delta^{(1)}(x)$ and $\bar{\Delta}(x)$ respectively (4) changes into

$$\frac{\partial K_{\mu\nu}}{\partial x^{\nu}} = -\delta(x)\frac{\partial \Delta^{(1)}}{\partial x^{\nu}} \tag{4a}$$

As $\frac{\partial \Delta^{(1)}}{\partial x^{\nu}}$ has a singularity of the type $\frac{x_{\nu}}{(x_{\lambda}x_{\lambda})^{2}}$ the expression $\delta(x)\frac{\partial \Delta^{(1)}}{\partial x^{\nu}}$ is (non-defined) and the condition (3) is *not an identity*. This fact cannot be changed by any "proof," however skillful it may be arranged, but only by additional assumptions, the mathematical side of which may appear as definitions.

From the standpoint exposed at the beginning of this letter the result that the left side of (3) is indetermined at the origin is neither surprising nor alarming because the expression (1) for $K_{\mu\nu}$ from which we started has already the same lack of determination on the whole light cone. To give integrals of the form (2) a meaning it seems to me inevitable to use a certain limiting process. In order to make this process lorentz-invariant it is convenient to use for it the dependence of the singular functions in (1) on the rest mass of which only the square m^{2} occurs. Given a function (x, m^{2}) with a singularity at the light cone $x_{\mu}x_{\mu} = 0$, we construct with help of a regularizing function $\rho(\kappa)$ where κ represents the square of the mass, the function

$$\tilde{f}(x) = \int_{-\infty}^{\infty} \rho(\kappa)f(x, \kappa)d\kappa \tag{I}$$

Of the function $\rho(\kappa)$ we require at first only

$$\int_{-\infty}^{\infty} \rho(\kappa)d\kappa = 0 \quad \text{and} \quad \int_{-\infty}^{\infty} \rho(\kappa)d\kappa = 0 \tag{II}$$

as these conditions are sufficient to make the functions $\tilde{\bar{\Delta}}, \tilde{\Delta}^{(1)}$ as well as their derivatives and products of them regular at the light cone.[142]

[1] I omit in the following all irrelevant constant factors, use natural units $\hbar = c = 1$ and write m instead of κ_{0}. On the other hand I use your notations $\bar{\Delta}(x), \Delta^{(1)}(x)$.

Pauli proceeded to outline his method of regulators to which we shall return in chapter 10. The letter also explained how to apply regulators in the photon self-energy problem so as to maintain gauge invariance, and also contained a discussion of the computation of the electron self-energy using regulators.

Schwinger never replied to Pauli. He handed the letter to Bryce DeWitt, who at the time was doing a thesis with Schwinger on the photon self-energy in the case of interacting elelctromagnetic and (quantized) gravitational fields. DeWitt recalls that Schwinger told him:

Why don't you reply to Pauli? So in fact I did. However, I wasn't quite sure that I could really speak for Schwinger and was some-what embarrassed to let Schwinger see my letter. I had an intu-ition that it was pretty naive arguing on my part. So I sent the let-ter but told Pauli that this had not been approved by Schwinger; it was strictly my own. My arguments essentially were that you have to get a gauge invariant result: Therefore you keep playing tricks to make the result gauge invariant and if you do this you cannot get any result but the one that is gauge invariant. It is a unique gauge invariant result. I believe this is quite true that it is unique, but this method of arguing certainly didn't please a lot of people at the time.[143]

DeWitt got a reply from Pauli, in which he chided him and said: "Well since [the letter] didn't have Schwinger's approval he could hardly take it very seriously. Then he started poking fun at Schwinger. He said that this idea of gauge invari-ance of Schwinger's—his arguments on gauge invariance—he and his group at the ETH in Zurich always referred to as 'the Revelation.' "[144] To Oppenheimer, Pauli wrote:

My discussion with Schwinger, in which he never participated himself, makes me think on "His Majesty's" (Julian's) psychol-ogy. (An Evening seminar on this subject—ladies admitted—would be very funny. I can also tell experimental material from earlier times). His Majesty permitted one of his pupils (B. Seligmann) [B. DeWitt] to break the "blockade" of the ETH/Zurich by Harvard and to write to me a letter, but he *re-fused* to *read* this letter himself! The content of this diplomatic note (it was a very long one) is only this, that his Majesty had a kind of revelation on some Mt. Sinai, to put always $\frac{\partial \Delta^{(1)}}{\partial x_\nu} = 0$ for $x = 0$ (in contrast to $\frac{\partial \delta(x)}{\partial x^\nu}$ which has same symmetry prop-erties) wherever it occurs. We are calling here this equation "the revelation," but it did not help our understanding. This B. Seligmann and also a Mr. Glauber want to come here next spring, but both are unable to obtain a scientific recommenda-tion from his Majesty who prefers to "sacrifice" both of them rather than to write to me. I am enjoying this situation very much.[145]

In a slightly later letter to Oppenheimer in which he outlined the regulator method that he had devised with Villars, Pauli stressed that "he had no doubts on the actual correctness of Schwinger's result" but added that "generally I found a much deeper reason in his result but noted that picture of the new theory he conveys to his readers. The latter seems to me false because he is concealing behind a screen of mathematical virtuosity the assumptions which he silently uses. *He must have strong psychological reasons for the very conservative appearance of his theory*" (emphasis added).[146]

Pauli's letters indicate how myths are made. Schwinger not answering Pauli was attributed to "His Majesty's psychology." But Schwinger had not refused to read DeWitt's letter; DeWitt had been embarrassed to show his letter to Schwinger. Nor did Schwinger refuse to write letters of recommendations. He did write to Oppenheimer and Glauber, and DeWitt went to the Institute of Advanced Study. At the beginning of September 1949, Schwinger attended the International Congress for Nuclear Physics, Quantum Electrodynamics and Cosmic Rays, sponsored by the Italian and Swiss physical societies and held in Basel. At Rabi's request, he took the occasion to make a trip to Zurich to soothe Pauli's "ruffled feelings" for not having answered his long letter. Pauli sat Schwinger down and again voiced his unhappiness with various aspects of QED II. To each of his complaints Schwinger would in effect reply, "Yes, but I don't do it that way any more." According to Schwinger, "this refusal to be a stationary target left Pauli utterly exasperated."[147]

Schwinger had indeed taken a new approach to quantum field theory.

7.9 The Quantum Action Principle

The Lamb shift experiment, the Nafe-Nelson-Rabi and the Kusch-Foley experiments had determined the structure of Schwinger's initial theoretical apparatus. He had concentrated on these *specific* problems and had elaborated a formalism and a viewpoint only wide enough to attack those problems.

Schwinger came to Pocono in April 1948 on the crest of his success in accounting for the anomalous magnetic moment of the electron, having just completed his explicitly relativistic and gauge-invariant formulation of quantum electrodynamics. He did not pay close attention to Feynman's approach to QED at the conference: "I was in no way antagonistic. It wasn't the way I was thinking about it. If you come and are feeling on top of things you don't pay that much attention to the competition. You feel perfectly happy with the way things are. . . . Feynman, on the other hand, had been thinking in more general terms, so he came with a general philosophy, but not the techniques of calculations."[148] Schwinger spent the fall and winter of 1948 finishing the specific calculations he had undertaken and preparing materials for publication. He sent a summary of the results of his calculations of the radiative corrections for relativistic Coulomb scattering to the *Physical Review* on January 21, 1949. His letter included an expression for the Lamb shift

$$\delta E = \left(\frac{\alpha}{3\pi}\right)\left[\log\left(\frac{mc^2}{2\Delta W}\right) + \frac{31}{120}\left(\frac{\hbar}{mc}\right)^2 < \nabla^2 V >\right.$$

$$\left. + \frac{\alpha}{2\pi}\left(\frac{\hbar}{2mc}\right)(-i\beta\boldsymbol{\alpha}\cdot\nabla V)\right], \tag{7.9.1}$$

which when applied to the $2^2 S_{1/2} - 2^2 P_{1/2}$ levels of hydrogen, yielded a relative displacement of 1051 mc/sec. In the note Schwinger also made a point of indicating that French and Weisskopf and Kroll and Lamb had obtained results in agreement with (7.9.1).[149] They in fact had done so before Schwinger.

Schwinger remembers that when he came to the Oldstone conference in April 1949 he was asked when he sat down at the opening session, "What is the Harvard group doing?" and answering, "What is the Harvard group thinking about? They are not thinking; they are writing!" Schwinger commented: "By the time of Oldstone the work had been done. Oldstone was the beginning of Feynman's ascent. What he had done was being recognized and its utility had been appreciated. Naturally I sat and listened."[150] Schwinger's elaborate formalism and lengthy calculations had gotten him the reputation for powerful but very complicated techniques. "The methods I developed before 1949 were more than powerful enough for the immediate problems and it was clear *even to me* they got too messy after a while."[151] Dyson's paper on the "Radiation Theories of Tomonaga, Schwinger and Feynman" (Dyson 1949a) was the stimulus for setting Schwinger thinking anew and getting him started on a new formulation of the quantum theory of fields.

> I was a little puzzled in reading Dyson. It is a very different language. I didn't recognize what I was doing in what Dyson said I was doing. I thought it was nice that what Feynman was doing could be derived by these more standard methods. So I said: "If that is the case I don't have to break my head to figure out what Feynman is doing because it is the same thing." So to that extent I took myself off the hook. The fact that everybody kept getting more and more interested in this [Feynman's and Dyson's work] and in a sense turning away from these powerful methods [i.e., QED II and QED III] rather baffled me. So sooner or later I had to find out what they were doing. But even if that hadn't occurred there would have been the pressure to continue the technique, to find a more effective way of doing it....I had gone through these gyrations of these successive canonical transformations, [and] it was clear I couldn't have done this to the next order....I was stimulated by what Dyson claimed to have done because here he said was a method. OK. I am not going to do it that way, but let's find a method. So all these

gyrations ultimately led to the quantum action principle—which to me was the answer—because all follows from that.[152]

In his Fermilab symposium lecture, Schwinger outlined the genesis of this new approach: "My retreat began at Brookhaven National Laboratory in the summer of 1949. It is only human that my first action was one of reaction. Like the silicon chip of more recent years, the Feynman diagram *was bringing computation to the masses* [emphasis added]. Yes, one can analyse experience into individual pieces of topology. But eventually one has to put it all together again" (Schwinger 1983a, p. 343). Like Feynman, Schwinger found his vision in Dirac's 1932 paper on the Lagrangian in quantum mechanics. Dirac's paper had led Feynman to his integral-over-path formulation of particle quantum mechanics (Feynman 1942, 1948c) and its generalization to Bose-Einstein fields, but he could not deal with fermions and Fermi fields. Schwinger was keen to treat Bose-Einstein fields and Fermi-Dirac fields on the same footing, and did not see how to do so in the Feynman-Dirac integral approach: "This was not my idea of a fundamental basis for the theory. And, as the history of physics and my own experience indicated, integral statements are best regarded as consequences of more basic differential statements. Indeed, the fundamental formulation of classical mechanics, Hamilton's principle, is a differential, a variational, principle" (Schwinger 1983a, p. 343). Moreover, the action is a relativistic invariant. The challenge for Schwinger therefore became: "What is the general quantum statement of Hamilton's principle in variational form?" The answer turned out to be rather simple and was the starting point of the subsequent developments. In quantum mechanics, the time development of a system is represented by a transformation function U_{12} relating the states of the system at two times, or more generally on two spacelike surfaces, σ_1 and σ_2:

$$(\xi_1', \sigma_1 \mid \xi_2'', \sigma_2) = (\xi_2', \sigma_2 \mid U_{12}^{-1} \mid \xi_2'', \sigma_2). \tag{7.9.2}$$

Here ξ_1' and ξ_2' represent the eigenvalues of complete sets of commuting operators ξ_1, ξ_2 constructed from the observables attached to surface 1 and 2, respectively. Any infinitesimal change in the quantities on which the transformation function depends induces a corresponding alteration in δU_{12}^{-1}

$$\begin{aligned} \delta(\xi_1', \sigma_1 \mid \xi_2'', \sigma_2) &= (\xi_2', \sigma_2 \mid \delta U_{12}^{-1} \mid \xi_2'', \sigma_2) \\ &= \frac{i}{\hbar} (\xi_1', \sigma_1 \mid \delta W_{12} \mid \xi_2'', \sigma_2). \end{aligned} \tag{7.9.3}$$

Schwinger's basic postulate is that δW_{12} is obtained by variation of the quantities contained in a hermitian operator W_{12}, which has the general form

$$W_{12} = \frac{1}{c} \int_{\sigma_1}^{\sigma_2} (dx) \mathscr{L}[x], \tag{7.9.4}$$

where $\mathscr{L}$, for a field system, is an invariant hermitian function of the fields and their coordinate derivatives. In conformity with their classical analogues, Schwinger called W and $\mathscr{L}$ the action integral and the Lagrange function operators. He emphasized in his Fermilab lecture that "it is the introduction of operator variations that cuts the umbilical cord of the correspondence principle and brings quantum mechanics to full maturity" (Schwinger 1983a, p. 343). Furthermore, the quantum action principle allowed Fermi-Dirac fields to appear naturally and on an equal footing with Bose-Einstein fields. During the spring semester of the academic year 1949/50, Schwinger lectured on this new approach to quantum mechanics and quantum field theory. A set of notes entitled "Quantum Theory of Fields. A New Formulation" was edited by M. L. Goldberger and issued in July 1950. In the course, Schwinger showed how the quantum action principle led to operator commutation rules, equations of motions, and conservation laws. In addition, when the action operator is chosen to produce first-order differential equations of motion, or field equations, it predicts the existence of two types of dynamical variables, with operator properties described by commutators and anti-commutators, respectively. A series of papers entitled "The Theory of Quantized Fields" developed the quantum action principle further. Effectiveness came with the introduction of sources, in two famous, brief papers that appeared in the *Proceedings of the National Academy of Sciences* in 1951 (Schwinger 1951d). In this extended scheme, the functional derivative of the transformation function with respect to a source is the matrix element of the associated field. This allowed all operator field equations to be represented by numerical functional derivatives. The method became the basis of a powerful and seminal formulation of quantum field theory. Its importance to Schwinger is indicated by the fact that his *entire* Nobel Prize lecture was devoted to an exposition of his quantum action principle (Schwinger 1966). Regarding his award of the Nobel Prize, Schwinger commented: "I wouldn't have thought that the electrodynamic work would be it. It's splashy and all that, but it was an obvious development of ideas that were in the air. Frankly, I much prefer the work centering on the action principle, spin statistics, TPC. These are fundamental results. The interplay between quantum fields and classical sources, the invention of non-classical sources, these have lasting values."[153]

7.10 Philosophical Outlook

In the early sixties, Schwinger delivered a popular, nontechnical lecture on the subject of quantum physics and philosophy.[154] The lecture was given during the period when the high energy theoretical physics community was divided into two camps: one committed to field theory and the other to Chew's particle-based S-matrix philosophy. At the time theorists were deeply split about "what they consider[ed] to be the proper modes of development, the proper means of finding a logical explanation" for the properties of the vast number of the so called "elementary particles" (pp. 2–3). In his lecture Schwinger explained why he was a committed field theorist. But Schwinger's presentation is of interest beyond this, because in it he also outlined his "general world view" and the "world picture" he then held. Even though the lecture was given almost a decade after the close of the period of interest to us, it nonetheless reflects rather accurately Schwinger's viewpoint in the early fifties, a viewpoint he had honed in the courses on quantum mechanics and advanced quantum mechanics he gave at Harvard in the late forties and throughout the fifties (Schwinger 1970c).

Schwinger's lecture was recorded and transcribed but never published. I have quoted at length from it—including, at times, some of his repetitions—not because it contains new materials (the viewpoint expressed regarding quantum mechanics is in fact a pragmatic variant of the standard Bohr interpretation) but because the unedited transcript exhibits the powerful logic of Schwinger's exposition and the remarkable clarity of his mode of expression, a testimony to his mastery of the subject matter. It also captures his characteristic style of lecturing, his distinctive use of language, and his particular rhetoric. The only thing the transcript does not indicate immediately is the pace at which Schwinger lectures, usually without ever looking at his notes. But the tempo can be inferred from the fact that his one-hour lecture required some 70 pages of double-spaced, typewritten text.

The lecture was delivered to an audience not versed in the technical details of quantum mechanics. One of the didactic devices used by Schwinger was to repeat himself. The repetitions, often using but slightly differing words, allowed the listeners to catch their breath. The rephrasings also conveyed to the members of the audience a sense of familiarity—they were hearing the same thing for a second time—and undoubtedly lulled some of them into believing that they were understanding the matter fully.

In his opening remarks Schwinger stated that he approached the idea of giving the lecture "with great diffidence because he was neither trained nor particularly eloquent in the traditional modes of philosophical discourse"(p. 1). But on thinking the matter over, it seemed to him that there were "deep philosophical lessons" to be learned from the way in which the practicing theoretical physicist thinks about the foundations of the subject; that is, "the manner in which he approaches the problems, [and] the general criteria that he brings to bear on what is a reasonable solution." He added that he considered the topic of the lecture to be of particular importance, because quantum mechanics—"the rational mode of understanding microscopic or atomic phenomena"—has had "the greatest impact of any of the developments of physics upon the mode of thinking or the world picture of the physicist and thereby indirectly of the general citizen" (p. 2).

Schwinger introduced the subject matter historically, for he believed that to understand quantum physics one must see how the stage had been set by classical physics. The latter he identified as the causal, deterministic, "essentially precise formulation of the properties of matter as they were finally expressed in their essentially perfect form at the beginning of the 20th century" (pp. 3–4). He characterized a theory as causal if when "the state of the system is given at a definite time, . . . then the state is completely determined at any other time. . . . Causality is inference in time." In addition, a theory is deterministic if "the knowledge of the state also determines all possible physical phenomena precisely" (p. 5). The classical picture accommodated two distinct viewpoints. The first, the extrapolation and idealization of Newtonian physics, was a description based on massive point particles interacting by means of forces that act instantaneously at a distance. The second was the field point of view, whose paradigm was Maxwell's field theory of electromagnetism. In the field point of view, interactions are propagated locally from one point of space to the contiguous points with finite velocity. Both these theories are causal and deterministic, but they are very different in their characterization of the specification of a state. In the case of Newtonian physics, a *discrete, finite* number of quantities—the positions and velocities of the particles at a given time—specify that state. In the field case, a *continuous, infinite* number of quantities—the field intensities at every point of space—are required to specify the state at a given time. Particles and fields are the models of the two limits of classical behavior: the discrete and the continuous.

Continuing his historical exposition, Schwinger stated his belief that the "very important developments associated with the name of Einstein—the special and general theory of relativity— . . . were not radically new developments in the sense [that] quantum mechanics [was]; they were rounding out the framework of classical physics" (pp. 11–12). What culminated with relativity was "a theory in which there are particles and fields, standing side by side, neither explained in terms of the other. A dualistic theory, but in which the strict Newtonian point of view has been modified because we now recognize that the interactions between particles are not instantaneous but are propagated through the mechanism of the field. The field is there to supply the dynamical agency by which the particles interact" (p. 13).

The legacy of classical mechanics and relativity incorporated a fundamental duality: the discrete point of view of particles coexisting side by side with the continuous view of field. One of Schwinger's aims in his lecture was to trace how these two distinct classical concepts became "transcended in something that has no classical counterpart, the idea of the quantized field," an idea that Schwinger considered to be "perhaps the deepest expression of what has been learned within the framework of microscopic phenomena" (p. 14). The necessity for transcending these classical concepts was made clear by recalling some of the paradoxes encountered in phenomena involving light, the paradigmatic example of a wave or field phenomenon. Under certain circumstances—for example, in the photoelectric effect, light behaved in a particlelike manner. Similarly, electrons, the standard examples of microscopic particles, exhibited wavelike properties under certain experimental conditions. For example, interference rings are observed when a collimated beam of monoenergetic electrons is scattered from a crystalline solid. Schwinger then recounted Bohr's struggle to understand the spectrum of hydrogen and emphasized the radical nature of the break with classical descriptions that followed from Bohr's insight that quantities which classically assumed continuous values—for example, the energy and the angular momentum—took on discrete values for microscopic systems. Part of Bohr's greatness, Schwinger asserted, was that he had recognized earlier than anyone else that microscopic phenomena had to be understood in terms of new laws that transcended those with which physicists were familiar. The microscopic domain was a new world.

One characteristic feature of this new microscopic world is that the phenomenon of atomicity is all-pervasive.

> Not only must we account for the very existence of atoms, which after all is not a classical conception,—classically, there should be no limit to the extent to which you could subdivide matter. The fact that this subdivision cannot be carried out indefinitely, but ceases when we reach the atomic scale is, of course, the most fundamental statement that something new is involved. The phenomenon of atomicity, not only in the mere existence of atoms but also in the laws of mechanical motion—that of the atomicity of action, to put it in the most general way—is the basic phenomenon of microscopic physics. (P. 22)

The other new fact of microscopic physics is the essential statistical nature of microscopic phenomena. "This is another fundamental feature which must be accepted as the way that the microscopic world operates" (p. 23). This conclusion follows inescapably from the analysis of how interference patterns come about in the scattering of electrons off crystals:

> Once we have this apparent duality of entities—electrons behaving under some circumstances like discrete entities, landing at definite places on the screen, but in other respects acting as waves, producing in their overall intensity an interference pattern characteristic of wave phenomena,—we must accept the fact that the interference pattern is not going to be repeated in miniature every time an electron lands. And therefore there must be an aspect of randomness about where the electrons do land. [We must accept the fact] that the interference pattern is merely finally the statement of relative probabilities....
>
> It is not possible, in general, to predict the outcome of a specific event. But what one can predict, and ... the purpose of microscopic physics or quantum mechanics [is] to predict, the average result, the statistical result, of the repetition of a sufficiently large number of the same experiment. (Pp. 27–28)

Schwinger then asked: What must be the mathematical features of a theory that incorporates these "bizarre" features of microscopic physics, namely, the atomicity and the statistical nature of microscopic events? Before this question could be answered, "we must go back and think a little more consciously about the fundamental principles, call them philosophical, if you like, which underlie classical physics—or shall I say macroscopic physics—because that's now the distinction" (p. 30). More specifically, what must be addressed is the theory of measurements. "We have to recognize . . . that physics is an experimental science and it is concerned only with those statements which in some sense can be verified by an experiment. And the purpose of [any] theory is to provide a unification, a codification, or however you want to say it, of those results which can be tested by means of some experiment. In a sense therefore, what is fundamental to any theory of a specific department of nature is the theory of measurement within that domain" (p. 31).

Now the characteristic feature of the theory of measurements in the macroscopic classical domain is the concept of a nondisturbing measurement. In any measurement in that domain one can either make the interaction between the measuring apparatus and the system so small that there is a negligible disturbance, or, if by the nature of the measurement the disturbance cannot be made arbitrarily small, one can still calculate the effect of the disturbance and correct for it to arbitrary precision. Thus, by virtue of the causal and deterministic structure of classical physics, there is no limit to the accuracy with which simultaneous measurements can be made of any number of physical properties.

> So, the point is, therefore, that the classical theory of measurement says that there is no limit to the accuracy with which we can assign numerical values to all the quantities that are needed to specify the state, and since all these are deterministic theories, that means to all physical properties at once. For this reason, since physical properties, so to speak, can be assigned numerical values, one has never in classical physics drawn any distinction between the physical properties and the numerical values which they have at any particular time because we are always in the position of being able to assign to the physical properties considered, if you like, as an abstract thing, a very concrete representation by means of numerical values which a non-disturbing measurement would assign to them at a particular time.
>
> So, here we have again restated the foundations of classical physics: the idealization of non-disturbing measurements and the corresponding foundations of the mathematical representation; the identification of physical properties with numbers because nothing stands in the way of the continual assignment of numerical values to these physical properties. (P. 34)

Schwinger went on to contrast the macroscopic situation with the microscopic one. He noted that the atomicity of microscopic entities imposes sharp constraints on microscopic measurements:

> There is no half an electron. The electron has a definite mass. It has a definite charge. . . . I must take into account the fact that the strength of the interaction, which must be present if I am to talk of measurements at all and, therefore, talk meaningfully of physical phenomena, the strength of the interactions that are necessarily present if a measurement is to take place at all, cannot in general be made arbitrarily small because the physical phenomena that interact, the atoms, the electrons, in general have relevant physical properties which come in certain units—quanta, the origin of the name of the subject that we are discussing.
>
> Now, that might seem as this were not an insupportable difficulty. . . . In classical physics, we said the situation may be such that the measurement interaction . . . cannot be made arbitrarily weak, but this does not upset the underlying philosophy of measurement because I can calculate with arbitrary precision what the effect of that interaction was and can compensate for it, correct for it. Can I still do that now? The answer is no, because this is where the second fundamental aspect of microscopic measurements comes into play; namely the phenomena of statistics. The fact is that we cannot predict in detail what each individual will do but only make predictions on an average or statistical scale. . . . If an act of measurement has produced a large disturbance, we might say we could, nevertheless, correct for it in each individual instance. But we cannot produce a control over what happens in each individual instance in any detail. We can only predict or control what happens on the average; never the individual instance, which is to say therefore that the program of computing what the effect of the disturbance was and correcting for it is, in general, impossible. Impossible because we cannot control precisely what will happen in each individual circumstance. Therefore the two basic tenets of the theory of macroscopic measurement are both violated. Either the interactions cannot be made arbitrarily weak because of the phenomenon of atomicity, or if we wish to accept this and correct for it, we cannot do so because we do not have a detailed, deterministic theory of each individual event; we have only the ability to anticipate or control what happens on the average. (Pp. 36–37)

Schwinger therefore concluded that in order to construct a theory for microscopic physics one needed a whole new theory of microscopic measurements, and that a whole new mathematics must go with this. The new mathematics must "represent or mimic" the basic properties of microscopic measurements and must reflect the fact that we can no longer speak meaningfully of the numerical values that physical properties have at a given time:

> If we once recognize that the act of measurement introduces in the [microscopic] object of measurement changes which are not arbitrarily small, and which cannot be precisely controlled ... then every time we make a measurement, we introduce a new physical situation and we can no longer be sure that the new physical situation corresponds to the same physical properties which we had obtained by an earlier measurement. In other words, if you measure two physical properties in one order, and then the other, which classically would absolutely make no difference, these in the microscopic realm are simply two different experiments....
>
> So, therefore the mathematical scheme can certainly not be the assignment, the association, or the representation of physical properties by numbers because numbers do not have this property of depending upon the order in which the measurements are carried out.... We must instead look for a new mathematical scheme in which the order of performance of physical operations is represented by an order of performance of mathematical operations. (Pp. 40–42)

Schwinger thus made plausible why in quantum mechanics physical properties are set in correspondence with noncommutative operators. He emphasized that quantum mechanics is still a causal theory—given the state at one time the state is uniquely determined at any other later time—but it is only a statistically determinate theory: the knowledge of the state enables one to predict only the statistical, the average outcome of the result of the measurement of any physical property, "never the result of any specific event" (p. 44). Schwinger went on to talk about Bohr's complementarity principle—"perhaps the widest philosophical principle that has emerged from the studies of microscopic systems" (p. 49). In the case of a single particle, Schwinger took its expression to be the fact that the definition of the state never referred to all the physical properties of the particle, for example, its position and its momentum, but to only half of them. He went on to note that the concept of identity or indistinguishability is given an "entirely new turn in microscopic phenomena" (p. 54). In any experimental situation involving several identical particles,

these basic physical phenomena, the atomicity, the statistical nature of things, the inability to control in detail individual events,
implies corrrespondingly the absence of an ability in the fundamental experimental sense to tell in detail which particle is
which at every stage of the interaction, which means that my
description must take into account, in a fundamental way, this
fundamental failure of being able to place a tag on every particle
because there is no experiment I can perform that gives reality
to that label. Because to do so would represent an intervention
into this experiment, the performance of a detailed microscopic
localization experiment which would change completely the nature of the experiment.... The net result of all of this is to recognize the description of states of several particles necessarily
could only be done in a way which would incorporate from the
beginning the fact that they were indistinguishable; that particular labels have no significance. And to state the results simply,
it means that when we have several indistinguishable particles,
the states can only be described in a way that is perfectly symmetrical among all the particles that contribute to it ... In other
words, [since] these arbitrary labels are deprived of any distinguishing significance ... no matter how the labels are given, the
wave function or the state is the same. But this can be done
either by making them completely symmetrical or completely
anti-symmetrical. And in this way we recognize the existence of
two very distinct systems of identical particles [namely, bosons
and fermions]. (Pp. 58–60)

Having elaborated what he considered to be the essential features of
the phenomena of microscopic physics and of the theoretical structures representing them, Schwinger turned to the idea of the quantized field which he
saw as "a deeper unification of what has been achieved by the new principles" [p. 61].

Let me imagine, and at the moment it is nothing but imagining,
an idealized, abstract physical situation in which I will create
a particle [at the position x at time t].... And let me describe
this act of creation ... by a creation operator [that is labeled by
the space-time point x, t and that operates on a state]. An operator
because it symbolizes a physical property but something beyond
what we are accustomed to thinking of, and an operator because
it acts on a state, the state being the state in which nothing is
present, or physically a vacuum. (P. 62)

A state with an arbitrary number of particles at a given time t can then be obtained by the repeated application of creation operators on the vacuum state, and furthermore, that the symmetry properties of the states representing the two classes of particles that are observed in nature—bosons and fermions—become replaced by an algebraic property of the operators.

> We now transfer our attention to this operator as the basic physical object. And this is what I mean by the quantized field because it is on the one hand a field; it is a mathematical quantity which varies continuously in time and space. On the other hand, it is certainly not a classical field, because these as operators are not things that can be measured simultaneously, and [also because] in the operator character . . . we have the elements of discontinuity which is essentially the particle concept.
>
> In fact, I have described, I have obtained, a field—and I have described it to you in terms of physical operations on particles. Namely this is the symbolized $\psi(\mathbf{x}, t)$, the operation of creating a particle at a certain point in time but since I can do this anywhere in space, at any time, a field conception is introduced. In other words, the two entirely unrelated classical conceptions of discreteness of particles, of continuity of the field, now are unified in this entirely new conception; if not unified, then transcended because the new conception, which is beyond either, because the two are after all incompatible in the classical sense because there is nothing that is both discrete and continuous. We point to something which has neither of these two properties but which, in limited domains, can be characterized in terms of either of these conventional concepts.
>
> So, here is, so to speak the fundamental unification, the idea of the quantized field. The fact that we can . . . speak meaningfully [and] think of processes in which particles are created, [implies] correspondingly that we must also have the inverse processes in which particles are destroyed at various points in space and time.
>
> This, as it arose historically, was simply a convenient way of summarizing the mathematical properties of indistinguishable particles but soon, through the ever broadening developments of experimental science, what was here conceived simply as a convenient mathematical idealization became reality. (Pp. 62–64)

Schwinger then proceeded to outline the developments since the early thirties, and in particular, the consequences of the realization that the direct correspondence and association of fields with particles "as we know them" could not persist. "A new level of abstraction had to be reached and was reached in the course of attempting to understand in more detail—more detail demanded by the experimental data— to understand more of the properties of atomic phenomena than were successfully accounted for in the first flush of the development of quantum mechanics" (p. 67). The recognition that in the case of electrons one is really concerned with the dynamics of two interacting fields—that of the electron-positron field and the photon field, that is, the electromagnetic field—implied that the identification of each field with these physical names had been only an approximate one. Only in those cases that the interaction between fields is weak—as is the case of electrodynamics— can physical names be used in relation to the operators, the mathematical objects used in the description. In other words, the physical object we call the photon is not what is created by the mathematical operator.

> Once you recognize this, you now say that we draw a distinction between two levels of physical description. There is the phenomenological level, in which we recognize the properties of electrons and photons, as we see them, of course, the enormously detailed analysis of microscopic experiments. [And] there is the attempt to deepen the understanding in terms of more primitive objects which are these fields, which are no longer placed in immediate correspondence but through a chain of dynamical development. (P. 69)

Schwinger cited the advances in quantum electrodynamics in the post-World War II period as illustrative of this process, and noted that "what were once considered anomalies, things that were unexpected, now became the predicted outcome of this deepening level of understanding of what the observed particles in nature were; that our level of understanding is not to be found in terms of what we actually see but something at a more fundamental level" (p. 69). He indicated that the process has continued in the submicroscopic realm, in the world of mesons, nucleons and hyperons. The strategy has been to extend the analysis of particles "as we know them" and their association with fields by introducing "yet more fundamental fields at a deeper level which have fewer properties . . . because this attempt at understanding is always to strive for a simpler level, to have deeper, more symbolic laws, with fewer arbitrary constants. In other words, unlike the experimental situation in which the charge of the electron, the mass of the electron, the magnetic moment of the electron, are all unrelated constants, the deeper understanding attemps to explain one or more of these in terms of a fewer number of fundamental things" (p. 70). But this process ran into difficulties at the subnuclear level in the early sixties not only because "the phenomena [are] bewilderingly complicated" but also because "we lack the mathematical means to draw the implications of any particular hypothesis about what is going on." As a consequence of this, a very deep schism has occurred in the theoretical physics community over what should be the fundamental nature of an explanation at the subnuclear level.

> Should it be the continuation of this point of view of the search-ing for deeper understandings in terms of ideally a very small number of fundamental fields, which in their dynamic interplay and as a result of the complexity of that dynamics, finally bring about the manifold nature of the world as we see it.... Or must we really abandon this attempt altogether and simply describe nature in terms of what happens when we take various mi-croscopic particles and perform experiments on them ... [and] make no attempt to describe what goes on [in the region where the particles interact] and simply attempt to characterize what emerges finally when the particles are separated again. Is the purpose of theoretical physics to be no more than a cataloguing of all the things that can happen when particles interact with each other and separate? Or is it to be an understanding at a deeper level in which there are things that are not directly ob-servable as the underlying fields are, but in terms of which we shall have a more fundamental understanding. (P. 73)

Schwinger clearly believed at the time that the approach based on quantized fields was the correct one and the one to follow. He believed "that the fundamental ob-jective is to find a simple mathematical scheme and labor very hard to draw the mathematical consequences, and in this way explain vast areas of empirical evi-dence" (p. 80). He still believes this.

A good deal of Schwinger's intellectual production has been concerned with drawing out "mathematical consequences." He has been called a formalist by virtue of his impressive formal and mathematical abilities. But he disagrees.

> I am not fundamentally a formalist—even though everyone thinks I am. The formalism is simply the language with which I express the physical ideas that have already been thought out. I don't know whether there are actual mental pictures. I suppose you can't avoid having mental pictures. If somebody talks about an electron I don't think I see Schrödinger's equation. Every-body pictures something; it's a point moving. There has to be some sort of picture.
>
> "All of this work [on QED from 1947 to 1950] was mo-tivated, I think, by physical ideas and to some extent philosoph-ical ideas on how a theory should be built. But from that initial

physical impetus—which to some extent is visual, but certainly
not in the elaborate way of Feynman—it's translated into mathematics, in which the mathematics attempts to simulate the physical ideas. And I think by and large from that point on it now
becomes mathematical manipulation—the physics only looking
at the answer to ask is this reasonable.[155]

Commenting on the advances in QED in the late forties, Schwinger noted that
"QED was necessary in restoring faith in field theory with the understanding that
field theory was to be trusted in one place but not trusted in another. One simply
had to separate the untrustworthy area. That was a fundamental difference with the
Dyson approach. He ends with a mathematical scheme which is electrodynamics
to all orders and all energies."[156]

For Schwinger, renormalization

is the clear separation of what we don't know—but which affects
our experiments in a very limited way—from what we do know
and where we can calculate in detail. In fact, I insist that all
theories are like this.—People may not want to face up to it, [but]
there is always an area beyond where the theory either breaks
down or where other phenomena come into play that you don't
know about. They do not upset everything in the area you can
control, and you isolate that from it: That's what renormalization
is really about. Not sweeping infinities away but isolating the
unknown part and recognizing its limited influence.

I am not sure that I was at all interested in the mathematical question of convergence to all order. I don't think that is
a physical question. [Similarly] I have a feeling even then that
I did not take renormalizability too seriously. If in fact the theory had been not renormalizable at the 27th stage or whatever
have you, I would have said "O.K. That's good" because here
is a place where what we don't know, namely what happens at
very high energies, enters the theory and will learn something.
It wasn't essential to me that the theory be renormalizable to
all orders. That was nice to get the theory going to lowest order. What would be even more interesting is if it didn't work.
I wasn't very caught up in all these all order questions. To that
extent I consider myself a physicist.

While a happy thing to have, I don't regard renormalization as fundamental—even in terms of selection of theories. It
may be right, but one has to recognize that there is an element of
hubris involved in saying that we are so close to the final answer
that this criterion becomes relevant—that there isn't a whole

world unknown to us about to be discovered in higher energies. I'd rather have it that way. It is more interesting. The philosophic approach which I stumbled into in those early days lasted with me.[157]

Schwinger has remained a phenomenologist. For him theories are always descriptions of but a part of nature.

7.11 Epilogue

As important as Schwinger's papers—in quantum electrodynamics, field theory, nuclear physics, scattering theory, and a host of topics in applied mathematics—during the decade from the mid-forties to the mid-fifties, was his influence on the Harvard and MIT graduate students and faculty members who attended his lectures on waveguides, nuclear physics, quantum mechanics, and field theory, and his impact on the generation of Harvard graduate students who did their dissertation under his supervision, often on problems that he had given them.

Bryce DeWitt, who was an undergraduate at Harvard from 1939 to 1943 and returned to Harvard in 1946 for his graduate studies in physics after serving in the navy as a pilot, has recorded his reminiscences of Schwinger's lectures:

[His] first lecture, often the only lecture of the day, would begin at noon. Occasionally, he would give an 11:00 o'clock class. If so he was invariably late, not arriving until 11:30. The room was always packed with the front row particularly crowded, everyone patiently awaiting Schwinger's arrival. Then at the appropriate moment, the door at the front of the room would open and a large head with a shock of black hair on it, would poke its way, followed by Schwinger's stooped body, come into the room, pick up a piece of chalk and almost invariably begin with the word at the end of the last hour. He would take up from where he had left off and start covering the board with equations. He never lectured from notes. It all came out of his head, except once, in a nuclear physics course he was teaching. He hadn't memorized some cross-section or other and he dug into vest pocket or a shirt pocket and pulled out a small slip of paper on which he had written it down. Of course everybody hissed him roundly for having recourse to actual material.

It was virtually impossible to follow the lectures in class. However, it was obvious that they were wonderful. I would simply write down everything that he wrote, which was about all I could do. I barely had a moment to interject a comment or something that might guide me in the philosophy of what he was doing, or explain a step he either had just finished making or was about to make. I would take these precious notes back to Kirkland House and in the evening I would rewrite them, reconstructing everything that Schwinger was trying to do, filling in the gaps, and trying to make it all logical. Usually one could piece together what Schwinger was doing. Of course, it was all very elegant and formal and on a number of occasions I remember throwing down my pencil in disgust and saying: "the s.o.b. has done it again." What I meant was [that] he made something sound very plausible, but behind it were many deep unanswered questions or at least questions that if you looked at it carefully ought to have been addressed. Schwinger was not addressing them and was merely being guided by a kind of intuition, led by the formalism itself. That is, the formalism would take a life of its own and just lead you even though it might not be completely legitimate to do so.... It was always a challenge to try to fill in the gaps in Schwinger's lectures.[158]

Commenting on the fact that Schwinger never mentioned Feynman or Dyson in any of his lectures, DeWitt observed that

all this formalism of functionally differentiating with respect to the spacelike hyperspace was terribly frustrating. You would go on for page after page of formalism, [but] how do you calculate with it? Well obviously, if we'd had momentum space techniques and diagrams readily taught to us at Harvard it would have helped immeasurably on computations. On the other hand, the way Schwinger presented things to us...he taught us to think in a more unconstrained way. It's perhaps paradoxical but the limitations imposed by Schwinger not mentioning Feynman or Dyson led us to adopt a broader perspective on physics. I think this is really true. It is true in my case.[159]

The elations and frustrations DeWitt relates was experienced by all the students who took Schwinger's courses at Harvard. All of them remember Schwinger's magnificent lectures. All of them spent countless hours clarifying and mastering their contents. All of them point with pride to the set of notes that emerged from their confrontation with Schwinger's courses. Interestingly, no one can remember anyone ever asking any questions during any of Schwinger's lectures.

Roger Newton, who obtained his Ph.D. with Schwinger in 1954, writes:

We all took it for granted that we were on top of the world, that is that QED was the most important thing in physics at the time and Schwinger was its king. I admired his lecturing style enormously and rarely noticed that its polished elegance made some rather questionable points more persuasive than was really justified. I still regard him as an outstanding teacher in so far as lecturing is concerned. He never gave the impression of communicating routine matters but what he presented was always new, or so it seemed.... [A] group of us (Stanley Deser, Dick Arnowitt, Chuck Zemach, Paul Martin and I forgot who else) wrote up lecture notes on his Quantum Mechanics course but he never wanted them published because he "had not yet found the perfect way to do quantum mechanics."[160]

Paul Martin put it this way:

During the late 1940's and the early 1950's Harvard was the home of a school of physics with a special outlook and a distinctive set of rituals. Somewhat before noon three times each week, the master would arrive in his blue chariot and in forceful and beautiful lectures, reveal profound truths to his Cantabridgian followers, Harvard and M.I.T. students and faculty. Cast in a language more powerful and general than any of his listeners had ever encountered, these ceremonial gatherings had some sacrificial overtones—interruptions were discouraged and since the sermons lasted past the lunch hours, fasting was often required. Following a mid-afternoon break, private audiences with the master were permitted and, in uncertain anticipation, students would gather in long lines to seek counsel. (Martin 1979, p. 70)

The queues Martin refers to were those of the students doing their thesis under Schwinger's supervision. Although a highly successful and influential lecturer, Schwinger was somewhat less effective in his one-to-one interactions with his students. Roger Newton notes that "perhaps that was because his personal shyness prevented him from overcoming the barrier between his students and him. We were, of course, in tremendous awe of him, and he did nothing to make himself more human and accessible to us." Newton indicates that he rarely saw Schwinger outside class and that "we talked about my thesis only a few times. There certainly was no real guidance in any personal sense, though his lectures provided a lot of it indirectly."[161] Most of Schwinger's students of that period tell the same story. DeWitt relates that he went to see Schwinger when he ran into some complications in his thesis—which dealt with the self-energy of a photon in the case that in addition to the electron-positron field the quantized gravitational field is also considered. Schwinger saw him briefly and told him to "cut out" the electron-positron field and to stick to pure photons and gravitation. "I probably saw [Schwinger] during my work on the thesis a total of 20 minutes."[162]

Nonetheless, Schwinger was undoubtedly the most important influence in his students' scientific life. Bryce DeWitt tells of his experience:

> I was a member of the Institute of Advanced Study in 1954 and was present when Schwinger came down and gave a marathon series of lectures, for a total of eleven hours. Seven of them on one day, and four the next day.... Following the custom I had adopted at Harvard I wrote down every single equation that Schwinger wrote on the board.... It was all extremely formal but extremely beautiful to me.... I can say that marathon lecture had an enormous impact on me and has affected my research life ever since, to this day.... The lectures were never published. I have never seen a paper of Schwinger's that included the material that he gave. They contained the effective action the way it is thought of nowadays, it contained the Legendre transform, the superdeterminant that is involved when you have simultaneously bosons and fermions.[163]

DeWitt speaks for all of Schwinger's students when he asserts that "Schwinger is a great teacher and had a great impact on many people."[164] And Schwinger counts many distinguished theoretical physicists among his students: K. M. Case, F. Rohrlich, R. J. Glauber, M. Karplus, B. DeWitt, A. Klein, B. Mottelson, E. Merzbacher, C. Zemach, R. Arnowitt, S. Deser, T. Fulton, P. Martin, J. J. Sakurai, R. Newton, J. Bernstein, K. Johnson, S. F. Edwards, I. Shapiro, C. Sommerfield, D. Kleitman, M. Baker, N. Beyers, S. L. Glashow, C. Baker, L. Brown, D. Boulware, to mention but a few of them. Over sixty students did their doctoral dissertations under his supervision during his stay at Harvard. Two of them, Mottelson and Glashow, subsequently won the Nobel Prize.

The contrast between Schwinger before and during the war and the later Schwinger merits comments. The warm and affectionate encomium of his prewar and wartime colleagues and acolytes is markedly different in tone from the criticisms of his students at Harvard. It is also interesting to note that while most of the papers Schwinger wrote before and during the war were collaborative efforts, the majority of his papers on work done after the war were written by himself. In his later work he also addressed "more general theoretical questions rather than specific problems of immediate experimental concern, which were nearer to the center of [his] earlier work" (Schwinger 1966, pp. 113–114). All this reflects his working style after the war: he becomes more and more a loner. There is a tragic aspect to Schwinger's life after 1950, for he becomes progressively more and more isolated from the physics community.

Schwinger's personality was undoubtedly a factor. David Saxon has observed that "Schwinger always wanted to do everything for himself, by himself.

And he would want to do it his own way. He insisted on doing it his own way."[165] Schwinger himself corroborates this. He confided that "when I was 14 or 15 or so, I read everything breathlessly. Then when I got my own guns working I would get interested, and say: "Gee. How would I do that," which meant then that the actual reading of the details of all the papers got to be less and less and less—to the point, I am sorry to say, I now [1982] don't read anything. I have to change that."[166] Contributing to this necessity to "do everything for himself by himself" was Schwinger's fear of being dominated. Recall what he said about Oppenheimer: "He [was] a dominating personality but I didn't want to be dominated. So to some extent I would back away and I would say I have to think about this my own way."[167] Schwinger expressed similar apprehension regarding Rabi, Breit, and Wigner.

Schwinger is also a perfectionist. Saxon, who worked closely with him at the Rad Lab during the war, notes that at that time "there was a gap between what he did and what he aspired to do."[168] The result then was that Schwinger never got to write down anything for publication. It was Saxon who wrote up the lectures Schwinger gave.

Harold Levine and Herman Feshbach tell the same story. Both of them assert, "I did the writing of our papers."[169] Levine is certain that when they were collaborating, Schwinger had solved many other diffraction problems "that didn't see the light of day." Levine and Schwinger at one point were thinking of writing a book on waveguide theory "but it was clear that Julian would not devote time to it."[170] Similarly, Marcuvitz and Schwinger at one time began collaborating on a book dealing with the general theory of electromagnetism—but the project did not come to fruition because Schwinger was not satisfied with what either he or Marcuvitz had written. His lectures on quantum mechanics during the early fifties were not published because he had not yet found the "perfect way" to do the subject.

It appeared to his colleagues at Harvard and MIT that after Schwinger got married in 1947 his interactions with other physicists were reduced somewhat: his social needs were evidently being satisfied at home and with a circle of friends outside the physics community. The awe in which he was held by that community, particularly after Shelter Island and his successes with QED, created further barriers. Rabi thought that the community treated him "like a god." All these factors combined to conspire to isolate him. And so a tragedy ensued. His need to do things his own way made him develop his own language, his own approaches and techniques. His need to prove constantly to himself that he could do it by himself[171] resulted at times in his not acknowledging the works of others—for indeed, he had done it by himself and in his own way. This in turn led to resentment by the community which pounced on him when he was vulnerable.[172]

When Schwinger was at Harvard his numerous students—many of whom were truly outstanding—learned and disseminated his language. But as he became more isolated, fewer people understood and spoke the newer languages he created—for example, sourcery (Schwinger 1970b, 1973)—contributing to his further isolation.

The tragedy is that Schwinger's creativity was affected by his isolation. His enormous talents and awesome technical powers would undoubtedly have been used differently—and perhaps more effectively—had he interacted differently with others and with his community. It was a mutual loss, for both Schwinger *and* the community were the losers. Yet for all this, the assessment of Schwinger given by three of his students and two of his closest colleagues on the occasion of his sixtieth birthday is surely right:

> His work during the forty-four years preceding his sixtieth birthday extends to almost every frontier of modern theoretical physics. He has made far reaching contributions to nuclear, particle, and atomic physics, to statistical mechanics, to classical electrodynamics, and to general relativity. Many of the mathematical techniques he developed can be found in every theorist's arsenal. . . . He is one of the prophets and pioneers in the uses of gauge theories. . . . Schwinger's influence, however, extends beyond his papers and books. His course lectures and their derivatives constitute the substance of graduate physics courses throughout the world, and in addition to directing about seventy doctoral theses, he is now the ancestor of at least four generations of physicists. . . . The influence of Julian Schwinger on the physics of his time has been profound.[173]

8. Richard Feynman and the Visualization of Space-Time Processes

He [Feynman] is a second Dirac, only this time human.

—*E. P. Wigner*[1]

8.1 Background

When I was a child I noticed that a ball in my express wagon would roll to the back when I started the wagon, and when I stopped suddenly it would roll forward. I asked my father why, and he answered as follows: "That, nobody knows! People call it inertia, and the general rule is that anything at rest tends to remain at rest, and a thing in motion tends to keep on moving in the same direction at the same speed. By the way, if you look closely you will see that when you start the wagon the ball doesn't really move backwards, but it just doesn't start up from rest as fast as does the wagon when you pull it, and it is the back of the wagon which moves toward the ball.[2]

Feynman had very vivid memories of his father. One of his earliest and most joyous recollections was that of his father taking him to the American Museum of Natural History in Manhattan and telling him about glaciers: "I can hear his voice, [as he] explained to me about the ice moving and grinding."[3] His father often played games with him and constantly challenged him by posing puzzles to him. Their interaction was primarily verbal, and solving problems by talking about them became a pattern with Feynman. In their discussions, his father stressed that facts per se were not important; what mattered was the process of finding things out. Skepticism and disrespect for authority were other traits that his father inculcated in him. But his father also got him the *Encyclopaedia Britannica*, and the young Feynman avidly read through many of its entries. Feynman recalled with sadness his shock as a young teenager when he discovered that his father's answers to his mathematical and scientific questions were no longer adequate.

Feynman's father, Melville, immigrated to the United States from Russia as a boy of 5 and grew up in Patchogue, Long Island. Upon graduating from high school he enrolled in a homeopathic medical institute, but chose not to practice. Feynman's mother, Lucille Phillips, came from a well-to-do family and attended the Ethical Culture School in New York and did not go to college thereafter. Feynman's father was involved in various business undertakings but never was very successful in any of them. Financial difficulties were responsible for the family's move from Far Rockaway to Cedarhurst and back again to Far Rockaway.

Richard ("Dick") Phillips Feynman was born on May 11, 1918. A younger brother, born when Feynman was 4 or 5, died shortly after birth. The other member of the family is a sister, Joan, some nine years younger than Richard. Feynman attended both junior and senior high school in Far Rockaway and was fortunate to have some very gifted teachers in chemistry and mathematics.[4] He recalled the "real pleasure" of doing chemistry experiments after school while in junior high school. During this same period, a lecture on heavy water by Harold Urey made a deep impression on him. Feynman had read about Urey, and the lecture was "good and technical, and it was fun. And that was my first contact with a real scientist."[5] Commenting on the experiments in chemistry and electricity that he performed as a teenager in his laboratory at home, Feynman pointed out that he "never played chaotically with scientific things":[6] he realized even then the importance of doing "things" in a controlled fashion, carefully, and watching what happened.

Feynman's mathematical talents manifested themselves early. Many things seemed obvious to him. Having learned the meaning of an exponent as a high school freshman, it was intuitively clear to him that the solution of $2^x = 32$ was $x = 5$. As a sophomore, in 1933, he worked hard on the problem of the trisection of an angle with only compass and ruler and had fantasies about the acclaim he would receive upon solving the problem. During that same year, Feynman taught himself trigonometry, advanced algebra, infinite series, analytic geometry, and the differential and integral calculus. His progress is recorded in notebooks he kept at the time.[7] What is noteworthy about their content is the thoroughness and the practical bent they display. Feynman was not content to master the formal, theoretical aspects of trigonometry: his notes contain a table of sines, cosines, and tangents from $0°$ to $90°$ in $5°$ steps that he computed himself by various ingenious schemes. Similarly, his mastering of the calculus is recorded in a special notebook—a green "Scribble-in-Book" marked on the cover "The Calculus" and given the title "The Calculus for the Practical Man" on the first page[8]—which contains extensive tables of integrals that Feynman had worked out. Already then he exhibited the need to recast what he had learned in his own language. Feynman's presentation of complex numbers, conic sections and other topics of advanced algebra are contained in a carefully typed manuscript dated November 1933.[9] Since his typewriter did not have keys for mathematical symbols such as plus, equal, multiplication, and integral signs, Feynman devised "typewriter symbols" for them enabling him to type out all his notes. The manuscript also contains a lengthy table of integrals that he had compiled for which he invented an elaborate notation. (See photographs 15 and 16 in illustration section.)

8.2 Undergraduate Days: MIT

Feynman entered MIT in the fall of 1935 a rather diffident but ambitious young man. He initially declared his major to be mathematics. During the fall semester of his freshman year he went to Franklin, the head of the mathematics department, and asked him, "What is the use of higher mathematics besides teaching more mathematics?" Franklin answered: "If you have to ask that, then you don't belong in mathematics."[10] So he switched to electrical engineering, but he soon found engineering too practical for his taste. As a freshman, helping two seniors who lived in his fraternity house with the problems in a graduate physics course they were taking from Slater, decided Feynman to major in physics. As a sophomore he enrolled in that same course—Physics 8.461: Introduction to Theoretical Physics—which that year was being taught by Stratton from the book of the same name by Slater and Frank (Slater and Frank 1933).

Feynman remembered coming to Stratton's course in his ROTC uniform— "a dead give away"[11] of his sophomore status—and filling out a pink registration card that was to be given to the instructor; seniors and graduate students had cards of different colors—green and brown, respectively. Although he was a little worried—he was only a sophomore, after all—he was proud to be there and felt "pretty good,"[12] since almost everyone else was filling out green and brown cards. The only other student in the class wearing an ROTC uniform and also filling out a pink card sat down next to him. His name was Theodore—Ted, for short—Welton. Welton and Feynman had met briefly the previous spring at the Annual Open House at the end of their freshman year.

Before coming to Stratton's first lecture, Welton had gone to the physics library and had gotten out Levi-Civita's *The Absolute Differential Calculus*, which he hoped would deepen his knowledge of differential geometry.

His interest in that field had been kindled after reading Eddington's *The Mathematical Theory of Relativity* the previous year. When Feynman noticed the books Welton had, he announced "in a somewhat raucous Far Rockaway version of standard English"[13] that he had been trying to get a hold of Levi-Civita and could he see it when Welton was finished with it. Welton for his part observed that Feynman's stack of books contained the library's copy of *Vector and Tensor Analysis* by A. P. Wills, which explained why he had been unable to locate it. Since they were the only two sophomores in that class, it apparently occurred to them simultaneously "that cooperation in the struggle against a crew of aggressive-looking seniors and graduate students might be mutually beneficial."[14] From that moment a deep friendship developed between the two.

Stratton quickly recognized Feynman as a truly superior student. Because of the pressure of other duties, at times Stratton would skimp on preparation and come to an embarrassed halt during his lecture. With only a moment's hesitation he would then turn to Feynman and ask for his help. Whereupon Feynman would walk somewhat diffidently to the blackboard and indicate how to proceed, "always correct[ly] and frequently ingeniously."[15]

Welton remembers an "amazing" quirk displayed by Feynman in Stratton's course: his "maddening refusal to concede that Lagrange might have something useful to say about physics. The rest of us were appropriately impressed with the compactness, elegance and utility of Lagrange's formulation, but Dick stubbornly insisted that real physics lay in identifying all the forces and properly resolving them into components."[16] Nonetheless, Feynman would always obtain the correct equations of motion using his physical intuition and his previously gained insights. The incident disclosed more than a quirk: it revealed Feynman's fierce independence, and his need to do and understand things his own way.

During their first conversation, on the afternoon of that memorable first class with Stratton, Feynman told Welton that he wanted to learn general relativity. Welton, "with proper superiority," announced that he already knew some general relativity, and wanted to learn quantum mechanics. Whereupon Feynman suggested to Welton that he take a look at Dirac, the "good book on the subject" he had read. Welton, upon reading Dirac, rapidly found himself over his head. Somehow they located a more appropriate text, Pauling and Wilson's *Introduction to Quantum Mechanics* and together "wandered through much of quantum mechanics"[17] during that fall semester.

During their sophomore year, while taking the courses they had enrolled in, Feynman and Welton taught themselves general relativity and together explored relativistic quantum mechanics, exchanging ideas, problems, and possible solutions in a notebook that went back and forth between them.[18] They rediscovered the Klein-Gordon equation, and Feynman remembered very distinctly that Welton asked whether one could calculate the energy levels of a hydrogen atom with this equation to see whether it agreed with experiments.[19] "Remember what I said about your equation having been tried and found wrong. I saw that in Dirac's book," Welton wrote to Feynman in their notebook during the summer of 1936. "Why don't you apply your equation to a problem like the hydrogen atom, and see what results it gives."[20] Feynman assessed the negative results they obtained a "terrible" and very important lesson. He learned from it neither to rely on the beauty of an equation nor on its "marvelous formality": the test was "to bring it down against the real thing"[21] (see photograph 17).

To get an indication of the caliber of these two 18-year-old sophomores, here is an excerpt from another of Welton's entries in their joint notebook during the summer of 1936:

> Here's something. The problem of an electron in the gravitational field of a heavy particle. Of course the electron would contribute to the field, but I think that could be neglected. Take your equation[*] ($K = 0$) and substitute for $g^{\mu\nu}$ the values in the field of the particle
>
> $$[g^{\mu\nu}P_\mu P_\nu + m^2 c^2]\psi = 0$$
>
> I wonder if the energy would be quantized? The more I think about the problem the more interesting it sounds. I am going to

try it. "Let's see in a grav. field of a particle

$$ds^2 = \left(1 - \frac{2m}{r}\right)dr^2 + r^2(d\theta^2 + \sin^2\theta\, d\phi^2) - \left(1 - \frac{2m}{r}\right)dt^2$$

I think that's right.
So,

$$\left[\left(1 - \frac{2m}{r}\right)P_r^2 + r^2 P_\theta^2 + r^2\sin^2\theta P_\phi^2 \right.$$
$$\left. - \left(1 - \frac{2m}{r}\right)P_t^2 + m^2\right]\psi = 0$$

$$c = \text{unity}$$

I'll work out the Christofell symbol and find explicit expressions for $P_r^2, P_\theta^2, P_\phi^2$ and P_t^2.

I'll let you know the result later. I'll probably get an equation that I can't solve anyway. That's the trouble with quantum mechanics. It's easy enough to set up equations for various problems, but it takes a mind twice as good as the differential analyser to solve them."[22]

[*]Welton, on a previous page, had commented on "Feynman's equation"
$$[(P_\mu - K_\mu)g^{\mu\nu}(P_\nu - K_\nu) + m^2 c^2]\psi = 0$$
$$K_\mu = \left(\tfrac{e}{c}A_1, \tfrac{e}{c}A_2, \tfrac{e}{c}A_3, -e\phi\right)$$

The entry concludes with Welton's comment: "I can't arouse much interest any more in classical quantum mechanics (Schrödinger's equation, etc.). Relativistic wave mechanics is the only stuff."[23]

The notebook also exhibits Feynman's mathematical virtuosity: normed matrix vector spaces are casually introduced, laborious tensor calculations are elegantly dispatched, and his affinity for useful notation is repeatedly displayed.

In 1937 the second half of "Introduction to Theoretical Physics," Physics 8.462, was taught by Philip Morse, who included in his course lectures on wave mechanics. Impressed by the problem sets Feynman and Welton were handing in and by the questions they were asking during and after the lectures, Morse invited Feynman, Welton, and a junior in the class, Al Clogston, to come to his office for one afternoon a week the next year to be properly exposed to quantum mechanics. They all accepted with alacrity.

During the fall semester of their junior year Feynman and Welton carefully studied Dirac's *Quantum Mechanics* with Morse. Clogston remembers "most vividly...the chastening encounter with Dick Feynman's quick mind"[24] and how hard he had to work to keep up. After they had gone through Dirac, Morse informed Feynman and Welton[25] that they were ready for a "little real research" and suggested some calculations of atomic properties, using a formulation of the variational method which he had published with L. Young (Morse and Young 1933).[26] This they did and in the process they learned a great deal about hydrogenic (Feynman called them "hygienic") wave functions and became nimble-fingered experts on Marchant calculating machines, the "chug-chug-ding-chug-chug-chug-ding..." hand-operated calculators of those prewar days.[27]

In his senior year Feynman took a metallurgy course to learn about the applications of physics and a laboratory course given by George Harrison. Welton recalls Harrison's lectures as "a pleasure to attend, with a wealth of ingenious applications of physical principles."[28] The laboratory required a project, and Feynman exhibited his great gadgeteering ingenuity there with a mechanism to obtain the ratio of the speeds of two rotating shafts.

Feynman and Welton in their last year at MIT also took a seminar given by Morse and Frank, in which they studied the review articles on nuclear physics that Bethe and Livingston had recently published (Bethe 1986).

By the time Feynman finished his undergraduate studies at MIT in 1939 he had mastered many of the fields of theoretical physics. In the tutorial Philip Morse had given him, he had learned quantum mechanics well enough to write a senior dissertation under Slater on "Forces and Molecules," in which his impressive formal and calculational talents are manifest. This research was published in the *Physical Review* and contains a result now known as the Hellmann-Feynman theorem (Feynman 1939).[29] He worked with Vallarta (Vallarta and Feynman 1939) on cosmic ray problems and with Harrison and Herring on aspects of solid state physics. He had also spent a great deal of time in the library reading a vast number of advanced texts: "I was very avid for reading and studying and learning"[30] is the way Feynman put it.

At MIT, Feynman's outstanding abilities were clearly recognized. He had done so well in all his courses that the department of physics "had taken the unusual step of proposing that he be granted his bachelor's degree at the end of three years instead of four" (Morse 1977, p. 126). While at MIT, Feynman changed from a somewhat shy, insecure, and timid teenager into a confident, brash young man. While there he also shed his fear of women.

In his autobiography, Morse recalls Feynman's father coming to MIT from New York in the fall of 1938 and telling him, "My son Richard is finishing his schooling here next spring. Now he tells me he wants to go on to more studying, to get still another degree. I guess I can afford to pay his way for another three or four years. But what I want to know, is it worth it for him.... Is he good enough...? Will it help him?" (Morse 1977, pp. 126–127). Morse reassured him that his son was the brightest undergraduate he had ever met, and yes "he really needed the extra schooling to be able to enter his chosen profession" (Morse 1977, p. 127). Feynman would have liked to stay on at MIT for his graduate studies, but

Slater insisted that he leave. Although he had not applied to Harvard, Feynman was offered a scholarship to study there because he had won the national Putnam contest in mathematics.[31] But he declined the invitation as he had already agreed to go to Princeton University and had accepted the physics department's offer to be Wigner's research assistant. As it turned out he was assigned to be Wheeler's assistant—a propitious event in retrospect. John Archibald Wheeler, who had just come to Princeton as a 26-year-old assistant professor in the fall of 1938, proved to be an ideal mentor for the young Feynman. Full of bold and original ideas, "a man who had the courage to look at any crazy problem, a fearless and intrepid explorer" (Wilson 1979, p. 213), Wheeler gave Feynman viewpoints and insights into physics that would prove decisive later on. Feynman recalls that when they first met, Wheeler indicated to him that they would have a limited and fixed amount of time to discuss things during their scheduled meetings. When Feynman appeared at Wheeler's office for their first conference, Wheeler took out his pocket watch and put it on the table so that he could see how much time had elapsed and how much time was left. After this initial meeting Feynman bought himself a dollar pocket watch. At their next scheduled conference, when Wheeler put his watch down on the table, Feynman also took out his watch and put it down on the table. Wheeler thereupon burst out laughing and so did Feynman. They both laughed so hard that they couldn't stop and couldn't get to work for a while.[32] This revealing incident marked the beginning of a deep and lasting friendship and of a seminal association between two minds that complemented each other. The boldness and brillance of Wheeler's apparently impossible ideas fell on fertile soil.

The first problem Wheeler assigned Feynman was to explain the shape of the Compton line in the scattering of X rays by atoms in order to learn what it revealed about the momentum distribution of the electrons in the atom. It was in this context that Feynman first learned of Wheeler's work on the scattering matrix (Wheeler 1937). Wheeler gave him lectures on scattering theory, and presented his views that all quantum-mechanical descriptions of physical phenomena could be construed as scattering processes. In particular he indicated to him how one could interpret the Schrödinger equation as describing a succession of scattering events.[33]

While working on these problems, Feynman continued to spend time on an idea he had fallen "deeply in love with" as an undergraduate at MIT, an idea on how to solve the difficulties plaguing quantum electrodynamics (QED). While at MIT Feynman had become acquainted with the leading problems in quantum electrodynamics by reading the books by Dirac and Heitler (Dirac 1935c; Heitler 1936). He knew that in classical theory the self-energy of a point charge was infinite, and he had studied various schemes that had been put forward to solve or bypass this difficulty. In his Nobel Prize lecture, he recalled that his understanding of the problem in the quantum theory was much hazier. It seems he believed that there the divergence arose from the fact that one was dealing with a *field* system with an infinite number of degrees of freedom, each degree having a finite zero-point energy (Feynman 1966a, p. 700).

To overcome both these difficulties, he put forth the following "quite evident" suggestions: first, that a charged particle does not act on itself, it acts only on other charged particles; second, that there was no electromagnetic field in order to eliminate the infinite number of degrees of freedom associated with the electromagnetic field. Since the field was completely determined by the motion of the charged particles, it could be expressed in terms of the particle variables. The field therefore did not have any independent degrees of freedom and the infinities he had associated with them would then be removed. "The general plan was first to solve the classical problem . . . and to hope that . . . [in] a quantum theory . . . everything would just be fine" (Feynman 1966a, p. 700).

8.3 Graduate Days: Princeton

By the time he came to Princeton as a graduate student, Feynman had noted "a glaringly obvious fault" with his theory (Feynman 1966a, p. 700). Self-interaction was necessary to account for the radiation resistance. He had learned that Lorentz had used the action of a charged particle on itself to explain the force of radiation resistance. More work is required to accelerate a charged particle than a neutral one because an accelerated charge radiates. A charged particle did seem to act on itself; and moreover this force was necessary to preserve the conservation of energy.

Feynman hoped that nonetheless he would be able to patch up his theory by considering the reaction back on the radiating particle from the induced motion of the other charges affected by the radiation. He presented his ideas to Wheeler, who immediately pointed out its flaws: "The answer you get for the problem with . . . two charges . . . unfortunately will depend upon the charge and the mass of the second charge and will vary inversely as the square of the distance, R, between the charges, while the force of radiation resistance depends on none of these things. . . . Finally when you accelerate the first charge, the second acts later, and then the reaction back here at the source would still be later . . . the action occurs at the wrong time (Feynman 1966a, p. 700). Wheeler went on to give a lecture on possible modifications of Feynman's approach. Suppose, Wheeler suggested, "that the return action by the charges in the absorber reaches the source by advanced waves as well as by the ordinary retarded waves of reflected light, so that the law of interaction acts backward in time, as well as forward in time" (Feynman 1966a, p. 700). Wheeler used advanced waves to get the reaction back at the right time, and then noted that if the advanced waves came back from the absorber phase shifted (but, by assumption unchanged in wavelength), then by a suitable adjustment at the index of refraction of the absorber, the action at the source of these advanced waves was completely independent of the properties of the charges of the absorber and moreover of the right character to represent radiation reaction. Wheeler asked Feynman to calculate how much advanced and how much retarded waves were needed to get reaction effects numerically right, and to "figure out what happens to the advanced effects that you would expect if you put a test charge . . . close to the source . . . [i.e.] why would that test charge not be affected by the advanced waves from the source?" (Feynman 1966a, p. 701).

At the time Feynman attributed Wheeler's insights to his natural brilliance. He was not aware that since coming to Princeton, Wheeler had been studying the action-at-a-distance formulations of electromagnetic theory of Schwartzschild (1903), Tetrode (1922), Frenkel (1925), and Fokker (1929a,b, 1932). While working with Breit in 1933 Wheeler became convinced

> that the great white hope of theoretical physics was the electron-positron theory and that people had been too early and too glib and too facile in ruling out the idea of the electron in the nucleus; that pair theory offers mechanisms for binding electrons in very small regions of space that never got a thorough discussion in these offhand comments of why there couldn't be electrons in the nucleus . . . I didn't leave the idea that electrons were the basic building materials until 1947. And the fanaticism with which I pursued that view is shown I guess not least by the fact that I felt that if electrons were the building blocks of atomic nuclei, the forces that were involved would not be the static electric forces but the radiation components of the forces. Therefore, it was of great importance to understand the influences set up by a rapidly accelerated electron.[34]

The interactions of highly accelerated relativistic electrons was thus a subject of great interest to Wheeler, and he had discussed these matters with Léon Rosenfeld during the latter's visit to Princeton in the spring of 1939. Wheeler had tried to meet Rosenfeld's objections "that electromagnetic radiation seemed to have no place in this picture" (Wheeler 1979, p. 258) and he recalls that "sometime later, reflecting quietly at home one Sunday afternoon on the back of an envelope, I suddenly recognized that if there were enough absorber particles around to absorb completely the radiation from an accelerated source, it would make no difference how numerous were these particles, nor what their properties. However, I failed by a factor of 2 to get the right result for the familiar force of radiative reaction. The next day I told Richard Feynman, then a graduate student, about my line of thought and about my results. Thanks to our usual lively discussion the factor two was cleared up along with many other ramifications" (Wheeler 1979, p. 258).

Feynman indeed discovered that one could account for radiation resistance if one assumed all actions are via half advanced and half retarded solutions of Maxwell's equation and that all sources are surrounded by material absorbing all the light that is emitted. Radiation resistance could then be explained as "a direct action of the charges of the absorber acting back by advanced waves on the source" (Feynman 1966a, p. 701).

Feynman remarked that when Wheeler first suggested using advanced waves so that the law of interaction also acts forward in time—a concept at first sight at variance with elementary notions of causality—he was enough of a physicist not to say, "No, how could that be?" Rather, he felt that it was like in the old days with Bohr. He had learned from the history of physics, in particular from Einstein's and Bohr's work, that "an idea which looks completely paradoxical at first, if analyzed to completion in all its details and in experimental situations, may in fact not be paradoxical" (Feynman 1966a, p. 700).

Wheeler and Feynman worked out the details of their action-at-a-distance theory in the fall of 1940 (Wheeler and Feynman 1945). Feynman gave a colloquium on their work that was attended by Einstein, Pauli, Wigner, and Henri Norris Russell, among others. Feynman recalls Wigner trying to reassure him before the lecture, and to convince him not to worry. "If Professor Russell falls asleep during your lecture," Wigner told him, "it does not mean it's no good, it's just because Professor Russell always falls asleep; but he is listening. And if Professor Pauli is nodding 'yes' during the entire lecture don't be too impressed, because . . . [he] has palsy and nods "yes" all the time."[35] Before his lecture, Feynman had filled all the blackboards in the room with equations. He remembers getting up to give the lecture and opening the envelope that contained his notes with a shaking hand: "I can see the shaking hand. Because it was quite a thing. And I started to talk about the subject. And then a thing happened that has happened ever since, and is just great: as soon as my mind got on the physics and trying to explain it, and organize the ideas, how to present it, there was no more worrying about the audience as personalities! It was all in terms of physics. I was calm, everything was good, I developed the ideas, I explained everything to the best of my abilities." [36]

Immediately after Feynman had finished, Pauli got up and criticized the theory. Neither Feynman, nor Wheeler,[37] nor Wigner[38] could remember Pauli's objections to the theory, but Feynman did remember Einstein's comments. Einstein noted that although the idea of action-at-a-distance involved in the Wheeler-Feynman theory was not consistent with the field views expressed by general relativity, he cautioned that general relativity was not as well established as electrodynamics. He, Einstein, would therefore not use that argument against the theory, because one could perhaps also develop a different way of doing gravitational interactions.[39]

Wheeler had been scheduled to give a lecture on how to quantize their action-at-a-distance theory at the next meeting of the colloquium. After Feynman's lecture, while walking back from Fine Hall to Palmer Labs with Feynman, Pauli asked him what Wheeler was going to say. Feynman replied that he didn't know. "Oh," said Pauli, "the professor doesn't tell his assistant how he has it worked out?

Maybe the professor hasn't got it worked out!" As it turned out, Wheeler had in fact overestimated his results, and he canceled the lecture. Feynman was impressed by Pauli's astuteness.[40]

In the spring of 1941 Feynman wrote up a 21-page draft of a paper entitled "The Interaction Theory of Radiation," which concisely summarized what had been worked out.[41] The assumptions of the theory were clearly spelled out.

II. PRINCIPLES OF THE INTERACTION THEORY

We make the following assumptions:

(1) The acceleration of a point charge is due only to the sum of its interactions with other charged particles (and to 'mechanical forces').* A charge does not act on itself.

(2) The force of interaction which one charge exerts on a second is calculated by means of the Lorentz force formula,** in which the fields are the fields generated by the first charge according to Maxwell's equations.

(3) The fundamental (microscopic) phenomena in nature are symmetrical with respect to interchange of past and future.[42]

(4) The limit of the velocity of each charge for increasingly remote (past or future) times is less than the velocity of light.

According to the second assumption alone, the force exerted by one charge on a second might be obtained from the field derived from the retarded potentials of Lienard and Wiechert. Thus, the second charge would be affected by an amount determined by the *previous* motion of the first charge. This is not the only possibility however; one could, for example, use the advanced potentials. In this case the second charge would be affected by an amount depending on the *later* motion of the first charge. The requirement that the effects be unchanged if one interchanges past and future removes the ambiguity and demands that one utilize one half the retarded plus one half the advanced potentials to calculate the force on the second point charge due to the first. This is exactly the law of interaction that one derives from the principle of least action of Fokker, and that principle may well have formed the starting point of this theory.

We shall now discuss the application of this law to some simple idealized situations in order to get an idea of its physical meaning.

In the first place, we notice that a single accelerating charge in otherwise charge free space will radiate no energy. There can be no radiative damping since there are no electrodynamic fields acting on the charge, no other charges being present to generate such fields.

*The present theory is one to describe those phenomena which are usually considered to be due to electromagnetic effects. Forces on charged particles such as nuclear forces on protons, or 'quantum forces' on electrons will be classified as 'mechanical' forces and will not be discussed further in this paper.

**Force $= e[E + \frac{v}{c} \times H]$

Next, Feynman presented the explanation he and Wheeler had given for the mechanism of radiation damping in their absorber model, and then addressed the problem of runaway solutions that Dirac had encountered. Feynman gave his manuscript to Wheeler, who reworked it and expanded it. Wheeler returned a new, unfinished manuscript to Feynman in the spring of 1942.[43] By then, Wheeler was deeply involved in the building of an atomic pile at the University of Chicago and could no longer devote any time to this project. The title of the new manuscript was "Action at a Distance in Classical Theory: Reaction of the Absorber as the Mechanism of Radiation Damping." Most of it is to be found in the paper Wheeler and Feynman submitted to the Festschrift celebrating Niels Bohr's sixtieth birthday (Wheeler and Feynman 1945). In this newer version, Wheeler gave an elegant explanation of how radiation damping arose. When there are n particles interacting, the field acting on particle 1 is

$$F_{\mu\nu} = \sum_{n \neq 1} F_{\mu\nu}^{(n)}, \tag{8.3.1}$$

where

$$F_{\mu\nu}^{(n)} = \frac{F_{\mu\nu\ ret}^{(n)} + F_{\mu\nu\ adv}^{(n)}}{2}. \tag{8.3.2}$$

$F_{\mu\nu\ ret}^{(n)}$ and $F_{\mu\nu\ adv}^{(n)}$ are the retarded and advanced solutions of Maxwell's equations generated by particle n.

The field on particle 1 can also be rewritten as

$$F = \sum_{n \neq 1} F_{ret}^{(n)} + \frac{1}{2} \sum_{n \neq 1} \left(F_{adv}^{(n)} - F_{ret}^{(n)} \right)$$

$$= \sum_{n \neq 1} F_{ret}^{(n)} + \frac{1}{2} \sum_{all\ n} \left(F_{adv}^{(n)} - F_{ret}^{(n)} \right) - \frac{1}{2} \left(F_{adv}^{(1)} - F_{ret}^{(1)} \right). \tag{8.3.3}$$

The term $\frac{1}{2}(F^{(1)}_{\mu\nu\,adv} - F^{(1)}_{\mu\nu\,ret})$ had been shown by Dirac (1938b) to give a force on particle 1 equal to

$$\frac{2}{3}e_{(1)}\left(\frac{d^2}{ds^2}v^{\mu}_{(1)}\right) + \left(\frac{dv_{(1)}}{ds}\right)^2 v^{\mu}_{(1)}, \tag{8.3.4}$$

which is just the Lorentz damping term. The term $\frac{1}{2}\sum_n\left(F^{(n)}_{adv} - F^{(n)}_{ret}\right)$ vanishes if one has absorbing walls: "If a source radiates for a time, at a very long time afterwards the total retarded field vanishes, for all the light is absorbed. But also the total advanced field vanishes at this time (for charges are no longer accelerating and the advanced field exists only at times previous to their motion). Hence the difference vanishes everywhere at this time and, since it is a solution of Maxwell's homogeneous equations, at all times" (Feynman 1948a, p. 941, n. 6). A particle thus effectively only interacts with the retarded fields of the other charged particles and experiences a Lorentz damping force due to its own acceleration. Although Wheeler and Feynman had started with a formulation symmetric with respect to past and future, they ended up with a solution that stressed the retarded character of the interaction—and which was the same as the one obtained by Dirac in 1938 using only retarded solutions of Maxwell's equation. Wheeler and Feynman attributed this to an asymmetry in the initial conditions with respect to time. The particles which constitute the absorber were assumed to be in random motion (or at rest) so that the sum of their retarded potentials has no effect on the acceleration of the source. The prevalence of retarded over advanced potentials was attributed to statistical considerations: the particles in the absorber tend to go from ordered to disordered states of higher entropy rather than vice versa. If the direction of time is reversed, one can inquire as to the initial conditions necessary for advanced potentials to play the dominant role that retarded potentials play in the usual picture. For this to be the case it would appear as if the chaotic motion of the absorber became ordered so that all the absorber particles were able to radiate in phase at precisely the right moment for the radiation to converge on the particle when it is accelerated. This latter set of initial conditions is much less probable than the first.[44]

The following statement of their views was given in Feynman's 21-page synopsis of their work:

> It might be worth while to make a few remarks at this point about the irreversibility of radiative phenomena. We must distinguish between two types of irreversibility. A sequence of natural phenomena will be said to be microscopically irreversible if the sequence of phenomena reversed in temporal order in every detail could not possibly occur in nature. If the original sequence and the reversed in time one have a

vastly different order of probability of occurrence in the macroscopic sense, the phenomena are said to be macroscopically irreversible.

The Lorentz theory predicts the existence of microscopically irreversible phenomena in systems which are not closed (for example, energy is always *lost* by the system to empty space as radiation). In our theory phenomena are microscopically reversible in any system. It seems at first sight, paradoxical that the two theories can ever lead to the same results, as they do in closed systems. The reason is that the phenomena predicted for closed systems are actually reversible even within the framework of the Lorentz theory which uses only retarded waves.[1] The apparent irreversibility in a closed system, then, either from our point of view or the point of view of Lorentz is a purely macroscopic irreversibility. The present authors believe that all physical phenomena are microscopically reversible, and that, therefore, all apparently irreversible phenomena are solely macroscopically irreversible.

[1]That this and the following statement are true in the Lorentz theory was emphasized by Einstein in a discussion with Ritz [Einstein and Ritz, *Phys. Zeits.* 10, p. 323 (1909)]. Our viewpoint on the matter discussed is essentially that of Einstein. (We should like to thank Prof. W. Pauli for calling our attention to this discussion.)

Feynman appended a handwritten note to the bottom of the page which pointed to the last paragraph and stated:

> ~~Prof~~ [*sic:* Feynman's erasure] Wheeler
> This is a rather sweeping statement. Perhaps you
> don't agree with it.

During the fall and winter of 1940 the theory had been formulated in many different versions. Its most elegant presentation was based on the observation that the equation of motion for the charged particles

$$m^{(i)}\frac{dv_\mu^{(i)}}{ds^{(i)}} = m^{(i)}\dot{v}_\mu^{(i)} = e^{(i)}v_\nu^{(i)}\sum_{j\neq 1}\frac{1}{2}\left(F_{\mu\text{ ret}}^{(j)\nu} + F_{\mu\text{ adv}}^{(j)\nu}\right) \tag{8.3.5}$$

could be derived from Fokker's variational principle: $\delta I = 0$, with

$$I = \sum_i \int m^{(i)}ds^{(i)} + \sum_{i\neq j}\int\int e^{(i)}e^{(j)}\delta\left(R_{ij}^2\right)dz_\mu^{(i)}dz^{(j)\mu}, \tag{8.3.6}$$

where

$$R_{ij}^2 = \left(z_\mu^{(i)} - z_\mu^{(j)}\right)\left(z^{(i)\mu} - z^{(j)\mu}\right)$$

and $ds^{(i)} = (dz_\mu^{(i)} dz^{(i)\mu})^{1/2}$ is the proper time along the path of particle (i). As Feynman later stated:

> We have in [Fokker's action principle] a thing that describes the character of the path throughout all of space and time. The behavior of nature is determined by saying her whole space-time path has a certain character. For an action like [Fokker's] the equations obtained by variation [of $z_\mu^{(j)}(s_j)$] are no longer at all easy to get back into Hamiltonian form. If you wish to use as variables only the coordinates of particles, then you can talk about the property of the paths—but the path of one particle at a given time is affected by the path of another at a different time. If you try to describe, therefore, things differentially, telling what the present conditions of the particles are, and how these present conditions will affect the future—you see it is impossible with particles alone, because something the particles did in the past is going to affect the future. (Feynman 1966a, p. 702)

It was during this same period, the fall of 1940, while working out the space-time picture of action-at-a-distance that Wheeler called Feynman up one Saturday evening and told him:

> "Feynman, I know why all different electrons have the same charge and the same mass."
> "Why?" asked Feynman.
> "Because they are all the same electron."

And Wheeler explained over the phone:

> "Suppose that the world lines which we were ordinarily considering before in time and space, instead of only going up in time, were a tremendous knot and then when we cut through all the knot, by the plane corresponding to a fixed time, we would see many, many world lines and that would represent many electrons—except for one thing. If in one section this is an ordinary electron world line in the section in which it reversed itself and is coming back from the future we have the wrong sign to the proper time—to the proper four velocity—and that's

equivalent to changing the sign of the charge and therefore that part of the path would act like a positron" (see Fig. 8.3.1).

"But," said Feynman immediately, "there aren't as many positrons as electrons. Where are all the positrons?"

"Well," answered Wheeler, "maybe they are hidden in the protons or something." (Feynman 1966a, p. 702)

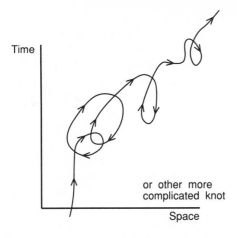

Time

or other more
complicated knot

Space

Figure 8.3.1

Feynman did not "take the idea that all the electrons were the same one ... as seriously as the observation that positrons could simply be represented as electrons going from the future to the past in a back section of their world lines" (Feynman 1966a, p. 702). The fact that Wheeler had a theory that could represent both electrons and positrons in classical physics in a very simple way "by letting the world lines go backwards and forwards in time" (Feynman 1966a, p. 702) made an indelible impression on Feynman.

It should be noted that the zigzag world line description of pair annihilation, unbeknownst to Wheeler, had also been put forward by Stueckelberg (1941b) at this same time.

An indication of Feynman's other interests and activities while at Princeton can be gleaned from a letter Wheeler wrote to Feynman in 1949. At the time Wheeler was working with John Toll, then a graduate student, on the general relation between dispersion and absorption in quantum electrodynamic processes. In the letter, Wheeler requested the title and author of an article Feynman had reported on in 1941:

> I'm writing you now because I remember you gave a report at Journal Club one Monday evening in 1941 on the relation between phase change and amplitude gain for a linear amplifier. The little black box had two input leads and two output leads. The magician was able to deduce all he needed from the requirement that energy shouldn't come out of the box on the right hand side before it had been put in on the left. You were reporting on a paper about which I remember neither the author nor the title nor the journal. . . .

I am having an interesting time trying to develop the theory of world lines, about which we once talked a little; a description of nature which makes no use of the concepts of space and time. But that is something else![45]

Soon thereafter, Feynman answered Wheeler's query:

The article which I reported at the Journal Club in Princeton in 1941 was "Relations Between Attenuation and Phase in Feed-Back Amplifier Design" by A. W. Bode in the Bell System Technical Journal, Vol. 19, page 421, 1940. Unfortunately this paper gives only the relation between attenuation and phase and does not describe how they may be obtained. It was my guess at the time I described the paper that they were simply a consequence of the assumption that signals could not come out of the amplifier before they were put in. In mathematical laws this corresponds to the assumption that all singularities in the impedance relationship must occur for frequencies with positive imaginary parts. The rest I had hoped would result from some maneuvering Cauchy's theorem. . . .[46]

The theory of world lines which we once spoke about has been subjected to investigation by somebody. I remember reading a long mathematical article on the subject in what I think was an english journal but I can not remember the author's name nor the journal unfortunately.[*] He seems to have gone somewhat further than I remember that we did but along almost exactly the same lines. He did not go far enough, however, to have something definite and interesting come of it. What do you think the quantum mechanical analogue of that picture is?[47]

[*]A. G. Walker, *Proc. Roy. Soc. Edinburgh,* **67**, 319 (1949).

8.4 Ph.D. Dissertation

After he solved the problem of expressing classical electrodynamics in a way that dispensed with the electromagnetic field, Feynman proceeded to formulate its quantum theory. The classical theory, when formulated in terms of Fokker's action principle, involved two different times which in turn meant that there was no Hamiltonian for the system. Thus the problem to be addressed was how to formulate the quantum theory of a system describable by an action principle of the form

$$S = \int L \, dt, \qquad (8.4.1)$$

where L is the Lagrangian of the system, but which did not admit of a Hamiltonian. Feynman's doctoral dissertation, "The Principle of Least Action in Quantum

Mechanics," which he presented to the Department of Physics in May 1942 just before joining the Manhattan Project, solved the problem of "finding a quantum description applicable to [nonrelativistic] systems which in their classical analogues are expressible by a principle of least action, and not necessarily by Hamiltonian equations of motion" (Feynman 1942, p. 6).

Wheeler, who also had been trying to quantize their absorber theory, at one stage told Feynman not to bother working on the quantization since he had already solved it. But since Wheeler hadn't shown Feynman his results, Feynman "still had to find out."[48] Feynman first considered simple models, such as a harmonic oscillator interacting with another harmonic oscillator with a time delay ("Put the essential in, but keep everything else simple"), and worked out the quantum-mechanical description of these toy models. But the results obtained from the analysis of such systems of interacting harmonic oscillators did not give "much of a clue as to how to generalize [them]...to other systems" (Feynman 1966a, p. 703).

Then in the spring of 1941 the essential clue in arriving at a general solution was provided by Herbert Jehle, who was then visiting Princeton. At a Nassau Tavern party Jehle, in answer to Feynman's query whether he knew of a way to go from a classical action to quantum mechanics without invoking a Hamiltonian, indicated that Dirac in 1932 had written a paper on how to go directly from a Lagrangian to the quantum theory (Dirac 1933a). In that discussion with Feynman, Jehle also had made the point "that the Lagrangian formulation permits a more simple, straightforward relativistically covariant approach than the Hamiltonian method."[49] The next day they studied Dirac's paper together, which pointed out that the transformation matrix

$$(q_t \mid q_T) \text{ "corresponds" to } \exp\left[i \int_T^t L \, dt / \hbar\right].^{50} \qquad (8.4.2)$$

In the infinitesimal case this becomes

$$(q'_{t+\epsilon} \mid q_t) \text{ corresponds to } e^{\frac{i}{\hbar}\epsilon L\left(\frac{q'-q}{\epsilon}, q\right)} = e^{\frac{i}{\hbar}S(q'_{t+\epsilon}, q_t)}, \qquad (8.4.3)$$

that is, $e^{\frac{i}{\hbar}S}$ is the "analogue" to the quantity which in quantum mechanics carries the wave function of a particle—classically described by the Lagrangian $L(q', q)$—from time t to $t+\epsilon$. To understand what Dirac had meant by "analogous,"[51] Feynman derived the Schrödinger equation for a particle whose Lagrangian was

$$L = \frac{1}{2}mv^2 - V(x) \qquad (8.4.4)$$

and concluded that by "analogous" Dirac had meant "proportional," that is, equal except for a proportionality constant. In his interview with Charles Weiner,

Feynman recalls that at the Princeton bicentennial celebration in the fall of 1946 he asked Dirac what he had meant by "analogous." "Did you know that they were proportional?" asked Feynman. "Are they?" Dirac inquired. "Yes," said Feynman. "Oh, that's interesting," was Dirac's final comment.[52]

Feynman assumed that by virtue of eq. (8.4.3) the wave function at time $t+\epsilon$ of a particle described by the (classical) Lagrangian eq. (8.4.4) is related to the wave function at time t by

$$\psi(q',t+\epsilon) = \int \frac{1}{A} e^{\frac{i\epsilon}{\hbar} L(\frac{q'-q}{\epsilon},q)} \psi(q,t)\, dq, \tag{8.4.5}$$

where A is a proportionality constant. In Jehle's presence he proved that eq. (8.4.5) yields Schrödinger's equation

$$\left(-\frac{h^2}{2m}\frac{\partial^2}{\partial q^2} + V(q)\right)\psi(q,t) = ih\frac{\partial}{\partial t}\psi(q,t), \tag{8.4.6}$$

provided

$$A = \sqrt{\frac{2\pi\hbar it}{m}}. \tag{8.4.7}$$

After Feynman had derived this result, the astonished Jehle commented: "This is an important discovery. You Americans are always trying to find out how something can be used. That's a good way to discover things." Indeed, Feynman thought he was finding out what Dirac had meant by "analogous," but in fact he had obtained the connection between the Lagrangian and quantum mechanics, albeit "still with wave functions and infinitesimal times" (Feynman 1966a, p. 703.) A few days later, lying in bed, Feynman "imagined" what would happen if he wanted to calculate the wave function at a finite time later.

In his paper Dirac had shown that $(q_t \mid q_T)$ could be written as

$$(q_t \mid q_T)$$
$$= \int \cdots \int (q_t \mid q_n)dq_N \cdots dq_m(q_m \mid q_{m-1})dq_{m-1}\cdots(q_2 \mid q_1)dq_1(q_1 \mid q_T), \tag{8.4.8}$$

where q_m denotes q at the intermediate time t_m. From the expression (8.4.8) one infers that

$$(q_t \mid q_T) \text{ corresponds to } \int e^{\frac{i}{\hbar}\int_T^t L dt} dq_N \cdots dq_1. \tag{8.4.9}$$

The finite time propagator could thus be written as

$$(q_t \mid q_T) = \lim_{\substack{N \to \infty \\ N\epsilon = t - T}} \int \cdots \int e^{\frac{i}{\hbar} \int_T^t L dt} \frac{dq_1}{A} \frac{dq_2}{A} \frac{dq_3}{A} \cdots \frac{dq_N}{A}, \qquad (8.4.10)$$

a result Feynman obtained by dividing the time interval $T \to t$ into a large number of small intervals of duration ϵ, $T \to t_1, t_1 \to t_2 \cdots t_N \to t$, $N\epsilon = t - T$, by introducing a sequence of intermediate times $t_m = T + m\epsilon$. Dirac had also indicated how in the limit as $h \to 0$ the only important contribution in the domain of integration of the q_k came from those q_k's for which a comparatively large variation produced only a very small variation in $\int L \, dt$—which corresponds to the set of points $q_1 \cdots q_N$ for which $\int L \, dt$ is stationary with respect to small variations in q_k, that is, the classical path.

Eq. (8.4.8) was, however, not interpreted by Dirac as an integral over paths: that interpretation is implicit in Feynman's thesis and becomes explicit when Feynman wrote up his formulation of quantum mechanics for the *Reviews of Modern Physics* in 1947.

Although no diagrams appear in that paper, the quantum-mechanical amplitude $(q't' \mid qt)$ is conceived as receiving a contribution from each path Γ that connects qt to $q't'$ (see fig. 8.4.1).

Each path Γ contributes a phase $e^{\frac{i}{\hbar}S(\Gamma)}$ to the total amplitude, where $S(\Gamma)$ is the action computed for the path Γ:

$$S(\Gamma) = \int_{\Gamma} \int_t^{t'} L(q, \dot{q}, t) \, dq. \qquad (8.4.11)$$

The total amplitude is given by

$$(q't' \mid qt) = \frac{1}{A'} \sum_{\substack{\text{over all paths } \Gamma \\ \text{from } qt \text{ to } q't'}} e^{\frac{i}{\hbar}S(\Gamma)} \qquad (8.4.12)$$

A' is a normalization constant, and expresses the idea that "the microscopic point particle makes its way from qt to $q't'$, not by a unique history, but by pursuing every conceivable history *with democratically equal amplitudes*" (Wheeler 1989; emphasis added).

These visual aspects of the integral-over-path approach to quantum mechanics were not stressed in Feynman's thesis, although it is clear that the conceptualization of the transformation function $(q't' \mid qt)$ in terms of space-time trajectories is central to the enterprise. In later years Feynman would stress the equivalence of his approach to that of Heisenberg, Dirac, and Schrödinger and would

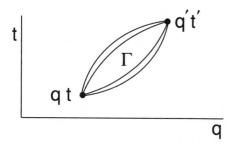

Figure 8.4.1

suggest that this multiplicity of possible description of quantum phenomena attests to our having captured key elements in our description of atomic phenomena—this multiplicity being an expression and "representation of the simplicity of nature" (Feynman 1966a, p. 702). In 1941, however, Feynman conceived and saw his approach as a generalization of the quantum mechanics of Dirac, Heisenberg, and Schrödinger, valid in circumstances when the usual approach presented unsurmountable obstacles or proved intractable or impotent. In point of fact, Feynman's formulation could handle situations where the ordinary concept of wave function was inadequate. Thus for a system of particles that interacted with a time delay, the concept of a wave function is not a convenient way of expressing information about the system. Feynman formulated his theory in terms of transition amplitudes (to go from one condition to another) and in this he was clearly influenced by the S-matrix viewpoint Wheeler had expounded to him.

The final section of his thesis summarized both the accomplishments and the difficulties of the formalism:

12. Conclusion

We have presented, in the foregoing pages, a generalization of quantum mechanics applicable to a system whose classical analogue is described by a principle of least action. It is important to emphasize, however, some of the difficulties and limitations of the description presented here. One of the most important limitations has already been discussed. The interpretation of the formulas from the physical point of view is rather unsatisfactory. The interpretation in terms of the concept of transition probability requires our altering the mechanical system, and our speaking of states of the system at times very far from the present. The interpretation in terms of expectations, which avoids this difficulty, is incomplete, since the criterion that a functional represent a real physical observable is lacking. It is possible that an analysis of the theory of measurements is required here. A concept such as the "reduction of the wave packet" is not directly applicable, for in the mathematics we must describe the system for all times, and if a measurement is going to be made in the interval of interest, this fact must be put somehow into the equations from the start. Summarizing: a physical interpretation should be sought which does not refer to the behaviour of the system at times very far distant from a present time of interest.

    A point of vagueness is the normalization factor, A. No rule has been given to determine it for a given action expression. This question is related to the difficult mathematical question as to the conditions under which the limiting process of subdividing the time scale, required by equations such as (45.1), actually converges.

    The problem of the form that relativistic quantum mechanics, and the Dirac equation, take from this point of view remains unsolved. Attempts to substitute, for the action, the classical relativistic form (integral of proper time) have met with difficulties associated with the fact that the square root involved becomes imaginary for certain values of the coordinates over which the action is integrated.

    The final test of any physical theory lies, of course, in experiment. No comparison to experiment has been made in the paper. The author hopes to apply these methods to quantum electrodynamics. It is only out of some such direct application that an experimental comparison can be made.

    The author would like to express his gratitude to Professor John A. Wheeler for his continued advice and encouragement. (Feynman 1942, pp. 73–74)

In order to confront the meaning of measurements in his more general version of quantum mechanics in those cases when the concept of wave function was inapplicable, Feynman had to fully understand the more conventional approaches of the quantum theory of measurement, and in particular von Neumann's formulation of the measurement problem (von Neumann 1932). He was dissatisfied with von Neumann's solution "that it does not make any difference where you made the cut, but you had to make some cut" because it left the possibility that there was a "vital force."[53] It seemed to him quite possible that there was no "vital force," and he did not want to decide that ahead of time from the principles of quantum mechanics. Moreover, he felt that it was unsatisfactory to have part of the world not described by quantum mechanics. While still a graduate student he discussed these problems with von Neumann at the Institute for Advanced Study. As Feynman recalled: "It didn't seem possible, really honestly to have decided that there must be a part that wasn't involved, just because I never really believed that philosophical things like that were really sound and even though you might want to use that to make an easy explanation, it was incomplete."[54] So he tried to find a way to give an objective definition of a measurement and arrived at the following statement: If you could correlate the position of a piece of the world with other pieces, then if the other pieces (or their action or energy or something) would go to infinity and the correlation approached a finite value, then that thing is measurable. In 1946, as he was preparing his article, he wrote himself a little note on the subject:

> When you start out to measure the property of one (or more) atom say, you get, for example, a spot on a photographic plate which you then interpret. But such a spot is really only more atoms & so in looking at the spot you are again measuring the properties of atoms, only now it is more atoms. What can we expect to end with if we say we can't see many things about one atom precisely, what in fact can we see. Proposal,
>
> Only those properties of a single atom can be measured which can be correlated (with finite probability) (by various experimental arrangements) with an unlimited no. of atoms. (Ie: the photographic spot is "real" because it can be enlarged & projected on screens, or affect large vats of chemicals, or big brains, etc., etc—it can be made to affect ever increasing sizes of things—it can determine whether a train goes from N.Y. to Chic.—or an atom bomb explodes—etc)[55]

More recently, the same was expressed in more vivid language:

> In other words, you tell me how much matter you want to get screwed up by correlating with something: you measure an electron, you turn a light green or red; not good enough for you to turn red or green, then let the light turn an atomic bomb off or on.... You give me any value no matter how large but not infinite, then I can define what it means to measure. So instead of putting the thing into the mind, or psychology, I put it into a number (I tried to take the idea of a mathematical limit in which the order of limits is reversed). There is no absolute definition of measurement but you tell me how accurate you want to be and I can show you that this thing would be measurable, and in practice the accuracy is fantastic for a small amount of excess matter getting screwed; once you get past a light bulb the rest of the correlations converge rapidly.[56]

While working on his thesis Feynman also realized that the answers to statistical mechanical problems were given by the same kind of path integrals except that the exponents were real. He could arrive at that result from the differential equation for the density operator $\rho = e^{-\beta H}$,

$$-\frac{\partial \rho}{\partial \beta} = H\rho, \qquad (8.4.13)$$

but it seemed to him that there must be another way "that was very beautiful, if only [he] could find it by which one writes quantum mechanics directly in terms of path integrals without going to the Schrödinger equation and then one goes directly across to statistical mechanics."[57] Feynman recounted:

> There I studied the question what do you mean by equilibrium. What do I have to *do* to a physical system to get it to equilibrium? So I would put the effective perturbations and couplings on my path integral as a dynamic object, and ask how does it transform into the correct answer for statistical results, hoping to do that in a fundamental way without ever descending, if I might put it that way, to the Schrödinger equation. The reason in my philosophy not to descend to the Schrödinger equation and to do as much of the physics as I could without doing that, is that I really believed at that time, in 1941–42, that this back action, this Wheeler-Feynman thing was really a forward step. That's why I was doing everything. That's why I found everything. I wanted to get the quantum mechanics of that, and that was in the form of a path integral; it had no Hamiltonian. So the general subject was how can we describe all quantum mechanics, all of physics indeed, when there is a principle of least action, but not a Hamiltonian. How does one get statistical mechanics if there is an action, but no Hamiltonian."[58]

It should, however, be stressed that Feynman was not trying to unify all of physics. He conceived of physics even then as many closely interrelated pieces: there were problems all about. The only things that he really "knew" were the algorithms to compute "something," and "that's what physics knowledge really is."[59] For Feynman the complete accurate statement of an algorithm *is* the theory. Path integrals represented a powerful algorithm for doing quantum mechanics, statistical mechanics and handling questions in the theory of measurements. The power— and great value—of the method lay in the fact that it allowed one to separate the system into pieces and "integrating out parts of it," something that was impossible to do with ordinary quantum mechanics stated in its differential form. Already in his thesis, Feynman had posed the following question: Consider the interaction of a harmonic oscillator of mass m, frequency ω, coupled to two systems Y and Z

described by Lagrangian L_y and L_z so that the total system is described by an action

$$S = \int dt \left\{ L_x + L_y + \left(\frac{m\dot{x}^2}{2} - \frac{m\omega^2 x^2}{2} \right) + \left(I_y + I_z \right) x \right\}. \tag{8.4.14}$$

Is it possible to find an action $\mathcal{A}$, a functional of $y(t)$, $z(t)$ only, such that, as far as the motion of Y and Z is concerned (i.e., for variations of $y(t)$, $z(t)$) the action $\mathcal{A}$ is a minimum? The answer is yes but with a specific proviso.

On the assumption that $I_y = I_y(y(t), t)$ and similarly $I_z = I_z(z(t), t)$, the equation of motion for $x(t)$,

$$\ddot{x}(t) + m\omega^2 x(t) = \left[I_y(t) + I_z(t) \right] = \gamma(t), \tag{8.4.15}$$

can be integrated. Feynman obtained the result that only if the specifications of $x(t)$ are in terms of $x(0)$ and $x(T)$, that is, in terms of initial and final positions (rather than initial position and velocity) does an $\mathcal{A}$ exist, for example,

$$\mathcal{A} = \int_{-\infty}^{+\infty} dt \, [L_y + L_z] \, dt + \frac{1}{2m\omega} \int_{-\infty}^{\infty} \int_{-\infty}^{t} \sin\omega(t - s)\gamma(t)\gamma(s) \, ds \, dt. \tag{8.4.16}$$

This is where matters stood in the spring of 1942 when Feynman wrote up his thesis at Wheeler's strong urging.[60] The completion of the dissertation terminated Feynman's discipleship with Wheeler. Wheeler had deeply influenced Feynman. They had discussed "many, many" physics problems together.[61] Wheeler had helped develop Feynman's "geometrical" way of thinking. When Wheeler explains things, he always makes use of pictures and diagrams. Wheeler says of himself: "I think geometrically rather than in words. I think in term of pictures."[62] From Wheeler, Feynman also learned to think in an abstract way. As Feynman put it, "His grandiose views were useful. Everything could be done this way. Everything could be looked at that way. I learned physics could be looked at from many different ways; each one makes an axiom system and they are all different. They start from the other guy's theorem as the axiom for the other. Wheeler was important."[63]

8.5 The War Years

Right after completing his dissertation, Feynman joined the group working with Robert ("Bob") Wilson on the electromagnetic separation of U^{235} and U^{238} using the "isotron," a device that accelerated beams of ionized uranium, and tried to separate the isotopes by bunching them by the application of a high-frequency voltage to a set of grids part way down a linear tube (Hewlett and Anderson 1962, p. 59). He thereafter became absorbed in the problems of making an atomic bomb.[64] He was well prepared for the task. He had taken Wheeler's course on nuclear physics and had edited a lucid, detailed set of notes based on that course (Wheeler 1940). He had learned a great deal about the properties of materials from Wigner's "very good course"[65] on solid state physics. And to prepare himself for the general examinations for the Ph.D. in the spring of 1940, he had reviewed all of physics and had written a notebook entitled "Things I Don't Know."[66]

Early in 1943 Feynman left for Albuquerque; he was one of the first people to arrive at Los Alamos. He had been invited by Oppenheimer, who personally arranged for the transfer of Feynman's wife, who was ill with lymphatic tuberculosis, from a hospital near Princeton to one in Albuquerque (Feynman 1976, p. 13).

Feynman has written about his experiences at Los Alamos (Feynman 1976, 1980). From his account, and those of others, it is clear that he was one of the most versatile, imaginative, ingenious, and energetic members of that community of outstanding scientists.[67] Feynman came to Los Alamos as a regular staff member and was quickly recognized by Bethe and Oppenheimer as one of the most valuable members of the theoretical division. Bethe recalls that Feynman

> was very lively from the beginning, . . . I realized very quickly that he was something phenomenal. The first thing he did since we had to integrate differential equations, and at that time only had hand computers, was to find an efficient method of integrating third-order differential equations numerically. It was very, very impressive. Then, within a month, we cooked up a formula for calculating the efficiency of a nuclear weapon. It is named the Bethe-Feynman formula, and it is still used. I thought Feynman perhaps the most ingenious man in the whole division, so we worked a great deal together. (Bernstein 1979, p. 61)

As early as November 1943, Oppenheimer wrote Birge, the chairman of the Department of Physics at Berkeley:

> As you know, we have quite a number of physicists here, and I have run into a few who are young and whose qualities I had not known before. Of these there is one who is in every way so outstanding and so clearly recognized as such, that I think it appropriate to call his name to your attention, with the urgent request that you consider him for a position in the department at the earliest time that is possible. You may remember the name because he once applied for a fellowship in Berkeley: it is Richard Feynman. He is by all odds the most brilliant young

physicist here, and every one knows this. He is a man of thoroughly engaging character and personality, extremely clear, extremely normal in all respects, and an excellent teacher with a warm feeling for physics in all its aspects. He has the best possible relations both with the theoretical people of whom he is one, and with the experimental people with whom he works in very close harmony.

The reason for telling you about him now is that his excellence is so well known, both at Princeton where he worked before he came here, and to a not inconsiderable number of "big shots" on this project, that he has already been offered a position for the post war period, and will most certainly be offered others. I feel that he would be a great strength for our department, tending to tie together its teaching, its research and its experimental and theoretical aspects. I may give you two quotations from men with whom he has worked. Bethe has said that he would rather lose any two other men than Feynman from his present job, and Wigner said, "He is a second Dirac, only this time human." (Smith and Weiner 1980, pp. 268–269)

Feynman became a group leader in the Theoretical Division under Bethe and worked on most aspects of the design and properties of the bombs.[68] He was sent to Oak Ridge, Tennessee, to help insure the safety of the isotope separation plants there,[69] and during the final phase of the bomb project at Los Alamos he was put in charge of computing, one of the most critical sections of the entire enterprise. A good deal of Feynman's initial work at Los Alamos dealt with problems in neutron diffusion.[70] The fast neutrons on which the U^{235} atomic bomb was to operate had a fission cross section σ_f almost as large as the scattering cross section σ_s. A differential diffusion equation approach was therefore not adequate, since it required $\sigma_f \ll \sigma_s$. A theory based on integral equations had to be devised. This approach, incidentally, was very close in spirit to the propagator formalism that Feynman developed after the war to deal with the Schrödinger equation. In both situations, a global viewpoint, based on integral equations, defines the approach.

Although the neutron diffusion integral equations could be solved for spherical geometries, the geometries involved in the U^{235} bomb assembly could not be solved analytically and a decision was made to use IBM machines to solve them. In those days computers were electromechanical machines: "The computation was done mechanically, much as in a desk calculator, the sensing was by an electrical contact through the holes in a punch card." Bethe remembers that

> Richard Feynman, Eldred Nelson and Stanley Frankel had studied the manuals carefully. So when the machines arrived, in big boxes and disassembled into a thousand pieces, they set

to work immediately to put them together. . . . The mechanical design was quite complicated, but Feynman, the mechanical wizard who had been repairing all the desk calculators in the laboratory whenever they went wrong, and the skillful Nelson and Frankel managed. Nobody outside IBM had ever assembled a machine successfully. When the official [IBM assembly] man arrived, he was most complimentary to our 3 physicist-mechanics.[71]

After von Neumann had suggested that an implosion be considered the favored method of assembling a plutonium bomb, Feynman, Metropolis, and Teller (1949) devised an equation of state for uranium and plutonium in order to determine how much compression could be expected. Feynman supervised all the computations involved in the assembly of the plutonium bomb. The success of the Trinity test is in no small measure attributable to Feynman's contributions in successfully expediting all the difficult calculations involed in the assembly. Los Alamos marked the beginning of Feynman's lifelong involvement with computers and computing (see Hillis 1989).

Because Feynman always explained the problems they were working on to the members of the group of whom he was in charge, even at the risk of violating security regulations—his deep respect for human rationality—he obtained from them a dedication that resulted in remarkable productivity (Groueff 1967, p. 121). His versatility is legendary. His genius at lockpicking, repairing Marchant and Monroe calculators, assembling IBM machines, solving puzzles and difficult physics problems, suggesting novel calculational approaches, and explaining theory to experimenters and experiments to theoreticians earned him the admiration of all those with whom he came in contact.

Feynman's love for solving puzzles merits comment. Welton observes:

Once presented with a clearly formulated physical paradox, mathematical result, card trick, or whatever [Feynman] would not sleep until he had the solution. Shortly after I had arrived [at Los Alamos] I presented Dick with a problem (later immortalized in the *Feynman Lectures*, Vol. II, section 17.4). I had gotten it from a friend at NRL and had immediately solved it. I stated the problem and Dick asked if I had solved it. I said yes, but (truthfully and a bit strangely) the answer had slipped away. He promptly set to work on it, with me steadily demolishing his attempted solutions but still not remembering my own solution. We parted to get some sleep (I thought), but the next morning Dick showed up at the office a bit the worse for wear but triumphant. This sort of thing happened over and over again with important matters rather than trivia.[72]

This love for solving problems—his father's nurturing brought to full bloom—was quintessentially Feynman: part of it was an obsessive need to "undo" what is "secret,"[73] part of it a need to constantly prove to *himself* that he was as good as anyone else, part of it was a fiercely competitive nature that converted challenges into creative opportunities. Another facet of the story Welton tells should be noted: after Welton had told Feynman that he had solved the puzzle (even though he had forgotten the answer), Feynman never doubted it.

It was at Los Alamos that Feynman first met Bethe. Bethe's unerring physical intuition, his awesome analytical powers, his sagacity, stamina, and erudition, his unaffected, straightforward demeanor, his "unflappability," his forthright collegiality, and above all his integrity impressed Feynman deeply. Bethe's personality and his sense of humor were such that Feynman got along exceedingly well with him. He came "to love this man."[74] Stephane Groueff, in his book on the Manhattan Project, has vividly described Feynman and Bethe's interactions at Los Alamos:

> Richard Feynman's voice could be heard from the far end of the corridor: "No, no, you're crazy!" His colleagues in the Los Alamos Theoretical Division looked up from their computers and exchanged knowing smiles. "There they go again!" one said. "The Battleship and the Mosquito Boat!"
>
> The "Battleship" was the division leader, Hans Bethe, a tall, heavy-set German who was recognized as a sort of genius in theoretical physics. At the moment he was having one of his frequent discussions with Dick Feynman, the "Mosquito Boat", who, from the moment he started talking physics, became completely oblivious of where he was and to whom he was talking. The imperturbable and meticulous Bethe solved problems by facing them squarely, analyzing them quietly and then plowing straight through them. He pushed obstacles aside like a battleship moving through the water.
>
> Feynman, on the other hand, would interrupt impatiently at nearly every sentence, either to shout his admiration or to express disagreements by irreverent remarks like "No, you're crazy!" or "That's nuts!" At each interruption Bethe would stop, then quietly and patiently explain why he was right. Feynman would calm down for a few minutes, only to jump up wildly again with "That's impossible, you're mad" and again Bethe could calmly prove it was not so. (Groueff 1967, p. 202)

"Bethe had a characteristic which I learned," Feynman recalled, "which is to calculate numbers. If you have a problem, the real test of everything—you can't leave [it] alone—you've got to get the numbers out; if you don't get down to earth with it, it really isn't much. So his perpetual attitude [is] to use the theory. To see how it really works is to really use it." Feynman wistfully added:

"What I wasn't able to learn is his personality. He is able to write page after page, quietly. Everything he presents is organized." Moreover, "he [Bethe] doesn't say something wrong and then gets it right," which is how Feynman sees himself working. And to prove his point, Feynman quoted the Los Alamos aphorism "If Feynman says it three times it is right."[75] The fact of the matter is, however, that on almost any subject Feynman did get it right, and usually the first time around.

Feynman's characteristic forthrightness was in evidence already then. His own description of his first encounters with Niels Bohr is revealing (Feynman 1980, pp. 129–130). Bohr had escaped from Denmark to Sweden in September 1943. The *New York Times* on October 9, 1943, reported:

> Dr. Niels H. D. Bohr, refugee Danish Scientist and Nobel Prize Winner for atomic research, reached London from Sweden today bearing what a Dane in Sweden said were plans for a new invention involving atomic explosions.
>
> The plans were described as of the greatest importance to the Allied War effort. Dr. Bohr, who arrived in London by plane escaped the Nazi persecution of Jews in Denmark by hiding in a fishing boat, arriving in Sweden September 28, according to the best information.

After a brief stay in London he received a Rockefeller Foundation grant for a presumed stay at the Institute for Advanced Study,[76] but, with his son Aage, he went to Los Alamos to work on the bomb.

Nicholas Baker's arrival at Los Alamos created quite a stir because "even to the big shots, Bohr was a great God" (Feynman 1980, p. 129). "Nicholas Baker" was the name Bohr had assumed for security reasons and he was affectionately known as Uncle Nick. Bohr had not been satisfied with his first conference with members of the Theoretical Division at which the problems of the bomb were discussed. Before the second conference, Bohr asked to see Feynman. When they met, Bohr indicated that his son Aage and he had been thinking of ways of making the bomb more efficient. He outlined his idea. After listening to him, Feynman immediately told Bohr why it would not work. Bohr then suggested a different approach, which Feynman once again found impractical. Feynman comments: "I was always *dumb* about one thing. I never knew who I was talking to. I was always worried about the physics. If the idea looked lousy, I said it looked lousy. If it looked good, I said it looked good. . . . I have always lived that way" (Feynman 1976, p. 28).

Feynman's discussion with Bohr went on for quite a while, and when it was finished Bohr said, "I guess we can call in the big shots now."[77] At that point Bohr assembled the leading members of the theoretical division and had a discussion with them. Aage Bohr later explained to Feynman what had transpired. After his first conference his father had told him that Feynman had been the only person at the meeting who was not afraid of him and who had been willing to say that an idea of his was "crazy." "So," said Bohr, "next time when we want to discuss ideas, we are not going to do it with these [big shots] . . . who say everything is yes, yes Dr. Bohr. Get [Feynman] and we'll talk with him first" (Feynman 1976, p. 29).

In his reminiscences of Los Alamos, Feynman also tells of the influence of von Neumann on him:

> Then there was Von Neumann, the great mathematician. We used to go for walks on Sunday. We'd walk in the canyons, and we'd often walk with Bethe, and Von Neumann, and Baker. It was a great pleasure. And Von Neumann gave me an interesting idea; that you don't have to be responsible for the world that you're in. So I have developed a very powerful sense of social irresponsibility as a result of Von Neumann's advice. It's made me a very happy man ever since. But it was Von Neumann who put the seed in that grew into my active irresponsibility. (Feynman 1976, p. 28)

Welton has adumbrated Feynman's interactions with his peers at Los Alamos: "We all saw him diplomatically, forcefully, usually with humor (gentle or not, as needed) dissuade a respected colleague from some unwise course. We all saw him forcefully rebuke a colleague less favored by his respect, frequently with definitely ungentle humor. Only a fool would have subjected himself twice to such an experience."[78]

The community's recognition of Feynman's talents can be gauged by the offers he received while at Los Alamos. As indicated by the previously quoted letter of Oppenheimer to Birge recommending an appointment for Feynman at Berkeley, Feynman had already received an offer for "a position for the post war period" at Cornell in November 1943. Birge, the somewhat pedantic, very formal, and very conservative chairman of the physics department at Berkeley was unwilling to make a commitment that far in advance. Oppenheimer expressed his disappointment to Birge in May 1944, and in his letter gave the following assessment:

> As for Feynman himself, I perhaps presumed too much on the excellence of his reputation among those to whom he is known. I know that Brode, McMillan, and Alvarez are all enthusiastic about him, and it is small wonder. He is not only an extremely brilliant theorist, but a man of the greatest robustness, responsibility and warmth, a brilliant and a lucid teacher, and an

untiring worker. He would come to the teaching of physics with both a rare talent and a rare enthusiasm. We have entrusted him here with the giving of a course for the staff of our laboratory. He is one of the most responsible men I have ever met. He does not regard himself as a privileged artist but as one of a group of hard working men for whom the development of physical science is an obligation, and the exposition both an obligation and a pleasure. He spends much of his time in the laboratories and is always closely associated with the experimental phases of the work. He was associated with Robert Wilson in the Princeton project and Wilson attributes a great part of the success of that project to his help. We regard him as invaluable here; he has been given a responsibility and his work carries a weight far beyond his years. In fact he is just such a man as we have long needed in Berkeley to contribute to the unity of the department and to give it technical strength where it has been lacking in the past. (Smith and Weiner 1980, pp. 276–277)

Feynman accepted Bethe's offer of a three-year appointment as an assistant professor at Cornell. This meant that he was formally on leave from Cornell University with Los Alamos paying 11/9 of his annual salary of $3,000 as set by his Cornell contract. In July 1945, at Oppenheimer's insistence, Feynman was finally offered an assistant professorship at Berkeley at the then considerable salary of $3,900 per year. Birge wrote him that "during the period that the department was being built up by the addition of men like Lawrence, Brode, Oppenheimer, Jenkins and White, and more recently by MacMillan and Alvarez, no one to whom we made an offer ever refused it. If you come to Berkeley, I am certain you will never regret the decision."[79] But Feynman did refuse.[80] His affection, respect, and esteem for Bethe were the decisive factors; furthermore, at Bethe's insistence, Cornell had immediately countered Berkeley's offer and set his "potential" salary at $4,000 per annum.[81] In recommending this increase of salary, Bethe indicated that his already high opinion of Feynman had further increased over the course of the year. "[Feynman] has been absolutely invaluable to this project. You can ask him to do anything at all from the most complicated theoretical calculations to the organization of a group of men to do machine computations. Everybody goes to him to have things explained and I think he is one of the best teachers our department ever had."[82]

In the fall of 1945, Feynman went to Cornell eager to assume his new professorial duties. He was one of the first persons to leave Los Alamos. The war's end in August 1945 had put great pressure on universities to accommodate the vast numbers of students that were expected to start or resume their studies. A year later, in the fall of 1946, H. D. Smythe, the chairman of the physics department at Princeton, wrote Feynman that the university and the Institute for Advanced Study would like to make him an offer "of a permanent position at a substantial salary"[83] whereby he would spend half of every academic year in the Department of Physics at the university and the other half as a member of the Institute, free of any teaching duty. Again Feynman decided to stay at Cornell.[84] The Princeton offer resulted in his being promoted to associate professor. An offer of an associate professorship at UCLA was also declined.[85] He did consider very seriously invitations to visit Berkeley—one for the academic year 1947/48 and one for the subsequent year—but Oppenheimer's acceptance of the directorship of the Institute for Advanced Study in the spring of 1947 convinced him to stay at Cornell.[86]

At the time, Feynman did not think that he merited these offers. Although he was giving interesting and stimulating graduate courses—in mathematical physics and electricity and magnetism—he felt that his research wasn't going anywhere. He was depressed; both his wife and his father had died during the previous year.[87] And he began to think that he was "burned out," that this was the end, and that he wouldn't accomplish anything. When he received the offer from Princeton, his reaction was that "they were absolutely crazy."[88]

It was during this period that Feynman accepted Wigner's invitation[89] to comment on the paper Dirac was presenting to the Nuclear Physics session of the Princeton Bicentennial conference on September 24, 1946 (Osgood 1951; Wigner 1947). Dirac had given Feynman a handwritten copy of his paper and he had studied it.[90] In his comments Feynman was rather critical of Dirac's work, believing him to be going "on the wrong track"[91] by working more and more with Hamiltonians and not coming to the central problems that quantum electrodynamics was facing. Since Feynman didn't feel very confident, he made many more than his normal quota of jokes during his comments and Weisskopf criticized him afterwards for what he considered a poor presentation. Bohr, however, came to Feynman's defense and agreed with him "that we have some important problems here to discuss."[92]

It should be pointed out that his close associates at the time—Philip Morrison with whom he shared an office, and Hans Bethe with whom he had frequent discussions—were unaware of his depression. Bethe explains "Feynman depressed is just a little more cheerful than any other person when he is exuberant."[93] On the other hand, Welton "was entranced as always by the flow of ideas" when he met Feynman during that time at Physical Society meetings, "but it was clear that his mind was not really where it properly belonged."[94] In the spring of 1947 Feynman revealed to Bob Wilson, who had just come to Cornell as the director of the newly founded Newman Laboratory for Nuclear Studies, his concerns that Cornell had made a "bad bet" with him. Wilson chided Feynman: "When we hire someone we take a risk and it is *our* risk." Feynman asserts that talking to Wilson "turned him around."[95]

8.6 Research, 1946

When Feynman came to Cornell, he resumed the investigations that the war had interrupted.

Although at Los Alamos he had worked intermittently on the problems of QED and statistical mechanics, primarily on the bus rides to and from Albaquerque to visit his wife in the hospital, he did not discuss these problems with his colleagues.[96] Moreover, less and less time was spent on these matters as the pace at Los Alamos intensified in late 1944 and 1945. The initial phase of his researches at Cornell was devoted to reconstructing and reevaluating the work he had carried out during the last stages of his dissertation and while at Los Alamos. One of the problems addressed was to understand the approach to thermodynamic equilibrium of a system whose charged constituents interacted via delayed interactions, or via 1/2 advanced + 1/2 retarded electromagnetic interactions. Among Feynman's papers at the Cal Tech archives are extensive notes of his calculations to prove the adiabatic theorem to various orders in perturbation theory.[97] These notes are interesting because in them Feynman resorts to diagrammatic mnemonics to keep track of the terms he encounters. However, the results were ambiguous and Feynman was not satisfied with them. The research was motivated by the belief that a system interacting via delayed interactions would reach equilibrium at some temperature; and Feynman sought a specification for the probability of a given motion of the system at a finite temperature directly in terms of the action that described the dynamics of the system.

Another problem Feynman worked on was how to describe a spin 1/2 particle in his integral over path of formulation of quantum mechanics—his "track theory"—and more particularly, how to describe a relativistic spin 1/2 particle, that is, a Dirac particle. Since the concept of the spin of a pointlike particle is lacking in classical theory and does not enter in the classical action, it is not immediately obvious what kinds of paths should be contemplated and a fortiori the amplitudes that should be assigned to them to give rise to the Dirac equation. Feynman's research activities at the beginning of 1947 are described in a letter he wrote to Welton: "I am engaged now in a general program of study—I want to understand (not just in a mathematical way) the ideas of all branches of theor[etical] physics. As you know I am now struggling with the Dirac Eq."[98] Feynman had been able to give a "derivation" of the Dirac equation for a particle of mass m in a world consisting of one space and one time dimension.[99] The particle is assumed to be able to travel in both the $+$ and $-x$-direction, and to move with the speed of light. It starts at $x = 0$ at $t = 0$, and ends at X at time T where $| X | \leq T$. The interval $[0,T]$ is divided into a large number n of small intervals of duration ϵ so that

$$T = n\epsilon. \tag{8.6.1}$$

If we suppose that in the entire interval $[0, T]$ the particle travels in the $+x$-direction n_1 times, and in the $-x$-direction n_2 times, then

$$n\epsilon = (n_1 + n_2)\epsilon = T \tag{8.6.2a}$$

$$(n_1 - n_2)\epsilon = X \tag{8.6.2b}$$

(the speed of light has been taken for convenience to be $c = 1$). A typical path will consist of null segments (since it travels with velocity c) meeting in sharp corners. The propagator to go from $(0,0)$ to (T, X) is obtained by summing over all paths the expression $\Sigma_r A(\Gamma, R)$, where $A(\Gamma, R)$ is the amplitude for the path Γ, with R corners. Feynman showed that one could derive the Dirac equation in one-space–one-time dimension if the amplitude for a path with R corners is taken to be $(im\epsilon)^R$. Stated differently, each time the electron reverses spatial direction, it acquires a phase factor $e^{i\pi/2}$ (see photograph 20).

Feynman encountered difficulties in extending the idea "of loading each turn thru θ by $e^{i\theta/2}$" which worked in one-space dimension to higher dimensions because in those situations the angles θ are in different planes.[100] He tried to use quaternions and "octonions" (quaternions representing euclidean 4-dimensional rotations) to represent wave functions, but he was not able to obtain a "natural" representation of the Dirac equation as an integral over path.

Although these researches had given him many insights into the Dirac equation in 1, 2, 3, and 4 spatial dimensions, Feynman's paper on the "Space-Time Approach to Non-Relativistic Quantum Mechanics," which he submitted in the summer of 1947 to the *Reviews of Modern Physics,* dealt only with spin "in a formal way." Feynman characterized his incorporation of spin and relativity into the formalism as "adding nothing to the understanding of these [Dirac and Klein-Gordon] equations" but added the statement, "There are other ways of obtaining the Dirac equation which offer some promise of giving a clearer physical interpretation to that important and beautiful equation" (Feynman 1948c, p. 387).

The "other ways" Feynman was alluding to were the ones he had outlined in his letter to Welton. He concluded that letter with some interesting philosophical remarks:

> Now I would like to add a little hooey. The reason I am so slow is not that I do not know what the correct equations, in integral or differential forms are (Dirac tells me) but rather that I would like to *understand* these equations from as many points of view as possible. So I do it in 1, 2, 3, & 4 dimensions with different assumptions etc....
>
> I find physics is a wonderful subject. We know so very much and then subsume it into so very few equations that we can say we know very little (except these equations—Eg. Dirac, Maxwell, Schrod). Then we think we have *the* physical picture with which to interpret the equations. But there are so very few equations that I have found that many physical pictures can give the same equations. So I am spending my time in study—in seeing how many new viewpoints I can take of what is known.
>
> Of course, the hope is that a slight modification of one of the pictures will straighten out some of the present troubles.

I dislike all this talk of others [of there] not being a picture possible, but we only need know how to go about calculating any phenomenon. True we only *need* calculate. But a picture is certainly a *convenience* & one is not doing anything wrong in making one up. It may prove to be entirely haywire while the equations are nearly right—yet for a while it helps. The power of mathematics is terrifying—and too many physicists finding they have correct equations without understanding them have been so terrified they give up trying to understand them. I want to go back & try to understand them. What do I mean by understanding? Nothing deep or accurate ∼ just to be able to see some of the qualitative consequences of the equations by some method other than solving them in detail.

For example I'm beginning to get a mild "understanding" of the place of Diracs α matrices which were invented by him "to produce an equation of first order in the differential coefficient in the time," but by me in order "to keep track of the result of a succession of changes of coordinate system...."

Why should the fundamental laws of Nature be so that one cannot explain them to a high-school student—but only to a quite advanced graduate student in physics? And we claim they are simple! In what sense are they simple? Because we can write them in one line. But it takes 8 years of college education to understand the symbols. Is there any simple *ideas* in the laws?[101]

Feynman did not answer the questions he posed. In his Reith Lectures, Oppenheimer did:

All this means that science is cumulative in a quite special sense. We cannot really know what a contemporary experiment means unless we understand what the instruments and the knowledge are that are involved in its design. This is one reason why the growing edge of science seems so inaccessible to common experience. Its findings are defined in terms of objects and laws and ideas which were the science of its predecessors. This is why the student spends many long years learning the facts and arts which, in the acts of science, *he will use and take for granted* [italics mine]—why this long tunnel, at the end of which is the light of discovery, is so discouraging for the layman to enter, to be an artist, scholar or man of affairs. (Oppenheimer 1953, p. 25)

During the summer of 1947 Feynman started writing up the results of his dissertation for the article "Space-Time Approach to Non-Relativistic Quantum

Mechanics," which he submitted to the *Reviews of Modern Physics* early in the fall of that year. The writing did not go well. Corben, with whom he spent part of the summer in Pittsburgh, recalls that "[Feynman] was articulate (as always) but had difficulty at first in putting his ideas down on paper. Along with Alfred Schild, who lived in the same house, we practically locked Dick in a room and told him to start writing. The first draft was poor but after three weeks the ... publication emerged."[102]

In a letter to a graduate student who was trying to obtain a copy of his Ph.D. thesis in 1949, Feynman pointed out the main differences between the thesis and the *RMP* article:

> In the thesis I was trying to generalize the idea [of using integral over paths as a quantization procedure] to apply to any action function at all—not just the integral of a function of values [*sic*; should read velocities] and positions (see Section 12). All the ideas which appear in the Review article were written in such a form that if any generalization is possible, they can be readily translated (in particular the important equation 45). The thesis contains a somewhat more detailed analysis of the general relation of the invariance properties of the actional [action] functional and constants of the motion. Also the problems of elimination of intermediate harmonic oscillators is done more completely than is done in Section 13.
>
> The reason I did not publish everything in the thesis is this. I met with a difficulty. An arbitrary action functional S produces results which do not conserve probability; for example, the energy values come out complex.[103] I do not know what this means nor was I able to find that class of actional functionals which would be guaranteed to give real eigen values for the energies.[104]

Important conceptual advances had also been made since 1942. There was now an unmistakable visual aspect to the formalism. Although no pictures appear in the paper, the text explicitly enjoins the reader to conceive of the amplitude for a particle to go from x at time t to x' at t' as receiving contributions from all the trajectories that can be drawn between xt and $x't'$. Feynman asserted that "clarity came from writing up the *RMP* article." Although he had amplitudes before, the "pictures" only came at this stage. When doing path integrals he now visualized paths, "I could see the path ... each path got an amplitude."[105]

Similarly, although propagators for finite time intervals were not explicitly introduced, it is clear that Feynman was conceiving problems in terms of their use. Thus in section 8 of that article, perturbation theory is presented in a fashion that makes evident that the higher-order terms are most simply expressed in terms of finite time propagators. Implicit in the *RMP* article is a perturbation theory for the kernel

$$K(2,1) = \int \ldots \int \mathscr{D}(\text{path}) \exp\left\{\frac{i}{\hbar} \int L dt\right\} \qquad (8.6.1)$$

when the action is of the form

$$S = S_0 - \int_{t_0}^{t} U(x(t)) dt, \qquad (8.6.2)$$

with U a perturbation, and S_o defining the "unperturbed" kernel

$$K_0(2,1) = \int \ldots \int \mathscr{D}(\text{path}) \exp\left\{\frac{i}{\hbar} S_0\right\}. \qquad (8.6.3)$$

Expanding the exponential yields the familiar result

$$K(2,1) = K_0(2,1) - \frac{i}{\hbar} \int d\tau_3 K_0(2,3) U(3) K_0(3,1) + \ldots \qquad (8.6.4)$$

(which is the equation following eq. 44 in Feynman 1948a, stated in a different notation). The steps by which the right-hand side of eq. (8.6.4) are obtained were of great importance for the subsequent developments. The expansion of $\exp\left\{-\frac{i}{\hbar} \int_{t_0}^{t_2} U(x(t)) dt\right\}$ yields the following first-order term:

$$-\frac{i}{\hbar} \int \ldots \int \mathscr{D}(\text{path}) E^{\frac{i}{\hbar} S_0} \int_{t_1}^{t_2} U(x(t)) dt, \qquad (8.6.5)$$

which is given meaning by the Feynman specification of a path $x(t)$ by the value of $x(t)$ at $t_1, t_1 + \epsilon, t_1 + 2\epsilon, \ldots t_2$ and

$$\mathscr{D}(\text{path}) \rightarrow \frac{dx_1}{A} \ldots \frac{dx_N}{A}. \qquad (8.6.6)$$

Thus eq. (8.6.5) is to be understood as

$$-\frac{i}{\hbar} \int \ldots \int \frac{dx_1}{A} \ldots \frac{dx_\ell}{A} \frac{dx_{\ell+1}}{A} \ldots \frac{dx_N}{A}$$

$$\exp\left\{\frac{i}{\hbar} \sum_{\ell+1}^{N} L(x_j x_{j+1}) \epsilon\right\} \cdot \epsilon \sum_{\ell} U(x_\ell) \exp\left\{\frac{i}{\hbar} \sum_{j}^{\ell} L(x_j x_{j+1}) \epsilon\right\}.$$

$$(8.6.7)$$

Fix x_ℓ: then the integration over x_1 to x_ℓ can be performed, yielding $K_0(\ell, 1)$. Similarly the integration from $\ell + 1$ to N can be performed and one is left with

$$-\frac{i}{\hbar} \int dx_\ell \sum_\ell \epsilon K_0(2, \ell) U(\ell) K_0(\ell, 1),$$

the first-order term in the right-hand side of eq. (8.6.4). Attached to eq. (8.6.4) is a visual picture that represents the integral as obtaining contributions connecting first the space-time point 1 to ℓ at $\ell\epsilon$, and then the path s connecting ℓ to 2. (Fig. 8.6.1)

The kernel $K(2,1)$ solves the problem

$$\psi(2) = \int K(2, 1)\psi(1)d^3(1). \tag{8.6.8}$$

In fact, the equivalence of the integral-over-path method to the standard version of quantum mechanics was demonstrated by showing that eq. (8.6.8), with K defined as an integral over path, yielded the Schrödinger equation when 2 was infinitesimally close to 1. Thus the integral-over-path representation of eq. (8.6.8) could also be looked upon as giving meaning to the expression

$$\mid \psi(t_2) >= e^{\frac{1}{\hbar} \int_{t_1}^{t_2} H(t')dt'} \mid \psi(t_1) >, \tag{8.6.9}$$

which is a *formal* solution of $H \mid \psi >= i\hbar\partial_t \mid \psi >$ when the exponential factor is properly interpreted. Feynman's calculus of ordered operators, which is closely linked to the meaning given to expressions such as eq. (8.6.9) by their integral-over-path formulation, was developed during the summer and fall of 1947 as a natural outgrowth of generalizing the results he obtained for nonrelativistic particles to the case when the particles are described by the Dirac equation.

8.7 Shelter Island and Its Aftermath

Feynman was one of the "young men" invited to the Shelter Island conference that took place in June 1947. It was his first "pure" physics conference with "big men."[106] Some twenty years later he commented that he never attended a more important conference.[107] On the final day of the conference, Feynman lectured on his space-time approach to nonrelativistic quantum mechanics and also talked about his unsuccessful attempts to give an integral-over-path formulation of the Dirac equation.[108]

Shelter Island was indeed the stimulus that made Feynman address once again the problems of quantum electrodynamics. But more precisely, it was Bethe

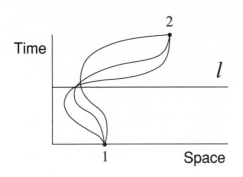

Figure 8.6.1

who got Feynman to work on these problems again. After Bethe had completed his famous train-ride calculation of the level shift, he called Feynman "excitedly" from Schenectady to tell him that he understood the Lamb shift (Feynman 1966a, p. 705). When Bethe returned to Cornell in early July, he gave a lecture explaining his nonrelativistic calculation. He indicated that he had encountered a logarithmic divergence for the Lamb shift—because in this non-relativistic theory the electron self-energy is linearly divergent—but he argued that in a relativistic calculation the Lamb shift would be finite, because the self-energy of an electron in hole theory is only logarithmically divergent. In concluding his lecture, Bethe stressed that if there were a way of making electrodynamics finite with a relativistic cutoff procedure, then it would be much simpler to carry out a relativistic quantum field theoretic calculation of the Lamb shift.

After Bethe's lecture, Feynman went up to him and told him, "I can do that for you. I'll bring it in for you tomorrow" (Feynman 1966a, p. 705). How to introduce a relativistically invariant cutoff into the *Lagrangian* of classical electro-dynamics was something Feynman knew how to do. Using his integral-over-path method, he could then quantize the theory (in terms of this altered but still invari-ant Lagrangian) without invoking either a Hamiltonian or equal-time commuta-tion rules that destroyed the manifest invariance. However, he didn't know how to compute a self-energy, since in Wheeler-Feynman theory charged particles don't interact with themselves. In fact, the elimination of self-interactions had been the initial motivation for his approach.

When they met the next day, Bethe showed Feynman how the expression for the self-energy is derived, and Feynman tried to apply his cutoff method to it. But for some reason, working together at the blackboard they found that Feynman's cutoff method didn't yield a finite answer for the self-energy. (Incidentally, neither Feynman nor Bethe were ever able to discover where they had gone wrong.) How-ever, Feynman never doubted that it would. And thus he got started on his epic researches. Feynman has described the sequel to that encounter:

> So, I went back to my room and worried about this thing and went around in circles trying to figure out what was wrong be-cause I was sure physically everything had to come out finite. I couldn't understand how it came out infinite. I became more and

more interested and finally realized I had to learn how to make a calculation. So, ultimately, I taught myself how to calculate the self-energy of an electron, working my patient way through the terrible confusion of those days of negative energy states and holes and longitudinal contributions and so on. (Feynman 1966a, p. 705)

In retrospect, it was a good thing that, working with Bethe, the initial attempt at rendering the self-energy finite failed. It forced Feynman to learn the "practical" aspects of quantum electrodynamics and to convince him that the cut-off method he had proposed "if carried out without making a mistake" was "all right"; it gave a finite answer for the self-energy and "nothing went wrong... physically" in other processes (Feynman 1966a, p. 706).

Feynman has outlined the genesis of his approach in his article, "Space-Time Approach to Quantum Electrodynamics," which he submitted to the *Physical Review* at the beginning of May 1949:

The conventional electrodynamics was expressed in Lagrangian form of quantum mechanics described in the *Reviews of Modern Physics* [Feynman 1948c]. The motion of the field oscillators could be integrated out (as described in Section 13 of that paper), the results being an expression of the delayed interaction of the particles. Next the modification of the delta-function interaction could be made directly from the analogy to the classical case [Feynman 1948a,b]. This was still not complete because the Lagrangian method had been worked out in detail only for particles obeying the non-relativistic Schrödinger equation. It was then modified in accordance of the Dirac equation and the phenomenon of pair creation. This was made easier by the reinterpretation of the theory of holes [Feynman 1949b]. Finally for practical claculations the expressions were developed in a power series in $e^2/\hbar c$. It was apparent that each term had a simple physical interpretation.

Feynman's outline, though accurate and corroborated by his notes and letters, does not convey the "trial-and-error" aspect of the synthesis, nor does it indicate how the diagrammatic component evolved. Also, Feynman's "Relativistic Cut-Off for Quantum Electrodynamics," a paper submitted to the *Physical Review* of July 12, 1948, and published in the November 15, 1948, issue, gives a misleading impression of what he had accomplished up to that time. That article, which presented some of his results on the radiative corrections to the properties of an electron in an external electromagnetic field, was written using old-fashioned computational methods, in part to make it comprehensible to the theoretical physics community. In fact, by the Pocono conference of April 1948 Feynman had obtained most of the results that were to be published much later: his version of positron theory; his operator calculus; closed expressions for the transition amplitudes; rules for calculating the contributions to the transition amplitudes to each order of perturbation theory, contributions that could be represented succinctly by diagrams—the famous Feynman diagrams; as well as invariant cutoff methods for dealing with the divergences arising from photon exchanges.

What remained to be clarified after Pocono were the problems associated with vacuum polarization and more generally the divergences connected with closed loops. After Pocono, Feynman spent a great deal of time and energy to develop ever more effective ways of computing more and more complicated diagrams; a synopsis of these methods is to be found in the appendixes of his "Space-Time Approach to QED" (Feynman 1949b).

The clarity and simplicity of the presentation in Feynman's two classic papers "The Theory of Positrons" (Feynman 1949a) and "The Space-Time Approach to Quantum Electrodynamics" (Feynman 1949b) belie the magnitude of the task that had been involved in their preparation.[109] Like Schwinger, Feynman had reworked all of quantum electrodynamics and had obtained a formulation that allowed one to bypass the divergence difficulties by using the renormalization procedures, and to obtain answers to problems that could not be addressed previously. But more importantly, the relativistically invariant computational techniques he had developed were so effective that in a few hours he could do calculations that would take (or had taken) months, and in some cases years, using conventional techniques. Moreover, these techniques could easily be generalized to apply to meson–theoretic problems.

8.8 The Genesis of the Theory

Classical Cutoffs

Feynman's notes and letters from the summer and fall of 1947 allow us to reconstruct the genesis of his version of quantum electrodynamics. His starting point was the Wheeler-Feynman statement of classical electrodynamics (Wheeler and Feynman 1945). Feynman adopted the formulation that Wheeler had given in his expanded version of Feynman's 1941 manuscript on the subject. The following exposition is taken from Feynman's notes written during the summer of 1947 entitled "Brief Description of Wheeler-Feynman Electrodynamics."[110] These notes later formed the basis for his article "A Relativistic Cut-off for Classical Electrodynamics" (Feynman 1948a). The action was taken to be

$$S = \sum_a m_a \int \sqrt{da_\mu da^\mu} + \sum_{\substack{a,b \\ a \neq b}} e_a e_b \int \dots \int \delta(s_{ab}^2) da_\mu db^\mu. \qquad (8.8.1)$$

The notation is as follows: particles a, b,... have mass m_a, charge e_a, and coordinates a_μ, $b_\mu (\mu = 1,2,3,4,)$ which may be considered as functions of parameters α, β ... on their paths.

$$\dot{a}_\mu \equiv da_\mu/d\alpha, \dot{b}_\mu \equiv db_\mu/d\beta, \text{etc.} \qquad (8.8.2)$$

The proper time of particle a is defined by $d\tau_a = \sqrt{da_\mu da^\mu}$;

$$x_\mu y^\mu = x_4 y_4 - x_1 y_1 - x_2 y_2 - x_3 y_3 \qquad (8.8.3)$$

$$s_{xy}^2 = (x - y)_\mu (x - y)^\mu \qquad (8.8.4)$$

$$\Box_x^2 = \frac{\partial}{\partial x^\mu}\frac{\partial}{\partial x_\mu} = \frac{\partial^2}{\partial x_4^2} - \frac{\partial^2}{\partial x_1^2} - \frac{\partial^2}{\partial x_2^2} - \frac{\partial^2}{\partial x_3^2}. \qquad (8.8.5)$$

Note that

$$\Box_x^2 \delta(s_{xy}^2) = 4\pi\delta(x_4 - y_4)\delta(x_1 - y_1)\delta(x_2 - y_2)\delta(x_3 - y_3). \qquad (8.8.6)$$

The equations of motion follow from the requirement that the action

$$S = \sum_a m_a \int \sqrt{\dot{a}_\mu \dot{a}^\mu} d\alpha$$

$$+ \sum_{a,b}{}' e_a e_b \int \delta(s_{a_\alpha b_\beta}) \dot{a}^\mu(\alpha) \dot{b}_\mu(\beta) d\alpha d\beta \qquad (8.8.7)$$

be a minimum for all variations of all paths $a^\mu(\alpha)$ of all particles, that is, varying $a^\mu(\alpha)$ to $a^\mu(\alpha) + \delta a^\mu(\alpha)$. The following equations result:

$$m_a \frac{d}{d\alpha}\left(\frac{da_\nu}{d\tau_a}\right) = e_a \dot{a}_\mu \sum_{b\neq a} e_b \int \delta'(s_{ab}^2)[2(a_\nu - b_\nu)\dot{b}^\mu - 2(a_\mu - b_\mu)\dot{b}^\nu]d\beta$$

$$= e_a \dot{a}^\mu \sum_{b\neq a} F_{\mu\nu}^{(b)}(a) \qquad (8.8.8)$$

or equivalently

$$m_a \frac{d}{d\tau_a}\left(\frac{da_\nu}{d\tau_a}\right) = e_a \frac{da^\mu}{d\tau_a} \sum_{b\neq a} F_{\mu\nu}^{(b)}(a), \qquad (8.8.9)$$

where $F_{\mu\nu}^{(b)}(x)$ is the field at x_μ due to particle b and is given by

$$F_{\mu\nu}^{(b)}(x) = \frac{\partial A_\mu^{(b)}(x)}{\partial x^\nu} - \frac{\partial A_\nu^{(b)}(x)}{\partial x^\mu}, \tag{8.8.10}$$

with

$$A_\mu^{(b)}(x) = e_b \int \dot{b}_\mu \delta(s_{xb}^2) d\beta. \tag{8.8.11}$$

By virtue of eq. (8.8.6),

$$\begin{aligned}
\Box_x^2 A_\mu^{(b)}(x) &= 4\pi e_b \int \dot{b}_\mu \delta(x_4 - b_4)\delta(x_3 - b_3)\delta(x_2 - b_2)\delta(x_1 - b_1)d\beta \\
&= J_\mu^{(b)}(x), \tag{8.8.12}
\end{aligned}$$

where the right–hand side of eq. (8.8.12) is the 4-current vector of a point charge e_b. Since

$$\delta(t^2 - r^2) = \frac{1}{2|r|}[\delta(t - r) + \delta(t + r)], \tag{8.8.13}$$

the $F_{\mu\nu}$ defined by eq. (8.8.10) satisfy Maxwell's equations, but the potential defined by eq. (8.8.11) are the retarded plus the advanced solutions of Lienard and Wiechert. The relation of this formulation to the usual theory that uses only retarded effects had been expounded at great length in Wheeler and Feynman (1945). If self-interactions are allowed, then the terms with $a = b$ are not excluded in the expression eq. (8.8.7) for the action. These terms give rise to an infinite contribution but allow the action to be written in the form

$$S = \sum_a m_a \int d\tau_a + \frac{1}{(4\pi)^2} \int d^4x \int d^4y\, j^\mu(x)\delta(s_{xy}^2) j_\mu(y). \tag{8.8.14}$$

Feynman then inquired what happens if one assumes that the action eq. (8.8.1) is of the form

$$S = \sum_a m_a \int d\tau_a + \frac{1}{2}\sum_{a,b} e_a e_b \int \int f(s_{ab}^2)\, da^\mu db_\mu, \tag{8.8.15}$$

where f is an invariant function of s_{ab}^2 that behaves like $\delta(s_{ab}^2)$ for large distances. Feynman assumed that $f(s^2)$ is such that interactions exist for s timelike and less than some small length a of the order of the classical electron radius $r_0 = \frac{e^2}{mc^2}$, that is, for $s^2 \leq a^2$, $a^2 > 0$, for example,

$$f(s) = \frac{1}{2a^2}e^{-|s|/a} \text{ for } s^2 > 0$$
$$= 0 \qquad\qquad \text{for } s^2 < 0. \tag{8.8.16}$$

Note that as long as f is a function of the interval s_{12} only, the covariance of the theory is maintained. The form

$$f(s^2) = \int d^4k\, e^{-ik\cdot(x-y)}\tilde{f}(k^2), \tag{8.8.17}$$

with $\tilde{f}$ a function of $k_\mu k^\mu$ only, is the most general one that will make f a function of s^2 only; $\tilde{f}(k_\mu k^\mu) = \delta(k^2)$ yields the original theory with interactions along the light cone only. For calculational purposes Feynman found it convenient to consider an $\tilde{f}$ of the form

$$\tilde{f} = \int_0^\infty [\delta(k^2) - \delta(k^2 - \lambda^2)]G(\lambda)d\lambda, \tag{8.8.18}$$

with

$$\int_0^\infty G(\lambda)d\lambda = 1. \tag{8.8.19}$$

The action eq. (8.8.1) with the $\delta(s_{xy}^2)$ function implied that interaction occurred between events x and y whose 4–dimensional interval vanishes, that is, between those points y that lie on the past and future light cone of x. The consequences of an f, which is different from 0 in the shaded region of fig. (8.8.1), for which $s^2 = t^2 - r^2 \leq a^2$, can be inferred as follows: $(t-r)(t+r) \leq a^2$ implies that when $r \gg a$, $t - r \approx a^2/2r$. Hence the velocity of propagation at large distances becomes closer and closer to the velocity of light c. Similarly, when t is large, since $\Delta s^2 = 2t\Delta t$, there is a spread in the time of arrival of an electromagnetic signal of an amount of the order $a^2/2t$. Thus the interaction between charges separated by a large distance remains essentially unchanged; there is a slight alteration of the interactions when particles are close to one another (i.e., when $r_{ij} \sim a$). There is, however, considerable modification for the action of a charged particle on itself: the infinite self-energy is reduced to a finite value. When the terms $a = b$ are

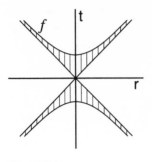

Figure 8.8.1

included in the action, the self-force is of the form

$$e_a \frac{da^\mu}{d\tau} F^{(a)}_{\mu\nu}(a)$$

$$= 2e_a^2 \int \dot{a}_\mu(\alpha)\dot{a}_\mu(\alpha')(a_\nu(\alpha)$$

$$- \alpha_\nu(\alpha'))f'(s^2_{a_\alpha a_{\alpha'}})d\alpha' \qquad (8.8.20)$$

$$- 2e_a^2 \int \dot{a}_\mu(\alpha)\dot{a}_\nu(\alpha')(a_\mu(\alpha)$$

$$- a_\mu(\alpha'))f'(s^2_{a_\alpha a_{\alpha'}})d\alpha'.$$

Feynman showed that when the acceleration of the particle is small, the right–hand side of eq. (8.8.20) reduces to the form $M \frac{d^2 a^\mu}{d\tau^2}$, with

$$M = e^2 \int_{-\infty}^{+\infty} \epsilon^2 f'(\epsilon^2) d\epsilon + e^2 \int_{\infty}^{+\infty} f(\epsilon^2) d\epsilon$$

$$= \frac{1}{2} e^2 \int_{-\infty}^{+\infty} f(\epsilon^2) d\epsilon.$$

The mass is positive and if desired, all the mass of the particle could be considered as electromagnetic in origin.

Feynman obtained another important result. He realized that the action eq. (8.8.15) allowed the possibility of describing pair production in an external field in this classical theory by considering positrons as electrons running backwards in time, the idea previously suggested to him by Wheeler (see Feynman 1948a, pp. 943–944). The purpose of Feynman's investigation of classical electrodynamics with a cutoff was to formulate a finite and consistent theory that included self-interactions, and that could be quantized using his integral-over-path formulation of quantum mechanics. The replacement of the δ function by a smooth function $f(s^2_{xy})$ in the interaction term of the Fokker action, that is, in the term $\int \int d^4x d^4y j_\mu(x)\delta(s^2_{xy})j^\mu(y)$, indeed yielded a finite theory. Furthermore, it was possible to establish the equivalence of this 1/2 (advanced + retarded) formulation of classical electrodynamics to the usual retarded formulation in a universe where all radiation is absorbed at infinity.[111]

A quantum-theoretic proof of the equivalence was more ambiguous. In any case, the immediate problem that Bethe had posed was to indicate how one introduces a cutoff in the quantized version of the usual *retarded* electrodynamics. Feynman knew from the research he had carried out for his thesis that the elimination of the transverse oscillators in the retarded formulation of quantum electrodynamics yielded, for the interaction action that entered into the transition amplitude, an expression very similar to eq. (8.8.14), the difference being that the δ function was replaced by another singular function, the function Feynman later called the $\delta_+(s^2)$ function.

Elimination of Radiation Oscillators

In a set of notes written during the summer of 1947 Feynman once again carried out the calculation of the elimination of the transverse oscillators in the expression for the transition amplitude.[112] The results obtained are those set forth in the first four sections of the article–"Mathematical Formulation of the Quantum Theory of Electromagnetic Interaction" (Feynman 1950, pp. 441–445). The Lagrangian was taken to be

$$L = L_p + L_{tr} + L_I + L_c, \tag{8.8.21}$$

where

$$L_p = \frac{1}{2} \sum_n m_n \dot{\mathbf{x}}_n^2 \tag{8.8.22}$$

is the Lagrangian of the particles;

$$L_{tr} = \frac{1}{2} \sum_k \sum_{r=1}^4 (\dot{q}_k^{(r)})^2 - k^2 (q_k^{(r)})^2 \tag{8.8.23}$$

is the Lagrangian of the transverse electromagnetic field

$$L_c = -\frac{1}{2} \sum_n \sum_m e_n e_m / r_{mn} \tag{8.8.24}$$

is the Coulomb interaction term; and

$$L_I = \sum_n e_n \dot{\mathbf{x}}_n \cdot \mathbf{A}^{tr}(\mathbf{x}_n) \tag{8.8.25}$$

is the interaction Lagrangian, with

$$\mathbf{A}^{tr}(x) = (8\pi)^{1/2} \sum_k \left[e_{1k} \left(q_k^{(1)} \cos k \cdot x + q_k^{(3)} \sin k \cdot x \right) \right.$$
$$\left. + e_{2k} \left(q_k^{(2)} \cos k \cdot x + q_k^{(4)} \sin k \cdot x \right) \right], \tag{8.8.26}$$

where e_{1k} and e_{2k} are two orthogonal unit polarization vectors perpendicular to the direction of propagation k. The sum over k means, for unit volume, $\int d^3 K / (2\pi)^3$, and each $q_k^{(r)}$ can be considered as the coordinate of a harmonic oscillator, since the transverse part has a Lagrangian given by eq. (8.8.23). The elimination of the transverse oscillators can be done one at a time. In the transition amplitude, for the matter system to go from a state $\chi_{t'}$ at t' to the state $\chi_{t''}$ at t'' with the radiation field remaining in its vacuum state ϕ_0 (i.e., with no photon emission, none being present initially), the result of this elimination is the following expression:

$$
\langle \chi_{t''} \phi_{t''} \mid \chi_{t'} \phi_{0t'} \rangle
$$

$$
= \int \int d \text{ (particle variables)}
$$

$$
\bar{\chi}_{t''} \exp \left[\frac{i}{\hbar} (S_p + S_{int}) \right] \chi_{t'} \mathcal{D} \text{ (path particle variables)}, \quad \textbf{(8.8.27)}
$$

with S_{int} given by

$$
S_{int} = \frac{i}{2} \sum_n \sum_m e_n e_m \int_{t'}^{t''} \int_{t'}^{t} (1 - \dot{\mathbf{x}}_n(t) \cdot \dot{\mathbf{x}}_m(s)) \delta_+ ((t - s)^2
$$

$$
- (\mathbf{x}_n(t) - \mathbf{x}_m(s))^2) dt ds, \quad \textbf{(8.8.28)}
$$

where the δ_+ function is defined by

$$
\delta_+(x) = \frac{1}{\pi} \int_0^{\infty} e^{-ikx} dk = \lim_{\epsilon \to 0} \frac{-i}{\pi(x - i\epsilon)}
$$

$$
= \delta(x) - \frac{i}{\pi x}. \quad \textbf{(8.8.29)}
$$

Feynman's notes contain extensive discussions of the transient terms (which have been omitted in the expression (8.8.27)) that occur because the amplitude is being evaluated from a (sharp) time t' to the (sharp) time t''. The interaction term is invariant. The invariance is made explicit by noting that

$$
1 - \dot{\mathbf{x}}_n(s) \cdot \dot{\mathbf{x}}_m(t) = \dot{x}_{n\mu}(s) \dot{x}_{m\mu}(t). \quad \textbf{(8.8.30)}
$$

For the case of a single charged particle, the amplitude for remaining in the initial state ψ_0 (the state at $t' = 0$) after a long time $t'' = T$ is given by

$$
\langle \psi_{0T} \mid \psi_{00} \rangle = \int_0^T \int_0^T \mathcal{D}(\text{path}) \, dx_T \, dx_0 \, \psi_0^*(x_T)
$$

$$
\exp \frac{i}{\hbar} \{ S_p(x(t)) + \frac{ie^2}{\hbar c} \int \int (1 - \dot{\mathbf{x}}(t) \cdot \dot{\mathbf{x}}(s)) \delta_+ ((t - s)^2
$$

$$
- (\mathbf{x}(t) - \mathbf{x}(s))^2) \} \, ds \, dt \, \psi_0(x_0). \quad \textbf{(8.8.31)}
$$

If the particle did not interact with the radiation field (and hence would not interact with itself), the time dependence of $< \psi_{0T} \mid \psi_{00} >$ would be $\exp(-iE_0T)$, with E_0 the energy of the unperturbed state ψ. In the presence of self-interaction,

$$\langle \psi_{0T} \mid \psi_{00} \rangle = e^{-iR_0T} e^{-iE_0T}, \qquad (8.8.32)$$

where R_0—the effect of interaction, is assumed to be small so that it can be expanded in powers of $\frac{e^2}{\hbar c}$. Writing $e^{-iRT} \approx 1 - iRT$, and expanding in powers of $\frac{e^2}{\hbar c}$, Feynman obtained the following expression for R_0:

$$R_0 = \lim_{T \to \infty} \frac{e^{iE_0T}}{T} \frac{e^2}{\hbar c} \int \int dx_0 \, dx_T \, \mathcal{D}(\text{paths})$$

$$\psi_0^*(x_T)e^{iS_P} \int_0^T \int_0^T \dot{x}_\mu(t)\dot{x}_\mu(s)$$

$$\delta_+((t-s)^2 - (x(t)-x(s))^2) \, ds \, dt \, \psi_0(x_0). \qquad (8.8.33)$$

This expression is symmetric in s and t. Assuming that s is later than t (and therefore multiplying the result by 2), and holding fixed the space-time points x_t, t and x_s, s, the integration over paths can be performed and yields the result

$$R_0 = \frac{2ie^2}{\hbar c} \lim_{T \to \infty} \frac{e^{iE_0T}}{T} \int dx_t \int dx_s \int_0^T \int_0^t ds$$

$$\psi_0^*(x_s, T - s)\dot{x}_\mu(s)K_0(x_s, s; x_t, t)\dot{x}_\mu(t)$$

$$\delta_+((t-s)^2 - (x_t - x_s)^2)\psi_0(x_t, t). \qquad (8.8.34)$$

Inserting into this expression the representation for δ_+ and K_0,

$$\delta_+(x_\mu^2) = -\frac{i}{2\pi} \int \frac{d^3k}{k} e^{i\mathbf{k}\cdot\mathbf{x}} e^{-ikt} \qquad (8.8.35)$$

and

$$K_0(2, 1) = \sum_n \psi_n(x_2)\psi_n^*(x_1)e^{-iE_n(t_2 - t_1)}, \qquad (8.8.36)$$

Feynman deduced that

$$R_0 = \frac{2e^2}{\hbar c} \sum_n \int \frac{d^3k}{2\pi k} \frac{(0 \mid e^{i\mathbf{k}\cdot\mathbf{x}} \dot{x}_\mu \mid n)(n \mid e^{-i\mathbf{k}\cdot\mathbf{x}} \dot{x}_\mu \mid 0)}{E_0 - E_n - \hbar ck}. \qquad (8.8.37)$$

The real part of R_0 gives rise to the infinite shift in energy levels; the imaginary part, which gives rise to a time dependence of the form $e^{+R_{im}T}$ in the transition amplitude, is the reciprocal lifetime of the state. The real part, upon neglecting the retardation factors $e^{i\mathbf{k}\cdot\mathbf{x}}$, is identical to Bethe's expression for the level shift. After mass renormalization, Feynman showed it to correspond to Bethe's formula for the Lamb shift. Formula (8.8.37) applies also to a Dirac particle, except in that case the velocity operators $\dot{x}(t)$ must be replaced by the Dirac $\boldsymbol{\alpha}$ matrices, the "velocity" operator for a Dirac particle, or equivalently $\dot{x}_\mu$ by γ_μ with the understanding that ψ^* goes over to the adjoint spinor. This correspondence was shown to be correct by noting that eq. (8.8.37) then yielded the expression for the second–order self-energy for a Dirac particle. Although the perturbation–theoretic formula Feynman obtained by the replacement $\dot{x}_\mu \to \gamma_\mu$ was anticipated, the precise structure depended on what was assumed for K in this situation. Stated differently, in the sum over intermediate states, different choices for K resulted in either the hole–theoretic result (all the negative energy states filled) or the result stemming from a one–particle theory.

With the hole-theoretical specification of the intermediate states, the formula (8.8.37) generalized to Dirac particles yielded the expression that Weisskopf (1939) had derived. In that expression the integral $\frac{d^3k}{k}$ over $\mathbf{k}$ space could be replaced by its equivalent $2\int d\omega\, d^3k\, \delta(\omega^2 - k^2)$, the integral being over all positive ω and all wave numbers $\mathbf{k}$. This step allowed Feynman to introduce his cutoff by replacing $\delta(\omega^2 - k)^2$ with $g(\omega^2 - k^2)$, with

$$g(\omega^2 - k^2) = \int_0^\infty [\delta(\omega^2 - k^2) - \delta(\omega^2 - k^2 - \lambda^2)]G(\lambda)d\lambda$$

$$\int_0^\infty G(\lambda)d\lambda = 1. \qquad \textbf{(8.8.38)}$$

The self-energy evaluated with $g(\omega^2 - k^2) = \delta(\omega^2 - k^2) - \delta(\omega^2 - k^2 - \lambda^2)$ (and with the intermediate states specified by hole theory) yielded a convergent result that depended logarithmically on the cutoff λ (Feynman 1948b).

These results were first obtained in a somewhat more haphazard fashion. Initially, Feynman derived his expression for the self-energy by eliminating the transverse photons and evaluating the contribution from the Coulomb interaction term—that is, of the longitudinal photons—separately. In his Nobel Prize speech, Feynman noted:

> I was very surprised to discover that it was not known at that time that every one of the formulas that had been worked out so patiently by separating longitudinal and transverse waves could be obtained from the formula for the transverse waves alone, if instead of summing over only the two perpendicular

polarization directions you would sum over all four possible directions of polarization. It was so obvious from the action $\left[\int j_\mu(x)\delta_+(x-y)^2 j_\mu(y)d^4x\, d^4y\right]$ that I thought it was general knowledge and would do it all the time. I would get into arguments with people, because I didn't realize they didn't know that. (Feynman 1966, p. 706)

Where matters stood at the end of the fall of 1947 can be gauged from a letter Feynman wrote to Bert and Mulaika Corben, with whom he had spent part of the summer in Pittsburgh: [113]

I have been working very hard recently so there has been no letter. But interesting things are piling up, so I thought I had better write some of them to you.

I sent my paper to the "Physical Review" and have not heard, as yet, about it but I have continued working with electrodynamics in the range of quantum mechanics which is described in the paper. You may remember I was able to eliminate explicit reference to the field oscillation in the equations of quantum mechanics. While I was working on this, there was so much talk around here about self-energy, that I thought it would be the easiest thing to calculate directly in my form. The result is exactly the same as one gets for ordinary perturbation theory (except for some nice simplification waves). It therefore also gives infinity. I then altered the delta function in the interaction to be a sum of less sharp function. This corresponds to a kind of finite electron. Then the self-energy of a non-relativistic particle is finite. Actually it comes out complex, the imaginary part represents the rate of radiation to the negative energy states. If I cause the negative energy states to be full, then the formation is no longer relativistically invariant and gives a finite self-energy to an electron, in fact all mass can be represented as electro-magnetic.

It therefore seems that I have guessed right, that the difficulties of electro-dynamics and the difficulties of the hole theory of Dirac, are independent and one can be solved before the other. I am now working on the hole theory, in particular, I now understand the Klein paradox, so that it is no longer a paradox and can tell you what an atom with a nuclear charge more than 137 would behave like, but I still haven't solved the whole problem. The main reason I am writing to you, is to tell you about this result which I feel is of very great significance.

It is very easy to see, that the self-energy of two electrons is not the same as the self-energy of each one separately.

That is because among the intermediate states which one needs in computing the self-energy of particle number 1, say, the state of particle 2 can no longer appear in the sum because a transition of 1 into the state of 2 is excluded by the Pauli exclusion principle. The amount by which the self-energy of two particles differs from the self-energy of each one separately, is actually the energy of their electrical attraction.[114] Therefore, the electro magnetic interaction between two particles can be looked upon as a correction to the self-energy produced by the exclusion principle. Thus Eddington is right in that it is a consequence of the exclusion principle.[115]

Finally, I have learned that the classical theory with a finite electron which is deduced from a principle of least action, can show the phenomenon of pair production. The action is made a minimum sometimes by a pair which reverses itself in time, in the way we have discussed often when I was there.[116]

On November 12, 1947, Feynman gave a seminar at the Institute for Advanced Study on "Dirac's Electron from Several Points of View." He had stopped in Princeton on his way to attend the tenth Washington Conference in Theoretical Physics which took place from Thursday to Saturday, November 13, 14, and 15, 1947. Arthur Wightman attended Feynman's lecture. His notes indicate that Feynman briefly presented the content of his *RMP* article and proceeded to derive eqs. (8.8.27) and (8.8.28) for the transition amplitude.[117] Explicit formulas were given for the S_{int} for the case of two interacting particles:

$$
S_{int}(x, y) = \frac{e_x e_y}{4\pi} \int_0^T \int_0^t (1 - \dot{\mathbf{x}}_{(t)} \cdot \dot{\mathbf{y}}_{(s)})
$$

$$
\left[\delta((t - s)^2 - (\mathbf{x}(s) - \mathbf{y}(t))^2 + \frac{i/\pi}{(t - s)^2 - (\mathbf{x}_t - \mathbf{x}_s)^2} \right] ds dt,
$$

$$
(8.8.39)
$$

and eq. (8.8.33) was derived "for the contribution to the energy shift," which Wightman recorded as

$$
\Delta = \frac{e^2}{4\pi T} \int_0^T \int_0^t \bar{\psi}_0 e^{iS'/\hbar}
$$

$$
(1 - \dot{\mathbf{x}}_t \cdot \dot{\mathbf{y}}_s)[\delta(t - s)^2 + \frac{i/\pi}{(t - s)^2}] \psi_0 \, dx_0 \dots dx_n. \qquad (8.8.40)
$$

During the rest of his lecture Feynman reviewed his various attempts to give an integral over path formulation for the Dirac electron. It is of interest to note that Feynman's original attempts to "derive" the Dirac equation involved space-time trajectories which had the property that the time parameter only increased along the paths. In his lecture at Princeton, the paths could now go backwards in time.

Dirac, who was visiting the Institute during that academic year, attended Feynman's seminar. Harish-Chandra, who at the time was a student of Dirac's, reported to Mrs. Corben that "Dirac is very impressed by Feynman and thinks he does some interesting things."[118]

From Princeton, Feynman went to Washington to attend the Washington Theoretical Physics Conference whose subject was "Gravitation and Electromagnetism." At the meeting Schwinger briefly lectured on the results he had obtained since the Shelter Island conference. Schwinger's comments were the only worthwhile report to the conference as far as Feynman was concerned. Writing to the Corbens on November 19, Feynman indicated to them:

> The meeting in Washington was very poor, don't quote me. The only interesting thing was something that Schwinger said at the end of the meeting. It was interesting because it got Oppy so excited but I did not have time to understand exactly what Schwinger had done. It has to do with the electro-magnetic self-energy problems. *One thing he did point out that was very interesting though, was that the descrepancy in the hyperfine structure of the hydrogen noted by Rabi, can be explained on the same basis as that of electro-magnetic self-energy, as can the line shift of Lamb* [emphasis mine]. The rest of the meeting was concerned with gravitation and the curvature of the universe and other problems for which there are very powerful mathematical equations—lots of speculation but very little evidence.
>
> I met Mrs. Schwinger and had hoped to come back to Princeton from Washington with them on the train. I was going to find out from Julie [*sic*; Julian] then, what he was trying to explain at the meeting.
>
> Unfortunately they did me dirt and did not come to Princeton. I stopped off at Princeton on my way back to Ithaca to talk to Pias [*sic*; Pais] and Bohm and used up all my time with Pias—unfortunately, because I also wanted very much to talk to Bohm.[119]

Schwinger's observation could readily be incorporated into Feynman's approach by considering S_p in eq. (8.8.31) the action for a charged particle in an external electromagnetic field. The transition matrix for radiationless scattering by the external field could then be calculated to first order in the external field and to order $\frac{e^2}{\hbar c}$ in the radiative corrections. This problem had just been reexamined by Lewis (1948). Lewis had redone Dancoff's calculation (Dancoff 1939) and had discovered that Dancoff had made an error (he had omitted certain matrix elements). Lewis found that after mass renormalization the amplitude for radiation-less scattering—although infrared divergent[120]—did not contain any ultraviolet divergences. Feynman calculated the amplitude for radiationless scattering using his photon cutoff and obtained the results to be found in his paper, "A Relativistic Cutoff for Quantum Electrodynamics" (Feynman 1948b).

By the middle of January 1949, just prior to the New York meeting of the American Physical Society, Feynman could report to the Corbens:

> In the last letter I wrote you, I made a mistake. As you know, I have been working with a theory of electricity in which the delta function interaction is replaced by a less sharp function. Then (in quantum mechanics) the self-energy of an electron in-cluding the Dirac hole theory comes out finite. The mistake in the last letter was to say that it is finite and not relativistically in-variant. Actually, the self-energy comes out finite and invariant and is therefore representable as a pure mass. The magnitude of the mass change is a fraction of the order of 1/137 times the log-arithm of the Compton wave length over the cutoff width of the delta function. Thus, all mass cannot be represented as electro-dynamic unless the cutoff is ridiculously short. The experimen-tal mass is of course the sum of inertia and this electromagnetic correction.
>
> I then turned to the problem of radiationless scattering which has always given such trouble in electrodynamics. I get the result that the cross section for scattering of a particle going past a nucleus without emitting a quantum is finite. If the cutoff is made to go to zero, the answer comes out infinite. If, however, the cross section is first expressed in terms of the experimental mass and then the cutoff is made to go to zero keeping the exper-imental mass as a constant, when the limit is taken, the result is finite. This therefore agrees with the result of Lewis and Oppen-heimer. I believe it also confirms the idea of Schwinger because I think that the terms which diverge logarithmically as the cutoff goes to zero are just the terms that Schwinger said one should subtract in a consistent electrodynamics.
>
> I have not computed the self-energy to second order,[121] I only hope it is also finite. If so, I think all the problems of elec-trodynamics can be unambiguously solved by this process: First compute the answer which is finite (but contains the cutoff log-arithm). Then express the result in terms of the experimental mass. The answer still contains the cutoff but this time not log-arithmically. Take the limit which now exists, as the cutoff goes to infinity.

I have not mentioned polarization of the vacuum for as yet, I do not completely understand the problems in which it appears. However, a calculation of the phenomenon also gives a finite answer for the polarizing of the electric charge. However, unfortunately for reasonable cutoff, the polarizability is very large as far as I can see, so that things do not looks as nice as they do for self-energy.

I am very excited by all this of course, because I think that the problem is at least solved either by my way or Schwinger's. I hope to prove the equivalent or at least to compare the two ideas shortly.[122]

At the New York meeting of the American Physical Society at the end of January 1948, Schwinger gave an invited lecture on "Recent Developments in Quantum Electrodynamics." This lecture had to be repeated a second time because of the vast number of people unable to hear him the first time. In his address, Schwinger reported on his calculation of the anomalous magnetic moment of a free electron and his initial results for the Lamb shift. He also indicated that he had encountered some difficulties: the calculated value of the anomalous magnetic moment of an electron in a Coulomb field did not agree with the value $\frac{\alpha}{2\pi}$ he had calculated for an electron in a magnetic field. The reason for the discrepancy was that Schwinger's calculational methods were not relativistically invariant (Schwinger 1983a, pp. 335–336).

After Schwinger's talk, Feynman got up and reported that he had computed all the things Schwinger had and that he agreed with Schwinger's results, except that he had found the magnetic moment of an electron in an atom also to be $(1 + \frac{\alpha}{2\pi})\frac{e\hbar}{2mc}$. If the calculation maintains proper covariance there was really no difficulty. Feynman "wasn't trying to show off"; he was trying to tell Schwinger that "he had caught up."[123] To those close to Feynman at the time it was evident that Schwinger was "at least his personal competitor."[124] Leonard Eyges, who was a graduate student in theoretical physics at Cornell from 1946 to 1949, recalls that "once when apparently he [Feynman] and Schwinger were doing the same calculation (magnetic moment of the electron?) he was with Bethe acting like a hyperactive child: (Do you suppose Schwinger has this? Do you suppose Schwinger is ahead?) Feynman asked Bethe, until even the patient and unperturbable Bethe had to tell him in effect, for God's sake, desist!"[125] On March 20, he wrote the Corbens:

I have been working on my little theory of electrodynamics in which the interaction is not exact on a delta function because there was some confusion in the Schwinger-Weisskopf-Bethe Camp as to what the correct answer was for the line shift. I worked that out in detail, my way. I find a shift in the magnetic

moment of an electron equal to $e^2/2\pi\hbar c$. The line shift in hydrogen has two terms, one a logarithmic one proportional to the
expected form of $\nabla^2 V$ and the other is a correction to the spin orbit interaction. This correction is exactly the same as the amount
that you would calculate from the change in the magnetic moment, that is, everything is nicely relativistically invariant. The
calculation took me four days and can be put neatly on four pages
of paper. Now that I understand it, it is really a very simple problem. What I did, was to compute the change in the Dirac Hamiltonian due to the fact that an electron can emit and absorb virtual
quanta when it is in a slowly varying external potential. If ϕ and
A are the scaler and vector potential in a problem, the correction
to the Hamiltonian is the

[no formula was inserted in the carbon copy of the letter] answer diverges for very low energy quanta, so I have expressed it
in terms of k_{min} which is the slowest momentum quanta which
have been included. This avoids the infrared catastrophe and the
low energy non-relativistic end can be worked out in a straightforward way, such as has already been done by Bethe. The actual
shift comes out around 1040 megacycles, I think.

   My theory of representing positrons as electrons going
backward, is working very well but nobody believes me because
I haven't got everything complete yet. I can only deal with pair
production and annihilations in a complete fashion. Polarization
of the vacumn still remains somewhat of a puzzle, it has not been
included in the above formula for ΔH.[126]

Feynman's letter indicates that by the spring of 1948 he had fully incorporated "Wheeler's old idea about going backward in time being positrons" into
his formulation of quantum electrodynamics.

Theory of Positrons

   There exists a set of notes, with the title "Theory of Positrons" and probably written in late 1947, which outlines Feynman's formulation of the theory as
of that time. The notes begin with the statement: "We shall consider that when
an electron travels along as proper time increases so does true time. For a positron
proper time increases as true time decreases. This is classically. Quantum mechanically the situation is that the wave function has a phase (in $e^{-i\phi}$ define ϕ as phase)
which increases as you move in positive true time for an electron, and increases in
negative true time for a positron."[127]

   Denote by $\psi(x)$, the amplitude for an electron to arrive at x; $\psi(x)$ contains
only positive energy components. "If a positron were to arrive at x, it would come

from a wave from the future of x & would give a $\psi(x)$ with only negative energy components."

Feynman then considered the action of a potential at ☒ and noted that "Whereas Dirac says ☒ sends an electron initially at A into states of positive and negative energy states both of which spread upward in time." We [Feynman] say instead that ☒ scatters a wave B toward future representing scattered electron, and a wave C toward past representing ... a positron with which the electron may have annihilated by action of potential at ☒ "[128] (see photograph 21).

The amplitude (scattered) arriving at (2) (electron) is (to first order in $\not{A}$)

$$\int \not{S}(2, x) \frac{i}{\hbar} \not{A}(x)\psi(x)d^4x$$

$$\not{A} = \gamma_\mu A_\mu.$$

The amplitude for the positron arriving at 3 is

$$\int \not{S}(3, x) \frac{i}{\hbar} \not{A}(x)\psi(x)d^4x,$$

where

$$\not{S}(A, B) = (\not{\nabla} + \mu)I(A, B); \not{\nabla} = \gamma_\mu \frac{\partial}{\partial x_\mu}, \tag{8.8.41}$$

with $I(A,B)$ a complex time-symmetrical solution of $(\Box^2 - \mu^2)I(A, B) = 0$ having the property that it only has positive frequency components for $t_A - t_B > 0$ and negative frequency components where $t_A - t_B < 0$.

Feynman then obtained the explicit representation for I:

$$I(1, 2) = \int d^4p\, e^{-ip\cdot x} \frac{1}{p_\mu p_\mu - \mu^2}, \tag{8.8.42}$$

"where μ^2 is considered to have an infinitesimal negative imaginary part."[129] Thus the Fourier transform of the $\not{S}$ operator is $\frac{\gamma_\nu p_\nu + \mu}{p_\mu^2 - \mu^2 + i\delta}$. It is clear that the approach is based upon the perturbative expansion

$$\not{S}_A(x, y) = \not{S}(x, y) + \int \not{S}(x, x') \frac{i}{\hbar} \not{A}(x') \not{S}(x', y)d^4x' + \dots \tag{8.8.43}$$

for what will be called the propagator. Analyzing the second-order contribution, Feynman noted that the term $A\!\!\!/S A\!\!\!/S$ includes the contribution from *both* of the diagrams indicated in figure 8.8.2.

Upon comparing the contribution of these two terms with those obtained from hole-theoretic perturbation theory, Feynman found that the contribution from diagram (b) correctly incorporated a minus sign which in hole theory was accounted for from the contribution of the diagram (fig. 8.8.3).

Feynman explained the result as follows:

> The minus sign on the last term arises as a result of the properties of $S\!\!\!/(y, x)$. In ordinary theory it is interpreted that at x a hole is made in sea, electron going to T, while hole is filled by electron S. This represents an exchange (relative to the first process where same electron S gets to T) of electron S and sea electron so by exclusion principle feeds with neg. amplitude.
>
> According to the present theory state T and S might be the same in 2nd term. Usual theory says no because then at time between t_y, t_y can't have 2 electrons in same state. We say

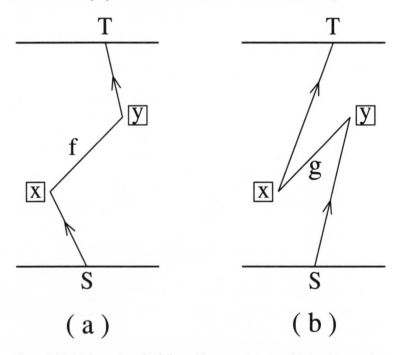

Figure 8.8.2 (a) Scattering of x followed by scattering at y. (b) A pair created at x, the positron of which annihilates the electron at y, and the electron of which is found at I.

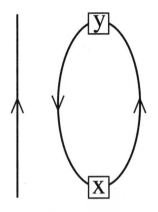

Figure 8.8.3 The vacuum diagram that must be included in hole theory.

it is the same electron so Pauli principle doesn't operate. Old theory has such a term anyway for it contemplates pair created by x annihilated by y (vac. polarization type) one of which is excluded if electron is in state S—namely the pair created at x (& destroyed at y) whose electron is in S. So term is subtracted relative to (infinite) vacuum if electron is at S. This is same term as we have so both theories give same result here.[130]

Feynman convinced himself of the validity of his formulation by comparing its results with those obtained from hole-theoretic perturbation theory. No justification was given—in late 1947—why "bubble diagrams" need not be considered. The fact that the "correct" expressions for the scattering amplitudes were obtained with the exclusion of "closed loops" reinforced Feynman's predilection not to consider "closed loops."

In January 1949, Feynman presented a paper at the New York meeting of the APS (Feynman 1949a) entitled "The Theory of Positrons." His paper was the fifth one given at Session T of the meeting. Feynman had carefully prepared his talk, and a manuscript entitled "T5. Theory of Positrons"[131] is to be found among his papers at the Millikan Library Archives. Although the paper was delivered in January 1949, its content reflects Feynman's thinking of one year earlier. The manuscript is of great interest because it contains, besides the famous "bombardier" metaphor[132] of his "Theory of Positrons" paper (Feynman 1949b), other metaphors (including the one about the letter N) to illustrate his notions of positrons as electrons moving backward in time. It also gives us a glimpse of Feynman in a more philosophical vein. After a brief historical sketch of Dirac's hole theory, Feynman elaborated:

> One of the disadvantages of this [hole] theory is that even the simplest processes become quite complicated in its analysis. One must take into account besides the limited number of real particles, the infinite number of electrons in the sea. The present work results from a reinterpretation of the Dirac equation so that this complexity is not required.
>
> It results from a different mode of representation of the phenomena of pair production. We can discuss it by a simple model.

Suppose a black thread be immersed in a cube of collodion, which is then hardened. Imagine the thread, although not necessarily quite straight, runs from top to bottom. The cube is now sliced horizontally into thin square layers, which are put together to form successive frames of a motion picture. In each frame will appear a black dot, the cross-section of the thread, which will move about in the movie depending on the waverings of the thread. The moving dot can be likened to a moving electron. How can pair production be visualized?

Suppose the thread did not go directly top to bottom but doubled back for a way (somewhat like the letter N with the straight parts extended to top and bottom). Then in successive frames first there would be just one dot but suddenly two new ones would appear when the frames come from layers cutting the thread through the reversed section. They would all three move about for a while when two would come together and annihilate, leaving only a single dot in the final frames. In this way new phenomena of pair creation and annihilation can be represented. They are similar to the simple motion of an electron (and are thus governed by the same equations), but correspond to a more tortuous path in space and time than one is used to considering. In common experience the future appears to us to develop out of conditions of the present (and past). The laws of physics have usually been expressed in this form. (Technically, in the form of differential equations, or "Hamiltonian Form"). The formulae tell what is to be expected to happen if given conditions prevail at a certain time. The author has found that the relations are often very much more simply analyzed if the entire time history be considered as one pattern: The entire phenomena is considered as all laid out in the four dimensions of time and space, and that we come upon the successive events. This is applied to simplify the description of the phenomena of pair production in the present paper. A bombardier watching a single road through the bomb-sight of a low flying plane suddenly sees three roads, the confusion only resolving itself when two of them move together and disappear and he realizes he has only passed over a long reverse switchback of a single road. The reversed section represents the positron in analogy, which is first created along with an electron and then moves about and annihilates another electron.

The relation of time in physics to that of gross experience has suffered many changes in the history of physics. The obvious difference of past and future does not appear in physical time for microscopic events (the connection of the laws of Newton and of statistical mechanics). Einstein discovered that the present is not the same for all people. (For those in motion it corresponds to cutting the same collodion cube at slight angle from the horizontal). It may prove useful in physics to consider events in all of time at once and to imagine that we at each instant are only aware of those that lie behind us.

The complete relation of this concept of physical time to the time of experience and causalilty is a physical problem which has not been worked out in detail. It may be that more problems and difficulties are produced than are solved by such a point of view. In the application to the description of positrons it should be emphasized that there still appear to be difficult unsolved problems and that the proposed viewpoint may eventually not prove to result in as much simplification as it appears to do at first sight.[133]

The propagator $\mathcal{S}$ was first introduced in the calculation of the self-energy. Feynman noted that the propagator defined as a sum over positive energy states

$$\mathcal{S}(2, 1) = \sum_{\text{pos } E_n} \phi_n(2)\bar{\phi}_n(1) \exp(-iE_n(t_2 - t_1)) \tag{8.8.44}$$

for $t_2 > t_1$ and the negative sum over negative energy states

$$= -\sum_{\text{neg } E_n} \phi_2(2)\bar{\phi}_n(1) \exp(-iE_n(t_2 - t_1)) \tag{8.8.45}$$

for $t_2 < t_1$ yielded the correct hole-theoretic expression for the self-energy. Feynman recognized that the use of $\mathcal{S}$ in place of the nonrelativistic propagator K_0 and the replacement of $\dot{x}_\mu$ by γ_μ yielded the correct hole-theoretic results in perturbation theory. The "Theory of Positron" notes of 1947 represent a formalization of these previously obtained results, applied to the case of charged spin 1/2 particle in an external field. Feynman confirmed this. He recalled that the negative energy states had always given him "trouble." He made a project imagining what would happen if an electron's space-time trajectory were like the letter N in time, "they would back up for a while and then go forward again" and found that he obtained the right formulas for the "positron end of the cases." He proceeded to make empirical rules about what sign should be given to particular terms, by doing more and more complicated problems and comparing the results with those of standard perturbation theory. He thus developed empirical rules for "computing every thing."[134]

Diagrams evolved as a shorthand to help Feynman translate his integral-over-path perturbative expansions into the expressions for transition matrix elements being calculated. In his interview with Charles Weiner in 1966, Feynman remembered that "I was working on the self-energy of the electron, and I was making a lot of these pictures to visualize the various terms and thinking about the various terms, that a moment occurred—I remember distinctly—when I looked at these, and they looked very funny to me. They were funny-looking pictures. And I did think consciously: Wouldn't it be funny if this turns out to be useful, and the *Physical Review* would be all full of these funny-looking pictures? It would be very amusing."[135] Feynman repeated these remarks in 1981. He then recalled "fantasizing" while drawing diagrams and saying to himself, "Wouldn't it be funny if these diagrams were to become really useful."[136]

8.9 Renormalization

As noted in chapter 2, the quantum electrodynamics of Dirac (1927b,c) and Heisenberg and Pauli (1929, 1930), had been very successful in describing experimental phenomena to lowest order in $e^2/\hbar c$. Higher-order corrections, however, gave rise to infinite divergent integrals. For example, the attempt to calculate the shift in the energy of an electron in a Coulomb field due to its intereaction with the quantized electromagnetic field led to an infinite answer (Oppenheimer 1930a). Even a free electron received an infinite correction, δm, to its rest energy, so that it would appear to have an infinite electromagnetic mass. In a consistent, finite theory the experimentally observed mass, m_{exp}, should be equal to the "mechanical" mass, m_{mech}, the mass that is put into the equation of motion of the electron, plus the correction δm, $m_{exp} = m_{mech} + \delta m$. (There is of course no experimental way to determine how much m_{exp} is m_{mech}.)

It is clear that if the electron has an infinite correction to the mass, other difficulties arise. For example, the energy of the ground state of hydrogen becomes $\frac{1}{2}(m_{mech} + \delta m)\frac{e^4}{\hbar^2} + \ldots$. Similarly, one would expect that the corrections to Rutherford scattering would result in a cross section which vanishes since no deflection is expected by an electron of infinite inertia.

It was Kramers' insight to note that if the single difficulty of the infinite self-energy of a free electron could be surmounted, all the other problems of the quantum electrodynamics of nonrelativistic charged particles would be solved at the same time. When Bethe (1947) calculated the Lamb shift, he assumed that it was due to the reaction of the electron to its own electromagnetic field. By comparing the divergent integrals he had calculated for an electron in a Coulomb field with those obtained in calculating the self-energy of a free electron, he was able to identify the divergent terms that were giving rise to the simple effect $m_{exp} = m_{mech} + \delta m$. These he discarded, because they had already been included when using m_{exp} instead of m_{mech} in calculating the Rydberg. Bethe argued that the remaining terms would be finite in a relativistic theory and obtained a value of about

1000 megacycles for the shift (after introducing appropriate cutoffs). Bethe's calculation suggested that Kramers' method might be general: all such divergences would be removed by renormalizing the mass.

The difficult problem of determining in an unambiguous manner which terms represented this mass effect in a fully relativistic treatment of quantum electrodynamics was first solved by Schwinger (1948b,e,f). He cast quantum field theories into a form in which the relativistic invariance of the terms was transparent. This reformulation had previously been independently invented by Tomonaga (1943b, 1946). From the clues suggested by the invariance properties of various terms, Schwinger was able to argue the elimination of certain divergent integrals in quantum-electrodynamical calculations because they represented *unobservable* mass and charge renormalizations (Schwinger 1948b,e,f, 1949a,b).

Feynman tried to solve these problems in a different way. He assumed that the laws of electrodynamics were altered at very short distances (or more precisely, at short proper times). The result of this modification is to produce a finite, relativistically invariant answer to all problems in quantum electrodynamics, including the rest mass of an electron.

The value of the mass correction, δm, depends on the type of alteration that is made (i.e., on the form of the function f in eq. 8.8.16). It depends logarithmically on the short distance, a, at which the effect of the modification becomes appreciable. All other processes also depend in this way on a if expressed in terms of the mechanical mass, m_{mech}. However, when expressed in terms of the experimental mass, m_{exp}, they are very insensitive to the exact value of a, if a is much less than the electron Compton wavelength. In fact, there is a definite limit for all observable processes as the cutoff length goes to zero. It was in this manner that Feynman computed the Lamb shift and the other results described in his letters to the Corbens and in his article "Relativistic Cut-Off for Quantum Electrodynamics" (Feynman 1948b). Vacuum polarization processes, however, could not be made finite by altering photon propagators. Vacuum polarization effects (to order $e^2/\hbar c$) arise because the potential, instead of scattering a charged particle directly (fig. 8.9.1a), can do so by first creating a pair that subsequently annihilates, creating a photon that does the scattering (fig. 8.9.1b).

The contribution from fig. 8.9.1b is divergent. How to cut off (and therefore circumvent) this divergence in a *gauge-invariant* fashion gave Feynman a great deal of difficulty until the beginning of 1949. (His views regarding vacuum polarization will be presented in section 8.11.) What he knew in early 1948 (Feynman 1948b,f) was that the divergent part of the correction from fig. 8.9.1b had the same structure as the contribution from fig. 8.9.1a and that their sum could be interpreted as though the potential were of another strength. Equivalently, if one thinks of the potential as being created by a charge, one can introduce a bare charge e_{bare}, which appears in the equation of motion and define Δe as the correction from fig. 8.11b. The "experimental" charge $e_{exp} = e_{bare} + \Delta e$ defines a charge "renormalization" in a manner analogous to mass renormalization. All observable quantities are finite when expressed in terms of e_{exp}.

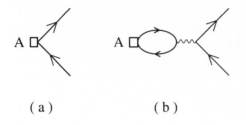

where matters stood in the spring of 1948 can be inferred from Feynman's lecture at the Pocono conference, which took place from March 30 to April 1, 1948. It is to this presentation that we now turn.

(a)   (b)

Figure 8.9.1

8.10 The Pocono Conference: March 30–April 1, 1948

Feynman's presentation of his "Alternative Formulation of Quantum Electrodynamics" at the Pocono conference followed Schwinger's extended exposition of his version of QED, which had lasted almost the whole day.[137] In after-dinner remarks, on the occasion of Schwinger's sixtieth birthday, Feynman recalled the meeting:

> Each of us [Schwinger and I] had worked out quantum electrodynamics and we were going to describe it to the tigers. He described his in the morning, first, and then he gave one of these lectures which are intimidating. They are so perfect that you don't want to ask any questions because it might interrupt the train. But the people in the audience like Bohr, and Dirac, Teller, and so forth, were not to be intimidated, so after a bit there were some questions. A slight disorganization, a mumbling, confusion. It was difficult. We didn't understand everything, you know. But after a while he got a good thing. He would say, perhaps it will become clearer if I proceed, so he continued.[138]

While Schwinger was lecturing, Bethe had noted that no one in the audience was giving him any difficulty as long as his presentation was formal and mathematical. However, as soon as he lapsed into a physical argument, he would be interrupted and heated discussions would ensue. Bethe therefore advised Feynman to present his formalism "mathematically" rather than "physically."[139] This Feynman proceeded to do with dismal consequences.

Feynman was prepared to present "this whole thing backward . . . not formally . . . with all physical ideas starting from path integrals." Instead he gave a presentation that emphasized the mathematical aspects based on a formalism that was totally unfamiliar to his audience. "I had too much stuff," Feynman points out. "My machines came from too far away." And so he felt that his lecture was a complete failure, "a hopeless presentation."[140]

Feynman began his exposition of his formulation of the laws of quantum electrodynamics with an outline of his calculus of ordered operators:

> A formal solution of the time-dependent Schrödinger equation may be written
>
> $$\psi(x,t) = e^{-\frac{i}{\hbar}Ht}\psi(x,0) \tag{8.10.1}$$
>
> when the operator H does not depend explicitly on time, or as
>
> $$\psi(x,t) = e^{-\frac{i}{\hbar}\int_0^t H(t')dt'}\psi(x,0) \tag{8.10.2}$$
>
> where the Hamiltonian H may depend on time.
> If there be a perturbing potential $V(t)$; i.e. the new Hamiltonian is $H + V(t)$ we have
>
> $$\psi(x,t) = e^{-\frac{i}{\hbar}(\int_0^t H(t')dt' + \int_0^t V(t')dt')}\psi(x,0) \tag{8.10.3}$$
>
> and this leads to the usual quantum mechanical perturbation theory.[141]

Feynman had devised a calculus to handle situations in which the order of the operators is of importance, and was thus able to give meaning to formal solutions such as those on the right-hand side of (8.10.1) and (8.10.3). The calculus was based on a simple notational device whereby the operators are given labels that indicate the order in which they are to be applied. Thus $A_2 B_1 \psi$ means B acts first on ψ, then A; that is, $A_2 B_1 \psi = AB\psi$, since the labels keep track of the order. Similarly $B_2 A_1 \psi = BA\psi$.

To determine the solution $\psi(t)$ at $t = T$ of $i\hbar\partial_t\psi(t) = H(t)\psi(t)$ given $\psi(0)$ at $t = 0$, break up the interval from 0 to T into n steps of length Δt over each of which $H(t)$ is nearly constant. Then clearly,

$$\psi(T) = e^{-iH(t_n)\Delta t_n}e^{-iH(t_{n-1})\Delta t_{n-1}}\ldots e^{-iH(t_1)\Delta t_1}\psi(0)$$
$$= \prod_{i=1}^{n} e^{-H(t_i)\Delta t_i}\psi(0). \tag{8.10.4}$$

The order in which the various $H(t_i)$ are to operate is specified by the value t_i, so that those with smaller t_i act on ψ before those with larger t_i: the $H(t_i)$ are ordered operationally as they are ordered temporally. Hence in the limit as $\Delta t \to 0$, the solution can be written in the form

$$\psi(T) = e^{-i\sum_{i=1}^{n} H(t_i)\Delta t_i}\psi(0)$$

$$= e^{-i\int_0^T H(t)dt}\psi(0), \tag{8.10.5}$$

where this last expression is to be interpreted in terms of the operational symbolism.

Applying this formalism to the Dirac equation in an external field $A_\nu(x)$,

$$\gamma_\nu(i\hbar\partial_\nu - \frac{e}{c}A_\nu)\psi = m\psi, \tag{8.10.6}$$

Feynman defined

$$\psi(x, w) = \exp\{+\frac{i}{\hbar}\int_0^w (i\hbar\slashed{\nabla}_{w'} - \frac{e}{c}\slashed{A}_{w'})dw'\}\psi(x, 0), \tag{8.10.7}$$

where x stands for the coordinates $xyzt$ of the particle and w is a parameter. The right-hand side of eq. (8.10.7) is understood in the notation of the ordered calculus, that is, the γ's, the $\frac{\partial}{\partial x_\nu}$, and the $A_\nu(x(w))$ have a label attached to them that orders sequentially their operation on $\psi(x, 0)$. The so-defined $\psi(x, w)$ satisfies the equation

$$-i\hbar\frac{\partial\psi(x, w)}{\partial w} = (i\hbar\slashed{\nabla} - \frac{e}{c}\slashed{A})\psi(x, w). \tag{8.10.8}$$

Since the Dirac equation for a particle with a well-defined mass is given by eq. (8.10.6), one must demand that

$$-i\hbar\frac{\partial\psi}{\partial w} = m\psi, \tag{8.10.9}$$

that is, that ψ be periodic in w:

$$\psi(x, w) = e^{+\frac{i}{\hbar}mw}\chi(x). \tag{8.10.10}$$

In other words, if eq. (8.10.8) is taken as a new relativistic equation for a spin 1/2 particle, the physical solutions are those which are periodic in w and will then correspond to a particle with mass m. Hence if $\psi(x, w)$ is any special solution of eq. (8.10.8), a solution of eq. (8.10.9) is obtained by finding

$$\int_0^\infty e^{-imw}\psi(x,w)dw = \int_0^\infty dw e^{-imw}\exp\{+\frac{i}{\hbar}\int_0^w(\frac{\hbar}{i}\nabla_w - \frac{e}{c}A_{w'})dw'\}\psi(x).$$

$$(8.10.11)$$

The right-hand side of eq. (8.10.11) will then be a solution of the Dirac equation eq. (8.10.6) for a particle with mass m. The integral in the right-hand side of eq. (8.10.11),

$$\int_0^\infty dw\exp\{+\frac{i}{\hbar}\int_0^w(i\hbar\nabla(w') - \frac{e}{c}A(w') - m)dw',$$

is the ordered operator representation of the Feynman propagator for a Dirac particle

$$\frac{1}{i(i\hbar\nabla - \frac{e}{c}A - m)} = \int_0^\infty \exp[i(i\hbar\nabla - \frac{e}{c}A - m)w]dw, \qquad (8.10.12)$$

where the mass m is assumed to have a small negative imaginary part.[142] For a system of n Dirac particles, the ordered operator $\exp -\frac{i}{\hbar}\int^t H(t')dt'$ in (10.5) thus corresponds to

$$\int_0^\infty dw_1 \ldots \int_0^\infty dw_n \exp\left[\frac{i}{\hbar}\sum_n \int_0^{w_n}(i\hbar\nabla(w'_n) - \frac{e}{c}A(w'_n) - m)dw'_n\right]$$

Feynman then wrote down the analogue of eq. (8.8.27) for a system of charged Dirac particles interacting via the quantized electromagnetic field when no real photons are present initially or finally. It is a formula for the wave function (considered as a function of the variables $w_1, w_2 \ldots w_n$) at "time" $w_1 \ldots w_n$ in terms of the wave function at "time" $w_1, \ldots w_n = 0$:

$$\psi(x_1, \ldots x_n, w_1, w_2, \ldots w_n)$$

$$= \exp\left\{ +\frac{i}{\hbar}\sum_n [-\int_0^{w_n}\frac{\hbar}{i}\nabla(w'_n)dw_{n'} - \int_0^{w_n}\frac{e}{c}A(x(w'_n))dw'_n]\right.$$

$$\left. -\frac{1}{2}\frac{i}{\hbar}\sum_{n,m}\frac{e_n e_m}{c}\int_0^{w_n}\int_0^{w_m}\gamma_\mu^{(n)}(w'_n)\gamma_\mu^{(m)}(w'_m)\delta_+([x_\mu^{(m)}(w'_m) - x_\mu^{(m)}(w'_n)]^2)dw'_m dw'_n\right\}$$

$$\cdot \psi(x_1, x_2, \ldots x_n, 0, \ldots 0) \qquad (8.10.13)$$

The right-hand side of eq. (8.10.13) is the generalization to the case of Dirac particles of Feynman's result for the elimination of the transverse photons in the expression for the transition amplitude. The first two terms in the exponential are the propagators for the particles in an external field A, and the third term is the interaction action obtained from eq. (8.8.28) when the velocity for the nth particle, $\dot{x}_{\mu}^{(n)}$ is replaced by the corresponding $\gamma_{\mu}^{(n)}$ matrix. Eq. (8.10.13) was essentially the first formula Feynman wrote down at Pocono. This same equation (8.10.13) integrated on both sides with

$$\int_0^\infty dw_1 e^{\frac{-i}{\hbar} mw_1} \dots \int_0^\infty dw_n e^{\frac{-i}{\hbar} mw_n},$$

so as to yield wave functions for Dirac particles of mass m, is the *last* formula in Feynman's *concluding* paper on quantum electrodynamics (eq. 75 in Feynman 1951a; see also n. 19 of that paper). It is derived there starting from the conventional formulation of quantum electrodynamics and can be considered the culmination and encapsulation of his formalism.

At Pocono, however, Feynman did not have a "complete formal derivation" for eq. (8.10.13). He recalled that Dirac interrupted him and asked him, "Is it unitary?" As he was explaining "how [he] was going to work out positrons and so on," Feynman remembered Schwinger's trick and said, "Perhaps it will become clear as we proceed." "But Dirac was not put off, and like the Raven kept saying, 'Is it unitary?'" Not being quite sure what Dirac meant, Feynman asked, "Is what unitary?" To which Dirac replied, "Is the matrix that carries you from past to future unitary?" Feynman wasn't quite sure what Dirac meant by "the matrix," nor exactly what Dirac understood by unitary in connection with eq. (8.10.13). "Since I had tracks going backwards and forwards in time I don't know whether it is unitary"[143] was Feynman's reply. He felt that probably his answer was not quite satisfactory.

Feynman then considered the case of a free electron ($A = 0$) with "the goal ... to find what is the permanent effect of the interaction."[144] To lowest order in $e^2/\hbar c$,

$$\psi(x, w) = e^{\frac{-i}{\hbar} \int_0^w \frac{\hbar}{i} \nabla(w)dw} \psi(x, 0)$$

$$+ \exp\left\{ -\frac{i}{\hbar} \int_0^w \frac{h}{i} \nabla(w')dw' \right\}$$

$$\dots \left\{ \frac{e^2}{i\hbar c} \int_0^w dw' \int_0^w dw'' \gamma_\mu(w')\gamma_\mu(w'')\delta_+(s_{w'w''}^2) \right\} \psi(x, 0).$$

$$(8.10.14)$$

The effect of the second term is proportional to the elapsed time. If it corresponds to a change in mass from m to $m + \delta m$, it will be proportional to

$$+iwe^{+imw}\delta m\psi(w,0)$$

and hence

$$e^{+\frac{i}{\hbar}wm}\delta m = \frac{i\hbar}{w}\{\exp\frac{i}{\hbar}\int_0^w \nabla(w')dw'\}$$

$$\bullet \frac{e^2}{i\hbar c}\int_0^w\int_0^w dw'dw''\gamma_\mu(w')\gamma_\mu(w'')\delta_+(s^2_{w'w''}). \qquad \textbf{(8.10.15)}$$

Upon writing

$$\int_0^w = \int_0^{w'} + \int_{w'}^{w''} + \int_{w''}^w, \qquad \textbf{(8.10.16)}$$

ordering the terms according to their chronological parameter and carrying out the integrations, Feynman obtained the result

$$\delta m = \int \frac{\gamma_\mu(\not p + \not k + m)\gamma_\mu}{p^2 + k^2 - (m - i\epsilon_1)^2}\frac{d\omega_k}{k^2 - i\epsilon_2}$$

$$d\omega_k = \frac{dk_4dk_3dk_2dk_1}{2\pi^2 i}, \qquad \textbf{(8.10.17)}$$

which "is the standard formula for the self-energy."[145] It diverges logarithmically. Upon introducing his cutoff, that is replacing the δ_+ function by

$$f_+(s^2) = \int e^{ik_\mu x_\mu}\frac{d\omega_k}{k^2 - i\epsilon}\int_0^\infty \frac{-\lambda^2}{k^2 - \lambda^2}G(\lambda)d\lambda, \qquad \textbf{(8.10.18)}$$

where $\int_0^\infty G(\lambda)d\lambda = 1$, δm is finite but depends logarithmically on the cutoff

$$\delta m = m\left(\frac{3}{2}\ln\frac{\lambda}{m} - \frac{1}{2}\right). \qquad \textbf{(8.10.19)}$$

Feynman also indicated the equivalence of his approach to the conventional one. He derived the usual expression for the self-energy from eq. (8.10.17) by integrating over k_4, the location of the poles being stipulated by the $i\epsilon$ factors.

Next, Feynman pointed out that the term that arises in the discussion of the self-energy

$$\int \gamma_\mu \frac{\not{p} + \not{k} + m}{(p^2 + \not{k}^2) - (m - i\epsilon)^2} \gamma_\mu \frac{d\omega_k}{k^2}$$

is the Fourier transform of

$$\bar{\psi}(2)\gamma_\mu f_+(2, 1)S_+(2, 1)\gamma_\mu \psi(1)$$

in which $S_+(2, 1)$ is the amplitude for arrival at 2 of an electron given out 1, $f_+(2, 1)$ is the electromagnetic disturbance at 2 due to charge at 1, and that this term could be represented diagrammatically as in figure (8.10.1). Feynman then gave an exposition of his theory of positrons, as outlined in the notes previously referred to. The diagrams to be found in these notes were used to illustrate the difference between Feynman's and Dirac's hole theory. Feynman stressed that his formalism was rooted in an approach that computed transition amplitudes where the data are specified at time 0 and time T: "We cannot find the amplitude for a negaton to be at x merely by knowing the amplitude for the negaton to be at a previous time; we have also to know the amplitude for a positron to be at a later time."[146] Feynman went on to explain how in his formalism one didn't have to worry about the Pauli principle in intermediate states. At this point Teller interrupted him and inquired, "You mean that helium can have three electrons in the S state for a little while?" To which Feynman replied, "Yes." "That was chaos," Feynman recalled.[147] The notetaker recorded, "The Exclusion Principle Comes Out Automatically," and also noted that "Bohr Has Raised The Question As To Whether This Point Of View Has Not The Same Physical Content as The Theory Of Dirac, But Differs in A Manner Of Speaking Of Things which Are Not Well-Defined Physically."[148]

Feynman then asserted that the effect of an external electromagnetic field could be included by taking the propagator in the transition amplitude to be

$$\exp \frac{i}{\hbar}[-\frac{\hbar}{i} \int \not{\nabla}_w dw - \frac{e}{c} \int \not{A}(w)dw + \frac{e^2}{c} \int \int \gamma_\mu(w')\gamma_\mu(w'')f_+(s^2_{w'w''})dw'dw''].$$

To lowest order in the external field, and to order $\frac{e^2}{hc}$, three terms contribute. The "straddle or central term"[149] for the transition between states $\not{p}_2$ and $\not{p}_1$ was shown by Feynman to be given by

$$\int \gamma_\mu \frac{1}{(\not{p}_2 + \not{k} - m)} \gamma_\nu \frac{1}{(\not{p}_1 + \not{k} - m)} \gamma_\mu \frac{d\omega_k}{k^2}.$$

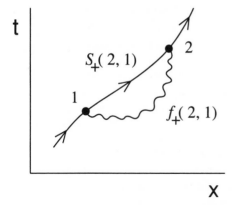

Figure 8.10.1 Self-energy diagram as drawn in Wheeler's Pocono notes.

Transformed to coordinate space it corresponded to $\bar\psi(2)\gamma_\mu \not{S}_+(3,2)$ $\not{A}(2)\not{S}_+(2,1)\gamma_\mu \psi(1)f_+(3,1)$ and could be represented by the diagram in figure 8.10.2a. This term diverges without the cutoff (i.e., with δ_+ instead of f_+). Upon the inclusion of two self-energy terms (figs. 8.10.2 b,c), the three terms combine to give a finite answer. "No convergence factor is needed."[150] The notes give an expression for the vertex function. They also report the results of Feynman's computation for the contributions to the Lamb shift of this radiative correction: 1000 m.c. (as compared to 1050 m.c. for Schwinger).

Feynman's calculation did not include closed loops—"paths which give rise to infinite polarizability of vacuum." The notes report that "Feynman Believes It Possible To Get A Consistent Theory Without Using Loops." The report on Feynman's lecture closes with the remark, "Still To Be Investigated in This New Feynman Theory is the Scattering of Light by Light."[151]

Feynman deemed his lecture at Pocono a failure. On the day he received the announcement that he had received the Nobel Prize, he gave an interview to the *California Tech,* the student newspaper at Cal Tech. That article gave the following account of the Pocono conference:

> Schwinger went first, giving a very mathematical presentation of his methods; whenever he tried to give a physical example, the audience threw so many questions at him that he

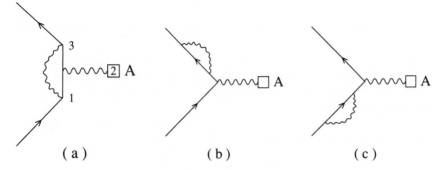

Figure 8.10.2

postponed the example and went back to the math. Then Feynman came to bat. His [Feynman's] ideas were greeted with even
less enthusiasm [than Schwinger's], largely because field theory
was then in vogue and his theory relied upon particle analysis.
He found it very difficult to explain his formulations because
they relied heavily upon physical arguments and intuition.

At each step he was asked to justify his procedure; instead he offered to work out a physical example to demonstrate
the correct results it produced. But the audience objected to the
time this would require and the hair involved, even though these
had been drastically reduced by his methods. The culmination
of his audience's feeling that Feynman was running amok without being rigorous came when Niels Bohr stood up, objected
to Feynman's use of trajectories for small particles, and started
reminding him about Heisenberg's uncertainty principle. Here
Feynman gave up in despair, realizing that he couldn't communicate the fact that his analysis was justified by its correct results.

Feynman then decided to publish what he had so far,
without waiting to remove completely the divergence difficulties, as he had originally planned. It turned out to be a good idea,
because the difficulties have yet to be removed, even after 17
years.[152]

Feynman also talked about the reception of his lecture at Pocono in his interview with Charles Weiner in the spring of 1966. He there recalled one of Bohr's
objections: his use of trajectories was not "a legitimate idea in quantum mechanics." Bohr reminded him that it was realized "already in the early days of quantum
mechanics" that the uncertainty principle rendered the "classical idea of a trajectory" invalid. Feynman also recalled that in the afternoon Bohr came up to him
and apologized. At lunch his son Aage had told him that he had misunderstood
what Feynman was saying and that in fact Feynman's viewpoint "was consonant
with the principles of quantum mechanics." To Charles Weiner, Feynman gave the
impression that although he was somewhat discouraged by the criticisms that had
been expressed during his talk—re unitarity, the Pauli principle in intermediate
states, the use of space-time trajectories—he was not "unhappy from that":[153] he
was merely resigned and felt that he had to publish. His interview with the *California Tech* probably gave a more accurate picture. He did give up in "despair."

Bethe also remembers that "Feynman was quite despondent" and that he
had lots of talks with him after Pocono to reassure him "that he [Bethe] believed
him to be right."[154]

There is one further aspect of the Pocono conference that Feynman has
stressed. He recounted that after he had finished his talk

[Schwinger and I] got together in the hallway and although
we'd come from the end of the earth with different ideas, we had

climbed the same mountain from different sides and we could check each other's equations.

I must explain [that] our methods [Schwinger's and mine] were entirely different. I didn't understand about those creation and annihilation operators. I didn't know how these operators that he was using worked and I had some magic from his point of view.

We compared our results because we worked out problems and we looked at the answers and kind of half described how the terms came. He would say, well I got a creation and then an annihilation of the same photon and then the potential goes. . . . Oh, I think that might be that, [and] I'd draw a picture. He didn't understand my pictures and I didn't understand his operators, but the terms corresponded and by looking at the equations we could tell, and so I knew, in spite of being refused admission by the rest, by conversations with Schwinger, that we both had come to the same mountain and that it was a real thing and everything was all right.[155]

Feynman also indicated that

we discussed matters at Pocono and later also over the telephone and compared results. We did not understand each other's method but trusted each other to be making sense—even when others still didn't trust us. We could compare final quantities and vaguely see in our own way where the other fellow's terms or error came from. We helped each other in several ways. For example, he showed me a trick for integrals that led to my parameter trick, and I suggested to him that only one complex propagator function ever appeared rather than his two separate real functions. Many people joked we were competitors—but I don't remember feeling that way.[156]

8.11 Vacuum Polarization

Aage Bohr, who attended the Pocono conference, recalls that "Feynman (who after Schwinger's lengthy review was left with rather little time to present his results) put a question mark on the need to include the vacuum polarization term, which appeared as a separate effect. This gave rise to criticism from many sides and I think my father [Niels Bohr] also raised objections to this view. The way the meeting developed, therefore, had the unfortunate effect of leaving some of the audience with the impression that Feynman's approach was less complete than Schwinger's.[157]

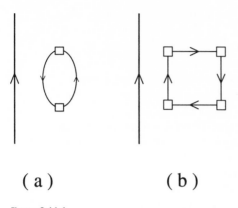

(a)   (b)

Figure 8.11.1

The problem of vacuum polarization always played a special role in Feynman's approach. Closed loops, which were responsible for the vacuum polarization phenomena (figure 8.9.1b), represented special paths—"unnatural" ones—in his integral-over-paths formalism: "From one point of view we are considering all routes by which a given electron can get from one region of space-time to another, i.e. from the source of electrons to the apparatus which measures them. From this point of view the close loop path leading to . . . [vacuum polarization] is unnatural. It might be assumed that the only paths of meaning are those which start from the source and work their way in a continuous path (possibly containing many reversals) to the detector. Closed loops would be excluded" (Feynman 1949b, p. 779).

Moreover, his successful reformulation of classical electrodynamics, which excluded self-interactions, had made a deep impression on him. Wheeler and Feynman's mechanism for radiation resistance—as stemming from the advanced interactions of the absorber particles with the charge emitting the radiation—had given him "a greater appreciation for the possibilities":[158] He was not convinced that the explanation given by convential QED of vacuum polarization, the scattering of light by a potential, or the scattering of light by light was unique and the only way to describe these phenomena.

When Bethe's explanation of the Lamb shift convinced him that self-interactions must be allowed—at least in conventional QED—Feynman proceeded to derive a closed expression for the transition amplitude from which he could compute the radiative corrections to scattering by an external field. It was not immediately obvious from this expression where the processes that are mediated through closed loops in conventional QED came from. The reason for this was that he had generalized a nonrelativistic theory (in which the number of charged particles is conserved and no pair production occurs) to one where particles could be created by virtue of his positron theory propagators. Also, in Feynman's (initial) formulation of his positron theory, disconnected diagrams did not occur. The Feynman propagator for a Dirac particle in an external field was given in a perturbative expansion as

$$S_{A+} = S_+ - \frac{i}{h}S_+AS_+ + \left(-\frac{i}{h}\right)^2 S_+AS_+AS_+ + \ldots, \qquad \textbf{(8.11.1)}$$

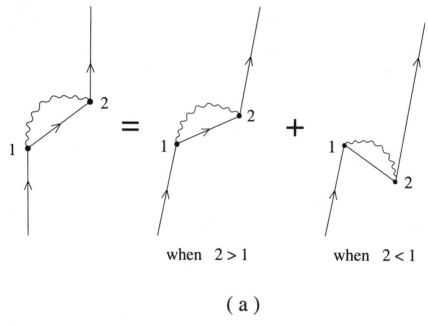

when 2 > 1   when 2 < 1

(a)

Figure 8.11.2(a)

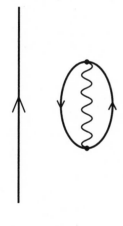

(b)

Figure 8.11.2(b)

and no diagrams such as those indicated in figure 8.11.1 appeared. That reformulation of positron theory strengthened his skepticism about closed-loop effects. His theory of positrons—using his S_+ propagator—automatically gave the correct perturbation theory result for the self-energy of an electron to order $e^2/\hbar c$ by calculating the contribution of (a) of figure 8.11.2 without having to invoke (b), which had to be taken into account in a hole-theoretic calculation to obtain the correct contribution. In Feynman's original conception the vacuum was a simple structure. Feynman of course knew the phenomena that closed loops gave rise to in conventional QED: the Uehling effect, the scattering of light by light, and so on. His letter to the Corbens in January 1948 indicated that although he did not completely understand the problems connected with the polarization of the vacuum, he could compute finite but non-gauge-invariant answers. No progress was made over the next few months because, in March 1948, he reiterated to the Corbens that "polarization of the vacuum still remains somewhat of a puzzle."[159]

In his "Theory of Positron" notes from the fall of 1947, Feynman had outlined how to calculate the polarization of the vacuum using his propagators. There he wrote:

> Vac. Polariz $t_{\nu\mu}$ = Spur $(\mathcal{S}(2,1)\gamma_\nu\mathcal{S}(1,2)\gamma_\mu)$ assume each $\mathcal{S}$ has different mass. Let us take q_μ component of Fourier transform:

$$t_{\nu\mu} = \int \frac{\text{Spur}[\gamma_\sigma(p_\sigma + q_\sigma) + m]\gamma_\nu[\gamma_\tau p_\tau + m]\gamma_\mu}{[(p_\mu + q_\mu)^2 - m_f^2][p_\mu^2 - m_0^2]} dp \tag{8.11.2}$$

> Numerators equals $p_\nu(p_\mu + q_\mu) + p_\mu(p_\nu + q_\nu) - g_{\mu\nu}[p_\sigma(p_\sigma + q_\sigma) - m^2]$.

The purpose of giving each propagator a different mass was so he could apply his cutoff method, or as Pauli was later to call it, to "regularize" each $\mathcal{S}$ separately. Although the method yielded a convergent answer, the result was not gauge invariant.[160] It is of interest to note that alternative ways to obtain gauge-invariant results were sought. In December 1949, writing to Wheeler, who had just informed him of his dispersion-theoretic approach to calculating the scattering of light by light (from the cross section for pair production in two-photon collisions), Feynman noted:

> I am very interested in the proposals that you have made with regard to the relation between absorption and dispersion, or in other words between real and virtual processes.... Professor Bethe suggested to me a couple years ago that all these problems of vacuum polarization, etc. could be studied by studying the real processes such as pair production to which they are related as absorption is to dispersion. The real processes represent the residues at the poles of some complex function. The virtual processes give the remainder of the description of the function, which should however be determined by the character of its poles. But neither of us has done anything in this direction and I would be very anxious to hear more details about your results.[161]

In March 1948 Feynman communicated to the Corbens that "my theory of representing positrons as electrons going backwards, is working very well but nobody believes me because I haven't got everything complete yet. I can deal with pair production and annihilation in a complete fashion. Polarization of the vacuum still remains somewhat of a puzzle, it has not been included in the above formula for ΔH."[162]

Freeman Dyson, who had gone to the Institute for Advanced Study after a year's residence at Cornell, visited Feynman in late October 1948. Dyson had just finished writing his paper on "The Radiation Theories of Tomonaga, Schwinger and Feynman"[163] proving the equivalence of Feynman's approach to that of Schwinger (Dyson 1949a). On this trip to Cornell, Dyson was accompanied by Cecile Morette. Dyson's account of his journey is related in a letter to his parents written shortly thereafter:

> Feynman himself came to meet us at the station, after our 10-hour train journey, and was in tremendous form, bubbling over with ideas and stories and entertaining us with performances on Indian drums from New Mexico until 1 A.M.
>
> The next day, Saturday we spent in conclave discussing physics. Feynman gave a masterly account of his theory, which kept Cecile in fits of laughter and made my talk at Princeton a pale shadow by comparison. He said he had given a copy of my paper to a graduate student to read, then asked the student if he himself ought to read it. The student said "No" and Feynman accordingly wasted no time on it and continued chasing his own ideas. Feynman and I really understand each other; I know that he is the one person in the world who has nothing to learn from what I have written; and he doesn't mind telling me so. That afternoon, Feynman produced more brilliant ideas per square minute than I have ever seen anywhere before....
>
> In the evening I mentioned that there were just two problems for which the finiteness of the theory remained to be established; both problems are well-known and feared by physicists, since many long and difficult papers running to 50 pages and more have been written about them, trying unsuccessfully to make the older theories give sensible answers to them. Amongst others, Kemmer and the great Heisenberg had been baffled by these problems.
>
> When I mentioned this fact, Feynman said "We'll see about this," and proceeded to sit down and in two hours, before our eyes, obtain finite and sensible answers to both problems.

It was the most amazing piece of lightning calculation I have ever witnessed, and the results prove, apart from some unforseen complication, the consistency of the whole theory.

　　The two problems were, the scattering of light by an electric field, and the scattering of light by light. After supper Feynman was working until 3 a.m. He has had a complete summer of vacation, and has returned with unbelievable stores of suppressed energy.

　　On Sunday Feynman was up at his usual hour (9 A.M.) and we went down to the Physics building, where he gave me another 2-hour lecture on miscellaneous discoveries of his.[164]

However, a week later Feynman was to write Dyson:

I hope you did not go bragging about how fast I could compute the scattering of light by a potential because on looking over the calculations last night I discovered the entire effect is zero. I am sure some smart fellow like Oppenheimer would have known such a thing right off.

　　Any loop with an odd number of quanta in it is zero. This is because among the various possibilities which must be summed there is one corresponding to the electron going around one way and another with the electron progressing around the loop in the opposite direction. The latter is the same as the former with reversal of the sign of the charge, thus all quanta and potential interactions change sign, so if there is an odd number of them the total result is zero.[165]

Feynman summarized the situation in the late fall of 1948 in a letter to his friend Ted Welton:

In regard to "Q.E.D." as you put it, I don't have the cold dope. I can calculate anything and everything is finite but the polarization of the vacuum is not gauge invariant when calculated. This is because my prescription for making the polarization integrals convergent is not gauge-invariant. If I threw away the obvious large gauge dependent term (a procedure which I can not justify legally, but which is practically un-ambiguous) the result is a charge renormalization plus the usual Uhling term. The amount of charge renormalization depends logarithmically on the cut-off. The Uhling [sic] terms are practically independent of the cut-off and give the usual $-1/5$ in the Lamb shift.

These terms come from closed loops, (in my way of talking, which I think you understand) in which two quanta are involved. Loops with a higher number of quanta always converge and in fact give definite answers practically independent of the cut-off, so that they could be computed by conventional Q.E.D. Incidentally, it is easy to show that all loops with an odd number of quanta of field interactions give zero. You know about these things. It is widely known that scattering of light by a potential only occurs with completed second order in the potential, i.e. probably fourth order in the potential. I think you told me it was so some time ago.

To me it has become clear that all the problems of Q.E.D. appear to be involved in the simplest problems, (self-energy and vacuum polarization) the more complicated ones always converge.[166]

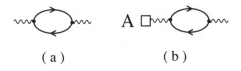

(a)   (b)

Figure 8.11.3

The last point was of course the thrust of Dyson's paper, and presumably had been expounded to Feynman by him on his visit.

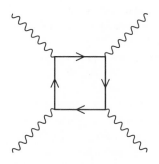

Figure 8.11.4

Thus by the end of 1948, Feynman had the following results regarding closed loops: The divergent lowest-order bubble diagrams—the ones that occurred in the self-energy of the photon—and the Uehling effect (figs. 8.11.3a,b) could not be cut off in a gauge-invariant way. All higher-order closed-loop effects, such as the one giving rise to the scattering of light by light (fig. 8.11.4) were finite.

At the end of January 1949, Bethe received a communication from Pauli with an important enclosure: a lengthy letter ("which became more similar to a smaller paper than to an ordinary letter") which Pauli had written to Schwinger.[167] In this "small paper" Pauli outlined his method of regulators by which he gave "definitional meaning" to the singular integrals encountered in Schwinger's approach. In particular, he indicated that the vacuum polarization divergence could be given a gauge-invariant regularization by calculating

$$t_{\mu\nu}^P = \int_0^\infty [t_{\mu\nu}(m^2) - t_{\mu\nu}(m^2 + \lambda^2)]G(\lambda)d\lambda, \qquad \textbf{(8.11.3)}$$

where $t_{\mu\nu}(m^2)$ is the expression eq. (8.11.2) with $m_f^2 = m_0^2 = m^2$, and imposing on $G(\lambda)$ the conditions satisfying $\int_0^\infty G(\lambda)d\lambda = 1$ and $\int_0^\infty G(\lambda)\lambda^2 d\lambda = 0$.

Bethe reported this regularization method to Feynman, who adopted it. In Feynman's "Space-Time Approach to Quantum Electro-dynamics," the self-energy of the photon and the divergent contribution to polarization of the vacuum were invariantly cut off using Pauli's regulator method. Feynman now knew how to circumvent the vacuum polarization difficulties, but his skepticism about the reality of closed-loop effects was not totally dispelled. In his "Space-Time Approach to QED" he commented:

> The closed loops are a consequence of the usual hole theory in electrodynamics. Among other things, they are required to keep probability conserved. The probability that no pair is produced by a potential is not unity and its deviation from unity arises from the imaginary part of $J_{\mu\nu}$ [what was called $t_{\mu\nu}$ above in eq. (8.11.2)]. Again, with closed loops excluded a pair of electrons once created cannot annihilate one another again, the scattering of light by light would be zero, etc. Although, we are not experimentally sure of these phenomena, this does seem to indicate that the closed loops are necessary. (Feynman 1949b, p. 779).

And in a footnote Feynman added: "It would be very interesting to calculate the Lamb shift accurately enough to be sure that the 20- megacycles expected from vacuum polarization are actually present" (Feynman 1949b, n. 18).

Settling this issue was of great interest and importance to Feynman, and he actively participated in the relativistic calculation of the Lamb shift (Baranger et al. 1953). This was "getting the numbers out" and thereby checking the theory. Developing better calculational tools "to get the numbers out" was something Feynman couldn't leave alone: a good deal of the second half of 1948 was devoted to this enterprise. The Feynman rules were also extended to apply to spin 0 and spin 1 particles, and during the spring of 1949 the rules for the various meson theories were obtained. The appendixes of his "Space-Time Approach to Quantum Electrodynamics" attest to the success of those efforts.

8.12 Evaluating Integrals

Feynman spent part of the summer of 1948 in Albuquerque and Santa Fe. Dyson has told the story of the trip out west in *Disturbing the Universe* (Dyson 1979, chapter 6). In New Mexico, "where love had drawn him," Feynman found "on arrival love dispersed,"[168] so he returned to work on improving the efficiency of his computational methods. Using "ever newer & more powerful methods" he checked again the radiationless scattering, and found agreement with his previous results. In a letter to Bethe he indicated:

> I am the possessor of a swanky new scheme to do each problem in terms of one with one less energy denominator. It is based on the great identity

$$\frac{1}{ab} = \int_0^1 dx \frac{1}{[ax + b(1 - x)]^2} \qquad (8.12.1)$$

> so 2 energy denominators may be combined to one—reserving the parametric x integration to the indefinite future (there's the rub, of course).[169]

Feynman's cutoff replaced the photon propagator $1/k^2$ by a new propagation kernel given by $\frac{1}{k^2} - \frac{1}{(k^2-L)}$, which he now conveniently represented as an integral,

$$\frac{1}{k^2} - \frac{1}{k^2 - \lambda^2} = -\int_0^{\lambda^2} \frac{dL}{(k^2 - L)^2}. \qquad (8.12.2)$$

Using his "swanky" integral representation for $1/ab$ Feynman, could then reduce all the integrations encountered in evaluating Feynman diagrams thus far to the following:

$$\int_{-\infty}^{+\infty} \frac{(1; k_\sigma)d^4 k}{(2\pi)^4 (k^2 + i\epsilon - L)^3} = \frac{1}{32\pi^2 iL}(1; 0). \qquad (8.12.3)$$

In eq. (8.12.3) the notation $(1; k_\sigma)$ means that either 1 or k_σ appears in the numerator, in which case on the right-hand side the $(1;0)$ is 1 or 0, respectively. The power of his new techniques were such that he believed he would be able to send Bethe the radiative corrections to the Klein-Nishina formula "in a few days." Also, if Bethe "were vitally interested in correction to Møller," Feynman thought he could deliver these "in short order (less than week?)."

Two weeks later, Feynman wrote Bethe from Santa Fe:

> I have been working on the Compton effect & the few days I promised the answer in turned into weeks. There are lots of integrals & terms to be added all together etc. & I kept looking

for a new & easy way because it was so complicated. But I think it is like calculating π to 107 decimal places—there is no short cut but to carry out the digits. So here I am beginning to believe that the answer is not much less simple than the steps leading to it—so I finally buckled down & did it....

I have set up & indicated how every integral can be reduced to transendental [*sic*] integrals in one variable, exactly. But I haven't done all the work of putting all the pieces all together & writing down the answers. I have however worked out a special limiting case in detail.[170]

The rest of the year was spent working hard preparing materials for his two papers, "Theory of Positrons" and "Space-Time Approach to QED." In late fall 1948 he informed Ted Welton, "I am very busy these days writing all my stuff on my paper...I am working like a demon."[171]

8.13 The January 1949 American Physical Society Meeting

The January 1949 meeting of the American Physical Society (APS) in New York proved to be another important landmark in Feynman's formulation of QED. As a result of a controversy between Slotnick and Case into which Feynman got drawn, he finally had to learn the formalism of second quantization. This proved to be of great value in writing up his "Theory of Positrons" and his subsequent papers (Feynman 1950, 1951a).

At the APS session dealing with nuclear scattering and neutron velocity spectrometer measurements, Rainwater, Rabi, and Havens reported on their recent measurements of the neutron-electron interaction as determined by the scattering of slow neutrons in lead and bismuth (Rainwater et al. 1949). These experiments essentially measured the neutron's electric form factor (for zero momentum transfer), a quantity of considerble theoretical interest. Several calculations of the neutron form factor had been performed in the past using various meson theories with various couplings, but the results were always plagued by the canonical divergence difficulties.[172] More recently, Murray Slotnick (Slotnick and Heitler 1949), in an impressive dissertation under Heitler and Bethe, had computed the interaction between a neutron and the electrostatic field of an electron in pseudoscalar meson theory. Although he had used old fashioned computational methods, he had made use of renormalization techniques and had obtained expressions for the "equivalent interaction potential"[173] in pure charged and in symmetrical pseudoscalar meson theory for both pseudoscalar coupling and pseudovector coupling. The result for pseudoscalar coupling was finite—"-7 kev for pure charged and -15 kev for the

Symmetrical theory" (Slotnick 1949)—whereas for pseudovector coupling he had obtained a logarithmically divergent result.

Slotnick had submitted an abstract of his work for the APS meeting (Slotnick 1949) and his presentation was scheduled to follow the paper of Rainwater, Rabi, and Havens.

Oppenheimer was in the audience, and after Slotnick's talk he commented that Slotnick's results must be wrong since they contradicted "Case's theorem." Oppenheimer pointed out that Case, who was a postdoctoral fellow at the Institute for Advanced Study, had just proved a theorem which stated that (to a certain approximation) pseudoscalar meson theories with pseudoscalar coupling were equivalent to ones with pseudovector coupling even in the presence of an external electromagnetic field. Since Slotnick's calculations violated "Case's theorem," they "must be" in error. Case was due to give a paper next day (Case 1949a). Since Slotnick did not know of Case's work—no paper or preprint had yet appeared—he was at a loss to reply to Oppenheimer's pointed criticism.

When Feynman arrived in New York that evening, he was told what had happened at the session. He received a report on the calculations of Slotnick, the "numbers" he had obtained after long and laborious computations and Oppenheimer's slashing criticism. He was then asked to comment on the validity of Slotnick's results in the light of "Case's theorem." Feynman had likewise not heard of this theorem. In fact, up to that point he had not interested himself in meson-theoretic calculations at all. However, between the results of a person who had calculated "numbers" and those of a formalist, the choice was clear. To corroborate his hunch that Slotnick was right, he got someone to explain to him what was meant by pure charged and symmetric meson theory, by pseudoscalar and pseudovector coupling, and he readily translated this information into the rules to compute the relevant matrix elements using his methods. He spent a few hours that evening calculating the difference between the proton and neutron electric form factor in various meson theories with both pseudoscalar and pseudovector couplings. The next morning he got a hold of Slotnick in order to compare his results with those that Slotnick had obtained, "because he wasn't quite sure that he had transcribed properly the usual formulation of meson theories into his rules."[174] When they compared their calculations, Slotnick asked him the meaning of the q^2 in Feynman's formulas. Feynman answered that it was the momentum transferred by the electron in the scattering. Feynman had calculated the full vertex function for arbitrary momentum transfer. "Oh," said Slotnick, "my results are only for $q^2 = 0$." "That's OK," Feynman indicated. "I can readily take the $q^2 = 0$ limit" (Feynman 1966a, p. 706), which he proceeded to do and then compared his answer with Slotnick's. They agreed. Slotnick was flabbergasted. He had spent close to two years on the problem and over six months on a calculation that took Feynman one evening. Even though Feynman had only calculated the difference between the neutron and the proton form factor while Slotnick had obtained the separate form factors, it was clear that with another few hours' work Feynman could easily get the separate pieces. Feynman was excited:

> This is when I really knew I had something. I didn't really know that I had something so wonderful as when this happened.... That was the moment that I really knew that I had to publish— that I had gotten ahead of the world.... That was the fire.
>
> That was the moment when I got my Nobel prize when Slotnick told me that he had been working two years. When I got the real prize it was really nothing, because I already knew I was a success. That was an exciting moment.[175]

After Case gave his paper, Feynman got up and commented, "But what about Slotnick's calculation? Your theorem must be wrong because a simple calculation shows that it's correct. I checked Slotnick's calculation and I agree with it."[176]

He was of course turning the tables on Oppenheimer for his arrogant dismissal of Slotnick's calculations. "I had fun with that," Feynman admits.[177]

Case sent Feynman a preprint of his paper, and Feynman felt obliged to find out "what is the matter with the damned thing."[178] Since it was written in the usual field-theoretic language using second quantized field operators, Feynman had difficulty reading it. Up to that time he had not studied second quantization! He later remembered that on a previous occasion, "when someone had started to teach me about creation and annihilation operators, that this operator creates an electron, I said 'how do you create an electron? It disagrees with the conservation of charge', and in this way I blocked my mind from learning a very practical scheme of calculation" (Feynman 1966a, p. 706). But this time he got Scalletar, then a graduate student at Cornell, to explain to him this formalism and proceeded to find the mistake that Case had made.[179]

Learning to express hole theory in the second quantized formalism turned out to be useful. It allowed Feynman to deal with vacuum processes in a way that had not been possible before. The appendixes in Feynman's "*The Theory of Positrons*" (Feynman 1949a), in which the equivalence of his approach with the second quantized version of positron theory is demonstrated (Appendix A) and the rules for handling vacuum processes are justified (Appendix B), are some of the fruits of this labor.

The other dividend from the Slotnick episode was that Feynman learned the different kinds of meson theories and formulated the rules for calculating with them. In less than two months, during the spring of 1949, he recalculated to order g^2 all the meson-theoretic calculations that had ever been performed up to that time—and many more. These efforts were summarized in the concluding paragraph of his "Space-time Approach to Quantum Electrodynamics": "Calculations are very easily carried out in this way to lowest order in g^2 for the various theories for nucleon interaction, scattering of meson by nucleons, meson production by nuclear collisions and by gamma-rays, nuclear magnetic moments, neutron-electron scatterings, etc. However, no good agreement with experiment[al] results, when these are available, is obtained. Probably all of the formulations are incorrect. An uncertainty arises since the calculations are only to first order in g^2, and are not valid if $g^2/\hbar c$ is large" (Feynman 1949b, p. 784).

By the spring of 1949 everything was in place. Simple rules for obtaining the contributions from the various orders of perturbation theory could be stated in terms of their associated Feynman diagrams, efficient calculational methods had been developed, and gauge-invariant cutoff methods were available for rendering finite the vacuum polarization, self-energy, and vacuum diagrams. Feynman could have proceeded by first publishing the equivalence of his approach with conventional quantum field theory. But Rabi at the Pocono conference had urged him to publish his rules and computational methods as soon as possible.[180] Feynman followed Rabi's advice. Also, Feynman's proof of the equivalence was predicated on his operator calculus, which would have given his presentation a mathematical and formal aspect he wanted to eschew. Finally, the order of appearance of the various papers—first the simple rules and efficient calculational methods, then the formal aspects—also reflected a latent hope: that he might yet be able to make quantum electrodynamics finite.

8.14 Retrospective

Commenting on what he had accomplished in the period from 1947 to the writing of his two classic papers in 1949, Feynman expressed some disappointment. He had come to quantum electrodynamics "from the desire to fix this problem," but he "didn't fix it." He had invented a more efficient way of calculating, but it wasn't "fixing it."[181] "I invented a better way to figure, but I hadn't fixed what I had wanted to fix. . . . I had kept the relativistic invariance under control and everything was nice . . . but I hadn't fixed anything. . . . The problem was still how to make the theory finite. . . . I wasn't satisfied at all."[182]

Feynman expressed these same feelings to the student newspaper that interviewed him on the day of the announcement that he had received the Nobel Prize: "It was the purpose of making these simplified methods of calculating more available that I published my paper in 1949, for I still didn't think I had solved any real problems, except to make more efficient calculations. But it turns out that if the efficiency is increased enough, it itself is practically a discovery. It was a lot faster way of doing the old thing."[183]

Feynman had the distinct recollection that he felt he was doing something "sort of temporarily"[184] while exploring the consequences of a patched-up retarded formulation of quantum electrodynamics, and that his real love lay in the $^1/_2$ (advanced plus retarded) formulation: "I was still expecting that I would some day come through the other end of my original idea . . . and get finite answers, get that self-radiation out and the vacuum circles and that stuff straightened out . . . which I never did."[185]

There is in fact a paragraph in his "Space-Time Approach to Quantum Electrodynamics" in which Feynman apologizes for publishing his theory prematurely because he couldn't make it finite:

> One can say . . . that this attempt to find a consistent modification of quantum electrodynamics is incomplete. . . . The desire to make the methods of simplifying the calculation of quantum electrodynamics processes more widely available has prompted this publication before an analysis of the correct form for f_+ is complete. One might try to take the position that, since the . . . discrepancies discussed vanish on the limit [that the cutoff] $\lambda \rightarrow \infty$, the correct physics might be considered to be that obtained by letting $\lambda \rightarrow \infty$ after mass renormalization. I have no proof of the mathematical consistency of this procedure but the presumption is very strong that it is satisfactory. (Feynman 1949b, p. 778)

Feynman added the further statement that the presumption that a satisfactory form for f_+ could be found "is [also] very strong." In retrospect, Feynman considered this paragraph "the one big mistake in the paper."

Feynman's hopes of being able to find a finite, consistent formulation of QED polarized his view of renormalization. He believed that he understood renormalizing "crudely":

> All I knew was that I had done a few problems, that I had noticed the obvious. As soon as the diagrams get more complicated the number of denominators increases and everything is OK. . . . That the only things that gave you trouble were these things [the self-energy diagrams]. I knew that. I knew that idea that was suggested by Weisskopf and Bethe, to me by Bethe, that [if] you correct the mass everything would be all right, that [if] you correct the charge everything will be all right. . . . That it was right, because these two diagrams are the only ones that make any difference. But I never proved it. If Dyson were to have come to me and told me that there is some difficulty with bubble diagrams or something, I would have been a little nervous because maybe there is some trouble with bubble diagrams.

In my world of physics, things were known better or worse, more sure or less sure. I knew a lot more than I could prove . . . as being extremely likely. So I would look at something like this [renormalization] and I considered I knew everything was all right. I didn't know it in the sense that I could prove it to anybody fully, carefully; but I knew . . . because I had done enough things that everything was OK. But if he Dyson would have come along with a diagram I didn't do, and discovered that there was something wrong there, I would say that there was a certain probability that he was right. He didn't actually! That's the kind of way I knew it. The odds were on it, the odds were for it. I realized how it worked and how it probably worked but I didn't really check it out.[186]

Feynman's views on renormalization were stated publicly in an invited paper he delivered in June 1949 at the APS meeting at the University of Washington in Seattle:

The philosophy behind these ideas [renormalization] might be something like this: A future electrodynamics may show that at very high energy our theory is wrong. In fact we might expect it to be wrong because undoubtedly high energy gamma rays may be able to produce mesons in pairs, etc., phenomena with which we do not deal in the present formulation of the electron-positron electrodynamics. If the electrodynamics is altered at very short distances then the problem is how accurately can we compute things at relatively long distances. The result would seem to be this: the only thing which might depend sensitively on the modification at short distances is the mass and the charge. But that all observable processes will be relatively insensitive and we are now in a position to be able to compute these real processes fairly accurately without worrying about the modifications at high frequencies. Of course it is an experimental problem yet to determine to what extent the calculations we are now able to make are in agreement with experience. In other words, it seems as though with these methods of mass and charge renormalization we have a consistent and definite electrodynamics for the calculation of all possible processes involving photons, electrons, and positrons.

I do not expect that this electrodynamics is correct at all energies, but in some sense it is modified at high energies. Of course, I do not expect that the particular modification I have chosen is correct. (It is, furthermore, completely unsatisfactory being at variance with energy conservation).

The statement that the electrodynamics is now definite, consistent, and free of divergences is my opinion. Although various partial proofs or convincing arguments may be and have been brought to bear on the subject, the actual situation is very poor from the point of view of mathematical rigor. I shall try to describe the situation in its most favorable light although my personal opinion is that in electrodynamics of electrons everything is all right.[187]

Feynman did not study Dyson's two papers on renormalization (Dyson 1949a, b). He accepted Dyson's statement that he knew what was in them. In any case, Feynman never felt order by order was anything but an approximation to the "thing" [and] the "thing" was the path integral.[188]

He "could write the expression which is supposed to be correct to arbitrary coupling and the expansion was only a demonstration of a practical calculation."[189]

However, "the rules I made [for the diagrams] were simpler than the way I got them. These rules were in fact equivalent to the field theory. A way of saying what quantum electrodynamics was, was to say what the rule was for the arbitrary diagram—although I really thought behind it was my action form."[190] Incidentally, these rules reflected an amalgamation of what Wheeler and Bethe had taught Feynman. The Feynman rules followed the algorithm Wheeler had given for carrying out calculations using the solutions of the 1/2 (advanced + retarded) formulation of classical electrodynamics in their absorber theory (Wheeler and Feynman 1945;, Galison 1983). But the calculations in that case were aimed at checking the theory's consistency and comparing its results with the conventional formulation using retarded potentials. The influence of Bethe is apparent in that the Feynman rules were designed to calculate observable phenomena and to "get numbers out" as efficiently as possible.

The 1949 papers contained the rules for the diagrams. The justification (and validity) of the rules appeared in two papers Feynman published in 1950 and 1951. These papers contain much of what Feynman had done in 1947 and 1948—for example, the derivation of the action when both the longitudinal and transverse radiation oscillators were integrated out—but now the results were *derived* in a rigorous fashion. Whereas previously, to handle Dirac particles, Feynman would replace the velocity v by α in the action, he could now justify these steps by using his calculus of ordered operators: "I knew the facts all the time, but I didn't know how I knew it." Now he proved all the rules that had seemed intuitively right or had been established by trial and error and checked with answers obtained by the more conventional approaches. The appendixes to these papers contain a wealth of deep insights. Feynman characterized them as "all my equipment being distributed—all the things I discovered on the way."[191]

The central result of Feynman's 1950 paper is a functional differential equation for the transition amplitude for a system of charges, each carrying the same charge e, and each interacting with the quantized electromagnetic field. In addition, each charge is assumed subject to an external (classically prescribed) electromagnetic field generated by a 4-vector potential B_μ, a function of space and time. Feynman made the assumption that the charge e and the potentials B_μ could be independently varied. If $T = T_{e^2}[B]$ is the probability amplitude for finding the system in a given final state at time t'', given the initial state of some earlier time t', then Feynman showed that T satisfies the functional differential equation

$$\frac{dT_{e^2}[B]}{de^2} = \frac{1}{2}i \int_{t'}^{t''} \int_{t'}^{t''} \frac{\delta^2 T_{e^2}[B]}{\delta B_\mu(1)\delta B_\mu(2)}\delta_+(s_{12}^2)\, d\tau_1\, d\tau_2, \qquad (8.14.1)$$

where the integration is over the two space-time points x_1, x_2 and $s_{12}^2 = (x_1 - x_2)_\mu(x_1 - x_2)_\mu$. The amplitude $T_0[B]$ describes the behavior of noninteracting particles in the external field B_μ, and can be computed exactly (in principle) in all cases of interest. The differential equation (8.14.1) determines $T_{e^2}[B]$ uniquely given the boundary value at $e^2 = 0$. Feynman then verified that when the equation is solved by a series expansion, the result is the calculational rules given in "Space-Time Approach to QED" (Feynman 1949b). In the final section of the paper, Feynman deduced additional rules that extended the results to processes in which real photons are present in the initial and final states. In Appendix A of the paper, Feynman gave a formally covariant treatment of particles satisfying the Klein-Gordon equation, making use of proper time as an independent variable. In Appendix B he discussed, in a general way, relations between real and virtual processes involving photons. In Appendix C he derived a wave equation for $\Phi_{e^2}[B, x]$, the probability amplitude for finding at x_μ a Dirac electron interacting with the quantized electromagnetic field and with an external field B_μ

$$(i\slashed{\nabla} - m)\Phi_{e^2}[B, x] = \slashed{B}(x)\Phi_{e^2}[B, x] + ie^2 \int \delta_+(s_{x1}^2)\frac{\delta\Phi_{e^2}[B, x]}{\delta B_\mu(1)}d\tau_1. \qquad (8.14.2)$$

This "equation contains in compact form the modification introduced into the Dirac theory of the electron by the interaction of the electron with its own field" (Dyson 1952d). It also encapsulates the advances made in the 20-year period from the time Dirac advanced his equation for a spin 1/2 particle and the post-World War II developments in QED. Feynman's final paper in the series, "An Operator Calculus Having Application in Quantum Electrodynamics" (Feynman 1951a) was the capstone on what he had accomplished in the 1947–1949 period.

A letter Feynman wrote to Wheeler in the spring of 1951 conveys the feeling that a chapter in Feynman's intellectual life had been terminated with the writing of these papers:

> I wanted to know what your opinion was about our old theory of
> action at a distance. It was based on two assumptions:
> (1)   Electrons act only on other electrons
> (2)   They do so with the mean of retarded and advanced poten-
>   tials
>
> The second proposition may be correct but I wish to deny the
> correctness of the first. The evidence is two fold. First there is
> the Lamb shift in hydrogen which is supposedly due to the self-
> action of the electron. It is true that we do not have a complete
> quantum theory of proposition one, so that we cannot be certain
> that the Lamb shift would not come from the net action on the hy-
> drogen atom of the atoms in the surrounding walls. That is why
> I am asking for your opinion. Do you believe that this reaction
> part of the energy could really be accounted for in this way?
>
> The second argument involves the idea that positrons
> are electrons going backwards in time. If this were the case, an
> electron and positron which were destined to annihilate one an-
> other would not interact according to proposition one, since they
> are actually the same charge. Thus, positronium could not be
> formed and subsequently decay, for if it were to decay it would
> mean that the electron and positron were the same particle and
> therefore should not have been exerting a force on one another.
> But they were exerting a Coulomb interaction in order to form
> the bound positronium state. . . .
>
> Finally, Deutsch[192] has . . . experimental evidence that
> positronium is formed in a stable state and subsequently decays.
>
> So I think we guessed wrong in 1941. Do you agree?[193]

Feynman himself states that "QED was over when I did the papers."[194]
These papers, on the space-time approach to nonrelativistic quantum mechanics
(Feynman 1948c), on quantum electrodynamics (1949a,b, 1950), and on his operator calculus (Feynman 1951a) must surely be placed near the top of any list of
the most seminal and influential papers in theoretical physics during the twentieth
century.

8.15 Style, Visualization, and All That

In a talk he gave to the Cal Tech YMCA in 1956 on "The Relation of
Science and Religion," Feynman characterized science as follows:

> Science can be defined as a method for, and a body of in-
> formation obtained by, trying to answer only questions which

can be put into the form: If I do this, what will happen? The technique, fundamentally is: Try it and see. Then you put together a large amount of information from such experiences. All scientists will agree that a question—any question, philosophical or other—which cannot be put into the form that can be tested by experiment (or, in simple terms, that cannot be put into the form: If I do this, what will happen?) is not a scientific question; it is outside the realm of science. (Feynman 1956b, p. 23)

In order to explain to his audience why it was difficult for many scientists to believe in "the kind of personal God, characteristic of western religions, to whom you pray and who has something to do with creating the universe and guiding you in morals" (Feynman 1956a, p. 21), Feynman indicated that

it is imperative in science to doubt; it is absolutely necessary, for progress in science, to have uncertainty as a fundamental part of your inner nature. To make progress in understanding we must remain modest and allow that we do not know. Nothing is certain or proved beyond all doubt. You investigate for curiosity, because it is *unknown*, not because you know the answer. And as you develop more information in the sciences, it is not that you are finding out the truth, but that you are finding out that this or that is more or less likely.

That is, if we investigate further, we find that the statements of science are not of what is true and what is not true, but statements of what is known to different degrees of certainty.... Every one of the concepts of science is on a scale graduated somewhere between, but at neither end of, absolute falsity or absolute truth. (Feynman 1956a, p. 21)

In his lecture Feynman stressed that it was necessary to accept this idea of uncertainty "not only for science, but also for other things." Scientists who are religious, and want to be consistent with their science, "can only say something like this to themselves: 'I am almost certain there is a God. The doubt is very small.' But this is quite different from saying 'I know that there is a God.'" Feynman went on to say that he did not believe that a scientist could ever obtain "that really religious understanding, that real knowledge that there is a God—that absolute certainty which religious people have." With the adoption of such a "skeptical" view, ethics becomes unhinged from its theological underpinnings, but Feynman proceeded to indicate "why the moral problems ultimately come out relatively unscathed: It is possible to doubt the divinity of Christ, and yet firmly believe that it is a good thing to do unto your neighbor as you would have him do unto you. It is possible to have both these views at the same time."

This notion of the uncertainty of our knowledge is central to Feynman's approach in his attempts to understand the physical world. Recall his statement: "In my world of physics, things [are] known better or worse, more sure or less sure."

Among Feynman's papers is an undated two-page manuscript on Fermat's theorem[195] (that $x^n + y^n = z^n$ is impossible for positive integer values of x, y, z, n if $n > 2$), which illustrates the meaning of "more sure, less sure" in Feynman's world. He first asks, "What is the prob[ability] N is a perfect n^{th} power?" To answer the question he observed that the spacing between $N^{1/n}$ and $(N + 1)^{1/n}$ is $\frac{1}{n} N^{1/n-1}$ (for large N). Therefore:

Probability that N is perfect n^{th} power is $\frac{N^{1/n}}{nN}$ [and] therefore probability that $x^n + y^n$ is perfect n^{th} power is $\frac{(x^n+y^n)^{1/n}}{n(x^n+y^n)}$. Therefore total probability any $x^n + y^n$ is perfect n^{th} power for $x > x_0$ and $y > y_0$ is equal to

$$\int_{x_0}^{\infty} \int_{x_0}^{\infty} \frac{1}{n} (x^n + y^n)^{-1+1/n} dx dy = \frac{2}{nx_0^{n-3}} c_n,$$

where

$$c_n = \frac{1}{2} \int_0^{\infty} \int_0^{\infty} (u^n + v^n)^{-1+1/n} du dv.$$

Feynman noted that if $n = 2$ the integral diverges badly—"i.e. [we] expect infinite no. of coincidences. For $n = 3$ the result is independent of x_0, but c_3 diverges logarithmically "so we cannot tell about $n = 3$ by chance. Must do theorem analytically. Restrict $n \geq 4$." For larger n, c_n is roughly $\simeq \frac{1}{2(n)}$, therefore "probability success for n, $x_0 = \frac{1}{n^2 [x(n)]^{n-3}}$ [and] probability success any $n > n_0 = \int_{n_0}^{\infty} \frac{d\mu}{\mu^2 [x_0]^{\mu-3}} = \pi_{n_0}$." Feynman then asked what is the smallest x_0 for which the theorem "has any chance at all." From the knowledge that the theorem is true for $n \leq 100$, and an estimate for π_{n_0} for $n_0 \leq 100$, he deduced that "probability (success) is certainly less than 10^{-200}, " and so he concluded, "For my money Fermat's theorem is true." Of course it would be very satisfying to have an elegant proof of the theorem, but as far as he was concerned he "knew" it was right even though he couldn't prove it rigorously.

The maxim, "The main job of theoretical physics is to prove yourself wrong as soon as possible," characterizes the way Feynman worked. He attributes the saying to "his friend Welton."[196] Very early on Feynman searched for ways to implement this strategy.

In a letter to his friend H. C. Corben written in the fall of 1947, Feynman elaborated on this approach to theoretical physics. Corben had asked him to comment on some work he had done on a relativistic theory of classical spinning particles, and Feynman replied, "I think the quickest way to find out whether there is anything really in your stuff would be this: Take some specific problem or problems, e.g., a single particle in an external field, if that means anything—or two interacting particles. Try to work the thing, if necessary, in one dimension (I mean four: space, time, momentum and energy). I have always found that it is when I try to do simple problems, that I find the main problems. This way you will find out just what the quantities mean or can mean."[197]

Feynman's genius combined great analytical skills, keen powers of visualization, and impressive physical intuition with almost unbounded physical energy and the ability to concentrate intensely on the demands of any task. He had a deep need to understand things his own way and to work out problems his own way. This obsession, combined with his immense powers, made it easier for him to derive the results of a paper on his own than to read it: "I have a lot of trouble reading papers. I have a lot of trouble understanding them. I don't have trouble working them out for myself. . . . It is easier working it out for myself than reading it; except a new idea somebody will *tell* me and I'll go home with the clever idea . . . and work out what he is trying to tell, and understand what he is trying to tell me; but if he writes the paper I have trouble understanding."[198]

In a revealing account of how his mind works, Feynman stated:

> I cannot explain what goes on in my mind clearly because I am actively confusing it and I cannot introspect and know what's happening. But visualization in some form or other is a vital part of my thinking and it isn't necessary I make a diagram like that. The diagram is really, in a certain sense, the picture that comes from trying to clarify visualization, which is a half-assed kind of vague, mixed with symbols. It is very difficult to explain, because it is not clear. My atom, for example, when I think of an electron spin in an atom, I see an atom and I see a vector and a ψ written somewhere, sort of, or mixed with it somehow, and an amplitude all mixed up with xs. It is impossible to differentiate the symbols from the thing; but it is very visual. It is hard to believe it, but I see these things not as mathematical expressions but a mixture of a mathematical expression wrapped into and around, in a vague way, around the object. So I see all the time visual things associated with what I am trying to do.
>
> It was always with visualization. There was a lot of visualization and a lot of analysis. Analysis is much more powerful when you can do it, especially when you want to publish

something or explain something; or when you want to be sure that what you have thought is clear and correct. Then the analysis, the mathematics is wonderful. That's why it looks like, when I write it, trying to do mathematics.

What I really am trying to do is to bring birth and clarity, which is really a half-assedly thought out pictorial semivision thing. OK?

I would see this jiggle-jiggle-jiggle, or the wiggle of the path or the influence of the other thing. Even when I talk now about the influence functional: I see the coupling and I try, I take this turn—like as if there is a big bag of stuff and try to collect it away and to push it. It's all visual. . . .

I see the character of the answer before me—that's what the picturing is.

Ordinarily I try to get the pictures clearer but in the end, the mathematics can take over and can be more efficient in communicating the idea than the picture.

In certain particular problems that I have done it was necessary to continue the development of the picture as the method before the mathematics could really be found.

One delightful example that I really got big pleasure out of, is the liquid helium problem that I like because I spent a long time picturing that damn thing and doing everything without writing any equations and it was one time in my life where I did an awful lot of physics with my hands tied behind my back because I didn't know how to write a damn thing and there was nothing I could do but keep on picturing . . . , and I couldn't get anything down mathematically. . . .

So the whole thing was worked out first, in fact, was published first as a descriptive thing . . . which doesn't carry much weight, but to me was the real answer. I really understood it and I was trying to explain it. It is a very difficult tool to explain pictures, and our methods of communications are nowhere near as good. But I really, honestly believed that paper with the description . . . would really be a very clear description [of] how it works, . . . and that anyone reading it could also visualize it without writing anything. Fortunately, in that case I then got the picture clear enough, I suggested the type of wave function, . . . it came out numerically and could show I am on the right track by having the analysis. But I knew I had the right answer in the visual thing. That was an answer where the weight was on the other side.[199]

When one listened to Feynman explain anything, it became evident that not only the "visual" but also the "acoustical" played an important role in "giving birth and clarity." The sounds—the jiggle-jiggle-jiggle, the swishing or fading sounds denoting exponential growth or decay; the modulation of his voice, the rapidity of his speech; the "verbal" trying to keep pace with the "mental"—all made clear that oral communication was more than translating the "visual" into sound. Verbal interaction—to explain, to clarify, to obtain information or criticism—was for Feynman the most efficient way to communicate.

What was immediately clear in any form of communication with him was that one was in contact with a remarkable human being. Salam and Wigner's assessment of Dirac applies equally well to Feynman. To paraphrase: "Posterity will rate Feynman as one of the greatest physicists of all time. The present generation values him as one of its greatest teachers.... Of those [who were] privileged to know him, Feynman has left his mark ... by his human greatness. He ... set the highest possible standards of personal and scientific integrity. He was a legend in his own lifetime and rightly so."[200]

8.16 A Postscript: Schwinger and Feynman

The two previous chapters presented separate biographical studies of the young Schwinger and of the young Feynman. Before turning to the contributions of Freeman Dyson, with which I deal in the next chapter, I want to make some brief observations regarding Schwinger and Feynman by noting some of the similarities and some of the differences between them. Incidentally, only by looking at Schwinger and Feynman *together* can a full account of the events under investigation be given, for part of the excitement and drama of the story derives from the fact that these two remarkable people, who were at the center of the developments, were seen by the community, and at times by themselves, as competing with each other.

Obvious similarities are apparent when we look at Feynman's and Schwinger's lives in parallel. Both grew up in New York and were raised in Jewish middle-class families. One should probably differentiate between the Jewish subculture of Far Rockaway (Feynman's), which had more in common with that of Brooklyn and the Lower East Side, and that of the upper Manhattan Jewish community (Schwinger's), which was in closer contact with the well-to-do, older, and more established German Jewish community. But I believe the salient features were to grow up Jewish in greater New York during the twenties and the Depression, and to be raised by immigrant parents. In an autobiographical narrative, Feynman made a point to indicate that he was "brought up in the Jewish religion— my family went to temple every Friday" (Feynman and Leighton 1988, p. 25) and he there also noted the trauma of his confrontation, at age twelve, with his Jewish faith and with religious belief in general. Schwinger grew up in a "typical Jewish home," with orthodox grandparents living in the apartment next door. Freud—that nonbeliever and alienated Jew—once remarked that "I owe only to my Jewish nature the two characteristics which had become indispensable in my difficult life's way. Because I was a Jew, I found myself free from many prejudices that limited others in the employment of their intellects, and as a Jew I was prepared to go into opposition and to do without the agreement of the 'compact majority' " (Gay 1987). To some degree, his statement is probably also applicable to Schwinger and Feynman. Schwinger asserts that he "never felt very Jewish." Nonetheless, the values that permeated the household of his parents—respect for learning and scholarly achievement, respect for craftsmanship and elegance, love of books—surely left their imprint on him and allowed certain of his traits to become fixed, in particular his shyness and his withdrawal into books.

Both Feynman and Schwinger remembered partaking of the intellectual and cultural riches that New York offered even during the Depression. Part of that legacy is a characteristic they shared: both were very musical. Recall that Schwinger once asserted that he would quit physics at age 40 and devote himself to composing.

The outstanding fact is of course that both were remarkably gifted; the physics community has called them "off-scale." Both displayed awesome powers of concentration, and both could work with relentless intensity and inexorable determination for long stretches of time. While engaged in his QED researches, Feynman described himself as "working like the devil." In a letter to James Watson, when the latter was embroiled in a heated controversy over the content of *The Double Helix*, Feynman corroborated Watson's description of what it felt like to do important research:

> Don't let anybody criticize that book who hasn't read it to the end. Its apparent minor faults and petty gossipy incidents fall into place as deeply meaningful and vitally necessary to your work....
>
> From the irregular trivia of ordinary life mixed with a bit of scientific doodling and failure, to the intense dramatic concentration as one closes in on the truth and the final elation (plus with gradually decreasing frequency, the sudden pangs of doubt)—that *is* how science is done. I recognize my own experiences with discovery beautifully (and perhaps for the first time!) described as the book nears its close. There it is utterly accurate.
>
> And the entire "novel" has a master plot and deep unanswered human question at the end: Is the sudden transformation of all the relevant scientific characters from petty people to great and selfless men because they see together a beautiful corner of nature unveiled and forget themselves in the presence of the

wonder? Or is it because our writer suddenly sees all his characters in a new and generous light because he has achieved success
and confidence in his work, and himself?

Don't try to resolve it. Leave it that way. Publish it with
as little change as possible. The people who say "that is not how
science is done" are wrong.[201]

Both Feynman and Schwinger had a need to do things their own way.
Each one created and developed his own language: Feynman, path integrals;
Schwinger, Green's functions and sources. Both were fiercely competitive. Each
needed to constantly prove to himself that he could do all the things others could,
and do so better. At some stage, both stopped reading the papers others wrote and
kept abreast by reading only abstracts of preprints and papers, and in Feynman's
case by talking to colleagues and attending seminars and conferences.

Both were under great pressure to do something out of the ordinary as
young men. Both labored under the myth that unless they did something outstanding before they were 30 they would not be able to do so later. Schwinger was
depressed on his twenty-fifth birthday as he felt that time was running out and
that he had not produced what had been expected of him. Feynman was deeply
depressed when he came to Cornell because he feared he was "burnt out."[202]

Both were perfectionists. It took Feynman a year to write his two classic
1949 papers. One of the (perhaps, unconscious) reasons for Schwinger's extensive
collaborations before World War II was that the research would be written up, even
if it did not quite meet his standards. Moreover, a cooperative effort implied that
this failure to bridge the gap between what he had aspired to do and what had been
accomplished was a shared responsibility.

Both men were independent, but their desire for independence manifested
itself in different ways. Feynman's avowal, "Now one of my diseases, one of my
things in life, is that anything that is secret I try to undo" is to be contrasted to
Schwinger's fear of being dominated. Both wanted to assert their independence,
but in Feynman's case the necessity to be in control became transformed into a
creative activity: a passion for unraveling nature's secrets and thereby acquiring a
mastery over it. In Schwinger's case, on the other hand, being in control became an
impediment. His nighttime working habits stemmed from his fear of being dominated by Breit and Wigner and were reinforced by his interactions with Rabi and
Oppenheimer. His fear contributed to his becoming socially isolated from his community; and social isolation encouraged intellectual isolation.

Both recognized they had "special brains." Both felt they had a unique
responsibility to engage in intellectual activities by virtue of their mental capabilities and that they had a special obligation to use their powers responsibly. For
both this sense of uniqueness and specialness translated itself into a decision not
to partake in the political process for most of their adult professional life. Neither one of them participated in departmental or university academic affairs. For

a brief period after World War II, Feynman was involved with the Federation of American Scientists. Schwinger lent his name to the organization but was never an active member. Neither ever worked as a consultant on defense projects. However, Feynman got deeply involved in educational matters and served on various committees charged with the improvement of the teaching of science and mathematics in the secondary schools of the state of California. Feynman's *Lectures on Physics* are further proof of his educational commitments. His archives give ample evidence of his deep concern with the intellectual growth of young people. Although many of the letters he received were but cursorily acknowledged—and many went unanswered—all the letters sent to him by students, whether they were attending elementary school, high school, college, or graduate school, were given thoughtful replies.

For both men the desire for independence was coupled with a strong sense of integrity. Two examples illustrate Feynman's sense of integrity. Feynman never accepted an honorary degree during his lifetime. In 1967 he was offered an honorary degree by the University of Chicago, which was to be bestowed at the special convocation commemorating the seventy-fifth anniversary of the founding of the university. Feynman's reply to Beadle, then the president of the university, tells much about the man:

> Yours is the first honorary degree that I have been offered, and I thank you for considering me for such an honor. However, I remember the work I did to get a real degree at Princeton and the guys on the same platform receiving honorary degrees without work—and felt that an "honorary degree" was a debasement of the idea of a "degree which confirms certain work has been accomplished." It is like giving an "honorary electrician's license." I swore then that if by chance I was ever offered one I would not accept it. Now at last (twenty-five years later) you have given me a chance to carry out my vow.[203]

In 1960 Feynman resigned from the National Academy of Sciences. In a brief letter to its president, Feynman noted that "I have found that I have little interest in the activities of the Academy, so would you please accept my resignation as a member."[204] The resignation was of course not accepted, but Feynman persisted and reiterated his desire to withdraw: "I should like to resign my membership in the Academy. I do not find myself with enough time or interest to actively participate."[205] The Academy, feeling that Feynman's resignation would tarnish its image, resorted to delaying tactics, and hoped that time would allow the matter to be forgotten. For a number of years the president of the Academy was successful in blunting Feynman's objective. However, Feynman had the last word. His resignation was finally accepted after he had made clear that "my request for resignation from the National Academy of Sciences is based entirely on personal psychological quirks. It represents in no way any implied or explicit criticism of the Academy, other than those characteristics that flow from the fact that most of the membership consider their installation as a significant honor."[206]

Other similarities between Schwinger and Feynman can be pointed out. The intellectual productions of both of them were deeply affected by their experiences during World War II. Their wartime activities—Schwinger working on the theory of radar devices at the Rad Lab, Feynman supervising computational activities at Los Alamos—helped develop and reinforce viewpoints and approaches they had been drawn to earlier: for Schwinger, the differential viewpoint; for Feynman, global approaches.

Yet for all their similarities, the differences between them should not be overlooked. It is Feynman's father who was the principal influence on him as a child; it is the mother in Schwinger's case. Feynman was a firstborn child, some nine years older than his sister. Schwinger grew up in the shadow of an accomplished and gifted older brother. Schwinger's mathematical mode of thinking is analytic and algebraic; Feynman's was visual and geometric. And the most striking difference between them was their personality.

In the final analysis, Goethe's aphorism that character is destiny haunts us. One is tempted to correlate Schwinger's and Feynman's styles in physics with their personalities. Schwinger's conservatism, as reflected in his attempts to synthesize the received dogma of quantum *field* theory and special relativity, is in character; so is his differential viewpoint. In general, he is of the opinion that one can extrapolate from what is known in but a *limited* fashion. Feynman was more daring—his global viewpoint reflects this—and in his work on QED, more revolutionary. Initially, he had conceived of his formulation of QED—which was based on the Wheeler-Feynman representation of electromagnetic phenomena in which only charged *particles* interacting via action-at-a-distance forces and no fields appear—as a *new* theory that would be rid of all divergence difficulties. One aspect of Feynman's persona was his brashness, his need to be at center of the stage, what has been called his performance personality. He needed to talk and he needed the give-and-take of verbal communication. He needed an audience and an intense interaction with his listeners, but he also needed his solitude. Schwinger, on the other hand, is socially very shy and retiring. Although he too needs an audience, the polished and dazzling performance is an end in itself. One has the sense that Feynman's psychological makeup was responsible, in part, for his enduring creativity. Feynman undoubtedly recognized this and always kept in touch with people. Appositely, Schwinger's personality was, in part, responsible for his drama. His marriage and his coming to Harvard created an environment in which his proclivities toward social isolation were nurtured. His sense of self-sufficiency was reinforced by the importance of his scientific contributions, by the impact of his brilliant lectures, and by his influence on the vast number of Ph.D.

students flocking to him. But what were strengths at one time became liabilities later on. As fewer people learned and spoke his language, he became intellectually isolated; and to shield his vulnerability he isolated himself further.

Personality and presentation of self helped shape not only the magnitude but also the character of Feynman's and Schwinger's scientific output. As Gleick's sensitive biography of Feynman makes clear (Gleick 1992), Feynman's relationship with Arline—his first girlfriend, his first and perhaps his only true love, his first wife—shaped his subsequent life. Arline was diagnosed as having lymphatic tuberculosis while Feynman was at MIT. Over the strenuous objections of his parents[207] they got married before Feynman went to Los Alamos, even though he knew she was dying—because Arline and he had promised one another to do so (Feynman and Leighton 1988). This was an assertion of Feynman's integrity. Throughout their stay there, Arline was in a hospital in Albuquerque and she died shortly before Trinity. In many ways Feynman never recovered from the loss. He seemingly came to the conviction that he would never again be fulfilled in love. Two years after she died, Feynman—the supreme rationalist—wrote Arline a wrenching, heart-rending love letter. His subsequent interactions with women were always affected by the scars incurred by the loss, scars that never healed. Until his marriage to Gweneth, the emptiness and yearning were filled by one-night stands and tempestuous, destructive love affairs (Gleick 1992). Similarly in physics, he early on came to accept the human limitations: unification, and in particular, a theory of everything was a fantasy with which the community deluded itself. Physics consisted of a set of algorithms that answered with a high degree of certainty questions of the form: "If I do this, what will happen?" Feynman searched for such algorithms with extraordinary courage and unshakable integrity.

To understand Feynman and the nature of his "genius" one has to apprehend his peculiar connection to his social world and his distinctive intellectual relationship to the scientific community in general and the physics community in particular. In intellectual matters Feynman was able to straddle the gulf between self and community. He could adhere to many of the tenets, assumptions, forms of thought and styles of reasoning that characterized the theoretical physics of his day and simultaneously transcend these limitations. One aspect of Feynman's "genius" was that he could make clear and explicit what was abstruse and obscure to most of his contemporaries. His doctoral dissertation and his 1948 *Reviews of Modern Physics* article that presented his path-integral formulation of nonrelativistic quantum mechanics helped clarify in a striking manner the assumptions that underlay the usual quantum-mechanical description of the dynamics of microscopic entities. And he did this in the very act of transcending the usual formulation with a startling innovation. It may well be that his reformulation of quantum mechanics and his integral-over-paths will turn out to be his most profound and enduring contributions. They have deepened considerably our understanding of quantum mechanics and have significantly extended the systems that can be quantized. And judging from the work of Atyah and Witten, Feynman's path integral has already substantially enriched mathematics and has provided new insights into spaces of infinite dimensions. Early on, Feynman also learned to walk the tightrope between the psychological needs of his self and the requirements imposed by belonging to a community. He came to accept and to appreciate the fact that the act of creation was for him also an act of consummate isolation. The latter was also true for Schwinger.

The creative act depends on private visions and on solitary constructions and always draws on the legacy and the resources of the community—be it in the arts, literature, technology, or the sciences. We call a person a genius because her or his ability to synthesize these communal resources overwhelms us, and if in addition the synthesis results in a startling outcome—as was the case of Schwinger and Feynman—we are astounded, amazed, and awed.

But I do not believe that it is helpful to create a category of persons we label geniuses, for we are then on a slippery slope that may lead us to believe that there is an innate quality attached to this categorization. Instead, we should focus on the social and cultural dimension of the attribution. Yet we should not forget that these people do give us a vision of what the human brain—and what the species—is able to understand, create, and accomplish.

9. Freeman Dyson and the Structure of Quantum Field Theory

Enough is good, but there is no use in satiety.
The bird in the forest can perch but on one bough,
And this should be the wise man's pattern.

—*Tso Ssu (translated by Arthur Waley)*

9.1 Family Background

In an autobiographical sketch that prefaces the reprinting of some of his articles (Dyson 1992), Freeman Dyson relates that when his mother died in 1974 at the age of 94, he found among her papers a long-forgotten manuscript of an unfinished novel with the title "Sir Phillip Roberts's Erolunar Collision," which he had written in 1932–1933 when he was 8 or 9.[1] In the story, the hero, Sir Phillip, having successfully predicted a collision between the asteroid Eros and the moon, organizes a scientific expedition to the moon to observe the event. Dyson speculates that the theme of his novel was suggested by the close approach of Eros to the earth in 1931, and by the account he had read of the expeditions to Africa and Brazil organized by Sir Frank Dyson, the Astronomer Royal,[2] to observe the gravitational bending of light during the solar eclipse of 1919. Dyson characterizes the literary style of the novel as borrowed from Jules Verne, and in particular Verne's story "From Earth to Moon and a Trip Round It." Dyson's passion for writing—and for space travel—developed early. His literary talents were nurtured by his parents, whose interests were artistic rather than scientific. His father, George Dyson, was teaching music at Wellington College, a boys' school near the Berkshire village of Crowthorne when Freeman was born. George Dyson was very talented; he had wide interests and possessed a quick and incisive mind: he could have done anything he wanted and chose a musical career. His writings reveal him to be literate, forthright and forceful (G. Dyson 1924). He was an accomplished composer and conductor, and over the years he acquired a considerable degree of recognition in the English musical scene. In 1924 he accepted the position of Master of Music at Winchester College, and in 1938 he was invited to become the director of the Royal College of Music, one of the two major conservatories of England, a position he held until 1952. Dyson's mother, Mildred Lucy Dyson, came from a fairly prosperous middle-class family; her father was a solicitor. She was a sensitive, extremely intelligent and well-read woman who "remembered everything about everybody."[3] Although she did not attend university, she had obtained a law degree, but never practiced. Freeman's parents had met in 1912 when George was teaching at Marlborough College. One of George's closest friends there was Freeman Atkey, Mildred's younger brother, a very gifted classicist who also taught at the college. George Dyson and Freeman Atkey—in addition to their common interest in literature and music—shared a passion for motor bicycles. Before World War I, they had ridden together on their motorcycles over the Alps to Italy. Freeman Atkey was killed in 1915 while serving at the front in the British army. Both Mildred and George were shattered by his death. They saw more and more of each other and finally married in 1916, when Mildred was 37 and George was 34. Their daughter, Alice, was born three years later. Their only son, born on December 15, 1923, was christened Freeman John in memory of his uncle and his paternal grandfather. Both Alice and Freeman recall their parents as "reserved Victorians"[4] who were "difficult to get to know." Neither Alice nor Freeman ever heard a cross word between them. It was a good, solid marriage between two close friends who had a great deal of respect for each other's abilities. Thus, even though Mildred was "totally unmusical,"[5] she helped George with the selection and the arrangement of the literary texts for his choral pieces. She would go to concert after concert not understanding "in the least" the music that her husband conducted, but she was intensely interested in the musicians and the students. She became very much liked as a result.

Both children felt very much closer to their mother than their father, even though she was the "serious" one and he the more lighthearted of the two. Both found it difficult to talk to their father. "He was always right,"[6] Alice recalls. Freeman remembers him as "ever cheerful" and with an ironic disposition. Because they seemed so old to him, Freeman thought of his parents as being more like grandparents: "They were somewhat remote." He felt much closer to Alice than to either parent. It was to Alice he would turn for comfort, entertainment, and companionship—and they have remained close.[7] Freeman characterizes his mother as very much a Victorian lady. Every afternoon she would retire for a rest "which was considered the unquestionable habit of a lady." Once a week she would put the children to bed. The needs of the house were attended to by a cook, a housemaid, a gardener, and a nursemaid. The family was well off since the father was on a fixed salary (of approximately £1000 a year) and their house was provided by the college. They were even better off during the Depression when prices fell. During Christmas and Easter recesses, the entire family would go off to their South Coast cottage in Lymington, a little house on 40 acres of water-logged land. (Keeping the acreage drained proved to be a constant challenge!) Summers would be spent traveling to Yorkshire, to Wales, and to France. Both of Freeman's parents were liberally inclined in politics and voted Labour. His father never forgot his humble working-class background. He was a product of the industrial north: Freeman's grandfather had been a blacksmith and his grandmother a weaver. During the thirties, Freeman's mother devoted a considerable amount of her energy and talents helping the Winchester birth control clinic—a cause none too popular at the time. The household was not religious. Although George faithfully discharged his duties as musician at the college chapel, he was a skeptic of orthodox belief, creed, and church dogma. He once told Alice that "my religion is my music."[8] Freeman's mother was a faithful churchgoer, but she too did not accept dogma. She was some-what of a mystic, who had discovered that the college-fostered muscular Christian-ity that filled the college chapel with singing boys satisfied a need within her.

Both Alice and Freeman remember their home in Winchester as a very happy one. Their house was spacious, and a well-kept garden abutted the college's playing fields. Alice describes it as "an idyllic setting, a warm home in beauti-ful surroundings." It was also a very stimulating home. Books on every subject packed the shelves of the house's ample library, and Freeman took advantage of their presence. Mildred had taught both children to read at a very early age and Freeman read omnivorously. Some of Alice's most vivid memories of the young Freeman are of him surrounded by encyclopedias and by sheets of paper on which he was constantly calculating one thing or another. His precocity and remarkable talents of assimilation were obvious to everyone in the household. At age six he computed the number of atoms in the sun and complained to Alice that he wished "Papa would get his facts right." On one occasion when asked, while immersed in his calculations, where his nanny was, Freeman replied "I expect her to be in the absolute elsewhere."[9]

Both George and Mildred cared deeply about intellectual matters—and with Freeman they got exactly the child they desired.

9.2 Early Education: Twyford and Winchester

Some of Dyson's earliest and most pleasant memories are of his nanny. She came into the house when he was five, and she looked after both him and Alice. Freeman started attending the day school of Miss Mary Scott at roughly this same age. "It was just marvelous. She was a wonderful lady who taught the ten boys who made up the school a tremendous lot."[10] Most of the time after school was spent reading, writing, and doing calculations—as had been the case already before he went to Miss Scott. As a child Freeman was not as friendly or outgoing as Alice. "He was more reserved, more shy; other children were not interested in the things he was."[11] He was considered a "strange bird"[12] because science and mathematics were his passion already then. His introduction to the sciences had come from the books he read. Freeman remembers lavishing much time at age 7 on *The Splendour of the Heavens*, a huge, beautifully illustrated compendium of popular astronomy.

Twyford College

Freeman's father was tremendously proud of his precocious son and he was very anxious to do all he could to nurture his son's great gifts. He believed that the best way to do so was to have his son attend a good prep school and thereafter matriculate at a good private school. Thus, when Freeman became 8 and 3/4 years old he was sent to boarding school. The school was Twyford, located some three miles from home. On the eve of his arrival in the United States to study at Cornell, which happened to coincide with the fifteenth anniversary of his entering Twyford, Dyson still vividly remembered being "deposited in the strange and forbidding environment" of that school.[13]

Although Dyson still says "Twyford was atrocious," on calmer reflection he concedes that Twyford wasn't all that bad—"all the boys were just too young for it." Already at Twyford, he had a sense of being different. Since grouping in English "public" (i.e., private) schools is by ability and not by age, Dyson found himself at Twyford, and later at Winchester, attending classes with boys much older than he was. He remembers the "enormous" fat knees of these older boys and "his own little wobbly ones" as he sat alongside them. Being somewhat small for his age, the discrepancy in size between his older classmates and him was always glaring. Nonetheless, Dyson recalls himself as easygoing, and being rather good at staying invisible and keeping out of trouble. Dyson also remembers that even though Twyford was only a short distance away, he never went home except on holidays. Nor did his parents ever visit him at Twyford.[14]

In the summer of 1936, at age 12, Dyson sat for the scholarship examination for Winchester College. The examination consisted then—as it does now—of three days of grueling written tests, followed by *viva voce* interview with the examiners. In the written part, the 12- or 13-year-old student is expected to be able to translate passages from Greek and Latin authors and to be familiar with all the niceties of the grammar of both languages. In former times, the contestant was further expected "to compose Ciceronian prose and Ovidian verse of his own with faultless accuracy and stylistic polish" (Sabben-Clare 1981, p. 56). There are also papers in English, algebra and geometry, French, a combined history and geography paper, and until recently a divinity paper.[15] Only in 1966 was science added to the schedule of papers.

Winchester is the intellectual summit of the English public school system and the hardest private school from which to obtain a scholarship. When Freeman was growing up, a Winchester scholarship was the highest ambition for a bright 12-year-old boy. Although he thought he would get a scholarship, he remembers being "absolutely astonished" to find himself first in all the separate papers and first on the list in order of merit. He never had thought of himself as anything that remarkable. His father was beside himself with joy. "It was a bigger event in our family than getting a Nobel prize."[16] Twyford thought so too. The school declared a holiday, for it considered Freeman's accomplishment a great triumph.

Winning first place was very much a turning point in Freeman's life. Thereafter people looked at him as somebody special.

Winchester College

Except for the youth of the students, there is little to differentiate Winchester College from the wealthiest and most beautiful Oxonian or Cantabrigian colleges. Both the old and the new buildings, the magnificent chapel, the well groomed athletic fields, the bright, alert faces of the students, the confident and self-assured presentation of self of the tutors and masters—all attest to Winchester's elite status. The school was founded in 1392 by William of Wykeham, Bishop of Winchester, to supply students—"well grounded in Latin"—for New College, the college he had established at Oxford in 1380 to maintain an educated priesthood (Sabben-Clare 1981, p. 2). Winchester College has been in continuous operation since 1394. Its students, who are called Wykehamists, have always been chosen from among the brightest "boys" in England. Even when they could not pay the tuiton, they were able to attend the college by being elected to a scholarship. "Scholars" can still be identified by the fact that they wear a gown at all times, whereas "commoners" do not. Even now Winchester remains "above all a place where boys of high intelligence, whatever their social backgrounds, can live among other likeminded pupils, and find the fullest encouragement to develop their particular talents" (Sabben-Clare 1981, p. 38). Able and dedicated teachers devote themselves to the nurturing of the intellectual, moral, and aesthetic sensibilities of the five hundred students in their charge. The responsibility of *in loco parentis* is taken seriously, and so are athletics and competitive sports.

Winchester was—and still is—an ideal environment for an intellectually gifted child. "It is such an intense emotional experience that many people regard the rest of their lives as an anticlimax." "It is so perfect," Dyson claims, "that it is hard to come out into the real world."[17]

When he entered Winchester in 1936, Freeman heeded the advice of his mother, to whom he had gotten close, not to immerse himself totally in mathematics and develop into a narrow specialist. In any case, he had become "a little disillusioned with mathematics."[18] His science teacher, Mr. Lucas,[19] got him excited about biology and during his first year at Winchester, Freeman decided to become a biologist. He carefully read H. G. Wells, Julian Huxley, and G. P. Wells' *The Science of Life* (1934), "that wonderfully broad view of biology,"[20] which gave him an appreciation of the range of biological inquiries. That winter, during the holiday recess at their seashore cottage, Freeman, in a burst of enthusiasm, brought a crayfish home to be dissected, only to discover that the procedure was more delicate and more involved than his readings had led him to believe. Freeman was also exposed to the other sciences at the school. Eric James, one of the science tutors, claims "that he learned his chemistry by teaching Dyson."[21] J. C. Manistry, who taught Freeman mathematics, recalls him as "both outstanding" and "remarkable for his humanity and thoughtfulness about the feelings of younger boys."[22]

Winchester has long had the reputation of being the best mathematical school in England. It has been the nursery of many wranglers and some of the best English mathematicians of the present century. G. H. Hardy and F. P. Ramsey were students there. It is thus not surprising that Dyson's mathematical interests should reassert themselves. At the time there were several other "boys" with outstanding mathematical abilities at the college: James Lighthill, who is exactly the same age as Freeman; and Christopher and Michael Longuet-Higgins, who are, respectively, one year older and two years younger than Freeman. The brilliance and genius for exposition of C. V. Durell,[23] the college's math tutor, helped stimulate these students' mathematical curiosity. On their own, they explored the mathematical worlds accessible to them. Having found in the library a copy of *Der Keplersche Körper* by Gerhardt Kowalewski, Freeman took an interest in polyhedra. He studied Coxeter, DuVal and Petrie's book on *The Fifty-Nine Icosahedra* and together with Christopher Longuet-Higgins constructed cardboard models of many of the stellations. These are still on display in one of the school's classrooms.

Michael Longuet-Higgins, who also became interested in making models of polyhedra, particularly the uniform polyhedra, recalls that "I owe my interest [in polyhedra] ultimately to [Dyson]. He was encouraging towards me, and arranged for me to give a talk on the subject to the school Mathematical Society which he and M. J. Lighthill had started."[24]

Together with Lighthill, Dyson worked through the three volumes of Camille Jordan's *Cours D'Analyse,* which they had found on the upper shelves of the library. It is quite possible that these volumes were donated to the library by G. H. Hardy. In his *Apology,* Hardy wrote of Jordan's *Cours*: "I shall never forget the astonishment with which I read that remarkable work, the first inspiration for so many mathematicians of my generation, and learned for the first time as I read it what mathematics really meant. From that time onwards, I was in my way a real mathematician, with sound mathematical ambitions and a genuine passion for mathematics" (Hardy 1968, p. 147).

The same was true for Freeman. His passion for analysis had been kindled by his readings in the famous eleventh edition of the *Encyclopaedia Britannica* (1910). Freeman had taught himself "all of the calculus and most of complex function theory" by studying its 28-page entry on "Functions," written by the geometer H. F. Baker. In that article, he first encountered Cauchy's theorem. Using the theorem, "which seemed like magic to him," he "could calculate impossible-looking integrals with astounding ease." The beauty and power of Cauchy's theorem turned him into a mathematician.[25]

In *Disturbing the Universe* (Dyson 1979), his revealing account of some chapters of his life, Dyson recounts spending Christmas vacation in 1938 working from six in the morning till ten in the evening with short breaks for meals, going through Piaggio's *Differential Equations* and solving its more than seven hundred problems. He manifests this same intensity, stamina, and power of concentration in all his undertakings.

Winchester has a tradition of handing out annual awards to the best students in the various subjects. The prize winners receive money with which they can buy books of their choosing. Freeman was an outstanding student. Among the many awards he won was the first prize in physical sciences *inter Wiccamicos Alumnos* in 1937 and the George Richardson Prize for the best student in mathematics in 1939. He built up an impressive small library of mathematical books with his prize monies. Many of the books he purchased then are still on his shelves: Whittaker and Robinson's *The Calculus of Observations*; Eddington's *The Mathematical Theory of Relativity*; *The Collected Papers of Srinivasa Ramanujan*; Hardy and Wright's *The Theory of Numbers*; E. T. Bell's *Men of Mathematics*; Felix Klein's *Lectures on the Icosahedron*; L. Wittgenstein's *Tractatus Logico-Philosophicus*. Each one was read carefully upon being received from the bookseller.

Another facet of Dyson's brilliance is his great facility for acquiring languages. When in 1938 he became interested in number theory, and was told of I. M. Vinogradov's classic introduction to the subject, then available only in Russian, he proceeded to teach himself Russian and translated Vinogradov. A neat manuscript of the full translation of Vinogradov's third edition of *An Introduction to the Theory of Numbers*, written in beautiful calligraphy, is still in the translator's possession.

Both Dyson and Lighthill had exhausted Winchester's resources in mathematics by the end of their third year. In their last year at the college, Durell arranged for Daniel Pedoe, an instructor at the nearby University College of Southampton, to come to Winchester once a week to teach his "two brilliant boys." Pedoe recalls that teaching the special math section of sixth form, the top class, as "an amazing experience." In Southampton, he taught an honours class for a university degree and used Durell's books. He could assign about three problems for homework at a time and hope that they would be done. Winchester naturally also used the Durell books, and the top class would do sixty problems at a time![26] After a time, Dyson approached Pedoe and asked him for a special assignment. As Pedoe had just published a nice little paper *On the Representation of Circles by Points in Space of Three Dimensions*, he gave Dyson the initial idea to work on. By Pedoe's next visit to Winchester, Dyson had written a paper as long as Pedoe's, "not quite exactly the same, but excellent."[27] Pedoe then, one Friday, lent Dyson the formidable *Vorlesungen über Algebräische Geometrie* by Francisco Severi, and on the following Monday Dyson turned up and told Pedoe that he had been making hay while the sun shines. Since it was a wonderful summer, Pedoe told him that he was glad he had been out of doors. But Dyson meant that he had worked right through the Severi book, and produced new theorems for Pedoe to look at.[28] "Pedoe is a real mathematician, a very brilliant geometer. Pedoe gave me my first glimpse of the abstract way of doing things."[29]

Pedoe also remembers that in 1946, when both he and Dyson were at Cambridge and Dyson wanted to switch from mathematics to theoretical physics, Dyson had asked him, "Shall I just go straight to Dirac, tell him I am rather good at pure mathematics, and would like advice on how to switch . . . ?" Pedoe comments: "One has to have a Winchester background to act in this way!"[30]

But a false impression is conveyed by singling out Dyson's mathematical abilities and prowess. Michael Longuet-Higgins, who was at Winchester with Dyson from 1937 to 1941 and in the same house, College West, remembers him "as a very bright, slightly built boy with the same piercing eyes and infectious, slightly sardonic laugh that he still has. He played the violin well and won the steeple chase, coming in far ahead of everyone else. Even at that time, he had unusually broad intellectual interests. For example he was one of the few people at school to study Russian. Above all he was supremely good at mathematics."[31]

In addition to his great intellectual gifts, what made Dyson unusual even as a teenager at Winchester was his keen awareness of the world around him. In *Disturbing the Universe*, Dyson wrote of growing up, constantly being reminded of the tragic consequences of the First World War. Winchester College had erected a monument, the War Cloister, for the five hundred Wykehamists who had been killed in World War I. Inscribed high on the walls inside the arcades that make up the cloister is a moving memorial to these fallen "boys." It encapsulates the trinitarian ethos of Winchester:

> Thanks be God for the Service of These Five hundred Wykehamists who were found faithful unto Death amid the manifold changes of the Great War, In the Day of Battle They forgot not God, who created them to do his will, nor their Country, the stronghold of Freedom, nor their School, the Mother of Goodness and Discipline, Strong in this threefold faith, They went forth from home and kindred to the battlefields of the world And Treading the path of duty and sacrifice Laid down their Lives for Mankind.
>
> Thou Therefore For whom they died Seek not Thy Own, But Serve as they Served.
>
> And in Peace or in War Bear Thyself Ever as Christ's Soldier Gentle in All things, Valiant in Action, Steadfast in Adversity.

Several times every day the young Dyson would walk through the War Cloister and would be reminded that many, if not most, of the best and the brightest of a whole generation of young Englishmen had died prematurely. This missing generation was the explanation for the mediocrity of English life. "Everywhere," in the government, in the professions, "tired men of 65 were doing the work that vigorous men of 45 should have done" (Dyson 1979). Already in 1937 Dyson was convinced that another blood bath was approaching, and that his chances of surviving the carnage of the coming war were slim. At best he had another five or six years to live. Moreover, he saw no possibility that a more hopeful future would emerge from the conflagration. At age 15, Dyson became a staunch pacifist. He had no illusions about the probable outcome of such a stand against Nazi Germany. But he harbored a fantasy of converting the German soldiers to massive disobedience. These goose-stepping soldiers would then return to Germany to use on their government the tactics pacifists had taught them. Hitler would be impotent, faced "with the refusal of his own soldiers to hate their enemies." This, in turn, would lead to the collapse of military institutions everywhere, and to an era of world peace and sanity. Gandhi had shown Dyson the way to this deeply moral way of life.[32]

World War II broke out in September 1939. The war did not change Freeman's pacifist views. He remembers well his lengthy discussions on the morality of pacifism with his father, who had welcomed Great Britain's entry into the war. His father saw war as the only means of stopping the spread of Nazism. The fall of France in the summer of 1940, and the establishment of the Laval-Pétain government, finally convinced Freeman of the futility of his position. Freeman considered the Pétain regime "in some sense a pacifist government,"[33] one which had abandoned violent confrontation and chosen the path of reconciliation. But unfortunately there was no way of distinguishing sincere French pacifists, who had committed themselves to nonviolent resistance to evil, from opportunists and collaborators like Laval. And the same would very probably happen in England if pacifist principles were put into practice there: pacifism would be destroyed as a moral force by its cooptation by collaborators.

9.3 Cambridge, 1941–1943

Dyson graduated from Winchester in the summer of 1941. In the fall of that year he went up to Cambridge, where he had been admitted to Trinity College on a scholarship. Physics was "probably" his first choice for his field of study. In his last year at Winchester he had studied Eddington's *The Mathematical Theory of Relativity* and had also worked through Joos's *Theoretical Physics,* and he was keen to pursue such studies. However, all the Cambridge physicists, except Dirac, had left to work on war-related projects. On the other hand, two of Cambridge's great mathematicians, Hardy and Besicovitch, were there.[34] Furthermore, since there were almost no advanced students in residence, they were teaching undergraduates. Dyson attended many of the courses given by them. He took Besicovitch's "beautiful course on integration," a class attended by three people. He also went to the lectures given by Dirac and by Hardy. He was disappointed with Dirac's course on quantum theory, which turned out to be essentially a verbatim repetition of his famous *Quantum Mechanics*. Dyson had read the book "without any understanding" at Winchester and had hoped that in his lectures Dirac would shed further light on the subject. For his part, Dirac had given considerable thought to the best way of presenting the subject, and believed that he had achieved this in his book. He knew no way of improving upon that exposition. Dyson found Hardy a distant figure and "not encouraging."[35] Later he learned from reading C. P. Snow's introduction to Hardy's *A Mathematician's Apology* that Hardy was in a deep depression at the time. But in Besicovitch, Dyson found not only a valuable teacher and mentor but also a good friend. They often went together on long walks and Dyson worked closely with him.

During his first year at Cambridge, Dyson studied mostly pure mathematics. His first article was written in 1941 but was only published in 1944 in *Eureka* ("The Journal of the Archimedeans"), the publication of the Cambridge undergraduates majoring in mathematics. The paper was a proof that every algebraic equation has a root (Dyson 1944a). Already here he exhibits a certain panache. The article begins with the statement that "there are so many proofs of the theorem that every equation has a root, that it seems almost criminal to produce another. I can however say two things in my defence: first, the proof I shall give is probably not a new one; second, if my proof is new it has a certain advantage over other proofs in using only the most elementary arguments."

The proof gives evidence of one of Dyson's great strengths: his insightful use of the growth of functions. Let

$$F(z) \equiv z^n + a_1 z^{n-1} + \ldots + a_n = 0 \qquad (9.3.1)$$

with

$$z = re^{i\theta} = x + iy. \qquad (9.3.2)$$

$F(z)$ can also be written in the form

$$F(z) = P(x, y) + iQ(x, y), \qquad (9.3.3)$$

where P and Q are polynomials in x and y with real coefficients. The terms of highest degree in $P(x, y)$ are

$$x^n - \binom{n}{2}x^{n-2}y^2 + \binom{n}{4}x^{n-4}y^4 + \ldots = r^n \cos n\theta. \qquad (9.3.4)$$

They vanish on n straight lines through the origin, making angles $\frac{\pi}{2n}$, $\frac{3\pi}{2n}$, ... $\frac{(2n-1)\pi}{2n}$ with the x-axis. Similarly, the graph of $Q = 0$ has n real asymptotes, making angles 0, $\frac{\pi}{n}$, ... $\frac{(n-1)\pi}{n}$. A sufficiently large circle C with center at the origin will cut the graph $Q = 0$, $P = 0$ at $4n$ points $Q_1, Q_2, Q_3, \ldots Q_{2n}$; $P_1, P_2, P_3, \ldots P_{2n}$ lying alternately around the circle. For a sufficiently large circle C, the value of P on the circle will change sign at points P_i and the value of Q at the points Q_i. By a simple topological argument, Dyson shows that Q has opposite signs at the ends of a connected arc inside C which joins P_i to P_{i+1}, and on which $P = 0$. Being a continuous function, Q vanishes at some point (x, y) on that arc. At this point $z = x + iy$, $P + iQ = 0$, hence z is a root of $F(z) = 0$.

Dyson's second publication also dates from his undergraduate days. He is still very pleased with the research which went into this paper, the title of which is "Some Guesses in the Theory of Partitions" (Dyson 1944b).

The properties of partitions constitute one the most interesting and fundamental areas in number theory. Euler's classic work on this topic initiated a long line of research. Rogers in 1894 made an important contribution by discovering two suprising identities that could be given a combinatorial interpretation. The identities were also discovered independently by Ramanujan around 1913 and by Schur in 1917. Together with a third formulated by Rogers in 1917, they have become known as the Rogers-Ramanujan identities. They insightfully illustrate the different methods of proof frequently used in combinatorial analysis. The reasoning of Rogers and Ramanujan was generalized by Selberg in 1936 and some new formulas were found. Dyson (1944b) in his paper pointed out that Selberg's result was already implicitly contained in Rogers' work, and he simplified the proofs considerably. Its introduction gives a revealing glimpse of its youthful, self-confident author:

> Professor Littlewood, when he makes use of an algebraic identity, always saves himself the trouble of proving it, he maintains that an identity, if true, can be verified in a few lines by anybody obtuse enough to feel the need of verification. My object in the following pages is to confute this assertion....
>
> ... After a few preliminaries, I state certain properties of partitions which I am unable to prove: these guesses are then transformed into algebraic identities which are also unproved although there is conclusive numerical evidence in their support; finally, I indulge in some even vaguer guesses concerning the existence of identities which I am not only unable to prove but unable to state. I think this should be enough to disillusion anyone who takes Prof. Littlewood's innocent view of the difficulties of algebra.

To place the matter in context: Littlewood was probably England's best mathematician at the time. Littlewood had worked with Hardy, and together they wrote some of the most important papers in mathematics during the first two decades of the twentieth century. The partnership between Hardy and Littlewood was probably the greatest and certainly the most famous collaboration in the history of mathematics. Hardy's judgment was that Littlewood was the more powerful mathematician of the two: he knew of "no one else who could command such a combination of insight, technique and power" (Hardy 1968, p. 29). He stated that he took huge satisfaction in being able to say, "[I] have collaborated with both Littlewood and Ramanujan on something like equal terms" (p. 148). Dyson had read Hardy's *A Mathematician's Apology* when it was first published in 1940; also, Hardy had often praised Littlewood during the lecture course that Dyson attended.

From reading Hardy and Wright's *Theory of Numbers,* Dyson had learned that Ramanujan had proved the divisibility property of partitions, which gave him the idea of working on the problem of rank congruences (Dyson 1944b). A partition of a number n is a representation of n as the sum of any number of positive integers (Hardy and Wright 1938, chapter 19). Thus $3 = 3 = 2+1 = 1+1+1$ has three partitions and $5 = 5 = 4+1 = 3+2 = 3+1+1 = 2+2+1 = 2+1+1+1 = 1+1+1+1+1$ has seven partitions. The number of partitions of n is denoted by $p(n)$. Thus $p(3) = 3$, $p(5) = 7$. The generating function of $p(n)$

$$F(x) = \sum_{n=0}^{\infty} p(n)x^n \qquad (9.3.5)$$

was found by Euler and is

$$F(x) = \frac{1}{(1-x)(1-x^2)(1-x^3)(1-x^4)} \cdots \qquad (9.3.6)$$

Ramanujan discovered and proved three arithmetical properties of $p(n)$ (Hardy and Wright 1938, p. 286):

$p(5n + 4) \equiv 0 \pmod{5}$
$p(7n + 5) \equiv 0 \pmod{7}$
$p(11n + 6) \equiv 0 \pmod{11}$.

The number which results from subtracting the number of parts in a partition from the largest part Dyson calls the "rank" of the partition. The ranks of partition clearly take on the values $1-n, 3-n, 4-n, \ldots -2, -1, 0, 1, 2, \ldots n-4, n-3, n-1$. Denote the number of partitions of n with rank m by $N(m, n)$. The number of partitions of n whose rank is congruent to m modulo q Dyson denotes by $N(m, q, n)$. Thus,

$$N(m, q, n) = \sum_{r=-\infty}^{\infty} N(m + rq, n). \qquad (9.3.7)$$

Dyson conjectured that

$$N(0, 5, 5n + 4) = N(1, 5, 5n + 4) = N(2, 5, 5n + 4)$$
$$= N(3, 5, 5n + 4) = N(4, 5, 5n + 4),$$

that is, the partitions of $5n+4$ are divided in five equally numerous classes according to the five possible values of the least positive residue of their ranks modulo 5. In the same way,

$$N(0, 7, 7n + 5) = N(1, 7, 7n + 5) = \ldots = N(6, 7, 7n + 5).$$

The corresponding conjecture with modulo 11 is, however, "definitely false." "The guesses were really an experimental discovery. It is a very simple problem to state, and also a very simple problem to find a solution of once you get the criterion." Although the problem was finally solved only five years later by Atkin and Swinnerton-Dyer, Dyson's paper formulated the problem and gave important insights (1953); see also Dyson 1969, 1972) on how to solve it. Dyson still vividly remembers "the indescribable joy of having found something. I was delighted with this partition stuff. It was beautiful, elegant, [but] old fashioned. It belonged to the past rather than the future. It was [work] in the 19th century Cayley, Sylvester style."[36] Dyson worked out several other problems at Cambridge and published his findings (Dyson 1943a, b, c, 1944a). Michael Longuet-Higgins recalls that when he went up to Trinity in 1943, "Freeman Dyson and James Lighthill had left behind a reputation as two of the most brilliant undergraduate mathematicians seen there for some years."[37] Even though his mathematical researches gave him much joy and he was very pleased with his published results, Dyson's stay in Cambridge from 1941 till 1943, when he joined Bomber Command at High Wycombe, was a "miserable time." Wartime Cambridge was a great anticlimax after Winchester. It was a "lonely life." Although, he got to know some of the other students in his classes and knew many of the Wykehamists who had come up with him to Cambridge, he had few friends. Two kept him going. Dyson reflects that he "would have been much more depressed than [he] was" had he not met them. The first of these "good" friends was Peter Sankey, an engineering student at Magdalene, who also was a Wykehamist. Freeman had known him at Winchester, where both of them used to climb the Winchester buildings at night.

Night climbing takes place on the face of buildings, on drainpipes, chimneys, and window sills—and on the spires and towers of the roof tops. It is done at night to avoid proctors. It is said that there is a kind of fear associated with night climbing which is very closely akin to love, and that this is the fear the climber enjoys. With every success there is a gain in confidence, and the successful climber acquires a certain fearlessness. Night climbing of buildings is an ancient tradition at English schools. Indeed, the most challenging college buildings usually have chockstones to prevent—or at least discourage—the more adventurous climbers. Sankey and Dyson got hold of *The Night Climbers of Cambridge* (Whipplesnaith 1937) and proceeded to climb many of the college buildings. The big Cambridge climbs, such as King's Chapel, proved to be too difficult, but they did most of the other climbs, including St. John's. "It was beautiful to go out at night, to be up there on top of the building . . . [and] listen to the bells chime."[38] During the winter recess of his first year at Cambridge, Freeman went rock climbing in Wales and was hit by a falling stone. A cut from the accident got infected and it took a two-week stay in a hospital and six weeks of further recovery at home to get him back on his feet. His mother at that time exacted from him a promise that he would no longer climb buildings and that he would go rock and mountain climbing only when properly equipped with ropes. Dyson sadly relates that Peter Sankey went to the army in 1942 and died in the summer of 1944 taking part in the attack on Arnhem.

Freeman's other good friend at Cambridge was Oscar Hahn. Oscar was the nephew of Kurt Hahn. Kurt Hahn had studied at Oxford before World War I and had been the private secretary of Prince Max von Baden. He had helped him found the Salem Co-educational School at Salem Castle in the south of Germany, a progressive school for the children of the wealthy elite. The school was closed when Hitler came to power. Kurt Hahn was arrested by the Nazis in March 1933, but was released through the intervention of Ramsey MacDonald. He immigrated to Great Britain, where he opened Gordonstoun School in the north of Scotland and operated it on the same principles as Salem. The Duke of Edinburgh was one of his "boys." Oscar was an undergraduate reading engineering when Freeman and he first became friends. At age 12 Oscar had contracted polio, which had left his legs paralyzed. He was thus confined to a wheelchair in which he would rush around Cambridge. Oscar's father had emigrated to England in 1933 and had started a business in Birmingham similar to the one he had had in Germany. Oscar's mother was a Warburg and ran an organization looking after refugee children who had gotten out of Germany without their parents. Dyson recalls her as a "tremendously effective person." Freeman visited them with Oscar. It was his first glimpse "of people of that type. They were a hundred times as energetic as anybody I had met before. They looked at the world as a challenge. In England everything was restricted and everything was seen as a problem of barely getting by and making do with as little as is possible. The Hahns, by contrast, were living in splendor. In the middle of the war it would never have occurred to them not to eat supper with the silver and the candles." Oscar's family was very warm and friendly, and every member looked after every other. "Our lives were somewhat subdued in comparison. They gave me a taste of what fun life could be."[39]

On one occasion Freeman and Oscar decided that they were going to "walk" from Cambridge to London. The challenge for Oscar was that he had been told that it never had been done in a wheelchair before. They trained for several weeks for the event. Together they "walked" 10 miles every morning. Then the great day came. They started out at three in the morning. They covered the 57 miles in 17 hours, arriving in London at Oscar's uncle's house at 10 P.M. From there Freeman walked to his parents' house near the Royal College of Music. In proper English fashion, Freeman did not tell them he had walked from Cambridge during the day. They of course assumed that he had come in on the train: he did not say anything to betray that impression.

Oscar Hahn played an important role in Dyson's life. He helped him overcome his sense of "being on the outside." Until he had met Oscar, Dyson felt left out of the social world of Cambridge. Oscar helped him shed his shyness with "girls." Wartime Cambridge was "swarming with girls" but Freeman "was scared" of them. He had lived a totally monastic life at Winchester, with essentially no contacts with members of the opposite sex; he thus had "no practice in getting to know girls." Dyson explains further: "Winchester has a homosexual culture and is famous for producing homosexuals. In Winchester you formed these intense emotional ties which are not quite homosexual, but platonic homosexual. . . . I was not sure at that point whether I was a homosexual. You don't know whether you ever become normal after you come out of that.... That was one of the delights about Oscar. He had some real girls. He was enormously attractive whether you were male or female." Freeman became enamoured of one of the "girls" that "he had floating around him." "I just fell in love with her. That was sort of my first discovery that I was capable of falling in love at all. That was very helpful." Freeman well remembers another incident involving Oscar. At the time it was absolutely forbidden for women to come in and eat supper in Hall. Nonetheless, Oscar succeeded in persuading one of his friends to cut off her "beautiful hair" and come and eat in Hall disguised as a boy. Thinking about it still makes Dyson chuckle.[40]

9.4 Bomber Command

During the early years of the war, Dyson had struggled with his pacifist beliefs, and he had had lengthy discussions with his father about the morality of fighting and killing. The "considerable courage and good humour" (Dyson 1979, p. 87) displayed by the British in their day-to-day struggle for survival convinced him that he too must help his country win the war. In 1943 he went to work as a civilian scientist for the Royal Air Force Bomber Command. He was sent to High Wycombe, the Bomber Command Headquarters, a base from which each night, weather permitting, squadrons of Lancaster bombers would be ordered to take off to bomb cities in Germany. He had come to High Wycombe after an interview with C. P. Snow, who was responsible for placing technically trained personnel into appropriate jobs. Dyson was 19 at the time; he had finished two years at Cambridge and was rated a trained mathematician. He was assigned to the Operational Research Section of the Bomber Command and his job was to analyze the factors that would increase the efficiency of the airplanes in their sorties and minimize their losses, and to make the appropriate recommendations. On his very first day at the base, he calculated the distribution of "window"—the aluminized paper strips the bombers released—to confuse enemy radars. He remembers being surprised at how rapidly he got "into the swing of things." He characterizes most of the work he did at Bomber Command as "problems of common sense." A report on "Loss Rates" written at High Wycombe can still be found among Dyson's papers. Its opening paragraph states: "Very frequently it is required to discover, by a comparison of loss rates, whether aircraft equipped with some piece of apparatus are in fact protected by it or endangered by it. It is often difficult to make a direct comparison of loss rates between equipped and unequipped aircraft owing to external factors such as Type of aircraft or Target which are different for equipped and non equipped aircraft" (Dyson 1944c).

The objective of Dyson's report was to suggest a method for obtaining significant results when the available data had to be divided into numerous small categories, "the figures within each category being strictly comparable but too small to be significant." Dyson's method had the merit that it did not introduce "any unwarrantable hypothesis concerning external factors" and did not waste any available information.

Often his findings went against the accepted wisdom, and "all the good advice we gave had no result." Dyson felt very inadequate. He felt that he was not making any dent in minimizing bomber losses or in bombardiers' lives; he also felt keenly that he "should have been fighting" like the other young men of his age. His awareness of the exorbitant price being paid —in lives, in planes, in bombs—to achieve objectives that were dubious to start with and his sense of moral failure for not having spoken out and disseminated this information much more widely have led him to compare himself to the Nazi bureaucrats who likewise "were calculating how to murder most economically. . . . The main difference was that they were sent to jail or hanged as war criminals, while I went free. I felt a certain sympathy for these men."[41] He felt that he had been a coward. Dyson's wartime experiences have haunted him. In a recurring dream he used to see an airplane crash and burst into flames while he was standing by in terror, unable to move as he watched the occupants of the plane burning. Dyson's son, George, remembers being woken up by him as a small boy in the middle of the night:

> I used to go to him when I had nightmares, but this time he came to me. He said he needed to talk to me. He had just had a dream where an airplane crashed. The plane was in flames. People were standing outside, and some of them were running into the flames to rescue the passengers. Freeman couldn't move. He was rooted to the spot.
>
> He told me it wouldn't mean anything to me now, but later it would. He told me that a father, though he seems powerful, is just a man, with weaknesses. He wanted me to remember, so that if I was ever in that situation, I would be able to move. (Brower 1978, p. 24)

For over a quarter of a century Dyson lived with the fear and self-hate of the dream. In *Disturbing the Universe,* he relates that in 1969, during the upheavals at American universities, a package containing a bomb was opened by the janitor of the faculty club at Santa Barbara where he was staying. When it exploded he acted out his dream and stood rooted to the floor. He then had to live with the guilt that perhaps he might have saved the caretaker, suffering from mortal wounds, if he had responded promptly. He discovered then that he was "after all, a coward." The dream did not reappear after that. But it was only in 1975 that he was finally redeemed by his son: George rescued two men at sea near Vancouver while Dyson was visiting him. Thereafter, a certain wholeness was restored (Brower 1978; Dyson 1979; Manuel 1979).

9.5 Imperial College and Cambridge University

Although Dyson worked quite hard at the Bomber Command base (from 9 A.M. to 7 P.M.), he found time to continue some of his mathematical researches and to do some readings in physics. He lived a few miles from the base in a "miserably cold" house, and he would ride a bicycle to get back and forth. These half-hour trips gave him the opportunity to think about mathematics; the solution of the "$\alpha\beta$ sequence" problem was effected on these bicycle rides (Dyson 1945). While at High Wycombe, Dyson also worked "very carefully" through Gödel's *The Consistency of the Axiom of Choice and of the Generalized Continuum Hypothesis with the Axioms of Set Theory*. A year earlier he had ordered the book from a bookseller and was quite surprised when it arrived in the mail: it had crossed the Atlantic, having been sent by its publishers from Princeton. He also had gotten hold of von Neumann's book on the mathematical foundations of quantum mechanics (von Neumann 1932) and became quite "frustrated" by it for it did not make quantum mechanics any clearer than Dirac had, and it gave "no clue on how to connect with the real world."[42] He also acquired the second edition of Heitler's *Quantum Theory of Radiation* when it was published in 1944. He found Heitler "enormously refreshing, because you could calculate with it."[43] In August 1945, after the surrender of Germany, the unit Dyson was attached to was all set to fly to Okinawa, to carry out bombing raids over Japan. Hiroshima and Nagasaki changed these plans.

With the war over, Dyson accepted a job at Imperial College as a demonstrator in mathematics. His duties consisted in correcting and grading papers and answering the questions of students. Most of his time was spent reading in the old science library. Once a week he went to Birkbeck College to meet Davenport, an outstanding number theorist with whom Dyson had gotten in touch. It was Davenport who suggested to Dyson that he tackle the Minkowski problem (Dyson 1947b). While at Imperial, he wrote a fellowship thesis for Trinity College, which consisted essentially of two previously solved problems—the $\alpha\beta$ problem and the Minkowski problem—together with an introduction setting these problems in context (Dyson 1946, 1947a,b). In the 1947a paper Dyson refined some results obtained by Borel in 1909 and by Bernstein in 1912 concerning the rational approximation of irrational numbers by continued fractions. The *Acta Mathematica* paper (Dyson 1947b) is concerned with rational approximations to algebraic numbers. It has proven to be very important and has greatly influenced research activities in number theory. The main content of Dyson's fellowship dissertation (1946) was published in Dyson 1945, 1948a,b. The 1948b paper gives a proof for $n = 4$ of Minowski's conjecture in the geometry of numbers, following the same general line as Remak's proof in 1923 for $n = 3$. The novelty of the paper lies in the fact that it is the first introduction of algebraic topology into the study of the geometry of numbers. The topological arguments used in the 1948b paper were separately published in Dyson 1948a. These are generalizations of results by Lebesgue, and of Phragmen and Brouwer, that relate the homology group of the intersection of two sets of points to those of the two sets taken separately.

Although he would write other papers on mathematics, the stay at Imperial College marked Dyson's final year as a mathematician. He had read the Smythe (1945) report, the official account of the development of the atomic bomb under the auspices of the United States government, and its content had much to do with his decision to become a theoretical physicist. He had been aware of the possibility of nuclear energy from his earlier readings in Eddington (Dyson 1979). Now it was a reality: the challenges were clearly in physics.

When he returned to Trinity in 1946 as a fully qualified fellow, his efforts were concentrated on the study of physics. Dyson obtained a desk at the Cavendish Laboratory and interacted with many of the young theoreticians there. His friendships with Tommy Gold and Herman Bondi date from that period.

Nicolas Kemmer, a former student of Gregor Wentzel and of Wolfgang Pauli, who before the war had done outstanding work in meson theory (Kemmer 1965, 1982, 1983) was a lecturer at Cambridge at the time. Dyson attended Kemmer's lectures on "Nuclear Physics" and on "Particle and Fields" and he calls Kemmer his "first real physics teacher." Kemmer introduced him to Wentzel's *Einführung in der Quantentheorie der Wellenfelder*, a copy of which had been sent to Kemmer by Wentzel. This copy was "treasured" and carefully studied by both Kemmer and Dyson. There is a story—which Kemmer corroborates as being true—that after one of Kemmer's lectures Dyson, Harish-Chandra, and Kemmer were walking to lunch. Harish-Chandra, who up to that time had been a physics student at the Cavendish, made the following remark: "Theoretical physics is in such a mess, I have decided to switch to pure mathematics," whereupon Dyson remarked, "That's curious. I have decided to switch to theoretical physics for precisely the same reason!"[44] From his readings it had become clear to Dyson that the United States was the country in which theoretical physics was flourishing. He went to G. I. Taylor at the Cavendish to ask him what he should do. Taylor, who had spent the war years at Los Alamos, told him, "You must go to Cornell to work with Bethe." He also said, "You might go talk to Peierls at Birmingham. He might be in a better position to advise you."[45] Dyson took G. I. Taylor's advice and wrote Peierls, indicating to him that he was a mathematician recently released from National Service who had just been elected to a fellowship at Trinity.

> I should like to enter the field of fundamental research into problems of electrodynamics and nuclear physics, and for this I need first of all personal contacts and guidance. Sir Geoffrey Taylor, with whom I have discussed this possibility, advised me to attach myself to one of the schools of research in the U.S.A., but recommended to me to consult you first as a specialist in the field. . . .

> Please can you be so kind as to give me an appointment
> for an interview? As I possess a motorcycle it will be quite easy
> for me to come and see you at Birmingham on almost any day
> you suggest.[46]

Peierls promptly answered Dyson, telling him that he will be glad to see him "and discuss all your problems."[47] When Dyson accepted the date Peierls had suggested for their appointment and indicated that he would meet him at "2:00 after lunch"[48] Peierls responded, "I suggest that if you can get here by 1 o'clock you look in and we have lunch together, otherwise I shall expect you at 2:00."[49]

On the appointed day Dyson traveled on his motorcycle to Birmingham to confer with Peierls to help him decide where he should study in the United States. The ride on the motorcycle turned out to be "an extraordinary hardship. It was raining cold [and] I could barely see anything through my goggles." Peierls told him that "Oppenheimer, who was at Berkeley, was the deepest thinker at present in the field of physics, but . . . still unofficially involved in a lot of secret work at Los Alamos, so that [Dyson's] position would be rather dubious if [he] went to work with him."[50] Bethe, the head of theoretical physics at Cornell, was the other person Peierls recommended. Dyson also consulted Kemmer, and he too advised him to go study with Hans Bethe. Dyson then applied for a Commonwealth Fellowship to study at Cornell. In the spring of 1947 he was informed that he had been awarded the fellowship.

Before going to the United States, Dyson spent three weeks in the summer of 1947 in Germany attending the University of Münster. The university had invited a group of foreign students to give German students an opportunity to become acquainted with the world outside Germany. The stay allowed Dyson to perfect his German and to familiarize himself further with German literature and culture. Winchester had inculcated in Dyson a sense of international responsibility. "At Winchester we were certainly concerned about the fate of the world."[51] It was therefore natural for him to go to and help Germany reestablish links with England. His trip would affirm that some Englishmen held out their hand in reconciliation. But another motivation for the visit was the desire to see at first hand the effect of the Allied bombings and to come to terms with his own involvement in these matters. Although two years had elapsed since the signing of an armistice, German cities were still a mass of rubbles. "The first and most striking thing," Dyson commented in a letter to his parents, "is the lack of strangeness when one walks round the town. Munster is indeed a 'Trummerstadt' about 90% of the buildings in the inner town are totally destroyed; but the ruins are all overgrown with grass and vegetation and one hardly notices them . . ."[52]

Dyson was impressed "by the liveliness and energy of people and by the superficial normality of things" despite the devastation all around. But he was also struck "by the deep psychological gulf which separates a country like Germany from one like England; here there is no academic and intellectual pessimism, but a pessimism which permeates the people's lives."[53]

In Münster, Dyson lived in the Borromaeum, the Catholic Theological College. The Borromaeum, a slate-faced structure made of reinforced concrete, had survived the bombings rather well: it was about two-thirds intact. Here is the description of one of his evenings while in Münster: "We had a party at the Borromaeum at which the University quartet played Mozart quartets, and I was sitting at an open window gazing out at a landscape of ruins and shrubs and flowers and trees and twilight and feeling distinctly mystical."

Dyson was a little taken aback to find the Germans so ready "to unburden their hearts without restraint" and to be "so genuinely and obviously happy" when reminiscing about the war years: "A description by the U-boat sailor of what happens when a petrol-tanker is torpedoed was given with the most single-minded enthusiasm. It reminded me vividly of the description we used to read at Bomber Command of successful incendiary attacks and of the elation we felt when such attacks succeeded. It is ironic that when enemies meet and come together as friends they should still be able to entertain each other with such stories."[54]

He was also surprised to find so many of the German students *tiefsinnig,* that is, having a philosophical bent. He commented to his parents that the "German character" is not as mythical "as he had believed."[55]

9.6 Cornell University

Dyson came to Cornell with a well-earned reputation. He had received glowing recommendations from his teachers at Cambridge and in particular from Nicolas Kemmer. Philip Morrison, who was then a young assistant professor in the Department of Physics, recalls Hans Bethe inviting him and Richard Feynman into his office in the spring of 1947 to go over the applications of the students interested in doing theory. Bethe informed them that there seemed to be one unquestionably outstanding application—that of Freeman Dyson, for whom G. I. Taylor had written essentially the following letter of recommendation: "Dear Bethe: You'll have received an application from Mr. Freeman Dyson to come work with you as a graduate student. I hope you will accept him. Although he is only 23 he is in my view the best mathematician in England. Very Sincerely yours, G. I. Taylor."[56] Indeed, the mathematics department at Cornell knew of an English mathematician named Dyson but refused to believe for a while that the Dyson who had come to Cornell was that same person. It invited Freeman to give a seminar on some number-theoretic topic. "It was very good," Phil Morrison recalls, and although they thereafter were convinced of his identity they remained puzzled by his shift of fields.

Dyson had been accepted as a regular graduate student working toward the doctorate. Ph.D. students in the physics department at Cornell are assigned a committee of two or more faculty members who advise them in their course of studies. The members of the committee are nominated by the student, and they are responsible for following the student's progress. The chairman of the committee is usually the student's thesis supervisor. The other members are chosen from fields different from that of the student's main field of interest, and see to it that he or she is exposed to other fields of research.[57] The chairman of Dyson's committee was Hans Bethe, and the other member of the committee was Robert Wilson, representing experimental physics.[58] At Bethe's recommendation Dyson registered in the "Advanced Quantum Mechanics" course that he was teaching and in the lecture course on the "Theory of Solids" given by Smith. He also attended Wilson's lectures on "Experimental Nuclear Physics," and two afternoons a week he took the laboratory course on modern experimental techniques required of all graduate students. Bethe also assigned him a research problem: the relativistic quantum electrodynamic calculation of the Lamb shift for a charged spin zero particle in a Coulomb field.

Dyson's first impression of Bethe is recorded in his weekly letter home: "Bethe himself is an odd figure, very large and clumsy and with an exceptionally muddy old pair of shoes. He gives the impression of being very clever and friendly, but rather a caricature of a professor; however, he was second in command at Los Alamos, so he must be a first-rate organizer as well."[59]

Dyson saw Bethe every day at lunch and had frequent discussions with him about his research problem. The difference in atmosphere between Cambridge and Cornell must have struck him. The Department of Physics at Cornell, and the Newman Laboratory in particular, was (and still is) an informal and very friendly place. A strong sense of community pervaded all of its activities—a legacy of Los Alamos where many of the faculty members had worked during the war.

Every day at lunch time, the contingent of Bethe's graduate students[60] could be seen walking with him to the nearby campus cafeteria. Everyone was— and is—on a first-name basis, and the luncheons provided a daily mechanism for relaxed exchanges on the progress of their own work, advances in physics, the state of the nation and that of the world. It was a very stimulating community. Jack H. Smith, who was one of Bethe's graduate students at the time, writes:

> Our weekly theory seminars were attended and contributed to by graduate students and faculty. Many papers and subjects, from nuclear physics to relativity, were discussed quite thoroughly.
>
> Lunch at Home Ec cafeteria was frequently a lively session, with Bethe at its center and a table of graduate students about him, discussing topics from current courses and/or research of the day. Evening meals at Home Ec similarly had their share of discussion amongst graduate students.[61]

Four weeks after his arrival at Cornell and his assignment of a problem by Bethe, Dyson was writing home:

I am seeing a lot of Bethe, far more in fact than I expected when I arranged this trip. I am more and more impressed with his intellect; he had the reputation of being a real terror when he was younger.... Nowadays he is mellowed, but still thinks and argues in a most forceful way, and is never caught in a mistake. He has given me a lengthy calculation to do which has considerable theoretical interest although it concerns an unobservable phenomenon. I am stimulated to work hard at it by frequent discussions with Bethe, and particularly by the feeling that I am on trial, and upon my success in this job will largely depend the amount of pull I shall have when new jobs come along. As a matter of fact, my own calculation is now almost finished and has given a clear answer to the problem.

Bethe has some six official research students whom he looks after in this thorough way, and this is a carefully picked bunch. He said at a seminar the other day that he considered the present situation in physics the most exciting there had been since the great days of 1925–1930; this attitude makes him an ideal disseminator of knowledge, and is highly contagious.[62]

Dyson, like all of Bethe's students, developed a deep admiration and a lasting affection for Bethe. Bethe's integrity, his marvelous common sense, his deep commitment to research and teaching, his overwhelming power of thought, his determination, his wry sense of humor, and above all his caring and concern for everyone around him were a model to emulate. Dyson was soon to tell his family that he "was bowled over by Bethe's complete generosity and unselfishness."[63]

What particularly struck Dyson about Bethe was that he always sat down and calculated things. "His view was to understand anything meant to be able to calculate the number. That was for him the essence of doing physics. When you came into his office he would always be calculating something. He would always talk to you in a very friendly fashion and when you got up to leave he would continue exactly where he had left off in his calculations."[64]

One thing Dyson was not aware of was how unsettling "his arrival and the revelation of his brilliance" had been on the other graduate students. For them Dyson's presence raised serious questions about their own abilities; in fact, one of Bethe's students may well have left physics because he believed that he could not compete with either Dyson or Feynman.[65] Although they found Dyson friendly and helpful, they were overwhelmed by his intellect and powers. Morrison remembers Dyson when he first came to Cornell as "enigmatic and private" and "extremely learned."[66]

At the end of October 1947, a little over a month after he had arrived at Cornell, Dyson was writing home:

> I am living a highly professionalized existence, without any pri-
> vate life to speak of, and wake up in the mornings thinking about
> mesons and photons.... I spend all the time at the Physics de-
> partment, doing my own research and reading other people's. To
> relieve this program, there are two courses of lectures which I
> go to, and two afternoons a week of lab work; also occasional
> seminars and a great deal more informal discussions; in ... the
> evenings I can stay here if nothing else is happening, but about
> one night in two there is a party of some kind or a show; other-
> wise I can go for a walk or in the browsing library. You will see
> from this that the life is in fact highly circumscribed; at present
> I have no desire for anything except this sort of life, and the only
> question is whether I shall be able to keep it up indefinitely.... I
> am however looking forward to a break when the skiing season
> starts and it is likely that I will then take whole days off.[67]

Dyson further commented that the most striking difference between the Cornell atmosphere and that of Cambridge is that "here it is taken for granted that one works hard; and this strengthens the natural tendency that I have to bury myself. For the first time in my life I am thinking about Physics continually and without effort, and I want to confirm the habit before I let it drop." He continued the letter by relating the progress he had made: "My work on the 'unobservable phenomenon' is now finished, and much to my surprise Bethe has asked me to write it up as a full-scale paper for the 'Physical Review,' the biggest journal on the subject." Dyson noted this was well worth doing "from the point of view of my personal advancement, if not for higher motives."[68] Bethe also asked him to give a short talk on the subject at the forthcoming meeting of the American Physical Society in New York in January (Dyson 1948a).

    In explaining to his parents what his calculation was about, Dyson in-
formed them that Bethe had earlier published a calculation of the level shift in a simplified model "ignoring the spin of the electron and also relativity." Although the answer "had come out to be infinity" by a plausible approximation, one could turn the infinite answer into a finite one which agreed with experiment. "Having had a good look at the frightfulness of the exact calculation," Bethe handed this problem to a graduate student, Scalettar by name, who had been struggling with it for the past few months and had not as yet been able "to get it down to a finite result, and it is not obvious that this is possible." Hence as an interim program, Bethe had assigned him "the problem of working out the answer with a rigorous relativistic theory but ignoring spin." Dyson found that with spin zero the exact relativistic calculation could be carried out without impossible complication, and gave a finite

answer precisely as Bethe had predicted, agreeing with experiment. "This result is useful, since it is unlikely that the spin in fact makes any essential difference to the behavior of the animal, whereas the same could not be said of relativity."[69]

Lamb Shift Calculation

Dyson's paper on "The Electromagnetic Shift of Energy Levels" appeared in the March 15, 1948, issue of the *Physical Review* (Dyson 1948c). It was Dyson's first research problem in physics. When Bethe gave him the problem he had handed him his paper (Bethe 1947) on the calculation of the Lamb shift for a nonrelativistic charged particle, and Dyson worked through it. His knowledge of quantum field theory allowed him to readily generalize Bethe's approach to the case of a relativistic spin zero particle. The system investigated consisted of a quantized complex scalar field interacting with a static Coulomb potential and with a quantized electromagnetic field. Dyson's mastery of the formalism of the quantum theory of fields as outlined in Wentzel (1943) is immediately apparent upon reading the paper. The complete Hamiltonian of the system is

$$H = H_1 + H_2 + H_3 + H_4, \qquad (9.6.1)$$

where H_1 is the Hamiltonian of the scalar field

$$H_1 = \int d\tau \left\{ \pi^* \pi + c^2 (\nabla \psi^* \cdot \nabla \psi) + \frac{m^2 c^4}{\hbar^2} \psi^* \psi \right\}, \qquad (9.6.2)$$

H_2 is that of the quantized electromagnetic field, H_3 is the term corresponding to the interaction between the matter field and the Coulomb potential

$$H_3 = \int \rho V d\tau, \qquad (9.6.3)$$

with

$$\rho = \frac{ie}{\hbar} (\pi^* \psi^* - \pi \psi) \qquad (9.6.4)$$

the charge density operator for the matter field. H_4 is the interaction Hamiltonian of the matter field with the radiation field

$$H_4 = \int \left\{ \phi \rho - \frac{1}{c} \mathbf{A} \cdot \mathbf{s} - \frac{e^2}{\hbar^2} \psi^* \psi A^2 \right\} d\tau, \qquad (9.6.5)$$

with $\mathbf{s} = i\frac{e}{\hbar}c^2(\nabla\psi^*\psi - \nabla\psi\psi^*) - 2\frac{e^2}{\hbar^2}c\mathbf{A}\psi^*\psi$, the charge current 3-vector in the presence of an electromagnetic field. The Hamiltonian $H_0 = H_1 + H_2 + H_3$ defines an unperturbed system in which the energy levels are given by the Klein-Gordon equation for the hydrogen atom[70]

$$\left\{ \left(E_s + \frac{e}{\hbar}V \right)^2 + c^2\nabla^2 - \frac{m^2c^4}{\hbar^2} \right\} \psi_s = 0. \tag{9.6.6}$$

The level shifts introduced by the interaction with the quantized radiation field are found by calculating the expectation value of the perturbing Hamiltonian H_4 in the stationary states of the unperturbed system.

The formalism for the quantized charged scalar field in the presence of an external field V is more complicated than that expounded in Wentzel's *Einführung* because the solutions of eq. (9.6.6) are not orthogonal, but satisfy the weaker relations

$$\int \psi_t^* \left(E_r + E_s + 2\frac{e}{\hbar}V \right) \psi_s d\tau = -\hbar\epsilon_s\delta_{st}, \tag{9.6.7}$$

where ϵ_s denotes $+1$ or -1 depending on whether ψ_s represents a state of negative or positive charge. The way to overcome this difficulty and obtain the correct quantized relativistic field theory had been shown by Snyder and Weinberg (1940) and by Blatt (1947), and Dyson adopted their formalism.[71]

The level shift is computed for initial states of the unperturbed system consisting of a negatively charged particle in a $2s$ and a $2p$ state, respectively. Dyson calculated the self-energy for these states by the method used by Weisskopf (1934, 1939) for the Dirac electron with appropriate modifications being made for the change from Fermi to Bose statistics as befits a spin zero particle. However, the mass renormalization was not without ambiguity because of the noncovariant integration methods used, and in the paper Dyson carefully explained the difference between the "single particle prescription" and the "wave packet prescription" that he used (Dyson 1948c). The difference between the results computed according to these two prescriptions is small but not negligible in comparison with the total line shift.

Dyson then proceeded to do the extensive calculations and obtained an expression for the displacement, D, of the $2s$ level relative to a $2p$ level:

$$D = \frac{1}{3\pi}Z^4\alpha^3\mathrm{Ry}\left[\text{``log''} + \frac{13}{36} - \frac{11}{15}\log 2 + \frac{1}{12}\epsilon - \frac{1}{40} \right], \tag{9.6.8}$$

where "log" is the Bethe logarithm occurring in the nonrelativistic calculation (Bethe 1947), ϵ is a calculable quantity but which takes on different values depending on the mass renormalization prescription used, and the factor $-1/40$ is the contribution due to vacuum polarization. Inserting numerical values, Dyson found $D = 1034$ and 1000 megacycles with the wave packet and single-particle subtractions, respectively.

As Dyson indicated in his paper:

> From the theoretical standpoint, the convergence of the present calculation is noteworthy and somewhat unexpected. The non-relativistic theory, in which the self-energy before subtraction diverges linearly, gives a logarithmically divergent level shift; and it was to be expected that the Dirac hole theory, in which the self-energy is only logarithmically divergent, would give a convergent level shift. In the case of the scalar particle the self-energy itself is quadratically divergent, and after the subtraction the transverse and longitudinal self-energies are still logarithmically divergent; the level shift finally converges only by an almost exact cancellation of the transverse and longitudinal contributions. The underlying significance of this cancellation is still obscure but it certainly argues in favor of the general correctness of Bethe's interpretation of self-energies. (Dyson 1948c, p. 618)

Dyson was pleased with this result. Since the system he had considered "is at least conceptually a physical system (which a nonrelativistic system is not), it is satisfactory to find that the theory gives a convergent expression for the level shift. Further, the values obtained are very close to the non-relativistic approximations and to the observed shifts" (Dyson 1948c, p. 617).

The striking features of Dyson's paper are the impressive computational abilities displayed, the maturity of the physical understanding indicated by the approximations made, and the mastery of the field-theoretic apparatus that was necessary to carry out the difficult calculations in record time. Dyson's assessment of his work when he had finished the calculations was that he "had done nothing in the last two months that you could call very clever or difficult; nothing one tenth as hard as my fellowship thesis; but because the problems I am now dealing with are public problems and all the theoretical physicists have been racking their brains over them for ten years with such negligible results, even the most modest contributions are at once publicized and applauded."[72] He also warned himself should he have the luck "to do something really clever in this field," he would have to be very careful not to have his head turned.[73]

It had become quite clear to Bethe as he watched Dyson's rapid progress and impressive performance in this Lamb shift calculation that Dyson was an unusually gifted and talented young physicist. In early December came the issue of the renewal of Dyson's fellowship and, in connection with this, the question what he should do the following year. Bethe indicated to Dyson that he would strongly recommend to the Commonwealth Fund that his fellowship be extended for another year, because "this was in the interests of science as well as your own interests."[74] He suggested to Dyson that he spend the year with Oppenheimer at the Institute for Advanced Study at Princeton. It was also evident that Bethe had already talked to Oppenheimer, because he could tell Dyson that Oppenheimer "would be very glad to look after him."[75] The letter of recommendation Bethe sent to the Commonwealth Fund was unequivocal:

> Mr. Dyson is absolutely unusual in his ability and accomplishments. I can say without reservation that he is the best I have ever had or observed. He is far superior, even to the best American graduate students we have at Cornell. In the short time here he has made very important contributions to theoretical physics....
>
> If Dyson is enabled to spend a year with Oppenheimer, he will broaden his knowledge of this field to which he has already contributed so much.[76]

After the Lamb shift calculation had been completed, Bethe asked him to look at the application of the newly developed renormalization methods to meson-theoretic problems: the problem of the nuclear forces, meson-nucleon scattering and meson production in γ-ray–nucleon collisions. The last two problems were of great interest, because the synchrotron that Wilson was building at Cornell was rapidly approaching completion and was due to come into operation during the spring of 1948. Once again within a short period Dyson obtained interesting results. At the end of February 1948 he submitted a note to the *Physical Review* on "The Interactions of Nucleons with Meson Fields" (Dyson 1948d), in which his results on the equivalence of certain kinds of couplings in meson theories were reported.

Actually, after Dyson had completed the Lamb shift calculation for a spin zero particle, it would have been logical for him to undertake the Lamb shift calculation for a Dirac particle. "That was the number one problem." Bethe, however, had already assigned that problem to Richard ("Dick") Scalettar during the summer of 1947. Dyson points out:

> The spin 1/2 case is really easier. Most of the difficulties I had were specific to spin 0. You really had ambiguity in how to define things which would have been easier in spin 1/2. The spin 1/2 case is more complicated in terms of the algebra but the formulation of the problem is in fact easier. It was sort of sad for Bethe that we didn't get the spin 1/2 done. It shouldn't have taken us a year to get that done. It was just accidental that Scalettar had the job and couldn't do it [quickly]. I couldn't grab it away from him.[77]

Scalettar and Dyson were in fact good friends and "very close." Both were bachelors and they always ate together. Not only couldn't he grab the problem away from him, but he had no wish "to push him aside." "So I did meson theory."[78]

During the spring of 1948 Dyson immersed himself ever more deeply in the formalism of the quantum theory of fields and the problems of quantum electrodynamics. In late February a package of physics journals had arrived at Cornell, including the first issue of *Progress of Theoretical Physics* that contained Yukawa's article on invariant quantization (Yukawa 1947) and Tomonaga's seminal paper on how to generalize Dirac's many time formalism so that the formalism could be applied to a quantum field (Tomonaga 1943b, 1946). Dyson was quick to recognize their importance and was genuinely elated that these Japanese workers "had caused rather a sensation because they had done such a lot of first rate work in isolation of the rest of the world." He wrote home to his family: "The reason that everyone is so enormously pleased with this work of Tomonaga is partly political. Long-sighted scientists are worried by the growing danger of nationalism in American Science.... Apart from these considerations, the flowering of physics in present day Japan is a wonderful demonstration of the resilience of the human spirit, and is admired and welcome for its own sake."[79]

In March 1948 preparations were underway for the forthcoming Pocono conference, and Bethe had asked Oppenheimer that Dyson be invited. The unfolding of the drama is related in one of Dyson's letters home:

> You can imagine that I was highly pleased and flattered when
> Bethe told me that he asked for me to be invited; though I thought
> right from the start that it was a little too good to be true, and that
> it would be rather invidious for me to be invited over the heads
> of other research students. In consequence I was more amused
> than sorry when Bethe came in a few days ago and announced
> awkwardly and apologetically that he had pleaded hard with Op-
> penheimer (who is acting as host), but that Oppenheimer insisted
> that if it was to be any good the conference must be kept small,
> and they could not begin inviting students.[80]

When Bethe received the Tomonaga letter that Oppenheimer circulated shortly after the conference—the letter which contained a summary of the Japanese effort in quantum field theory—Dyson was once again overjoyed: "It seemed like a miracle that they were able to do anything at all. We had thought of them as down and out and here they were doing something exciting."[81]

But there was another reason why Dyson reacted so favorably to the Japanese contribution. The difference between Tomonaga's approach and Schwinger's was that Tomonaga's presentation was simple. Dyson had found Schwinger's theory, as outlined in the Pocono conference notes, complicated and shrouded in difficulties that somehow only Schwinger could manage to overcome. Tomonaga's papers *demystified* relativistic quantum field theory.

Dyson was very impressed by the clarity of Tomonaga's presentation. Although Tomonaga and Schwinger had essentially the same idea on how to formulate a quantum field theory in a relativistically invariant manner, "Tomonaga expressed his in a simple, clear language so that anybody could understand it and Schwinger did not. When you read Schwinger you had the impression it was immensely complicated from the start. Tomonaga set the framework in a very beautiful way. To me that was very important. It gave me the idea that this was after all simple."[82]

9.7 The Michigan Symposium, Summer 1948

In early March 1948 Weisskopf visited Cornell and presented an account—"rather garbled in transit" Dyson told his parents[83]—of Schwinger's new version of quantum electrodynamics, work that he had announced at the New York meeting but had not finished at that time (Schwinger 1948b,c). "This new theory is a magnificent piece of work," Dyson wrote home, "difficult to digest but with some highly original and undoubtedly correct ideas." Dyson was working hard at "understanding it thoroughly."[84] In this same letter Dyson remarked that Feynman had been attacking the same problem as Schwinger from a different direction, "and has now come out with a roughly equivalent theory, reaching many of the same ideas independently; this makes it pretty clear that the theory is right." He concluded his letter by indicating that "Feynman is a man whose ideas are as difficult to make contact with as Bethe's are easy; for this reason I have so far learnt much more from Bethe, but I think if I stayed here much longer, I should begin to find that it was Feynman with whom I was working more."[85]

Already in the fall semester Dyson had had some discussions with Feynman during which Feynman had drawn many diagrams on the board. Dyson had been "tremendously impressed" by Feynman's calculational abilities. "He could calculate all these things much quicker than we could. He had these magic tricks. Everything came out in 2 lines instead of 10 pages."[86] In November 1947, just after getting to know him—they had driven together to Rochester to the first of the Joint Rochester-Cornell conferences on theoretical physics—Dyson wrote of his admiration for Feynman: "[He is] the brightest of the young theoreticians here, and . . . the first example of that rare species, the native American Scientist." Feynman "is always sizzling with new ideas . . . hardly any of which get very far before some newer inspiration eclipses them." Moreover, as a sustainer of morale, Feynman was invaluable: "When he bursts into the room with his latest brainwave and proceeds to expound it with the most lavish sound effects and waving about the arms, life . . . is not dull."[87]

By the spring of 1948, Dyson and Feynman were friends. They had not only talked physics, but they had shared with one another some of their war experiences. During the spring Feynman explained at length to Dyson his own approach to quantum mechanics, and to quantum electrodynamics. "He would write things straight down instead of solving equations." This bothered Dyson.

> To me it was magic. I felt all the time that there must be some way to understand this differently. It wasn't clear that what he was doing was right. Nothing had really been firmly checked. [By the spring of 1948] there were still some discrepancies between Feynman's Lamb shift [calculation] and other people's Lamb shift and it wasn't clear that Feynman would turn out to be right.
>
> So I had a sort of open mind about that. He was just guessing. Maybe it's true, maybe it isn't true. Also he [Feynman] didn't believe in vacuum polarization and that was something that could be checked experimentally. I was skeptical about his approach.[88]

It should be stressed that during the time that Dyson was at Cornell, Feynman never gave a coherent exposition of his work. "Everything I learned from Feynman was in his office. He would talk to me at the blackboard. He never gave a seminar."[89]

Dyson had studied the notes that Wheeler issued after the Pocono conference containing some of the details of Schwinger's and Feynman's presentations. It was clear to him that the problem of the relationship between Feynman's and Schwinger's formulation of quantum electrodynamics was something that had to be addressed.

The spring term passed rapidly. In the beginning of June, Bethe left Cornell for his summer travels in Europe. The separation was a difficult one for Dyson and came "with unexpected suddenness." Bethe gave him "his parting blessing" and warned Dyson not to expect too much at Princeton. "You can imagine that it is sad to lose such a uniquely patient and helpful teacher," Dyson wrote his parents, "and the building sounds empty deprived of his ringing laughter."[90]

Dyson was pleased with what he had accomplished during his year at Cornell, and he had every reason to be satisfied. He had become recognized as one of the most promising of the upcoming young theoreticians, and had carried out some important research. Although he could write to his sister Alice that he believed himself "a success as a physicist" and that "life [was] generally busy and happy enough," he did have real regrets. "I live now more than ever for myself alone and seem further than ever from social usefulness." The prospect of living the life of a succesful physicist was quite satisfactory and suited him "very well." But physics is "a fundamentally competitive game, and it is not good to get too buried in it." This, and the example of his married friends around him, "make me look more and more to marriage as the necessary and desirable corrective."[91]

Having worked very hard since coming to the United States, Dyson considered spending the summer in Mexico. A trip to Mexico was being sponsored by the Friends Relief Committee and was to last from July 1 to August 15. "The idea of the trip was to combine a holiday with a gesture of international goodwill, and they want to collect people from as many countries and religions as possible; the students will spend much of the time on manual labour, in particular in building a hospital in some rather inaccessible part of the country, and the rest of the time, I suppose, trying to fraternize with the Mexicans....I should certainly be foolish to miss such an opportunity."[92] But Dyson eventually changed his mind and decided instead to go to the Michigan Summer Symposium in Ann Arbor because Schwinger was going to lecture there on his most recent work.

His plans were greatly helped "by an offer of a ride across the country by Feynman." Feynman was driving to Albuquerque at the end of the semester and had offered to meet Dyson in Cleveland and take him along. Since Dyson had two weeks to kill before the start of the summer school he gladly accepted Feynman's invitation. Dyson has told the story of this ride in *Disturbing the Universe* (Dyson 1979).

At the summer school Dyson attended Schwinger's "excellent" lectures. He "worked a lot" and "worked hard to understand all the details of the lectures to his own satisfaction" and took every opportunity to discuss with Schwinger points that were not quite clear. Dyson found Schwinger "not an easily approachable person" but discovered that he was friendly and easy to talk to in private and "not all as forbidding as his reputation would have it." Dyson found satisfaction in the fact that by the time Schwinger departed, he knew "at least" who he was.[93]

Dyson came away from Ann Arbor with the feeling that Schwinger's theory was "unbelievably complicated." His impression was that this couldn't "be the right way to do it." In his lectures, Schwinger had dressed the theory up with details, and had displayed "tremendous virtuosity in doing integrals." The work was presented as "something which needed such skills that nobody besides Schwinger could do it. If you listened to the lectures you couldn't see the motivation; it was all hidden in this wonderful apparatus."[94]

From Michigan, Dyson communicated to Bethe, who was vacationing in Switzerland, the results he had obtained since he had last seen him, and Bethe urged him to write them up and publish them.[95] Dyson was surprised at Bethe's suggestion, for he had not intended "to publish any of it, without having bothered to think the matter out." In reporting this incident to his family, Dyson noted that upon analyzing his motives for not wishing to publish, "I found that unconsciously I am expecting to do much more important things soon, which is an encouraging sign. Anyway to publish this old stuff would mean a lot of work and it is all being done by other people too."[96]

This letter from Dyson to his family was written from San Francisco on August 26. After the Michigan summer school was over, Dyson had taken a Greyhound bus to Berkeley, where he was to spend a week meeting with people there. He was restless during most of his stay in Berkeley. "I cannot find much to do here," he wrote home, "one cannot just go around bothering people all day. So I go to meetings and spend a lot of time reading in the library. I shall have had quite enough of it by the time one week is over."[97]

On September 2, Dyson boarded the bus to go back East.

9.8. Princeton: The Institute for Advanced Study

On arriving in Princeton, Dyson wrote his family:

> On the third day of the journey a remarkable thing happened; going into a sort of semi-stupor as one does after 48 hours of bus-riding, I began to think very hard about physics, and particularly about the rival radiation theories of Schwinger and Feynman. Gradually my thoughts grew more coherent, and before I knew where I was, I had solved the problem that had been in the back of my mind all this year, which was to prove the equivalence of the two theories. Moreover, since each of the two theories is superior in certain features, the proof of the equivalence furnished incidentally a new form of the Schwinger theory which combines the advantages of both.[98]

The same process of solving a difficult problem occurred in Cambridge when Dyson had addressed the rank congruence problem. Then too, he thought "very hard" about it, had seen what needed to be done, and after some preliminary calculations, had established how long it would take. He then sat down and wrote out his findings in essentially the time he had believed it would take.

To his family, Dyson characterized his work as "neither difficult nor particularly clever, but it is undeniably important," especially if nobody else had done it in the meantime. He was particularly elated because he would now "encounter Oppenheimer with something to say which will interest him" and thus be able "to gain at once some share of his attention."[99]

A week later, Dyson wrote his family that his

> work consisted of a unification of radiation theory, combining the advantageous features of the two theories put forward by Schwinger and Feynman. Now it happened that Feynman and Schwinger talk such completely different languages, that neither of them is able to understand properly what the other is doing. It also happened that I was almost the only young man in the world who had worked with the Schwinger theory from the beginning and had also had long personal contact with Feynman at Cornell, so I had a unique opportunity to put the two together.[100]

What Dyson realized on the bus was that he could write a solution of the Tomonaga-Schwinger equation

$$i\hbar \frac{\delta}{\delta\sigma(x)} \Psi(\sigma) = \mathcal{H}_I(x)\Psi(\sigma) \tag{9.8.1}$$

as a power series expansion

$$\Psi(\sigma) = \sum_{n=0}^{\infty} \left(-\frac{i}{\hbar c}\right)^n \int_{\sigma_0}^{\sigma} d^4x_1 \dots \int_{\sigma_0}^{\sigma} d^4x_n P\left(\mathcal{H}_I(x_1) \dots \mathcal{H}_I(x_n)\right)\psi(\sigma_0), \tag{9.8.2}$$

where P is a formal operator, the "chronological ordering operator" that rewrites the factors $\mathcal{H}_I(x_1) \dots \mathcal{H}_I(x_n)$ according to their time label, that is,

$$P\left(\mathcal{H}_I(x_1) \dots \mathcal{H}_I(x_n)\right) = \mathcal{H}(x_{i_1}) \dots \mathcal{H}(x_{i_n})$$
$$\text{if} \quad x_{i_{10}} > x_{i_{20}} > \dots x_{i_{n0}}. \tag{9.8.3}$$

The idea of chronologically ordering operators had been introduced by Feynman in his calculus of ordered operators and Dyson certainly knew of Feynman's calculus. However, the *explicit* introduction of the P operator was psychologically important. It greatly simplified the perturbative calculus and made transparent why eq. (9.8.2) was a solution of eq. (9.8.1).

In Chicago, where Dyson stopped for a few days on his way east from San Francisco, working intensely, he convinced himself that with this time-ordered perturbative solution, he could derive the entire Feynman theory with the Feynman Green's functions for the electron and photon propagators appearing naturally: the latter were essentially the vacuum expectation values of $P(\bar{\psi}_\gamma(x)\psi_\beta(y))$ and $P(A_\mu(x)A_\nu(y))$, respectively.

Although he didn't have enough time in Chicago to work out all the details of his insight, after some calculations he felt confident that he had indeed solved the problem. In great excitement he wrote Bethe a letter "announcing the triumph":

> I have succeeded in re-formulating the Schwinger method, without any changes of substance, so that it gives immediately all the advantages of Feynman theory. This means, one can now write down 2nd and 4th order radiative corrections in a very concise form and with a minimum of labour, and there are general formulae for the radiative corrections of $2n^{th}$ order. The method is also a little more foolproof than the Feynman theory, since it gives the signs of the various terms automatically; also it includes vacuum-polarization terms on an equal

footing with the others. Incidentally, the complete equivalence
of Schwinger and Feynman is now demonstrated.

I think it may now not be impossibly difficult to prove
by general arguments the finiteness of all radiative corrections.[101]

But Dyson added the cautionary remark, "However, that is as it may be."

What had to be done was outlined in his letter to his parents at the beginning of September. First, to write up the connection between Schwinger and Feynman "and get that straight." Second, to look at the higher orders.[102] This Dyson set out to do when he arrived in Princeton. He comments:

> The big change in my thinking, what I didn't understand on
> the bus, was that the thing to calculate is the S-matrix. On the
> bus I was still thinking in the Schwinger style, calculating by
> transforming the perturbations away and getting new Hamilto-
> nians and translating that into Feynman propagators. Only after
> I came to Princeton and started writing did it become obvious
> that much the simpler approach was to say: "Look we are cal-
> culating scattering and it is just a matter of writing down the
> S-matrix. Then everything is much simpler." Then you can re-
> ally start doing the higher order. Then I started working on the
> S-matrix.[103]

Dyson was familiar with S-matrix theory. Pauli and Stueckelberg had talked about Heisenberg's work on the S-matrix at the International Conference on Fundamental Particles and Low Temperature that had been held at the Cavendish in July 1946. Although Dyson had not attended the conference, Kemmer did, and the importance of Heisenberg's program had been impressed on Kemmer. Thus, after he arrived in Cambridge, Dyson read Heisenberg's papers (Heisenberg 1943a,b, 1944). He recalls that at that time the accepted view was that quantum field theory was totally inadequate and that a radically new approach was needed to overcome the difficulties. One need only note that the title of Bohr's opening address at the Cavendish conference had been "*Problems* of Elementary-Particle Physics," that of Pauli's paper "*Difficulties* of Field Theories and of Field Quantization" and that of Dirac "The *Difficulties* in Quantum Electrodynamics" to corroborate that impression. During his year at Cambridge, Dyson had also studied the different suggestions that were being made to overcome the divergence difficulties in the higher orders of perturbation theory. He remembers giving a seminar on Heitler's and Ma's works that eviscerated higher-order perturbation theoretic calculations (Heitler 1941; Ma 1943), and criticizing that approach as faulty and unconvincing. "It came as a great shock to me in Princeton to find that the work we had been doing was calculating the S-matrix: Heisenberg was only *talking* about it; that we could do all that Heisenberg wanted to do and do it right. Heisenberg's work was always presented as being new physics so the surprising thing was to find the old physics led to it in a natural way."[104]

Neither Feynman nor Schwinger had been "talking S-matrix." At the Michigan summer school, Schwinger had computed the radiative corrections to the one-particle current operator and had obtained an effective *Hamiltonian* for a one-particle system with which to calculate the Lamb shift. It should also be stressed that in proving the equivalence between Schwinger's and Feynman's radiation theories, Dyson simultaneously solved the problem of formulating Feynman's approach as a *field theory*. Recall that until 1949 Feynman had thought of electrons as "particles." The idea of writing down a field equation for the "electron-positron" field and treating the interaction between this matter field and the Maxwell field by a term $\bar{\psi}\gamma_\mu A^\mu \psi$ in the Hamiltonian, with the $\bar{\psi}$, ψ, and A all field operators, was completely foreign to Feynman (see chap. 8, sec. 13). Dyson stresses this point. "Nobody at Cornell understood that the electron field was a field like the Maxwell field. That was something that was in Wentzel but was nowhere else. That was what was lacking in the old fashioned way of calculating. The electron was a particle, the photon was a field, and the two were just totally different. This notion of just two interacting fields with this simple interaction term $\bar{\psi}\gamma_\mu A^\mu \psi$ was essentially what I brought to Cornell with me from England out of Wentzel's book."[105] This was also the approach that Schwinger had stressed at Pocono and Ann Arbor.

Dyson's proof of the equivalence of the theories of Feynman and Schwinger was an important advance. It had been clear, to Dyson and to others, that the Schwinger formalism was too cumbersome to permit higher-order calculations to be performed routinely, and thus precluded the possibility of answering rigorously the question whether the theory could be made finite to every order of perturbation. Dyson recognized what was most valuable in Schwinger's and Feynman's approaches and his synthesis kept the advantages of the respective methods.

But to make his mark and prove his mettle he would have to identify and solve a problem that neither Feynman nor Schwinger had addressed and make *it* his own. The proof that after mass and charge renormalization the S-matrix for quantum electrodynamics was finite when calculated to any order of perturbation theory became that problem. He made it *his* problem and with great virtuosity and insight conquered it.

9.9 The "Radiation Theories" Paper

The intensity with which Dyson worked upon his arrival to Princeton is conveyed in his letter to his parents:

> After a brief visit to Cornell, to collect my various belongings, I settled down to work at writing up the physical theories I mentioned in the last letter. I found as one often does when one comes to write, that the job was even bigger than I had imagined,

and I was for about five days stuck in my rooms, writing and thinking with a concentration which nearly killed me. On the seventh day, the paper was complete and with immense satisfaction I wrote the number 52 at the bottom of the last page. Having emerged from that, I feel I shall not do any more thinking for the rest of the year. However, there are still a number of rather delicate calculations which have to be checked, references to be found, and the whole job of duplication lies ahead; so I am keeping on working for some days longer; I am naturally impatient to get the thing off my hands.

While I was struggling to get these ideas into shape, I thought they were so difficult I should never make them properly intelligible; however reading the paper through after it was done, it seemed so simple and clear as hardly to be worth the effort expended on it. It is, in fact, impossible for me to judge at present whether the work is as great as I think it may be. All I know is, it is certainly much the best thing I have done yet.[106]

The primary aim of Dyson's paper was to simplify the Schwinger theory for those who wanted to use it and to prove that it yielded Feynman's version. The abstract of the paper succinctly conveyed its content:

A unified development of the subject of quantum electrodynamics is outlined, embodying the main features of both the Tomonaga-Schwinger and of the Feynman radiation theory. . . .

The chief results obtained are (a) a demonstration of the equivalence of the Feynman and Schwinger theories and (b) a considerable simplification of the procedure involved in applying the Schwinger theory to particular problems, the simplification being the greater the more complicated the problem.

Dyson started his paper by sketching Tomonaga's and Schwinger's devistions of their eponymous equation, which describes the evolution of interacting quantized fields. If $\mathcal{H}(\mathbf{r})$ denotes the energy density of the interacting matter and radiation fields,

$$\mathcal{H}(\mathbf{r}) = \mathcal{H}_0(\mathbf{r}) + \mathcal{H}_I(\mathbf{r}), \tag{9.9.1}$$

where $\mathcal{H}_0$ is the energy density of the free electromagnetic and electron-positron fields and $\mathcal{H}_I(\mathbf{r})$ is that of their interaction with each other and any external (classical prescribed) fields that may be present, then the Tomonaga-Schwinger equation for the system is

$$i\hbar\frac{\delta\Psi(\sigma)}{\delta\sigma(x_0)} = \mathcal{H}_I(\mathbf{r}, x_0)\Psi(\sigma), \quad (9.9.2)$$

where

$$\mathcal{H}_I(\mathbf{r}, x_0) = e^{-\frac{i}{\hbar}H_0 x_0}\mathcal{H}_I(r)e^{+\frac{i}{\hbar}H_0 x_0} = \mathcal{H}_I(x)$$

$$= -\frac{1}{c}\bar{\psi}(x)\gamma_\mu\psi(x)[A^\mu(x) + A^{\mu e}(x)]$$

$$= \mathcal{H}^i(x) + \mathcal{H}^e(x) \quad (9.9.3)$$

with

$$H_0 = \int \mathcal{H}_0(\mathbf{r})d^3\mathbf{r}. \quad (9.9.4)$$

Dyson noted that both sides of eq. (9.9.2) are relativistic invariants. The Tomonaga-Schwinger equation thus avoids one of the most unsatisfactory features of the old theories, namely that "the invariant $\mathcal{H}_I$ was added to the non-invariant $\mathcal{H}_0$." The Schwinger perturbative solution of eq. (9.9.2) was then written down,

$$U(\sigma, -\infty) = 1 - \frac{i}{\hbar c}\int_{-\infty}^{\sigma}\mathcal{H}_I(x_1)d^4x_1$$

$$+ \left(-\frac{i}{\hbar c}\right)^2\int_{-\infty}^{\sigma}d^4x_1\int_{-\infty}^{\sigma}d^4x_2\mathcal{H}_I(x_1)\mathcal{H}_I(x_2) + \dots, \quad (9.9.5)$$

and Dyson pointed out that $U(\infty, -\infty)$ is identical with the Heisenberg S-matrix. Mass renormalization was formally effected by writing

$$\mathcal{H}(r) = (\mathcal{H}_0(r) + \delta mc^2\bar{\psi}\psi(r))$$

$$+ (\mathcal{H}_I(r) - \delta mc^2\bar{\psi}\psi(r)) \quad (9.9.6)$$

and redefining $\mathcal{H}_0(r) + \delta mc^2\bar{\psi}\psi(r)$ as the energy density of the electromagnetic field and that of the matter field with the electron having its *observed* rest mass. The value of δm is to be adjusted so as to cancel the self-energy effects in $U(\infty, -\infty)$. No such adjustment is needed for the photon self-energy since the photon self-energy must be identically zero by gauge invariance.

In the absence of external fields (i.e., when $A^e_\mu(x) = 0$) the vacuum, Ψ_0, and one particle states, Ψ_1—that is, one-photon states, one-electron states, and one-positron states—have the property of being "steady," that is, they satisfy

$$U(\infty, -\infty)\Psi_0 = \Psi_0 \qquad (9.9.7)$$

$$U(\infty, -\infty)\Psi_1 = \Psi_1. \qquad (9.9.8)$$

In the presence of an external field, the effect of radiative corrections is incorporated in the unitary operator $S(\sigma, -\infty)$ that satisfies

$$i\hbar\frac{\delta S}{\delta\sigma(x)} = \mathcal{H}^i(x_0)S$$

$$\mathcal{H}^i(x) = -\frac{1}{c}j_\mu(x)A^\mu(x), \qquad (9.9.9)$$

whose perturbative expansion is given by

$$S(t, -\infty) = 1 - \frac{i}{\hbar c}\int_{-\infty}^{\sigma} d^4x_1 \mathcal{H}^i(x_1)$$

$$+ (\frac{-i}{\hbar c})^2 \int_{-\infty}^{\sigma} d^4x_1 \int_{-\infty}^{\sigma_1} d^4x_2 \mathcal{H}^i(x_1)\mathcal{H}^i(x_2) + \dots. \quad (9.9.10)$$

The state vector $\Omega(\sigma)$ defined by

$$\Psi(\sigma) = S(\sigma)\Omega(\sigma) \qquad (9.9.11)$$

then obeys the equation

$$i\hbar\frac{\delta\Omega}{\delta\sigma(x_0)} = S(\sigma)^{-1}\mathcal{H}^e(x_0)S(\sigma)\Omega$$

$$= \mathcal{H}_T(x_0)\Omega. \qquad (9.9.12)$$

Substituting into eq. (9.9.12) the expression (9.9.10) for $S(\sigma)$ and $S(\sigma)^{-1}$ yields

$$\mathcal{H}_T(x_0) = \sum_{n=0}^{\infty}(\frac{i}{\hbar c})^n \int_{-\infty}^{\sigma(x_0)} d^4x_1 \int_{-\infty}^{\sigma(x_1)} d^4x_2 \dots \int_{-\infty}^{\sigma(x_{n-1})} d^4x_n$$

$$[\mathcal{H}^i(x_n), [\dots [\mathcal{H}^i(x_2), [\mathcal{H}(x_1), \mathcal{H}^e(x_0)]\dots]. \qquad (9.9.13)$$

The repeated commutators in eq. (9.9.13) are characteristic of the Schwinger theory, and "their evaluation gives rise to long and rather difficult analysis" (Dyson 1949a, p. 491). Schwinger had obtained the $\frac{\alpha}{2\pi}$ correction to the magnetic moment of an electron and his value for the Lamb shift from $\mathcal{H}_T$, eq. (9.9.13), evaluated to second order.

In Feynman's theory, "the basic principle is to preserve symmetry between past and future" (Dyson 1949a, p. 492), and the matrix elements of the operator $\mathcal{H}_T$ are evaluated in a "mixed representation." If in the Schwinger theory the initial state and final states are specified by the vectors Ω_1 and Ω_2, respectively, then Dyson claimed that in the Feynman theory they are characterized by the state vectors Ω_1 and $S(\infty)^{-1}\Omega_2$, respectively. Since

$$
\begin{aligned}
(\Omega_2, \mathcal{H}_T\Omega_1) &= (S(\infty)^{-1}\Omega_2, \mathcal{H}_T\Omega_1) \\
&= (\Omega_2, S(\infty)\mathcal{H}_T\Omega_1),
\end{aligned}
\tag{9.9.14}
$$

the operator which replaces H_T in the mixed representation is

$$
\begin{aligned}
\mathcal{H}_F(x_0) &= S(\infty)\mathcal{H}_T(x_0) \\
&= S(\infty)S(\sigma)^{-1}\mathcal{H}_T(x_0)S(\sigma).
\end{aligned}
\tag{9.9.15}
$$

Dyson then derived the "fundamental" formula of Feynman by making use of the chronological operator. If $F_1(x_1)\ldots F_n(x_n)$ are any operators defined respectively at the points $x_1,\ldots x_n$ of space-time, then

$$
P(F_1(x_1)\ldots F_n(x_n))
$$

denotes the product of these operators taken in their chronological order, reading from right to left, in which the times $x_{10},\ldots x_{no}$ occur, that is, if

$$
x_{i_{1}0} > x_{i_{2}0} \ldots > x_{i_{n}0}
\tag{9.9.16}
$$

then

$$
P(F_1(x_1)\ldots F_n(x_n)) = F_{i_1}(x_{i_0})\ldots F_{i_{n}0}.
\tag{9.9.17}
$$

Since in most applications $F_i(x_i)$ commutes with $F_j(x_j)$ as long as x_i and x_j are outside each other's light cone, the right-hand side of eq. (9.9.17) can be generalized to mean ordering in which spacelike surfaces $\sigma(x_{i_1}), \sigma(x_{i_2})\ldots$ occur in time. Dyson then proved that by virtue of the symmetry of the integrand,

$$
\begin{aligned}
\mathcal{H}_F(x_0) &= S(\infty)\mathcal{H}_T(x_0) \\
&= S(\infty)S(\sigma)^{-1}\mathcal{H}^e(x_0)S(\sigma)
\end{aligned}
\tag{9.9.18}
$$

can be written as

$$\mathcal{H}_F(x_0) = \sum_{n=0}^{\infty} \left(-\frac{i}{\hbar c}\right)^n \frac{1}{n!} \int_{-\infty}^{+\infty} d^4 x_1 \dots \int_{-\infty}^{+\infty} d^4 x_n P\left(\mathcal{H}^e(x_0)\mathcal{H}^i(x_1) \dots \mathcal{H}^i(x_n)\right),$$
(9.9.19)

where the range of the integrations—from $-\infty$ to $+\infty$—is achieved by the replacement of the commutators with time-ordered products. When $\mathcal{H}^e(x_0)$ is replaced by the unit matrix, one obtains

$$S(\infty) = \sum_{n=0}^{\infty} \left(-\frac{i}{\hbar c}\right)^n \frac{1}{n!} \int_{-\infty}^{+\infty} d^4 x_1 \dots \int_{-\infty}^{+\infty} d^4 x_n P\left(\mathcal{H}^i(x_1) \dots \mathcal{H}^i(x_n)\right).$$
(9.9.20)

On the basis of eq. (9.9.20), Dyson then derived the Feynman rules for the class of problems involving initially and finally one charged particle and no photons (initial state A, final state B). The derivation of these rules "from what is fundamentally the Tomonaga-Schwinger theory constitutes the proof of the equivalence of the two theories" (Dyson 1949a, p. 493). Dyson showed that the matrix element of

$$P\left(\bar{\psi}(x_0)\gamma^{\mu_0}\psi(x_0)A^e_{\mu_0}(x_0)\bar{\psi}(x_1)\gamma_1^\mu\psi(x_1)A_{\mu_1}(x_1) \dots \bar{\psi}(x_n)\gamma_n^\mu\psi(x_n)A_{\mu_n}(x_n)\right)$$

between two such one-particle states is a sum of contributions, each contribution arising from a specific way of dividing the factors $\bar{\psi}(x_0)\psi(x_0)$ $\dots \psi(x_n)\psi(x_n)A_{\mu_1}(x_1) \dots A_{\mu_n}(x_n)$ into two single factors containing a single ψ and a single $\bar{\psi}$ and paired factors $\bar{\psi}(x_j)\psi(x_{r_j})$ and $A_{\mu_l}(x_1)A_{\mu_m}(x_m)$. The pairing of two A operators gives rise to a contribution that corresponds to the vacuum expectation of the P product:

$$\langle P(A_\mu(x)A_\mu(y)) \rangle_0 = \hbar c g_{\mu\nu} D_F(x-y),$$
(9.9.21)

while that of ψ and $\bar{\psi}$ pairing contributes

$$\langle P(\psi_a(x)\bar{\psi}_\beta(y)) \rangle = \frac{1}{2}\epsilon(x,y)S_F(x-y),$$
(9.9.22)

where S_F and D_F are the Feynman propagators for a photon and electron, respectively. The contribution of the nth order term eq. (9.9.20) to the transition matrix element from A to B is thus a sum of terms of the form

$$M = \epsilon' \prod_{i \neq k} \frac{1}{2} S_F\left(x_i - x_{r_i}\right) \prod_j \frac{1}{2}\hbar c D_F\left(x_{s_j} - x_{t_j}\right) \bar{\psi}(x_k)\psi(x_{r_k}),$$
(9.9.23)

where ϵ' is + or −1 depending only on the type of matrix element and not on the points $x_0 \ldots x_n$; ϵ' can readily be determined from the pairings involved. Each such matrix element M can be represented by a "graph" as follows: The points $x_0, \ldots x_n$ are represented by $n + 1$ points. For each paired factor $\bar{\psi}(x_i)\psi(x_{r_i})$ with $i \neq k$ (to which correspond a factor $S_F(x_i - x_{r_i})$), a directed "electron" line is drawn from x_i to x_{r_i}; for the single factor $\bar{\psi}(x_k)$, a directed "electron" line is drawn leading from x_k to the edge of the diagram; and for the single factor $\psi(x_{r_k})$, a directed line from the edge of the diagram to the point x_{r_k}; for each pair $(A(x_{s_i})A(x_{t_i}))$, an undirected "photon" line from x_{s_i} to x_{t_i}.

There exists clearly a one-to-one correspondence between types of matrix elements and graphs. Dyson's way of thinking is analytical and in his paper "graphs" were not drawn explicitly. I shall draw them for the convenience of the reader. Thus, for example, the graphs arising for $n = 2$ are shown in fig. 9.9.1.

Dyson pointed out that in Feynman's theory "the graph corresponding to a particular matrix element is regarded, not merely as an aid to calculation, but as a picture of the physical process which gives rise to that matrix element." In Feynman's hands this pictorial interpretation "has been used as the basis for the derivation of most of the results of the present paper" (Dyson 1949a, p. 496).

Dyson next proved that the matrix elements giving rise to the disconnected graphs such as those drawn in figures 9.9.1e and 9.9.2 do not contribute. Their contribution sums to $(\Psi_0, S(\infty)\Psi_0)$; but $(\Psi_0, S(\infty)\Psi_0) = 1$. Dyson then showed that the contribution of the diagrams indicated by figure 9.9.3a,b,c . . . can

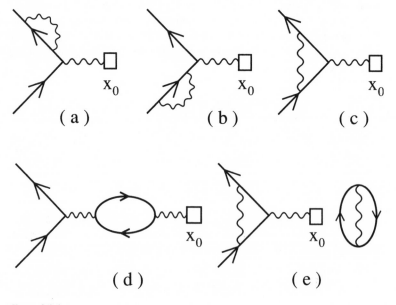

Figure 9.9.1

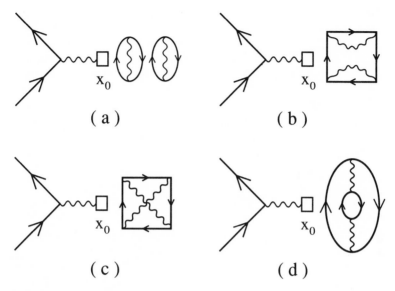

Figure 9.9.2

be obtained by replacing $\langle B|\bar{\psi}(x_0)|0\rangle$ in the lowest-order matrix element by

$$\langle B|\sum_{n=0}^{\infty}(-\frac{i}{\hbar c})^n\frac{1}{n!}\int d^4y_n P(\bar{\psi}(x_0)\mathcal{H}^I(y_1)\ldots\mathcal{H}^I(y_n))|0\rangle.$$

Dyson then indicated that relativistic invariance dictates that this matrix element reduces to a constant multiple of that obtained from $\bar{\psi}(x_0)$, that is, the sum is obtained by the replacement $\bar{\psi}(x_0) \rightarrow R_1\bar{\psi}(x_0)$, where R_1 an absolute constant (albeit infinite!). Similarly, the contributions of the diagrams of figure 9.9.4 can be obtained by the replacement $\bar{\psi}(x_0) \rightarrow R_1\bar{\psi}(x_0)$. Analogously, all the diagrams that arise by inserting into a directed electron from x_1 to x_2 all possible "self-energy" insertions (i.e., the contribution indicated by figures 9.9.5a,b,c) are obtained by the replacement of $S_F(x_2 - x_1)$ by $S'_F(x_2 - x_1)$, where

$$\frac{1}{2}\epsilon(x_2 - x_1)S'_F(x_2 - x_1) = \langle\sum_{n=0}^{\infty}(-\frac{i}{\hbar c})^n\frac{1}{n!}\int d^4y_1\ldots\int d^4y_n$$
$$P(\bar{\psi}(x_2)\psi(x_1)\mathcal{H}^I(y_1)\ldots\mathcal{H}^I(y_n))\rangle_0. \quad \textbf{(9.9.24)}$$

The self-energy insertions into a photon line joining the points x_1 and x_2 (fig. 9.9.6) are obtained by replacing $D_F(x_1 - x_2)$ by $D'_F(x_1 - x_2)$, where

$$\frac{1}{2}g_{\mu_1\mu_2}D_F(x_1 - x_2) = \langle\sum_{n=0}^{\infty}(-\frac{i}{\hbar c})^n\frac{1}{n!}\int d^4y_1\ldots\int d^4y_n$$
$$P(A_{\mu_1}(x_1)A_{\mu_2}(x_2)\mathcal{H}^I(y_1)\ldots\mathcal{H}^I(y_n))\rangle_0. \quad \textbf{(9.9.25)}$$

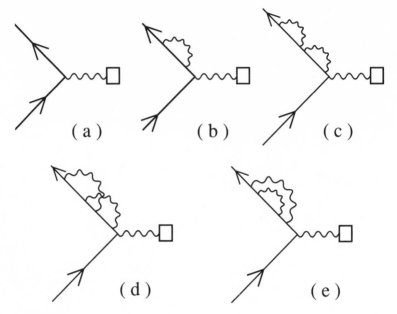

Figure 9.9.3

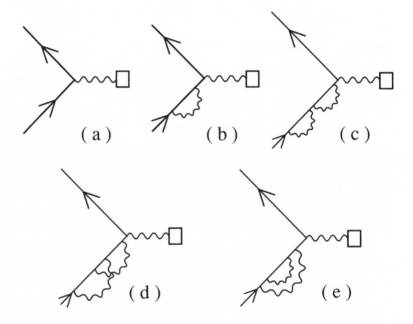

Figure 9.9.4

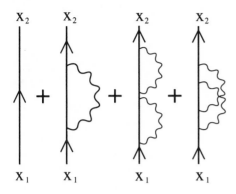

Figure 9.9.5

Dyson pointed out that as a result of these observations, the contribution of graphs with self-energy parts can always be obtained from certain "irreducible" graphs containing no self-energy parts, by replacing the S_F and D_F factors in the latter by S'_F and D'_F functions. This "elimination of graphs with self-energy parts is a most important simplification of the theory" (Dyson 1949a, p. 498). Dyson conjectured that S'_F has a series expansion of the form

$$S'_{F\beta\alpha}(x) = (R_2 + a_1(\Box + \kappa_0^2) + a_2(\Box + \kappa_0^2)^2 + \ldots)S_{F\beta\alpha}(x) \qquad (9.9.26)$$
$$+ (b_1 + b_2(\Box + \kappa_0^2) + \ldots b_2(\Box + \kappa_0^2)^2 + \ldots)(i\gamma\partial + \kappa_0)_{\beta\rho}S_{F\rho\alpha}(x),$$

where the a, b are numerical coefficients, and that D'_F has the following series representation:

$$D'_F(x) = (R_3 + c_1\Box + c_2\Box^2, + \ldots)D_F(x). \qquad (9.9.27)$$

Dyson also indicated that "it is believed (this has been verified only for second order terms)" that all the nth order matrix element of (9.9.20) will involve the factors R_1, R_2, R_3 only in the form of a multiplier $(eR_2R_3^{1/2})^n$. "Now the only possible experimental determination of e is by means of measurements of the effects described by various matrix elements of (31), and so the directly measured quantity is not e but $eR_2R_3^{1/2}$. Therefore, in practice the letter e is used to denote this measured quantity and the multipliers R no longer appear explicitly in the matrix elements of [eq. 9.9.20]; the change in the meaning of the letter e is called 'charge renormalization,' and is essential if e is to be identified with the observed electronic charge. As a result of the renormalization, the divergent coefficients R_1, R_2, and $R_3 \ldots$ are to be replaced by unity, and the higher coefficients $a, b,$ and c by expressions involving only the renormalized charge e."

Finally the "external vacuum polarization" induced by the potential A^e_μ can be treated in precisely a similar manner. Graphs describing external vacuum effects are those where the point x_0 is connected with the rest of the graph by a single photon line (fig. 9.9.7).

Effectively, these graphs are obtained by a replacement $A^e_\mu(x) \to A^{e'}_\mu(x)$. After a renormalization of the unit of potential "similar to the renormalization of

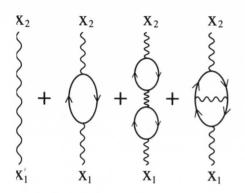

Figure 9.9.6

charge," the modified potential $A^{e'}$ takes the form

$$A_\mu^{e'}(x) =$$
$$(1 + c_1\square + c_2\square^2 + \ldots)A_\mu^e(x),$$
$$(9.9.28)$$

where the coefficients are the same as in eq. (9.9.27).

The penultimate section of the paper summarized its content by restating the rules for writing down the matrix elements of $\mathcal{H}_F(x_0)$. The concluding section illustrated the advantage of Feynman's approach by deriving the expression from which the second-order radiative corrections could be computed and contained the only diagram in the paper. After some delay in the typing of the manuscript—the secretary had gotten ill—the paper was sent to the *Physical Review* on October 4, 1948. The hurry in which it was written allowed several mistakes in it to go unnoticed. In "Notes added in Proof" Dyson had to warn the reader that some of its conclusions were wrong. But the basic insight that the divergences could be absorbed into multiplicative renormalizations of the charge was correct. It was to be the cornerstone for the analysis of the S-matrix of QED in the general case.

9.10 The Institute for Advanced Study: Oppenheimer

In the fall of 1929, a month or so before the crash of the stock market, Louis Bamberger and his sister Carrie Bamberger Frank Fuld sold their Newark,

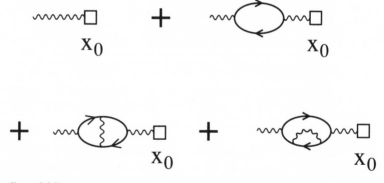

Figure 9.9.7

New Jersey, business to R. H. Macy in order to devote their energies and fortunes to philanthropy. Their own interest in education, and the advice of their friends and counselors, led them to explore the possibility of establishing a postdoctoral research and teaching institution. They sought out Abraham Flexner, who had argued for the need for such an institution and had outlined a model for an institute for advanced studies. Although initially the Bambergers had thought of endowing a "graduate school, in which the students would have an opportunity to pursue work qualifying them for a higher degree, and in which the Faculty, unburdened by the teaching of undergraduates, would be free to devote themselves to their researches, and the training of graduate students,"[107] economic factors, among other things, were responsible for their acceptance of Flexner's suggestion for an Institute for Advanced Study. In a memorandum to the Bambergers, Flexner sketched his ideas. His institute was designed so that it "should be small, that its staff and students or scholars should be few, that administration should be inconspicuous, inexpensive, subordinate, that members of the teaching staff, while freed from the waste of time involved in administrative work, should freely participate in decisions involving the character, quality, and direction of its activities, that living conditions should represent a marked improvement over contemporary academic conditions in America, that its subjects should be fundamental in character and that it should develop gradually."[108]

Flexner conceived of the permanent faculty of his institute as "contributors to the progress of knowledge and the solution of problems" and also as teachers "choosing a few competent and earnest disciples engaged in mastery of a subject." In addition, "The institute will be neither a current university, struggling with diverse tasks and many students, nor a research institute, devoted solely to the solution of problems. It may be pictured as a wedge inserted between the two—a small university, in which a limited amount of teaching and a liberal amount of research are both to be found."[109] Flexner envisioned the growth of the institution as follows:

> It should, one by one, as men and funds are available—and only then—create a series of schools or groups—a school of mathematics, a school of economics, a school of history, etc. The "schools" may change from time to time; in any event, the designations are so broad that they may readily cover one group of activities today, quite another as time goes on. Thus, from the outset the school of mathematics might well contain the history or philosophy of science.... Each school should conduct its affairs in its own way; for neither the subjects nor the scholars will all fit in one mold.[110]

In 1930 the Bambergers contributed $5 million to endow the Institute for Advanced Study in Princeton. The following year Flexner was appointed its first director and Oswald Veblen was invited to become the Institute's first professor of mathematics. The acceptance of a permanent position by Albert Einstein in 1933 and the appointment of James Alexander, Marston Morse, John von Neumann, and Hermann Weyl to the faculty of the School of Mathematics established the Institute's eminence and status. For the first few years after its founding, the Institute was housed in Fine Hall on the campus of Princeton University. With the completion of Fuld Hall in 1939, it moved from its temporary quarters to its present Olden Farm tract of 500 acres.

The Institute's reputation grew during the thirties, and it achieved distinction not only in mathematics. Because of the presence of Einstein and by virtue of its distinguished visitors, among them Bohr, Dirac, and Pauli, theoretical physics acquired equal fame. Furthermore, each year a number of younger theoreticians came as visiting members and helped establish the Institute's repute as an outstanding center in theoretical physics.

When in the fall of 1947 Oppenheimer accepted an appointment as professor of theoretical physics in the School of Mathematics and director of the Institute for Advanced Study at a salary of $20,000 a year, his contract stipulated that he was expected "with the advice and consent of the Trustees to determine an academic policy for the Institute as a place of learning and study."[111] Under Oppenheimer's leadership the Institute expanded. In his report to the Board of Trustees covering the first five years of his directorship, Oppenheimer noted that in the years from 1948 through 1953 "the work of the Institute for Advanced Study has grown, not dramatically, but steadily. Our faculty has become larger; our annual membership has increased; we have added new buildings."[112]

When Dyson arrived at the Institute in September 1948, these new buildings were just being completed. Until they could move into the new quarters, the visiting fellows in physics had desks in the large office of the director. This was a legacy of Oppenheimer's California days, when his graduate students and postdoctoral fellows similarly shared space and desks in his office. This had not caused much difficulty during Oppenheimer's first year at the Institute since there were only four fellows working with him, and he spent most of his time in Washington. Nor did it create too much havoc in the fall of 1948, even though the number of new postdoctoral fellows was much larger, as Oppenheimer was traveling in Europe until the new buildings were ready for occupancy.

For the academic year 1948/49, the visiting fellows in physics were Kenneth Case, Daniel Feer, Robert Karplus, Joseph Lepore, Fritz Rohrlich—all students of Schwinger—Bruria Kaufman, Norman Kroll, Cecile Morette, Sheila Power, Jack Steinberger, and Kenneth Watson. All were outstanding and all went on to distinguished careers in physics. Also visiting that year was George Uhlenbeck from the University of Michigan, a close friend of Oppenheimer, and Hideki Yukawa, the original proponent of the meson theory of nuclear forces, who had come from Kyoto University in Japan. Another member of the Institute's physics circle was Abraham Pais, who had just received a long-term appointment.

Dyson's first impression of the community was very favorable: "From what I have seen of them, they are a very nice crowd."[113] Yukawa he found "most friendly and approachable." He hoped to establish a friendship with Pais, "a man of very wide interests and culture."[114] Most of these theorists were struggling to understand the Schwinger radiation formalism, and Dyson did not tell them that he had been striving to supersede it. "That would be bad manners." His aim was to publish "his bombshell as soon as possible, preferably before Oppenheimer comes to pull it to pieces."[115]

Oppenheimer was expected to return to the Institute in the middle of October 1948. He had been visiting Europe since the beginning of September and had attended the Birmingham conference and the eighth Solvay Congress. Dyson depicted the mood before his arrival in his weekly letter home (see also Dyson 1979, pp. 70–83): "On Wednesday Oppenheimer returns. The atmosphere at the Institute during these last days has been rather like the first scene in 'Murder in the Cathedral' with the women of Canterbury awaiting the return of their archbishop."[116] When Oppenheimer came back Dyson gave him a copy of his paper. During their first conversation, Oppenheimer "shocked [Dyson] by being unreceptive to new ideas in general, and Feynman [theory] in particular." Oppenheimer's cool reception of Dyson's work had a chilling effect on their relationship initially. Dyson wrote home: "It is this general attitude of hesitation which I now see I shall have to fight with in the next few months; I am sure I shall have no difficulty in the long run, and the great thing at present is to avoid antagonising people by being impatient at their conservatism."[117]

Dyson did become irritated by Oppenheimer's "semi-defeatist attitude to the whole business, and . . . [his] complete lack of enthusiasm for a lot of the things that I consider most hopeful of fruitful advances."[118] He also became very annoyed by the fact that Oppenheimer's "lethargy" was making it difficult for him or anybody else "to go ahead with it."

"What I want to do now," Dyson complained to his parents, "is to get some large-scale calculations done to apply the theory to nuclear problems, and this is too big a job for me to tackle alone."[119]

Shortly after Oppenheimer had come back he handed Dyson a copy of the report he had submitted to the eighth Solvay Congress, which had been held in Brussels from September 27 to the October 2.[120] In it Oppenheimer had given an account of the recent advances in quantum electrodynamics, focusing on the logical and procedural aspects of the developments, and he also had appended some remarks and questions "on [the] applications of these developments to nuclear problems and on the question of the closure of electrodynamics" (Oppenheimer 1950).

The report was typical Oppenheimer vintage: literate, subtle, involuted, and difficult. It was a fair account of what had been accomplished, but it was not an enthusiastic one; and in the conclusion Oppenheimer downplayed the importance and cast doubt on the novelty of the recent progress:

> It is tempting to suppose that these new successes of electrody-
> namics, which extend its range very considerably beyond what
> had earlier been believed possible, can themselves be traced to
> a rather simple general feature: [namely that] electrodynamics
> is an almost closed subject; changes limited to very small dis-
> tances, and having little effect even in the typical relativistic
> domain $E = mc^2$, could suffice to make a consistent theory; in
> fact, only weak and remote interactions appear to carry us out
> of the domain of electrodynamics, into that of the mesons, the
> nuclei, and the other elementary particles.

The report ended on a somewhat pessimistic tone: "That electrodynamics is also
not quite closed is indicated, not alone by the fact that for finite e^2/hc the present
theory is not, after all, consistent, but equally by the existence of those small interactions with other forms of matter to which we must in the end look for a clue,
both for consistency, and for the actual value of the electron's charge" (Oppenheimer 1950). Dyson seized upon the report to vent his irritation and frustration.
He drafted a memorandum to which was appended a short note:

THE INSTITUTE FOR ADVANCED STUDY
SCHOOL OF MATHEMATICS
PRINCETON, NEW JERSEY

From   October 17th 1948

Mr. F.J. Dyson.

Dear Dr. Oppenheimer:

As I disagree rather strongly with the point of view expressed in
your Solvay Report (not so much with what you say as with what
you do not say), and as my own opinions are not firmly enough
based for me to put them up against you in public discussion,
I decided to send you a short written memorandum. This is a
statement of aims and hopes, and I would be glad if you would
read it before starting on the arid details of my long paper on
radiation theory.

MEMORANDUM

    I.   As a result of using both old-fashioned quantum electro-
           dynamics (Heisenberg-Pauli) and Feynman electrody-
           namics, on problems in which no divergence difficulties

arise, I am convinced that the Feynman theory is considerably easier to use, understand, and teach.

II. Therefore I believe that a correct theory, even if radically different from our present ideas, will contain more of Feynman than of Heisenberg-Pauli.

III. I believe it to be probable that the Feynman theory will provide a complete fulfillment of Heisenberg's S-matrix program. The Feynman theory is essentially nothing more than a method of calculating the S-matrix for any physical system from the usual equations of electrodynamics. It appears as an experimental fact (not yet known for certain) that the S-matrix so calculated is always finite; the divergencies only appear in the part of the theory which Heisenberg would in any case reject as meaningless. This seems to me a strong indication that Heisenberg is really right, that the localisation of physical processes is the only cause of inconsistency in present physics, and that so long as all experiments are interpreted by means of the S-matrix the theory is correct.

IV. The Feynman theory exceeds the original Heisenberg program in that it does not involve any new arbitrary hypothesis such as a fundamental length.

V. I do not see any reason for supposing the Feynman method to be less applicable to meson theory than to electrodynamics. In particular I find the argument about "open" and "closed" systems of fields irrelevant.

VI. Whatever the truth of the foregoing assertions may be, we have now a theory of nuclear fields which can be developed to the point where it can be compared with experiment, and this is a challenge to be accepted with enthusiasm.[121]

After he had finished the memorandum, he was unsure whether to give it to Oppenheimer. The resolution of the dilemma was reported in a letter home:

On Sunday night I went for a walk into a field outside the town, where the sky was unobscured by lights, and sat down on the grass to make up my mind whether I should send the letter off. After some time I had finally decided to do it, and then suddenly the sky was filled with the most brilliant northern lights I have ever seen. They lasted only about five minutes, but were a rich blood-red and filled half the sky. Whether the show really was staged for my benefit I doubt, but certainly it produced the same psychological effect as if it had been. I sent the letter off.[122]

On the next day, Oppenheimer ran into Dyson and told him he was delighted with his letter and had arranged for Dyson to present a seminar to give him an opportunity to expound his views publicly.

Oppenheimer in fact had scheduled Dyson for a series of seminars to allow for a full presentation of his recent work. That work was progressing "splendidly."[123] Dyson was extending the proof of the renormalizability of the S-matrix he had given in his radiation theory paper for one-particle systems to arbitrary n-particle systems.

Dyson's first seminar took place in the last week of October. "It went very well and I am to continue twice a week until it is finished."[124] It was actually not Dyson's first presentation of his work. Earlier that month, Bethe, who was visiting Columbia University during the fall semester, had invited him to New York to give a complete account of what he had done. Starting at 2, Dyson had gone "steadily on" until 5. And when he finished, Bethe had said to him that "he understood everything and liked it very much."[125]

Although at the conclusion of Dyson's first Institute seminar, Oppenheimer had professed to the audience that "this is the first new idea I have heard for 6 months,"[126] he had constantly interrupted the presentation with comments, criticisms, and questions. As the seminars proceeded, matters grew steadily worse. "I have been observing rather carefully his behavior during seminars," Dyson wrote home.

> If one is saying, for the benefit of the rest of the audience, things that he knows already, he cannot resist hurrying on to something else; then when one says things that he doesn't know or immediately agree with, he breaks in before the point is fully explained with acute and sometimes devastating criticisms, to which it is impossible to reply adequately even when he is wrong. If one watches him one can see that he is moving around nervously all the time, never stops smoking, and I believe that his impatience is largely beyond his control.[127]

Relations between Dyson and Oppenheimer could not have been helped by a letter Dyson received from Leonard Eyges: "You should be told Oppenheimer's remark on hearing of your results. Bethe announced them at a session on field theory at the B'ham conference and told how you had done it on a 50 hr. ride etc. Then Oppenheimer after a short silence said: 'There wasn't enough room in the margin to write down the proof.'"[128] In mid-November Dyson informed his family: "On Tuesday we had our fiercest public battle so far, when I criticized some unwarrantably pessimistic remarks he had made about the Schwinger theory. He came down on me like a ton of bricks, and conclusively won the argument so far as the public was concerned. However, afterwards he was very friendly and even apologized to me."[129] It was Bethe who saved the day. At the end of November Dyson wrote his parents:

> At last on Wednesday of this week Bethe came to my rescue. He came down to talk to the seminar about some calculations he was doing with the Feynman theory. He was received in the style to which I am accustomed with incessant interruptions and confused babbling of voices, and had great difficulty in making even his main point clear; while this was going on he stood very calmly and said nothing, only grinned at me as if to say "Now I can see what you are up against." After that he began to make openings for me, saying in answer to a question "Well, I have no doubt Dyson will have told you all about that," at which point I was not slow to say in as deliberate a tone as possible "I am afraid I have not got to that yet." Finally Bethe made a peroration in which he said explicitly that the Feynman theory is much the best theory and that people must learn it if they want to avoid talking nonsense; things which I had begun saying but in vain.[130]

After his seminar Bethe went off and had dinner with the Oppenheimers and Dyson did not see him again. But the next day he discovered that three extra seminars had been arranged for him, all in one week. "My triumph was complete . . . Bethe is a great and good man, and I wrote to him and told him so," Dyson informed his parents. "The tact and strategy which he used, to pull the opinion of the Institute onto my side, could not have been more effective."[131]

In fact, Oppenheimer not only listened to Dyson and did not interrupt him but "was wonderfully cooperative."[132] In order to get all his results presented in time, Dyson, at the suggestion of Uhlenbeck, had filled the blackboards with formulas before starting the lecture, and by "ruthlessly going on talking for 1 3/4 hours on the first two occasions and for 2 hours on the last"[133] he felt that he had put across at least all the main ideas. At the end Oppenheimer made a short speech, saying: "It is not possible to say on the basis of these talks that the consistency of the theory [has been] rigorously proved, but at least we all have learnt a very great deal and shall have plenty to argue about from now on."[134] Dyson noted that the exhausted audience departed quietly "in no mood to start an argument." On the morning after the final seminar, Dyson found in his mailbox a small piece of paper with the words " 'Nolo Contendere. R.O.' scribbled on it"[135] (see photograph 27). The victory was sweet (Dyson 1979)!

After the series of seminar talks was over, Dyson went to Oppenheimer to review them. Earlier, Dyson had communicated to his parents his belief that once Oppenheimer understood and believed in what had been accomplished, "he will certainly have a great deal of useful advice and experience to offer us in applying it. Also he may be able to help me decide what I should do next, though I am fairly determined already on a thorough going attempt to prove the whole theory consistent."[136] During his meeting with Dyson, Oppenheimer was extremely pleasant and indicated to Dyson his agreement with all his main contentions. But he had no very concrete proposals to make for pushing the theory further and advised Dyson to "follow his destiny"[137] and to go on thinking about it until he had squeezed all he could out of it. At the end of their conversation Oppenheimer asked Dyson what he intended to do the following year. When Dyson told him he was returning to England, Oppenheimer warmly approved and urged him to resist the temptation to settle permanently in the United States. He informed him that "Dirac and Bohr both felt that their proper place is in England and Denmark," and that they had an arrangement whereby they can visit the Institute one year in three or thereabouts so that they can keep in touch with people and developments in the United States. "Certainly we shall be able to do something of the kind for you."[138] Although Dyson felt that Oppenheimer had no evidence on which he could place him in the same class as Dirac and Bohr—and he doubted that he would ever achieve their distinction—he confided to his parents that "this kind of talk is of course vastly satisfying to my ego." He was grateful for this arrangement to return periodically to the Institute, which would enable him to keep abreast of what was happening in physics. Concluding his letter, Dyson also observed: "I think [Oppenheimer's] remarks are chiefly interesting as a key to his own character, and an explanation of his previous behavior. It is just this sudden and exaggerated enthusiasm which he showed when Schwinger first produced his theory, and the sudden exaggerated lack of enthusiasm with which he viewed Schwinger and Feynman when I began my talks. He is a curious mixture so cool and accurate in his speech and appearance, and so nervous and unstable inside."[139]

Reminiscing some thirty-five years later, Dyson felt that "Oppenheimer was a great disappointment. He hadn't time for the details. As compared to Hans Bethe, Oppenheimer was completely superficial. To talk to Oppenheimer was interesting. It was like meeting some very famous person who had interesting things to say but I just never got anything that you could really call guidance. I wasn't needing much guidance. . . . For that year it didn't matter. I had plenty to do. . . . [But] he had a bad effect on other people who needed the guidance more than I did."[140]

After his "triumph", life at the Institute became more pleasant. Under Dyson's guidance several major projects were initiated using Feynman's methods and Dyson's renormalization techniques: Karplus and Kroll started on their ground-breaking fourth-order calculation of the magnetic moment of the electron, and Watson and Lepore began a computation of the nuclear forces to fourth order in the meson-nucleon coupling constant.

By Thanksgiving, Dyson's fame had spread to such an extent that he had been offered over half a dozen jobs. One was for the position of chief assistant at Greenwich Observatory, "with excellent prospect of being Astronomer Royal by about 1965."[141] Mott invited him to Bristol as a lecturer "with expectation of a professorship soon."[142] Rabi offered him a position at Columbia, which Dyson declined because his Commonwealth Fellowship stipulated that at the end of his tenure he had to return to Great Britain or go to one of the Commonwealth countries. "With bitterness in [his] heart" Dyson lamented that "there is no place in the world which would be better for me than Columbia. They have the finest experimental department in the world, and it is just in the contact with experimentalists that I have the most to learn."[143] When he received Rabi's offer, Oppenheimer had not yet offered him the visiting appointment to the Institute, and Dyson dreaded the isolation that was awaiting him in England: "It is a grim prospect to be cut off without more than rumours and months-old reports of what Feynman or Schwinger or Columbia or Berkeley is doing."[144] He even considered obtaining a position in Toronto in order to be closer to Bethe and Feynman. The prospect of intellectual isolation was not the only thing which made Dyson apprehensive: he had also begun to appreciate the American academic life-style. He had befriended many members of the Princeton physics community, although he did feel lonely and isolated at times.[145] Life at the Institute had many charms and it suited him well. The *New Yorker* described its ambience in one of its columns:

> The little vine-covered cottage atmosphere ... is just right for most of the Institute's geniuses, who seem to be in their early twenties, and still breathless from taking their Phds on the run ... We peeked in on a dozen offices in one of the Institute's three red brick college buildings, and were introduced to several thinkers occupied with their labors. One was doodling on a scratch pad, presumably developing a momentous equation; one was staring out the window at four crows in a field; and a third was writing a letter to his mother on a portable typewriter. [Dyson!]
>
> That evening, we watched a square dance in the common room at the Institute. The participants were mainly Oppenheimer's young atomic physicists and mathematicians. It was like a square dance at any small college, except that more languages were in use. The boys [*sic*] were in shirtsleeves, slacks or jeans, and sneakers; the girls [*sic*]—some of them wives, some of them secretaries, some of them scientists—were mostly wearing peasant blouses, dirndls, and either saddle shoes or ballet slippers.[146]

9.11 The *S*-Matrix in QED

Shortly before Christmas 1948, Dyson was "suddenly seized with an impulse to get another big paper written," covering the work he had been doing since the one on the radiation theories of Tomonaga, Schwinger, and Feynman had been finished. On Christmas Day, Dyson informed his parents:

> The second paper turned out to be even more formidable in length and difficulty than the first; however I took it a good deal easier since I was by this time a lot more familiar with the main ideas I had to get across. This morning, after ten days of fairly intensive work, I had covered 58 pages and had come to the end of the major difficulties....
>
> On the whole I am very well satisfied with the new paper. Roughly, what it does is to prove by a frontal attack on the most complicated problems imaginable that the Feynman theory can be consistently carried through for any kind of problem, and not just for the simplest ones so far considered....It turns out that with the Feynman theory...there are very serious difficulties in applying the theory to the simplest problems, as the history of the Lamb shift shows; but once these difficulties are got over, then no new ones appear at all in more complicated situations. This seems to me a most remarkable property of the theory, and a strong argument for thinking that it may even be the right theory.

To clarify further his results, Dyson explained to his parents:

> As you may know, the Newtonian theory of gravitation can deal exactly with the problem of two bodies, but not with that of three. Einstein then improved on this by producing a theory that could deal exactly with the problem of one body but no longer with two. So you see it is really something, when one finds that with Feynman one can deal satisfactorily with a problem involving n bodies, and also the creation and annihilation of any number of additional particles. In dealing with these complicated problems, the main difficulties, always supposing the theory makes sense, are mathematical. And so I found when writing this last paper, that a familiarity with the technique of handling masses of mathematics came in very useful. I think the work has an air of finality about it, for nobody in his senses is likely to ever want to do anything quite as complicated again.

Dyson was wrong in his prediction. A vigorous, international community of researchers devotes itself to the problems of renormalization and have done calculations (like eighth-order radiative corrections in QED and two-loop processes in quantum gravity) that Feynman and Dyson would not have dreamed of actually doing. And all the work in the field is an outgrowth of Dyson's (1949b) paper on "The S-matrix in Quantum Electrodynamics." In his assessment of this seminal work, C. N. Yang observed:

> The papers of Tomonaga, Schwinger and Feynman did not complete the renormalization program since they confined themselves to low order calculations. It was Dyson who dared to face the problem of high orders and brought the program to completion. In two magnificently penetrating papers, he pointed out and resolved the main problems of this very difficult analysis. Renormalization is a program that converts additive subtractions into multiplicative renormalization. That it works required a highly nontrivial proof. That proof Dyson supplied. He defined the concept of primitive divergences, skeleton graphs, and overlapping divergences. Using these concepts, he pushed through an incisive analysis and completed the proof of renormalizability of quantum electrodynamics. His perception and power were dazzling. (Yang 1983, p. 65)

Renormalization theory is an enormously complicated subject. In commenting on the early work in renormalization, Salam and Matthews noted that one of "the difficulties . . . in all this work, is to find a notation which is both concise and intelligible to at least two people of whom one may be the author" (Matthews and Salam 1951). Renormalization theory has a history of errors by distinguished theorists. It also has a reputation for perversity: a particular method can be shown to work up to thirteenth order in the perturbation theory series but to fail in the fourteenth order. Arguments that seem plausible do not hold up (Mills and Yang 1966; see also Yang 1983, pp. 64 and 383; Ward 1951).

Dyson's paper on the renormalizability of the S-matrix (Dyson 1949b) has been expounded in all the textbooks on the subject (see, for example, Jauch and Rohrlich 1955; Schweber 1961; Akhiezer and Berezetski 1963; Bjorken and Drell 1965; Itzykson and Zuber 1980). Rather than recapitulate the content of Dyson's paper, I will here sketch how "renormalization converts additive [divergent] subtractions into multiplicative renormalization." I shall base my exposition on the lectures on "Renormalization" that Dyson delivered at the 1954 summer school in Les Houches (Dyson 1954). Dyson's presentation was based, in part, on Gupta's (1951) and Takeda's (1952) approach to renormalization (see also Matthews and Salam 1951; Matthews 1954). My aim is not to give a proof of the renormalizability of QED, but rather to indicate how renormalization works, and to point to the difficulties that Dyson encountered in his proof of the renormalizability of the S-matrix in quantum electrodynamics.

The Lagrangian density from which the equations of motion of quantum electrodynamics are derived is

$$\mathcal{L} = -\frac{1}{4}\hat{F}^{\mu\nu}\hat{F}_{\mu\nu} - \frac{\lambda}{2}(\partial_\mu \hat{A})^2 - (-\frac{1}{2}i\hat{\bar{\psi}}\gamma_\mu \overset{\leftrightarrow}{\partial}{}^\mu \hat{\psi} - \hat{m}\hat{\bar{\psi}}\hat{\psi})$$
$$- \hat{e}\hat{\bar{\psi}}\gamma_\mu \hat{A}^\mu \hat{\psi} - \frac{1}{4}\hat{F}^e_{\mu\nu}\hat{F}^{e\mu\nu} - \frac{1}{2}\hat{F}_{\mu\nu}\hat{F}^{e\mu\nu}$$
$$- \hat{e}\hat{\bar{\psi}}\gamma_\mu \hat{A}^{e\mu} \hat{\psi}. \tag{9.11.1}$$

This Lagrangian describes the interactions of the electron-positron field with both the quantized electromagnetic field, $\hat{A}$, and an external electromagnetic field, $\hat{A}^e$. As was indicated in our exposition of Dyson (1949a) in section 9, the Heisenberg operators $\hat{A}^H(x)(\hat{\bar{\psi}}^H(x), \hat{\psi}^H(x))$ derived from eq. (9.11.1) do not have finite matrix elements between the vacuum state and one-particle states. The perturbative expansion of $< 0|\hat{A}^H_\mu(x)|1\text{photon} >$ is represented in the diagrams of Figure 9.11.1. and the contributions corresponding to (b), (c),... are all divergent. Relativistic and gauge invariance allowed Dyson to infer that $< 0|\hat{A}^H_\mu(x)|1\text{photon} >= Z_3^{1/2} < 0|A_\mu(x)|1\text{photon} >$, where $A_\mu(x)$ is a free field and Z_3 a (gauge-invariant) divergent constant.

Based on this observation, Dyson supposed that the field operators that have finite matrix elements between physical states are not those appearing in (9.11.1) but field operators proportional to them,

$$A_\mu(x) = Z_3^{-1/2}\hat{A}_\mu(x) \tag{9.11.2}$$

$$A^e_\mu(x) = Z_3^{-1/2}\hat{A}^e(x). \tag{9.11.3}$$

These operators were assumed to have the same Lorentz and gauge transformations as $\hat{A}_\mu$ and $\hat{A}^e_\mu$. Similarly Dyson conjectured that

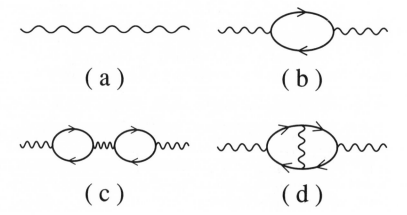

(a)   (b)

(c)   (d)

Figure 9.11.1

$$\psi(x) = Z_2^{-1/2}\hat{\psi}(x) \tag{9.11.4}$$

$$\bar{\psi}(x) = Z_2^{-1/2}\hat{\bar{\psi}}(x). \tag{9.11.5}$$

The quantities Z_3 and Z_2 are found to be gauge dependent, real (divergent) constants. The new fields $A_\mu(x), \psi(x), \bar{\psi}(x)$ are called the renormalized fields. The charge and the mass parameters entering in (9.11.1) are likewise renormalized:

$$\hat{e} = Z_1 Z_2^{-1} Z_3^{-1/2} e \tag{9.11.6}$$

$$\hat{m} = m(1 - Z_2^{-1}Y), \tag{9.11.7}$$

where Y is defined by eq. (9.11.7). These definitions of the renormalized fields and of the renormalized mass and charge imply that

$$\hat{e}\hat{\bar{\psi}}\gamma_\mu \hat{A}^\mu \hat{\psi} = Z_1 e\bar{\psi}\gamma_\mu A^\mu \psi \tag{9.11.8}$$

and

$$\hat{\bar{\psi}}(i\gamma_\mu \partial^\mu - \hat{m})\hat{\psi} = Z_2 \bar{\psi}(i\gamma_\mu \partial^\mu - m)\psi - Ym\bar{\psi}\psi \tag{9.11.9}$$

Eq. (9.11.9) gives the interpretation of Y. The Lagrangian (9.11.1) can now be rewritten in the form

$$\mathcal{L} = \mathcal{L}_0 + \mathcal{L}_I, \tag{9.11.10}$$

$\mathcal{L}_0$ being the Lagrangian of the "free" renormalized fields

$$\mathcal{L}_0 = \frac{1}{4}F_{\mu\nu}F^{\mu\nu} - \frac{1}{2}(\partial_\mu A^\mu)^2 - (-\frac{1}{2}i\bar{\psi}\gamma_\mu \overleftrightarrow{\partial}^\mu \psi - m\bar{\psi}\psi)$$

$$-\frac{1}{4}F_{\mu\nu}^e F^{e\mu\nu} - \frac{1}{2}F_{\mu\nu}F^{e\mu\nu} \tag{9.11.11}$$

and $\mathcal{L}_I = \mathcal{L} - \mathcal{L}_0$ is given by

$$
\begin{array}{cc}
(1) & (2) \\
\end{array}
$$
$$\mathcal{L}_I = e\bar{\psi}\gamma_\mu A^\mu \psi - e\bar{\psi}\gamma_\mu A^{e\mu}\psi$$
$$
\begin{array}{cc}
(3) & (4) \\
\end{array}
$$
$$-eL\bar{\psi}\gamma_\mu A^\mu \psi - eL\bar{\psi}\gamma_\mu A^{e\mu}\psi$$
$$
\begin{array}{cccc}
(5) & (6) & (7) & (8) \\
\end{array}
$$
$$-\frac{1}{4}CF_{\mu\nu}F^{\mu\nu} - \frac{1}{2}CF_{\mu\nu}F^{e\mu\nu} - \frac{1}{4}CF_{\mu\nu}F^{e\mu\nu} - mY\bar{\psi}\psi$$
$$
\begin{array}{c}
(9) \\
\end{array}
$$
$$-B\bar{\psi}(-i\gamma_\mu \partial^\mu - m)\psi, \tag{9.11.12}$$

where

$$C = Z_3 - 1 \qquad\qquad (9.11.13)$$
$$B = Z_2 - 1 \qquad\qquad (9.11.14)$$
$$L = 1 - Z_1. \qquad\qquad (9.11.15)$$

The (arbitrary) constant λ in (9.11.1) has been chosen equal to Z_3^{-1}, $\lambda = Z_3^{-1}$, so as to set equal to zero the term $\frac{1}{2}(\lambda Z_3 - 1)(\partial_\mu A^\mu)^2$. Note there are now nine terms in the interaction Lagrangian instead of the original two. Renormalizability consists in showing that the coefficients C, B, L, Y of these supplementary terms can be so chosen that in the computation of any scattering process to any given order of perturbation theory, the divergences that stem from the two first terms in $\mathcal{L}_I(x)$ are eliminated, and only finite contributions remain. To lowest order these nine terms give rise to the diagrams in figure 9.11.2. One readily verifies that the S-matrix element M for the scattering of an electron by an external potential to second order in e, and to first order in A^e, obtains contributions from the diagrams indicated in table 9.11.1.

Note that the terms labeled (3) and (5) in eq. (9.11.12) do not contribute because C and L are at least of order e^2, and that the term labeled (7) does not contribute because it is of order $(A^e)^2$. The meaning of the entries in the third column in table 9.11.1 is as follows: If the lowest-order matrix element is denoted by

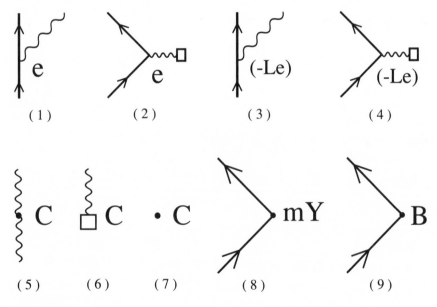

Figure 9.11.2

Table 9.11.1

	Terms of L	Diagram	Contribution to $\mathbf{M}$
1.	$(1, 2)$		$M_2^1 + L_2 M_0$
2.	(2)		M_0
3.	$(1, 2)$		$M_2^2 + C_2 M_0$
4.	$(1, 2)$		$(\partial m M_0) + B_2 M$
5.	(4)		$-L M_0$
6.	$(8, 2)$		$-Y M_0$
7.	$(9, 2)$		$-B M_0$
8.	$(1, 6)$		$-C M_0$

$M_0 \propto e\bar{u}(p')\gamma_\mu a^\mu(q)u(p), p' - p = q$, the contribution from diagram (1) is given by

$$M_{(1)} \propto e^3 \bar{u}(p') \int \frac{d^4 k}{k^2 + i\epsilon} \gamma_\alpha \frac{1}{p - \not{k} - m + i\epsilon} \gamma_\mu \frac{1}{p' - \not{k} - m + i\epsilon} \gamma^\alpha u(p) a^e_\mu(q)$$

$M_{(1)}$ decomposes into a term proportional to M_0 and a finite contribution M_2^1,

$$M_{(1)} = L_2 M_0 + M_2^1.$$

L_2 diverges logarithmically and the finite part is made definite by the requirement that $M_2^1 = 0$ when $p = p'$ with $p^2 = p'^2 = m^2$. The contribution from (4) is analogously derived. It gives rise to a contribution

$$M_{(4)} \propto e^3 \int \frac{d^4 k}{k^2 + i\epsilon} \bar{u}(p')\gamma^\nu \frac{1}{p + \not{k} - m + i\epsilon} \gamma_\nu \frac{1}{p - m + i\epsilon} \gamma_\mu u(p) a^{e\mu}(q)$$
$$= (\delta m M_0) + B_2 M_0.$$

The other entries in the third column of table 9.1 are similarly derived. One notes that all divergences to second order will be eliminated if one supposes that the coefficients L_1, B_1, Y, C have a power series development in e^2, such that their lowest-order contributions are $L_2, B_2, \delta m$, and C_2, respectively.

In order to investigate the possibility of renormalization to all orders, Dyson considered an arbitrary matrix element M_G of

$$S_{m+n} = (\frac{-ie}{\hbar c})^{n+m} \frac{1}{n!} \int \frac{1}{m!} \ldots d^4 x_1 \ldots d^4 x_{m+n}$$

$$P(\bar{\psi}\gamma^\mu A_\mu \psi(x_1)\ldots \bar{\psi}\gamma_\mu A^\mu \psi(x_n)\bar{\psi}\gamma_\mu A^{e\mu}\psi(x_{n+1})\ldots \bar{\psi}\gamma_\mu A^{e\mu}\psi(x_{m+n})),$$

$$(9.11.16)$$

that is, arising from the interaction term $e\bar{\psi}\gamma_\mu A^\mu \psi + e\bar{\psi}\gamma_\mu A^{e\mu}\psi$ to order e^{n+m}. M_G will correspond to a diagram with $m + n$ vertices, n of which are vertices with two electron lines and a photon line joining at the vertex, and m of which will have two electron lines and an external potential meeting there.

The integrations $\int d^4 x_i$ will result in momentum space in a δ function $\delta(p' - p + k)$ for each vertex x_i, corresponding to the conservation of energy momentum of the momenta, p', p, k, carried by the lines (or external potential) meeting at the vertex. In momentum space, M_G will be given by an expression involving integrations over the momenta of all the *unconstrained internal* lines appearing in the diagrams (one of the δ functions arising from integrations over $d^4 x_1 \ldots d^4 x_{m+n}$ will give rise to the overall energy momentum conservation for the external lines of the diagram). Dyson designated by

I_F the number of internal electronic lines

I_B the number of internal photonic lines

E_F the number of external electron lines

E_B the number of external photon lines,

so that the number of variables of integrations that remain after the elimination of all the δ functions is

$$F = I_F + I_B - m - n + 1, \tag{9.11.17}$$

and M_G has the form

$$M_G \propto \int d^4 k_1 \ldots \int d^4 k_F \prod_{I_B} \frac{1}{k_i^2 + i\epsilon} \prod_{I_B} \frac{1}{k_j - m + i\epsilon}. \tag{9.11.18}$$

Dyson did not consider the divergences of M_G that may occur because of the small k behavior of the integrand (infrared red divergences) but only those that arise from the large k behavior. In order to make his analysis, Dyson rotated the paths of integration so that instead of integrating from $k_{j0} = -\infty$ to $k_{j0} = +\infty$ the paths go from $-i\infty$ to $+i\infty$. That it is always possible to do so (modulo some residues) is not a trivial matter to demonstrate and was justified by (Dyson 1949b, p. 1744). Thereafter a change of the variables of integration from k_0 to ik_0' transform the Minkowski denominators $k_0^2 - \mathbf{k}^2$ into $-(k_0'^2 + \mathbf{k}^2)$, and they become negative definite.

Dyson could then formally determine the convergence for large k of the integral (9.1.18) by counting powers in the numerator and denominator. The difference in the "degree" of the numerator and denominator is $4F - (I_F + 2I_B)$. In general, a sufficient condition for the *divergence* of this integral is that the degree of the numerator be greater than the denominator

$$4F \geq IF + 2I_B. \tag{9.11.19}$$

Using eq. (9.11.17) this becomes

$$4(m + n) \leq 3I_F + 2I_B + 4 \tag{9.11.20}$$

Now, since every vertex has two electron lines meeting at it, and every internal electron line connects two vertices,

$$2I_F + E_F = 2(m + n). \tag{9.11.21}$$

Similarly, the total number of photon lines is related to the number of photon vertices, n, by

$$2I_B + E_B = n. \tag{9.11.22}$$

Hence, eq. (9.11.20) can be rewritten as

$$\frac{3}{2}E_F + E_B + m \le 4, \tag{9.11.23}$$

so that the divergence is controlled by the number of external lines (m is the number of times A^e acts), and not by the number of internal lines! Call

$$\omega(G) = 4 - \frac{3}{2}E_F - E_B - m \tag{9.11.24}$$

the degree of divergence. The integral will be divergent if $\omega(G) \ge 0$, and as the number of external lines is increased an integral representing a Feynman diagram becomes more convergent. This remarkable feature of $\omega(G)$—that it is independent of the number of internal lines and hence of the internal details of the Feynman diagram—is the key to the success of Dyson's renormalization program. It is a consequence of the fact that the coupling constant $e/\sqrt{hc}$ is dimensionless and that m, the electron mass, is the only dimensional constant in the theory. As one goes to higher orders in the calculation of an amplitude (by inserting more internal lines), its dimension does not change. Thus $\omega(G)$ is determined by overall dimensional arguments. For theories containing derivative coupling in $\mathcal{L}_I$, for example, $\partial \bar{\psi} \gamma_\mu \psi \partial^\mu \phi$, and dimensional coupling constants, $\omega(G)$ depends on the number of internal lines and the perturbative renormalization program fails. Such theories are called nonrenormalizable.

The criterion $\omega(G) \ge 0$ is an assertion about the divergence of the integral over the whole set of variables. It is possible for the overall integration to be (superficially) convergent but for some of the subintegrations to be divergent. Thus, for example, in the diagram of figure 9.11.3 for the total integral $E_F = 4$, $E_B + M = 0$ so that $\omega(G) = -2$ and the integral converges superficially. For the integration over k_1, g_1 has $E_F = 2$ and $E_B = 1$, hence $\omega(g_1) = 0$ and the integral diverges; for the integration over k_2 alone, g_2 has $E_F = 4$, $E_B = 2$, hence $\omega(g_2) = -4$ and the integration converges. Note the Feynman diagram of figure 9.11.3 corresponds to the insertion of a (divergent) vertex part into the diagram of figure 9.11.4.

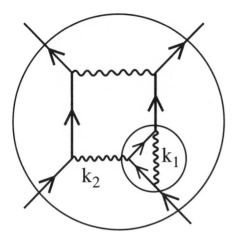

Figure 9.11.3

In order to determine the possible types of divergent diagrams, Dyson introduced the concept of a primitively divergent diagram. He called a divergent M primitively divergent if whenever one of the momentum 4 vectors in its integrand is held fixed, the integration over the remaining variables is convergent. Equivalently, a diagram is primitively divergent if upon the opening of any one of its internal lines (into two external lines) it becomes a convergent diagram. Since for a primitive divergent the integral converges when one of the k_i is held fixed, the integral is divergent *if and only if*

$$\frac{3}{2}E_F + E_B + m \le 4. \tag{9.11.25}$$

There are thus only a finite number of types of primitively divergent diagrams corresponding to the values of E_F, E_B, m compatible with the relation (9.1.25). These are indicated in figure 9.11.5.

Furry's theorem (Furry 1937) implies that diagrams with $E_B = 3, E_F = 0, m = 0$, (diagrams d and j) vanish, as do those with $m = 1, E_F = E_B = 0$. The bubble diagrams $E_F = E_B = 0, m = 0$ (diagram c) are eliminated by noting they are a common multiplicative factor to all diagrams and correspond to a renormalization of the vacuum energy. Dyson was thus left with the four basic types of primitively divergent graphs indicated in table 9.11.2.

The scattering of light by light is in fact not divergent because of gauge invariance. Dyson considered next the primitively divergent diagram of Fig. 9.11.6 corresponding to a photon self-energy contribution. If $\{s\}$ denotes the set of variables associated with the internal integration, then its contribution can be represented as

$$M_{\mu\nu} \propto \int ds W_{\mu\nu}(s, k),$$

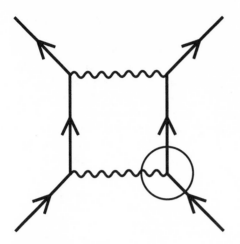

Figure 9.11.4

where k corresponds to the momentum of the external photon line (k^2 need not, however, be equal to 0). W is a rational function of the momenta of the internal lines, such that $\int ds W_{\mu\nu}(s, k)$ is superficially quadratically divergent. Dyson extracts a finite part for this integral by making a Taylor expansion of the integrand about $k = 0$:

$$W(s, k) = W(s, 0) + k_\alpha \left[\frac{\partial W}{\partial k_\alpha}\right]_{k=0}$$
$$+ \frac{1}{2!} k_\alpha k_\beta \left[\frac{\partial^2 W}{\partial k_\alpha \partial k_\beta}\right]_{k=0} + R(s, k). \qquad (9.11.26)$$

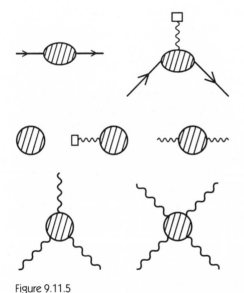

Figure 9.11.5

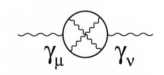

Figure 9.11.6

Table 9.11.2

Type	E_F	E_B	Lowest-Order Diagram	$\omega(G)$	Effective Divergence
Photon self-energy	0	2		2	Logarithmic because of gauge invariance
Scattering of light by light	0	4		0	Convergent because of gauge invariance
Electron self-energy	2	1		0	Logarithmic
Vertex part	2	1		0	Logarithmic

Then

$$\int ds \left\{ W(s, k) - W(s, 0) - k_\alpha \left[\frac{\partial W}{\partial k_\alpha} \right]_{k=0} + \frac{1}{2!} k_\alpha k_\beta \left[\frac{\partial^2 W}{\partial k_\alpha \partial k_\beta} \right]_{k=0} \right\}$$

is a convergent integral; this because the remainder in eq. (9.11.26), $R(s, k)$, has three powers of s more in the denominator than does $W(s, k)$.

Thus $M_{\mu\nu}$ has been separated into a finite and a divergent part:

$$M_{\mu\nu} = M_{\mu\nu}^{\text{(finite)}} + \int ds \{ W_{\mu\nu}(s,0)$$
$$+ k_\alpha \left[\frac{\partial W_{\mu\nu}}{\partial k} \right]_{k=0} + \frac{1}{2} k_\alpha k_\beta \left[\frac{\partial^2 W_{\mu\nu}}{\partial k_\alpha \partial k_\beta} \right]_{k=0} \}. \qquad (9.11.27)$$

Relativistic invariance and symmetry allowed Dyson to deduce that

$$M_{\mu\nu} = M^{\text{(finite)}} + A g_{\mu\nu} + \frac{1}{2} C k^2 g_{\mu\nu} + D k_\mu k_\nu. \qquad (9.11.28)$$

Finally, for $M_{\mu\nu}$ to be gauge invariant, $A = 0$ and $D = -\frac{1}{2}C$, and $M^{\text{(finite)}}$ therefore has the form

$$M_{\mu\nu}^{\text{(finite)}} = F(k^2)[g_{\mu\nu} k^2 - k_\mu k_\nu] \qquad (9.11.29)$$

with $F(0) = 0$ because $M^{\text{(finite)}} = 0(k^3)$ by construction. Hence for this particular graph W,

$$M_{\mu\nu}(W) = [k^2 g_{\mu\nu} - k_\mu k_\nu][\frac{1}{2} C_W + F_W(k^2)]. \qquad (9.11.30)$$

Now the term $-\frac{1}{4} C F_{\mu\nu} F^{\mu\nu}$ in the interaction Lagrangian will give rise to a contribution $-\frac{1}{2} C(g_{\mu\nu} k^2 - k_\mu k_\nu)$ which by an appropriate choice of C to that order will cancel C_W in $M(W)$ and leave only the finite remainder. Similar procedures can be used to extract the divergent parts of other primitively divergent graphs.

Dyson next examined whether it was possible to define a procedure for subtracting all the divergences to a given order arising from the subdiagrams of the given diagram. To adumbrate the procedure and the difficulties that are encountered, Dyson considered first some simple cases in which the divergences are entirely contained one within the other. Thus, for example, in figure 9.11.7 the divergence corresponding to subdiagram 1 is eliminated by the diagram in figure 9.11.8. The divergence corresponding to 2 is thereafter eliminated by the diagram of figure 9.11.9 and the last corresponding to 3 by the diagram in

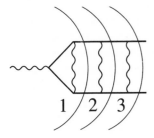

Figure 9.11.7

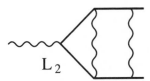

Figure 9.11.8

Figure 9.11.9

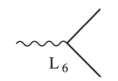

Figure 9.11.10

figure 9.11.10. With each subtraction the integrals remain logarithmically divergent. The reason for the success of the procedure is of course that the radiative corrections are the same, no matter what the diagram considered, and these are rendered finite by the counter terms with constant $L, B, \ldots$. Incidentally, it should be obvious that the entire analysis could not have been carried out without the use of diagrams. Similarly, in the diagram of figure 9.11.11, before removing the divergence (3) one must remove the self-energy correction to the photon line. But since this self-energy subdiagram is entirely contained within (3), and thus separated from the divergences (1) and (2), it does not create new difficulties for (3): one begins by eliminating it using the C counterterm in figure 9.11.12. But it is not always the case that the separation of divergent parts is as straightforward as in the case of entirely nested or separated divergences. For example, in the diagrams of figure 9.11.13 the vertex part insertion in one corner can also be looked upon as a vertex part insertion for the other corner: this is a simple case of what Dyson called a *b*-divergence. (These have become known as *overlapping divergences*.) A more complicated case of an overlapping divergence is illustrated in figure 9.11.13(b), where the overall integral is quadratically divergent, and the contributions for the various subintegrations are logarithmically divergent. An unambiguous procedure must be formulated to treat such problems. Two methods have been devised for treating overlapping divergences. The first method is due to Salam and works only on renormalizable theories. The second, by Bogoliubov and Parasiuk (1957), is of wider scope and is effective for any polynomial Lagrangian theories. That it can be made to work for these cases was shown by Hepp (1966) (who corrected some errors in Bogoliubov and Parasiuk). It consists in giving an unambiguous subtraction method for an arbitrary integral of order n (see also Hepp 1969; Velo and Wightman 1976; Manoukian 1983).

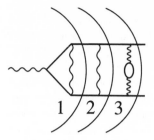

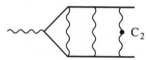

Figure 9.11.11

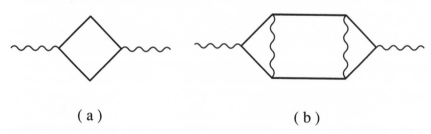

Figure 9.11.12

The story of the resolution of the problem of overlapping divergences by Abdus Salam is of interest. Salam was a research student at Cambridge early in 1950. He had asked Kemmer to be his thesis supervisor, but Kemmer allowed him to work with him only "peripherally" since he already had eight students doing their Ph.D. with him. Kemmer suggested to Salam that he contact Paul Matthews—who was just finishing a Ph.D. on the application of Dyson's methods to meson theories—to see whether he had any problems left (Salam 1987). In his thesis, Matthews had come to the conclusion that no meson theory with derivative coupling could be renormalized, and that among the direct coupling theories only the spin zero meson theory and the neutral spin 1 meson theory with conserved currents were renormalizable in the one-loop approximation. Furthermore, Matthews had shown that in the spin zero case an additional $\lambda\phi^4$ counterterm had to be added to the Lagrangian, where ϕ is the meson field. He could not go beyond one loop because overlapping divergences made their appearances, and this problem had to be solved before a proof of the renormalizability to all orders could be given. Dyson had been the external examiner at Matthew's Ph.D. defense. In the Ph.D. viva, as the defense is known in Cambridge, Dyson asked Matthews whether he had come across overlapping divergences. "And if so, how do you resolve the problems posed by these?" Matthews replied "You have claimed in your paper on QED that these infinities—which occur in the self-energy graphs—can properly be taken care of. I am simply following you." Evidently no further question on these infinities was asked on this embarrassing subject.

(a)

Figure 9.11.13(a)

(b)

Figure 9.11.13(b)

Overlapping divergences, as has been indicated, had indeed reared their ugly head in Dyson's QED paper. He had noted that a general self-energy graph could be regarded as an insertion of a modified vertex at either end of the lowest order self-energy graph. Insertion of modified vertices at both ends would correspond to double-counting. "But Dyson, in his paper, while discussing these, had recommended precisely this—that one should subtract the vertex—part sub-infinities twice before subtracting the final self-energy infinity" (Salam 1951a). But no proof of this assertion was given.

Salam decided to make the overlapping divergences his problem. He thought that the best way to solve the problem "would be to ask Dyson's direct help" (Salam 1987). So he rang Dyson up and said, "I am a beginning research student; I would like to talk with you. I am trying to renormalize meson theories, and there is this problem of overlapping divergences which you have solved. Could you give me some time?" Dyson indicated that he was leaving for the United States on the next day, so if Salam wanted to talk to him he had to come to Birmingham "tonight." This Salam did. The two of them got together the next morning. This was the first time that Salam met Dyson. He asked him, "What is your solution to the overlapping infinity problem?" Dyson answered, "But I have no solution. I only made a conjecture" (Salam 1987). Salam recalled that for a young student who had just started on research, "this was a terrible shock. Dyson was our hero. His papers were classics. For him to say that he had only made a conjecture made me feel that my support of certainty in the subject was slipping away." But Salam notes that Dyson "was being characteristically modest about his own work. He explained to me what the basis of his conjecture was. What he told me was enough to build on and show that he was absolutely right" (Salam 1987).

During the summer of 1950, "at Cambridge amid the summer roses at the Backs of the Colleges," Salam tackled the overlapping divergences problem, and "using a generalization of Dyson's remarks," was able to show that QED and spin zero meson theories were indeed renormalizable to all orders (Salam 1951a, b, 1987).

A second method to handle overlapping divergences, which reduced the treatment of proper self-energy parts (in which overlapping divergences occur) to that of proper vertex parts (in which no overlaps occur), was devised by John Ward (1951). In Ward's method great simplications ensue because of the gauge invariance of the theory. By doing a lowest-order calculation, Dyson noted that $L_2 + B_2 = 0$. Ward gave a proof of this identity that is valid to all orders of perturbation theory. Gauge invariance requires that the Lagrangian (9.11.1) be invariant under the transformation

$$\hat{\psi} \rightarrow \exp\left\{ -\frac{i}{\hbar c} \hat{e} \hat{A} \right\} \psi \qquad (9.11.31)$$

$$\hat{A}_\mu \to \hat{A}_\mu + \partial_\mu \hat{\Lambda}, \tag{9.11.32}$$

and similarly (9.11.11) must be invariant under

$$\psi \to \exp\left\{ -\frac{i}{\hbar c} eA \right\} \psi \tag{9.11.33}$$

$$A_\mu \to A_\mu + \partial_\mu \Lambda, \tag{9.11.34}$$

and hence

$$\hat{e}\hat{A} = eA. \tag{9.11.35}$$

Now since $\hat{e} = Z_1 Z_2^{-1} Z_3^{-1/2} e$ and $\hat{A} = Z_3^{+1/2} A_\mu$, it follows that $Z_1 = Z_2$, or equivalently $L + B = 0$, and $\hat{e}/e$ depends only on the factor Z_3. Now Z_3 obtains contributions only from closed loops. If one has several kinds of charged particles, the Z_3 for the different charged particles will all be the same by including the closed-loop contributions from all the charged particles. All the renormalized charges will then be the same.

Dyson's proof of the renormalizability of QED consists in showing inductively that if all divergences have been removed in order e^{2n}, then the outlined procedure removes them to order e^{2n+2}. Dyson's original proof that the renormalized theory is finite was based on integral equations satisfied by the S_F', D_F' and the vertex operator $\Gamma_\mu'(p^1, p)$. It also relied on a conjecture later proved by Weinberg (1960) that a Feynman integral converges if the degree of divergence of the diagram as well as the degree of divergence associated with each possible subdiagram is negative. The reader is referred to Bjorken and Drell (1965, pp. 330ff) and Itzyckson and Zuber (1980) for presentations of the proof.

9.12 The *S*-Matrix Paper: Retrospective

Dyson recalls that he was not confident that the higher orders could really be handled by renormalization until he had almost finished the paper. An important simplification had occurred when he learned from Feynman, during his trip to Ithaca in October 1948, that the scattering of light by light was convergent: "Obviously I had been frightened about the scattering of light by light. It looked like a mess and I was not prepared to say I could do it. The scattering of light by light was legendary as a difficult problem, as attested by Euler and Kockel's work during the thirties. It came as a tremendous relief to find from Feynman that after all one could do it and that it came out easily."[147] Dyson adds that Feynman's fearlessness was the important thing in overcoming this problem. He also remembers: "I had very much the feeling that I was skating on thin ice when it came to overlapping divergences. I know I thought the paper was simply wrong when I first realized there were overlapping divergences. Then I patched it up as best I could. I was well aware that it was a matter of faith that the thing would actually work."[148] The paper was thus less rigorous than a mathematician like Dyson must have been used to. In a note to Peierls to whom he sent a preprint of this paper, Dyson commented, "It is very long, and not very well written, and some of the statements in it ought to be qualified with a 'perhaps' at the beginning and a 'maybe' at the end. However, I take as my motto 'Ογεγραφα γεγραφα.' "[149]

Dyson took much more trouble with the writing of the S-matrix paper than with the one on radiation theory. He was much more concerned about doing a careful job of presentation because it was so much more complicated: "It was a level of sophistication which was unusual for physicists at the time. I also had more time. I was much more relaxed. I felt I had no competitors. When this second paper was three quarters written, I went off skiing in Canada and finished it when I came back, which implied feeling confident and being in the comfortable position of not being scooped."[150]

The S-matrix paper was finished in mid-January. It was typed and circulated just before the January 1949 American Physical Society meeting in New York. "It is very nice to have it sent out so quickly," Dyson wrote his parents, "because it answers so categorically a lot of questions which were raised in the first paper, about which people are always worrying me."[151]

The paper's conclusions were immediately assented to, almost uncritically so. "I was surprised by the universal acceptance of the paper. Perhaps it had so little trouble attracting attention because it was so mathematical and so complicated."[152] Dyson therefore welcomed the scrutiny that his paper was being subjected to by the "younger" members of the theoretical physics community. When Peierls forwarded him some comments, he replied:

> I am glad you have had a talk with Pauli, who seems to be the one member of the "old gang" who takes the trouble to thoroughly understand the new methods. We are all very pleased with his regulators; especially the programme of systematic segregation of renormalization outlined in my S-matrix paper. The regulators make it clearer why the rules of procedure I have proposed are sensible, and vice versa. To demonstrate the equivalence of my rules with Pauli's, it is only necessary to show that all the convergent operators which appear in my method as the physical effects after separation from the renormalizations, actually tend to zero with increasing rest-mass. That this is so seems to follow just from dimensional arguments; since the real effects always begin by being proportional to the particle

momenta to some positive power, they must also have some
positive power of the electron mass in the denominator. However, I have not yet tried to make a rigorous argument out of
this.[153]

Although most of the paper was mathematical and complicated, the last
section entitled "Discussion of Further Outlook" was of a more philosophical bent.
This last section had been added by Dyson after the main body was completed, because he had anticipated "some trouble in getting the paper past Oppenheimer."[154]
The paper had consisted of eight "long and involved" sections, to which the short
ninth section was appended "to purposely distract Oppenheimer."[155] In its original
version, this appendage dealt with the general meaning of the whole work and was
"full of provocative statements."[156] Oppenheimer objected very strongly to it and
Dyson and he had some informative discussions deciding what should be put in its
place. In the end, Dyson drafted a new version to which Oppenheimer "agreed."
After writing it, Dyson also felt that it was a great deal better than the original one.
Moreover, "since this is the only part of the paper that the average reader will ever
read, it was worthwhile to spend some trouble on it."[157]

In section 9 of his paper, Dyson pointed to a surprising feature of S-matrix
theory: its success in avoiding difficulties even though it is based on an infinite
interaction Hamiltonian density which is devoid of physical meaning.

> Starting from the methods of Tomonaga, Schwinger and Feyn-
> man, and using no new ideas or techniques, one arrives at an
> S-matrix from which the well-known divergences seem to have
> conspired to eliminate themselves.... This automatic disappear-
> ance of the divergences is an empirical fact, which must be given
> due weight in considering the future prospects of electrodynam-
> ics. Paradoxically, opposed to the finiteness of the S-matrix is the
> second fact, that the whole theory is built upon a Hamiltionian
> formalism with an interaction function
>
> $$H_I(x) = -eA_\mu(x)\bar{\psi}(x)\gamma^\mu\psi(x) - \delta mc^2\bar{\psi}(x)\psi(x)$$
>
> which is infinite and therefore physically meaningless. (Dyson
> 1949b, p. 1754)

The stress is on the fact that no new ideas or techniques have been used.
Indeed, it had been Dyson's intention to prove the soundness of the conservative
position. Taking the received formulation of quantum mechanics and of special
relativity, his aim had been to see how far the synthesis could be pushed through

without tampering with the basic tenets. The analysis of the paper supported the position that all the difficulties could be circumvented by renormalization, and that as far as the S-matrix was concerned, no novel or revolutionary ideas need be introduced. Dyson emphasized this point in a letter to Peierls written in the spring of 1949:

> The hope, which is expressed in the last section of the S-matrix paper, and which I believe is a promising hope, is that electro-dynamics can now be put into a consistent and divergence-free shape by a *purely mathemetical reformulation, without any additions to its physical content* [italics mine].
>
> I certainly envisage the necessity later on of making alterations in electrodynamics, of a more fundamental and physical kind. But I think the nature of such alterations cannot be guessed at.... [You are willing to make] such a guess. I am less ambitious and confine myself to squeezing all I can out of the existing theory.[158]

And the same theme was reiterated two years later in a popular article Dyson wrote for *Physics Today*:

> We are proud of our new quantum electrodynamics ... it is a triumph of ingenuity, and it succeeds in reconciling all the contradictions of the old theory without abandoning anything of value. [But] the whole success of the theory is based on an unexplained miracle.... In QED the starting equations involve the unobservable and mathematically meaningless symbols e_0 and m_0. There is [then] a complicated mathematical cancellation, so that in calculations of observable quantities the final results are independent ... of the meaningless symbols. Why these miraculous cancellations occur, the theory does not explain. (Dyson 1952a)

To explain the discrepancy between the finite renormalized S-matrix and the divergent Hamiltonian density, Dyson, in the last section of his S-matrix paper, referred to Bohr and Rosenfeld's (1933) analysis of the measurements problem in quantum electrodynamics:

> We interpret the contrast between the divergent Hamiltonian formalism and the finite S-matrix as a contrast between two pictures of the world, seen by two observers having a different choice of measuring equipment at their disposal. The first

picture is of a collection of quantized fields with localizable interactions, and is seen by a fictitious observer whose apparatus has no atomic structure and whose measurements are limited in accuracy only by the existence of the fundamental constants c and h. This ["ideal"] observer is able to make with complete freedom on a sub-microscopic scale the kind of observations which Bohr and Rosenfeld employ . . . in their classic discussion of the measurability of field-quantities. The second picture is of collection of observable quantities (in the terminology of Heisenberg) and is the picture seen by a real observer, whose apparatus consists of atoms and elementary particles and whose measurements are limited in accuracy not only by c and h, but also by other constants such as α and m. (Dyson 1949b, p. 1755)

A "real observer" can measure energy levels, and perform experiments involving the scattering of various elementary particles—the observable of S-matrix theory—but cannot measure field strengths in small regions of spacetime. The "ideal" observer, making use of the kind of "ideal" apparatus described by Bohr and Rosenfeld, can make measurements of this last kind, and the commutation relations of the fields can be interpreted in terms of such measurements. The Hamiltonian density will presumably always remain unobservable to the real observer whereas the ideal observer, "using non atomic apparatus whose location in space and time is known with infinite precision" is presumed to be able to measure the interaction Hamiltonian density. "In conformity with the Heisenberg uncertainty principle, it can perhaps be considered a physical consequence of the infinitely precise knowledge of location allowed to the ideal observer, that the value obtained by him when he measures Hamiltonian density is infinity" (Dyson 1949b, p. 1755). If this analysis is correct, Dyson speculated, the divergences of QED are directly attributable "to the fact that the Hamiltonian formalism is based upon an idealized conception of measurability" (p. 1755).

If this notion of measurability is accepted, the correlation between expressions which are unobservable to a real observer and expressions which are infinite "is a physically intelligible and acceptable feature of the theory." "The paradox," Dyson continued, "is the fact that it is necessary . . . to start from infinite expressions in order to deduce finite ones." What may be therefore looked for in "a future theory" is not necessarily a modification of the present theory which will make all infinite quantities finite "but rather a turning around of the theory so that the finite quantities shall become primary and the infinite quantities secondary." Dyson concluded his paper with the statement: "The purpose of the foregoing remarks is merely to point out that there is no longer, as there has seemed to be in the past, a compelling necessity for a future theory to abandon some essential features of the present electrodynamics. The present electrodynamics is certainly incomplete, but is no longer certainly incorrect" (Dyson 1949b, p. 1755).

In his S-matrix paper Dyson regarded the Hamiltonian framework as an auxiliary tool. He believed that "the future theories" which would probably still be based on a modified Lagrangian approach, would be able to address more local and detailed questions than scattering matrix elements did. Dyson turned to these questions in a series of important papers that we will consider in section 9.16. Although Dyson did not introduce cutoffs in his S-matrix paper, Feynman's approach (using a cutoff) gave rise to the hope that a quantum field theory based on a Hamiltonian could be made mathematically well defined by introducing cutoffs. Renormalizability then simply meant that the bare mass and the bare charge would converge to their desired value as the cut-off was removed. With the advent of axiomatic and constructive quantum field theory this hope was converted into mathematical theorems (Wightman 1956, 1976, 1986; Velo and Wightman 1973, 1976). Using constructive methods it has been shown that mathematically well-defined quantum field theories exist in $2+1$ space-time, and rigorous proofs have been supplied that demonstrate that the cutoff theory converges to a cutoff-free theory satisfying all the Wightman axioms. However, there exists as yet no proof of the existence of a cutoff-free QED in four dimensions.

Dyson's concern with measurement theory in the last section of his S-matrix paper had come from his discussions with Oppenheimer. Dyson recalls that Oppenheimer had taken the mathematical analysis for granted. "He wanted to talk about what it meant for Physics. Oppenheimer had this extreme reverence for Niels Bohr. He had a very dogmatic view that everything in the paper of Bohr and Rosenfeld on measurability of field operators was the revealed truth and everything had to be consistent." The analysis of Bohr and Rosenfeld in 1934 was carried out for free fields (or equivalently for the renormalized fields to 0th order in the fine structure constant.) In 1951 they extended their analysis to consider the measurability of the renormalized fields to order α and showed that there were no inconsistencies. What Dyson was suggesting is that it is unlikely that the unrenormalized interaction picture was well defined. Oppenheimer had not seen the point and he had thought that Dyson had shown that Bohr's definition of measurability was inadequate.

Dyson returned to these measurability questions when he addressed the problem of whether renormalization would render well-defined, local operators such as the current operator, $j_\mu(x)$ and the energy density, $\mathcal{H}_I(x)$ (Dyson 1951a, b).

9.13 The S-matrix Paper: Aftermath

Dyson's fame spread by word of mouth and also from the reading of the preprint of his "Radiation Theory" paper. By the time of the APS meeting in New York at the end of January 1949, he had become a celebrity. To his parents he acknowledged:

> The New York meeting...was from my point of view a fantastic affair....On the first day the real fun began. I was

sitting in the middle of the hall and in the front, with Feynman beside me, and there rose to the platform to speak a young man from Columbia whom I know dimly. The young man had done some calculations using methods of Feynman and me, and he did not confine himself to stating this fact, but referred again and again to the "beautiful theory of Feynman-Dyson" in gushing tones. After he said this the first time, Feynman turned to me and remarked in a loud voice, "Well, Doc, you're in." Then, as the young man went on, Feynman continued to make irreverant comments, much to the entertainment of the audience near him.[159]

On the next day, Oppenheimer delivered his presidential address on the subject "Fields and Quanta" and gave a "very good historical summary of the vicissitudes of field theory." At the end of his address Oppenheimer spoke in glowing terms of the work Dyson had been doing and said it was pointing the way for the immediate future, "even if it did not seem deep enough to carry us farther than that." Dyson was "overwhelmed."

Thereafter Dyson was constantly being pursued by one person after another asking to be told all about it. "I am really becoming a Big Shot with a vengeance" Dyson confided to his parents.[160] He also told them that he didn't understand why Oppenheimer should have been so "indiscreet" as to reveal the content of his as-yet-unavailable preprint. "Perhaps," Dyson mused "it is partly because I am English, and he is so much concerned with impressing upon the nationalistic war generation the fact that science is an international activity. However, that may be, I believe I am wise enough to enjoy this sort of success without having been taken in by it; if I were not, I have the example of Feynman to instruct me."[161]

But Dyson was somewhat "taken in by it." People remember him as being at times arrogant during that period. And a little after the APS meeting he wrote Bethe: "I am sorry our conversation at the New York meeting became so contentious; when I come to Cornell I will be able to listen to your arguments more calmly."[162]

It would have been very difficult not to have been affected by the adulation showered upon him by the physics community. He received invitations to present colloquia from all the major universities on the eastern seaboard as well as invitations for week-long visits from Rochester, Chicago, and Toronto. K. K. Darrow invited him to be one of the featured speakers at the Washington meeting of the American Physical Society. And once the radiation theory paper appeared in the *Physical Review*, correspondence began to pour in upon him to the point that he joked to his family, "Soon I shall have to engage a secretary."[163]

At the end of March 1949 he informed Rudolf Peierls in Birmingham: "I have spent the last month, and shall spend the next travelling around from place to place and giving talks and consuming excessive quantities of food and liquor. While this sort of life is good for acquiring general knowledge about what is going on, it is certainly not conducive to serious thinking."[164]

After a week of being wined, dined, and "partied" in Chicago, he informed his parents that "travelling will be a welcome change and rest" and added "since last summer, I have had a nostalgic feeling for long distance rides in Greyhound Buses."[165]

There were also new offers for positions in England. When Dyson asked Oppenheimer to help him choose between Birmingham, Bristol, and Cambridge, Oppenheimer replied without a moment's hesitation: "Well, Birmingham has much the best theoretical physicist to work with, Peierls; Bristol has much the best experimental physicist, Powell; Cambridge has some excellent architecture; you can take your choice."[166] Dyson chose Birmingham and accepted Peierls' offer of a research fellowhip.

The little note that Bethe appended to his letter of recommendation for Dyson's appointment as a fellow at Birmingham is indicative of the community's assessment of his talents and accomplishments.[167] In it Bethe confided to Peierls, "I think the enclosed letter on Dyson is strong enough but the strange thing is that it is all true and sincere. He is really incredibly good." Bethe's recommendation stated:

> In my opinion Dyson is one of the very best theoretical physicists now living, regardless of age. He is probably the best English theorist since Dirac. In this country I would rate him on the same level with Schwinger who is a full professor at Harvard University and is considered the leading theorist under 40 years.... Many universities including Columbia, have offered him an associate professorship which would carry a salary in the neighborhood of $6000 or $7000. He has refused these offers because he feels under obligation to return to England.[168]

Later that spring Peierls was influential in Dyson being awarded a prestigious Royal Society Warren Research Fellowship.[169] Peierls wrote Dyson: "This [fellowship] has several advantages since it would give you a somewhat more recognized status and it would make it easier for you if one day you decided that some other place in this country would interest you more than Birmingham. From our point of view it would save the University some money."[170]

When in June 1949 Cambridge University offered Dyson a lectureship, Peierls promptly wrote him that he should not consider himself bound by his commitment to Birmingham. Dyson confessed to his family: "A permanent job of this sort at Cambridge—is a rare opportunity not likely to recur often. Also, it is in a way the thing that in the long run I want to have more than anything else. Nevertheless;

I again refused it."[171] Once again the quality of theoretical physics at Birmingham was the decisive factor in arriving at the decision.

On the whole Dyson was pleased with his newly acquired fame, but there was one aspect of it which disturbed him somewhat. He was at times concerned that he had deprived Feynman of some of the credit that was due him: "I sometimes feel a little guilty for having cut in front of him with his own ideas," Dyson wrote home. But he assuaged his guilt feelings with the observation: "However, he is now at last writing up two big papers, which will display his genius to the world; and it is possible that I have helped to make him do this by making him just a shade conscious of being cut in on, which if it is true is a valuable service on my part."[172]

Dyson did take every opportunity to communicate to the world what Feynman had done. Thus for his invited talk in September 1949 at the International Congress on Nuclear Physics and Quantum Electrodynamics, his first to the European physics community, he delivered a lecture on "The Radiation Theory of Feynman." In it he stressed that "the FEYNMAN theory is not to be regarded as a theory in competition with the SCHWINGER theory. It is rather a collection of ideas, of a somewhat intuitive character, which create a deeper understanding, on the one hand of the physical assumptions underlying existing electrodynamics, and on the other hand of the possibilities which exist for new theoretical developments" (Dyson 1950b, p. 242).

9.14 Oldstone

There is no better indication of Dyson's standing in the theoretical physics community than the fact that Oppenheimer invited him to attend the Oldstone conference. As had been the case at the two previous Oppenheimer-run conferences, Shelter Island and Pocono, the pace at Oldstone was intense. Dyson's impression of the conference was summarized in a letter to his parents written on Easter Sunday, 1949:

> Oldstone is a country hotel, about 50 miles north of New York, in a splendid situation overlooking the Hudson, and with hills behind it and hills facing it across the river. We had lovely weather for the conference, and could sit outside whenever we were not conferring. However, since the conference was run by Oppenheimer that was not often. One of the things which simply amazes me about Oppenheimer is his mental and physical indefatigability; this must have had a lot to do with his performance during the war. There was no fixed program for the conference, and so we just talked as much or as little as we liked; nevertheless Oppenheimer had us in there everyday from

ten A.M. till seven P.M. with only short breaks, and on the first day also after supper from eight till ten, this night session being only dropped on the second day after a general rebellion. And all through these sessions Oppenheimer was wide awake, listening to everything that was said and obviously absorbing it.[173]

Dyson remembers feeling very ashamed of himself for falling asleep while Lamb was talking and thereafter thinking to himself, "Gosh, isn't this terrible to have the great experimenter Mr. Lamb himself talking and you can't even stay awake."[174] Dyson was the featured speaker at session 5 on "Electrodynamics and Mesodynamics." In his weekly letter to his parents, Dyson reported: "On the second day I stood up and talked for about an hour. I found this very easy, as I did not have to prepare anything but only to answer questions and summarize what the other young people at the Institute have been doing."[175]

The ease with which he could discharge his assignment reflects not only his mastery of every facet of the subject, but also his great self-confidence. The notes of the conference relate that Dyson dealt with "problems susceptible to expansion type of treatment" and that he reviewed his S-matrix paper. He also reported on Karplus and Kroll's calculation of the magnetic moment of the electron to order α^2. "Result not yet reached, but looks as if it will be reasonable. Difficulty of calculations shows that electrodynamics will not be easy to apply to complex problems."[176] Reporting on the nuclear magnetic moment calculations that were being done, Dyson remarked that the quantum electrodynamics of particles with an intrinsic moment of the Pauli type whose interaction Hamiltonian is $H_R = e j_\mu A_\mu + \mu_0(\bar\psi \sigma_{\mu\nu}\psi)F_{\mu\nu}$ is not "satisfactory": the theory is not renormalizable. The notes report that Feynman suggested that this criterion "can perhaps not be taken as a decisive argument against such theories." Dyson felt differently and argued that renormalizability should be adopted as a criterion for selecting theories and that, in particular, this criterion should be used for selecting meson theories. He wrote down the condition for the renormalizability of theories of nucleons interacting with scalar or pseudoscalar mesons via couplings of the form $\bar\psi\Gamma\psi\phi$ or $\bar\psi\Gamma_\mu\psi\partial^\mu\phi$ or with vector mesons via a coupling of the form $\bar\psi\Gamma_\mu\psi V^\mu$ or $\bar\psi\Gamma_{\mu\nu}\psi\partial^\mu V^\nu$. A table in the notes gives a characterization of the various meson-nucleon theories.

Meson	*Coupling*	*Remarks*
Scalar	Scalar	Renormalizable
Scalar	Vector	Diverges—Non-renormalizable
Pseudoscalar	Pseudoscalar	Renormalizable
Pseudovector	Pseudoscalar	Non-renormalizable
Vector	Vector	Diverges (Feynman)[177]

Furthermore, Dyson reported that on the basis of this selection criterion, computations were being done at the Institute on the nucleon-nucleon interaction in a PS(PS) theory.

Dyson asserts, "I always believed renormalizable theories were the ones to work on."[178]

Shortly after the Oldstone conference, Dyson delivered a 40-minute invited talk on "Recent Developments in the Quantum Theory of Fields" at the Washington, D.C., meeting of the American Physical Society. Dyson remembers saying in that lecture:

> We have the key to the universe. Quantum electrodynamics works and does everything you wanted it to do. We understand how to calculate everything concerned with electrons and photons. Now all that remains is merely to apply the same to understand weak interactions, to understand gravitation and to understand nuclear forces.
>
>   It was a very gung-ho sort of talk. I felt it was all in the bag. We just have to do the other parts of nature in the same fashion and we will understand the whole lot. The task at hand was to clean up everything we know how to calculate—we had to get the various (then disparate) parts of nature properly formulated and afterwards see whether you can get an overall pattern. Maybe we can unify but let us not do this first.[179]

That invited paper was the high point of Dyson's faith in field theory. He believed QED was incomplete, but not incorrect; incomplete mathematically because of the divergence difficulties, and incomplete physically because it did not deal with all of nature. And he remembers having an argument with Bryce de Witt at the time about whether the inclusion of general relativity might make the theory convergent. "I believed then that if you could bring in gravity, perhaps everything would converge."[180] He was right about QED, wrong about gravity.

9.15 Return to Europe

The year at the Institute passed very rapidly. It was exceedingly successful both at the professional and at the personal level. Dyson had become a "big shot." His personal life had also taken a dramatic turn for the better. During the last few weeks of his stay at the Institute he had met and fallen in love with Verena Haefeli. He was going back to England with all his expectations and some of his fantasies more than fulfilled.

He was fully aware of this. In a letter home he wrote:

> It is queer how things move. I remember a long conversation with Alice in her room at Guilford, in which I said how I would

like to be a cementer of international science, and to travel round and organise conferences and learn to speak a lot of languages. And I said this was all very well, but one had first to concentrate on being a good physicist, so that one could have some excuse for one's conferences and one's travelling around. So I said I was going to the U.S. to talk nothing but English for two years, and do nothing but learn some physics.

It has certainly worked out better than I then expected.[181]

Upon his return to Europe, Dyson was one of the invited speakers at the International Congress on Nuclear Physics and Quantum Electrodynamics which convened in Basel in early September 1949 (Dyson 1950b). Two weeks later he attended the Como conference of the Italian Physical Society in that same capacity. In Basel and at Lake Como he met all the leading European physicists, including Pauli and Heisenberg. Pauli, after a canonical rude introductory remark, was very friendly to him. Heisenberg was deeply impressed by Dyson. After their meeting in Como, Heisenberg "studied carefully" Dyson's S-matrix paper and thereafter wrote him, asking his opinion "on a few questions concerning the current density operator." In his letter Heisenberg told Dyson:

You replace the ordinary expression $\psi^+\gamma_\mu\psi$ by the new effective current density $\psi^+\Gamma_\mu\psi$, where Γ_μ is a rather complicated integral operator. In the old theory, one could prove easily that the integral $\int d\sigma_\mu\psi^+\gamma_\mu\psi$, taken over any finite space-like volume, has only integer eigenvalues, in other words that the charge contained in any finite volume is always an entire multiple of the electronic charge. This situation is changed completely if you replace γ_μ by Γ_μ and there seem to be 3 possibilities with respect to the eigenvalues of $\int d\sigma_\mu\psi^+\Gamma_\mu\psi$ taken over a finite volume σ.

1. The eigenvalues are integers for any value of the real electronic charge.

2. The eigenvalues are generally not integers.

3. The eigenvalues are integers only in the special value $hc/e^2 = 137$ for the electronic charge.[182]

On the basis of some perturbative calculations, Heisenberg argued against the correctness of 1 and 2, leaving alternative 3, "certainly the most interesting one." Heisenberg inquired of Dyson whether he had "so-detailed knowledge of the operator Γ_μ, that [he] can already disprove alternative 3." In a lengthy reply to Heisenberg, a month later, Dyson summarized his conclusion by saying that he "believes there is no reason to doubt that alternative 2 is the correct one..."

and [moreover] that [he] does not think that this result is in any way physically paradoxical."

Dyson actually proved that alternative 1 is impossible and adduced strong reasons why alternative 3 could not be taken seriously. He further remarked that "so far as the methods of expanding in powers of e^2/hc is justified, the present electrodynamics appears to be a consistent and physically satisfactory theory irrespective of the particular value of e^2/hc. Therefore the number 137 can only be understood either (i) from a treatment of electrodynamics not based upon expansion in powers of e^2/hc, or much more probably (ii) from arguments lying altogether outside electrodynamics."[183]

The general impression that Dyson obtained from meeting the European physicists was that they "took it very much for granted that everything interesting was happening in America." He perceived his function was "to teach these backward Europeans" how to do physics.[184] When Dyson arrived in Birmingham in the fall of 1949 to assume his Warren Fellowship, he found Peierls similarly eager to learn the new developments. Peierls felt very strongly that England had been left behind and he encouraged Dyson to teach everybody as much as he could. While in Birmingham that year, Dyson was constantly giving seminars and lecturing on Feynman's and his own work.

As Dyson had not gotten a Ph.D. while at Cornell, Peierls initiated steps for him to obtain an advanced degree while at Birmingham. He had Dyson register as a research student. As there is a tuition charged to such students, Peierls wrote the registrar a note suggesting that Dyson "be exempted from tuition fees. He is a man of very senior standing in research and it would be more correct to say that we are receiving tuition from him than vice versa."[185] However, Dyson never obtained his D.Sc. degree because there is a two-year residence requirement at Birmingham University which Dyson didn't satisfy (Peierls 1985). He visited the Institute for Advanced Study during the fall semester of 1950 and left for Cornell in the fall of 1951.

Dyson lived with the Peierls during the first year he spent in Birmingham. The Peierls' children used to make fun of Dyson's slow and meticulous way of eating, which they claimed included carefully peeling each pea (Peierls 1985, p. 235). Peierls also recounts that Dyson "had periods where he felt that he had run out of imagination, and that he would never again have an original idea in physics. At such time he would seriously talk about changing his profession, perhaps to study medicine. But each time, the next challenging idea appeared before long, and all such doubts were forgotten" (Peierls 1985, p. 235).

9.16 Heisenberg Operators

No sooner had the "S-matrix in QED" paper been finished than Dyson turned to meson theory. "This week I began once more to think in a fundamental way, having got the second paper off my mind," he wrote home at the end of January 1949.

The second paper was still concerned with the physics of outside the nucleus; it was in fact a general demonstration of the effectiveness of the Schwinger-Feynman methods in dealing with a very general class of extra-nuclear problems; the main achievement of it was the building up of a technique which made it possible to use the theory almost automatically. Now, at last, having got the theory into such a handy and powerful shape, I have decided to sally forth and apply these methods to the nucleus. Almost at once I began to get encouraging results and I believe the thing will really begin to go now.[186]

Experiments had established that π mesons were pseudoscalar particles and Dyson had verified by his power-counting argument that pseudoscalar meson theories with pseudoscalar couplings to the nucleon field, $\mathcal{L}_I = g\bar{\psi}\gamma_5\tau_i\psi\phi_i$, were renormalizable. The comparison of the prediction of PS(PS) theory with experiments thus constituted an important problem. But since the coupling constant g was expected to be large, a perturbative expansion in powers of g^2/hc could not be trusted. Deriving results which could be believed was the challenge. As Oppenheimer remarked to Peierls, "It is turning out to be a very tough thing to digest these developments of the last two years and maintain any sort of perspective, the more so because at the moment one can neither get any sensible results with the mesons, nor devise methods sufficiently powerful to justify expecting them."[187]

Dyson devised a scheme which allowed the high frequency part of the interaction, which presumably is ineffective except in producing renormalization effects, to be treated separately from the low frequency part. On the basis of his experience with quantum electrodynamics, he argued that the high frequency interaction produces only small physical effects and that an expansion in powers of the high frequency interaction should be rapidly convergent, and therefore convenient for practical calculations.

Interestingly, Dyson's new approach abandoned Feynman's S-matrix viewpoint and was based on the Hamiltonian formalism. It was in fact patterned after Schwinger's procedure of making contact transformations to eliminate virtual effects. However, it eliminated making an adiabatic assumption, a central feature of Schwinger's approach.

If H is the Hamiltonian of the interacting system

$$H = H_0 + H_R + H_e \tag{9.16.1}$$

with H_0 the Hamiltonian for the free Dirac and radiation field, and

$$H_R = -e \int \bar{\psi}(x)\gamma_\mu A^\mu(x)\psi(x)d^3x \tag{9.16.2}$$

$$H_e = -e \int \bar{\psi}(x)\gamma_\mu A^{e\mu}(x)\psi(x)d^3x, \tag{9.16.3}$$

then the interaction representation state vector satisfies the equation

$$i\hbar\frac{\partial \Psi(t)}{\partial t} = (H_R + H_e)\Psi(t). \tag{9.16.4}$$

Formally all virtual effects are eliminated by making the contact transformation

$$\Psi(t) = S(t)\Phi(t) \tag{9.16.5}$$

such that

$$i\hbar\frac{\partial S(t)}{\partial t} = H_R S(t). \tag{9.16.6}$$

Φ then satisfies the equation

$$i\hbar\frac{\partial \Phi}{\partial t} = S^{-1}(t)H_e S(t)\Phi(t). \tag{9.16.7}$$

The power series solution of eq. (9.16.6) is given by

$$S(t,t_0) = \sum_{n=0}^{\infty} \frac{1}{n!}(\frac{-i}{\hbar})^n \int_{t_0}^{t}\int_{t_0}^{t} P(H_R(u_1)\ldots H_R(u_n)du_1\ldots du_n. \tag{9.16.8}$$

The essence of Schwinger's program had consisted in the computation of $H_e' = S^{-1}H_e S$. This canonical transformation could be interpreted as the replacement of a real particle, interacting with the radiation field and external potential $A^e(x)$, by a bare particle moving in a modified external potential described by H_e'. Schwinger obtained his result by imagining the interaction between matter and radiation to have been adiabatically switched on, so that in the remote past the particle appeared without interaction. Mathematically this was accomplished by extending the initial time t_0 backward to $t_0 = -\infty$ and averaging out all terms that depended periodically on t_0. The prescription, however, destroyed the unitarity of $S(t,t_0)$.

To eliminate the adiabatic assumption, Dyson introduced a new representation, which he called the smoothed interaction representation (SIR), defined to be such that at any time t the Hamiltonian functions $H_R(u)$ in eq. (9.16.6) are transformed to

$$H'_R(u) = e^{\gamma(u-t)}H_R(u) \qquad (\gamma > 0). \qquad (9.16.9)$$

One readily verifies that $\Phi_\gamma(t)$ defined by

$$\Psi(t) = S_\gamma(t)\Phi_\gamma(t) \qquad (9.16.10)$$

with

$$S_\gamma(t) = \sum_{n=0}^{\infty}(-\frac{i}{\hbar})^n\frac{1}{n!}\int\int du_1\ldots\int du_n e^{-\gamma(nt-u_1-u_2-\ldots u_n)}P(H_R(u_1)\ldots H_R(u_n))$$

$$(9.16.11)$$

satisfies the following equation of motion:

$$i\hbar\frac{\partial\Phi_\gamma(t)}{\partial t} = \left\{S^{-1}(t)\left(H_e - i\hbar\frac{\Delta}{\Delta t}\right)S_\gamma(t)\right\}\Phi_\gamma(t), \qquad (9.16.12)$$

where $\frac{\Delta}{\Delta t}$ is defined as the time differentiation which acts only on the integrand in expressions such as eq. (9.16.8). To lowest order in $H_R(u)$,

$$-i\hbar\frac{\Delta}{\Delta t}S = \gamma\int_{-\infty}^{t} du H_R(u)e^{-\gamma(t-u)}. \qquad (9.16.13)$$

If $H_R(u)$ has matrix elements of the form $ae^{i\omega t}$, then

$$i\hbar\frac{\Delta}{\Delta t}S = \frac{ae^{i\omega t}}{1+i\frac{\omega}{\gamma}} \approx \begin{cases} H_R(t) & \text{for } \omega << \gamma \\ 0 & \text{for } \omega >> \gamma, \end{cases} \qquad (9.16.14)$$

so that to this order the right-hand side of (9.16.12) is given by H_e plus the low frequency part of H_R. Dyson's method thus separates the high and low frequency effects, high and low being scaled by the parameter γ. Moreover, for finite γ the limit $t_0 \to -\infty$ maintains the unitarity of S. Dyson verified that with the addition of suitable counterterms to (9.16.12), the operator $S^{-1}(H_e - i\hbar\frac{\Delta}{\Delta t})S$ is finite and yields a finite renormalized Hamiltonian to the order concerned. One therefore obtains a finite Schrödinger equation with which bound states can be calculated.

Dyson lectured on this formulation of quantum field theory at the 1950 Michigan summer school (Dyson 1950) and illustrated the theory with a calculation of the Lamb shift. The method has several advantages: The integrals one encounters in computing S_γ to lowest order are just as easy to handle and analyze as the integrals from the same graphs in the calculation of the S-matrix. Moreover, whereas the S-matrix integrals involve three distinct types of divergences, namely (1) infrared divergences, (2) singularities arising from several poles of the integrand running together, and (3) divergences at high frequencies, in Dyson's SIR all divergences of types (1) and (2) are eliminated. Dyson's method also automatically resulted in the cancelation of the renormalization divergences that appear in diagrams of figure 9.16.1 if the mass renormalization counterterm $\delta m \bar{\psi} \psi$ carries the damping factor $e^{-2\gamma(t-u)}$.

In pushing the investigation further, Dyson discovered that the method not only yielded a finite Hamiltonian but could also be applied to more general operators such as the current operators $S_{\gamma}^{-1} j_{\mu}(x) S_{\gamma}$. In the middle of July 1950 he communicated to Bethe: "I have got a new method of doing calculations in electrodynamics which is very pretty and makes the calculation of 'modified current operators' and such things in the Schwinger theory as simple and general as the calculation of the S-matrix. Thus it should now be possible to carry through the renormalization program formally to all orders in α including bound states."[188]

At Michigan Dyson proved that the one-loop expansion of $S_{\gamma}^{-1} j_{\mu}(x) S$ was rendered finite by renormalization. The notes ended with the statement that "The great question now is whether the operators of the SIR such as H_{γ} or $j_{\mu\gamma}(x) = S_{\gamma}^{-1} j_{\mu}(x) S_{\gamma}$ can be analyzed quite generally in the same way as the S-matrix was analyzed.... I do not know. If the answer is yes, then we shall presumably be able to extend the systematic removal of renormalization effects from the S-matrix to all operators occurring in the SIR. But this is a rash claim, and we have a long way to go to fulfill it" (Dyson 1950c).

The order-by-order analysis of Heisenberg operators such as $S_{\gamma}^{-1} j_{\mu} S_{\gamma}$ is exceedingly complex. At the end of the summer Dyson wrote Peierls that "My own work has not progressed any further since we left Ann Arbor. I am held up by some mathematical difficulties which may not be easy to overcome."[189] But Dyson continued to work on the problem. During the fall he informed his parents:

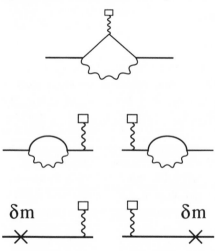

Figure 9.16.1

"I am these days working very hard at my physics and getting somewhere with it. But it is slow work. I have chosen a very tough problem to attack."[190] In the middle of November 1950 he could report to them that he had "started at last the job of writing a big paper for the *Physical Review*."[191] However, "a serious crisis" with his big paper occurred at the end of that month. He had "found an important mistake which took 24 hours of frantic work to repair. Now everything is in order again, this time, I am sure permanently."[192]

Just before Christmas he wrote excitedly to his parents:

> Yesterday I managed unexpectedly to think out the solution of
> one of the outstanding problems which remained in my physics
> researches. This is a wonderful piece of luck. The whole thing
> is now ready, so I can hand the results over to the students and
> other people who will make practical use of them. Now I do not
> ·have to think about these problems any more. I will just sit down
> and write another long paper. I am immensely pleased with all
> this piece of work. It is really a breakthrough which will get the
> whole of radiation theory moving forward again after a hold-up
> which has lasted 2 years. But it will take 6 months or so before
> this is effective, the new method is very sophisticated and people
> need quite a long time before they can assimilate it. People who
> have read my recent paper are respectful but also bewildered.
> This is true even of Oppy.[193]

The achievement was summarized in a communication entitled "The
Renormalization Method in quantum electrodynamics" (Dyson 1951). In his letter to the *Royal Society* recommending its publication, Peierls noted that "this
paper seems to me of such outstanding importance that it would deserve some
preferential treatment."[194]

With this paper the programme that Dyson had outlined in "Heisenberg
Operators I" (Dyson 1951d) was widened "so that the objective is a proof of the
total disappearance of all divergences from quantum electrodynamics after the dynamical variables have been transformed by a suitable contact transformation. . . .
Furthermore, it is intended that the new divergence-free formulation of the theory
shall be practically useful and adaptable to approximate calculations."

In the paper the physical and heuristic ideas underlying the transformation S_γ were briefly explained, and a physical picture to intuitively explain the
success of the renormalization technique was adumbrated. The electron-positron
and the electromagnetic field, Dyson explained, are to be thought of as two characteristic properties of a single "fluid" which fills the whole of space-time. This fluid
is in a state of violent quantum-mechanical fluctuation, the fluctuations becoming
more and more pronounced as the region over which they are observed is made
smaller. The fluctuations have the property that at sufficiently high frequencies
and in sufficiently small regions of space they are essentially isotropic and uniform over the whole of space-time, just like the fluctuations of a classical fluid in
a state of isotropic turbulence. Statements concerning the behavior of the fluid at
a particular space-time point are observationally meaningless: they translate into
the mathematical divergences that obtain when the theory is expressed in terms
of operators defined at sharp space-time points. However, because of the isotropy
and homogeneity of the fluctuations, the macroscopic properties of the fluid are

observable and well defined. A divergent-free description of the fluid is obtained as soon as the behavior is expressed entirely in terms of new dynamical variables that are averages of the original instantaneous variables over finite intervals of space and time.

According to Dyson, renormalization is always associated with an averaging-out of high frequency fluctuations of the fields. The averaging-out is achieved by integrating the equation of motion of the fields explicitly with respect to the time. In the Schwinger theory the state vector is transformed so that the new state vector reflects the behavior of the system in the infinite past. In the Feynman theory the description is directly in terms of an overall space-time picture in which localization of processes in space and time is abandoned. It is precisely the averaging over an infinite time interval that introduced the two fundamental limitations of the formalisms that had been developed thus far: (1) the inapplicability of the S-matrix apparatus to bound state problems, and (2) the nonconvergence of the power series expansions in the radiation interaction. "It is [therefore] reasonable to hope that the theory can be freed from both limitations if the averaging over infinite time-intervals is dropped and the removal of high frequency fluctuations is accomplished by integrating the equations of motions over finite intervals" (Dyson 1951d).

Dyson then proceeded to indicate how this is to be done. The original state vector Ψ of the interaction picture is to be replaced by a new one, Φ, according to

$$\Psi(t) = S(t)\Phi(t), \tag{9.16.15}$$

where Φ is to be the smoothed-out average of Ψ over a finite time interval; $S(t)$ therefore is to follow accurately the high frequency fluctuations of Ψ but not the slow long-term variations. If $H_1(e_1, t)$ is the interaction Hamiltonian appearing in the Tomonaga-Schwinger equation (including mass and charge renormalization counterterms all expressed in terms of the renormalized charge), Dyson took $S(t)$ to satisfy

$$i\hbar \frac{\partial S(t)}{\partial t} = h_g(t, t')S(t), \tag{9.16.16}$$

where

$$H_g(t, t') = H_1(e_1 g(t - t'), t') \qquad \text{for} \qquad t' \leq t, \tag{9.16.17}$$

with $g(a)$ a continuous function of a with the property $g(0) = 1$, $g(a) \to 0$ as $a \to \infty$. This is of course nothing but a more formal and general version of the smoothed interaction representation that he had introduced in the Michigan lectures. The function $g(t - t')$ is scaled by a constant with the dimension of time, T; T^{-1} is then the frequency defining the division into high and low frequencies. In the limit $T \to 0$ $g(a) = 0$ and S is the identity operator; in the limit $T \to \infty$ all frequencies are considered as high, and S satisfies the same equation as Ψ so that Φ is the (constant) Heisenberg state vector describing the system.

In two difficult papers Dyson (1951b, c) proved that his intermediate representation provides a complete and divergence free formulation of quantum electrodynamics and yields a divergence-free Schrödinger equation that accurately describes the behavior of any physical system.

Dyson's paper (1951c) in which the finiteness of the Hamiltonian operator for the Schrödinger equation of the field system is proved, ended with the statement: "Since the equation is divergence-free, all the well-known approximate methods, trial wave functions, variation methods, iteration methods, and if necessary numerical integrations, may be applied to it freely. The techniques developed in this series of papers may thus be instrumental in restoring to the theoretical physicist working in quantum field theory the freedom of action which he enjoys when solving partial differential equations in other branches of physics" (p. 1216)

The papers were met with total silence. Dyson thinks back on that series of papers melancholically:

> To me there is a certain sadness about that whole program of extending [the S-matrix methods] to Heisenberg operators. The whole point about that whole program was to be able to do something about strong coupling theories, to do meson theories with strong interactions. In fact, this whole apparatus that I worked out of separating the high and low frequencies was for me a tremendous triumph. I really thought that it would be as important as anything else I had done because I thought it would enable us finally to get to grips with the meson theories and actually to calculate things with large coupling constant. It looked like if you did the high frequency part separately then the theory would converge. This was not philosophically driven at all: it was intended as a practical tool.[195]

The reason for the silence was that the formalism was exceedingly complicated and complex. Simpler methods for dealing with Heisenberg operators were then being advanced by Gell-Mann and Low (1951) and by Schwinger (1951c). Also, Salpeter and Bethe had discovered a covariant way of treating bound state problems within the Feynman scheme (Salpeter and Bethe 1951).

Furthermore, as Dyson notes "the [meson] theories weren't any good in the first place."[196] But in retrospect these papers were pioneering in the effort to relate all p-space vacuum expectation values by a single p-space analytic function and led naturally to the work of Lehmann, Symanzik, and Zimmermann on advanced and retarded functions (1955, 1957). Equally important were Dyson's ideas regarding scale dependence within the framework of quantum field theory, as expounded in his work on the smoothed interaction representation (Dyson 1951). He there argued that when g is varied, some modification had to be made in the definition of the g-dependent interaction, in order to compensate the effect caused by the change of the g-dependent charge. In line with this idea of Dyson, Landau and his collaborators developed a similar concept of a smeared-out interaction in a series of influential papers (1954a, b, c, d; 1956). Both Dyson and Landau argued that the magnitude of the interaction should not be regarded as a constant but a function of the radius of interaction which must fall off rapidly when the momentum exceeds a critical value $P \sim = 1/a$, where a is the range of the interaction. As a decreases, all the physical results tend to finite limits. Correspondingly, the electron's charge must be regarded as an as yet unknown function of the radius of interaction. Both Dyson and Landau had the idea that the parameter corresponding to the charge of the electron was scale dependent. In addition, Dyson hinted, though only implicitly, that the interaction should be scale independent. Landau, more explicitly, suggested that the physics of QED might be asymptotically scale invariant. In later works, in particular in those of Stueckelberg and Petermann (1953) and of Gell-Mann and Low (1954), Dyson's varying parameter g and Landau's range of interaction were further specified as the sliding renormalization scale, or subtraction point. In Gell-Mann and Low, the scale-dependent character of parameters and the connection between parameters at different renormalization scales were elaborated in terms of renormalization group transformations, and the scale-independent character of the physics was embodied in renormalization group equations. However, these elaborations were not appreciated until much later—in the late sixties and early seventies—when a deeper understanding of the ideas of scale invariance and of renormalization group equations was gained, mainly through the researches of K. G. Wilson, the result of fruitful interactions between QFT and statistical physics (K. Wilson and Kogut 1974; K. Wilson 1983).

9.17 Divergence of Perturbative Series

After completing his paper on the smoothed interaction representation, Dyson wrote Bethe:

> My own work has now come to an end for the time being. I have written 2 more papers... and this completes the proof that the equations of motion in the Intermediate Representation

are free of divergences. The...two papers are long but quite straightforward. ...The next problem is to prove the convergence of the perturbation theory expansion. I am convinced now that this can be done, and that the convergence will occur even for large values of the coupling constant. But the proof will be a long and very difficult piece of analysis, and I am not in any hurry to start on it. It needs to be thought about at leisure.[197]

During the summer of 1951, which he spent in Zurich visiting the ETH, Dyson discovered a simple argument which suggested that the S-matrix series diverges. His reasoning was the following: Suppose one were to calculate a physical observable in a power series in the coupling constant, e^2. If this series is convergent for some positive value of e^2, it must converge in some circle of radius e^2 centered around the origin in the complex e^2 plane. The series must therefore also converge on some interval of the negative real axis, that is, for e^2 negative. Now e^2 negative corresponds to a world in which like charges would attract one another and opposite charges would repel each other. However, if e^2 is negative, then a state which contains a large number N, of electron-positron pairs in which the electrons are clustered together in a region V_1 of space and the positrons are clustered together in another region V_2 of space far from V_1, would have an energy lower than that of the vacuum for N large enough; in other words, the binding energy of a large collection of particles would exceed the energy necessary to create them. The vacuum state is therefore unstable relative to such states. Since the larger the number of pairs is, the more pronounced the effect becomes, the higher order terms in the power series expansion must become more and more important so that the series cannot converge. At best, it can only be an asymptotic series.

QED was over for Dyson in the summer of 1951, when he found this heuristic argument that the perturbation theory diverges. Until that moment he had thought that one could get a convergent theory. "All my efforts up to that point had been directed toward building a complete convergent theory. Finding out that after all the series diverged convinced me that was as far as one could go....That was of course a terrible blow to all my hopes. It really meant that this whole program made no sense." And in a philosophical vein Dyson adds, "The humour of it is: that was the only part that did make sense."[198]

Dyson had always taken the view that the quantum electrodynamics was nothing more than the perturbative series. The so-called exact equations were mathematically so ill-defined that he couldn't really "put any trust in them." "I had this rather positivistic view that all QED was the perturbative series. So if that failed you didn't really have a theory. In that way I was somewhat different from Feynman. I felt path integrals were an intuitive guide, that they weren't anything one could use. [At that time] I never found a way that I could make a real use of path integrals."[199]

9.18 Closure

In May 1950 Dyson accepted a professorship at Cornell to replace Feynman, who was leaving for Cal Tech. In the letter making the offer, Bethe told Dyson:

> When Feynman's leaving had become definite there was a unanimous opinion among the staff... that there was only one man in the world who could replace him, namely you. Everyone of the professors whom I asked for suggestions mentioned your name immediately and spontaneously. The rank we are offering you, i.e. a full professorship, and the salary which goes with it [$8,000 a year] are meant to express how much we want you to come. The offer has the enthusiastic approval of the Dean and the President of the University.[200]

In the summer of 1950, Freeman Dyson and Verena Haefeli became engaged. Later that year they were married. The course of Dyson's personal and intellectual life seems to run along parallel tracks thereafter.

The first few years of his marriage witnessed the birth of his daughter Esther and son George, and the publication of his papers on the renormalizability of Heisenberg operators, perhaps his most impressive work in quantum field theory. But disappointments were soon to follow. During the summer of 1951 he convinced himself that the perturbative series upon which most of his work had been based did not converge. In the fall of 1951 Dyson came to Cornell, only to discover that the professorial life did not suit him well. He enjoyed teaching, even though it was hard work because he was always teaching new things. His impressive set of lecture notes on "Advanced Quantum Mechanics" was one of the fruits of this labor (Dyson 1952c). But taking care of graduate students proved "very painful." There was a mismatch between their needs and his. "Graduate students need stability. They require a problem which is stable for two or three years. My attention span is too short for that. I didn't fit into the Ph.D. system. That was the main reason for going to the Institute."[201]

In 1953 Dyson accepted Oppenheimer's invitation to join the Institute as a permanent member. In January 1957 his marriage broke up when his wife left him. A chapter in Dyson's life closed with these events.

The paper on the divergence of perturbation theory in QED marked the end of Dyson's involvement with QED. Although during 1952/53 Dyson was deeply involved in the Cornell project analyzing meson-nucleon scattering using field-theoretic methods (Dyson 1954), Dyson confesses "my heart was not in this Tamm-Dancoff work. There were no grand hopes."[202] Dyson had been deeply hurt by the reception of his extended hard work in QED. The community had not appreciated what he had done and "being a practical person, [he] didn't feel like going on all by [himself] into the wilderness and wait for the world to catch up." The difference between Feynman, Schwinger, and Dyson was that Feynman and Schwinger had "stopped at the crest of the wave." Dyson had been more courageous, only to find "the river disappearing into sand."[203]

Thereafter he never invested the same amount of energy and commitment into any fundamental physics program. "I wasn't so much disappointed that these papers were not noticed," Dyson claims. "I thought to myself: well really it was silly of me to have worked so hard on something which turned out not to be important."[204]

9.19 Philosophy

Replying to remarks by his parents comparing the Dyson-Schwinger-Feynman theory to the Athanasian creed, Dyson indicated that they would be disappointed if they harbored any hopes that it might prove an effective substitute: "The central idea of the theories, is to give a correct account of experimental facts while deliberately ignoring certain mathematical inconsistencies which come in when you discuss things that cannot be directly measured; in this there is a close similarity to the Athanasian Creed. However, there is the important difference that these theories are expected to last only about as long as no fresh experiments are thought up, so they will hardly do as a basis for a Weltanschauung."[205]

For Dyson the function of theories is to account for experimental phenomena. Should a theory fail to do so, it will be replaced. Moreover, "The nature of a future theory is not a profitable subject for theoretical speculation. [A] future theory will be built, first of all upon the results of future experiments" (Dyson 1949b, p. 1755).

Dyson is a positivist. His positivism was clearly articulated in a letter to his family written in the spring of 1949, in which he compared Oppenheimer's views with his own:

> We are both agreed that the existing methods of field theory are not satisfactory, and must ultimately be scrapped in favour of a theory which is physically more intelligible and less arbitrary. We also agree that the final theory should explain why there are the various types of particles which we see and no more. However, Oppenheimer believes that the nature of the nuclear forces will itself give us enough information on which to build the new theory; in other words, the nuclear forces will not be describable at all except in terms of a theory which explains the existence of elementary particles. I on the other hand take a more pessimistic view, that we shall be able to give a complete account of the nucleus on the basis of the present field theory, and that when this is done the theory will work equally well

for a proton whose mass is 1835 electron masses as for a proton whose mass is 1836. If I am right, the discovery of a finally satisfactory theory of elementary particles will be a much deeper problem than those we are tackling at present, and may very well not be achieved within the framework of microscopic physics.

Probably the reason for our disagreement is largely a matter of history; Oppenheimer has spent all his life seeing the field theory fail on one problem after another, whereas I have grown up during one of its brief periods of success.[206]

QED had assumed for Dyson a privileged role—and this because of his positivism. As he explained in a popular account of the theory:

> Quantum electrodynamics occupies a unique position in contemporary physics. It is the only part of our science which has been completely reduced to a set of precise equations. It is the only field in which we can choose a hypothetical experiment and predict the result to five places of decimals confident that the theory takes into account all the factors that are involved. Quantum electrodynamics gives us a complete description of what an electron does; therefore in a certain sense it gives us an understanding of what an electron is. It is only in quantum electrodynamics that our knowledge is so exact that we can feel we have some grasp of the nature of an elementary particle. (Dyson 1953a)

And he adds:

> I always felt it was a miracle that electrons actually behaved the way the theory said. To me it was always an amazing experimental fact that this perturbation series was somehow real, and everything the perturbation series said turned out to be right. I never felt that we really had understood the theory in the philosophical sense—where by understood I mean having a well defined and consistent mathematical scheme. [Nonetheless] I always felt that it was obviously true, true even with a big T. Truth to me, means agreeing with the experiments.... For a theory to be true it has to describe accurately what really happens in the experiments.[207]

This in contrast to a model which is essentially a "toy." A model is not supposed to be a description of reality; it creates an imaginary world where things are simple enough to calculate. It has some aspect of the real world built into it, but it is not supposed to be true in any sense. Quantum electrodynamics captures something essential and true about the furniture of the world and it is "one of the few theories that seems to be very closely in contact with reality."[208]

9.20 Style

When working on a problem, "I always sit and calculate, and fill great numbers of pages of calculations as the first stage. You can't really understand anything unless you can calculate it. The only way I know how to work is to calculate. I can't think about things very much in the abstract. So I try out various things."[209]

This has always been the way Dyson tackles a problem, whether in mathematics or in physics. He recollects that after his bus ride from Berkeley to the Midwest in the summer of 1948, "I scribbled a lot in Chicago and got things fairly well understood. There was a lot of pages of stuff which later were organized and put into the Radiation theory paper." It was by doing second-order calculations in Chicago that Dyson had obtained his insight of grouping all the divergences into multiplication factors. And it was by doing perturbative calculations that he noted the equality of Z_1 and Z_2 to second order.

The real work is always "this blind exploring," the final writing of the paper is "really more or less a question of literary composition." The task there is to try to present things "in a logical and clear fashion and to be persuasive."[210] Dyson obtains a great deal of satisfaction from writing up his results. When he was at work on his "Heisenberg Operators I" paper (Dyson 1951a), he wrote his family: "I have now started at last the job of writing down a big paper for the Physical Review. It is very pleasant to be finally writing down my ideas; I enjoy the labour of getting everything clear and into shape."[211]

Dyson's calculational abilities are awesome. The ease with which difficult computations are undertaken and carried through is revealed in a letter to his family written shortly after he had completed his S-matrix paper:

> During the last week I have started some heavy calculations which are restful as they do not require any serious thinking for their performance. They are a fairly straightforward, but important application of the theory Feynman and I have been developing. If all goes well, they will be finished in about 2 weeks and 200 pages, and the result will be one little number, the fourth-order correction to the electron's magnetic moment. ... This little number is the most near to being observable of all the effects the theory predicts, beyond the original effects calculated by Schwinger. Probably it will be too small to be detected for some years, even with Rabi devoting himself to the job; it will probably be about one part in a million, whereas the exper-

iments are so far good only to about one in forty thousand. It will be a fine thing in one way when this calculation is done, for it will be the first time for many years that the theory has been a jump ahead of the experiment; usually the experiments come first, and then it is not so convincing when the theory gets them right.[212]

Baranger and Salpeter tell identical stories corroborating Dyson's computational prowess.[213] During the summer of 1952, the three of them had undertaken to calculate the fourth-order vacuum polarization contribution to the magnetic moment of the electron. The computed diagrams are indicated in figure 9.20.1.

They had decided to do the computations independently and had agreed that the correct answer would be taken to be the result obtained by a majority of the calculators. Dyson finished his calculation after a week of concentrated effort, having invented clever ways to reduce the complexity of the problem. Several weeks later Baranger and Salpeter completed theirs. The numbers obtained by Dyson and Salpeter agreed with each other, but their result was different from that obtained by Baranger. Baranger eventually found a mistake in Dyson's calculation, proving that Dyson is not immune from making mistakes. But the speed with which he can do such difficult calculations allows him to readily repeat them and weed out any mistakes.[214]

Dyson's impressive memory, his vast erudition, his great powers of concentration, and the acuity with which he can construct tentative solutions and explore their validity allow him to complete difficult tasks in prodigiously short amounts of time.

In one of his letters to his parents he casually remarked: "Meanwhile I had four days of holiday from physics in which I did quite an interesting piece of pure mathematics. This happened quite suddenly. I had a good idea and worked it out in 4 days. This will probably lead to a paper quite soon."[215]

All of Dyson's important papers—whether in physics or mathematics—exhibit his great analytical skills. If a mathematician's "nature" can be characterized by either an analytical or a geometrical approach and by the ability to pick the flaws in an argument, then Dyson's nature is that of an analyst: "I never adopted Feynman's visual way. I found it alien, like a foreign language to me. That explains why there is but one diagram in the radiation theory paper. I had worked it out combinatorially."[216]

A consequence of Dyson's "nature" (and the legacy of his training as a mathematician—the influence of Hardy and Besicovich) is that aesthetics plays an important role in the problems he addresses. There are two independent criteria for judging science: the criterion of elegance, and the criterion of usefulness. Dyson's scientific productions are always elegant. He has always felt that he should conform to his own nature and not try to be somebody else. That is why he has tended to gravitate toward problems where well-defined mathematical questions could be answered and "where elegance was useful." "All my work has been like that. It

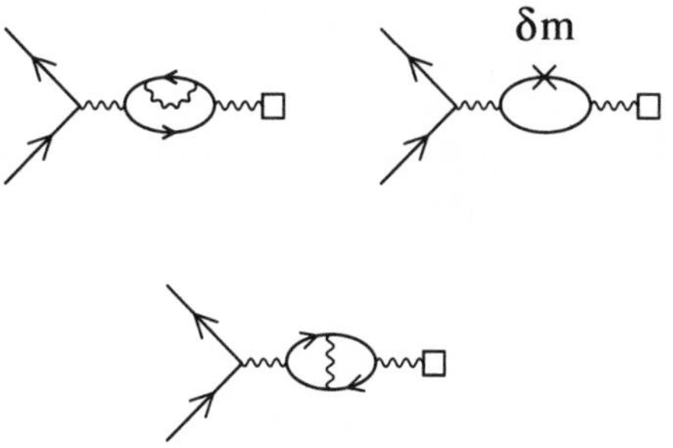

Figure 9.20.1

can happen on rare occasions that the work is also physically important. Quantum electrodynamics is elegant and useful. That was a lucky accident—essentially a matter of chance—and was not expected to happen again. Elegance comes first; if the problem that is solved is physically interesting or important that's a bonus!"[217]

"My nature . . . ," Dyson muses, "the only strength I have is being an artist with mathematical tools. The same thing is true of my writing. I enjoy the craftsmanship of writing the same way I like the craftsmanship of doing mathematics." And after pondering, he adds, "I never really set out to solve the great problems of nature. I am not a dedicated scientist in the proper sense." To illustrate this last point, Dyson points to the eigenvalue correlation problem (Dyson 1962; see Mehta 1967) on which he worked for a while. To him that was something "exceedingly beautiful." It turned out that he could actually compute in closed form the n level correlation function and obtain an elegant formula for it. "It is not easy to do, and it is not even easy to prove when you know the answer." The solution of that problem gave Dyson exactly the same kind of pleasure as the S-matrix paper. "You really can do it to all orders and it works! Everything that happens at second order goes all the way. It is a beautiful piece of architecture. It is more like architecture than anything else. But it happens that this piece of work is absolutely useless. Nobody actually needs to know how to calculate a three level correlation function or a five level correlation function. It has no relevance to anything except to itself. And I did it with the same loving care that I did the S-matrix calculation and for me it was as satisfying."

Dyson avers that there is another pleasure that comes with the craftsmanship of doing mathematical physics: "It is like writing a novel where you as author have complete control over the characters. It is a self-contained world where you understand everything, the parts and the whole."

9.21 Epilogue

Dyson has written extensively on unifiers and diversifiers in science. He commented: "The passion to understand is, I think, characteristic of unifiers, whereas the typical diversifier is somebody who is in love with the object itself, and I belong to that category. I gazed at the stars as a young boy. That's what science means to me. It's not theories about stars; it's the actual stars that count"(Dyson 1981, pp. 122–3).

Actually, Dyson was a unifier during his QED phase. He devoted himself to the solution of that problem to the exclusion of almost everything else. His great achievement was to *clarify* the structure of field theories and in doing so, he paid homage to science as a luciferous enterprise. After his triumph and his disappointment—*true to his nature*—he became a diversifier. And much more so than before he came to doubt the worth of science unless it is fructiferous also. He has become engineer, astrophysicist, applied mathematician, theoretical biologist while yet remaining the pure theoretician. He has also become a prophet while still nourishing his priestly sensibilities by concerning himself with origins and ends, of worlds and of life.

And he found his niche at the Institute. He epitomizes the faculty Flexner wanted to bring there: "These men know their own minds; they have their own way; the men who have throughout human history meant most to themselves and to human progress have usually followed their own inner light; no organizer, no administrator, no institution can do more than furnish conditions favorable to the restless prowling of an enlightened and informed human spirit" (Flexner 1940).

9.22 A Postscript: Tomonaga, Schwinger, Feynman, and Dyson

In the summer of 1980 Julian Schwinger delivered a memorial lecture for Tomonaga Sinitiro at the Nishina Memorial Foundation in Tokyo. In his eulogy Schwinger noted that among the three "partners" who had shared the Nobel Prize for their researches on quantum electrodynamics, his work "was most akin in spirit to that of Tomonaga." He indicated that he had been struck, while preparing his lecture, how much Tomonaga's scientific life "had in common" with his own (Schwinger 1983b).[218]

The parallels are indeed striking. Both were deeply influenced by Dirac's work. Both are by nature conservative. Understanding the phenomena was the driving force for both of them. Both had "read everything" during the thirties—Tomonaga by virtue of his geographical isolation, Schwinger by virtue of his social isolation. The paths that led to their strikingly similar contributions to quantum electrodynamics from 1946 to 1949 had indeed many common landmarks. Dirac's papers, and in particular Dirac, Fock, and Podolsky on the multi-time formulation of quantum electrodynamics (Dirac et al. 1932a) was, for both of them, the starting point for their research in physics. In the late thirties, nuclear physics and meson theory were the central focus of their research interests, and both extended Wentzel's formulation of strong coupling methods in meson theory using canonical tranformations. During the war both made important contributions to the electrodynamics of microwaves, and both developed an S-matrix approach to microwave circuitry.[219]

Schwinger and Tomonaga's contributions are examples of the momentum of internal factors in the advances of physics. That so remarkably similar solutions were advanced by Tomonaga and by Schwinger, working in complete isolation from one another, in two remarkably dissimilar cultures under radically differing working conditions—war-ravaged Tokyo and affluent Harvard—is surely proof that physicists advance explanatory schemes that transcend national boundaries and national styles.[220] Schwinger's and Tomonaga's contributions to QED also illustrate the impressive efficacy and instrumental power of modern physical theories.

If we enlarge the cast of actors, what stands out is not only the momentum of internal factors, but the inordinate influence of some individuals. Dirac, in particular, towers above almost everyone else. All the theorists who started their career in the late twenties and in the thirties—Stueckelberg, Tomonaga, Yukawa, Schwinger, Feynman, Dyson—are "students" of Dirac. They all learned quantum mechanics either from Dirac's papers or from his classic treatise on the subject. Except for Darwin's *Origin of Species*, no book since Newton's *Principia* explained so much of so wide a realm of nature. It is difficult to think of another physics text that conveys more effectively the power of a simple, logical presentation. Probably no other book has ever given its readers a greater appreciation of the aesthetic dimension of theoretical physics. If we turn to Dirac's scientific papers, rarely have the articles of a single individual proved so influential. Dirac's papers from 1927 to 1933 molded the subsequent synthesis of special relativity and quantum mechanics. Let me corroborate my assertion by looking at just two of the papers Dirac published in the early thirties: that on the "Lagrangian in Quantum Mechanics" and the one on quantum electrodynamics written in collaboration with Fock and Podolsky. The former paper is the point of departure for Feynman's path integral formulation of quantum mechanics, Schwinger's quantum action principle, and Yukawa's and Tomonaga's attempts to give a manifestly covariant formulation of quantum field theory. The paper by Dirac, Fock, and Podolsky is the starting point for Wentzel's limiting procedure, for Stueckelberg's investigations of the interactions between particles, and for Schwinger's and Tomonaga's covariant formulation of quantum field theory during the 1940s. If one adds to these two papers the ones on quantum electrodynamics, on the relativistic spin 1/2 wave equation, on hole theory and vacuum polarization, on magnetic monopoles, the result is staggering.

Even though there is clearly continuity with earlier contributions in the post-World War II period, the break with the intellectual approaches of the thirties is striking. Many of the problems in quantum field theory that had been worked out during the thirties were forgotten, or at best only lip service was paid to them. The ease with which one could calculate with Feynman methods implied that it was much easier to do the computations *de novo* rather than try to decipher what people had done in earlier times with their cumbursome notations and recondite hole-theoretic approaches. The coming of age of a new generation of young American physicists who had been introduced to quantum field theory through Wentzel's book and the papers of Schwinger, Dyson, and Feynman added a generational component to the discontinuity. Many of these younger physicists had been trained in the war time laboratories; these activities and the "atmospherics" under which they were carried out were important factors in causing the rupture in style and outlook.[221]

When we look at Tomonaga, Schwinger, Feynman, and Dyson, some of the necessary conditions for making theoretical contributions of such magnitude become apparent. All are endowed with powerful memories, great powers of concentration, and remarkable computational abilities. All knew their strengths and were confident of their powers. Each one of them was aware of the problems that had to be solved. Dyson had heard Feynman and Schwinger lecture about their work. Bethe was the source of information for Feynman. Schwinger had read everything; so had Tomonaga and Dyson. They were all trained by, or came into contact with, the masters of the older generation: Tomonaga with Heisenberg; Schwinger with Rabi and Oppenheimer; Feynman with Wheeler and Bethe; Dyson with Bethe. Yet each one of them had to free himself from their influence to find his own voice.

It is interesting to note that Tomonaga, Feynman, and Dyson (but *not* Schwinger) each had as teenagers a close friendship with another talented youth of his own age: Tomonaga with Yukawa, Feynman with Welton, Dyson with Lighthill and Longuet-Higgins. These friendships played an important role. Undoubtedly there was rivalry between the friends, and this competition nurtured their erudition, stimulated their productivity, and goaded them to exhibit the depth of their understanding and the novelty of their presentation. It helped them assert their uniqueness and their creativity but in a setting that also fostered trust, cooperation, and caring.

Though there are features common to all four, the uniqueness of the individual contributions is also striking. And here Feynman stands out. The genuine novelty of his approach is in marked contrast with the more conservative and traditional methods of Tomonaga and Schwinger. It would seem that Feynman's more pragmatic and skeptical attitude, his greater irreverence, his being less sensitive to the weight of tradition allowed him to be or perhaps made him more innovative. In some sense, Tomonaga and Schwinger knew too much. The chasm between the approach of Feynman and that of Tomonaga and Schwinger was bridged by Dyson's work, but there remained a divergence in the outlooks. Influenced by the Pauli-Fierz-Wentzel tradition, Tomonaga and Schwinger's formulation gave a local description that allowed inferences to be made about the results of local measurements in small regions of space-time. The essence of the Tomonaga-Schwinger method for quantum elctrodynamics was to recast the theory so that a single charged particle travels with its accompanying Coulomb and Biot-Savart fields, carries a total charge e (its experimentally measured charge), and has a mass m (its experimentally measured mass). This was achieved by making successive canonical transformations (as had been done by Pauli and Fierz, and by Wentzel). The transformed Tomonaga-Schwinger equation described the interaction between such dressed particles, and hence was rather complicated. Not even Schwinger could carry out the program beyond order e^2.

The Feynman approach, on the other hand, is global. It is an S-matrix theory, and its simplicity derives from the fact that it only attempts to predict the results of collision processes. The states it deals with are asymptotic states of freely moving particles. In contrast to the Tomonaga-Schwinger approach, instead of making the single particle description more and more complicated it describes the collisions with more and more precision.

As noted earlier, the overall triumph was a conservative one. Feynman's formulation was based on his intuitive yet radical integral-over-path approach. After Dyson had indicated how Feynman's formulation could be derived from the standard version of quantum field theory, Feynman's derivation was (temporarily) ignored. For a while at least, the triumph of Dyson enthroned field theory. And because field theory became the gospel, it was not fully appreciated at the time that Feynman was committed to a particle point of view.

For reasons that are not entirely clear, Dyson's achievement was not rewarded by the community. Some have suggested that his contributions were primarily mathematical; others have noted that his proof of the renormalizability of the S-matrix in quantum electrodynamics was imcomplete—that the full prescription for the correct handling of overlapping divergences was formulated by Salam (1950) and that the rigorous proof of the counting theorem and of the renormalizability of QED were given later by Weinberg (1959), Hepp (1966), and others. Yet I would suggest that Dyson should have shared the Nobel Prize, that his contribution recast the way field theories were thought of and dealt with, and that his criterion of renormalizability for selecting theories had great import on the subsequent evolution of the subject. Without minimizing in any way Tomonaga's accomplishments, it seems to me that the developments in the period from 1947 to 1950 would not have been substantially different without him, yet one cannot conceive of the subsequent developments without Dyson. [222]

10. QED in Switzerland

10.1 Field Theory in Switzerland: Stueckelberg

The history of theoretical physics in Switzerland during the twentieth century merits a detailed exposition. Jungnickel and McCormmach (1986) have sketched part of the story in their magisterial account of *Theoretical Physics from Ohm to Einstein*, but only for the German-speaking Swiss universities and only for the first two decades of the century. A full account has yet to be written.

According to Jungnickel and McCormmach, the University of Zurich was "behind the times" at the beginning of the twentieth century, when compared to the universities in Basel, Bern, Geneva, and Lausanne. Whereas each of the other Swiss universities had two physics professorships, one in experimental physics and one in theoretical physics, Zurich had only an experimental chair. The first appointment in theoretical physics was that of Albert Einstein as an extraordinary (associate) professor in 1909. Einstein stayed in Zurich for only two years, as did Max von Laue, his successor. Peter Debye, who replaced von Laue, departed for a professorship in Holland after only a year's stay, and the position remained vacant during World War I. The position was elevated to an ordinary (full) professorship in 1921, and Schrödinger was its first incumbent (Jungnickel and McCormmach 1986). When Schrödinger left for Berlin in 1928, the faculty called on Gregor Wentzel to fill the position.[1] Wolfgang Pauli was appointed to the chair in theoretical physics at the Eidgenössische Technische Hochschule (E.T.H.) in Zurich at the same time. Pauli and Wentzel made Zurich a world center of theoretical physics.[2] Among Wentzel's Ph.D. students were some of the most distinguished mathematical and theoretical physicists of their generation: Valentine Bargmann, Nicolas Kemmer, Markus Fierz, Res Jost, Felix Villars; and all of Pauli's assistants became outstanding theorists: Casimir, Bloch, Peierls, Weisskopf, Kemmer, Fierz.

In 1933 E.C.G. Stueckelberg von Briedenbach came to the University of Zurich as a *Privatdozent*.[3] "He was a tall and, in his younger days, [an] elegant man who took pride in being an officer in the Swiss army and a baron in Germany" (Enz 1986). Stueckelberg was a brilliant, eccentric, and difficult individual who became afflicted with a mental illness that required increasingly frequent hospitalization. Many of his important contributions to physics went unappreciated, partly because of the difficulty of his presentations. His articles were cumbersome to read because he often insisted on creating his own novel notation; furthermore, they were frequently published in relatively obscure and inaccessible journals, such as the *Archives des Sciences Physiques et Naturelles* and the *Compte rendu des Séances de la societé de physique et d'histoire naturelle de Genève*. But they usually made insightful and valuable contributions and always displayed uncommon originality. In 1934, at roughly the same time as Yukawa, Stueckelberg advanced the hypothesis that the nuclear forces were generated by the exchange of a massive

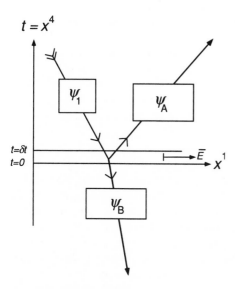

Fig 10.1.1 Scattering of a wave packet (from Stueckelberg 1941b).

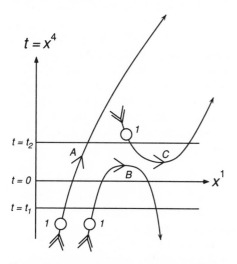

Figure 10.1.2 Pair production and annihilation (from Stueckelberg 1941b).

boson between the nucleons, but he did not publish his idea—in contrast to Yukawa—because of Pauli's opposition. In 1937 Stueckelberg was quick to suggest the identification of the newly discovered mesotron with Yukawa's meson, and in the late 1930s he made important contributions to the theory of nuclear forces. Independently of Wheeler and Feynman, Stueckelberg pointed out in 1941 that pair production could be described classically by considering positrons as electrons running backwards in time (Stueckelberg 1941b, 1942c), and he illustrated these concepts with graphs of space-time trajectories similar to the diagrams Feynman began drawing in the summer of 1947 (figs. 10.1.1 and 10.1.2).

When Stueckelberg came to the University of Zurich in 1933, Wentzel was working on eliminating the divergences in relativistic field theories (Wentzel 1933a,b, 1934) and he got Stueckelberg interested in quantum field theory. Stueckelberg began working on the field-theoretical description of the interaction between particles. It is to be noted that Dirac, Fock, and Podolsky's (1932a) article on quantum electrodynamics was the point of departure for both Wentzel's and Stueckelberg's research programs (Stueckelberg 1934b, 1935a, 1936). In 1938, again taking the Dirac, Fock, Podolsky paper as his starting point, Stueckelberg gave a formulation of quantum electrodynamics and of various meson theories in what later became known as the interaction picture. He stressed its advantages, namely, the manifest covariance of the "Schrödinger" equation in that picture, and the possibility of writing covariant commutation rules for the field operators at different times (Stueckelberg 1938a,b). The following year he suggested that the space-time coordinates appearing in the quantities representing physical phenomena be analytically continued into the complex plane, and he investigated the properties of the wave equations on such complex manifolds (Stueckelberg 1939b). Stueckelberg was instrumental in developing the method of contact transformations for extracting the physical content of field theories (Stueckelberg 1938c); this technique was then applied to strong coupling theories by Wentzel (1940, 1941). The divergences of quantum field theory became the focus of Stueckelberg's research efforts during the forties. The problems he addressed ranged from the nature of the singularities of the nuclear potential that could be derived from the various meson theories that were popular at the time, to the structure of the singularities of the matrix elements of the Heisenberg S-matrix (Stueckelberg 1940, 1944, 1947). In the fifties and thereafter, religion and metaphysics came to assume ever greater importance in Stueckelberg's life (he had converted to Catholicism) and he began extended investigations on the nature of time, entropy, and cosmology.

In the late thirties, Stueckelberg, together with Dirac and Kramers, believed that the correct strategy to overcome the divergence difficulties was to try to resolve them initially at the classical level. He first investigated classical field theories in which charged point particles were coupled to a neutral scalar field in addition to the electromagnetic field (Stueckelberg 1938). But he later abandoned such mixed field theories and accepted Dirac's (1938b) philosophy that considered electrons pointlike, permanent and immutable "elementary particles." He extended Dirac's Lorentz invariant subtraction procedure (Dirac 1938b) that guaranteed that the resulting theory would be relativistically invariant and that the conservation laws were obeyed, and derived divergence-free equations of motion for point charges. He also analyzed carefully the reaction of the electromagnetic field on the motion of the charges (radiation damping) in both classical and quantum theory (Stueckelberg 1941b).

The papers that Heisenberg published during the war in which he laid the foundation of S-matrix theory made a deep impression on Stueckelberg. Heisenberg had formulated his research program in order to overcome the divergence difficulties encountered in the usual Hamiltonian formulation of the quantum field theory. His aim was to sever the connection between the S-matrix and the Hamiltonian and its associated Schrödinger equation. He wanted to establish correlations between initial and final states, but without giving a description in space and time of the actual evolution of the initial state because he presumed this process to be unobservable. Although Heisenberg retained the general properties of the S-matrix that had been derived from the canonical formalism, for example, unitarity and relativistic invariance, his theory was very "incomplete . . . as long as no other rules [were] added as a substitute for the Schrödinger equation; it [was] like an empty frame for a picture yet to be painted" (Wentzel 1947). Stueckelberg immediately set about to find a "connecting link" between Heisenberg's approach and the usual formulation of quantum field theory. He observed that the S-matrix could be defined as the unitary operator that connects the state vectors $\Psi(T)$ and $\Psi(-T)$,

$$\lim_{T\to\infty} \Psi(+T) = S\Psi(-T), \tag{10.1.1}$$

where the vector Ψ is related to the Schrödinger state vector $\psi(t)$ in the following manner:

$$\Psi = \exp\left\{+\frac{i}{\hbar}H_o t\right\} \psi. \tag{10.1.2}$$

He also derived a power series expansion of S in terms of the Hamiltonian when it was assumed that $\psi(t)$ obeys the usual Schrödinger equation (Stueckelberg 1944b,c). Somewhat later Stueckelberg realized that the theory is much simpler, more transparent, and, in addition, maintains explicit relativistic invariance, if it is expressed in the interaction picture, that is, if formulated in terms of the vector $\Psi(t)$ and the equation it obeys. He reported on this work at the Cambridge conference of 1946 and was instrumental in bringing Heisenberg's work to the attention of the theoretical physics community (Stueckelberg 1947).

Like Heisenberg, Stueckelberg was intent on giving up the connection between the S-matrix and the Schrödinger equation. For Stueckelberg the initial motivation was to circumvent the acausal effects he had encountered in his theories of finite-sized charged particles. He too wanted a time translation operator for finite time intervals without basing himself on the Hamiltonian formalism; and like Heisenberg, he wanted to retain some of the general properties of such operators—in particular, unitarity and relativistic invariance. To gain further insights he analyzed the properties of the S-matrix for some of the simple one-particle systems he had previously analyzed at the classical level. To fill "at least partly" Heisenberg's "empty frame," Stueckelberg initially postulated that in a certain limit the S-matrix must yield a particular formulation of the classical theory of radiation damping (Stueckelberg 1944b,c; Wentzel 1947). A year later, in a general analysis of "asymptotic mechanics," he further refined this prescription by requiring that the ordinary wave-mechanical theory must result if radiation damping and other reaction forces are neglected (Stueckelberg 1945; Wentzel 1947).

Stueckelberg also investigated the field-theoretic case. In the standard formulation, the infinities of quantum field theory are generated by intermediate processes in which a quantum is emitted and eventually reabsorbed. Heisenberg, in his S-matrix paper, had observed that such divergent contributions never arise if the S-matrix is exhibited in a form where all creation operators are written to the left of all annihilation operators. In his initial attempts to circumvent the divergences, Stueckelberg was inspired by Heitler's heuristic prescription for avoiding the divergences encountered in the field-theoretic description of scattering processes, an approach that had become known as Heitler's theory of radiation damping (Heitler 1941; A. H. Wilson 1941). Stueckelberg (1945) tried to introduce a relativistically invariant subtraction procedure for the higher-order terms in the power series expansion of the S-matrix by imposing rearrangement rules for ordering the creation and annihilation operators. But his prescription had a great deal of arbitrariness (Wentzel 1947). During the course of these researches Stueckelberg noted an additional constraint that had to be imposed on the S-matrix besides unitarity and relativistic invariance. This was a requirement imposed by causality (Stueckelberg 1946). It translated itself into a requirement that only a certain "causal" Green's function, which Stueckelberg called the D^c, appear in the matrix elements of the S-matrix.[4]

In the course of this work Stueckelberg elaborated a formalism that allowed him to construct the S-matrix using field theory as an input, but without being bound by its particulars. The S-matrix elements were exhibited as a power series expansion in the coupling constant of the interaction term in the Lagrangian, and only D^c functions appeared in them. The formalism was similar to the one Feynman devised; Stueckelberg, however, could not give it secure foundations. Stueckelberg often presented his work at the theoretical seminar in Zurich. Felix Villars, who heard many of these talks, recalls that Markus Fierz was always very critical of Stueckelberg's approach";[5] Pauli much less so. Stueckelberg's presentations were always very difficult and their physical content and usefulness were not immediately apparent. According to Villars, "Stueckelberg never gave a clear account of how to calculate and why you introduce the D^c function.... His causality argument was not very coherent."[6] It was Dominique Rivier who assimilated all of Stueckelberg's insights and gave them a formulation for the field-theoretical case that was both lucid and understandable.[7] In a thesis presented to the university of Lausanne in 1949 he succinctly summarized Stueckelberg's program:

> In the form given by Stueckelberg, the theory rests on the integral equation
>
> $$\psi[\tau''(\ )] = S[\tau''(\ ), \tau'(\ )]\psi[\tau'(\ )]$$
> $$S^+S = 1$$
>
> The criteria for the validity of the theory are thus applied to the operator S. Relativistic *covariance* demands that S be a scalar invariant.... The probabilistic character of entrains the *unitarity* of S. Finally *causality*, which is no longer a priori guaranteed by the existence of a differential equation, requires that the operator S has the following particular structure: in the power series development of S [in terms of the Hamiltonian density] only one type of Green's function can appear, namely, the causal functions D^c.

This last point can be justified axiomatically: if by causal one understands a theory that can formulate a principle of causality, and if one notes that a necessary condition for such a principle is the exclusion of states of negative energy, one can show that the only Green's function of a quantum theory that is invariant and causal, is precisely the function

$$D_\kappa^c(x) = \frac{i}{2} \left[\theta_\kappa^+(x^4) D_\kappa^+(x) + \theta^-(x^4) D_\kappa^-(x) \right]$$

that contains only positive frequency waves in the future ($x^4 > 0$) and only negative frequency waves in the past ($x^4 < 0$)....But one can also demonstrate the necessary presence of the Green's function D^c directly, by constructing the operator S starting from a scalar hamiltonian density representing a definite type of interaction. (Rivier 1949)

The application of the formalism to the calculation of the magnetic moment of the neutron and of the proton "starting from . . . a definite type of interaction hamiltonian density" was carried out by Rivier and reported in a letter to the *Physical Review* (Rivier and Stueckelberg 1948). This letter contained an explicit representation of the causal function D^c that was central to the perturbative approach to Stueckelberg's program as a function of the space-time coordinates, $D^c(x)$, as well as of its Fourier transform, $D^c(k)$. In addition, Rivier and Stueckelberg noted that the various matrix elements appearing in any calculation would be rendered finite by the replacement of

$$D^c(k) = \frac{1}{k^2 + m^2 + i\epsilon}$$

with

$$D^c(k) = \frac{1}{k^2 + m^2 + i\epsilon} + \sum_i \frac{1}{k^2 + m_i^2 + i\epsilon},$$

and the limit $m_i \to \infty$ taken at the end of the calculation. The source of this method of rendering the calculated expressions finite was Stueckelberg's earlier investigations of mixed field theories (Stueckelberg 1939a), and the subsequent work of Pais (1945, 1946).

Pauli was very positive about this aspect of Rivier and Stueckelberg's work.[8] His own approach was predicated on the fact that if one had a formalism that is covariant, gauge invariant, and finite, one would then be able not only to calculate in a particular Lorentz frame, but one would also not have to worry about gauge invariance because if the calculated quantities are finite they must come

out with the right covariance properties. He was therefore very interested in seeing whether it was possible to regard the usual formalism as a limit of a finite
formalism. Stueckelberg and Rivier's procedure for rendering the S-matrix elements finite partially satisfied Pauli's criteria. Pauli, at the time also "entertained
the idea—at least for awhile—that maybe a problem was that we operated with
bare objects which had a specified mass and he had sort of a vague idea that if we
had a system with a mass spectrum things might be less singular."[9] The subtraction
procedure of Rivier and Stueckelberg made it apparent that the auxiliary masses
that were brought in, if considered as field quanta, must come into the formalism as
fields with imaginary coupling constants. Such a procedure could, therefore, only
be considered as a formal procedure. Pauli hoped that such a formal procedure
would nonetheless express some real formal physics. These considerations were
the point of departure for Pauli's work on "regularization" procedures in quantum
field theory, to which we turn in the next section.

10.2 Quantum Field Theory in Zurich: Pauli's Seminar, 1947–1950

Wolfgang Pauli was born on April 25, 1900, the son of the distinguished
Viennese physician and physiologist, Wolfgang Joseph Pauli.[10] Pauli's father was
a close friend of Ernst Mach, and at the time of Wolfgang junior's birth he "was
intellectually [*geistig*] entirely under his influence."[11] Wolfgang junior attended
the Doblinger Gymnasium and graduated from that institution in 1918 in a class
that became known as "the class of geniuses."

Pauli published his first paper while still in high school. The subject:
the energy momentum tensor in Einstein's theory of general relativity. He entered
Sommerfeld's seminar in the fall 1918. Two papers published during 1919 dealt
with Weyl's new theory of general relativity. In them, he already displayed his
sharp critical faculties, his great mathematical abilities, and his cautious, conservative, and positivistic inclinations. A year later he completed his classic presentation of Einstein's theory of relativity for the *Encyklopädie der mathematischen
Wissenschaften*, which Einstein reviewed in glowing terms: "No one studying this
mature, grandly conceived work would believe that the author is a man of twenty-
one. One wonders what to admire most, the psychological understanding for the
development of ideas, the sureness of mathematical deduction, the profound physical insight, the capacity for lucid, systematical presentation, the knowledge of the
literature, the complete treatment of the subject matter [or] the sureness of critical
appraisal" (quoted in Enz 1973, p. 768).

Pauli obtained his doctorate—summa cum laude—in the summer of
1921. His dissertation, carried out under Arnold Sommerfeld's supervision, was
an investigation of the structure of the hydrogen molecule ion in the old quantum
theory. During the winter semester of 1921/22 Pauli was in residence in Göttingen

as Max Born's assistant. He spent the summer of 1922 in Hamburg working with Wilhelm Lenz, and from there went to Copenhagen to spend a year in Bohr's institute. Bohr later remarked: "When Pauli . . . came to Copenhagen in 1922, he at once became, with his acutely critical and untiringly searching mind, a great source of stimulation to our group. Especially he endeared himself to us all by his intellectual honesty, expressed with candour and humour in scientific discussions as well as in all other human relations" (Bohr in Fierz and Weisskopf 1960, p. 2).

Pauli returned to Hamburg in 1924, and he remained there—first as Lenz's assistant and then as an associate professor—until 1928, when he was called to the E.T.H. in Zurich. Pauli stayed there until his death in 1957. In his eulogy for Pauli, Bohr noted: "At the same time as the anecdotes around his personality grew into a veritable legend, he more and more became the very conscience of the community of theoretical physicists" (Bohr in Fierz and Weisskopf 1960, p. 4).

Throughout the twenties Pauli was at the center of all the developments in both the old and the new quantum theory[12] and "[in] the heroic period during the 1930's, . . . Pauli's own work and constant critical vigilance were decisive in the conception of the basic ideas of quantum mechanics" (Fierz and Weisskopf 1960, p. vii).

Quantum electrodynamics and quantum field theory became Pauli's main concern after 1927. As indicated in chapter 2, Pauli and Heisenberg formulated the canonical approach to the quantization of field systems in 1929. During the thirties many of the seminal papers in quantum field theory were either written by Pauli, had Pauli as one of its authors, or were written under the acknowledged guidance of Pauli. In 1936 Pauli gave a preliminary statement of the spin-statistic theorem.[13] The theorem was refined over the next few years and became the content of the paper Pauli was to present to the 1939 Solvay Congress. Its essential point was that if one requires that two local observables commute if their space-time separation is spacelike, and that the energy be positive semi-definite, then integer spin-field theories must be quantized using commutation rules and odd half-integer spin fields using anticommutation rules. The state of affairs in quantum field theory at the end of the 1930s was summarized by Pauli in a famous article in the *Reviews of Modern Physics* (Pauli 1941).

Pauli spent the war years, from 1940 to 1946, at the Institute of Advanced Study in Princeton.[14] Most of his efforts during that period were devoted to meson theory. He returned to the E.T.H. early in 1946 and once again took charge, with Wentzel, of activities in theoretical physics in Zurich.[15] When Pauli was informed of the developments at Shelter Island during the summer of 1947, he shifted his attention to quantum electrodynamics. He was well prepared to jump on the bandwagon. Working with Pauli were some very gifted and able young theorists, among them Jost, Villars, and Luttinger. Jost and Villars had done dissertations with Wentzel on strong coupling methods in meson theory. Luttinger had obtained his Ph.D. at MIT and had come to Zurich in the fall of 1947 to work with Pauli.

He remained there until 1949 when, together with Pauli and Jost, he visited the Institute for Advanced Study in Princeton for a year. During his stay in Zurich, Luttinger and Jost made several important calculations that clarified the renormalization procedures in higher orders (Luttinger 1949a,b; 1950). The following account of Pauli's seminars from 1947 to 1949 is based on Luttinger's and Villars' reminiscences.

Joaquin ("Quin") Mazdak Luttinger was born in Manhattan, New York, on December 2, 1923, the youngest of the three children in the household. His mother, Shirley, was born in the United States shortly after her family had emigrated from Vilna, Lithuania, which at the turn of the century was one of the most renowned centers of Jewish learning in the diaspora. His father, Paul, came to the United States when he was fifteen years old. He was a man of enormous energy who was self-educated and had worked his way through New York University, becoming a well-known and prominent physician. He was also an able scientist, who at one time was the research director of the New York City Board of Health and had helped develop a vaccine against whooping cough. All his life he held radical political views that had evolved from anarchism in the 1910s to communism during the 1920s and 1930s. When he died, shortly before World War II, he was a Trotskyite. A prolific writer, he edited a medical magazine and during the thirties wrote a "Health Advice" column in the *Daily Worker*, the official newspaper of the Communist party in the United States. Throughout his life he was a staunch advocate of socialized medicine and was constantly battling the American Medical Association (AMA). His particular nemesis was Morris Fishbein, the editor of the *Journal* of the AMA and its chief spokesman in its relentless drive against socialized medicine and compulsory health insurance.

When Quin Luttinger was growing up, his family lived in Greenwich Village and was well off—the household was looked after by "servants." He was sent to the Evergreen School, a nursery school and kindergarten run on Montessori principles,[16] and later to the Hudson Park School, a private school that offered classes in the first six years of elementary education. He went on to Stuyvesant High School, which was within walking distance of his home. Quin was "very much a home scientist as a kid." His father nurtured his scientific proclivities by having him build crystal radio sets, encouraging him to experiment with chemistry kits, and stimulating him to read about science. His brother, who was some four years older than he and "very precocious," got him interested in "philosophy and the universe." As a result, when he was twelve or thirteen years old, Luttinger became deeply interested in astronomy and cosmology. He had the "good judgment" to recognize that he had to learn physics and mathematics in order to understand what he was reading. He thus began studying introductory college physics texts; these in turn required him to learn the calculus and differential equations. He became fascinated with the subject and "abandoned everything for mathematics." One of his friends at Stuyvesant was Irving Reiner,[17] and the two of them went through Whittaker and Watson's *Modern Analysis*, doing all the problems in the treatise. "I loved Whittaker and Watson. It was like a world opening for me. It has been invaluable to me in my later career." During his senior year at Stuyvesant, Luttinger discovered Lindsay and Margenau's *Foundations of Physics*. "That book had an incredible influence on me. . . . It had a great deal of clarity." Luttinger had read widely in physics, but only after reading this book did he "really understand for the first time a lot of stuff that people were doing." "I had tried to read quantum mechanics before but I didn't like recipes. From Lindsay and Margenau I realized that quantum mechanics was a systematic thing. Lindsay and Margenau had a profound influence on me."

Although Luttinger was accepted by MIT, the death of his father made it impossible for the family to meet the cost of tuition and of room and board there. He therefore went to Brooklyn College, where he came under the tutelage of William Rarita, who "appreciated" him and "was very encouraging." Luttinger never withdrew his application from MIT and he kept on applying for financial aid. His negotiations bore fruit and eventually he was awarded a tuition scholarship. In addition, Slater, the head of the physics department at MIT, offered him a job grading papers and instructing undergraduate recitation sections, and so in 1943 Luttinger transferred to MIT. With the scholarship, the remuneration from grading and teaching and some additional tutoring in mathematics and physics, he was able to make ends meet. Luttinger found MIT "very practically oriented": as a physics major he had to take several courses in electrical engineering. He satisfied many of the requirements for graduation without taking the courses involved by passing written examinations, and wrote a senior thesis under Laslo Tisza's supervision; for it he calculated the "Energies of Cubical Dipole Arrays."[18] Tisza found him "very talented and very promising."[19]

Luttinger graduated from MIT in 1944 and took a job with the Kellex Corporation in Jersey City. The Kellex Corporation, a division of the Kellogg Corporation, was a unit of the Manhattan Project that worked on the design of diffusion barriers for separating uranium isotopes (Smythe 1945, pp. 180–181). Luttinger "got very into that." He had an idea for a model of a barrier and worked intensely on it. The head of the theoretical division at Kellex was Elliot Montroll, who had obtained his Ph.D. only a few years earlier.[20] Even though they worked long hours at Kellex, Montroll and Luttinger organized a seminar on the Ising model that met weekly at Columbia University in the evening.[21] "We were very young and energetic," Luttinger commented. His stay at Kellex did not last very long: in the spring of 1945 he was drafted into the army and sent to Los Alamos. Working at Los Alamos required security clearance, which was denied because Luttinger's father had been a member of the Communist party. He was sent instead to Fort Crowder, a Signal Corps base in Missouri, and assigned to the New Equipment Introductory Detachment (NEID). The first project he worked on dealt with moving target radar and in connection with it he spent some time at the Rad Lab at MIT. While in Cambridge he talked to Tisza about some ideas he had had for solving the statistical mechanics of lattice of electric dipoles, and Tisza thought that the problem was a good topic for a Ph.D. dissertation. When the moving radar target project aborted, he returned to Missouri and spent the remainder of his tour of duty working with Leon Lederman and Jack Steinberger designing fast triangulation methods for mortar detection, testing equipment, and writing field manuals. But he also found time in the evenings and on weekends to work on his thesis problem. He was discharged in 1946 and that fall returned to MIT for graduate studies. Luttinger took his Ph.D. general examinations that winter and completed his dissertation under Tisza in the spring 1947.[22] He obtained his degree in June 1947 since there were no course requirements for the Ph.D. then. He did take some courses in the mathematics department. He recalls attending lectures by Norbert Weiner that he didn't understand and taking a course from R. H. Cameron on complex variables that he found disappointing because he knew everything in it. But he also remembers Witold Hurewicz's course on topology as one of "the great moments" in his life. "It was an incredible course. He made topology almost like physics. Hurewicz had a complete intuitive grasp of everything and could communicate it." It was quite common for MIT students, research associates, and faculty members to attend lectures at Harvard, and during his year in Cambridge, Luttinger went faithfully to the courses Schwinger was teaching that year. Luttinger characterized Schwinger's lectures on nuclear physics and on the theory of wave guides as "a phenomenon." The first semester of Schwinger's nuclear physics course was devoted to quantum mechanics. "The theory was just wonderful." Luttinger took copious notes and "the rest of his life" he has used the materials he learned in Schwinger's course in his researches and in the courses he has taught. He also attended Van Vleck's lectures on group theoretical methods. Although he thought that Van Vleck was not a particularly good lecturer, the content of the course proved to be very informative and valuable.

When it became clear, in the spring of 1947, that he would be finished in June, Luttinger thought about whom to work with as a postdoctoral fellow. He knew that he wanted "to do something very fundamental" and that he wanted to go to Europe. He felt "European" because of his upbringing and his cultural background. He consulted Weisskopf, who suggested that he go work with Pauli in Zurich. The suggestion proved attractive, and he applied for and obtained a Swiss-American Exchange Fellowship. Pauli's seminar consisted of young staff members—Jost, Villars, Corinaldesi, and Troesch—and a half dozen or so graduate students—among them Bleuler, Källén, Schaffroth, Telegdi, and Thellung.[23] Almost all of them were working with him on field-theoretic problems. Luttinger spent that summer studying Wentzel's *Einführung in die Quantentheorie der Wellenfelder*, going "through it very carefully" and "thoroughly."

When Luttinger first met Pauli in September 1947, he found him somewhat "reluctant." Pauli told him: "I only took you because Weisskopf said you were good. I find Americans never very serious." But Pauli soon discovered that Luttinger was very serious.

Much had happened that summer. The results of Shelter Island had been communicated to Pauli and it was clear that quantum electrodynamics was once again at the center of the stage. Pauli was kept informed of all developments because of the great respect the physics community had for him. Luttinger recalls that

> Pauli was in correspondence with everyone—Oppenheimer, Rabi, Bethe, Weisskopf—[and] knew what was going on. He got all the papers [that were being written] and was constantly telling us the most recent developments. Pauli wasn't doing any work at that moment [and] had no project he drew people into [but] he suggested things and oversaw everything in great detail. He listened to you—you had to talk to him all the time—and he was very, very good. He was the heart and soul of this [enterprise]. But for really day-to-day help it was Jost, who was very good and I worked mostly with him. . . .
>
> The first interesting thing I did had nothing to do with experiments but later we found there were experimental results.

Pauli had told Luttinger: "The Lamb shift is so complicated. See what happens in a magnetic field."

> That became my problem. I worked on the problem of the relativistic Dirac electron in a homogeneous magnetic field which can be solved exactly. I noticed that there was one state in which the orbital and spin energy cancelled out,[24] so that for that state the energy is simply $E = mc^2$. The change in the energy of this state due to the change in the mass and the charge of the electron that arise from the interaction of the electron with the vacuum fluctuations of the radiation field and [those] of the electron-positron field, is then the same as for a free electron $[E_{self}(0)]$, independent of the external field; so that $E = E_{self}(0)$ simply. Furthermore, it was well known that a homogeneous magnetic field gave rise to no polarization of the vacuum,[25] so that in fact, to order e^2, there could not be any effects due to charge renormalization. This meant that when one calculates the energy and finds terms which depend on the external field strength, these terms must represent the true change in the energy, and they must converge. The important point was that I realized that the mass renormalization for that state would be the same as for a free electron at rest, and I could make use of Weisskopf's results. The calculation was a little hard for me[26] . . . but I did it and got the $\alpha/2\pi$ correction to the g factor of the electron before we knew that Schwinger had also obtained this result. . . . We thought this very small correction was immeasurably small, [and that] it was not interesting. We then got the letter that Schwinger had published and became aware of Rabi's and Kusch's experiments.

An incorrect version of Schwinger's result had been communicated to Pauli by Oppenheimer early in December 1947. In his letter Oppenheimer commented:

> As to physics, I think the essential result of our meeting last spring at Ram Island was that (a) we saw that we did not know enough about the meson systems to be able to draw clear conclusions even from the failures of past theories, and that we needed to watch the sky and the sea while the evidence about mesons became a little more definite, and (b) that on the other hand, in electrodynamics, very considerable progress should be possible within the framework of the *present* theory, even though some questions would undoubtedly transcend it.[27]

Oppenheimer also informed Pauli of the content of the lecture Schwinger had given at the theoretical physics conference held in Washington in late November.

He then reported, without mentioning Dyson or Lewis by name, the results of Dyson's calculation of the Lamb shift for a spin zero particle and those of Lewis on the finiteness of the radiative corrections to the Rutherford formula for the scattering of a spin 1/2 particle by a Coulomb potential. All these calculations, Oppenheimer indicated,

> (a)... seem to give the Lamb shift in hydrogen in accord with experiment.... (b) They give, according to Schwinger, a *reduction* in the electron's g value of relative value $-1/2\pi \cdot e^2/\hbar c$ [emphasis added; note minus sign]. (c) They give finite radiative corrections to the problem of the scattering of an electron by a field.... I should add all three results hold, *irrespective of the spin* of the charged particle.... This seems to me particularly satisfactory.... Many of us are now attempting to simplify and extend these formal procedures and also to bring about, wherever possible, a good comparison with experiments. But, however that may be, it seems likely that all this constitutes a sort of discovery about the structure of quantum electrodynamics which points the way towards a real advance. In particular, there should no longer be difficulty in calculating the reactive corrections for energy levels and collision processes of much more complicated systems. Where and how far such methods can help the situation with meson fields is also being studied in an exploratory way.[28]

Shortly after receiving Oppenheimer's letter, Pauli got a short note from Rabi informing him of Kusch's experiment corroborating Breit's suggestion that the electron might have an anomalous magnetic moment. Rabi noted that "this increased value of the electron moment explains the divergence of our results from the hyperfine structure completely" and added that "Schwinger has made some relativistic calculations on the problem, à la Bethe, but relativistically, he finds indeed that the spin moment is increased by the fraction $\alpha/2\pi$ which is exactly our result."[29] While Luttinger was performing his calculation of the radiative corrections to the energy levels of an electron in a homogeneous magnetic field, he and Jost had also undertaken a calculation of the Lamb shift for a charged spin zero particle, but Oppenheimer's letter informed them that they had been "scooped" by Dyson. The state of affairs in Zurich at the end of 1947 is conveyed by the letter Pauli wrote in answer to Oppenheimer:

> Many thanks for your letter of December 9th. . . . My talks with Weisskopf in Copenhagen on the theory of the Lamb-shift gave me already the impression, that we are at the beginning of a new development in quantum electrodynamics. As Weisskopf this autumn wrote me, that "he had no time" to decide the question whether this shift is finite also for Bose particles with spin 0, we made some calculations about it in Zurich. My first improvisation of a rule for subtraction worked already well and in the course of two weeks we obtained the result, that for this kind of particles, too, the shift of the energy levels is finite. Later we heard from Weisskopf, that meanwhile he had derived the same result.
>
> All you say on the problem of the radiative corrections for the scattering cross section of an electron by a field [Dancoff 1939] is very plausible to me. The general problem, with which we are now confronted is this:
>
> 1) Can all the divergences in the present form of quantum electrodynamics be interpreted as due to a change of the mass of the charged particle?
>
> 2) Can the separation of the infinite from the finite terms be done with the help of a relativistically invariant and unique rule? Remark ad 1): Besides the infinities due to the self-energy there are also those due to the polarization of the vacuum. They can be described as due to a change of charge (instead of mass) and can possibly be treated in an analogous way.
>
> In order to study these questions, we have now decided to treat here in Zurich as an example the Compton effect in the approximation next higher (in powers of e^2/hc) to the usual one. (For Bose particles this is not very complicated.) I want to learn, how the above questions 1) and 2) can be answered in this example. Moreover there is an interesting connection of this problem with the emission of low frequency photons (compare

Jost's paper in Phys. Rev.). If the new method of "separation of singular inertial terms" works, it may be possible to improve the original method of the Bloch-Nordsieck-transformation considerably. The final goal must be to get rid of the use of the development in powers of e^2/hc in the definition of a new formalism. I write to you our plans of working, in the hope of a reasonable partition of labour between our group in Zurich and the groups in Princeton and in M.I.T.

Schwinger's result on the change of the magnetic moment of a free electron... seems less plausible (at first glance) than the other effects you mentioned, but may be true nevertheless. If so the result is very fundamental. (Meanwhile I got a letter of Rabi on experimental verifications of this change of the electron's g-value. Differently from you Rabi speaks of an increase of the value of the electron's g-value, both experimentally and theoretically, while you speak of a reduction of this value). I hope that Schwinger will find a collaborator to write down his calculation for print; as I know him, he never writes a paper alone.[30]

Thus by the end of 1947 Pauli was quite clear and explicit about the problems to be addressed. Pauli took as his problem the question of the uniqueness of the radiative corrections after mass and charge renormalization had been performed.

Pauli had been very excited by Luttinger's calculation of the magnetic moment of an electron and he asked him to "do the mesons by the same techniques." So Luttinger proceeded to calculate the magnetic moment of the neutron and the proton in different meson theories. Pauli had been invited to the Eighth Solvay conference that was to be held in Brussels in late September 1948, and he wanted to report on these calculations. Luttinger recalls that "Pauli drove me crazy by coming every five minutes to my office to ask me whether the calculation was finished." Later that year Pauli wrote Oppenheimer: "I am sending you a copy of the paper of Luttinger (in print in *Helvetia Physica Acta*) on magnetic moments of the nucleons derived with the method of the 'patent-state.' He did not send me the result to Brussels (what was silly) because they disagree with experiment. The reason for the discrepancy can of course be the use of perturbation theory with respect to the coupling constant but I do not think that the interplay of some heavier meson is entirely excluded. What do you think about it."[31]

The shift of Pauli's views with respect to the introduction of new particles should be noted. His suggestion that some "heavier meson" may be responsible for the discrepancy is in marked contrast to his reticence about his own neutrino, Dirac's positron, and Stueckelberg's meson during the early thirties.

Judging from his letters to Fierz during 1948, Pauli was not particularly excited by any of the developments during that year. He kept abreast of all developments—for example, Feynman's cutoff method, the calculations of the Lamb shift by Kroll and Lamb and by French and Weisskopf, Wentzel's criticism of Schwinger's treatment of the photon self-energy—and followed with great interest the activities of Rivier and Stueckelberg. After obtaining a copy of the proceedings of the Pocono conference in late summer 1948, the members of the Pauli seminar studied the presentations that Schwinger and Feynman had given there. A considerable amount of time was spent by Jost and others in proving that the expression that Feynman had written down for the self-energy of an electron, which involved only the D_F Green's function, was identical to Schwinger's that was expressed in terms of $D, \overline{D}$, and $D^{(1)}$ Green's functions. Jost also investigated whether it was possible to obtain a set of rules for writing down the terms that contribute to a given order in perturbation theory in Schwinger's formalism.[32] Luttinger reports that Pauli was "very down on Feynman until Dyson's first paper reached Zurich." That paper and Schwinger's "(59 pages)" manuscript of "QED. Part II" arrived around Christmas. Writing to Fierz after reading them, Pauli expressed the hope "that with the help of Jost he would be able to understand the "Feynman theory" better.[33] In a letter to Oppenheimer he characterized Schwinger's paper "as the first authentic text of himself available to me (after the only formal Part I)" and indicated "that of course I studied it with great care." He went on to note that "in some respect I was, however, disappointed. Wentzel's paper was not even mentioned and the questions raised in it were not cleared. Moreover he pretended to have 'proofs' at places, where assumptions in the form of defining rules are necessary due to the lack of uniqueness of the resulting mathematical expressions. This is particularly so for the gauge invariance of the additional current describing the vacuum-polarization. . . . But it also holds for the problems connected with the electron self energy."[34]

During the next few months the entire group—Pauli, Villars, Jost, Luttinger, Rayski, Schaffroth—worked hard not only to understand the connections between the various formulations (those of Rivier and Stueckelberg, Feynman, Schwinger, Dyson), but also to clarify the efficacy and the limitations of the computational methods used by these various authors. The fruits of these efforts were contained in the long letter Pauli wrote Schwinger at the end of January 1949, parts of which were presented in chapter 7. In it Pauli stressed that what had been accomplished during the past year by Schwinger, Feynman, and Stueckelberg depended in an essential manner on the fact that perturbation theory had been formulated in a Lorentz invariant fashion, and this made it possible to isolate the infinities and to identify and calculate finite radiative corrections. A critical issue for him was how to guarantee the uniqueness of the physical predictions of the theory after the infinities had been isolated.

The source of the ambiguity in the renormalization procedure was the singular nature of the Green's functions $\triangle(x; M)$ and $\triangle^{(1)}(x; M)$ on the light cone. Pauli was intent on finding a consistent prescription for dealing with the "undefined mathematical expressions" that the confluence of singularities in products of singular functions gave rise to. Pauli believed that such singularities would not appear in a "future theory." As a step in this direction Pauli, influenced by the work of Feynman (1948b) and Rivier and Stueckelberg (1948), suggested replacing the single mass associated with each field by a mass spectrum, which was adjusted so that the new, "regularized," Green's functions are free of singularities on the light cone. Thus, with a boson field, Pauli associated the propagator

$$\triangle_R(x) = \sum_i c_i \triangle(x; M_i) \tag{10.2.1}$$

and

$$\triangle_R^{(1)}(x) = \sum_i c_i \triangle^{(1)}(x; M_i), \tag{10.2.2}$$

with the c_i's satisfying the conditions

$$\sum_i c_i = 0 \tag{10.2.3a}$$

and

$$\sum_i c_i M_i^2 = 0. \tag{10.2.3b}$$

The δ-function type singularity and first order poles $[1/(x_\mu x_\mu)^{1/2}]$ are eliminated by virtue of eq. (10.2.3a), and finite jumps and logarithmic singularities are eliminated by condition (10.2.3b). It is the meaning of the regularization prescriptions that the first term in the series (10.2.1) represents the non-regularized function itself, that is, that

$$c_0 = 1 \qquad M_0 = m \tag{10.2.4}$$

and that all the $M_i (i > 0)$ should finally tend to ∞, and $c_i/M_i^2 \to 0$. The auxiliary masses are regarded only as a "formal" mathematical aid for the computation.[35] Pauli then outlined how the invariant regularization procedure was to be applied when calculating the self-energy of the photon and that of the electron.

In the concluding paragraph of his "small paper," Pauli told Schwinger the reason for his devoting so much energy to elucidating the computational procedures of the various investigators:

> It is my endeavor, on the one hand to clarify the logical situation to such a degree, that the danger of longer polemical controversies between you and other physicists is prevented; on the other hand, to preserve all new results and new methods developed

both by yourself and other physicists and to bring them in closer connection.

You and others may object against my point of view that it is only a formal prescription for calculations without a clear formulation of new concepts. This is perfectly true, but it is supposed to be only a preparation for further progress. I believe that in a future theory neither the functions, which are singular on the light cone, or the three dimensional-function in the commutators of the canonical formalism will occur and that the present theory can only describe averages over small but finite space-time regions approximately correct. The dimensions of these critical regions do not seem to be an "universal length," but must depend on the masses of the particles involved. As a substitute for concepts not yet precisely known appear now auxiliary masses and it seems to me only a formal preliminary measure to let them finally tend to infinity.[36]

Pauli and Villars smoothed out the further intricacies of the regularization procedure by applying the method to a host of problems, including the calculation of the electron's anomalous magnetic moment. Pauli "sent a joint paper with Villars, containing [their] calculations on regularization for print in the Einstein-issue of the *Reviews of Modern Physics*, edited by Pais" (Pauli and Villars 1949) in order to continue his role as "His Majesty's [Schwinger's] opposition."[37]

By the end of 1948, several members of Pauli's seminar were investigating the renormalization procedure in higher order of perturbation theory. Källèn (1949) proved that in the higher approximation of vacuum polarization in an external field, only the terms in e^2 and e^4 diverge, and that "the higher terms are both convergent and gauge invariant." Jost and Luttinger undertook an investigation of charge renormalization to order e^4 in QED. In late February 1949 Pauli wrote Oppenheimer: "While we made considerable progress . . . in the treatment of particular problems (higher approximation for vacuum-polarisation and for the magnetic moment of the electron and the nucleon . . .) no progress has been made regarding the understanding of the principles of the quantized field theory. We consider at present the 'self-charge' (resulting theoretically in the vacuum-polarisation) as the 'top-nonsense' of the present theory."[38]

When Dyson's papers were received in Zurich, they became the object of a very careful study.[39] Jost and Luttinger's investigation of charge renormalization to order e^4 was reformulated in Dyson's language. Besides trying to understand how the renormalizations were to be performed in higher orders—and therefore how to deal with overlapping divergences[40]—Jost and Luttinger were intent on clarifying two other issues. If one considers charge renormalization not merely as a formal elimination of divergences but ascribes to it some element of reality, one must then explain the universality of the renormalization, that is, why all charges are renormalized the same amount so that the proton, the positively charged muon, and the positron all have the same renormalized charge. Jost and Luttinger (1950) showed how this comes about to order e^2.[41] The second problem they addressed was a conjecture by Pauli. Pauli had entertained the hope that the cancelation of the divergences of order e^2 with those of order e^4 in the expression for the self-charge might fix the value of the fine structure constant.[42] The conjecture was proven wrong. Jost and Luttinger showed that the renormalized charge could be expressed as a power series in the unrenormalized charge

$$e^2 = e_0^2 c_1 + e_0^4 c_2 + \ldots,$$

and found that c_1 and c_2 had the same sign.[43]

Important calculations of the radiative corrections to various electrodynamic processes were also carried out in Zurich. Schaffroth (1949, 1950) computed the radiative corrections to Compton scattering; Jost and Luttinger (1950) those for pair production. These papers played an important role in indicating how the procedures outlined by Dyson, Feynman, Tomonaga, and Schwinger could actually be carried out in the first nontrivial order: namely, to fourth order in the coupling constant. However, the purpose of these calculations was to demonstrate the operation and the consistency of the renormalization program, not to get the numbers out.[44] This reflected Pauli's attitude toward experimental physics. He did follow the experiments that were of interest to him—for example, the Lamb experiments, the Kusch experiments—but when Luttinger was in Zurich he took no interest in any experimental activities in solid state physics or nuclear physics. He did become enormously excited about mesons and sent one of his assistants, Telegdi, to Powell to find out whether they were real. By this time Pauli was more of a mathematical physicist than a theorist. Luttinger indicates that Pauli "talked about experiments very nobly" but also cautioned "never believe what experimenters say first." One of his mottos was: "Never work too closely with experimenters. Allow the results to settle."[45]

As important as the postwar contributions of Pauli were, it is, I believe, fair to say that they were not core contributions. I would suggest that Pauli's emphasis on the foundational rather than on the instrumental aspects of quantum electrodynamics impeded him from making contributions that would be central to the postwar developments. To some extent Pauli gave a repeat performance of the role he assumed in the 1925–1927 period. Though he played a critically important role, Pauli was not the driving force in the developments of quantum mechanics. He was not as original a mind as the other founding fathers. "Pauli needed too much clarity. Whereas people like Bohr and Heisenberg could function in an atmosphere of very great uncertainty and have ideas, Pauli needed certainty. He always had to make it lucid—if he couldn't make it lucid, he couldn't think about it very well."[46]

Epilogue: Some Reflections on Renormalization Theory

Renormalization theory is a complicated conceptual system.[1] The renormalization procedure can be viewed as a technical device for circumventing—that is, isolating and discarding—the infinite results that occur in perturbative calculations in quantum field theory. The concept of renormalization has also helped to clarify the conceptual basis of quantum field theory and to establish its consistency. Furthermore, renormalizability can also be considered as a regulative principle, guiding theory construction and theory selection within the general framework of quantum field theory.

As we have seen, the emergence of renormalization theory in the late 1940s was a response to the divergence difficulties of quantum electrodynamics. In its original formulation, renormalization theory was technical and conservative in character, and it is important to keep these initial characteristics in mind when trying to understand its subsequent developments. It was technical because it involved a series of algorithmic steps for obtaining numerical results from the theory, numbers that could be compared with experimental data—for example, the Lamb shift and the anomalous magnetic moment of the electron. It was conservative because it took the framework of *local* quantum field theory as given, and made no attempts to alter its foundations, in particular the locality assumption. In fact, Freeman Dyson took the conservative character of renormalization theory as one of its endearing features.

The local coupling among fields and the fact that the application of field operators on the vacuum results in strictly local excitations imply that in (local) quantum field theoretical calculations one has to consider virtual processes involving arbitrarily high energy. However, except for the consequences imposed by such general constraints as unitarity, there exists essentially no empirical evidence for believing the correctness of the theory at these energies. Mathematically, the inclusion of these virtual processes at arbitrarily high energy results in infinite integrals. Thus the divergence difficulties are internal in the very nature of local quantum field theory: they are constitutive within the canonical formulation of the theory. In this sense the occurrence of the divergences clearly pointed to an inconsistency in the conceptual structure of QFT.

If we disregard the various proposals for radically altering the foundations of QFT by giving up locality, two different responses were advanced to overcome this inconsistency. The first was developed independently by Pais (1945) and by Sakata (1947), and was in the spirit of Poincaré's solution to the stability problem of the Lorentz electron. It put forth the idea of compensation: fields of unknown particles were introduced in such a way as to cancel the divergences produced by the known interactions. The second response was the renormalization program.

As was noted in chapter 2, Dirac, Heisenberg, Weisskopf, Kramers, Dancoff, and others had already put forth the idea of renormalization in terms of subtractions in the 1930s. But it required the precise experimental findings on the spectrum of hydrogen and deuterium obtained using techniques and instruments developed during World War II to stimulate the further elaboration of this idea. The explanation of the accurate and reliable data obtained by Lamb and Retherford (1947) in their measurements of the fine structure of hydrogen, and of Rabi's results on the hyperfine structure of hydrogen and deuterium, became an outstanding challenge for theoretical physicists. In the process they developed algorithms for obtaining finite numbers for the measured quantities in their QFT-based calculations, and put forth suggestive ideas for justifying the algorithms. The ideas and algorithms developed by Kramers, Bethe, Lewis, Schwinger, and Tomonaga can be summarized as follows:

> **1.** The divergent terms that occur in quantum electrodynamical calculations are identifiable in a Lorentz and gauge-invariant manner, and can be interpreted as modifying the mass and charge parameters that are introduced in the original Lagrangian.

> **2.** By identifying the modified, or renormalized, mass and charge parameters with the physically observable masses and charges of physical particles, all the divergences are absorbed into the mass and charge renormalization factors, and finite results in good agreement with experiments are obtained. Thus the measurements of Lamb and Rabi could be explained within the framework of (renormalized) QED.

A crucial assumption underlying the whole renormalization program was expressed succinctly by Lewis (1948): "The electromagnetic mass of the electron is a small effect and . . . its apparent divergence arises from a failure of present day quantum electrodynamics above certain frequencies." It should be clear that only when a physical parameter is actually finite and small, can its separation and amalgamation into the "bare" parameters be regarded as mathematically justifiable. However, when calculated perturbatively in quantum field theory, the parameters often turned out to be divergent.

The failure of quantum field theory at ultra-relativistic energies, as indicated by the divergences in perturbation theory, implied that the region in which the existing framework of quantum field theory is valid should be separated from the region in which it is not valid and in which new physics would become manifest. It is impossible to determine where the boundary is, and one does not know what theory can be used to calculate the small effects that are not calculable in quantum field theory. However, this separation of knowable from unknowable, which is realized mathematically by the introduction of a cutoff, can be schematized by using the phenomenological parameters that must include these small effects.

Neither Lewis nor Schwinger nor Tomonaga made explicit use of a cutoff. They directly identified the divergent terms with corrections to the mass and the charge, and removed them from the expressions for real processes by redefining the masses and charges. By contrast, Feynman's efficient calculational algorithm was based on the explicit use of a relativistic cutoff. The latter consisted of a set of rules for regularization, which made it possible to calculate physical quantities in a relativistically and gauge-invariant manner, but still resulted in divergent expressions in the limit as the cutoff mass goes to infinity. With a finite cutoff this artifice transforms essentially purely formal manipulations of divergent quantities, that is, the redefinition of parameters, into quasi-respectable mathematical operations. If after the redefinition of mass and charge other processes are insensitive to the value of the cutoff, then a renormalized theory can be defined by letting the cutoff go to infinity. A theory is called renormalizable if a finite number of parameters are sufficient to define it as a renormalized one.

Physically, Feynman's relativistic cutoff is equivalent to introducing an auxiliary field (and its associated particle) to cancel the infinite contributions due to the ("real") particles of the original field. But Feynman's approach differs from realistic theories of regularization or compensation. In the latter, auxiliary particles with finite masses and positive energies are assumed to be observable in principle and are described by field operators that enter the Hamiltonian explicitly. Feynman's theory of a cutoff is formalistic in the sense that the auxiliary masses are used merely as mathematical parameters which finally tend to infinity and are nonobservable in principle. "Realistic" approaches were proposed and investigated by Sakata (1947, 1950), Sakta and Hara (1947), Umezawa et al. (1948), Umezawa and Kawabe (1949a,b), and other Japanese physicists, and by Rayski (1948). Among the "formalists" are to be found, in addition to Feynman, Rivier and Stueckelberg (1948) and Pauli and Villars (1949). The renormalized version of QED has enjoyed great success because of its astonishing predictive power, both in calculating the anomalous magnetic moment of the electron and the Lamb shift in hydrogen, and in estimating the radiative corrections to high energy electron-electron and electron-positron scattering.

In the late forties and early fifties it was hoped that by successfully circumventing the divergence difficulties a consistent framework of quantum field theory could be constructed, and moreover, as Pauli suggested, that it might fix the masses and charges of the particles that appear in the theory. This turned out to be too optimistic although some advances toward a rigorous proof of the renormalizability of quantum electrodynamics were made. Crucial among these were:

1. The solutions to the overlapping divergences given by Salam (1951a,b), Ward (1950a,b, 1951), Mills and Yang (1966), and by Hepp (1966); and

2. The convergence theorem of Weinberg (1960), which is necessary for the proof that in renormalizable theories all the ultraviolet divergences do cancel to all orders of perturbation theory, despite the occurrence of complicated divergent subgraphs.

Conceptually, however, even though renormalizability had been accepted as a property of quantum electrodynamics, why the renormalization program actually worked remained quite unclear. The question can be divided into two parts:

1. Why do the apparent divergences arising from a failure of unrenormalized quantum electrodynamics above certain energies actually give rise to small effects?

2. More generally, why are the representations of nature by renormalized theories stable, and more specifically, why are they so very insensitive to whatever happens at very high energy?

While some progress toward an answer to the second part of the question has been made during the past two decades, no real insight has been obtained toward being able to give an answer to the first part during the more than four decades since Lewis first stated the smallness assumption. An equally fundamental question regarding renormalization theory was whether all the interactions in nature are renormalizable.

Dyson was aware that the answer to this last question was negative and so reported to the Oldstone conference. Thus the question arose: "Should nature be described only by renormalizable theories?" For physicists such as Bethe who had elevated renormalizability from a property of quantum electrodynamics to a regulative principle guiding theory selection, the answer was affirmative (see Schweber et al. 1955). They justified their position in terms of predictive power. They argued that since the aim of fundamental physics is to formulate theories that possess considerable predictive power, "fundamental laws" must contain only a finite number of parameters. Only renormalizable theories are consistent with this requirement. While the divergences of nonrenormalizable theories could possibly be eliminated by absorbing them into appropriately specified parameters, an infinite number of parameters would be required, and such theories would initially be defined with an infinite number of parameters appearing in the Lagrangian.

Most field theorists ignored the consistency problem in renormalization theory, and argued that if meaningful calculations could be carried out only within the framework of renormalized perturbative theory, then in fact renormalizability should be taken as a decisive constraint on theory construction. It is a historical fact that the further developments of quantum field theory beyond the scope of quantum electrodynamics have been accomplished using the principle of renormalizability as a guideline. The most convincing case in point is Weinberg's unified field theory of electroweak interactions. As Weinberg (1980a) remarked in his Nobel lecture, if he had not been guided by the principle of renormalizability his theory of electroweak interactions would have received contributions not only from SU(2) × U(1)-invariant vector boson exchanges—which were believed to be renormalizable, though not proven to be so until few years later by 't Hooft (1971a,b, 1972) and others (Lee and Zinn-Justin 1972; Becchi et al. 1974)—but also from SU(2) × U(1)-invariant four-fermion couplings, which were known to be non-renormalizable, and the theory would have lost most of its predictive power.

According to the renormalizability principle, the interaction Lagrangian of a charged spin 1/2 particle interacting with the electromagnetic field cannot contain a Pauli moment. Similarly, a pseudovector coupling of the pion to the nucleon was excluded. By the same reasoning, Fermi's theory of weak interaction lost its status as a fundamental theory. A more complicated application of the renormalization constraint was the rejection of the pseudoscalar coupling of pions to nucleons in the strong interactions. Formally, the pseudoscalar coupling was renormalizable. Yet its renormalizability was not realizable because the radiative corrections it produces are too large to be able to justify the use of perturbation theory, which is the only framework within which the renormalization procedure works. This paved the way for the popularity of the dispersion relations approach and for the adoption of Chew's S-matrix theory approach that rejected the whole framework of quantum field theory by a considerable number of theorists (cf. Cushing 1990; Cao 1991).

The case of the Yang-Mills field (1954a,b) deserves special attention. Physicists were interested in the original Yang-Mills theory in part because it was conjectured to be renormalizable, even though the massless bosons required by the gauge invariance could not be responsible for short-range nuclear forces. The massive version of Yang-Mills theory was unacceptable because the massive gauge bosons spoiled not only the gauge invariance but the renormalizability of the theory as well. Gell-Mann was attracted by the theory and tried to find a "soft-mass" mechanism which would allow the renormalizability of the massless theory to persist in the presence of gauge boson masses, but he did not succeed (cf. Gell-Mann 1987, 1989). Where Gell-Mann failed, Weinberg (1967) and Salam (1968) succeeded with the help of the Higgs mechanism. After 't Hooft's apparent proof of the renormalizability of a Yang-Mills theory whose vacuum state vector had its symmetry broken by the Higgs mechanism, the Yang-Mills theory became the paradigmatic case of quantum field theory.

It would not be too great an exaggeration to claim that the most substantial advances in quantum field theory that have been achieved in the past four decades have been guided and constrained by the renormalizability principle. It is certainly the case that renormalization theory saved quantum field theory, made it manipulable and allowed one to calculate with it, and thus revived the faith of theorists in quantum field theory. Be that as it may, no unanimous consensus has ever been reached as to whether renormalizability is an essential characteristic of quantum field theory, or a universal principle constraining all the possible descriptions of nature. As a matter of fact, since the early fifties serious arguments have been advanced challenging the consistency of renormalization theory and casting doubts on the foundations of quantum field theory. The debate has led to a deeper understanding of the physics and the philosophy of renormalization, and has helped to clarify the foundations of quantum field theory. This can be viewed as another way in which renormalization theory has advanced quantum field theory into a new phase.

An unrenormalized theory is inconsistent due to the presence of ultraviolet divergences and by virtue of the infinities that stem from the infinite volume of space-time. The latter have to be disentangled from the former and exclude the Fock representation as a candidate for the Weyl form of the canonical commutation relations. The occurrence of these two kinds of infinities makes the definition of a (mathematically respectable) Hamiltonian operator much more recondite, and the whole scheme of canonical quantization of quantum field theory collapses (Wightman 1976).

The consistency problem in renormalized theories is very different from that of unrenormalized ones. The ultraviolet divergences are supposed to be circumventable by the renormalization procedure. Some of the remaining difficulties were solved in a rigorous fashion by axiomatic field theorists with the help of distribution theory. Yet new problems created by renormalization theory invited other criticisms. Thus, Dirac (1968a,b) criticized renormalization theory for neglecting infinities instead of infinitesimals, a procedure radically at odds with the usual custom in mathematics. Lewis's smallness assumption anticipated and seems to invalidate Dirac's criticism, but the assumption itself has to be justified in the first place. At the physical level, Heitler (1961) and others noted that the mass differences of particles (such as the pions and the nucleons) which are identical except for their electric charge could not be calculated using renormalization theory. It is not difficult to establish that if the mass differences are of electromagnetic origin, then the divergent electromagnetic self-energy will lead to infinite mass differences. This difficulty clearly indicated that renormalization theory could not fulfill Pauli's hope that it would provide a general theory to account for the mass ratios of the "elementary particles." In addition, renormalization theory was criticized as being too narrow a framework to accommodate the representations of such important phenomena as the CP-violating weak interactions and the gravitational interactions.

The internal consistency of renormalization theory at the foundational level was seriously challenged by Dyson, Källén, Landau, and others not long after Dyson outlined his proof of the renormalizability of quantum electrodynamics. In 1953 Källén claimed to be able to show that, starting with the assumption that all renormalization constants are finite, at least one of the renormalization constants in quantum electrodynamics must be infinite. For several years this contradictory result was accepted by most physicists as evidence for the inconsistency of quantum electrodynamics. However, as was later pointed out by some critics (e.g., Gasiorowicz et al. 1959), his results depended upon some notoriously treacherous arguments involving interchanges of the orders of integration and summation over an infinite number of states, and was thus inconclusive. Källén himself later acknowledged this ambiguity (1966).

More serious arguments challenging the consistency of renormalization theory were expressed in terms of the breakdown of perturbation theory. Dyson's renormalization theory was formulated only within the framework of perturbation theory. The output of perturbative renormalization theory is a set of well-defined formal power series for the Green functions of a field theory. However, it was soon realized that these series, in particular the one for the S-matrix, were most likely divergent. Thus theorists were thrown into a state of confusion and could not give an answer to the question: "In what sense does the perturbative series of a field theory define a solution?" As noted in Chapter 9, the first theorist to be disillusioned by perturbative renormalization theory was Dyson himself. In 1952 he gave an ingenious argument that suggested that after renormalization all the power series expansions were divergent. The subsequent discussion by Hurst (1952), Thirring (1953), Peterman (1953a,b), and by Jaffe (1965) and other axiomatic and constructive field theorists added further weight to the assertion that the perturbative series of most renormalized field theories diverge, even though there is still no complete proof in most cases.

A divergent perturbative series for a Green function may still be asymptotic to a solution of the theory. In the mid-seventies the existence of solutions for some field-theoretical models was established by constructive field theorists, and these indicated a posteriori that the solution is uniquely determined by its perturbative expansion (cf. Wightman 1976). Yet these solutions were exhibited only for field-theoretic models in space-time continua of two or three dimensions. As far as the more realistic 4-dimensional quantum electrodynamics is concerned, already in 1952 Hurst had suggested that the excellent agreement of quantum electrodynamics with experiments indicated that the perturbative series is an asymptotic expansion. However, the investigations of the high-energy behavior of quantum electrodynamics by Källén, Landau, and especially that by Gell-Mann and Low (1954) showed that the perturbative approach in quantum electrodynamics unavoidably breaks down, ironically, as a consequence of the necessity of charge renormalization. Landau and his collaborators argued further that remaining within the perturbative framework would lead either to no interaction (zero renormalized charge), or to the occurrence of ghost states rendering the theory apparently inconsistent (Landau 1955, Landau and Pomeranchuck 1955). Both results demonstrated the inapplicability of perturbative theory in renormalized quantum electrodynamics.

After the discovery of asymptotic freedom in a wide class of non-Abelian gauge theories, especially in quantum chromodynamics, the hope was expressed that perturbative quantum chromodynamics would get rid of the Landau ghost and would thus eliminate most doubts as to the consistency of quantum field theory. However, this expectation did not last long. It was soon realized that the ghost which disappeared at high energy reappeared at low energy (cf. Collins 1984). Thus field theorists were reminded—forcefully and persistently—of the limits of applicability of perturbative theory. As a result, the consistency problem of quantum field theory in general, and of perturbative renormalization theory in particular, was in a state of uncertainty. One of the gravest defects of renormalization theory was recognized around 1970, namely, that it is in direct and irreconcilable conflict with the chiral and trace anomalies that occur in high orders of quantum field theory (Jackiw 1972).

The attitude of theoretical physicists toward the issue of consistency differed sharply. For most practicing physicists, consistency is just a pedantic problem. As pragmatists, they are guided only by their scientific experiences and have little interest in speculating about the ultimate consistency of a theory. For Dirac (1963b, 1973a, 1983), however, renormalization theory with the cutoff going to infinity was illogical and nonsensical physically. In his opinion, what was required were new forms of interaction and new mathematics, such as possibly the use of an indefinite metric (1942), or of nonassociative algebra (1973a), or perhaps something even more esoteric. The positions adopted by Landau and Chew were more radical and drastic (cf. Cao 1991). What they rejected were not merely particular forms of interactions and perturbative versions of quantum field theory, but the general framework of quantum field theory itself. For them the very concept of a local field operator and the postulation of any detailed mechanism for interactions in a microscopic space-time region were unacceptable because these were too speculative to be observable, even in principle. Their position was supported by the presence of divergences in quantum field theory and by the lack of a proof of the consistency of renormalization theory, even though Landau's arguments for the inconsistency of renormalized quantum electrodynamics could not claim to be conclusive.

The most positive attitude was taken by the axiomatic field theorists, who later became called constructive field theorists. In the spirit of Hilbert's tradition they tried to settle the question of the internal consistency of quantum field theory by axiomatization, and took this as the only way to give clear answers to conceptual problems. While Hilbert tried to establish the existence of mathematical entities with a proof of the consistency of a formal system consisting of these entities, the axiomatic field theorists went the other way around. They tried to prove the internal consistency of quantum field theory by constructing nontrivial examples whose existence is a consequence of the axioms alone. Without radically altering the foundations of quantum field theory, they tried to overcome the apparent difficulties with its consistency step by step. Although many of the important problems remain open, nowhere did they find any indication that quantum field theory contained basic inconsistencies.

The axiomatic field theorists took the fields to be operator-valued distributions defined with infinitely differentiable test functions of fast decrease at infinity or with test functions having compact support. Essentially, this was a mathematical expression of the physical idea of modifying the exact point model (Wightman 1981b). However, a thus defined theory may still be nonrenormalizable in the sense of perturbation theory.

Since the mid-seventies, there have been major efforts using the approach of constructive field theory[2] to understand the structure of nonrenormalizable theories and to establish the conditions under which a nonrenormalizable theory can make sense. One of the striking results of this enterprise is that the solutions of some nonrenormalizable theories have only a finite number of arbitrary parameters. This is contrary to their description in terms of the perturbative series. It has been speculated that the necessity for an infinite number of parameters to be renormalized in perturbation theory may come from an illegitimate power series expansion (Wightman 1986). It is certainly the case that in these efforts the axiomatic and constructive field theorists have exhibited an open and flexible frame of mind. Yet future developments in understanding the foundations and proving the consistency of renormalization theory may involve changes in some assumptions that have not been challenged up to the present, or which have not been captured by any axiomatization of the present theory. It is also possible that the present view of the consistency problem, including its function and significance in theory building, theory appraisal, and theory selection, may change radically as it is grounded in Hilbert's formalist tradition. In any case, the failure to construct a soluble 4-dimensional field theory, despite intensive efforts for close to four decades, indicates that the axiomatic and constructive field theorists are meeting considerable difficulty in solving the consistency problem of quantum field theory in Hilbert's sense. It also has dampened their initial optimism somewhat.

The finitists, among whom are Salam (1970), Salam and Strathdee (1973) and the various advocates of supergravity (e.g., Hawking 1980) and superstrings (e.g., Green 1985, Green et al. 1987), are more optimistic than the axiomatic and constructive field theorists. Their hope is that by including gravitational interactions in the existing formulations of quantum field theory systems, it will be possible to construct a finite theory without any infinite renormalizations. They thus hope to avoid the consistency problem of the renormalization theory. But hopes come and go, and all of them seem to be short lived.

Two additional contrary views on renormalization, advanced respectively by Sakata (1950, 1956), Sakata and Umezawa (1950), Sakata et al. (1952) and by Schwinger (1970, 1973, 1983a), were expressed in terms of their concerns about the structure of elementary particles. For Sakata, renormalization theory was only an abstract formalism, behind which lay hidden the concrete structure of elementary particles. His position was that when renormalization theory would encounter a defect (and its limitations would be exposed), it would become necessary to look for and analyze more closely the structure of the elementary particles. Under Sakata's influence, more efforts were invested in Japan in model-building of constituents of elementary particles than in the analysis of the theoretical structure of quantum field theory. As a result, little emphasis was placed on renormalization theory as an essential conceptual ingredient of quantum field theory (cf. Takabayasi 1983; Aramaki 1989).

Schwinger's views of renormalization are of particular interest, not only because he is one of the founders of renormalization theory, but also because he has given penetrating analyses of the philosophy of the renormalization program and is one of its most incisive critics. According to Schwinger, the un-renormalized description, which adopts local field operators as its conceptual basis, contains speculative assumptions about the dynamic structure of the physical particles that are sensitive to details at high energy. However, we have no reason to believe that the theory is correct in that domain. In accordance with Kramers' precept that quantum field theory should have a structure-independent character, which Schwinger accepted as a guiding principle, the renormalization procedure that he elaborated removed any reference to very high energy processes and the related small distance and inner structure assumptions. He thus transferred the focus from the hypothetical world of localized excitations and interactions to the observed world of physical particles. But Schwinger found it unacceptable to proceed in this tortuous manner of first introducing physically extraneous structural assumptions, only to delete them at the end in order to obtain physically meaningful results. This constitutes a rejection of the philosophy of renormalization. But renormalization is essential and unavoidable in a local operator field theory if the latter is to make any sense. In order to bring his criticism to its logical conclusion, Schwinger introduced numerically valued (nonoperator) sources and numerical fields to replace the local field operators. These sources symbolize the interventions that constitute measurements of the physical system. Furthermore, all the matrix elements of the associated fields, the operator field equations, and the commutation relations can be expressed in terms of the sources. An action principle gives succinct expression to the formalism. According to Schwinger, his source theory takes finite quantities as primary, and it is thus free of divergences. This theory is also sufficiently malleable to be able to incorporate new experimental results, and to extrapolate them in a reasonable manner. Most important, it can do so without having to extend the theory to arbitrarily high energies—regions that constitute unexplored domains where new unknown physics is sure to be encountered.

Thus in Schwinger's approach the ultimate fate of renormalization theory is to be removed and excluded from any description of nature. He tried to implement this by abandoning the concept of a local operator field, and this constitutes a drastic alteration of the foundations of quantum field theory. The radical character of Schwinger's approach, whose foundations were laid in 1951 and elaborated in the 1960s and 1970s, was not recognized until the mid-1970s when the renormalizability principle began to be challenged. By that time new important insights into renormalization and renormalizability had been gleaned from studies using renormalization group methods, resulting in a new understanding of renormalization, of quantum field theory, and also in novel attitudes toward scientific theories in general. Thus a renewed interest in nonrenormalizable theories began to be manifested, and what has become known as the "effective field theory" approach began to gain its popularity (Georgi 1989b). In this changed conceptual context, the most perspicacious theorists—for example, Weinberg (1979)—began to realize that Schwinger's ideas were essential in the radical shift of outlook in fundamental physics.

During the mid-1970s the fundamental nature and essential character of the renormalizability principle began to be challenged. As a result of two decades of fruitful interactions between QFT and statistical mechanics, the understanding by theoretical physicists of certain foundational aspects of renormalization theory underwent a radical transformation. At the heart of the transformation was the emergence of the new concept of "broken scale invariance" and the related renormalization group approach. Weinberg (1978) was one of the first to assimilate the physical insights developed principally by K. G. Wilson (1975; Wilson and Kogut 1974) in the context of critical phenomena—for example, the existence of the fixed-point solutions of renormalization group equations and the conditions for trajectories in coupling-constant space passing through fixed points—and to apply them within the context of quantum field theory. His intention was to explain or even replace the renormalizability principle by a more fundamental guiding principle that he labeled "asymptotical safety." Yet this program was soon to be overshadowed by another program, that of "effective field theory," also initiated by Weinberg (1979, 1980b). At first, effective field theory was a less ambitious program than that encompassed by asymptotically safe theories because effective field theory still takes renormalizability as its conceptual basis. Effective field theory, however, has led to a radical change of outlook, together with a thorough examination of the very concept of renormalizability and a clarification of the ontological basis of quantum field theory (cf. Georgi 1989b; Cao and Schweber 1993).

Effective field theory has given tentative and suggestive answers to the following two questions: How can one account for the great empirical success of renormalized quantum electrodynamics, seemingly a conceptually inconsistent theory? And why is the notion of a local field so effective? As David Gross (1985) has remarked: "Who would have expected that the concept of a local field, which originated in Faraday's attempt to vizualise macroscopic magnetic fields, would be so fruitful?" And to put the matter in perspective: With the establishment during the 1970s of successful gauge theories of the electroweak and the strong interactions, physicists believe that they are able to account for all physical phenomena from the macroscopic structure of the universe down to the structure of matter at distances of at least 10^{-15} cm in terms of local quantum field theories. In fact, the only place where physicists express doubt about the adequacy of local quantum field theories is in the description of physics at the Planck scale of 10^{-33} cm.

Whatever the future may bring, it is safe to assert that the theoretical advances made in the unraveling of the constitution of matter since World War II comprise one of the greatest intellectual achievements of mankind. They were based on the ground secured by the contributions of Tomonaga, Bethe, Schwinger, Feynman, and Dyson to quantum field theory and renormalization theory in the period from 1946 to 1951.

NOTES

Abbreviations and Archives

AHQP	Archives for the History of Quantum Physics. Niels Bohr Library, American Institute of Physics, College Park, Maryland
AIP	American Institute of Physics, Niels Bohr Library, College Park, Maryland
Bethe Papers	Archives, Olin Library, Cornell University, Ithaca, N.Y.
Breit Papers	Archives, Yale University Library, New Haven, Conn.
BSC	Bohr Scientific Correspondence. Bohr Library, Nordita, Copenhagen, Denmark. Also AHQP
Darrow Papers	Niels Bohr Library, AIP
DSC	Dirac Scientific Correspondence. Florida State University, Tallahassee or Churchill College, Cambridge, U.K.
Dyson Papers	Institute for Advanced Study, Princeton, N.J.
Fierz Papers	E.T.H. Library, Zurich, Switzerland
FJD, S^3	F. J. Dyson. Taped interview with S. S. Schweber, November 18–19, 1984
MacInnes Papers	Rockefeller University Archives, New York
NAS Archives	National Academy of Sciences Archives, Washington, D.C.
Oppenheimer Papers	National Archives, Library of Congress, Washington, D.C.
Peierls Papers	Bodleian Library, Oxford
Rabi Papers	National Archives, Library of Congress, Washington, D.C.
RPF, CIT	R. P. Feynman Papers. Millikan Library Archives, California Institute of Technology, Pasadena
RPF, CW	Richard P. Feynmann. Oral interview by Charles Weiner, AIP
RPF, S^3	R. P. Feynman. Taped interview with S. S. Schweber, November 13, 1984
TOM	Tomonaga Pamphlet. Tomonaga Memorial Room, University of Tsukuba, Japan
Van Vleck Papers	Niels Bohr Library, American Institute of Physics, College Park, Maryland
Weisskopf Papers	Archives, MIT Library, Cambridge, Mass.

Chapter 1
The Birth of Quantum Field Theory

1. My presentation in chapters 1 and 2 has relied heavily on secondary sources. Particular mention should be made of Olivier Darrigol's thesis and his articles on the early history of quantum field theory (Darrigol 1982, 1984, 1986). Similarly, Joan Bromberg's article on Dirac's quantum electrodynamics and the wave-particle dualism (Bromberg 1977) and Marcello Cini's on cultural traditions in the development of quantum electrodynamics (1925–1933) (Cini 1982) were very helpful. Finally, the appearance of Pais's *Inward Bound* (Pais 1986) rendered my task much easier: it allowed me to shorten my presentation and I refer the reader to his insightful account of these developments.

2. Quoted in Heilbron (1985), p. 221. In that article Heilbron has given an insightful and trenchant account of Bohr's early apostles: Heisenberg, Pauli, and Jordan. A particularly incisive portrait of Jordan, including his activities as a Nazi party member during the thirties, emerges from Heilbron's account. Beyerchen indicates that Jordan was one of the younger conservative physicists who during the 1930s believed that "they could deradicalize the Nazi movement and temper its excesses by joining it" (Beyerchen 1977, p. 228). In a letter to Bohr, written shortly after the war, Jordan indicated that he had "thought it to be a thrilling sport to give a book pleading for relativity and quanta the title *Die Physik des XX. Jahrhunderts* as an answer to A. Rosenberg's ill-famed *Mythos des XX. Jahrhunderts*." The latter was the standard Nazi primer. Jordan to Bohr, May 1945, BSC; quoted in Heilbron (1985), p. 222.

3. The most complete biographical account of Jordan is to be found in Mehra and Rechenberg (1982), vol. 3 of their *The Historical Development of Quantum Theory*, pp. 44ff. See also Jordan's interview with T. S. Kuhn (AHQP) upon which their presentation is based.

4. In a letter to Kronig later that year, Pauli reiterated his disdain for the formal mathematics of Göttingen: "Heisenberg's mechanics has again given me *zest for life* and hope. Although it does not give the solution of the puzzle, I believe that it is now again possible to make progress. One must next try to further liberate the Heisenberg mechanics from the torrent of Göttingen formal "learnedness" [*Gelehrsamkeitschwall*] and to better reveal its physical core" (Pauli 1979, p. 100).

5. A translation of Born et al. (1926) (which however omits chap. 4) can be found in van der Waerden (1967), pp. 277–306. I have used this translation. The book also contains a letter of Jordan to van der Waerden indicating who did what, and when.

6. It can be readily be shown that in classical electromagnetic theory a fluctuation of the energy in the chamber of volume v arises from a kind of beat phenomenon. Its means square $< \Delta E^2 >$ satisfies

$$< \Delta E^2 > = \frac{< E^2 >}{Z_v(v)dv}, \qquad (n6.1)$$

where $Z_v(\nu)$ is the number of proper oscillations of frequency between ν and $\nu + d\nu$ in the volume v:

$$Z_v(v)dv = \frac{8\pi v^2 dv}{c^3}. \qquad (n6.2)$$

Eq. (*n*6.1) is precisely the result one would obtain if the energy distribution in the cavity behaved according to the Rayleigh-Jeans formula:

$$< E > = \frac{8\pi}{c^3} v^2 dv \, v \, kT. \qquad\qquad (n6.3)$$

On the other hand, if the cavity radiation obeyed the Wein formula

$$< E > = \frac{8\pi}{c^3} v^2 dv \, v \, hv e^{-hv/kT}, \qquad\qquad (n6.4)$$

then

$$< \Delta E^2 > = hv < E >. \qquad\qquad (n6.5)$$

For a lucid analysis of the classical case, see Tomonaga (1962), vol. 1, pp. 299–305.

7. Jordan's calculation is also sketched in Heisenberg's Chicago lectures (1930a), pp. 95–101. It was Heisenberg who in 1931 convinced himself that Jordan's result was not right (Heisenberg 1931).

8. See Mehra and Rechenberg (1982) for an account of these investigations.

9. Bohr, upon receiving Dirac's book in August 1930, wrote him that he believes "it will be standing as a monument for a most interesting epoke [*sic*] in physical science, and one which you have contributed so eminently yourself to create." Bohr to Dirac, August 29, 1930, BSC.

10. For a more extensive biographical sketch of Dirac, see Mehra and Rechenberg (1982), vol. 3. I also found Dalitz and Peierls's *Memoir* (1986) very valuable. Some of the material of the section can also be found in Schweber (1985).

11. In an interview with Mehra, Dirac recalled that after working hard over Heisenberg's manuscript "During a long walk on a Sunday it occurred to me that the commutator might be the analogue of the Poisson bracket" (Mehra and Rechenberg 1982, vol. 4, p. 5). Dirac's commitment to the Hamiltonian formalism as well as his interpretation of quantum mechanics are analyzed in a lucid paper by de Maria and La Teana (1981, 1982). Incidentally, as late as 1982 Dirac insisted that the correct approach to relativistic quantum field theory was along Hamiltonian lines: "Physicists should not be working with a falsification of the Heisenberg equations. I have spent many years looking for a good Hamiltonian to put into the theory and have not found it. I shall continue to work on it as long as I can, and other people, I hope, will follow along the same lines. Some day people will find the correct Hamiltonian, and then there will be some new degrees of freedom, something we cannot understand according to classical ideas, playing a role in the foundations of quantum mechanics" (Dirac 1984b).

12. A postcard to Dirac in 1928 from Oppenheimer, who was aboard the *R.M.S. Franconia* on his return to the United States, gives an indication of both Dirac's standing and the intellectual climate:

> Dear Dirac,
>
> ...Jordan has just finished the fourth of his new papers but as yet he does not understand it well enough himself to see whether it is important.

I asked him to write you about it.

> With the best wishes
> Robert Oppenheimer.

(DSC 3/1, Churchill College). The Dirac Archive materials are now at Florida State University.

In the Dirac Archive at Churchill College (Box 3/1) there is a beautiful handwritten letter by Lorentz dated June 9, 1927, inviting Dirac to Leiden: "Many dutch physicists, in the University of Leiden and elsewhere are much interested in the development of quantum mechanics and in your contributions to it. So it has occured to them, that it would be greatly to their benefit if you could stay for a certain time in Leiden, for the purpose of discussing your theory and questions connected with it with those who work on the subject."

13. At the symposium in theoretical physics of the University of Michigan in the summer of 1929, Dirac gave a course called "Advanced Quantum Mechanics." Its content was listed in the catalog as follows: "General transformation theory with deduction of the wave equation. Applications to collisions and emission and absorption. The many electron problem with reference to the spin. Quantization of the electromagnetic field. Relativity theory of the electron."

14. The amazingly logical character of Dirac's mind and the precision of his verbal utterances is well known. For stories illustrating this facet of Dirac's personality, see Dalitz and Peierls (1986). Let me here quote only one such story reported by Condon and Shortley (1934): "When Dirac visited Princeton in 1929 he gave a seminar report on his paper showing the connection of the exchange energy with the spin variables of the electrons. In discussion following the report, Weyl protested that Dirac had said that he would derive the results without the use of group theory but, as Weyl said, all of Dirac's arguments were really applications of group theory. Dirac replied 'I said I would obtain the results without any previous knowledge of group theory.'"

15. The idea of going together on a trip to the Far East had been suggested by Heisenberg to Dirac in 1928. "It is very probable that I will go to Chicago from April 1929 to Sept. 1929; I have decided, not to go in this year; the journey for next year is not quite certain yet. Of course I would be extremely glad, if we could work together those 6 months in Chicago and bring European life into the American hurry. If you go to Chicago, I will certainly go. Perhaps we could have some pleasure from seeing beautiful parts of the country, f.e. from seeing California, which I would probably visit in July or June. Or we could go back to Europe via Japan—India or China etc. But of course you ought to do what you like best" (February 13, 1928, DSC 3/2). Dirac met Heisenberg in Yellowstone National Park and they traveled together to Japan. The ship stopped for a few hours in Honolulu and they were shown some of the island. Dirac to Van Vleck, June 7, 1973, AIP Niels Bohr Library. Some of the details of the trip, particularly their stay in Japan, are related in Brown and Rechenberg (1985). For Dirac's recollection of a visit with Heisenberg to a Japanese temple, see Dirac's introduction to Heisenberg's lecture in Trieste in 1971. The trip was not without its drama. Originally Dirac was supposed to travel through southern Manchuria, but tension between Russia and China over control of the Eastern Chinese Railway made it more advisable to travel through central Siberia. Nishina reserved a train seat for Dirac on the Trans-Siberian Railway (DSC 3/1).

16. DSC 3/3. It should be mentioned that chess was taboo in the Sommerfeld circles. Sommerfeld had explicitly forbidden the young Pauli to engage in such activities. Evidently

Dirac and Heisenberg learned to play "Go" while in Japan. In July 1930 Heisenberg wrote Dirac: "Recently I played 'Go' several times with a friend in Leipzig. I hope to beat you in Brussels in Go." The letter ended with a note stating, "Thinking night and days about quantization of space" (Heisenberg to Dirac, July 14, 1930, DSC 3/3).

17. The Royal Society had a policy that one of its Fellows (FRS) could submit papers by someone else and vouch for their accuracy and importance, obviating the need of an additional referee and thus expediting their publication. At the Bad Neuheim meeting in 1920, the German Physical Society had adopted a policy that any "reputable" physicist could submit his or her paper to the *Zeitschrift für Physik* and have it published without refereeing. This accounts for the rapid publication in this journal.

18. Incidentally, Dirac was fully aware that although he used the notation $e^{i\theta_r}$ and $e^{-i\theta_r}$, the product $e^{i\theta_r} e^{-i\theta_r}$ is not equal to 1, but is equal to

$$\begin{pmatrix} 0 & 0 & 0 & 0 & 0 & 0 \\ 0 & 1 & 0 & 0 & 0 & 0 \\ 0 & 0 & 1 & 0 & 0 & 0 \\ 0 & 0 & 0 & 1 & \ldots & \end{pmatrix}$$

Thus $e^{i\theta_r}$ and $e^{-i\theta_r}$ are not really reciprocal quantities, as the notation implies. This difficulty always occurs when one has action variables whose characteristic values do not extend from ∞ to $+\infty$.

The reason why this difficulty does not matter is that, as we shall find, $e^{i\theta_r}$ occurs in the analysis only with a factor $N_r^{1/2}$ just in front, equivalent to $(N_r + 1)^{1/2}$ just behind, thus $N_r^{1/2} e^{i\theta_r} (N_r + 1)^{1/2}$, and similarly $e^{-i\theta_r}$ occurs always in $(N_r + 1)^{1/2} e^{-i\theta_r}$ or $e^{-i\theta_r} N_r^{1/2}$. Now $N_r^{1/2} e^{i\theta_r} \cdot (N_r + 1) e^{-i\theta_r} = N_r$ which is the same as one would get if one assumed $e^{i\theta_r} e^{-i\theta_r} = 1$. Thus the notation of $e^{i\theta_r}$ and $e^{-i\theta_r}$ will not lead one into error when only the combinations $N_r^{1/2} e^{i\theta_r}, e^{-i\theta_r} N_r^{1/2}$ occur in the analysis.
(Dirac 1928f)

19. Eqs. 1.3.20a,b are of course equivalent to the familiar equations,

$$b_r \psi(\ldots N_r' \ldots) = \sqrt{n_r'} \; \psi(\ldots N_r' - 1 \ldots)$$

$$b_r^* \psi(\ldots N_r' \ldots) = \sqrt{N_r' + 1} \; \psi(N_r' + 1 \ldots).$$

20. Dirac to Bohr, February 19, 1927, BSC.

21. P. Jordan to E. Schrödinger, AHQP, reel no. 18.

22. Heisenberg to Dirac, February 13, 1928, DSC 3/2. See also Pauli's letter to Dirac of February 17, 1928, in DSC 3/11, which amplifies on Heisenberg's statements (Pauli 1979, [187], pp. 435–438).

23. The lessons learned from his study of "Volterra Mathematik" first appeared in the paper written by Pauli with Jordan during the summer of 1927 on the quantization of the electromagnetic field in the absence of charges (Jordan 1929). In Pauli's Nachlass at CERN there is a 17-page manuscript entitled "Zur Funktionalmathematik und der

'Hamilton-Jacobischen' Theorie fur Variationsprobleme, die aus die mehrfachen Integralen entspringen" (Pauli 1979, p. 386a).

24. Note that the invariance of the Hamiltonian seems to depend on a special structure of the Hamiltonian, "and that looked very suspicious" to Dirac. This was the point of departure of Rosenfeld's (1930a,b) investigation. Rosenfeld discovered an algebraic error in Pauli and Heisenberg's work. In 1932 Dirac again studied the Heisenberg-Pauli paper and wrote to Rosenfeld on June 5 that "I find it difficult to understand why their formalism is invariant under a Lorentz transformation. I can follow all their arguments except the last sentence in §5 on page 180 of Zeits. für Phys. vol 59. I should be very glad if you could explain the sentence a little more fully." (Dirac to Rosenfeld, June 5, 1932, BSC).

25. Pauli in his 1932 *Handbuch* article worked out this approach. There exists an extensive correspondence between Pauli and Dirac on the gauge-invariant formulation of quantum electrodynamics. The Pauli lectures and manuscript are in the Churchill College Archives and are reprinted in the correspondence of W. Pauli, vol. 2 for the year 1932.

26. This should not be interpreted as Dirac downgrading Bohr's enormous contribution to the development of the quantum theory from 1913 on. Dirac held Bohr in great respect. Upon receiving Bohr's letter congratulating him on the award of the Nobel Prize in 1933, Dirac responded: "Many thanks for your very nice letter—so nice that I find it a little difficult to answer. I feel that all my deepest ideas have been very greatly and favourably influenced by the talks I have had with you, more than with anyone else. Even if this influence does not show itself very clearly in my writings, it governs the plan of all my attempts at research" (Dirac to Bohr, November 28, 1933, BSC). In his Varenna lectures, Dirac stated: "I admired Bohr very much. [While I was in Copenhagen in 1926] we had long talks together, long talks in which Bohr did practically all the talking" (Dirac 1977c, p. 109).

27. In a letter to Rosenfeld, Dirac made the point that his paper was intended "to give a theory that is more closely connected with the results of observation than the preceding ones" (Dirac to Rosenfeld, June 6, 1932, BSC).

28. In 1934/35 Dirac was a member of the Institute for Advanced Study. During that year he visited Harvard and gave a lecture on "Quantum Electrodynamics" that elaborated the viewpoint he expressed in his 1932 paper. While QED is a "rather elaborate theory, which has never yet led to a practical result which had not already previously been obtained by more elementary methods, without the help of QED.... Yet Q. [uantum] Eld [electrodynamics] forms a great unifying principle, which collects together and puts in the most concise form all that is known about the interaction of charged particles with E. M. field" (DSC 2/7, Churchill College).

29. This collaboration came into being as follows: Fock and Podolsky, upon receiving a reprint of Dirac's "Relativistic Quantum Mechanics," generalized the "toy model" to three spatial dimensions and obtained "in first approximation (neglecting all relativity corrections) the Coulomb interaction with the correct sign. But there are other terms, namely infinite constants multiplied by ϵ_1^2, and ϵ_2^2 that are superfluous." They also calculated the interaction for the case of charged particles described by the Dirac equation, but obtained additional terms to the Breit interaction. The starting point of their calculation was the deduction of the commutation relations between the electromagnetic potentials. It was in clarifying these, and deducing the correct subsidiary condition in the presence of charges (in the interaction picture), that Dirac entered into the collaboration (DSC 3/4, Churchill College: Fock to Dirac, July 7, 1932; DSC 3/11: Dirac to Podolsky, November 11, 1932).

30. Bloch (1933) gave $\Phi(q_1 t_1 \ldots q_n t_n)$ a physical meaning when its argument lies in the region given by eq. (1.5.40). Namely, "$W(q_1 t_1 \ldots q_n t_n) = |\Phi(q_1 t_1 \ldots q_n t_n)|^2$ gives the relative probability that one finds the value q_1 in the measurement of the position of the first particle at the instant t_1, the value q_2 in the measurement of the particle at the instant $t_2 \ldots$"

31. See also Dirac 1977b. In that lecture Dirac is quoted as saying, "The transformation theory had become my darling, I wasn't interested in considering any theory which would not fit with my darling."

32. For that history, see Kragh (1990).

33. It is interesting to note that Dirac was not consistent as to the location of the conference. In his interview with T. S. Kuhn of May 7, 1963, Dirac stated, "I remember when I was in Copenhagen quite early Bohr asked me what I was working on, and I told him." In lectures delivered during a visit to Australia and New Zealand in August–September 1975, Dirac remembered "in particular an incident in the Solvay conference in 1927. During the interval before one of the lectures, Bohr came up to me and asked me: 'What are you working on now?' (Dirac 1978).

34. Heisenberg to Dirac, February 13, 1928, DSC 3/3.

35. Incidentally, it was Jordan who had pointed out these matrices to Pauli and who also drew his attention to their connection with quaternions (Pauli 1927b, n. 2, p. 607).

36. In his letter of February 13, 1928 (DSC 3/3), Heisenberg had asked Dirac: "Do you get the Sommerfeld formula in all approximation?"

37. The formulation of Hartree-Fock methods in terms of density matrices was another of his research programs of that period (Dirac 1930c, 1931a).

38. In 1983 Dirac stated his position as follows: "When I first got this idea, it seemed to me there ought to be symmetry between the holes and the ordinary electrons, but the only positively charged particles known to me at that time were the protons; so it seemed to me that these holes had to be protons. I lacked the courage to propose a new kind of particle . . . . In those days the climate of opinion was very much against the idea of proposing new particles. I certainly did not dare do it; so I published my idea as a theory of electrons and protons." (Dirac 1983, p. 42).

39. Bohr to Dirac, November 24, 1929, DSC. For further discussions of β-rays in this context, see the correspondence between Pauli and Heisenberg and Pauli and Klein 1930–1932 in Pauli (1985).

40. Dirac to Bohr, BSC.

41. Bohr to Dirac, December 5, 1929, DSC.

42. Dirac to Bohr, BSC.

43. Bohr to Dirac, December 23, 1929, DSC.

44. Heisenberg to Dirac, December 7, 1929, DSC.

45. Heisenberg to Dirac, December 28, 1929, DSC.

46. 3/3, Igor Tamm to Dirac, October 11, 1930, DSC.

47. The calculated frequency of occurrence of annihilation of protons (considered as "holes") with electrons must clearly be very small under ordinary conditions, "as these processes have never been observed in the laboratory." Dirac's own calculations (Dirac 1930b) as well as those of Tamm and Oppenheimer gave results much too large to be true. In fact, the order of magnitude was altogether wrong.

48. Upon his arrival to the United States in 1929, Heisenberg had communicated to Dirac that "Weyl thinks to have the solution of the $\pm$ e-difficulty" and asked Dirac to tell him "the main point of Weyl's work" (DSC). Weyl in the second edition of his book on group theory commented that according to Dirac's theory, "the mass of a proton should be same as the mass of the electron; furthermore, no matter how the action is chosen (so long as it is invariant under interchange of right and left), this hypothesis leads to the essential equivalence of positive and negative electricity under all circumstances—even on taking the interaction between matter and radiation rigorously into account" (Weyl 1930, p. 263).

49. Address to the British Association for the Advancement of Science (1929); see also DSC 3/3, Churchill College; Dirac 1930a,d. In his lecture accepting the Nobel Prize in December 1933, Dirac stated, "From general philosophical grounds one would at first sight like to have as few kinds of electromagnetic particles as possible, say only one kind, or at most two, and to have all matter built up of these elementary kinds." However, he went on, "To get an interpretation of some modern experimental results one must suppose that particles can be created and annihilated. Thus if a particle is observed to come out of another particle, one can no longer be sure that the latter is composite. The former may have been created. The distinction between elementary particles and composite particles now becomes a matter of convenience. This reason alone is sufficient to compel one to give up the attractive philosophical idea that all matter is made up of one kind, or perhaps two kinds of bricks" (*Nobel Lectures, Physics*, 1922–1941, p. 320).

50. For an informative, thorough, and insightful history of the discovery of the positron, see de Maria and Russo (1985). See also Hanson (1963) for an earlier important and influential account of the discovery.

51. The name "positron" was suggested to Anderson by Watson Davis, the director of *Science Service*. See de Maria and Russo (1985), pp. 271–272.

52. AHQP, Max Delbrück to Thomas S. Kuhn, March 13, 1962, AHQP.

53. E. C. Kemble to G. Birkhoff, March 27, 1933 (Kemble correspondence, Harvard University Archives; also AHQP, Mf 53,5).

54. See also Dirac (1952).

55. See the speeches given in honor of Dirac's seventieth birthday in Mehra (1973). See also Mehra (1972), Rohrlich (1985).

56. Ibid.

57. Dirac added that it may turn out that the work may have an application: "Then one has good luck."

58. In Fermi's first note on QED, the literature referred to is Jordan and Pauli's paper on the relativistically invariant quantization of the free electromagnetic field (Jordan and Pauli 1928) and Dirac's (1927a,b) two papers.

Chapter 2
The 1930s

1. The chapter is patterned after the presentation in my Les Houches lectures (Schweber 1985).

2. Other names could readily be added to these lists. Millikan was another member of the older generation who opposed identifying positrons with holes. The *New York Times* on November 26, 1933, featured an article by W. Kaempfert with the title "Millikan Denies Dirac Theory" (quoted in de Maria and La Teana 1981, p 276, n. 121). Uhlenbeck and Fermi were two of the younger physicists who very early accepted the identification (Fermi 1934).

3. The Γ_i are various combinations of products of Dirac matrices which differentiate between scalar ($\Gamma = 1$), pseudo scalar ($\Gamma = \gamma_5$), vector ($\Gamma = \gamma_\mu$), pseudo vector ($\Gamma = \gamma_5\gamma_\mu$), and tensor ($\Gamma = \gamma_\mu\gamma_\nu$) coupling. Fermi's original theory took the Γ_i to be $\beta\gamma_\mu$ in analogy with the electromagnetic interaction.

4. The point is that one cannot create a single fermion. Moreover, I have been slightly anachronistic. Fermi took the basic process to be $n \rightarrow p + \ell^- + \nu$, with a neutrino (rather than an antineutrino) being created in the reaction.

5. "The striking confirmation which this [Klein-Nishina] formula has obtained became soon the main support for the essential correctness of Dirac's theory when it was apparently confronted with so many grave difficulties." Bohr to Nishina, January 26, 1934, BSC; quoted in Brown (1985).

6. See Anderson's article (1983) and the other articles on the origin of cosmic ray research in Brown and Hoddeson (1983). Also Hanson (1963), Cassidy (1981), Galison (1983), and de Maria and Russo (1985). Chao was a Chinese student working in Millikan's laboratory at Cal Tech. His researches stemmed from Millikan's need to understand the absorption of cosmic rays—which he believed to be photons—in their passage through the atmosphere. Anderson's cloud chamber investigations, and his subsequent discovery of the positron, was part of that same research program. Anderson was also a student of Millikan's and he was complying with Millikan's request that he use a cloud chamber to study the stopping power of γ-rays.

7. See, however, the cautionary remarks about Dirac's recollections some fifty years after the fact in de Maria and Russo (1985), p. 267.

8. De Maria and Russo (1985, p. 282) indicate that in 1930 one article and three letters on cosmic rays were published in the *Physical Review* by four authors. By 1934 the count is eighteen articles and seventeen letters written by forty-one authors on the subject of cosmic rays and positrons.

9. Bohr to Dirac, August 29, 1930, DSC, Churchill College.

10. Ibid.

11. Throughout the decade, Pauli was obsessed with the problem of explaining the "value" of the fine-structure constant. At the end of the 1930s he suggested that its magnitude could perhaps be fixed by the requirement that the divergences of the self-energy of an electron in second and fourth order of perturbation theory cancel one another. Racah (1946) proved that this was not the case. In 1948 Pauli made a similar suggestion with respect to the vacuum polarization divergences. Jost and Luttinger (1950) disproved that conjecture.

12. For a fuller statement of Oppenheimer's pessimistic views during the 1930s, see Galison (1983).

13. This even though Dr. R. Serber "pointed out that corrections to the scattering cross section of order α resulting from the Coulomb interaction...should be considered." See footnote on page 962 of Dancoff's paper in which he stated that "the conclusions drawn below are unaffected by the presence of the Coulomb interaction."

14. Dancoff was actually the first one to dub the infinite subtraction associated with the wave function normalization a "renormalization."

15. Dancoff (1939), p. 961, n. 3.

16. A summary of the spectroscopic observation on the H_α line to 1945 is given in Lamb (1951) and will be reviewed in chapter 6. That Pasternack's work was taken very seriously can be inferred from the fact that Sommerfeld noted it in his revised edition of *Atombau und Spectrallinien* (Sommerfeld 1939, p. 820; see also Sommerfeld 1941).

17. V. Weisskopf, interview with author.

18. Abraham Pais in 1946 explored the possibility of explaining the 2S-2P difference as an observable consequence of his f-field hypothesis. The scalar f field had been introduced by Pais as a way to make the electron's self-energy finite to lowest order in perturbation theory. The static electron-proton potential in that theory was given by

$$ -\frac{e^2}{r} + \frac{f^2}{r}e^{-Kr}, $$

with $f^2 = 2e^2$ required to make the self-energy finite. The contribution of the f field thus raised the S levels of hydrogen. Pais found that with $1/K$, the range of the field (about six times the classical electron radius), he could account for the Pasternack value of 0.03 cm^{-1} for the S level shift (Pais 1946).

19. See Hacking (1983).

20. The Institute of Cinematography, the International Organization of Work, and the Institute in Private Rights were set up in Rome, and several institutes were located in Geneva (*Institut* 1947, pp. 48–49).

21. See Henri Bonnet's address to the conference in *New Theories* (1939), pp. xviii–xix.

22. The other conferences held under the aegis of the Institut, besides the four referred to, were:

> V. "The Foundations and the Method of the Mathematical Sciences," December 6–9, 1939, at the ETH in Zurich. The proceedings were published in 1941 by the Institute for Intellectual Cooperation.
>
> VI. "Magnetism," May 21–25, 1939, in Strasbourg. The proceedings were published in three volumes by the Institute.
>
> VII. "The Measurement of Ionizing Radiation," May 30–June 1, 1939, University of Groningen, Holland. The proceedings were published by the Institute for Intellectual Cooperation.
>
> VIII. "The Application of the Theory of Probability," July 12–15, 1939, University of Geneva, Zurich. The proceedings were published in 1946 by the Institute for Intellectual Cooperation.
>
> IX. "The Nomenclature and Terminology of Genetics and Cytology: August 15, 1939, Linnean Society, London. The proceedings were published in 1947 by UNESCO (*Institut* 1947, pp. 362–372).

23. Neither Russia nor Germany were members of the League. In the spring of 1936, the German education ministry had forbidden German scholars to have any contacts with any organization or event connected with the League of Nations. A similar rule was in effect for Italian scholars, in view of the sanctions imposed by the League on Italy following its invasion of Ethiopia and Eritrea in October 1935.

24. See Casimir (1983) for his recollections of these conferences, p. 88.

25. S. Goudsmit to A. Establier, August 14, 1938, Goudsmit Papers, AIP Niels Bohr Library.

26. S. Goudsmit to Bob and Jean Bacher, August 6, 1938, ibid.

27. Abstracts from letters received from invited members of the third Washington Conference on Theoretical Physics, February 15–20, 1937. Bethe Papers, 14/22/1976. Karl Hufbauer has collected the relevant materials on these conferences; they are in the AIP Niels Bohr Library.

28. Merle Tuve Papers, Box 4, Manuscript Division, Library of Congress. Also in the AIP Niels Bohr Library.

29. Ibid.

30. Science. Teller et al. 1941.

31. In the *Dreimännerarbeit*, the velocity of propagation on the string, $v = \sqrt{T/\rho}$, was set equal to 1.

32. Note that the zero point energy $1/2 \Sigma \hbar \omega_k$ is proportional to h, and therefore adding $-1/2 \Sigma \hbar \omega_k$ to H still yields the same classical Hamiltonian in the limit as $h \to 0$.

33. Stimulated by the Bohr and Rosenfeld paper and by his own investigations of the charge fluctuation in a volume v due to the normal creation and annihilation of pairs (Heisenberg 1934a), Heisenberg in March 1935 sent Pauli a 12-page memorandum about questions of observability in field operators and suggested that such studies might lead to fundamental innovations (Pauli 1985 [407] p. 386). For a criticism of the Bohr-Rosenfeld procedure, see Peierls (1963). See also Morrison (1939), Bohr and Rosenfeld (1950), Kalckar (1971).

34. The original text in English by Dirac is in DSC, Churchill College. It was subsequently translated by him and it is the latter version that appears in the proceedings of the Solvay Congress.

35. In the published version of the article, the vacuum polarization correction is given as $-\frac{4}{15\pi} \frac{e^2}{hc} \left(\frac{h}{mc}\right)^2 \Delta^2 \rho$. The correct numerical factor is $\frac{1}{15\pi}$, and was first calculated by Heisenberg.

36. Dirac to Bohr, September 10, 1933, BSC. See also the long letter Peierls wrote to Dirac.

37. Dirac to Bohr, November 10, 1933, BSC. This letter was written just as Dirac had heard that he had won the Nobel Prize. "I hope to be able to see you in December on my way to or from Stockholm." On November 28, 1933, Peierls wrote him: "Bethe told me that in London you reckoned to have an invariant method for the treatment of the polarization method in a fortnight, and as this was three weeks ago, I suppose your results are already in print." In his long letter, Peierls informed Dirac of his own results on the vacuum polarization problem that were to appear in Peierls (1934). Peierls to Dirac, November 28, 1933, DSC 3/11. See also Peierls to Dirac, December 8, 1933, DSC 3/5.

38. See Wightman (1986) and the references cited therein for this "modern" statement. It should however be noted that many of these developments were foreshadowed by Pauli, Heisenberg, and Weisskopf in the mid-thirties. See the Pauli-Heisenberg-Weisskopf correspondence for 1933–1935 in Pauli (1985).

39. A history of the treatment of the vacuum polarization problem during the 1930s that encompasses the contributions of Pauli, Heisenberg, Peierls, and Weisskopf and their students, of the theorists around Oppenheimer at Berkeley and those in Russia, remains to be written.

40. The $A(xt)$ operator in the Heisenberg picture has the following plane wave decomposition:

$$A(xt) = \sum_{k_\gamma} \epsilon_{k\gamma} \sqrt{\frac{h}{2kV}} \, e^{i(k \cdot x - kt)} \, a_{k_\gamma} + \text{h.a.},$$

where V is the normalization volume, and where λ runs over the two states of polarization $\lambda = 1, 2, k \cdot \epsilon_{k\lambda} = 0$:

$$[a_{k\lambda}, \, a^*_{k'\lambda'}] = \delta_{\lambda\lambda} \cdot \delta(k - k').$$

41. My presentation here is anachronistic. Bear in mind that until 1931 or so, calculations involving the Dirac equation made use of explicit representations of the Dirac α, β matrices. Traces were taken by multiplying out all the matrices and taking the sums of the diagonal elements!

42. For the corresponding reinterpretation of other processes, e.g., bremsstrahlung, pair production, etc., see Heitler (1945), pp. 186ff.

43. W. H. Furry to V. Weisskopf, June 21, 1934, Weisskopf Papers, MIT Archives. After Furry had examined Weisskopf's paper and had discovered Weisskopf's mistake, he went to Oppenheimer and asked him what to do. Oppenheimer told him, "You can either publish or do the noble thing." Furry wrote to Weisskopf (W. H. Furry, interview with J. Ovgaard and S. S. Schweber June 17, 1979).

44. V. Weisskopf to W. H. Furry, July 19, 1934, Furry Papers, Harvard University Archives.

45. W. Heisenberg to V. Weisskopf, November 2, 1934, Weisskopf Papers, MIT Archives.

46. R. Peierls to S. S. Schweber, May 31, 1985.

Chapter 3
The War and Its Aftermath

1. See in particular Heilbron's (1985) stimulating article on the propagation of the "Kopenhagener Geist der Quantenphysik."

2. See Brown and Hoddeson (1983) and references therein. See also Weiner (1974).

3. DuBridge (1946, pp. 52–53) commented:

> The war brought about the greatest flowering of large cooperative labo-
> ratories in history. Laboratories numbering many hundreds or even sev-
> eral thousand employees grew up for the development of radar, prox-
> imity fuses, nuclear energy and other fields. Their achievements were
> so astounding that they at once raised the question of why peacetime
> research could not be carried out in the same way.
>
> In the first place it is necessary to emphasize the fact that
> these huge war laboratories were not research laboratories...in pure
> research. They were applied physics laboratories; they were built and
> organized to develop specific weapons of war.
>
> They were built up under conditions peculiar to a war.

4. P. Morrison, "The Laboratory Demobilizes." Speech delivered at the Atomic Energy session of the *New York Herald Tribune* forum, October 9, 1946. Bethe Papers, 14/22/97b.

5. K. T. Compton, *Technology Review*, November 1945, p. 41. Quoted in Greber (1987).

6. Ibid., June 1941, p. 347.

7. In fact, V. Bush at the same convocation gave an address entitled "The Case for Biological Engineering."

8. K. T. Compton had been president of MIT since 1930. Within a few years of assuming the presidency, he had transformed MIT into a leading scientifically based technical institution.

9. The history of the Office of Scientific Research and Development is summarized in Baxter (1948). Details about the different parts of the organization are presented in a series of volumes with the common title, *Science in World War II,* published by Little, Brown and Co. in 1948. See also Dupree (1972) and Pursell (1979).

10. Various personal accounts of the Radiation Laboratory are available. See Rigden's (1987) biography of I. I. Rabi, Alvarez's (1987) autobiography, and the reminiscences of Pollard (1982).

11. H. A. Bethe to R. C. Gibbs, November 20, 1944, Bethe Papers, 14/22/97b.

12. Kroll's interview with Joan Bromberg (Kroll 1987) was in connection with her project on the history of the laser, and the one with Finn Aaserud in connection with researches on the relation between scientists and the military after World War II. I thank both of them for the opportunity of reading their interviews and for their permission to quote from them.

13. Summer sessions for advanced instruction and research in mechanics began being held at Brown in the summer of 1940 "as a contribution to the defence program of the nation by assisting essential American Industries." See "Advanced Instruction and Research Mechanics," Brown University. In summer 1943 the session lasted from June 15 to August 29, and J. D. Tamarkin and Willy Feller lectured on "Differential and Integral Equations of Mathematical Physics," and Louis Brillouin on "Principles of Mechanics," "Advanced Dynamics," and "Partial Differential Equations."

14. For a description of the apparatus, see Lamb (1946) and P. Morrison (1947).

15. But in August 1945 Oppenheimer could still write Charles Lauritsen at Cal Tech: "I proposed twice getting Rabi to the institute thinking it a good thing generally, and for us in particular a great source of strength. Has this fallen through? If so, is it a lack of money, is it a reluctance to add another jew to the faculty, is it a general feeling that he would not fit in?" (Smith and Weiner 1980, p. 299).

16. Robin S. Allan, Royal Society of New Zealand, to Jewett, October 28, 1946. NAS Archives.

17. K. K. Darrow's entry in his diary for Friday, November 16, 1945, recorded the following: "This was the 'big day' of the atomic-energy joint meeting of A. Phil. S. & N.A.S. in the U. of Pa. museum. The high spot, by any reckoning, was R. S. Stone's presentation of the 'health' activities of the projects, and J. Viner's incisive speech on political prospects & policies. JRO delivered a beautifully phrased speech of little practical consequence" (Darrow Papers).

18. The abstract of Wheeler's paper states: "Now that the broad principles of the science of nuclear transformations have become clear, and this field has found important practical application, it has become the task of science to explore the physics of the elementary particles. The purpose of this paper is to survey the outstanding problems and to discuss possible lines of investigation" (Wheeler 1946).

19. Forman (1971).

20. Let me indicate its flavor by quoting some brief passages:

> Each culture has its own possibilities of self-expression which arise, ripen, decay and never return. There is not one sculpture, one painting, one mathematics but many, each in its deepest essence different from the other, each limited in duration and self-contained."

> Western European physics—let no one deceive himself—has reached the limit of its possibilities. This is the origin of the sudden and annihilating doubt that has arisen about things that even yesterday were the unchallenged foundations of physical theory, about the meaning of

the energy principle, the concept of mass, space, absolute time, and causal natural laws generally.

Today in the sunset of the scientific epoch in the stage of victorious skepsis, the clouds dissolve and the quiet landscape of the morning reappears in all distinctness.... Weary after its striving, the Western Science returns to its spiritual home.

21. "There was...a strong tendency among German physicists and mathematicians to reshape their own ideology toward a congruence with the values and mood of that environment—a repudiation of positivistic conceptions of the nature of science, and in some cases, of the very possibility of the scientific enterprise....I am convinced and... endeavor to demonstrate that the movement to dispense with causality with physics, which sprang up so suddenly and blossomed so luxuriantly in Germany after 1918, was primarily an effort by German physicists to adapt the context of their science to the values of their intellectual environment (Forman 1971, p. 7).

22. See in this connection Serwer (1977), MacKinnon (1977), Hanle (1977, 1979), and the article by Hendry (1980) that takes issue with Forman's thesis.

23. C. Morette-deWitt, pers. comm., summer 1983.

24. The proceedings of the conference are contained in *Report of an International Conference on Fundamental Particles and Low Temperatures* held at the Cavendish Laboratory, Cambridge, July 22–27, 1946. 2 vols. (London: The Physical Society, 1947).

25. This requires an analytic continuation of S to complex values of the energy. This had been pointed out to Heisenberg by Kramers (Dresden 1988). For the early history of the S-matrix, see Cushing (1982, 1986, 1990); Dresden (1988); Rechenberg (1989).

26. In his review of Pauli's book Furry noted that beginning in the fall of 1944 a series of "rehabilitation" lectures were arranged at MIT, primarily for the staff of the Rad Lab. "There was such eagerness for 'rehabilitation' that the lectures, each of which took the heart out of a Saturday afternoon precious to people still heavily engaged in war work, drew large audiences." *Meson Theory of Nuclear Forces* reproduced the notes prepared by F. J. Carlson and A.J.F. Siegert on the first series of five lectures given by W. Pauli (Furry 1947).

27. A set of notes by Sneddon on the lectures, "Recent Developments in Relativistic Quantum Theory," which C. Møller delivered at the University of Bristol during the spring term 1946, are extant.

28. For a history of the S-matrix program, see the important work of Cushing (1986), (1990).

Chapter 4
Three Conferences

1. MacInnes Papers, 450M189, Box 9, K. K. Darrow to D. MacInnes, January 16, 1948. These comments should be contrasted with the remarks that I. I. Rabi had made shortly before the conference at a luncheon with K. K. Darrow and Willis Lamb: "The last 18 years has been the most sterile of a century in theo. [i.e., theoretical physics]." Darrow Diaries, entry for Monday, April 14, 1947.

2. R. P. Feynman to C. Weiner, oral interview, 1966, AHQP.

3. Until 1950, the presidency of the National Academy of Sciences was a nonpaying, part-time position, with the president usually spending one day a week in Washington. Frank

Jewett had been the president of Bell Telephone Laboratories, and in 1945 he was a vice-president of AT&T in charge of all research and development in the Bell System, with offices at 350 Fifth Avenue in Manhattan. For biographical information on Frank Baldwin Jewett, see Cochrane (1978), p. 382.

4. D. MacInnes to F. Jewett, October 6, 1945, MacInnes Papers, 450M189, Box 5, Folder J.

5. F. Jewett to D. MacInnes, October 9, 1945, MacInnes Papers, Box 5, Folder J.

6. D. MacInnes to F. Jewett, October 24, 1945, MacInnes Papers, Box 5, Folder J. In a memorandum that MacInnes prepared in November 1947 for Richards, the president of the NAS, he reiterated his views: "There is a need for such gatherings because the meetings of the larger scientific societies have become so large that the key men cannot readily get together and since the programs are crowded there is little or no time for discussion. Furthermore the papers and the discussions cannot be of high intellectual caliber or they will not be intelligible to most of the large audiences assembled. Such meetings usually deal with only one science, whereas many subjects under investigation involve several branches of science." NAS Post-War Conferences, D. MacInnes to Richards, November 7, 1947, NAS Archives.

7. D. MacInnes, "Autobiographical Sketch." The sketch was written in 1963 at the request of W. James King, the director of the project on the History of Recent Physics in the United States. It is on deposit in the "MS Biographies at the APS," AIP Niels Bohr Library. The sketch formed the basis of the obituary that Longworth and Shedlovsky (1970) wrote for the *Biographical Memoirs of the NAS*.

8. Longworth and Shedlovsky (1970), p. 305.

9. MacInnes Diaries, 450M189, Box 2, entries for October 20, 1944, p. 3; October 30, 1944; and January 19, 1945, p. 65.

10. NAS Post-War Conferences, F. Jewett to W. Weaver, April 23, 1946, NAS Archives. In this letter Jewett describes his meeting with MacInnes.

11. F. Jewett to W. Weaver, April 23, 1946, NAS Archives.

12. MacInnes had been studying Rojanski's book (1938) on wave mechanics, partly to master Fourier series, a topic which was of importance to his experimental studies of diffusion phenomena. MacInnes Diaries, October 18, 1944, and December 21, 1945. See, for example, entry for November 1, 1944. In a typed entry on May 5, 1945, MacInnes recorded: "The reading I have been doing in Kemble's book [Kemble 1937] is getting pretty deep. So I had better make another detour. However, I am getting the main ideas. The whole business seems to be largely a mathematical construct, but may have vaility [*sic,* validity] of a kind all the same." MacInnes Diaries, October 18, 1944–December 21, 1945, p. 110.

13. MacInnes recorded in his diary: "I told him [Darrow] about the idea I and Jewett have concerning the conferences and he agreed with some hesitation. He suggested that I write to Pauli about it." Entry for January 24, 1946, covering previous month's activities. MacInnes Diaries, 450M189, Box 2, Folder 3, p. 7.

14. D. MacInnes to F. Jewett, January 4, 1946, NAS Archives. In an entry in his diary for January 2, 1946, MacInnes recorded that "[Brillouin] said that what we are planning under the National Academy Auspices isn't far from what the Solvay Congresses did, and he made some excellent suggestions, one being that the group shouldn't be over thirty, and another that the discussion should be recorded. Both Darrow and he thought that 'The Postulates of Wave Mechanics' would be an excellent subject for a conference." León Brillouin, the respected French theoretical physicist, had become general director of the French National Broadcasting Society at the beginning of World War II and had taken refuge in the

United States after the fall of France in the spring of 1941. Since 1943 he held a professorial appointment at Columbia, and was a consultant on the development of magnetrons at the Radiation Laboratories at Columbia and MIT (see Thomas 1985).

15. K. K. Darrow to W. Pauli, January 4, 1946, Darrow Papers.

16. Ibid.

17. F. Jewett to D. MacInnes, January 14, 1946, NAS Archives.

18. *Report of an International Conference on Fundamental Particles and Low Temperatures* (London Physical Society, 1947). The conference was held in Cambridge from July 22 to July 27, 1946.

19. MacInnes Diaries, January 2, 1946–December 30, 1947, p. 6, entry for January 19, 1947.

20. Ibid., January 24, 1946, p. 6.

21. Ibid.

22. D. MacInnes to W. Pauli, January 25, 1946, MacInnes Papers.

23. See John Wheeler, Transcript of an Interview, taken on a tape recorder by Charles Weiner and Gloria Lubkin, April 5, 1967, Center for History and Philosophy of Physics, AIP. Also J. A. Wheeler's article, "Some Men and Moments in the History of Nuclear Physics: The Interplay of Colleagues and Motivations," in Stuewer (1979).

24. See Wigner (1946); Groueff (1967), pp. 304–308; Hewlett and Anderson (1962), pp. 175–180.

25. Both MacInnes and Darrow attended the NAS-APS meeting. Darrow's entry in his diary for November 16, 1945, reads: "This was the 'big day' of the atomic-energy meeting of A Phil Soc & NAS in the U of Pa museum." For Darrow's reaction see n. 17, chap. 3.

26. Darrow Papers. A copy of the letter is also in the MacInnes Papers.

27. W. Pauli and J. A. Wheeler to D. MacInnes, February 16, 1946, MacInnes Papers. See also MacInnes Diaries, February 20, 1946. The contrast between MacInnes' entry in his diary and the actual content of Pauli and Wheeler's letter is of interest. Whereas MacInnes speaks of a conference on "wave mechanics," Pauli and Wheeler were referring to "a conference of fundamental physical problems."
The letter from Bohr had been addressed to Wheeler, and in it Bohr indicated that he hoped "very much indeed that you will be able to come at that time [early 1947] and stay here for a while sufficiently long to allow us to have proper discussions about the fundamental problems in which we share so deep an interest." W. Pauli and J. Wheeler to D. MacInnes, February 16, 1946, MacInnes Papers, Box 7, Folder 8.

28. W. Pauli and J. A. Wheeler to D. MacInnes, February 16, 1946, MacInnes Papers.

29. MacInnes Diaries, entry for February 6, 1946.

30. K. K. Darrow to J. A. Wheeler, February 26, 1946 [copy], MacInnes Papers, the Wheeler letters.

31. There are other indications that some concerted efforts were being made at the time to "get out" the younger American physicists. In a letter to Van Vleck, Wigner indicated that he could not go along with the election to the NAS of some of the younger men if this meant that some distinguished and worthy older (foreign born) physicists would be bypassed. E. Wigner to Van Vleck, November 4, 1946, Van Vleck Papers, Box 10. See also n. 49 for a letter of MacInnes to Gamow written in June 1946 in which he stressed that it was very desirable that the "*younger American scientists* [italics mine] take part actively in preparing the papers for the Conference."

32. D. MacInnes to J. A. Wheeler, February 27, 1946, MacInnes Papers.

33. D. MacInnes to J. A. Wheeler, March 14, 1946, MacInnes Papers, the Wheeler letters.

34. D. MacInnes to Jewett, March 18, 1946, NAS Archives.

35. Peter Havas studied physics at the Technische Hochschule in Vienna from 1934 to 1938. After the Anschluss he went to France and worked with Thibaud and Guido Beck in Lyons. After the fall of France he came to the United States, and obtained his Ph.D. from Columbia in 1944. He taught at Columbia during the war, and went to Cornell in the spring of 1945 and stayed there for a year. His thesis (Havas 1944, 1945) was on quantum electrodynamics.

36. Washington Conference on Theoretical Physics. Bethe Papers, 14/22/1976.

37. The list is included in a letter of MacInnes to Darrow dated July 2, 1946, in which he tells him that Wheeler had sent him the names on March 19. MacInnes Papers.

38. D. MacInnes to J. A. Wheeler, March 28, 1946, Darrow Papers. A copy of this letter is also in the MacInnes Papers.

39. J. A. Wheeler to D. MacInnes, April 4, 1946, MacInnes Papers. Wheeler wrote a similar letter to Bohr.

40. D. MacInnes to J. A. Wheeler, April 16, 1946, Darrow Papers.

41. D. MacInnes to F. Jewett, April 16, 1946, Darrow Papers.

42. F. Jewett to D. MacInnes, April 22, 1946, NAS Archives.

43. MacInnes to Wheeler, May 7, 1946, Darrow Papers. The letter formalized the results of a conference between Wheeler, MacInnes, and Darrow held on May 5. MacInnes Diaries, entry for May 6.

44. J. A. Wheeler to D. MacInnes, June 7, 1946, Darrow Papers.

45. Ibid.

46. D. MacInnes to K. K. Darrow, June 18, 1946, MacInnes Papers.

47. Ibid.

48. K. K. Darrow to D. MacInnes, June 21, 1946, MacInnes Papers.

49. D. MacInnes to F. Jewett, June 20, 1946, NAS Archives. On July 3 MacInnes wrote Gamow that "although the point that a number of European workers in that field [i.e., wave mechanics] will be in this country during the coming autumn . . . carries a great deal of weight, it has appeared to Dr. Darrow and me that the original purpose of the conference *i.e.* a careful study of the foundations of wave mechanics, could not be carried out in the short interval remaining. *It has also appeared very desirable that the younger American scientists take part actively in preparing the papers for the conference without too much regard for authority and big names* [italics mine]." MacInnes to Gamow, July 3, 1946, MacInnes Papers, the Wheeler letters.

The ninth Washington Conference on Theoretical Physics was held on October 31, November 1 and 2, 1946. "The Physics of Living Matter" was the subject of discussion and was chosen in response to widespread interest among theoretical physicists in biological problems. Among the participants were G. Beadle, N. Bohr, M. Delbruck, M. Demerec, J. Edsall, G. Gamov, J. Kirkwood, F. London, S. Spiegelmann, W. Stanley, L. Szilard, E. Teller, J. von Neumann, and H. Weyl.

50. The draft for the announcement read as follows:

[encl 7 Nov 1946]

Dear

We take pleasure in inviting you to take part in a Conference on the Foundations of Quantum Mechanics, to be held at   on

>This is intended to be a Conference of an unusual and exper-
>imental type, possibly destined to be the forerunner of others of similar
>character. Our major wish is to promote free conversation and discus-
>sion of the theme of the Conference. There will be no formal papers,
>excepting three or four very brief ones which are to be so planned as to
>invite and excite discussion. Apart from these papers there will be no
>agenda, and it is not contemplated to make any record of the proceed-
>ings. Some will construe the theme of the Conference as being identical
>with Theories of Elementary Particles, others may not: we allow any
>reasonable construction. During certain hours of each day, discussions
>may take place under the chairmanship of some member of the Confer-
>ence serving as moderator; during the rest of the time, it is contemplated
>that the conferees will act quite as they please, remaining together or
>breaking up into smaller groups which may be as small as one member
>each.
>
>A list of the people invited to this Conference is given on
>another page. These will all be guests of the National Academy of Sci-
>ences at   from   to   . For people coming from dis-
>tances greater than 500 miles, the railway and lower-berth Pullman fare
>for the excess of the distance over 500 miles will be refunded by the
>Academy.

The announcement is in the MacInnes Papers, Box 9, NAS Conferences.

51. The participants included W.J.V. Osterhout, K. Cole, L. Blinks, Lorente de No—
who presented prepared papers—and D. Bronk, R. Cox, P. Debye, F. Jewett, J. G. Kirkwood,
I. Langmuir, L. Onsager, F. O. Schmitt, T. Shedlovsky, J. von Neumann, and T. Feng. For
the proceedings of the conference, see MacInnes Papers, 450M189, Box 9. The papers by
Osterhout, Blinks, and Cole, revised in the light of the discussions, were later published in
the *Proceedings of the National Academy of Sciences* 35 (1949): 547–575.

52. Darrow made the following entry in his diary for January 6, 1947: "APS morning:
λ [lunch] with Duncan MacInnes at B-P: we decided to postpone till May the conf'ce on
found'ns of wave mechanics and (in conformity with a request or hint of Jewett) to ask
MS of 3 speakers (tentatively Kemble, Opp. [Oppenheimer], one other) for future publ'
after post-conference revision & perh. [perhaps] some discussion added." Darrow Papers.
MacInnes for his part entered the following in his diary for January 6, 1947: "I had luncheon
with Darrow this noon. . . . Since a monograph is desired by Jewett we decided that it would
be well to put off the conference until late May so as to have reprints. He accepted my idea
of having Kemble of Harvard for the introductory paper, and he suggested Oppenheimer for
another. The monograph might be a real service to science as a whole." MacInnes Diaries.

53. D. MacInnes to F. Jewett, January 14, 1947, NAS Archives.

54. Darrow invited Oppenheimer on January 10, 1947. He indicated that MacInnes had
"enlisted the financial and moral support of the National Academy of Sciences through Dr.
F. B. Jewett" but that the latter felt "that to justify the support of the Academy something
printed should emerge from the deliberations. We therefore propose to ask not more than
three participants to write papers which shall be circulated in advance to all the rest, and
published later on. We would like you to be one of those three." Darrow to Oppenheimer,
Oppenheimer Papers, Box 72.

55. Darrow to MacInnes, February 4, 1947, MacInnes Papers.

56. Kemble was a respected professor of theoretical physics at Harvard, and the author of a treatise on quantum mechanics (Kemble 1937). Weisskopf is an eminent theoretician who was trained at Göttingen with Franck and Wigner, and had been Pauli's assistant from 1934 to 1936 and Bohr's in 1936/37. He had come to the United States in 1937 to a position at the University of Rochester, and during the war he was a group leader at Los Alamos. In the fall of 1945 he joined MIT. Kramers was the professor of theoretical physics at Leiden. He had succeeded Ehrenfest in 1933. He was spending the academic year 1946/47 in the United States visiting Columbia and serving as the chairman of the Scientific and Technical Committee of the United Nations' Atomic Energy Commission in New York. In spite of great difficulties, he succeeded in having his committee submit a unanimous report indicating that the control of nuclear bombs is technologically feasible (see Hewlett and Anderson 1962, pp. 593ff). He was a first-rate theoretical physicist with an international reputation, much respected by the physics committee and held in high esteem by everyone (see Wheeler's obituary of Kramers in Wheeler 1953; the *Dictionary of Scientific Biography* entry for Kramers; and Dresden 1987.) It was probably Rabi who suggested to Darrow that Kramers be asked to be one of the discussion leaders.

57. V. Weisskopf to K. K. Darrow, February 18, 1947 [copy], MacInnes Papers.

58. MacInnes Diaries, entry for February 28, 1947, p. 96.

59. MacInnes Diaries, entries for February 19, 1947, p. 90, and March 11, 1947, p. 101.

60. MacInnes Diaries, entry for March 26, 1947, p. 108.

61. K. K. Darrow to Kramers; NAS Archives; Darrow to Kramers, Oppenheimer, and Weisskopf, April 1, 1947 [copy], in RUA. When Darrow wrote Oppenheimer on March 18, 1947, to inform him of the date and location of the conference, he included a tentative list of participants and asked him "to make such additions (and subtractions) as you deem fitting." On the list were: Oppenheimer, Weisskopf, Kramers, Wigner, von Neumann, Bethe, Teller, Breit, Kusaka, E. B. Wilson, Feynman, Wheeler, Critchfield, Furry, Christy, Schiff, Hill, Schwinger, Serber, Rabi, Van Vleck.

62. K. K. Darrow to D. MacInnes, April 7, 1947, MacInnes Papers.

63. MacInnes Papers, Box 9. In the final letter sent by Darrow to the members of the conference on May 24, 1947, the participants were asked for "at least four choices as to roommates." MacInnes Papers.

64. K. K. Darrow to D. MacInnes, April 17, 1947, MacInnes Papers.

65. D. MacInnes to K. K. Darrow, April 18, 1947, ibid.

66. Ibid.

67. V. Weisskopf to K. K. Darrow, April 25, 1947, ibid.

68. K. K. Darrow to D. MacInnes, May 16, 1947, ibid. Of those who accepted, only Fermi did not attend. Although he came to New York to attend the conference, Fermi "had returned to Chicago having developed eye trouble." MacInnes Diaries, entry for June 6, 1947, pp. 128–129. Incidentally, Fermi in his acceptance letter of May 12 to Darrow informed him that Teller had not received an invitation. Teller evidently had also been invited to another conference at the same time and some confusion had ensued. On May 18 Darrow wrote Teller: "All's well that ends well—but I shall as Chairman, allot you ten minutes to explain the Great Teller Mystery, or How a conference on Quantum Mechanics in Long Island can be Indistinguishable from a Conference on Spectroscopy in Ohio." K. K. Darrow to E. Teller, May 18, 1947, MacInnes Papers.

69. K.K. Darrow to R. Serber, MacInnes Papers, Box 9, Correspondence to Serber.

70. In the fall of 1947 Oppenheimer left California under strained conditions (see, for example, Smith and Weiner 1980, pp. 298–304) to become the director of the Institute for Advanced Study. Serber was thus the only conferee from the West Coast to attend the Pocono conference (to which all the Shelter Island conferees were invited); Serber and Christy shared this distinction at the Oldstone conference.

71. Darrow's diaries record his meetings with MacInnes to discuss the preparation for the conference. On March 31, 1947, at a luncheon meeting they put the "finishing touches on the plans for the q-m conference," and on Tuesday, May 27, 1947, Darrow had "δ[dinner] with D. A. MacInnes final details of conf'ce" Darrow Papers. On May 28, 1947 MacInnes recorded in his diary: "All that remains to be done, at least by me, with regard to the conference is to see that enough transportaton is provided to get the men there and back, with the addition of two ladies. Mrs. Pauling and Mrs. Von Neumann, both of which have been added by means of polite blackmail . . . Feynmann [sic] has offered his car, but not enough address to call him up to accept." MacInnes Diaries, May 28, 1947, p. 127.

72. MacInnes Diaries, entry for May 20, 1947.

73. The bus that had been ordered developed a broken spring and had to be replaced, causing an hour and a half delay in the departure time. MacInnes Diaries, June 6, 1947. MacInnes in this same entry noted: "Also Oppenheimer, Fermi, and Rabi were conspicuously absent. Being responsible I got quite agitated, but couldn't think of anything to do about it. . . . During the meal [later in the evening at Greenport] Oppenheimer and Rabi arrived. They had come by plane."

74. MacInnes Diaries, entry for June 6, 1947, p. 128.

75. K. K. Darrow to Deming, June 9, 1947 [copy], MacInnes Papers.

76. Pers. comm., H. Feshbach, taped interview, October 1981.

77. "I have some reason to believe that your share was greater than we knew at the time . . . you may have been personally our host." K. K. Darrow to J. C. White, June 9, 1947, MacInnes Papers.

78. *New York Herald Tribune;* Stephen White, June 3, 1947. Incidentally, it was Oppenheimer who called Stephen White, possibly at Rabi's recommendation, to suggest that he come to Ram's Head Inn to cover the conference for the *Herald Tribune* (Stephen White, pers. comm.). MacInnes recorded that "we got the ten thirty ferry to the Rams Head Inn and arrived some minutes later to find the inn all lighted up to welcome us." MacInness Diaries, entry for June 6, 1947, p. 129. The following account of these proceedings was entered by Darrow in his diary probably very shortly after the conference was over:

> I went to AIP to meet the group convoked for the Shelter Island conference. At ca 3.50 I went off in a car with Van Vleck, Messrs. Worsdell & Deming of the NY State Dept of Commerce. The others followed in bus provided by the town of Greenport, at which we were given a dinner by some one—not yet clear whom!—and there were speeches of welcome by Mayor (Otis Burt), Worsdell and Chester White of the local Chamber of Commerce. To these I responded at some length (15')* and Oppenheimer briefly. Then to the Ram's Head Inn, brand-new white & lovely like an ocean liner newly put in service, and the analogy borne out by the smell of the sea. To bed at ca 12 in a double bed, rare pleasure!

John C. White, who was toastmaster and may have paid out
of his own pocket for the dinner, was an earnest, somewhat shy & somewhat religious who explained (and more than once to more than one)
that as Captain of Marines on Okinawa he had been in 5 invasions of
as many islands, expected to be in the invasion of Japan, expected to be
blotted out with his men & believed the A bomb had saved them and
him, wherefore he was grateful to us.
* JRO characterized this as a gay little speech.

MacInnes in his diary noted that throughout the speeches "the juke box [was] going
loudly all the time as is our peasant American custom." MacInnes Diaries, June 6, 1947,
p. 128.

79. MacInnes Diaries, entry for June 6 (2), 1947, p. 129.

80. Darrow to Katavolos, June 9, 1947, MacInnes Papers.

81. *New York Herald Tribune*, June 3, 1947.

82. H. Feshbach, taped interview, October 1981.

83. MacInnes Diaries, entry for June 17, 1947, p. 129.

84. MacInnes entered into his diary on June 17, 1947: "One evening [the discussions]
continued until eleven thirty."

85. Diary entry for Tuesday, June 3, 1947, Darrow Papers. This entry begins with the
following account: "At one of the meals, sitting w [with] JRO [Oppenheimer], IIR [Rabi],
HAB [Bethe] & others I remarked that while I could listen to un-understood papers on
eg radar, and circuits with a comfortable feeling of 'oh, this is just another specialty' I felt
humiliated at not following those of this conf ce, to which JRO retorted that this exemplified
the 'hierarchy' of science."

86. V. Weisskopf to S. S. Schweber, taped interview, February 14, 1981.

87. J. R. Oppenheimer to F. Jewett, June 4, 1947, NAS Archives.

88. J. A. Wheeler to D. MacInnes, June 6, 1947, in ibid.

89. J. A. Wheeler to F. Jewett, July 18, 1947, in ibid.

90. G. Breit to D. MacInnes, June 6, 1947; G. Breit to F. Jewett, June 16, 1947, in ibid.

91. MacInnes Diaries, entry for June 17, 1947, p. 129.

92. Ibid., entry for June 6, 1947, p. 128. On June 24, he recorded that he thought "the
conference was a real success."

93. J. R. Oppenheimer to Richards, December 1, 1947, Oppenheimer Papers.

94. F. Jewett to J. R. Oppenheimer, June 11, 1947, in ibid. Jewett's reaction should
also be placed in a wider context. From 1945 to 1948 the scientific community was deeply
involved in the politics connected with the establishment of a National Science Foundation.
Jewett was opposed (see Penick et al. 1972, pp. 132–134) to all the various bills that had
been introduced (Kilgore-Magnuson S 1850; Smith S 526; Thomas S 525; Mill HR 1830)
partly because he wanted a larger role for the National Academy of Sciences and the National Research Council. On August 6, 1947, Truman withheld his approval of S 526, which
had passed both the Senate and the House, and thus the bill died by a pocket veto. Frantic efforts were made to have a bill passed that would meet Truman's insistence on a presidential
appointment of the foundation's director. Jewett was one of the few representatives of the
scientific community who objected to the salvage operation. His testimony before Congress
included a reprint of Samuel Johnson's *Rambler No 91* (1751) "on the hazards to scientific
research of dependence upon government support": "The Sciences, after a thousand indig-

nities, retired from the palace of Patronage, and having long wandered over the world in grief and distress, were led at last to the cottage of Independence, the daughter of Fortitude; where they were taught by Prudence and Parsimony to support themselves in dignity and quiet" [quoted in Cochrane 1978, p. 469). See also Kevles (1978), chapters 21 and 22.

95. D. MacInnes to A. N. Richards, November 7, 1947, NAS Archives.

96. Ibid. Incidentally, MacInnes should have added to his list Weisskopf's letter to the editor, "On the Production Process of Mesons," which appeared in the same issue of the *Physical Review* as Marshak and Bethe's paper (Weisskopf 1947).

97. A. N. Richards to J. R. Oppenheimer, December 2 and December 4, 1947; Oppenheimer to Richards, December 1, 1947, Oppenheimer Papers.

98. D. MacInnes to J. R. Oppenheimer, December 6, 1947, Oppenheimer Papers.

99. The list of those attending the Pocono Manor Conference on Theoretical Physics held at the Pocono Manor Inn, Pocono Manor, Pa., March 29, 30, 31, and April 1, 2, 1948, is as follows:

H. A. Bethe, Cornell University, Ithaca, N.Y.

D. Bohm, Princeton University, Princeton, N.J.

A. Bohr, Institute for Advanced Study, Princeton, N.J.

N. Bohr, Institute for Advanced Study, Princeton, N.J.

G. Breit, Yale University, New Haven, Conn.

K. K. Darrow, Bell Telephone Laboratories, New York, N.Y.

P.A.M. Dirac, Institute for Advanced Study, Princeton, N.J.

Enrico Fermi, University of Chicago, Chicago, Ill.

H. Feshbach, Massachusetts Institute of Technology, Cambridge, Mass.

R. P. Feynman, Cornell University, Ithaca, N.Y.

W. Heitler, Columbia University, New York, N.Y.

W. E. Lamb, Columbia University, New York, N.Y.

R. E. Marshak, Institute for Advanced Study, Princeton, N.J.

A. Nordsieck, University of Illinois, Urbana, Ill.

R. Oppenheimer, Institute for Advanced Study, Princeton, N.J.

A. Pais, Institute for Advanced Study, Princeton, N.J.

I. I. Rabi, Columbia University, New York, N.Y.

B. Rossi, Massachusetts Institute of Technology, Cambridge, Mass.

J. Schwinger, Harvard University, Cambridge, Mass.

R. Serber, University of California, Berkeley, Calif.

E. Teller, University of Chicago, Chicago, Ill.

G. E. Uhlenbeck, University of Michigan, Ann Arbor, Mich.

J. von Neumann, Institute for Advanced Study, Princeton, N.J.

V. F. Weisskopf, Massachusetts Institute of Technology, Cambridge, Mass.

G. Wentzel, University of Chicago, Chicago, Ill.

J. A. Wheeler, Princeton University, Princeton, N.J.

E. P. Wigner, Princeton University, Princeton, N.J.

100. J. R. Oppenheimer to A. N. Richards, April 2, 1948, NAS Archives.

101. Ibid.

102. A. N. Richards to Council of NAS, January 7, 1949, NAS Archives.

103. Ibid.

104. Here is a list of these conferences:

Subject	Chairman	Date
Bioelectric Potentials	D. A. MacInnes	December 13–14, 1946
Quantum Mechanics	K. K. Darrow	June 2–4, 1947
Theoretical Physics	J. R. Oppenheimer	March 29–April 2, 1948
Low Temperature Physics	J. C. Slater	May 31, June 1–2, 1948
Fundamental Physics	J. R. Oppenheimer	April 11–14, 1949
The Gene	B. P. Kaufmann	May 30, June 1–2, 1949
The Ultracentrifuge	D. A. MacInnes	June 13–15, 1949
Evolution of the Earth	L. B. Slichter	January 23–25, 1950
Statistical Mechanics of Irreversible Processes	J. C. Kirkwood	June 6–8, 1950
Mechanics of Immunity	M. Heidelberger	June 13–15, 1950
Wave Machanics Theory of Valence	R. S. Mulliken	September 8–10, 1951

From MacInness Papers, Box 9, NAS Conferences.

In his report on the conference on bioelectric potentials, which introduced the papers by Osterhout, Blinks, and Cole, that were published in vol. 35, no. 10, of the *Proceedings of the National Academy of Sciences* in October 1949, MacInnes indicated that five conferences had taken place thus far and gave the following assessment:

> The conferences have been organized so as to bring together active workers in order to consider current research problems in a selected field of science. Formal presentation of papers has been avoided, the emphasis being on leisurely and informal discussion. The number of participants in each conference has been limited to 25. After the first conference, which was held in New York City, the remaining ones have been held at country inns.
>
> It has been found that much of value may be accomplished in two or three days of close association by key workers, who are frequently widely scattered geographically. Concentration on a given topic, the absence of distracting interests and the opportunity of free intimate discussions which is made possible by a small group, are particularly helpful.

105. J. R. Oppenheimer to A. N. Richards, January 4, 1949, NAS Archives.
106. Ibid.
107. The participants of the Oldstone Conference on Theoretical Physics were:

H. A. Bethe, Cornell University, Ithaca, N.Y.
Aage Bohr, Columbia University, New York, N.Y.
Gregory Breit, Yale University, New Haven, Conn.
Robert F. Christy, California Institute of Technology, Pasadena, Calif.

Freeman J. Dyson, Institute for Advanced Study, Princeton, N.J.

R. P. Feynman, Cornell University, Ithaca, N.Y.

Herman Feshbach, Massachusetts Institute of Technology, Cambridge, Mass.

Willis E. Lamb, Columbia University, New York, N.Y.

Robert E. Marshak, University of Rochester, Rochester, N.Y.

Robert Oppenheimer, Institute for Advanced Study, Princeton, N.J.

Abraham Pais, Institute for Advanced Studies, Princeton, N.J.

George Placzek, Institute for Advanced Study, Princeton, N.J.

I. I. Rabi, Columbia University, New York, N.Y.

Bruno Rossi, Massachusetts Institute of Technology, Cambridge, Mass.

Julian Schwinger, Harvard University, Cambridge, Mass.

R. Serber, University of California, Berkeley, Calif.

Edward Teller, University of Chicago, Chicago, Ill.

G. E. Uhlenbeck, Institute for Advanced Study, Princeton, N.J.

John von Neumann, Institute for Advanced Study, Princeton, N.J.

V. F. Weisskopf, Massachusetts Institute of Technology, Cambridge, Mass.

Gregor Wentzel, University of Chicago, Chicago, Ill.

J. A. Wheeler, Princeton University, Princeton, N.J.

Eugene P. Wigner, Princeton University, Princeton, N.J.

Hideki Yukawa, Institute for Advanced Study, Princeton, N.J.

Invited to the conference, but not attending:

Felix Bloch, Stanford University, Stanford, Calif.

Enrico Fermi, University of Chicago, Chicago, Ill.

C. Møller, Purdue University, Lafayette, Ind.

108. J. R. Oppenheimer to A. N. Richards, April 15, 1949. NAS Archives.

109. Copies of the "abstracts" that Kramers, Oppenheimer, and Weisskopf prepared are in the Oppenheimer papers, and in the MacInnes Papers, Box 9.

110. R. Serber, taped interview by S. S. Schweber, March 2, 1981. It is of interest to note that on Tuesday, October 2, 1945, Darrow made the following entry into his diary: "I spoke briefly with JRO [Oppenheimer] over the telϕ, greeting & polite questions only. On my asking whether the severe overwork of these years had affected him, he replied that he was 'mentally ruined but physically sound' adding 'aren't we all?' or words to that effect." Darrow Papers.

111. H. W. Lewis, pers. comm.; taped interview by S. S. Schweber, January 16, 1982.

112. For an overview of the history of meson physics in the 1930s, see Brown and Hoddeson (1983), "Introduction." See also the proceedings of the *International Colloquium on the History of Particle Physics* in the supplement to *Journal de Physique*, fasc. 12, tome 43, December 1982.

113. The Coulomb repulsion between a slow positively charged meson and the positively charged nucleus would prevent it from approaching the nucleus and being absorbed. A positive meson slowed down in matter would therefore wander around until it decayed. For a negatively charged meson the Coulomb attraction would favor its close approach to the positively charged nucleus and thus enhance the possibility of nuclear absorption.

114. V. Weisskopf to H. A. Bethe, May 19, 1947, Bethe Papers, 14/22/76.

115. V. Weisskopf to H. A. Bethe, May 1, 1947, Bethe Papers.

116. H. A. Bethe to V. Weisskopf, May 14, 1947, in ibid., Box 12.

117. V. Weisskopf, tape-recorded interviews, February 14 and March 10, 1981. Schwinger in Brown and Hoddeson (1983), p. 331.

118. J. Blatt to S. S. Schweber, September 27, 1984. John Blatt, who was a postdoctoral fellow at MIT from 1946 to 1952, asserts "very firmly" that "Weisskopf deserves very much more credit than he is accorded usually. He had a student (Bruce French) working on the actual calculation of the Lamb shift when *I got to MIT*, well BEFORE Lamb got any experimental results . . . Weisskopf . . . had developed an entirely correct method for calculating it. . . . Before the Lamb experiment, there seemed to be no reason to put pressure on a graduate student to do a quick calculation, so Bruce French was left to go at his own speed (nor has Viki Weisskopf ever been noted as a fast and accurate calculator himself)."

119. Breit Papers, Notebook, 1946, Box 18, pp. 72–87. I have interviewed or requested information from all the participants who were alive in 1983. Their memory of the event, however, is vague and hazy. Although everyone has vivid general impressions, no one remembers exact details. H. Feshbach, who did take notes, destroyed them at some later stage since he didn't think they would be of any further interest to him. R. Serber and H. Bethe could not locate the notes they had taken, and they are presumably lost.

120. *New York Herald Tribune*, June 3, 1947.

121. Breit Papers, Notebook, 1946, p. 72.

122. Darrow Diaries, entry for June 2, 1947, Darrow Papers.

123. MacInnes recorded in his diary: "After a night's sleep and a good breakfast the conference got down to work, and went off without any delay or explanation after a halting speech from me. It was immediately evident that Oppenheimer was the moving spirit of the affair and the preliminary agenda were hardly followed at all. The subjects seemed to be all of the difficulties of modern physics. The main discussions were by Kremers, Fineman, Lamb, and by Oppenheimer. One evening they continued until eleven thirty. It was all pretty much above my head by I got a bit here and there that was valuable, at least enough to justify my doubts about the quantum theory in its present form." MacInnes Diaries, June 17, 1947, p. 129.

124. Breit Papers, Notebook, 1946, p. 82.

125. Lewis, Oppenheimer, and Wouthuysen (1948), p. 1271, indicate that "it was recognized at the Shelter Island Conference that the most natural interpretation lay in the existence of 'structure' either for the nucleons or the mesons, or both. Thus Weisskopf suggested the existence of long-lived metastable states for the nucleons formed during the primary collision and subsequently decaying with a long lifetime to yield the mesons; whereas Marshak proposed the equally satisfactory view that the mesons originally created were metastable with regard to disintegration into those actually observed, as might, for example be the case if they had a higher mass."

126. See part 3 of Brown and Hoddeson (1983) and in particular Hayakawa's contribution to that volume detailing the development of meson physics in Japan.

127. This is to be compared with an estimated absorption lifetime of about 10^{-19} sec for a meson responsible for nuclear forces (i.e., for a π-meson).

128. The details of Feynman's lecture will be presented in chapter 8.

129. Darrow Papers, diary entry for June 4, 1947.

130. See, for example, "Inside AIP," vol. 6, no. 26, December 19, 1967, where a letter of Bethe to Darrow of November 15, 1967, is reprinted in which he informs him that "one problem . . . I did solve on a train; that was the Lamb shift. It was done between New York and Schenectady after the Shelter Island conference in 1947." See also Bethe's lecture

"Pleasure from Physics" given in June 1968 at the International Centre for Theoretical Physics in Trieste and published in *From a Life of Physics* (Salam 1969), where he states: "There was ... one paper which I did do on a train, and that was the Lamb shift."

131. In his paper, Bethe acknowledged that during the discussions at Shelter Island, "Schwinger and Weisskopf, and Oppenheimer ... suggested that a possible explanation might be the shift of energy levels by the interaction of the electron with the radiation field" (Bethe 1947, n. 6, p. 339).

132. W. Lamb, interview with S. S. Schweber, January 18, 1982.

133. Bethe to Oppenheimer, June 9, 1947, Oppenheimer Papers.

134. Bethe (pers. comm.) regrets not mentioning Kramers' contribution to the conference in his 1947 paper. Kramers is acknowledged in his unpublished report to the 1948 Solvay Congress on the Lamb shift. It should be recalled that Oppenheimer in 1930, using a pre-hole-theoretic formulation of quantum electrodynamics, had investigated whether energy level *differences* might be finite. The result was negative because of the linear divergence of the electron self-energy in that theory. Lamb and Kroll's calculation of the Lamb shift in 1947–1948 is in fact an updated hole-theoretic version of Oppenheimer's paper. Incidentally, Oppenheimer's work in 1946 with his students Wouthuysen and Lewis, on multiple productions of mesons, was an adaptation of the approach of Pauli and Fierz (1938) to (pseudoscalar) meson theory. The mass renormalization procedure used by Bethe (1947) in his calculation of the Lamb shift is closely related to that used by Pauli and Fierz in their classic paper.

135. V. Weisskopf to H. A. Bethe, June 17, 1947, Bethe Papers. "Es möchte doch schön sein," is an expression that Bohr often used. It is idiomatically incorrect. A native speaker would say "Es würde doch schön sein."

136. Nonetheless, as late as August 29 Weisskopf would write Oppenheimer, who was then vacationing at Ram's Head: "I didn't like Hans' first version and I hope that he did not send it to the P. R. It was too *half cooked.* And his conclusions *ubereilt.*" V. Weisskopf to J. R. Oppenheimer, August 29, 1947, Oppenheimer Papers.

137. V. Weisskopf to J. R. Oppenheimer, August 19, 1947, Oppenheimer Papers. Weisskopf had visited Ann Arbor that summer and had learned of Kramers' derivation from him. Weisskopf's further conclusions in his letter to Oppenheimer that "the log x shows, that an expansion in powers of x would never give a finite result. *This,* I think, is the explanation of Bethe's divergence" were wrong.

138. Lamb, in Brown and Hoddeson (1983), p. 324.

139. Schwinger, in Brown and Hoddeson (1983), pp. 332–334. M. L. Goldberger to H. A. Bethe, October 13, 1947; H. A. Bethe to M. L. Goldberger, October 20, 1947. Bethe Papers, Box 10.

140. Feynman Papers.

141. For a masterly nontechnical overview of the problem of the divergences in quantum field theory, see Weinberg (1977). See also Schwinger's "Preface" in Schwinger (1958), and Weisskopf (1983) for a more personal account. Other accounts—of a more technical nature—are in Pais (1947, 1972), Wentzel (1960), and Jost (1972). See particularly Weisskopf (1949) for an *anschaulich* account of the 1947–1949 developments.

142. Schwinger, in Brown and Hoddeson (1983), pp. 334–337.

143. Ditto notes, titled "Conference on Physics—Pocono Manor, Pennsylvania, 30 March–1 April 1948 sponsored by the National Academy of Sciences," were prepared informally by J. A. Wheeler. On the cover sheet he indicated: "There has been no opportunity to check them with the participants for accuracy or completeness." They are dated Princeton, April 2, 1948. The dittoed notes of the Pocono conference were widely circulated. The letter Jauch wrote Wheeler was characteristic: "A few days ago I received a set of notes which you took at the Pocono Manor Conference on physics. You can hardly imagine how much I appreciate having them." J. M. Jauch to J. A. Wheeler, May 17, 1948. Oppenheimer Papers, Box 72. The issuance of a set of notes at Pocono was a reaction to the fact that no proceedings had been published of the Shelter Island conference. As Jauch wrote Wheeler: "I would like to ask you whether it would not be possible to publish proceedings of such meetings in the future. I am sure that the cost of such publication would be gladly paid by the receivers of these proceedings." I obtained a set of the Pocono conference notes while I was a graduate student at Princeton in 1950. A xerox copy has been placed on deposit at the AIP Niels Bohr Library. For a summary of the Pocono conference, see Feynman (1948d).

144. Oppenheimer Papers.

145. The Lamb-Retherford experiment was featured in a front page article by William L. Laurence in the *New York Times* on September 21, 1947. The headline announced "RADAR FORCE IN ATOM." Laurence reported that the experiment changed "fundamental ideas about the nature and motion of electrons, ideas forming the keystone of current atomic theory." The official announcement describes the discovery as "the most significant advance in fifteen years in knowledge of the atom." This statement is attributed to Rabi in the *Newsweek* report on the Lamb-Retherford experiment. Both *Newsweek* and *Time* magazine featured the experiment in their September 29, 1947, issue. The *Newsweek* article (p. 60), however, was the only one that gave a quantitative indication of the magnitude of the shift. It explained that "according to present formulas, two of the many 'energy levels' for the electron in a hydrogen atom were supposed to be exactly equal. The Columbia experiment showed they were not, the difference being evidenced in the response of the atoms when 'tuned in' to the radio microwaves on a wave length of 2.74 centimeters." See chapter 8 for an extensive discussion of the theoretical work carried out in Japan and Switzerland.

146. See Schwinger's lecture, "Two Shakers of Physics: Memorial Lecture for Sin-Itoro Tomonaga," in Brown and Hoddeson (1983), pp. 336, 354–375, and Tomonaga's Nobel lecture in 1966, reprinted in Mehra (1973), pp. 404–412.

147. This letter is in the Oppenheimer Papers, as is Tomonaga's letter to Oppenheimer, which was dated April 24, 1948.

148. J. R. Oppenheimer to Tomonaga, night letter cable, April 13, 1948, Oppenheimer Papers, Box 72.

149. Tomonaga to J. R. Oppenheimer, May 14, 1948; R. E. Graves to J. R. Oppenheimer, May 26, 1948, Oppenheimer Papers.

150. F. J. Dyson to his parents, April 11, 1948.

151. A set of "rough notes" by J. A. Wheeler and E. P. Wigner, on the Oldstone conference, April 11–14, 1949, were prepared by J. A. Wheeler and written up by A. S. Wightman. They are among Oppenheimer's papers in the National Archives. A separate card in the file has a typed note (presumably from Oppenheimer to Wheeler) which states: "Pais has seen these; and would like to talk to you about them. He is not impressed with them; and he thinks it is a little late for publishing them." They were never issued nor circulated.

152. Even at Pocono, Bohr commented after Schwinger's presentation: "One may not be able to treat all physical problems without a fundamentally new idea." Pocono Notes, p. 25, AIP Niels Bohr Library.

153. For Dyson's account of these developments, see Dyson 1979. See also chapter 9.

154. Heisenberg (1938b), p. 99: "The passage from the pre-relativistic to the relativistic theory or from the pre-quantum to the quantum theory was not so much a correction of the older theories but the recognition that upon ascribing finite values to c and h the possibility of visualising physical phenomena in terms of the concepts of daily life must in part be abandoned. Thus in the relativity theory the finite velocity of light precludes the introduction of an absolute time independent of the state of motion of the observer while in the quantum theory the finite reaction of the observer upon the objects observed makes illusory every attempt of measuring the position and velocity of a particle simultaneously with any desired accuracy."

155. With characteristic hubris, physicists could declare a decade later—after super-conductivity had been understood—that all of atomic and macroscopic physics, except for gravity, was explained "in principle."

Chapter 5
The Lamb Shift

1. Dirac in his interview with T. S. Kuhn recalled:

> I might tell you the story I heard from Schrödinger of how when he first got the idea for his equation, he immediately applied it to the behavior of the electron in the hydrogen atom and then got results that did not agree with experiment. The disagreement arose because at that time it was not known that the electron has a spin. That of course was a disappointment to Schrödinger, and it caused him to abandon the work for some months. Then he noticed that if he applied the theory in a more approximate way, not taking into account the refinements required by relativity, to this rough approximation his work was in agreement with observation.
>
> I think there is a moral to this story, namely that is more important to have beauty in one's equations than to have them fit experiment. If Schrödinger had been more confident in his work, he could have published it some months earlier, and he could have published a more accurate equation.

2. For example, a molecule with no net electronic angular momentum but with a nucleus of spin I and a moment μ_N has a resonance frequency $f_N = g_N \mu_o H / h$ where $|g_N| = \mu_N / \mu_o I$. An atom whose nucleus has spin 0, whose electronic state is a $^2S_{1/2}$ state, has a resonance frequency $f_s = g_s \mu_o H / h$ where $|g_s| = 2\mu_s / \mu_o$.

3. W. E. Lamb, oral interview with S. S. Schweber, January 18, 1982. For other biographical materials on W. E. Lamb, see Franken (1978) and Lamb (1983).

4. Lamb recalculated the 2s lifetime and detected a small error in Breit and Teller's computation. The error is as follows: "One has $\vec{p} \cdot \vec{A}$ in the interaction energy, and uses the equations of motion to replace the momentum $\vec{p}$ by $m\vec{r}$ times a frequency. Now this frequency is determined by the energy difference between atomic states $E_{n'} - E_{n''}$, and not by the frequency of the photon emitted." Lamb found a value of 1/10 second for the lifetime, in comparison to the 1/7 quoted by Breit and Teller. W. E. Lamb to G. Breit, August 4 and August 8, 1947, Breit Papers, Yale University Archives. See also Wheeler (1947).

5. Oral interview, W. E. Lamb with S. S. Schweber, 1982.

6. Ibid.

7. N. Ramsey, oral interview with S. S. Schweber, December 2, 1986.

8. See "The Reminiscences of Norman F. Ramsey," Oral History Research Office, Columbia University, 1962.

9. G. Breit to I. I. Rabi, May 27, 1947, Breit Papers.

10. I. I. Rabi to G. Breit, July 18, 1947. In a postscript to his letter to Breit, Rabi indicated that "L. H. Thomas had taken a look at the problem and he claims that he is stuck and had practically given it up."

11. G. Breit to I. I. Rabi, February 20, 1948, Breit Papers.

12. Ibid., September 24, 1947.

13. I. I. Rabi to G. Breit, February 24, 1948, in ibid. The exchange of letters between Breit and Rabi occurred in early February 1948.

14. I. I. Rabi to G. Breit, December 12, 1947, in ibid. A handwritten footnote tells that "Schwingers result is $\mu = \mu_o(1 + \frac{\alpha}{2\pi})$."

15. G. Breit to I. I. Rabi, February 20, 1948, Breit Papers.

16. I. I. Rabi to G. Breit, February 24, 1948, in ibid.

17. H. A. Bethe, "Notes for Advanced Quantum Mechanics" (1946). Bethe Papers, Box 1.

18. Ibid.

19. Tape-recorded interview, H. A. Bethe with S. S. Schweber, 1983

20. Equivalently, assume that the Hamiltonian is reexpressed in terms of the observed mass so that it becomes

$$H = \frac{p^2}{2m_o} + H_{rad} + \frac{e}{m_o c}p \cdot A$$

$$= \frac{p^2}{2(m - \Delta m)} + H_{rad} + \frac{e}{(m - \Delta m)c}p \cdot A$$

$$\approx \frac{p^2}{2m} + H_{rad} + \frac{e}{mc}p \cdot A + \frac{p^2}{2m}\frac{\Delta m}{m} + 0(e^3),$$

where $\Delta m = \mu$ is the self-energy of a free-electron. To order e^2 (eq. 5.6.3) is the contribution of the $\frac{e}{mc}p \cdot A$ term; to this must be added the first-order contribution of the $\frac{p^2}{2m}\frac{\Delta m}{m}$ term.

21. Note the level shift is of order $\alpha^3 Ry$, where α is the fine structure constant, $e^2/\hbar c$. According to eq. (6.11) the effect should depend on the principal quantum number as $1/n^3$, while the dependence on the nuclear charge is more complicated. There is a Z^4 dependence modified by decrease of the Bethe logarithm with increasing Z. The average excitation $(E_n - E_o)_{AV}$ was found by Stehn and Stewart to be about 16.7 Ry for hydrogen and 16.72^2 Ry for a nucleus of charge Z. Bethe's formula predicts that a state of He^+ should have an effect 13 times greater than the state of the same n in hydrogen. Mack and Austern (1947) found a ratio of 11.5 ± 2.

22. E. Fermi to E. A. Uehling, September 26, 1947, Uehling Papers, University of Washington Archives. See also E. A. Uehling to E. Fermi, September 16, 1947, in ibid.

23. Bethe's calculation, including recoil, can be found in his unpublished report to the Solvay Congress of 1948. Kroll and Lamb, in their paper on the Lamb shift, pointed out in a footnote that "if the non relativistic theory is taken seriously to such an extent that retardation and recoil energy in the denominators are taken seriously the dynamic self-energy diverges only logarithmically and the $S - P_{1/2}$ level shift converges, and in fact with K determined to be $\overline{K = 2mc^2}$. The resulting shift of 1134 mc is in disagreement with the observations." Kroll and Lamb (1949), p. 338, n. 1

24. E. Fermi to E. A. Uehling, September 26, 1947. Uehling Papers.

25. M. L. Goldberger to H. A. Bethe, October 13, 1947, Bethe Papers.

26. H. A. Bethe to M. L. Goldberger, October 20, 1947, in ibid. On October 21, 1947, Bethe wrote Nathan Rosen: "I have done the relativistic calculation, but have not quite finished it. The calculation shows that the level shift is indeed finite and that the most important term is the one I gave in my short paper in the Physical Review.

"There are, however, a large number of additional terms which will add some numbers to the log in my formula. These terms have not been calculated but are being calculated by one of my students. Several other people have confirmed the result that the expression converges." H. A. Bethe to N. Rosen, October 21, 1947, in ibid.

27. H. A. Bethe to V. Weisskopf, October 28, 1947, in ibid.

28. Notes on Cosmic Ray Conference, September 10–12, 1947, compiled and edited by W. T. Scott, BNL-I-4, Brookhaven National Laboratory, Associated Universities, Upton, L.I., N.Y. The foreword by W. T. Scott notes that "a session devoted to a discussion by H. A. Bethe and W. E. Lamb, Jr. of the electromagnetic energy shift was omitted from these notes because it was beyond the main subject of the conference." Although a magnetic wire recording of the entire proceedings was made, it evidently is no longer extant.

29. "Lennox has finished the first calculation on the self-energy and has shown that the finite terms are not invariant. But they can be made invariant by a shift of the center of the sphere in momentum space over which the integration is taken." H. A. Bethe to V. Weisskopf, November 5, 1947, Bethe Papers.
Bethe also told Weisskopf: "Feynman and I are having a lot of fun on these problems. He has shown that the usual elimination of the longitudinal waves makes the resulting expression less invariant. If one does not eliminate them, one gets a much more plausible looking and simpler result for the self-energy which can be made invariant by using my cut-off procedure."

30. H. A. Bethe to V. Weisskopf, November 13, 1947, Bethe Papers.

31. V. Weisskopf to H. A. Bethe, November 17, 1947, in ibid.

32. "Rabi writes me that Schwinger has the theory connecting line shift and hyperfine structure. Is it true?" H. A. Bethe to V. Weisskopf, November 13, 1947, in ibid.

33. V. Weisskopf to H. A. Bethe, November 17, 1947, in ibid.

34. H. A. Bethe to H. J. Bhabha, December 12, 1947, in ibid.

35. H. A. Bethe to V. Weisskopf, November 21, 1974, in ibid.

36. T. A. Welton to S. S. Schweber, January 17, 1984.

37. V. Weisskopf to J. R. Oppenheimer, December 22, 1947, Oppenheimer Papers. In an undated letter written roughly at the same time to Bethe, Weisskopf indicated that his Coulomb calculation yields an anomalous moment of the electron of $(1 + \frac{e^2}{6\pi\hbar c})$. "I am still very unhappy that the g-value correction, which I get, is $\frac{1}{3}$ of the one Julian gets in a pure magnetic field." V. Weisskopf to H. A. Bethe, undated, but very likely late December 1947. Bethe Papers.

38. V. Weisskopf to J. R. Oppenheimer, December 22, 1947, Oppenheimer Papers.

39. French and Weisskopf, however, correctly incorporated the "exchange" contributions that Uehling had not considered (Schwinger and Weisskopf 1948).

40. "I define, more or less arbitrarily, the line shift as the difference between the self-energy W_s of the actual state in question, and the self-energy W_{free} of a wave packet ψ_k of a *free* electron, which is [at a given time t_0] identical with the eigenfunction ψ_s of the actual state. Thus ψ_s and ψ_{free} differ only in their time dependence at t_0 not in their space dependence. This calculation can be carried out straightforwardly by using *Hole* theory. It is slightly cumbersome but is definitely finite." V. Weisskopf to R. J. Oppenheimer, August 29, 1947, Oppenheimer Papers.

41. F. J. Belifante to V. Weisskopf, December 27, 1948. I thank Dr. French for providing me with a copy of this letter.

42. H. A. Bethe, "Electromagnetic Shift of Energy Levels." Institut International de Physique Solvay, Huitieme Conseil, February 9–10, 1948. The report was submitted by Bethe for the Solvay Congress. But since he did not attend the meeting, it was not included in the published proceedings of that conference.

43. V. Weisskopf to J. R. Oppenheimer, December 14, 1948. Oppenheimer Papers.

44. R. P. Feynman to V. Weisskopf, December 17, 1948, Weisskopf Papers.

45. J. B. French to R. P. Feynman, January 6, 1949; R. P. Feynman to T. A. Welton, December 15, 1952, Feynman Papers.

46. The results of Kroll and Lamb were in agreement with those of French and Weisskopf. Their treatment differs from FW in that they do not introduce a mass operator (to do the mass renormalization) and instead omit in the computation of W all terms independent of the external electrostatic field. Their treatment of the ambiguities (due to the lack of relativistic computational methods) differed from FW in that they did not use their own methods to calculate the energy shift in a magnetic field but used the criterion that the anomalous magnetic moment in a Coulomb field should be equal to that computed by Schwinger, namely $\alpha/2\pi$.

47. J. Blatt to S. S. Schweber, September 27, 1984.

48. Ibid.

49. The footnote reads as follows: "This result [for the value of the Lamb shift] agrees with that obtained by an earlier method [Schwinger 1948b], and announced at the January 1948 meeting of the American Physical Society. However, in the previous method the contribution of the additional magnetic moment to the energy in an electric field had to be artificially corrected in order to obtain a Lorentz invariant result. This difficulty is attributable to the incorrect transformation properties of the electron self-energy obtained from the conventional Hamiltonian treatment, and is completely removed in the covariant formulation now employed. Independent calculations by J. B. French and V. F. Weisskopf [1949], as well as N. M. Kroll and W. E. Lamb, Jr. [1949], are also in agreement with Eq. (8)."

50. Lewis had been an undergraduate at New York University. He came to Berkeley in 1943 to work with Oppenheimer, but Oppenheimer was already at Los Alamos. After a year in Berkeley working with David Bohm and Joseph Weinberg, Lewis went to the navy. In 1946 he returned to Berkeley and worked with Oppenheimer on a problem in multiple meson production (Lewis 1948). Epstein, Foldy, and Wouthuysen were also graduate students at the time. Robert Finkelstein was a postdoctoral fellow.

51. S. Epstein to S. S. Schweber, October 9, 1987.

52. Ibid.

53. Interview with R. Finkelstein, December 11, 1980.

54. S. Epstein to S. S. Schweber, October 9, 1987.

55. Ibid.

56. J. R. Oppenheimer to A. Nordsieck, January 6, 1948. Oppenheimer Papers.

57. Ibid.

Chapter 6
Tomonaga

1. See Ito (1988). The hut was an auxiliary building of the Ballistic Research Institute. It was given to the Tokyo University of Education after the war. Because of the extreme housing shortage that had been brought about by the heavy U.S. bombing raids, the offices of faculty members and graduate students became their living quarters.

2. I am not familiar with the political, social, and intellectual history of Japan. Moreover, the language barrier makes it impossible for me to use primary sources and I have had to rely exclusively on secondary sources. I have found Laurie Brown's important researches on Japanese physics particularly helpful. See Brown (1981, 1985, 1986), Brown et al. (1980, 1981, 1988), and the articles dealing with Japanese physics in Brown and Hoddeson (1983). For my account of Tomonaga I have relied on Tomonaga's Nobel lecture (Tomonaga 1971, p. 712), on Schwinger (1983b), and on Hayakawa (1988b); and especially on Darrigol's brief but insightful scientific biography of Tomonaga (Darrigol 1988a).

3. See, for example, the account of Taketani (1974) of his experiences during the 1930s in Nakayama et al. (1974).

4. The colonization policy had already been promulgated before the turn of the century. It was implemented by expeditions into Manchuria, Korea, and elsewhere, and was formalized by Japan's annexation and harsh occupation of Korea in 1910.

5. Existentialism, pragmatism, etc., were rejected "because they appeared to be still too characteristically Western for application to Japan's problems." Swain, in Nakayama et al. (1974), p. xvii.

6. See Taketani (1974), and especially Low (1984).

7. Nagaoka and Ishiwara both made notable contributions to atomic physics and relativity during the first two decades of the century and had won international recognition. See the entries for Nagaoka and Ishiwara in the *Dictionary of Scientific Biography*, and Koizumi (1975) on Japan's first physicists.

8. The Institute of Physical and Chemical Research was organized during World War I to undertake "creative research in physics and chemistry ... to contribute to world progress ... to enhance national prestige ... to further the sound development of industry," and "to help the nation become independent in military supplies hereafter and to acquire self-sufficiency in industrial materials" (Itakura and Yazi 1974, p. 184). Riken opened its door on March 20, 1917. It was supported by grants from the government and by contributions from industry. It also derived some income from patents, including one for the manufacture of saké. For an account of the early history of the Institute, see Itakura and Yagi (1974).

9. When first organized in 1931, the Nishina laboratory consisted of four people: Nishina, Sagane, Tekeuchi, and Tomonaga. By 1942 it had grown to about one hundred members. See Itakura and Yagi (1974).

10. This was also true at Osaka University, where Kikuchi led an active experimental group in cosmic ray research. Yukawa had gone there in 1932 and Sakata joined him in 1934. Later in the decade a cyclotron was built at Osaka University and experiments on proton-proton scattering were carried out. Kyoto University also had a group working in nuclear physics during the thirties.

11. Its title is "Myself and Experiments in Physics" and can be found in TOM #1 issued by the Tomonaga Memorial Room of the University of Tsukuba. This pamphlet is devoted to the personality and accomplishments of Tomonaga. "TOM" was Tomonaga's informal signature. I thank Dr. Odagaki for translating this material for me.

12. From biographical material Tomonaga submitted to the Nobel Foundation on the occasion of receiving the Nobel Prize in 1965. This high school was famous for its tradition of "liberalism," in contrast to the First Higher School that cultivated a tradition of "autonomy." This legacy of the Taisho democracy came under attack by the militarists and was gradually eliminated after Hirohito became emperor in 1926, the beginning of the Showa era.

13. TOM #1 contains some of Tomonaga's drawings: of cats, trees, as well as a self-portrait. They give proof of his great talent and sensitivity.

14. Nobel biographical material.

15. Tomonaga (1978).

16. Tomonaga described the disintegration of the mesotron in terms of its making a virtual transition into a neutron-proton pair, and the latter converting into an electron-neutrino pair via the Fermi interaction.

17. It is interesting to compare this entry with the letter Darwin wrote to his wife from Moor Park in 1858, during a break while writing the *Origin*: "Yesterday, . . . , I strolled a little beyond the glade for an hour and a half, and enjoyed myself—the fresh yet dark green of the grand Scotch firs, the brown of the catskins of the old birches, with their white stems, and a fringe of distant green from the larches, made an excessively pretty view. At last I fell asleep on the grass, and awoke with a chorus of birds singing around me, and squirrels running up the trees, and some woodpeckers laughing and it was as pleasant and rural a scene as I ever saw, and I did not care one penny how the beasts or birds had been formed" (Darwin 1958, pp. 194–195).

18. Tomonaga later recalled: "When I was in Germany I had wanted to stay another year in Europe, but once I was aboard a Japanese ship I became eager to arrive in Japan." Quoted in Schwinger (1983b), p. 359. Yukawa, who had come to Europe to attend the 1939 Solvay Congress, sailed on that same vessel when the conference was canceled because of the outbreak of hostilities. The ship docked in New York, and during the stopover Tomonaga visited the World's Fair. He later remarked: "I found that I was speaking German rather than English, even though I had not spoken fluent German when I was in Germany" (quoted in Schwinger 1983b, pp. 359–360). Yukawa disembarked in New York and made his way to the West Coast, lecturing at various universities; but Tomonaga stayed aboard the ship for the voyage home through the Panama Canal.

19. For an account of Tomonaga's intermediate coupling methods, see Maki (1988). Tomonaga's original papers are reprinted in Tomonaga (1971).

20. H. Fukuda, "Professor Tomonaga and I," in TOM #1.

21. Tokyo College of Science and Literature later became Tokyo University of Education (Bunrika Daigaku).

22. Fukuda, "Professor Tomonaga and I."

23. In his article Yukawa emphasized that "the existing formalism of quantum field theory is not perfectly relativistic" (Yukawa 1942). This work was the point of departure for his researches on nonlocal theories (Yukawa 1950).

24. In the spring of 1943, after he had completed his researches, Tomonaga gave a talk at Riken entitled "Relativistically Invariant Formulation of Quantum Field Theory" for which he furnished the following abstract: "In the present formulation of quantum fields as a generalization of ordinary quantum mechanics such non-relativistic concepts as probability amplitude, canonical commutation relations and Schrödinger equation are used. Namely these concepts are defined referring to a particular Lorentz frame in space-time. This unsatisfactory feature has been pointed out by many people and also Yukawa emphasized it recently.

I made a relativistic generalization of these concepts in quantum mechanics such that they do not refer to any particular coordinate frame and reformulate the quantum theory of fields in a relativistic invariant manner" (quoted in Schwinger 1983b, p. 362).

25. Yukawa in his paper had made use of this generalized transformation function (g.t.f.).

26. In his Nobel acceptance speech Tomonaga stated: "I was recalling Dirac's many-time theory which had enchanted me ten years earlier."

27. Stueckelberg (1938b) had previously used this device.

28. T. Miyamiza, "Magnetrons," in TOM #1. I thank Dr. Odagaki for translating this article. Tomonaga's papers on the theory of split-anode magnetrons and ultra-shortwave circuits are reprinted in vol. 2 of his *Scientific Papers* (Tomonaga 1976).

29. From biographical material submitted by Tomonaga to the Nobel Foundation.

30. During the war, high school studies were compressed to two and a half years in Japan. Hayakawa was thus rather young when he entered Tokyo University in the fall of 1943 to study physics. The same was true of the other students in the class.

31. "Monologue of the Decade," reprinted in TOM #1. I thank Dr. Odagaki for translating this article for me.

32. Many of Japan's most gifted students were attracted to Tokyo University by virtue of its prestige. Among its graduates with a Bachelor of Science degree in physics were: 1939—Miyazima; 1940—Takabayashi, Kubo, Kato; 1942—Hayashi, Nambu; 1943—Ona; 1945—Hayakawa, Koba, Miyamoto; 1946—Takeda, Tani, Fukuda; 1947—Kinoshita, Suura, Fujimoto, Yamaguchi; 1948—Nishijima. (The Bachelor of Science degree was roughly equivalent to the American Master's degree. Usually a student was about 21 years old at that stage). After the war the faculty in theoretical physics at Tokyo included Kubo, who had a group of young people working with him on problems in statistical mechanics; and Nakamura, Ochiai, and Kodaira in elementary particle physics. Ochiai had worked with Heisenberg before the war. Kodaira had a degree in both mathematics and physics. During the 1946 spring semester Kodaira ran a weekly 8-hour seminar on quantum field theory. Among the students attending were Kinoshita, Hayakawa, Nishijima, and Yamaguchi; some assistants and faculty members also participated. Although sometimes the seminar heard reports of original work by the participants, most of the time was spent reading Heitler's *Quantum Theory of Radiation* and the recent literature. Hayakawa remembers reading the papers by Bloch and Nordsieck (1937), Dirac (1938b), and Pauli (1940). Heisenberg's *S*-matrix papers were also read and discussed, as Kodaira was working on the mathematics of the *S*-matrix connected with the Schrödinger theory. Kodaira had formulated a general theory of eigenfunction expansions and had communicated his results to Hermann Weyl in Princeton. Weyl by return mail sent him a copy of the book Titchmarsh had published on the subject during the war and invited Kodaira to visit the Institute of Advanced Study for a year.

33. The English translation of Tomonaga (1943a).

34. See in particular Pauli (1941).

35. *C* for cohesive field.

36. In his Nobel speech, Tomonaga stated that "the first information concerning the Lamb shift was obtained not through the *Physical Review*, but through the popular science column of a weekly U.S. magazine." According to Hayakawa (1988b), p. 57, the magazine was *Newsweek*.

37. "The method of covariant contact transformation, by which we did Dancoff's calculation over again [was found] also [to] be useful for the problem of performing the relativistic calculation of the Lamb shift." Tomonaga (1971), p. 723.

38. Oppenheimer received the letter shortly after the Pocono conference and circulated it among the conferees. The letter is reprinted in chapter 5. For an assessment of the difficulties Tomonaga and his associates had encountered in the photon self-energy problem, see Tomonaga and Koba (1947), and the note appended to that letter by Oppenheimer. The letter from Pais to Tomonaga is dated April 13, 1948, and is reprinted in *Soryūshiron Kenkyū* (1948), no. 2, Kenkyu.

39. Nambu independently carried out a Lamb shift calculation "along the line of the ordinary perturbation theory" (Nambu 1949).

Chapter 7
Julian Schwinger

1. In 1925, 1.5 million dollars was appropriated by the Board of Trustees of Columbia University for the construction and equipment of a new building to meet "the probable needs of the Department [of Physics]."It was completed and formally opened in 1928. Michael Pupin, who had been at Columbia since 1879 as both student and teacher, in 1928 made substantial gifts to the university "to support research in the physical sciences." In 1935, after Pupin's death, the laboratories housed in the building—in which research in physics, astronomy, psychology, engineering, biophysics, and optometry was carried out—were named in his honor.

2. The proceedings were printed privately. *Celebration of the Fiftieth Anniversary of the Pupin Laboratories*, Department of Physics, Columbia University, undated.

3. J. Schwinger, typewritten manuscript of "Physics in the Future—A View from the Past," a lecture delivered, on November 5, 1977, at a session entitled "Physics in the Future" on the occasion of the fiftieth anniversary of the Pupin Laboratories. Some of Schwinger's papers have been deposited in the UCLA archives. Schwinger's lecture is reprinted in the *Celebration* volume.

4. Ibid.

5. Harold Schwinger to S. S. Schweber. Taped interview, October 28, 1984. Harold is Julian's older brother.

6. Ibid.

7. I. I. Rabi at J. Schwinger's sixtieth birthday celebration at UCLA; mimeographed transcription of speeches and lectures given on that occasion. The tapes and their transcription are available at the Center for the History of Physics, AIP, New York.

8. The high school was opened in 1849 as a one-year preparatory school for the predecessor to the City College of New York, the Free Academy, that had been founded by Townsend Harris. In 1906 it became a full-fledged city high school, offering a rigorous program for intellectually gifted students. In 1942 Mayor Fiorello La Guardia ordered the closing of Townsend Harris as a "nonessential educational unit." "Many of those who went to Townsend Harris say the school provided one of the most stimulating times of their lives." Samuel Weiss, "The New Townsend Harris High School," *New York Times,* June 10, 1985, p. B5.

9. J. Schwinger. Biographical material submitted to Nobel Foundation upon receipt of Nobel Prize in 1965.

10. William Rarita, "My Early Recollections of Julian S. Schwinger," September 9, 1977, Lawrence Radiation Laboratory, LBL-6287, Berkeley, Calif.

11. Bernard Feld, speech on the occasion of Julian Schwinger's sixtieth birthday, UCLA. A copy of the speeches delivered on that occasion are on deposit at the AIP. I thank R. Finkelstein for making them available to me.

12. For example, in physics: Bernard T. Feld, Herman Feshbach, Morton Hamermesh, Robert Herman, Robert Hofstadter, William Nierenberg, Arno Penzias, Mark Zemansky; in economics: Kenneth Arrow; in biochemistry: Julius Axelrod, Ernest Borek, Philip Handler, Clifford Grobstein, Arthur Kornberg, Abraham Mazur, Harold Scheraga, Sol Spiegelman; in mathematics: Jesse Douglass, P. Franklin, Edward Kasner, Jacob T. Schwartz; in the social sciences: Daniel Bell, Lewis Mumford; in literature: Paddy Chayefsky, Alfred Kazin, Bernard Malamud; and many more equally distinguished physicians, engineers, and lawyers.

13. Taped interview of Lloyd P. Motz with S. S. Schweber, 1987.

14. B. T. Feld, speech at J. Schwinger's sixtieth birthday celebration, UCLA, 1978.

15. See J. Schwinger. In Brown and Hoddeson (1983), Schwinger lists this manuscript as no. 1 in his list of publications. The manuscript is still in the possession of its author. In the autobiographical note attached to his Nobel Prize speech, Schwinger remarked that "to judge by a first publication, I *debuted* [emphasis added] as a professional physicist at the age of sixteen."

16. Morton Hamermesh is a nuclear physicist, the author of a well-known text on *Group Theory* (Hamermesh 1962). For many years he was chairman of the physics department at the University of Minnesota.

17. I do not know whether Schwinger acquired the book later, but in any case a copy of Graustein's *Introduction to Higher Geometry* (1935) can be found on his bookshelf in his office at UCLA.

18. M. Hamermesh, "Reminiscences" delivered on the occasion of J. Schwinger's sixtieth birthday celebration, UCLA, 1978.

19. Herman Feshbach, interview with S. S. Schweber, October 12, 1988.

20. J. Schwinger, taped interview with S. S. Schweber, November 14, 1982.

21. Ibid.

22. L. Motz, taped interview with S. S. Schweber 1987.

23. Ibid.

24. While working for his Ph.D. at Columbia, Rabi had been a tutor at City College from 1924 to 1928. Schwinger evidently was failing English, and Rabi was to have said to Schwinger, "Here you are flunking English. You speak very well and you sound like an educated person." Rabi, testimonial speech at Schwinger's sixtieth birthday celebration, UCLA, 1978. In another version of this same story, Rabi has Schwinger not only flunking English, but "just about everything else" too. Bernstein (1975).

25. Besides Bethe, Uhlenbeck was also asked to write a letter on behalf of Schwinger's admission to Columbia. Bernstein (1975).

26. Rabi, testimonial speech, J. Schwinger sixtieth birthday celebration, UCLA, 1978.

27. N. F. Ramsey, "Recollections of the Thirties." Videotape of discussion among I. I. Rabi, S. Millman, N. Ramsey, J. Schwinger, and J. Zacharias; J. Goldstein, moderator. Sloan Foundation, March 29, 1984. A slightly differing account was given by Ramsey in a taped interview with S. S. Schweber, June 1987.

28. Rabi, "Recollections of the Thirties." Transcription of videotaped interview by Jack Goldstein of I. I. Rabi, J. Schwinger, N. Ramsey, S. Millman, and J. Zacharias, March 29, 1984, poll.

29. Ramsey, "Recollections of the Thirties," p. 13.

30. Ramsey, "Recollections of the Thirties," pp. 12–13. For a different version of the same story, see Jeremy Bernstein's profile of Rabi in the *New Yorker* (Bernstein 1975). The story about Schwinger is found in the October 13 article, p. 105.

31. Rabi, at Schwinger's sixtieth birthday celebration, UCLA, 1978.

32. Rarita, LBL 6187.

33. M. Hamermesh, recollections delivered at Schwinger's sixieth birthday celebration, UCLA, 1978.

34. Bethe's main objection to "Schwinger's paper in its present form [was], however, the neglection of the atomic form factor." Bethe had worked on the same problem and had formulated a "phenomenological calculation" that had been published (Bethe et al. 1986). Bethe Papers. The date of Bethe's letter, January 1, 1936, merits comment. Bethe's referee report was written four days after the paper had been received by the editor of the *Physical Review*!

35. Schwinger included the paper on the scattering of neutron by ortho and para hydrogen in his "Selected Papers" "because I and not my distinguished colleague, wrote it" (Schwinger 1979, p. xxiii].

36. Oral interview of Lloyd P. Motz with S. S. Schweber, 1987.

37. S. Devons. Opening remarks on the university convocation on the occasion of the celebrating of the fiftieth anniversary of the Pupin Physics Laboratories. Privately printed by Columbia University.

38. M. Hamermesh, Schwinger's sixtieth birthday celebration, UCLA, 1978.

39. J. Manley to S. S. Schweber, October 15, 1984. In that same letter Manley also notes: "Julian was shy in those days, but I think only with strangers and perhaps about social situations. He was not shy in talking physics, in explaining items or about his own abilities. I'd say quiet but delightfully effective. He was considerate, thoughtful, and always willing to explain his role and the physics involved. He was the opposite of egotistical, and my impression of him then [was of] a very genuine, sincere, knowledgeable person devoid of any airs which his ability might have justified."

40. Rabi, at Schwinger's sixtieth birthday celebration, UCLA, 1978.

41. J. Schwinger, taped interview with S. S. Schweber, November 14, 1982.

42. Rabi, at Schwinger's sixtieth birthday celebration, UCLA, 1978, and "Recollections of the Thirties." Upon hearing this story from Rabi, Schwinger immediately commented that "these are compatible statements." "Recollections of the Thirties," p. 19. Nor are these statements incompatible with the fact that in a paper entitled "On the Spin of the Neutron," which was submitted from Madison on November 17, 1937, Schwinger (1937c) expresses "his deep gratitude to Professors Breit and Wigner for the benefit of stimulating conversations on this and other subjects."

43. Schwinger, "Recollections of the Thirties," p. 19.

44. Oral interview with S. S. Schweber, 1982.

45. Breit Notebooks, Log of Seminars for 1937–38, Box 18, Breit Papers, Yale University.

46. Taped interview with S. S. Schweber, November 1982.

47. Rabi, "Recollections of the Thirties."

48. Hamermesh, recollections delivered at Schwinger's sixtieth birthday celebration, UCLA, 1978.

49. Rabi, "Recollections of the Thirties."

50. Schwinger, taped interview with S. S. Schweber, November 1982.

51. L. P. Motz, taped interview with S. S. Schweber. March 1983.

52. Ibid.

53. J. H. Van Vleck to I. I. Rabi, December 5, 1938, Van Vleck Papers, Box 9.

54. J. H. Van Vleck to I. I. Rabi, January 7, 1939, in ibid.

55. Schwinger, taped interview with S. S. Schweber, November 1982.

56. Rabi, "Recollections of the Thirties."

57. Schwinger, taped interview with S. S. Schweber, November 1982.

58. "I was a great reader of the literature and I was always telling [Schwinger] about interesting problems and unfortunately one day I mentioned the absorption of sound in gases and that started him off on an enormous amount of work which I don't think he ever published, as far as I can tell. But he did all sorts of calculations on this." M. Hamermesh, after-dinner speech, Schwinger's sixtieth birthday celebration, UCLA, 1978.

59. Schwinger, taped interview with S. S. Schweber, November 1982.

60. Ibid.

61. Ibid.

62. H. C. Corben to S. S. Schweber, October 15, 1984.

63. In addition to a central potential $V(r)$ between nucleons, a spin dependent potential of the form $\boldsymbol{\sigma}_1 \cdot \boldsymbol{\sigma}_2 V_2(r)$ (where $\boldsymbol{\sigma}_1$ and $\boldsymbol{\sigma}_2$ are the spins of the two nucleons and r their separation) was invoked in order to explain the differences between the triplet and single states of the nucleon system. A tensor potential is of the form $(3\boldsymbol{\sigma}_1 \cdot \hat{\mathbf{r}} \boldsymbol{\sigma}_2 \cdot \hat{\mathbf{r}} - \boldsymbol{\sigma}_1 \cdot \boldsymbol{\sigma}_2)V_3(r)$. With this potential the predominant $^3S_1$ state of the deuteron is mixed with a $^3D_1$ component, and the combination gives rise to a quadrupole moment.

64. W. Rarita to S. S. Schweber, October 24, 1988.

65. I. I. Rabi to F. D. Fackenthal, September 30, 1946, Rabi Papers, Box 1/25, folder "Columbia University, 1947." Rarita, who accompanied Schwinger on his trip east from Berkeley to Ann Arbor, tried to get him "to start early in the day to go across country in his car, but it was late, way past sundown before we were on our way. . . . One advantage of travelling at night was that we could pass the deserts comfortably cool." Rarita remembers "crossing the Sierra Nevada about two o'clock in the morning, and at one point, we almost slipped off the side of the mountain on some loose gravel. . . . At noon the next day Julian got up and we were on our way again. We arrived in Ann Arbor, Michigan, with no major accident. Julian gave his lectures, and then drove through Canada to visit his cousins. Our trip ended in New York City." To have a position of this type at the age of 23 is impressive. William Rarita, "My Early Recollections of Julian S. Schwinger," LBL 6187, September 9, 1977.

66. R. G. Sachs to S. S. Schweber, August 15, 1988.

67. Ibid.

68. Ibid.

69. See also the interview of Charles Weiner with H. A. Bethe, AIP, 1967.

70. H. A. Bethe to T. Von Karman, October 29, 1940, and November 30, 1940. Bethe Papers.

71. H. A. Bethe to T. Von Karman, November 30, 1940, in ibid.

72. Bethe to DuBridge, February 12, 1942, in ibid., Box 2, "Calculations for MIT Radiation Lab 1942."

73. Ibid.

74. "Proposal of a Defense Research Project on the Theory of Electromagnetic Microwaves," in ibid.

75. Ibid.

76. Interview with S. S. Schweber, January 15, 1982.

77. B. T. Feld at Schwinger's sixtieth birthday celebration, UCLA, 1978. Among Schwinger's papers still in his possession is a folder containing some of his work for Wigner on "slowing down of neutrons, and cooling of pile (Winter 1943)."

78. Nathan Marcuvitz, interview with S. S. Schweber, October 20, 1988.

79. Harold Levine, interview with S. S. Schweber, October 19, 1988.

80. Ibid.

81. David Saxon, interview with S. S. Schweber, October 19, 1988.

82. Schwinger, biographical material submitted in connection with Nobel lectures (Schwinger 1972, p. 173).

83. Nathan Marcuvitz, interview with S. S. Schweber, October 20, 1988. The preface to the *Waveguide Handbook* succinctly summarizes what Schwinger had accomplished:

> This book endeavors to present the salient features in the reformulation of microwave field problems as microwave network problems. The problems treated are the class of electromagnetic "boundary value" or "diffraction" problems descriptive of the scattering properties of continuities in waveguides. Their reformulation as network problems permits such properties to be calculated in a conventional network manner from equivalent microwave networks composed of transmission lines and lumped constant circuits. A knowledge of the values of the equivalent network parameters is a necessary prerequisite to quantitative calculations. The theoretical evaluations of microwave network parameters entails in general the solution of three-dimensional boundary-value problems and hence belongs properly in the domain of electromagnetic field theory. In contrast, the network calculations of power distribution, frequency response, resonance properties, etc., characteristic of the "far-field" behavior in microwave structures, involve mostly algebraic problems and hence may be said to belong in the domain of microwave network theory. The independence of the roles played by microwave field and network theories is to be emphasized; it has a counterpart in conventional low-frequency electrical theory and accounts in no small measure for the far-reaching development of the network point of view both at microwave and low frequencies. (Marcuvitz 1950, p. vii)

84. M. Hamermesh to S. S. Schweber, August 29, 1988.

85. Schwinger interview with S. S. Schweber, 1982.

86. R. P. Feynman at Schwinger's sixtieth birthday celebration, UCLA, 1978.

87. Van Vleck to Dean Paul H. Buck, May 3, 1945. Rabi Papers, Box 17, folder "Harvard University Visiting Committee"; Van Vleck Papers, Correspondence AIP.

88. J. H. Van Vleck to E. P. Wigner, December 28, 1944; Van Vleck to W. Pauli, January 13, 1945, Van Vleck Papers, Box 9, Pauli folder. M. Hamermesh to S. S. Schweber, August 29, 1988.

89. W. Pauli to J. H. Van Vleck, January 30, 1945, and April 8, 1945; E. P. Wigner to Van Vleck, March 23, 1945; L. A. DuBridge to J. H. Van Vleck, January 1, 1945; G. Breit to J. H. Van Vleck, January 14, 1945; F. Bloch to Van Vleck, January 5, 1945; J. R. Oppenheimer to P. W. Bridgman, March 19, 1945; E. Fermi to Van Vleck, January 19, 1945, and March 16, 1945. Rabi Papers, Box 17, folder "Harvard University Visiting Committee." Pauli was enormously impressed by Schwinger. In his first letter to Van Vleck, he wrote: "Dr. Schwinger is in a position to be very helpful in a laboratory, in the theoretical interpretation of experiments. In research work he is entirely independent in setting problems, having initiative of his own and originality. Moreover he is especially gifted as a teacher and speaker, and he seems to me a modest and friendly person. All these qualities make him an outstanding figure among the theoretical physicists of his age, and therefore I would know of nobody among the younger physicists whom I could recommend more or as much as Dr. Schwinger." Corben, who knew both Schwinger and Pauli, wrote in 1940/41: "When I read of [Pauli's] Nobel Prize (in 1946) I wrote from Melbourne to congratulate him. His reply indicated his disappointment that Julian Schwinger had not also written. It seemed to me that he held Julian in such high esteem that he wanted his praise more than anyone else's." H. C. Corben to S. S. Schweber, October 15, 1984.

90. J. H. Van Vleck to P. Buck, May 3, 1945. Rabi Papers. J. H. Van Vleck to W. Pauli, June 27, 1945, Van Vleck Papers, Box 9.

91. I. I. Rabi to F. D. Fackenthal, September 30, 1946. Rabi Papers, Box 1, folder "Columbia University, 1947."

92. J. S. Schwinger, taped interview with S. S. Schweber, January 15, 1982.

93. Ibid.

94. Ibid. See also Schwinger (1983a).

95. Ibid. It should also be noted that in the winter of 1946 Schwinger had carefully gone through Wentzel's *Einführung in der Quantentheorie der Wellenfelder* (1943), which had been reissued in a lithographed edition by Edwards Brothers in the fall of 1946. Wendell Furry told the author that the speed with which Schwinger had plowed through the volume and had assimilated its content had left him (Furry) depressed and very discouraged (W. H. Furry, interview with J. Ovgaard and S. S. Schweber, January 17, 1979).

96. Interview with S. S. Schweber, January 15, 1982.

97. Ibid.

98. M. Hamermesh to S. S. Schweber, August 29, 1988.

99. The notes for these calculations are extant, and have been deposited among Schwinger's papers in the UCLA Library.

100. A simple way to see this is to note that if $| i >, | f >$ are two eigenstates of H_0, with $H_0 | i >= E_i | i >$ and $E_f \neq E_i$, then $< f | [H_0, S] | i >= (E_f - E_i) < f | S | i >$ and

$$< f | S | i > = \frac{i < f | H_{rad} | i >}{E_i - E_f} = -i < f | H_{rad} \frac{1}{E_f - H_0} | i >$$

$$= \lim_{\epsilon \to 0} \int_0^\infty d\lambda < f | H_{rad} e^{-i(E_f - H_0 + i\epsilon)\lambda} | i >$$

$$= \lim_{\epsilon \to 0} \int_0^\infty d\lambda < f | e^{iH_0\lambda} H_{rad} e^{-iH_0\lambda} | i > e^{-\epsilon\lambda},$$

hence

$$S = \lim_{\epsilon \to 0} \int_0^\infty d\lambda \, e^{iH_0\lambda} H_{rad} \, e^{-iH_0\lambda} \, e^{-\epsilon\lambda},$$

from which eq. (7.4.49) follows.

101. Taped interview, N. Ramsey with S. S. Schweber, 1987.

102. The announcement for the tenth Washington Conference on Theoretical Physics was sent out on October 9, 1947. The subject was "Gravitational and Electromagnetism" and it was to be held on November 13, 14, and 15, 1947, at the George Washington Library Building. The announcement indicated that "the meeting will have no fixed agenda, in accordance with usual custom, but in general problems of the expanding universe will occupy the first day and the Schrödinger and Blackett proposals regarding the relations between gravitation and electromagnetism will be discussed along with other topics relating to unified field theory the second day." Invitations were sent out to the following:

Albert Einstein, Princeton
H. P. Robertson, Cal Tech
Hans Bethe, Cornell
R. Feynman, Cornell
E. Teller, Chicago
C. Critchfield, Minnesota
E. Fermi, Chicago
J. Schwinger, Harvard
R. C. Tolman, Cal Tech
Edwin Hubble, Mount Wilson Observatory

John Wheeler, Princeton
Robert Oppenheimer, Princeton
Horace Babcock, Mount Wilson
  Observatory
W. M. Elsasser, University of
  Pennsylvania
Gregory Breit, Yale
M. Schwarzschild, Princeton
J. von Neumann, Princeton
L. Infeld, Toronto

Those attending included Breit, Critchfield, Feynman, Wheeler, Schwinger, Teller, Oppenheimer, and Weyl. *Science*, December 12, 1947.

103. R. P. Feynman to H. C. Corben, November 1947, Feynman Papers.

104. I. I. Rabi to J. Schwinger, December 12, 1947, Rabi Papers, RICC, folder "Physics. Technical Letters."

105. I. I. Rabi to H. A. Bethe, December 2, 1947; H. A. Bethe to I. I. Rabi, December 4, 1947. Bethe Papers.

106. I. I. Rabi to J. R. Oppenheimer, December 12, 1947. Oppenheimer Papers, Box 72.

107. Darrow Diaries, entry for January 31, 1948. Darrow Papers.

108. Official Register of Harvard University, vol. 47, May 16, 1950, p. 49.

109. F. J. Dyson to his parents, February 4, 1948. In his letter Dyson had also told his parents: "I was very interested to see Schwinger at close quarters, in view of his great reputation.... He is ... a young man aged 30, ... fat, and extremely uninterested in anybody except himself. He annoyed everyone as much as possible in his talk by failing to mention that Bethe or anyone else had had anything to do with the developments leading up to his theory; probably this was not done deliberately, but just from mental habit. Fortunately Bethe is not a man to squabble over priorities.... I spoke to one of his students, who said that as director of research he is absolutely useless and drives all his research students crazy, but he makes up for it by giving the most inspiring courses of lectures." Regarding Dyson's description of Schwinger as "fat," Schwinger confides that he was deeply affected by Pauli's death in 1957. He lost thirty pounds and "determined to be healthy." Interview with S. S. Schweber, January 15, 1982.

110. The entry in K.K. Darrow's diary for Wednesday, March 31, 1948 reads: "All this day was occupied by Schwinger." Breit in his notebook recorded Fermi's talk, indicating at which point Schwinger had finished the first part of his lecture and where he resumed. Breit Papers, Box 18.

111. Pocono notes taken by J. A. Wheeler. A copy is on deposit at the AIP Niels Bohr Library.

112. In his notebook Breit recorded: "will show $H^{(2)}_{PSI} = 0$ [Formally only actually $0 \cdot \infty$]." Breit Papers.

113. Breit notebook, ibid., p. 123.

114. The entries in Breit's notes and Wheeler's notes differ slightly. Eq. (7.5.51) is the one found in Breit's notes. The entry in the Pocono notes had the factor e/κ^2 in the first line replaced by i/κ.

115. Breit notebook, p. 129. Schwinger also reported that, using these covariant methods, Case had calculated the magnetic moment of the neutron and proton in meson theory "with no cut offs and Dirac theory for nucleons." The completely relativistic result he obtained was

$$\mu_p = 1 + \frac{1}{16\pi} \frac{g^2}{\hbar c} \left(\frac{M}{\mu}\right)^2 C_1$$

$$\mu_N = -\frac{1}{16\pi} \frac{g^2}{\hbar c} \left(\frac{M}{\mu}\right)^2 (C_1 + C_2).$$

116. J. Schwinger, oral interview with S. S. Schweber, January 15, 1982.

117. Biographical material on J. Schwinger, Schwinger Papers, UCLA.

118. William L. Lawrence, "Schwinger States His Cosmic Theory," *New York Times*, June 25, 1948, pp. 38–39.

119. Feshbach (1950) reports having these notes in June 1948.

120. J. Schwinger, interview with S. S. Schweber, November 12, 1982.

121. F. J. Dyson to his parents, July 20, 1948.

122. F. J. Dyson to H. A. Bethe, August 9, 1948, Bethe Papers.

123. Ibid.

124. William J. Robbins to E. V. Condon, May 6, 1943. ORG: NAS: National Science Fund: Project: Mayer Nature of Light Awards: Advisory Committee Appointments: Members: 1943–48. NAS Archives.

125. K. K. Darrow to W. J. Robbins, April 6, 1944, NAS-NSF Mayer Nature of Light Awards, NAS Archives.

126. Announcement for the Charles L. Mayer Nature of Light Awards, NAS Archives.

127. H. H. Sargent to members of Advisory Committee, January 24, 1946, April 24, 1946, Mayer Nature of Light Awards, NAS Archives.

128. H. H. Sargent to C. J. Eliezer, December 10, 1946. NAS Archives.

129. C. Mayer to H. H. Sargent. NAS Archives.

130. K. K. Darrow to H. H. Sargent, June 17, 1947; H. H. Sargent to K. K. Darrow, June 17, 1947. NAS Archives.

131. H. H. Sargent to R. P. Feynman, August 4, 1947; R. P. Feynman to H. H. Howland, September 4, 1947. NAS Archives.

132. R. A. Millikan to H. H. Sargent, April 6, 1948; H.H. Sargent to H. Shapley, April 22, 1948; H. Shapley to P. S. Epstein, October 5, 1948; P. S. Epstein to H. Shapley, October 12, 1948. NAS Archives.

133. Eva G. Rosensteel to the Members of the Advisory Committee, Charles L. Mayer Nature of Light Awards, March 30, 1949; H. Shapley to J. Schwinger, March 30, 1949; J. Schwinger to H. Shapley, April 11, 1949; H. Shapley to E. G. Rosensteel, April 14, 1949. NAS Archives.

134. H. H. Sargent to I. I. Rabi, April 4, 1949; I. I. Rabi to H. H. Sargent, April 8, 1949. *Boston Globe*, May 10, 1949.

135. The typewritten manuscript is in the archives of the NAS, Mayer Nature of Light Awards. Among Schwinger's papers is a handwritten manuscript entitled "Quantum Electrodynamics III. Radiative [Corrections to, crossed out] Modification of Particle Electromagnetic Properties. Vacuum Fluctuations and Particle...," which, except for some minor differences, is an earlier version of the manuscript submitted for the Mayer Award.

136. Schwinger, "QED III, Modification...," Mayer Nature of Light Awards, NAS Archives.

137. In the manuscript submitted to the Mayer competition, the notation for the infrared divergent term is $\log u_0 - 1$. Also the factor 17/40 is incorrectly recorded as 7/40.

138. Gregor Wentzel (1898–1978) was a student of Sommerfeld's. He received his doctorate in 1921 and stayed in Munich until 1926 as a *Privatdozent*. In 1926 he moved to Leipzig as an extraordinary professor of theoretical physics, and in 1928 he became Schrödinger's successor at the University of Zurich. At about that same time, Pauli came to the ETH. For many years the Pauli-Wentzel team represented theoretical physics in Zurich.

139. W. Pauli to J. Schwinger, January 21, 1949. A copy of the letter was also sent to Bethe, who forwarded mimeographed copies to Wentzel, Oppenheimer, and others. A copy of the letter can be found in the Oppenheimer Papers in the National Archives. Pauli's original letter and the mimeographed version differ trivially.

140. W. Pauli to H. A. Bethe, January 25, 1949, Bethe Papers.

141. Ibid.

142. W. Pauli to J. Schwinger, January 21, 1949.

143. Ibid.

144. B. DeWitt, taped communication to S. S. Schweber, August 15, 1984.

145. Pauli to Oppenheimer, undated but sometime in late February 1949. Oppenheimer Papers.

146. Pauli to Oppenheimer, February 22, 1949, in ibid.

147. J. Schwinger, taped interview, 1982.

148. Ibid.

149. The order of the citations is of importance. French and Weisskopf had obtained an expression for the Lamb shift before Schwinger, Feynman, and Kroll and Lamb. They had postponed publishing their correct result because it disagreed with Schwinger's and Feynman's result. Kroll and Lamb's paper therefore appeared in print before that of French and Weisskopf.

150. J. Schwinger, oral interview, 1982.

151. Ibid.

152. Ibid.

153. Ibid.

154. The typescript of the lecture is among the papers that Schwinger has deposited at UCLA. The page numbers throughout section 10 refer to this untitled and undated manuscript.

155. J. Schwinger, taped interview with S. S. Schweber, 1982.

156. Ibid.

157. Ibid. See also Schwinger (1973) for further comments on renormalization, especially pp. 418–419, 428–429.

158. B. DeWitt, taped communication to S.S. Schweber, August 15, 1984.

159. Ibid.

160. R. Newton to S. S. Schweber.

161. Ibid.

162. B. DeWitt, taped communication, 1984.

163. Ibid.

164. Ibid.

165. D. Saxon, interview with S. S. Schweber, October 19, 1988.

166. Taped interview with S. S. Schweber, 1982.

167. Ibid.

168. D. Saxon, interview with S. S. Schweber, October 1988.

169. Both of them made this identical statement. H. Levine, interview with S. S. Schweber, October 19, 1988. H. Feshbach, interview with S. S. Schweber, October 16, 1988.

170. H. Levine, interview with S. S. Schweber, October 1988.

171. The reader may have noted that Schwinger's rhetoric reflects a view of the world that is competitive.

172. See for example, the account of Källén's criticism of Schwinger at the 1959 Rochester conference in Wightman (1981a).

173. Deser, Feshbach, Finkelstein, Johnson, and Martin, in preface to *Physica* 96A (1979), nos. 1 and 2, *Essays in Honour of Julian Schwinger's 60th Birthday.*

Chapter 8
Richard Feynman

1. As quoted by J. R. Oppenheimer in a letter to Birge. Smith and Weiner (1980), p. 269.

2. Lecture by R. P. Feynman in 1954 entitled "Inertia." RPF, CIT 2.5.

3. Richard P. Feynman, CW. Transcription of a taped oral interview of Feynman by Charles Weiner, March 1966, for the Center for History and Philosophy of Physics. The transcript is on deposit at the AIP Niels Bohr Library. I thank R. P. Feynman and C. Weiner for permission to quote from it.

4. Ibid. See Feynman's lecture on "The Principle of Least Action" in Feynman et al. (1964), vol. 2, Chap. 19, 1–14.

5. RPF, CW.

6. Ibid.

7. These notebooks have been deposited in the Feynman Papers in the AIP Niels Bohr Library. Xerox copies are in RPF, CIT 15.3.

8. The notebook is in the Feynman Papers, AIP Niels Bohr Library. A xerox copy is in RPF, CIT 15.2.

9. Feynman Papers, AIP Niels Bohr Library; RPF, CIT 15.3.

10. R. P. Feynman to S. S. Schweber, January 28, 1985.

11. RPF, CW.

12. Ibid.

13. Welton, "Memories," a manuscript by T. A. Welton of his reminiscences of his interactions with R. P. Feynman. A copy has been placed on deposit at the AIP Niels Bohr Library.

14. Ibid, p. 3.

15. Ibid.

16. Ibid.

17. Ibid, pp. 3, 4.

18. The notebook is in the Feynman Papers, AIP Niels Bohr Library. See also RPF, CIT 15.4.

19. RPF, CW.

20. Feynman-Welton notebook, p. 56.

21. RPF, CW.

22. Feynman-Welton notebook, p. 52.

23. Ibid.

24. A. Clogston to S. S. Schweber, October 3, 1984.

25. Clogston was a senior at the time. He was doing research for a senior thesis under Philip Morse at MIT and Leo Goldberg at the Harvard Observatory.

26. See also RPF, CIT 6.7.

27. Welton, "Memories," p. 6.

28. Ibid. Martin Deutsch, who was an undergraduate at MIT at this same time. In the brief "Scientific Autobiography" he appended to the articles on "Discovery of Positronium" (chap. 2 in Maglich, *Adventures in Experimental Physics*, vol. 4, p. 125), Deutsch remarks: "It was Harrison's junior course in atomic physics that had the greatest influence on me [as an undergraduate at MIT]. The course was totally disorganized, and seemed to consist of a series of scientific anecdotes or vignettes. Somehow this style kindled my enthusiasm, and I still charge many of the insights into physics and the creative process which I acquired there to this influence." Incidentally, Deutsch remembers Feynman as standing out among his fellow students.

29. The dissertation is on deposit in the archives of the MIT Library. See also RPF, CIT 15.6. For a history of the Hellmann-Feynman theorem, see Musher (1966). For the relation of the Hellmann-Feynman and the viral theorems, see Epstein (1981).

30. RPF, CW.

31. When the mathematics department at MIT had discovered that they didn't have the four people needed to enter a team for the Putman contest, they asked Feynman to join the competition. Looking over their records, they had found that Feynman had been in mathematics. Feynman recalled that he "was unsure, but they gave me old exams to practice on." R. P. Feynman to S. S. Schweber, January 28, 1985.

32. RPF, CW.

33. Ibid. See also Wheeler (1989).

34. J. A. Wheeler, transcript of an interview by Charles Weiner and Gloria Lubkin on April 5, 1967. Center for History and Philosophy of Physics, AIP.

35. RPF, CW.

36. Ibid.

37. Oral interview with S. S. Schweber, October 15, 1980.

38. Oral interview with S. S. Schweber, August 15, 1981.

39. RPF, CW. See also Wheeler (1989).

40. RPF, CW.

41. RPF, CIT 6.1.

42. The original typewritten statement read: "(3) The fundamental equations are to be invariant with respect to interchange of the sign of the time in them (symmetrical with respect to interchange of past and future), and was changed by Feynman to the form indicated in the text."

43. The manuscript is among the Feynman Papers in the AIP Niels Bohr Library.

44. For an extended discussion of those matters, see Gold (1967). Mr. X in that volume is R. P. Feynman. See T. Gold to R. P. Feynman, October 1965. RPF, CIT 5.8.

45. J. A. Wheeler to R. P. Feynman, November 10, 1949. RPF, CIT 3.10.

46. Feynman wrote that "in high school, I had a very able friend, Herbert Harris, who when we graduated, went to Rensselaer Polytech to become an electrical engineer, while I went to MIT. One summer [probably at the end of their freshman year], he returned to Far Rockaway, we friends took a walk and he told me about the then new feedback amplifiers. He tried to design them in different ways avoiding oscillations and said he was convinced that there was some law of nature that made it impossible to make the impedance fall off too fast without introducing a large phase shift. I proposed it might be a reflection into the frequency response domain of the fact that signals cannot come out before they come in. But neither of us was apparently sophisticated enough to work this out mathematically." This is the reason why somewhat later Feynman had found Bode's paper "so interesting." R. P. Feynman to S. S. Schweber, January 28, 1985. Feynman worked once again on the mathematical theory of feedback amplifiers while at Los Alamos. Hoddeson and Baym (1979), p. 39.

47. R. P. Feynman to J. A. Wheeler, December 8, 1949. RPF, CIT 3.10. Another facet of Feynman's activities while a graduate student can be gleaned from Henry Barshall's 1940 experimental thesis on neutron scattering in which he expresses his great indebtedness to Mr. Richard Feynman for his help in computing the corrections to his data coming from scattering from the chamber's walls.

48. RPF, CW. After Feynman had obtained the alternative formulation of quantum mechanics that he had developed for his thesis, Wheeler again tried to quantize the Wheeler-Feynman theory. His approach consisted in linearizing the action using Dirac's trick of replacing $\sqrt{-da_\mu da^\mu} \to \Gamma_\mu da^\mu$. He sent his notes—consisting of some thirteen pages, dated November 1941—to Feynman at Los Alamos in 1945. They are in RPF, CIT 3.10.

49. Letter from H. Jehle to S. S. Schweber, September 10, 1980. Herbert Jehle was a student of Schrödinger's in the early thirties in Berlin. He was interned in a concentration camp in Nazi Germany as a Quaker and a pacifist but made his way to France, only to be interned in a concentration camp there as an enemy alien. He came to the United States in 1940 and taught at Harvard, and later at the University of Pennsylvania and the University of Nebraska.

50. Since $L(t)$ is a time-dependent operator (which does not commute with $L(t')$ for $t \neq t'$), the expression $\exp \frac{i}{\hbar} \int_T^t L dt$ must be defined. The calculus of ordered operators which Feynman invented in 1947 gives explicit meaning to such expressions.

51. RPF, CW.

52. Ibid.

53. RPF, S^3.

54. Ibid.

55. RPF, CIT 13.3. Box 44 of Feynman's papers contains his writeup of his attempts to understand measurement theory while working on his thesis. His intent was to understand where uncertainty enters the quantum mechanical description and to elucidate what is meant by a measurement and how irreversibility enters the process. "A correlation can be set up between an unlimited number of particles such that the removal (?) of a finite number of such particles at any time from the set—either before or after they have been reproduced— will not limit the number which can be so correlated. This we define as a measurement. . . . This is very close but require a better definition from pt of view of rigor. I'll mull it over and someday explain it to me again. Note that, and study why, reproduction and amplification and throwing away boxes is irreversible."

56. RPF, S^3.

57. Ibid.

58. Ibid.

59. Ibid.

60. Wheeler wrote to Feynman on March 26, 1942, from Chicago (where he was working with Fermi and Wigner on the first atomic pile) that he felt Feynman "had done more than enough for a thesis" and urged him "very strongly to write up what you have in the remaining few weeks before you get into the situation in which I now find myself." RPF, CIT 3.10.

61. RPF, S^3.

62. J. A. Wheeler, oral interview with S. S. Schweber, August 15, 1981.

63. RPF, S^3.

64. RPF, CW. See also Feynman (1976, 1980).

65. RPF, S^3.

66. RPF, CIT 15.1.

67. See, for example, the account in Groueff (1967), pp. 202–203. In 1979 Lillian Hoddeson and Gordon Baym interviewed Feynman to obtain his recollections of his activities at Los Alamos. They contain an extensive account of his technical contributions to the project. Hoddeson and Baym (1979).

68. The group was known as T-4. Its original members were Julius Ashkin, Frederick Reines, and Dick Ehrlich. Theodore Welton joined it in the early spring of 1944. For some of the problems the group worked on, see Welton, "Memories," pp. 11-15. See also Groueff (1967), p. 212.

69. The question arose whether the gas diffusion plant for separating nuclear isotopes could lead to an accumulation of U^{235} and cause a nuclear explosion. Compare Feynman's account (Feynman 1980) and that given by Teller in Blumberg and Owens (1976), p. 457.

70. Feynman worked with Fermi, de Hoffman, and Serber on many aspects of "pile theory." See, for example, the declassified Los Alamos reports LADC 256, LADC 269. Interestingly, the report on intensity fluctuations of a neutron chain reactor by Feynman, de Hoffman, and Serber (LADC 256) contains a diagram very suggestive of later Feynman diagrams.

71. H. A. Bethe, introduction to "Computers and Their Roles in Science." Bethe Papers, 14/22/1976.

72. Welton, "Memories," p. 10.

73. "Now, one of my diseases, one of my things in life, is that anything that is secret I try to undo." Feynman (1976), p.19.

74. RPF, S^3.

75. RPF, S^3. In his interview with Hoddeson, Feynman tells that Bethe taught him how to compute quickly the square, cube, square root, cube root, logarithm of any number. "From then on, we had a kind of fun competition, which he would almost always win, in which I tried to do arithmetic faster than he would." Hoddeson (1979), pp. 41–44.

76. Frank Aydelotte, Director, Institute for Advanced Study, to Warren Weaver, Rockefeller Foundation, October 11, 1943, and Aydelotte to Frank Blair Hanson, Rockefeller Foundation, December 20, 1943. Also, Oswald Veblen to V. Bush, November 8, 1943. Rockefeller Archive Center 200-1.1, 143, 1763, Pocantico Hills, North Tarrytown, N.Y.

77. The development was gradual. After he came to Cornell in the fall of 1945, Feynman was a member of the Association of Scientists of Cornell University, Rockefeller Hall, and gave public lectures on various aspects of the problems of atomic power and atomic weapons. In April 1946 he gave the first radio talk on these issues in a series called "The Scientists Speak." A copy of the talk is in the Bethe Papers.

78. Welton, "Memories," p. 10.

79. M. E. Deutsch to R. P. Feynman, July 3, 1945; R.T. Birge to R. P. Feynman, July 5, 1945. RPF, CIT 1.33.

80. Feynman to Deutsch, July 27, 1945, and August 8, 1945. RPF, CIT 1.33.

81. R. C. Gibbs to C. de Kiewiet, July 31, 1945; R. C. Gibbs to R. P. Feynman, August 3, 1945. RPF, CIT 1.33.

82. H. A. Bethe to R. C. Gibbs, July 25, 1945. RPF, CIT 1.33. This letter is among the Feynman papers at Cal Tech because Bethe's secretary at Los Alamos, "in the process of cleaning out Hans' file before he leaves ran across some rather interesting correspondence that [she] thought [Feynman] might like to keep." M. Bradeur to R. P. Feynman, December 29, 1945. RPF, CIT 1.33.

83. H. D. Smythe to R. P. Feynman, October 23, 1946. RPF, CIT 2.48.

84. R. P. Feynman to F. Aydelotte, January 6, 1947. RPF, CIT 2.48.

85. Telegram of J. Ellis to R. P. Feynman, February 18, 1947; R. P. Feynman to J. Ellis, February 19, 1947. RPF, CIT 1.33.

86. R. P. Feynman to R. J. Oppenheimer, November 6, 1947; January 6, 1947. RPF, CIT 2.31. R. P. Feynman to E. O. Lawrence, and to E. Segrè, July 15, 1947. RPF, CIT 1.33.

87. For a moving account of Feynman's relationship with his first wife, Arline Greenbaum, see Feynman and Leighton (1988) and Gleick (1992).

88. RPF, CW.

89. Wigner invited Feynman on February 26, 1946. RPF, CIT 3.11.

90. The title of Dirac's paper was "Elementary Particles and Their Interactions." It dealt with the Hamiltonian approach to quantum mechanics and quantum electrodynamics. A copy of it is in the University Archives, Seeley G. Mudd Manuscript Library, Princeton University, among the materials dealing with the Princeton Bicentennial.

91. RPF, CW.

92. Ibid.

93. Taped interviews of H. A. Bethe with S. S. Schweber, October 15, 1980, and P. Morrison with S. S. Schweber, August 9, 1984.

94. Welton, "Memories," p. 14.

95. RPF, CW.

96. Feynman's wife was ill with tuberculosis and was hospitalized in Albuquerque. On weekends Feynman would visit her (see Feynman and Leighton 1988). Nonclassified work was the only thing he could work on while taking the bus to and from the hospital. Welton in his "Memories" recalls that when he arrived at Los Alamos in 1944, Feynman met him at the train station. "After giving me a thorough briefing on the work of the project and of his group [T4], the talk degenerated to a description of our interest in non-military physics. . . . He showed me how . . . his later to-be-famous formulation in terms of a summation over all space time trajectories of the system . . . worked by a simple illustration." But no reference is made to any further talk along these lines later on. Welton, "Memories," p. 10.

97. These notes are labeled Assay, February 25, 1946. RPF, CIT 13.3.

98. R. P. Feynman to T. A. Welton, 1947. RPF, CIT 3.9.

99. This approach is outlined in his Nobel Prize speech, Feynman (1966a), p. 704.

100. R. P. Feynman to T. A. Welton, February 10, 1947. RPF, CIT 3.9.

101. Ibid.

102. B. Corben to S. S. Schweber, September 11, 1984.

103. For example, the quantization procedure applied to the action $S = \int x(t)\,x(t+a)dt$ for which there is neither a canonical momentum nor a Hamiltonian, and which gives rise to negative probabilities and imaginary energies.

104. R. P. Feynman to C. Kelber, February 21, 1949. RPF, CIT 2.12.

105. RPF, S^3.

106. RPF, CW.

107. Ibid.

108. Breit Notebooks, 1947, pp. 86, 87, Breit Papers.

109. In a letter to Cecille Morette in 1950, Feynman indicated that "in general I am very careless in my lectures, so I would appreciate it if they weren't quoted in detail and specific points argued. On the other hand, the papers that I write each take me about a year to get everything straightened out. So I give you full permission to quote from a published paper and argue with the equations." R. P. Feynman to C. Morette, June 5, 1950. RPF, CIT 2.26.

110. The notes are in RPF, CIT 12.9.

111. See the appendix of Feynman (1948a).

112. The notes are entitled "Elimination of Field Oscillators." RPF, CIT 12.9. Much of this duplicated calculations he had previously carried out for his thesis (Feynman 1942).

113. Feynman had met Bert Corben in Princeton in 1941 when Corben was studying with Pauli at the Institute for Advanced Study. Pauli had come to the United States in 1940 because he had feared that Switzerland would be overrun by Germany, and that he would be vulnerable because his mother was Jewish. The Rockefeller Foundation supported Pauli's stay at the Institute for Advanced Study in Princeton during the war. Corben had returned to the United States from Australia in July 1946 to accept a position at the Carnegie Institute of Technology in Pittsburgh. The Corbens had met Feynman at the New York meeting of the APS in January 1947 and had invited him to Pittsburgh for part of the summer. B. Corben to S. S. Schweber, September 11, 1984.

114. A set of notes, "Self-interactions of 2 Particles," written during the summer of 1947, gives the mathematical details. RPF, CIT 12.9.

115. Feynman had read Eddington's *Fundamental Theory* with Welton in 1937. Welton, "Memories," p. 4.

116. R. P. Feynman to Bert and Mulaika Corbin [*sic*: Corben], November 6, 1947. RPF, CIT 1.23. Feynman also reported to David Bohm his recent results: "I changed the delta function interaction in classical electrodynamics by another function. The function is arbitrary, but must have a short range of the order of the electron radius. The result is that all mass can be represented as electromagnetic. It also shows the interesting possibility of describing pair production classically. Some solutions of the least action principle exist which correspond to pairs running backwards and forwards in time." R. P. Feynman to David Bohm, November 19, 1947. RPF, CIT. Bohm at the time was working on a relativistic description of extended charges (Bohm and Weinstein 1948) and had noted that in his theory the charge distribution could oscillate. Bohm had told Feynman of this result during Feynman's visit to Princeton in early November. Feynman communicated to Bohm that "I looked to see if could get oscillation of my charge. The result depended upon the exact choice of the function for the interaction."

117. A. S. Wightman, "Notes: Quantum Field Theory 1947." They are in Wightman's possession at Princeton University. I thank him for permission to study them.

118. Mulaika Corben to R. P. Feynman, November 24, 1947. RPF, CIT 1.23.

119. R. P. Feynman to Mr. and Mrs. Corbin [*sic:* Corben], November 19, 1947. RPF, CIT 1.23.

120. Bloch and Nordsieck (1937) had shown the origin and solution of that particular difficulty.

121. By second order, Feynman meant to order $(e^2/hc)^2$.

122. R. P. Feynman to Prof. and Mrs. H. M. Corben, January 15, 1948. RPF, CIT 1.23. Feynman thought that the story of the "mistake" referred to at the beginning of his letter was "interesting": "As near as I can remember it, I first got a relativistic result (we were only working to order v^2/c^2). A student found an error in an early line and concluded it would not be invariant—when I wrote the first letter. But later on, several pages later, he found another error where I cancelled two equal complicated terms that I should have added. The original answer I had gotten was right—it was relativistic. The miracle of two cancelling errors was probably the result of a mixture of having a strong feeling for what the answer must be and algebraic carelessness." Feynman to Schweber, January 28, 1985. This "miracle of two cancelling errors" applies also to the proof of the gauge invariance given by Feynman in section 8 of Feynman (1949b). The "two cancelling errors" in that case were the omission of the contributions from the vanishing denominators $m - \not{p}$ in the amplitudes containing the self-energy corrections and the omission of the wave function renormalization constants, whose variation under gauge transformations exactly cancels the first contribution. The net result of Feynman's analysis—the gauge invariance of QED—was so obviously true that the flow in the derivation went unnoticed until pointed out by Bialynicki-Birula (1967, 1970).

123. R. P. Feynman to Prof. and Mrs. Corben, March 20, 1948. RPF, CIT 1.23.

124. L. Eyges, pers. comm., September 14, 1984.

125. Ibid. This incident prompts a general remark. The theoretical physics community and intellectual communities in general accept a wide latitude of social behavior from its most distinguished members. The more outstanding the person, the wider the deviations from the generally accepted norms that remain unchallenged. Part of the reason that these members can act in this manner is that they enter the community very young and need not go through the process of socialization. Landau is a famous case in point (see, for example, Casimir 1983, pp. 104–116).

126. R. P. Feynman to Prof. and Mrs. Corben, March 20, 1948. RPF, CIT 1.23.

127. Handwritten notes entitled "Theory of Positrons." RPF, CIT 13.1.

128. Ibid.

129. Ibid.

130. Ibid.

131. "T. S. Theory of Positrons." Manuscript for APS lecture in January 1949. RPF, CIT 6.13.

132. RPF, CW. The metaphor "the space time trajectory being like the letter N" is also used in the introduction of "The Theory of Positrons" (Feynman 1949a). There the bombardier metaphor is stated as follows: "Following the charge rather than the particles corresponds to considering this continuous world line as a whole rather than breaking it up into pieces. It is as though a bombardier flying low over a road suddenly sees three roads and it is only when two of them come together and disappear again that he realizes that he has simply passed over a long switchback in a single road." Feynman has indicated that: "The 'bombardier metaphor' was suggested to me by some student at Cornell (who had actually been a bombardier during the war) when I was writing up my paper and was asking for opinions of how to explain it and only had poor or awkward metaphors." R. P. Feynman to S. S. Schweber, January 28, 1985.

133. RPF, CIT 6.13.

134. RPF, CW.

135. Ibid.

136. RPF, S^3.

137. Breit Notebooks, 1947, Breit Papers, Yale University Archives.

138. R. P. Feynman at Schwinger's sixtieth birthday celebration. AIP.

139. RPF, CW.

140. RPF, S^3.

141. Pocono Conference Notes, p. 46, AIP.

142. Since $i/x = lim_{\epsilon \to 0} i/x + i\epsilon = \int_0^\infty \exp(iw(x + i\epsilon))dw$.

143. RPF, CW.

144. Pocono Conference Notes, pp. 48–49, AIP.

145. Ibid., p. 50.

146. Ibid., p. 53. Wheeler, who was the note taker, was trying to conform to the recommendation of the International Union of Physics by calling electrons the generic name of particles with the mass of an electron and charge $\pm e$, and giving the name "negaton" to a negatively charged electron and "positon" to the positively charged antiparticle.

147. RPF, S^3.

148. Pocono Conference Notes, p. 53, AIP.

149. Ibid., p. 54.

150. Ibid., p. 55.

151. Ibid.

152. *California Tech*, vol. 67, no. $5\frac{1}{2}$, Extra Edition, October 22, 1965, "Dr. Richard Feynman Nobel Laureate."

153. RPF, CW. A concise statement of visualization in the early days of quantum mechanics was given by Dirac in 1930:

> The classical tradition has been to consider the world to be an association of observable objects (particles, fluids, & c.) moving about according to definite laws of force, so that one could form a mental picture in space and time of the whole scheme. This led to a physics whose aim was to make assumptions about the mechanism and forces connecting these observable objects, to account for their behaviour in the simplest possible way. It has become increasingly evident in recent times, however, that nature works on a different plan. Her fundamental laws do not govern the world as it appears in our mental picture in any very direct way, but instead they control a substratum of which we cannot form a mental picture without introducing irrelevancies. The formulation of these laws requires the use of the mathematics of transformations. The important things in the world appear as the invariants (or more generally the nearly invariants, or quantities with simple transformation properties) of these transformations. The things we are immediately aware of are the relations of these nearly invariants to a

certain frame of reference, usually one chosen so as to introduce special simplifying features which are unimportant from the point of view of general theory. (P.A.M. Dirac 1930e).

154. H. A. Bethe to S. S. Schweber, taped interview, October 15, 1980.

155. Feynman, after-dinner talk at Schwinger's sixtieth birthday celebration, 1978. AIP.

156. R. P. Feynman to S. S. Schweber, January 28, 1985.

157. A. Bohr to S. S. Schweber, August 18, 1980. The letter continues: "But I do not believe my father held this view and I remember many good and humorous discussions between the three of us after the session."

158. RPF, S^3.

159. Feynman to B. and M. Corben, January 15, 1948; Feynman to B. and M. Corben, March 20, 1948. RPF, CIT 1.23.

160. "Theory of Positrons." RPF, CIT 13.1. See also Feynman (1949b), p. 246, and in particular n. 20.

161. R. P. Feynman to J. A. Wheeler, December 8, 1949. RPF, CIT 3.10.

162. R. P. Feynman to the Corbens, March 20, 1948. RPF, CIT 1.23.

163. A full account of Dyson's contributions is given in chap. 9.

164. F. J. Dyson to his parents, November 1, 1948.

165. R. P. Feynman to F. J. Dyson, October 29, 1948. RPF, CIT 1.30.

166. R. P. Feynman to T. A. Welton, October 30, 1940. RPF, CIT 3.9.

167. W. Pauli to H. A. Bethe, January 25, 1949. Bethe Papers.

168. R. P. Feynman to H. A. Bethe, July 7, 1948 (from Casa Grande Lodge, Albuquerque, New Mexico). Bethe Papers, Cornell University Library, 14/22/1976. Dyson, who had accompanied Feynman on his trip west, wrote to his parents in summer 1948: "Yesterday we were discussing his [Feynman's] new problem, this time again a girl in New Mexico with whom he is desperately in love. This time the problem is not tuberculosis, but the girl is a Catholic. You can imagine all the troubles this raises and if there is one thing Feynman could not do to save his soul it is to become a Catholic himself. So we talked and talked, and sent the sun down the sky, and went on talking in the darkness. I am afraid that at the end of it poor Feynman was no nearer to the solution of his problems, but at least it must have done him good to get them off his chest."

169. Ibid.

170. R. P. Feynman to H. A. Bethe, July 22, 1948 (from Frijoles Canyon Lodge, Santa Fe, New Mexico). Bethe Papers.

171. R. P. Feynman to T. A. Welton, October 30, 1948. RPF, CIT 3.9.

172. See the references given in Slotnick and Heitler (1949).

173. The value of $\int V(x) d^3x$, where V is the neutron-electron interaction potential.

174. RPF, S^3. See also Feynman (1966a), p. 706.

175. RPF, S^3. In his Nobel acceptance speech, Feynman put it thus: "That was a thrilling moment for me, like receiving the Nobel Prize, because it convinced me, at last, I did have some kind of method and technique and understood how to do something that other people did not know how to do. That was my moment of triumph in which I realized I really had succeeded in working out something worthwhile" (Feynman 1966a, p. 707).

176. RPF, S^3. Francis Low (pers. comm.) points out that it is interesting that Feynman got the same answer as Slotnick, since a controversy arose over the calculation of the electron-neutron interaction. There were two calculations using standard perturbation theory, one by Slotnick and Heitler (1949) and the other by Dancoff and Drell (1949), which agreed with each other. There were two others using the Schwinger formalism, one by Case (1949a) and the other by Borowitz and Kohn (1949), which agreed with each other but not with the calculations of Slotnick and Drell. The situation was resolved by Foldy (1952) who found Slotnick and Drell to be right.

177. RPF, S^3.

178. Ibid.

179. The paper that Case (1949b) submitted to the *Physical Review* contains an acknowledgment stating "Thanks are due to Dr. R. P. Feynman for pointing out an error in the original manuscript."

180. RPF, S^3.

181. Ibid.

182. Ibid.

183. *California Tech*, vol. 67, $5\frac{1}{2}$, Extra Edition, October 22, 1965.

184. RPF, S^3.

185. Ibid.

186. Ibid.

187. R. P. Feynman. *Phys. Rev.* 77 (1949): 584. The manuscript is in RPF, CIT 13.1.

188. RPF, S^3.

189. Ibid.

190. Ibid.

191. Ibid.

192. Deutsch (1951a,b); see also M. Deutsch, "Discovery of Positronium," in B. Maglich, *Adventures in Experimental Physics,* vol. 4, pp. 63–129.

193. R. P. Feynman to J. A. Wheeler, May 4, 1951. RPF, CIT 3.10.

194. RPF, S^3. RPF, CIT.

195. The power of the method was, however, found to be limited. Professor Morgan Ward indicated to Feynman that "the same argument would show that an equation like $x^7 + y^{13} = z^{11}$ (powers prime to each other) would be unlikely to have integer solution—but that they do, an infinite number of them!" R. P. Feynman to S. S. Schweber, January 28, 1985.

196. R. P. Feynman to the Corbens, November 19, 1947. RPF, CIT 1.23.

197. Ibid.

198. RPF, S^3.

199. Ibid.

200. Salam and Wigner in Fierz and Weisskopf (1960).

201. R. P. Feynman to J. Watson, October 2, 1967. RFP, CIT.

202. Recall that it was Robert Wilson who got Feynman back on track by reintroducing the element of play in his research. He got him to investigate the wobble and spin of a plate as it is tossed across a room.

203. Beadle to Feynman, January 4, 1967; Feynman to Beadle, January 16, 1967. Feynman Papers, CIT.

204. R. P. Feynman to P. Handler, November 9, 1960. RPF, CIT.

205. R. P. Feynman to P. Handler, February 20, 1961. RPF, CIT.

206. R. P. Feynman to P. Handler, July 15, 1969. RPF, CIT.

207. His parents' objection derived from their fear that he too might contract the desease.

Chapter 9
Freeman Dyson

1. The complete story is reprinted in Dyson (1992), pp. 3–7. Philip is, incidentally, the name of the hero of Edith Nesbit's *The Magic City*, one of Dyson's favorite storybooks when he was eight. See the opening chapter of *Disturbing the Universe* (Dyson 1979).

2. Sir Frank Dyson, although not related to the Dysons, was born in Halifax, the same Yorkshire town as Dyson's father. "His exalted glory" was responsible for turning the young Dyson's thoughts toward astronomy.

3. FJD, S^3.

4. Alice became a medical social worker and presently lives in the house that George and Mildred retired to in Winchester in 1952. Palmer (1984), in his biography of George Dyson, notes that she "took to heart her mother's favourite maxim from Terence: 'I am human and I let nothing human be alien to me.' "

5. Taped interview with Alice Dyson, June 1985.

6. Ibid.

7. Ibid.

8. Ibid.

9. Ibid. Some of the letters of Dyson to his family during his stay at Cornell have been reprinted in full under the title "Feynman in 1948," in Dyson (1992), pp. 322–337.

10. FJD, S^3.

11. Interview with Alice Dyson, June 1985.

12. Ibid.

13. Dyson to his parents, September 15, 1947, on board the *Queen Elizabeth*. These letters are in Freeman J. Dyson's possession. I thank him for permission to read them and to quote from them.

14. FJD, S^3.

15. Typical divinity questions: "What do you know about Melchizedek, Ghaazi, Beth-horon, Mount Ebal?" or "Quote any passages in Scripture, not from the book of Numbers, in which the name of Balaam occurs." Sabben-Clare (1981), p. 56.

16. FJD, S^3.

17. Ibid.

18. Ibid.

19. After World War II, Lucas went to Uganda and was responsible for building up the fine program in biology at Makarere University College. His idealism was fairly representative of many on the teaching staff.

20. FJD, S^3.

21. Interview with S. S. Schweber, May 18, 1984.

22. Manisty.

23. Durell.

24. M. S. Longuet-Higgins to S. S. Schweber, November 13, 1984.

25. FJD, S^3.

26. Daniel Pedoe to S. S. Schweber, July 21, 1986.

27. Ibid.

28. Ibid.

29. FJD, S^3. The copy of *Disturbing the Universe* Dyson presented to Pedoe is inscribed as follows: "In grateful memory of those happy days in 1941 when you opened my eyes to the beauties of algebraic geometry. You were the first real mathematician I ever met, and I have never forgotten what you taught me."

30. Pedoe to S. S. Schweber, July 21, 1986.

31. M. Longuet-Higgins to S. S. Schweber, November 13, 1984.

32. FJD, S^3.

33. Ibid.

34. Hardy was a pacifist, and Besicovitch, not being a born Englishman, could not be employed at Bletchley Park where many British mathematicians were working to break German codes.

35. FJD, S^3.

36. Ibid. At the end of May 1948, Dyson had written his father:

> Today I had an exciting and incoherent letter from Oliver Atkin. We have been corresponding for the last few months on the subject of some discoveries of mine in the theory of numbers which you may remember my talking about; I made these discoveries 5 years ago, but was unable to prove them; two years ago I decided to make a determined attack on them, and spent a month of concentrated effort on it, at the end of which I admitted final defeat. After this I spoke to Atkin and strongly urged him to try his wits on the problem. Some three months ago he and a Cambridge friend called Swinnerton-Dyer became seriously interested, and quite soon made a decisive advance. . . .
>
> I am very pleased about this, especially because it was largely my active encouragement that started them off; it is the first time that work of mine has borne fruit in other people's brains. Actually what Atkin and Dyer have done is a first-rate piece of work, the best thing either of them have yet done, and much more than my contribution to the problem; and oddly enough it is just this that makes me happy. (F. J. Dyson to his father, May 24, 1948).

Since the rank criterion was only valid for the modulo 5 and modulo 7 cases but not for the modulo 11 case, Dyson had guessed that there existed some analogue of the rank, which he called crank, that would explain the last case. After more than forty years, this guess was confirmed and the crank was discovered by Andrews and Garvan in 1987. (An elementary proof for the modulo 11 case had already been given by Winquist in 1969.) The generating function for partitions of a given rank reported by Dyson in a 1948 paper was used only as a practical tool for numerical calculations but was given no combinatorial meaning. However, generalizing the concept of partitions so that they are defined as parsing into nonnegative parts, instead of into positive parts only, led to a new symmetry of partitions and gave a combinatorial understanding to the generating function. This, incidentally, clarified some special feature of Euler's classic formula for the partition function. Dyson (1987), which consists mainly of an exposition of his previous interests in Ramanujan's work, also records his results in generalizing Ramanujan's function.

37. M. Longuet-Higgins to S. S. Schweber, November 13, 1984.

38. FJD, S^3.

39. Ibid.

40. Ibid. Oscar Hahn later became chairman of British Industries, and was the head of numerous charities. He and Dyson remained friends throughout his life.

41. FJD, S^3. For further remarks concerning Dyson's stay at the Wyton Air Force Base, see his talk, "Strategic Bombing in World War II and Today: Has Anything Changed?" reprinted in Dyson (1992), pp. 70–90.

42. Ibid. The Dover edition of von Neumann's *Mathemathische Grundlagen* is still in Dyson's possession. Its annotations and marginalia give proof of how carefully Dyson worked through the volume. Incorrectly numbered footnotes are corrected, as are printing errors. For example, Dyson noted that on page 211, the beginning of the second paragraph should read "Für $T \to \infty$" rather than "Für $T \to 0$" (as it appears in the text), and he added in the margin, "But this paragraph is distinctly skating on thin ice. See p. 183 especially."

43. FJD, S^3.

44. N. Kemmer to S. S. Schweber.

45. FJD, S^3.

46. F. J. Dyson to R. E. Peierls, October 20, 1946, Peierls Papers, MS Eng b 205 c sac 52/6/77.

47. R. E. Peierls to F. J. Dyson, October 24, 1946, Peierls Papers.

48. F. J. Dyson to R.E. Peierls, October 25, 1946, Peierls Papers.

49. R. E. Peierls to F. J. Dyson, October 29, 1946, Peierls Papers.

50. FJD, S^3.

51. Ibid.

52. F. J. Dyson to his family from Münster, August 8, 1947.

53. Ibid.

54. F. J. Dyson to his family from Münster, August 13, 1947.

55. Ibid.

56. P. M. Morrison, oral interview with S. S. Schweber.

57. Usually there is a member of the mathematics department on the committee, but since Dyson was recognized as a full-fledged mathematician, no mathematician was assigned.

58. Every theoretician had to fulfill a "minor" in experimental physics, and it was therefore usual for an experimentalist to be on the committee.

59. F. J. Dyson to his family, September 25, 1947.

60. The group consisted of Richard Scaletter, P.V.C. Hough, Ed Lennox, Walter McAfee, Leonard Eyges, and Jack Smith.

61. J. H. Smith to S. S. Schweber, October 26, 1984.

62. F. J. Dyson to his family, October 29, 1947. It is interesting to note that Dyson added: "Nevertheless the whole investigation is highly theoretical because it is unknown whether there exist in the universe any mesons with spin zero; and if they do they certainly disintegrate much too rapidly to become constituents of atoms." Thus the news of the discovery of π mesons had not reached Dyson by that time. However, by early 1948 Dyson was explaining to his parents Powell's discovery and its import for nuclear physics. F. J. Dyson to his family, January 14, 1948.

63. F. J. Dyson to his parents, October 29, 1947.

64. FJD, S^3.

65. Letter of Leonard Eyges to S. S. Schweber, September 14, 1986.

66. P. M. Morrison, oral interview with S. S. Schweber.

67. F. J. Dyson to his parents, October 29, 1947.

68. Ibid.

69. Ibid.

70. See, for example, Schiff (1949), pp. 309–311.

71. Jack H. Smith recalls Dyson "dealing with non-orthogonal functions and evidently felt it quite a challenge." Letter of J. H. Smith to S. S. Schweber, October 26, 1984.

72. F. J. Dyson to his parents, December 7, 1947.

73. Ibid.

74. F. J. Dyson to his family, December 7, 1947.

75. Ibid.

76. H. A. Bethe to E. K. Wickman, February 29, 1948. Bethe Papers, 14/22/976.

77. FJD, S^3.

78. Ibid.

79. F. J. Dyson to his parents, April 11, 1948.

80. Ibid. In early March, Bethe wrote Oppenheimer asking him whether it would be possible to invite Dyson to the forthcoming Oldstone conference, and commented: "He is the one who has done most about radiative corrections here. I think he knows more about it than either Feynman or I." H. A. Bethe to J. R. Oppenheimer, March 6, 1948, Oppenheimer Papers.

81. FJD, S^3.

82. Ibid.

83. F. J. Dyson to his parents, March 18, 1948.

84. Ibid.

85. Ibid.

86. FJD, S^3.

87. F. J. Dyson to his parents, November 19, 1947.

88. FJD, S^3.

89. Ibid.

90. F. J. Dyson to his parents, June 2, 1948.

91. F. J. Dyson to Alice Dyson, June 1948.

92. F. J. Dyson to his parents, April 11, 1948.

93. Ibid., August, 1948.

94. FJD, S^3.

95. F. J. Dyson to H. A. Bethe, August 9, 1948, Bethe Papers.

96. F. J. Dyson to his parents, August 26, 1948.

97. Ibid.

98. F. J. Dyson to his parents, September 18, 1948.

99. Ibid.

100. F. J. Dyson to his parents, September 30, 1948.

101. F. J. Dyson to H. A. Bethe, September 8, 1948, Bethe Papers. Bethe in fact reported on the results that Dyson had obtained at the Birmingham conference that took place in late summer 1948. The proceedings of that conference contain the following abstract of Bethe's report: "Dyson is working on a reformulation of Schwinger's theory which includes features of simplicity of Feynman's prescription and which enables discussion of the higher terms."

102. F. J. Dyson to his parents, September 3, 1948.

103. FJD, S^3.

104. Ibid.

105. Ibid.

106. F. Dyson to his parents, September 26, 1948.

107. The Institute for Advanced Study, pamphlet entitled "Some Introductory Information," Princeton, N.J. (1951). Oppenheimer Papers, Box 233.

108. For a brief history of the Institute for Advanced Study, see the foreword by Harry Woolf to *A Community of Scholars, The Institute for Advanced Study Faculty and Members*, 1930–1980 (Princeton: The Institute for Advanced Study, 1980). A more detailed history of the founding of the Institute is given in Porter (1988). See also Regis (1988). The quotation from Flexner's memorandum is in Woolf's foreword. See also Gertrude Samuels, "Where Einstein Surveys the Universe," *New York Times Magazine*, November 19, 1950, p. 14.

109. *A Community of Scholars.*

110. Ibid.

111. Oppenheimer Papers, Box 233, AIS—Director Appointment. Letter from Herbert H. Maass to J. Robert Oppenheimer, September 8, 1947.

112. Report of the Director, 1948–1953 (The Institute for Advanced Study, 1954). A copy of the report is in the Oppenheimer Papers, Box 233.

113. F. J. Dyson to his parents. Rabi made the following comments on the Institute in a letter to his friend Otto Halpern: "The Oppenheimer 'Cheder' is now operating. Thus far it has been interesting but not as effective as the rival Schwinger-Weisskopf 'Yeshiva' which while lacking the social elegance of the first makes up for it by displaying a more indigeneous (*sic*) and less eclectic art." Rabi to Otto Halpern, February 18, 1948. Rabi Papers, Library of Congress, RIIC, Folder "Physics-Technical Letters."

114. F. J. Dyson to his parents, October 16, 1948.

115. Ibid.

116. F. J. Dyson to his parents, October 10, 1948.

117. Ibid., October 16, 1948.

118. Ibid.

119. Ibid., November 1, 1948.

120. A copy of the draft of Oppenheimer's report can be found among his papers in the National Archives.

121. A carbon copy of the memorandum is among F. J. Dyson's papers.

122. F. J. Dyson to his parents, October 19, 1948.

123. Ibid., November 1, 1948.

124. Ibid.

125. Ibid.

126. Ibid.

127. Ibid., November 14, 1948.

128. L. Eyges to F. J. Dyson, October 26, 1948. L. Eyges had been a fellow graduate student with Dyson at Cornell. He had gone to Birmingham on a fellowship to work with Peierls. Oppenheimer's allusion is of course to Fermat's statement concerning his proof that no integral values of x, y, z can be found to satisfy the equation $x^n + y^n = z^n$ if n is an integer greater than 2.

129. F. J. Dyson to his parents, November 14, 1948.

130. Ibid.

131. Ibid., November 21, 1948.

132. Ibid.

133. Ibid.

134. Ibid.

135. Dyson papers.

136. Ibid., F.J. Dyson to his parents, December 4, 1948.

137. Ibid.

138. Ibid.

139. Ibid.

140. FJD, S^3.

141. F. J. Dyson to his parents, September 30, 1948.

142. Ibid., October 4, 1948.

143. Ibid., November 4, 1948.

144. Ibid., November 4, 1948.

145. Dyson's mood in mid-October is conveyed by his letters to his parents. He was very lonely, eating by himself and keeping to himself: "I believe I should find it much easier to live through this time if I had somebody to talk to; it is hard to be so much alone as I am here and to keep sane; thank goodness I can always make a dash and get relaxed by talking things over with my faithful friends Bethe at New York and Feynman at Cornell." F. J. Dyson to his parents, October 16, 1948.

146. *The New Yorker*, April 30, 1949, pp. 23–24.

147. FJD, S^3.

148. Ibid.

149. F. J. Dyson to R. E. Peierls, February 23, 1949, Peierls Papers.

150. FJD, S^3.

151. F. J. Dyson to his parents, January 22, 1949.

152. FJD, S^3.

153. F. J. Dyson to R. E. Peierls, March 31, 1949, Peierls Papers.

154. F. J. Dyson to his parents, January 15, 1949.

155. Ibid.

156. Ibid.

157. Ibid.

158. F. J. Dyson to R. E. Peierls, March 31, 1949, Peierls Papers. Peierls was working with McManus on a classical, but relativistically invariant, theory that attempted to give the electron an extended charge distribution. McManus (1949).

159. F. J. Dyson to his parents, January 30, 1949.

160. Ibid.

161. Ibid.

162. F. J. Dyson to H. A. Bethe, February 10, 1949, Bethe Papers.

163. F. J. Dyson to his parents, April 5, 1949.

164. F. J. Dyson to R. E. Peierls, March 31, 1949, Peierls Papers.

165. F. J. Dyson to his parents, April 5, 1949.

166. F. J. Dyson to his parents, January 22, 1948.

167. Dyson had obtained an offer for a fellowship from Peierls. R. E. Peierls to F. J. Dyson, January 16, 1949.

168. H. A. Bethe to R. E. Peierls, February 2, 1949, Peierls Papers.

169. The Warren Research Committee wrote Dyson that he had been awarded the fellowship "for your work on mathematical physics . . . . The stipend offered is £800 per annum (plus superannuation allowance of £80). The period of the appointment is for four years in the first instance. Warren fellowships are renewable. It is expected that a fellow devote his or her whole time to research." D. C. Martin to F. J. Dyson, May 17, 1949, Peierls Papers.

170. R. E. Peierls to F. J. Dyson, March 21, 1949, Peierls Papers.

171. F. J. Dyson to his parents, June 26, 1949.

172. F. J. Dyson to his parents, February 28, 1949.

173. F. J. Dyson to his parents, Easter Sunday, 1949.

174. FJD, S^3. In his letter to his parents written on April 17, 1949, Dyson told them: "For my part, knowing my limited capacity for attention, I chose a comfortable chair and allowed nature to take its course; three times during the conference I slept soundly and unnoticed and the rest of the time I was the better for it."

175. F. J. Dyson to his parents, April 17, 1949.

176. Dyson himself had made extensive calculations of radiative corrections. In a letter to Bethe in early February 1949, Dyson indicates that he had calculated the vacuum polarization correction to the vertex diagram: "As [it] is the easiest I started with that and it came out very nicely. It gives an increase of the magnetic moment by a relative amount

$$\frac{\alpha^2}{6} \left[\frac{119}{12\pi^2} - 1 \right]$$

...The way the (-1) comes in and nearly cancels the whole thing is very peculiar and beautiful." F. J. Dyson to H. A. Bethe, February 10, 1949, Bethe Papers. The Oldstone notes were written up by A. S. Wightman from the notes taken at the conference by J. A. Wheeler. A copy is in the Oppenheimer Papers.

177. Oldstone notes, Oppenheimer Papers.

178. FJD, S^3.

179. Ibid.

180. Ibid.

181. F. J. Dyson to his parents, March 11, 1949.

182. W. Heisenberg to F. J. Dyson, December 19, 1949. Dyson Papers.

183. F. J. Dyson to W. Heisenberg, January 23, 1950. Copy in Dyson Papers.

184. FJD, S^3.

185. R. E. Peierls to the registrar, Birmingham University, November 3, 1949, Peierls Papers.

186. F. J. Dyson to his parents, January 22, 1949.

187. J. R. Oppenheimer to R. E. Peierls, March 16, 1949.

188. F. J. Dyson to H. A. Bethe, July 13, 1950, from Ann Arbor, Michigan.

189. F. J. Dyson to R. E. Peierls, September 23, 1950.

190. F. J. Dyson to his parents, November 2, 1950.

191. Ibid., November 14, 1950.

192. Ibid., November 14, 1950.

193. Ibid., December 14, 1950.

194. R. E. Peierls to Royal Society, January 19, 1951.

195. FJD, S^3.

196. Ibid.

197. F. J. Dyson to H. A. Bethe, April 15, 1951, Bethe Papers.

198. FJD, S^3.

199. Ibid.

200. "You may have heard that Feynman is leaving Cornell. The California Institute of Technology has offered him a very high-class professorship. We tried our best to hold him here but we did not succeed. His main reason for going to Cal Tech seems to be that it is in a big city which will offer him more entertainment, and Cornell University could not quite manage to build a city of a million here at Ithaca; his second reason was that it was time for him to change and that Cal Tech was one of the very few places which he would consider at all. It took him very long to decide because he liked it here very much." H. A. Bethe to F. J. Dyson, May 5, 1950. Bethe Papers.

Officially Dyson refused the offer from Bethe, because the terms of his Commonwealth fellowship had stipulated that he must not return to the United States for a period of two years after its completion; However, his letter clearly intimated that he would welcome an offer later (F. J. Dyson to H. A. Bethe, May 10, 1950, Bethe Papers). Bethe immediately replied that he would keep the position open for one year and invited him to come to Cornell in the fall of 1951. H. A. Bethe to F. J. Dyson, May 17, 1950, Bethe Papers. Dyson then accepted the professorship. F. J. Dyson to H. A. Bethe, May 23, 1950, Bethe Papers. Actually the Commonwealth Foundation allowed Dyson to come to the United States during the summer and fall of 1950, to lecture at Michigan and to visit the Institute, because the "2-year rule is to prevent the fellowship being used as a stepping stone to permanent appointments" (F. J. Dyson to H. A. Bethe, May 23, 1950, Bethe Papers).

201. FJD, S^3.

202. FJD, S^3.

203. Ibid.

204. Ibid.

205. F. J. Dyson to his parents, October 4, 1948. The Athanasian creed asserts belief in the Trinity, as opposed to Arianism. The doctrines of Arius declared that Jesus was not of the same substance as God, but only the best of created living beings.

206. Ibid., Spring 1949.

207. FJD, S^3.

208. Ibid.

209. Ibid.

210. Ibid.

211. F. J. Dyson to his parents, November 14, 1950.

212. Ibid., February 15, 1949.

213. Personal interviews with S. S. Schweber, 1985.

214. FJD, S^3.

215. F. J. Dyson to his parents, November 2, 1950.

216. FJD, S^3.

217. Ibid.

218. In his lecture, Schwinger also pointed out that the Japanese character *shin* that appears in Tomonaga's name meant "to shake" or "to wave," and so did his own name; in German, *schwingen* means "to swing" or "to shake." For this reason he gave his lecture the title, "Two Shakers of Physics." Interestingly, Schwinger has also written a biographical sketch of Feynman, the other partner in the Nobel Prize award (Schwinger 1989).

219. Note that after the war, both of them also won prizes for contributions on the nature of light.

220. Such cases of simultaneous explanations given in widely differing contexts are usually adduced to lend support to the notion that physics captures some "objective" features of the physical world.

221. It is worth noting that the age of the contributors manifests itself in other ways. The contributions of the somewhat older physicists—Tomonaga, Bethe, Pauli, Peierls—are primarily group efforts. This, to some extent, is a reflection of the more senior positions they occupied and of their responsibilities toward graduate students and postdoctoral fellows; and in the case of the Americans, of their time-consuming roles as consultants to the government and to industry.

222. Only three people can share in the award, which made the assignments particularly difficult.

Chapter 10
QED in Switzerland

1. Gregor Wentzel was born in Düsseldorf, Germany, on February 17, 1898. After studying mathematics at Greifswald and Freiburg, Wentzel went to Munich in the fall of 1920 to work with Sommerfeld. It was there that he first met Pauli and Heisenberg, who at the time were students in Sommerfeld's seminar. Wentzel saw a good deal of Pauli then, and they remained close friends until Pauli's death. By the time of his appointment to the post in Zurich, Wentzel had made important contributions to the quantum theory of atomic structure and atomic spectra and to the quantum theory of scattering. He joined the faculty of the University of Chicago in 1948. He retired in 1970 and moved to Ancona, Switzerland, where he died on August 12, 1978. See Wentzel (1970).

2. Students would attend both Pauli's and Wentzel's courses. Wentzel was a brilliant lecturer, whereas Pauli was somewhat disorganized. Joint seminars were held every month at which the latest developments in theoretical physics were presented. During the thirties Pauli maintained close contact with Heisenberg; students and assistants moved freely between Zurich and Leipzig. Pauli also established strong ties with the Zurich mathematics community.

3. Stueckelberg was born in Basel on February 1, 1905. His father's family could trace their ancestry in the canton of Basel back to the fourteenth century, and his mother was the baroness of Breidenbach at Breidenstein and Melsbach, a tiny dukedom in central Germany. He had studied in Munich and had obtained his Ph.D. with Sommerfeld in 1927. He spent the next few years in Princeton, first as a research associate and then as an assistant professor. In collaborations with J. G. Winans and P. M. Morse, he made important contributions to the quantum theory of molecular structure and to scattering theory. In 1935, after a two-year stay in Zurich, he accepted the professorship of theoretical physics at the University of Geneva. From 1942 until 1947 he also held an appointment as professor of physics at the University of Lausanne. For further background material regarding Stueckelberg, see Crease and Mann (1987) and Enz's obituary (1986).

4. See the interesting discussion given by Wentzel (1947, p. 16), who reached a conclusion at variance with later findings.

5. According to Felix Villars, relations between Stueckelberg and Markus Fierz were very strained. On one occasion Fierz accused him of "poisoning the young." Interview with S. S. Schweber, March 1981. Fierz was born in Zurich on June 20, 1912. His father was a professor of chemistry at the E.T.H. He studied physics in Göttingen and at the University of Zurich and received his Ph.D. with a dissertation under Wentzel in 1936. He was Pauli's assistant in the late 1930s. He went to the University of Basel as a *Privatdozent* in 1940, became an associate professor of theoretical physics there in 1943 and a full professor in 1945. During the mid-forties he also was the editor of the *Helvetica Acta Physica*. Fierz regularly attended the monthly theoretical seminar that was held in Zurich. Stueckelberg would usually also come in from Geneva.

6. F. Villars, interview with S. S. Schweber, 1981. The most complete and succinct statement of Stueckelberg's views on the connection between causality and the S-matrix is to be found in Stueckelberg and Rivier (1950).

7. Rivier was born on November 12, 1918. He was educated at the University of Lausanne and Geneva and did his dissertation with Stueckelberg.

8. See correspondence between Rivier and Pauli, 1948. I thank D. Rivier for showing me these letters.

9. F. Villars, interview with S. S. Schweber.

10. Wolfgang Pauli senior made important contributions to colloid chemistry. In 1907 he was appointed an associate professor at the University of Vienna and in 1922 he became the director of the Institute for Medical Colloid Chemistry there.

11. Quoted in Enz (1973), p. 767. Mach was Wolfgang's godfather and the "fin-de-siècle style" silver cup that Mach gave as a christening present is among Pauli's mementos at CERN.

12. For an account, see Hendry (1984), Fierz and Weisskopf (1960), Enz (1973), Pais (1986). For an insight into Pauli's momentous contributions, see vols. 1 and 2 of his remarkable correspondence (Pauli 1979, 1985), and Pauli (1964).

13. E.g., Pauli and Weisskopf (1934), Pauli and Rose (1936), Pauli and Fierz (1938); e.g., Fierz's (1939) formulation of the theory of free fields of arbitrary spin is an instance of work written under the aegis of Pauli.

14. Pauli had come to the United States in 1940 because he had feared that Switzerland might be overrun by Germany. In any case, he was vulnerable because his mother had been Jewish, and he still carried an Austrian passport and was considered a German citizen after the Anschluss in 1938. When approached about working on the atomic bomb, he indicated that he was uncertain whether he should go into research directly connected with the war. From Los Alamos, Oppenheimer convinced him that it would be "a waste and an error" for him to do that, he being "just about the only physicist in the country who can help to keep those principles of science alive which do not seem immediately relevant to the war, and that is imminently worth doing." In the same letter Oppenheimer informed him that he was worried about the fact that "none of the people in our field are publishing work in the Physical Review. . . . It must be apparent to the enemy that these physicists are being put to good use and that this is in itself a piece of information on the nature of the work we are doing." "We have often wondered," Oppenheimer continued, "whether your great talents for physics and for burlesque could not appropriately be put to use by your publishing some work in the names of a few of the men who are now engaged in things that they cannot publish. It would give you a chance to express in the most appropriate way possible your evaluation of their qualities and you would have a delicious opportunity to argue with yourself in the public press without any interference. I think that you should not undertake this without getting the permission of your victims, but I know that Bethe, Teller, Serber and I would be delighted to grant you that and I do not doubt that there would be many others. Do not dismiss this thought too lightly." J. R. Oppenheimer to W. Pauli, May 20, 1943, in Smith and Weiner (1980).

15. Pauli's decision to return to Switzerland after World War II is described in Jost (1984).

16. Luttinger reports that "the school has disappeared . . . I saw a report that my mother had which showed that the kids from that school had been remarkably successful even though the school was totally undisciplined and unformed." Interview with J. Luttinger, October 1984. Unless otherwise indicated all the quotations in the present section are taken from that taped interview.

17. Reiner later became a professor of mathematics at the University of Illinois.

18. The senior thesis was submitted in February 1944 and is on deposit in the MIT Archives.

19. L. Tisza, telephone interview, January 2, 1989.

20. Elliot Montroll was born on May 14, 1916, in Pittsburgh. He obtained his B.S. degree in 1937 and a Ph.D. in mathematics in 1940, both from the University of Pittsburgh. During the academic year 1939/40 he was at Columbia as a research assistant in theoretical chemistry; he spent the following year working with Lars Onsager as a Sterling resident fellow at Yale. After a year at Cornell and one in Princeton, he joined the Kellex Corporation in 1943 as head of mathematical research.

21. The seminar was attended by Willis Lamb, Julius Ashkin, and interested graduate students. Luttinger first heard of Onsager's solution of the 2-dimensional Ising model and of Kramers and Wannier's researches on that model in the seminar.

22. The title of Luttinger's Ph.D. dissertation was "Dipole Interactions in Crystals"; it dealt with the classical and quantum statistical mechanics of dipoles on cubic lattices interacting with one another through dipole-dipole forces.

23. Luttinger found Källén "very smart and quite remarkable." Although only a young graduate student at the time, "he was already a fully established personality, who argued with everybody, Pauli included."

24. This is a direct consequence of the fact that the Dirac equation predicts a g factor exactly equal to 2 for the electron. See Luttinger (1948).

25. Weisskopf had shown this in 1936.

26. See Luttinger (1948) for the details of this elegant calculation.

27. J. R. Oppenheimer to W. Pauli, December 9, 1947. Fierz Papers.

28. Ibid.

29. I. I. Rabi to W. Pauli, December 19, 1947. Fierz Papers.

30. W. Pauli to J. R. Oppenheimer, January 6, 1948. Oppenheimer Papers. See also Pauli's letter to Fierz, January 7, 1948. Fierz Papers.

31. W. Pauli to J. R. Oppenheimer, October 29, 1948. Oppenheimer Papers. The report of Pauli (1948) to the Solvay conference indicated only that Luttinger had obtained finite answers for the magnetic moments that did not agree with observed values.

32. F. Villars, interview with S. S. Schweber.

33. W. Pauli to M. Fierz, January 3, 1949. Fierz Papers.

34. W. Pauli to J. R. Oppenheimer, February 22, 1949. Oppenheimer Papers.

35. What the invariant regularization does is easily seen for the case that $i = 3$. The conditions (10.2.3.a) and (10.2.3b) then imply that the Fourier transform of the regularized Green's function has the form

$$\sum_i \frac{c_i}{(k^2 - M_i^2)} = \frac{\text{constant}}{(k^2 - M_1^2)(k^2 - M_2^2)(k^2 - M_3^2)} \tag{A}$$

and will converge much more rapidly for large k. Pauli and Villars regarded the auxiliary masses as "formalistic." It was observed by Rayski (1948) and by Umezawa et al. (1948) that it might be possible to give a "realistic" interpretation to the masses, if the corresponding particles obey Bose statistics. In other words, the auxiliary masses would "compensate" the vacuum polarization divergences in the way the C-mesons did for the self-energy of the electron. See also Jost and Rayski (1949). An exploration of the possibility of interpreting the regularization procedure realistically in higher orders was made by Källén (1949). If one considers $\triangle_{R(x)}$ as the actual Green's function of a boson field, the latter satisfies the following interaction picture equation:

$$\prod_i \left(\Box^2 + M_i^2\right) \triangle_R (x) = \delta(x). \tag{B}$$

A multimass equation such as (B) implies that the theory must be quantized with an indefinite metric in Hilbert space (Pais and Uhlenbeck 1950). In fact, the work of Pais and Uhlenbeck demonstrated that it was evidently impossible to satisfy simultaneously the requirement that the theory be (1) finite, (2) causal, and (3) that the metric in Hilbert space be positive definite. See also Villars (1960), p. 84.

36. W. Pauli to J. Schwinger, January 24, 1949, Oppenheimer Papers.

37. W. Pauli to J. R. Oppenheimer, February 22, 1949, Oppenheimer Papers.

38. Ibid. Oppenheimer sent to Pauli a copy of Dyson's second paper in January 1949, and Dyson himself wrote Pauli on January 17, 1949. On February 22 Pauli informed Oppenheimer that "he was very glad about Dyson's letter . . . because its content agree[d] very well with my own ideas."

39. Interview with J. Luttinger, October 1984.

40. In their paper, Jost and Luttinger acknowledge Dyson's help in explaining to them how to handle ovelapping divergences (Jost and Luttinger 1950, p. 211).

41. In a footnote to their paper, Jost and Luttinger indicate that Schwinger in a seminar in Zurich in September 1949 was able to demonstrate the universality to all orders by considering charge renormalization as a renormalization of the electromagnetic field.

42. Racah (1946) has dispelled a similar conjecture by Pauli for the self-energy of the electron. Racah showed that the e^4 self-energy contribution for an electron at rest has the same sign as the order e^2 contribution.

43. Schwinger later proved that

$$\frac{e^2}{e_o^2} = \frac{1}{1 + C},$$

with $C > 0$. Similarly Jost and Rayski (1949) checked that the multiple ambiguities in Schwinger's calculation of the photon self-energy are satisfactorily resolved by the Pauli-Villars regulator method.

44. J. Luttinger, interview with S. S. Schweber.

45. Ibid.

Epilogue
Some Reflections on Renormalization Theory

1. The following is an adaptation from the article by Cao and Schweber (1993).

2. Wightman (1978) takes constructive quantum field theory as an offspring of axiomatic field theory, with some difference between them. While the concern of axiomatic field theory is the general theory of quantum fields, the constructive field theory starts from specific Lagrangian models and constructs solutions satisfying the requirements of the former. For early development of axiomatic field theory, see Jost (1965) and Streater and Wightman (1964); for constructive field theory, cf. Velo and Wightman (1973).

BIBLIOGRAPHY

Akhiezer, A. 1937. Über die Streuung von Licht an Licht. *Phys. Zeits. Sowjetunion* 11: 263–276.

Akhiezer, A., and Berezetski, V. B. 1963. *Quantum Electrodynamics*. 2d ed. New York: Wiley.

Akhiezer, A., and Pomeranchuck, I. 1937. Über die kohärente Streuung von γ-Strahlen durch Kernen . *Phys. Zeits. Sowjetunion* 11:478–498.

Allison, D. K. 1981. *New Eye for the Navy: The Origin of Radar at the Naval Research Laboratory*. Washington, D.C.: Naval Research Laboratory.

Alvarez, L. W. 1969. Developments in particle physics. *Science* 165:1071–1091. Nobel lecture.

Alvarez, L. W. 1987. *Alvarez: Adventures of a Physicist*. New York: Basic Books.

Anderson, C. D. 1932a. The apparent existence of easily deflectable positives. *Science* 76:238–239.

Anderson, C. D. 1932b. Energies of cosmic-ray particles. *Phys. Rev.* 41:405–412.

Anderson, C. D. 1933. The positive electron. *Phys. Rev.* 43:491–494.

Anderson, C. D. 1961. Early work on the positron and muon. *Am. J. Phys.* 29:825–830.

Anderson, C. D.; Millikan, R. A.; Needermeyer, S.; and Pickering, W. 1934. The mechanism of cosmic-ray counter action. *Phys. Rev.* 45:352–363.

Araki, G., and Tomonaga, S. 1940. Effect of the nuclear Coulomb field on the capture of slow mesons. *Phys. Rev.* 58:90–91.

Aramaki, S. 1987. Formation of the renormalization theory in quantum electrodynamics. *Historia Scientiarum* 32:1–42.

Aramaki, S. 1989. Development of the renormalization theory in quantum electrodynamics. II. *Historia Scientiarum* 37:91–113.

Aris, R.; Davis, H. T.; and Stuewer, R., eds. 1983. *Springs of Scientific Creativity*. Minneapolis: University of Minnesota Press.

Auger, P., and Ehrenfest, P., Jr. 1935. Clichés de rayons cosmiques obtenus avec une chambre de Wilson-Blackett dans des condition spéciales. *J. de Phys.* 6:255–256.

Bacon, F. 1942. *Essays and New Atlantis*. New York: Walter J. Black.

Badash, L.; Hirshfelder, J. O.; and Broida, H. P. 1980. *Reminiscences of Los Alamos: 1943–45*. Dordrecht, Holland: Reidel.

Baranger, M. 1951. Relativistic corrections to the Lamb shift. Ph.D. diss., Cornell University, Ithaca, New York.

Baranger, M.; Bethe, H. A.; and Feynman, R. P. 1953. Relativistic corrections to the Lamb shift. *Phys. Rev.* 92:482–501.

Bargmann, V., and Wigner, E. P. 1946. Group theoretical discussion of relativistic wave equations. *Proc. Natl. Acad. Sci. U.S.A.* 34:211–223.

Barut, A. O.; van der Merwe, A.; and Vigier, J.-P. 1984. *Quantum, Space and Time. The Quest Continues: Studies in Honour of Louis de Broglie, Paul Dirac and Eugene Wigner*. Cambridge, U.K.: Cambridge University Press.

Baxter, J. P. 1948. *Scientists against Time*. Boston: Little, Brown and Co.

Beasley, W. G. 1974. *The Modern History of Japan*. 2d. ed. New York: Preager.

Becchi, C.; Rouet, A.; and Stora, R. 1974. The Abelian Higgs-Kibble model, unitarity of the *S*-operator. *Physics Letters* 52B:344–346.

Becker, G., and Autler, S. 1946. Water vapor absorption of electromagnetic radiation in the centimeter wavelength range. *Phys. Rev.* 70:300–307.

Belloni, L. 1978. A note on Fermi's route to Fermi-Dirac statistics. *Scientia* 113:421–429.

Bernal, J. D. 1946. Lessons of the war for science. *Rep. Prog. Phys.* 10:418–436.

Bernstein, J. 1975. Rabi. *New Yorker,* October 13 and October 20, 1975.

Bernstein, J. 1979. *Hans Bethe: Prophet of Energy.* New York: Basic Books.

Bethe, H. A. 1932. Bremsformel für Elektronen relativistischer Geschwindigkeit. *Zeits. Phys.* 76:293–299.

Bethe, H. A. 1933. Quantenmechanik der Ein- und Zwei-Electronprobleme. In H. Geiger, K. Scheel, and A. Smekal, eds., *Handbuch der Physik.* Vol. 24, part 1, chap. 3, pp. 273–551. Berlin: Springer-Verlag.

Bethe, H. A. 1947. The electromagnetic shift of energy levels. *Phys. Rev.* 72:339–341. (Reprinted in Schwinger 1958, pp. 139–141.)

Bethe, H. A. 1948. On the electromagnetic shift of energy levels. *Phys. Rev.* 73:1272A.

Bethe, H. A. 1968. J. Robert Oppenheimer, 1904–1967. *Biog. Mem. Fellows Roy. Soc. London,* 14:391–416.

Bethe, H. A., and Fermi, E. 1932. Über die Wechselwirkung von Zwei Electronen. *Zeits. Phys.* 77:296–306.

Bethe, H. A., and Heitler, W. 1934. On the stopping of fast particles and on the creation of positive electrons. *Proc. Roy. Soc. London* A146:83–112.

Bethe, H. A., and Oppenheimer, J. R. 1946. Reaction of radiation on electron scattering and Heitler's theory of radiation damping. *Phys. Rev.* 70:451–458.

Bethe, H. A., and Salpeter, E. E. 1957. *Quantum Mechanics of One- and Two-Electron Atoms.* Berlin: Springer-Verlag.

Bethe, H. A.; Bacher, R. F.; and Livingston, M. S. 1986. *Basic Bethe. Seminal Articles on Nuclear Physics, 1936–1937.* New York: American Institute of Physics.

Betz, O. 1932. Über die Absorption kurzer elektrischer Wellen in ionisierten Gasen, ein Ursuch zum Nachweis der langwelligen Strahlung des Wasserstoff Atoms. *Ann. der Phys.* 15:321–344.

Beyer, H. J. 1978. Lamb-shift and fine-structure measurements on one-electron systems. In Hanle and Kleinpoppen, pp. 529–605.

Beyerchen, A. D. 1977. *Scientists under Hitler: Politics and the Physics Community in the Third Reich.* New Haven, Conn.: Yale University Press.

Bhabha, H. J. 1936. The scattering of positrons by electrons with exchange on Dirac's theory of the positron. *Proc. Roy. Soc. London* A154:195–206.

Bhabha, H. J. 1940. Elementary heavy particles with any integral charge. *Proc. Indian Acad. Sci.* 11:347–368. Errata 11:467.

Bhabha, H. J. 1945. Relativistic wave equations for the elementary particles. *Rev. Mod. Phys.* 17:200–216.

Bhabha, H. J., and Heitler, W. 1937. The passage of fast electrons and the theory of cosmic showers. *Proc. Roy. Soc. London* A159:432–458.

Bialynicki-Birula, I. 1960. On the gauge properties of Green's functions. *Nuovo Cimento* 17/6:951–955.

Bialynicki-Birula, I. 1970. Renormalization, diagrams and gauge invariance. *Phys. Rev.* D2:825–838.

Biedenharn, L. C. 1984. The "Sommerfeld puzzle" revisited and resolved. In A. O. Barut et al. 1984, pp. 258–279.

Birr, K. 1979. Industrial laboratories. In N. Reingold, ed., *The Sciences in the American Context,* pp. 193–208. Washington, D.C.: Smithsonian Institution Press.

Bjorken, J. D., and Drell, S. 1965. *Relativistic Quantum Fields.* New York: McGraw-Hill.

Blackett, P.M.S. 1933a. The craft of experimental physics. In H. Wright, ed., *University Studies: Cambridge 1933.* London.

Blackett, P.M.S. 1933b. The positive electron. *Nature* 133:917–918.

Blackett, P.M.S. 1969. The old days of the Cavendish. *Riv. Nuovo Cimento,* Special Issue 1, p. xxxiii.

Blackett, P.M.S. 1972. Rutherford. *Notes and Records Roy. Soc.* 27:57–72.

Blackett, P.M.S., and Occhialini, G.P.S. 1933. Some photographs of the tracks of penetrating radiation. *Proc. Roy. Soc. London* A139:699–720.

Blackett, P.M.S., and Wilson, J. G. 1937. Energy loss of cosmic ray particles in metal plates. *Proc. Roy. Soc. London* A160:304–323.

Blatt, J. M. 1945. On the meson charge cloud around a proton. *Phys. Rev.* 72:227–228.

Blatt, J. M. 1947. On Heitler's theory of radiation damping. *Phys. Rev.* 72:466–477.

Bloch, F. 1933. Bremsvermögen von Atomen mit mehreren Elektronen. *Zeits. Phys.* 81:363–376.

Bloch, F. 1934. Die physikalische Bedeutung mehrerer Zeiten in der Quantenelectrodynamik. *Phys. Zeits. Sowjetunion* 5:301–315.

Bloch, F. 1936. On the continuous γ-radiation accompanying the β-decay. *Phys. Rev.* 50:272–278.

Bloch, F., and Nordsieck, A. 1937. A note on the radiation field of the electron. *Phys. Rev.* 52:54–59. (Also reprinted in Schwinger 1958, pp. 129–134.)

Blumberg, S. A., and Owens, G. *The Life and Times of Edward Teller.* New York: G. P. Putnam's Sons.

Bogoliubov, N. N., and Parasiuk, O. 1957. Über die Multiplikation der Kausalfunktionen in der Quantentheorie der Felder. *Acta Math.* 97:227–266.

Bogoliubov, N. N., and Shirkov, D. V. 1959. *Introduction to the Theory of Quantized Fields.* New York: Interscience.

Bohm, D., and Weinstein, M. 1948. The self-oscillation of charged particles. *Phys. Rev.* 74:1789–1798.

Bohr, N. 1939. The causality problem in atomic physics. In *New Theories in Physics, 1939.* Reports to the Congress of the Institute of Intellectual Cooperation held in Warsaw, September 1938. Paris: Institute of Intellectual Cooperation.

Bohr, N., and Rosenfeld, L. 1933. Zur Frage der Messbarkeit der electromagnetischen Feldgrossen. *Kgl. Danske Vidensk. Selskab. Mat.-Fys. Medd.* 12. (English trans. in Cohen and Stachel 1979.)

Bohr, N., and Rosenfeld, L. 1950. Field and charge measurements in quantum electrodynamics. *Phys. Rev.* 78:794–798.

Bohr, N., and Wheeler, J. A. 1939. The mechanism of nuclear fission. *Phys. Rev.* 56:426–450.

Bopp, F. 1940. Eine lineare Theorie des Elektrons. *Ann. der Phys.* 38:345–384.

Born, M. 1952. Arnold Johannes Wilhelm Sommerfeld. *Obit. Not. Fellows Roy. Soc. London* 8/21:275–296.

Born, M. 1933. On the quantum theory of the electromagnetic field. *Proc. Roy. Soc. London* A143:410–437.

Born, M. 1977. *My Life: Recollections of a Nobel Laureate*. New York: Charles Scribner's Sons.

Born, M., and Infeld, L. 1934a. Foundations of the new theory. *Proc. Roy. Soc. London* A144:425–451.

Born, M., and Infeld, L. 1934b. On the quantization of the field equations of the new theory. *Proc. Roy. Soc. London* A147:522–546.

Born, M., and Infeld, L. 1935. On the quantization of the field equations of the new theory. II. *Proc. Roy. Soc. London* A150:141–166.

Born, M., and Jordan, P. 1925a. Zur Quantentheorie aperiodischer Vorgänge. *Zeits. Phys.* 33:479–505.

Born, M., and Jordan, P. 1925b. Zur Quantenmechanik. *Zeits. Phys.* 34:858–888.

Born, M., and Oppenheimer, J. R. 1927. Zur Quantentheorie der Molekulen. *Ann der Phys.* 84:457–484.

Born, M.; Heisenberg, W.; and Jordan, P. 1926. Zur Quantenmechanik. II. *Zeits. Phys.* 35:557–615.

Borowitz, S., and Kohn, W. 1949. On the electromagnetic properties of nucleons. *Phys. Rev.* 76:818–827.

Borowitz, S.; Kohn, W.; and Schwinger, J. 1950. On the self-stress of the electron. *Phys. Rev.* 78:345.

Braunbeck, W., and Weinmann, E. 1938. Die Rutherford-Streuung mit Berücksichtigung der Ausstrahlung. *Zeits. Phys.* 110:360–372.

Breit, G. 1929. The effect of retardation on the interaction of two electrons. *Phys. Rev.* 24:553–573.

Breit, G. 1947a. Relativistic corrections to magnetic moments of nuclear particles. *Phys. Rev.* 71:400–402.

Breit, G. 1947b. Does the electron have an intrinsic magnetic moment? *Phys. Rev.* 72:984.

Breit G., and Teller, E. 1940. Metastability of H and He levels. *Astrophys. J.* 91:215–238.

Bromberg, J. 1971. The impact of the neutron: Bohr and Heisenberg. In R. McCormmach, ed., *Historical Studies in the Physical Sciences*, vol. 3, pp. 307–331. Princeton, N. J.: Princeton University Press.

Bromberg, J. 1976. The concept of particle creation before and after quantum mechanics. In R. McCormmach, ed., *Historical Studies in the Physical Sciences*, vol. 7, pp. 161–191. Princeton, N.J.: Princeton University Press.

Bromberg, J. 1977. Dirac's quantum electrodynamics and the wave-particle equivalence. In C. Weiner, ed., *History of Twentieth Century Physics,* pp. 147–157. New York: Academic Press.

Brower, K. 1978. *The Starship and the Canoe*. New York: Holt, Rinehart and Winston.

Brown, L. M. 1978. The idea of the neutrino. *Physics Today* 31/9:23–28.

Brown, L. M. 1981. Yukawa's prediction of the meson. *Centaurus* 25:71–132.

Brown, L. M. 1985. How Yukawa arrived at the meson theory. *Prog. Theor. Phys.* (suppl.) 85:13–19.

Brown, L. M. 1986. Yoichiro Nambu: The first forty years. *Prog. Theor. Phys.* (suppl.) 86:1–11.

Brown, L. M., and Cao, T. Y. 1991. Spontaneous breakdown of symmetry: Its rediscovery and integration into quantum field theory. *Hist. Stud. Phys. Bio. Sciences* 21/2:211–236.

Brown, L. M., and Hoddeson, L. 1983. *The Birth of Particle Physics*. Cambridge, U.K.: Cambridge University Press.

Brown, L. M. and Konuma, M., eds. 1986. Proceedings of the Japan–U.S. collaborative workshops on history of particle theory in Japan, 1935–1960. Research Institute for Fundamental Physics, University of Kyoto.

Brown, L. M., and Moyer, D. M. 1984. Lady or tiger?—The Meitner-Hupfeld effect and Heisenberg's neutron theory. *Am. J. Phys.* 52:130–136.

Brown, L. M., and Rechenberg, H. 1987. *Paul Dirac and Werner Heisenberg—A partnership in science.* In Kursunoglu and Wigner 1987.

Brown, L. M.; Dresden, M.; and Hoddeson, L., eds. 1989. *Pions to Quarks: Particle Physics in the 1950s.* Cambridge, U.K.: Cambridge University Press.

Brown, L. M.; Kawabe, R.; Konuma, M.; and Maki, Z. 1988. Elementary particle theory in Japan, 1935–1960. *Proceedings of the Japan–USA Collaborative Workshops on the History of Particle Theory in Japan, 1935–1960.* Yukawa Hall Archival Library. Research Institute for Fundamental Physics, University of Kyoto.

Brown, L. M.; Konuma, M.; and Maki, Z., eds. 1980. *Particle Physics in Japan, 1930–1950.* Vols. 1, 2. Research Institute for Fundamental Physics, University of Kyoto.

Brown, L. M.; Konuma, M.; and Maki, Z., eds. 1981. *Particle Physics in Japan, 1930–1950.* Vol. 3. Research Institute for Fundamental Physics, University of Kyoto.

Brown, L. S., and Gabrielse, G. 1986. Geonium physics: Physics of a single electron or ion in a Penning trap. *Rev. Mod. Phys.* 58:233–311.

Buber, M. 1958. *Tales of the Hasidim.* New York: Shocken Books.

Burchard, J. 1948. *Q.E.D.: M.I.T. in World War II.* New York: Technology Press.

Burrill, E. A. 1948. The accelerator conference. *Phys. Today* 1/5:15–18.

Bush, V. 1945. *Science: The Endless Frontier.* A report to the President. Washington, D.C.: U.S. Government Printing Office.

Bush, V. 1960. *Science: The Endless Frontier.* A report to the President on a program for post-war scientific research, July, 1945. (Reprinted, National Science Foundation, Washington, D.C.)

Butterfield, H. 1931. *The Whig Interpretation of History.* London: G. Bell. (Reprinted, New York: Norton, 1965.)

Cao, T. Y. 1991. The Reggeization program, 1962–1982: Attempts at reconciling quantum field theory with S-matrix theory. *Arch. Hist. Exact Sci..* 41:239–283.

Cao, T. Y., and Schweber, S. S. 1993. The conceptual foundations and philosophical aspects of renormalization theory. *Synthèse* (in press).

Carlson, F. J., and Oppenheimer, J. R. 1937. On multiplicative showers. *Phys. Rev.* 51: 220–231.

Cartwright, N. 1983. *How the Laws of Physics Lie.* Oxford: Clarendon Press.

Case, K. M. 1949a. Equivalence theorems for meson-nucleon coupling. *Phys. Rev.* 75:1506

Case, K. M. 1949b. On the neutron moment and the neutron electron interaction. *Phys. Rev.* 76:14–17.

Case, K. M. 1949c. Equivalence theorems for meson-nucleon coupling. *Phys. Rev.* 76:1–14.

Casimir, H. 1983. *Haphazard Reality: Half a Century of Science.* New York: Harper and Row.

Cassidy, D. C. 1981. Cosmic ray showers, high energy physics, and quantum field theories: Programmatic interactions in the 1930's. *Hist. Stud. Phys. Sci.* 12(1):1–40.

Chadwick, J. 1932a. Possible existence of a neutron. *Nature* 129:312–313.

Chadwick, J. 1932b. The existence of a neutron. *Proc. Roy. Soc. London* A136:692–708.

Chadwick, J,; Blackett, P.M.S.; and Occhialini, G.P.S. 1933. New evidence for the positive electron. *Nature* 131:473.

Chao, C. Y. 1930a. The absorption coefficient of hard γ-rays. *Proc. Natl. Acad. Sci. U.S.A.* 16:431–433.

Chao, C. Y. 1930b. Scattering of hard γ-rays. *Phys. Rev.* 36:1519.

Chew, G. 1966. *The Analytic S-Matrix.* New York: W. A. Benjamin.

Cini, M. 1980. The history and ideology of dispersion relations. The pattern of internal and external factors in a paradigmatic shift. *Fundamentae Scientiae* 1:157–172.

Cini, M. 1982. Cultural traditions and environmental factors in the development of quantum electrodynamics (1925–1933). *Fundamenta Scientiae* 3:229–253.

Cobas, A., and Lamb Jr., W. E. 1944. On the extraction of electrons from a metal surface by ions and metastable atoms. *Phys. Rev.* 65:327–337.

Coben, S. 1971. The scientific establishment and the transmission of quantum mechanics to the United States, 1919–1932. *Amer. Hist. Rev.* 76:442–466.

Coben, S. 1976. Foundation officials and fellowships: Innovation in the patronage of science. *Minerva* 14:225–240.

Cochrane, R. C. 1978. *The National Academy of Sciences. The First Hundred Years, 1863–1963.* Washington, D.C.: National Academy of Sciences.

Cohen, R. S., and Stachel, J. J., eds. 1979. *Selected Papers of Leon Rosenfeld.* Dordrecht, Holland: Reidel.

Cohen, V. W.; Goldsmith, H. H.; and Schwinger, J. 1939. The neutron-proton scattering cross section. *Phys. Rev.* 55:106.

Collins, G. B., ed. 1948. *Microwave Magnetrons.* New York: McGraw-Hill.

Collins, J. C. 1984. *Renormalization.* Cambridge, U.K.: Cambridge University Press.

Compton, A. 1945. Science and our nation's future. *Science* 101:207–209.

Compton, A. 1956. *Atomic Quest: A Personal Narrative.* New York: Oxford University Press.

Compton, K. T. 1942. Scientists face the world of 1942. In *Scientists Face the World of 1942: Essays by K. T. Compton, R. W. Trullinger and V. Bush.* New Brunswick, N.J.: Rutgers University Press.

Compton, K. T. 1946. Forword. In *MIT, Five Years at the Radiation Laboratory.* Cambridge, Mass.: MIT Press.

Condon, E. U. 1938. Mathematical models in modern physics. *Franklin Inst.* 225:255–261.

Condon, E. U., and Shortley, G. H. 1934. *The Theory of Atomic Spectra.* Cambridge, U.K.: Cambridge University Press.

Conversi, M. 1983. The period that led to the 1946 discovery of the leptonic nature of the mesotron. In Brown and Hoddeson 1983, pp. 242–250.

Conversi, M.; Pancini, E.; and Piccioni, O. 1947. On the disintegration of negative mesons. *Phys. Rev.* 71:209–210.

Corben, H., and Schwinger, J., 1940a. The electromagnetic properties of mesotrons. *Phys. Rev.* 58:191.

Corben, H., and Schwinger, J., 1940b. The electromagnetic properties of mesotrons. *Phys. Rev.* 58:953–968.

Corben, M. 1956. *Not to Mention the Kangaroos.* London: Hammond.

Crease, R. P., and Mann, C. C. 1987. *The Second Creation.* New York: Macmillan.

Critchfield, C.; Teller, E.; and Wigner, E. P. 1939. The electron-positron theory of nuclear forces. *Phys. Rev.* 56:530–539.

Crussard, J., and Leprince-Ringuet, L. 1937. Étude dans le grand électro-aimant de Bellevue de traversées d'écrans par des particules du rayonnement cosmique. *Compt. Rend.* 204:240–242.

Cushing, J. T. 1982. Models and methodologies in current theoretical high-energy physics. *Synthèse* 50:5–101.

Cushing, J. T. 1986. The importance of Heisenberg's *S*-matrix program for the theoretical physics of the 1950s. *Centaurus* 29:110–149.

Cushing, J. T. 1990. *Theory Construction and Selection in Modern Physics: The S-Matrix.* Cambridge (U.K.) and New York: Cambridge University Press.

Dalitz, R. H., and Peierls, R. E. 1986. Paul Adrien Maurice Dirac, 1902–1984. *Biog. Mem. Fellows Roy. Soc.* 32:138–185.

Dancoff, S. M. 1939. On radiative corrections for electron scattering. *Phys. Rev.* 55: 959–963.

Dancoff, S. M., and Drell, S. 1949. Electrostatic scattering of neutrons. *Phys. Rev.* 76: 205–212.

Darrigol, O. 1982. *Les débuts de la théorie quantique des champs* (1925–1948). 3d cycle Ph.D. diss., University of Paris (Partheon-Sorbonne).

Darrigol, O. 1984. La genèse du concept de champ quantique. *Ann. Phys. Fr.* 9:433–501.

Darrigol, O. 1986. The origin of quantized matter waves. *Hist. Stud. Phys. Sci.* 16/2: 198–253.

Darrigol, O. 1988a. Elements of a scientific biography of Tomonaga Sin-itiro. *Historia Scientiarum* 35:1–29.

Darrigol, O. 1988b. The quantum electrodynamical analogy in early nuclear theory, or the roots of Yukawa's theory. *Rev. d'Histoire Sci.* 41:225-297.

Darrow, K. K. 1934. Discovery and early history of the positive electron. *Scientific Monthly* 38:5–15.

Darwin, C. 1958. *The Autobiography of Charles Darwin, and Selected Letters.* Edited by Francis Darwin. New York: Dover Publications.

Davis, N. P. 1968. *Lawrence and Oppenheimer.* New York: Simon and Schuster.

de Boer, J.; Dal, E.; and Ulfbeck, O., eds. 1986. *The Lesson of Quantum Theory.* Amsterdam: Elsevier.

de Broglie, L. 1943. *Théorie générale des particules à spin. Méthode de fusion.* Paris: Gauthier-Villars.

de Broglie, M. 1927. *Rechèrche sur la théorie des quanta.* Leipzig.

de Broglie, M. 1951. *Les Premiers Congrès de Physique Solvay et l'orientation de la physique depuis 1911.* Paris: Editions Albin Michel.

de Broglie, M., and Langevin, P. 1912. *La théorie du rayonnement et les quanta.* Rapports et discussions de la reunion venues à Bruxelles du 30 Octobre au 3 Novembre 1911. Sous les auspices de M. E. Solvay. Paris: Gauthier-Villars.

Dehmelt, H. G. 1981. Invariant frequency ratios in electron and positron geonium spectra yield refined data on electron structure. In D. Kleppner and F. M. Pipkin, eds., *Atomic Physics 7*, pp. 337–372. New York: Plenum Presss.

Dehmelt, H. G. 1983. Electron. In *Yearbook of Science and Technology*, pp. 204–206. New York: McGraw-Hill.

de Maria, M., and La Teana, F. 1981. Dirac's "unorthodox" contribution to orthodox quantum mechanics (1925–1927). Nota Interna no. 768, Instituto di Fisica G. Marconi, University of Rome.

de Maria, M., and La Teana, F. 1982. Schrödinger's and Dirac's unorthodoxy in quantum mechanics. *Fundamentae Scientiae* 3/2:1–20.

de Maria, M., and Russo, A. 1985. The Discovery of the positron. *Riv. Stor. Sci.* 2(2): 237–286.

Deutsch, M. 1951a. Evidence for the formation of positronium in gases. *Phys. Rev.* 82:455(L).

Deutsch, M. 1951b. Three-quantum decay of positronium. *Phys. Rev.* 83:866–867(L).

Deutsch, M., and Dulit, E. 1951. Short-range interaction of electrons and fine structure of positronium. *Phys. Rev.* 84:601–602(L).

Deutsch, M.; Evans, R. D.; Feld, B. T.; Friedman, F.; and Goodman, C. 1948. *The Science and Engineering of Nuclear Power.* Cambridge, Mass.: MIT Press.

Dewey, J. 1931. *Philosophy and Civilization..* New York.

Dirac, Margit. 1987. Thinking of my darling Paul. In Kursunoglu and Wigner 1987, pp. 6–9.

Dirac, P.A.M. 1924a. Dissociation under a temperature gradient. *Proc. Camb. Phil. Soc.* 22:132–137.

Dirac, P.A.M. 1924b. Note on the relativity dynamics of a particle. *Phil. Mag.* 47: 1158–1159.

Dirac, P.A.M. 1924c. Note on the Doppler principle and Bohr's frequency condition. *Proc. Cambridge Phil. Soc.* 22:432–433.

Dirac, P.A.M. 1924d. The conditions for statictical equilibrium between atoms, electrons and radiation. *Proc. Roy. Soc. London* A106:581–596.

Dirac, P.A.M. 1925a. The adiabatic invariants of the quantum integrals. *Proc. Roy. Soc. London* A107:725–734.

Dirac, P.A.M. 1925b. The effect of Compton scattering by free electrons. *Mon. Not. Roy. Astr. Soc. London* 85:825–832.

Dirac, P.A.M. 1925c. The adiabatic hypothesis for magnetic fields. *Proc. Cambridge Phil. Soc.* 23:69–72.

Dirac, P.A.M. 1925d. The fundamental equations of quantum mechanics. *Proc. Roy. Soc. London* A109:642–653.

Dirac, P.A.M. 1926a. Quantum mechanics and a preliminary investigation of the hydrogen atom. *Proc. Royal Soc. London* A110:561–579.

Dirac, P.A.M. 1926b. The elimination of the nodes in quantum mechanics. *Proc. Roy. Soc. London* A111:281–305.

Dirac, P.A.M. 1926c. Relativity quantum mechanics with an application to Compton scattering. *Proc. Roy. Soc. London* A111:405–423.

Dirac, P.A.M. 1926d. *Quantum Mechanics.* D. Phil. diss., Cambridge University.

Dirac, P.A.M. 1926e. On quantum algebra. *Proc. Cambridge Phil. Soc.* 23:412–418.

Dirac, P.A.M. 1926f. On the theory of quantum mechanics. *Proc. Roy. Soc. London* A112:661–677.

Dirac, P.A.M. 1926g. The Compton effect in wave mechanics. *Proc. Cambridge Phil. Soc.* 23:500–507.

Dirac, P.A.M. 1927a. The physical interpretation of the quantum dynamics. *Proc. Roy. Soc. London* A113:621–641.

Dirac, P.A.M. 1927b. The quantum theory of the emission and absorption of radiation. *Proc. Roy. Soc. London* A114:243–265. (Reprinted in Schwinger 1958, pp. 1–23.)

Dirac, P.A.M. 1927c. The quantum theory of dispersion. *Proc. Roy. Soc. London* A114:710–728.

Dirac, P.A.M. 1927d. Über die Quantenmechanik der Stossvorgänge. *Zeits. Phys.* 44: 585–595.

Dirac, P.A.M. 1928a. The quantum theory of the electron. *Proc. Roy. Soc. London* A117:610–624.

Dirac, P.A.M. 1928b. The quantum theory of the electron. Part II. *Proc. Roy. Soc. London* A118:351–361.

Dirac, P.A.M. 1928c. Discussion. In *Electrons and Photons* (Cinquième Conseil de physique de l'Institut International de Physique, Solvay, October 24–29, 1927, Brussels), pp. 258–263. Paris: Gauthier-Villars.

Dirac, P.A.M. 1928d. A theory of electrons and protons. *Proc. Roy. Soc. London* A117: 360–365.

Dirac, P.A.M. 1928e. Über die Quantentheorie des Elektrons. *Phys. Zeits.* 29:561–563. (Report on Dirac's lecture at the Leipziger Universitätswoche, June 18–23, 1928.)

Dirac, P.A.M. 1928f. Lectures on Modern Quantum Mechanics. Handwritten notes for lectures given in the fall of 1928. Churchill College, Cambridge, U.K.

Dirac, P.A.M. 1929a. The basis of statistical quantum mechanics. *Proc. Cambridge Phil. Soc.* 25:62–66.

Dirac, P.A.M. 1929b. Quantum mechanics of many-electron systems. *Proc. Roy. Soc. London* A123:714–733.

Dirac, P.A.M. 1930a. A theory of electrons and protons. *Proc. Roy. Soc. London* A126: 360–365.

Dirac, P.A.M. 1930b. On the annihilation of electrons and protons. *Proc. Cambridge Phil. Soc.* 26:361–375.

Dirac, P.A.M. 1930c. Note on exchange phenomena in the Thomas atom. *Proc. Cambridge Phil. Soc.* 26:376–385.

Dirac, P.A.M. 1930d. The proton. *Nature* 126:605–606.

Dirac, P.A.M. 1930e. *The Principles of Quantum Mechanics.* 1st ed. Oxford: Clarendon Press.

Dirac, P.A.M. 1930f. Approximate methods. (Manuscript in English at Churchill College Archives, Cambridge, U.K. This is the text of chap. 11A of the Russian edition of Dirac 1930e [trans. M. P. Bronshtein and D. D. Iwanenko], pp. 243–257, with the note "Author's addition to the Russian edition.")

Dirac, P.A.M. 1931a. Note on the interpretation of the density matrix in the many electron problem. *Proc. Cambridge Phil. Soc.* 27:240–243.

Dirac, P.A.M. 1931b. Quantized singularities in the electromagnetic field. *Proc. Roy. Soc. London* A133:60–72.

Dirac, P.A.M. 1931c. Quelques problèmes de mécanique quantique. *Ann. Inst. Henri Poincaré* 1(4):357–400.

Dirac, P.A.M. 1931d. Lectures on quantum mechanics. Reported by B. Hoffmann, Princeton, N.J. (Mimeographed, Library, Institute for Advanced Study, Princeton, N.J., and elsewhere.)

Dirac, P.A.M. 1932a. Relativistic quantum mechanics. *Proc. Roy. Soc. London* A136: 453–464.

Dirac, P.A.M. 1933a. The Lagrangian in quantum mechanics. *Phys. Zeits. Sowjetunion* 3:64–72.

Dirac, P.A.M. 1933b. Homogeneous variable in classical dynamics. *Proc. Cambridge Phil. Soc.* 29:389–400.

Dirac, P.A.M. 1933c. Statement of a problem in quantum mechanics. *J. London Math. Soc.* 8:274–277.

Dirac, P.A.M. 1934a. *Théorie du positron.* Septième Conseil de Physique Solvay: Structure et propriétés des noyaux atomiques, October 22–29, 1933, pp. 203–230. Paris: Gauthier-Villars.

Dirac, P.A.M. 1934b. Teoriya pozitrona. In *Atomnoe yadro, sbornik dokladov i vsesoyoznoi yadernoi konferentsii,* ed. M. P. Bronshtein, V. M. Dukelśkii, D. D. Ivanenko, and Yo. B. Khariton, pp. 129–143. Leningrad and Moscow: Izdatelśtvo, Akademii Nauk USSR.

Dirac, P.A.M. 1934c. Theory of electrons and positrons. In *Nobel Lectures—Physics, 1922–41,* pp. 320–325. Amsterdam, 1965.

Dirac, P.A.M. 1934d. Discussion of the infinite distribution of electrons in the theory of the positron. *Proc. Cambridge Phil. Soc.* 30:150–163.

Dirac, P.A.M. 1935a. The electron wave equation in deSitter space. *Ann. Math.* 36: 657–669.

Dirac, P.A.M. 1935b. Lectures on quantum electrodynamics (1934–1935). Notes by Dr. Boris Podolsky (1st semester) and Dr. Nathan Rosen (2d semester). Institute for Advanced Study, Princeton, N.J.

Dirac, P.A.M. 1935c. *The Principles of Quantum Mechanics.* 2d ed. Oxford: Clarendon Press.

Dirac, P.A.M. 1936a. Does conservation of energy hold in atomic processes? *Nature* 137:298–299.

Dirac, P.A.M. 1936b. Relativistic wave equations. *Proc. Roy. Soc. London* A155: 447–459.

Dirac, P.A.M. 1936c. Wave equations in conformal space. *Ann. Math.* 37:429–442.

Dirac, P.A.M. 1937a. The cosmological constants. *Nature* 139:323.

Dirac, P.A.M. 1937b. Physical science and philosophy (a reply to Dr. H. Dingle). *Nature* 139:1001–1002.

Dirac, P.A.M. 1937c. Complex variables in quantum mechanics. *Proc. Roy. Soc. London* A160:48–59.

Dirac, P.A.M. 1937d. The reversal operator in quantum mechanics. *Izv. Acad. Nauk SSSR,* nos. 4–5:569–575.

Dirac, P.A.M. 1938a. A new basis for cosmology. *Proc. Roy. Soc. London* A165:199–208.

Dirac, P.A.M. 1938b. Classical theory of radiating electrons. *Proc. Roy. Soc. London* A167:148–169.

Dirac, P.A.M. 1939a. The relation between mathematics and physics. James Scott Prize Lecture. *Proc. Roy. Soc. Edinburgh* 59:122–129.

Dirac, P.A.M. 1939b. A new notation for quantum mechanics. *Proc. Camb. Phil. Soc.* 35:416–418.

Dirac, P.A.M. 1939c. La théorie de l'électron et du champ électromagnetique. *Ann. Inst. Henri Poincaré* 9:13–49.

Dirac, P.A.M. 1940. The physical interpretation of quantum mechanics. *Proc. Roy. Soc. London* A80:1–40.

Dirac, P.A.M. 1941. The theory of the separation of isotopes by statistical methods (n.d. or p.). Manuscript.

Dirac, P.A.M. 1942. The physical interpretation of quantum mechanics. Bakerian Lecture, 1941. *Proc. Roy. Soc. London* A180:1–40.

Dirac, P.A.M. 1943. Quantum electrodynamics. *Comm. Dublin Inst. Advanced Studies,* A1:1–36.

Dirac, P.A.M. 1945a. Unitary representations of the Lorentz group. *Proc. Roy. Soc. London* A183:284–295.

Dirac, P.A.M. 1945b. On the analogy between classical and quantum mechanics. *Rev. Mod. Phys.* 17:195–199.

Dirac, P.A.M. 1945c. Applications of quaternions to Lorentz transformations. *Proc. Roy. Irish Acad.* A50:261–270.

Dirac, P.A.M. 1945d. *Quelques développements sur la théorie atomique.* (Conference faite au Palais de la Découverte, December 6, 1945.) Paris: Université de Paris, 1948.

Dirac, P.A.M. 1946. Developments in quantum electrodynamics. *Comm. Dublin Inst. Advanced Studies* A3:1–33.

Dirac, P.A.M. 1947a. *The Principles of Quantum Mechanics.* 3d ed. Oxford: Oxford University Press.

Dirac, P.A.M. 1947b. The difficulties in quantum electrodynamics. *Rept. Intl. Conf. on Fundamental Particles and Low Temperatures, July 1946,* 1:10–14. London: The Physical Society.

Dirac, P.A.M. 1948a. On the theory of point electrons. *Phil. Mag.* 39:31–34.

Dirac, P.A.M. 1948b. Quantum theory of localizable dynamic systems. *Phys. Rev.* 73: 1092–1103.

Dirac, P.A.M. 1948c. The theory of magnetic poles. *Phys. Rev.* 74:817–830.

Dirac, P.A.M. 1949a. Forms of relativistic dynamics. *Rev. Mod. Phys.* 21:392–399.

Dirac, P.A.M. 1949b. La séconde quantification. *Ann. Inst. Henri Poincaré* 11(1):15–47.

Dirac, P.A.M. 1950a. A new meaning for gauge transformations in electrodynamics. *Nuovo Cimento* (9)7:925–938.

Dirac, P.A.M. 1950b. Generalized Hamiltonian dynamics. *Can. J. Math.* 2:129–148.

Dirac, P.A.M. 1950c. Field theory. In *Proc. of the Harwell Nuclear Physics Conference, September 1950* (AERE report no. 68, 114–115).

Dirac, P.A.M. 1951. A new classical theory of the electron. *Proc. Roy. Soc. London* A209:291-296.

Dirac, P.A.M. 1952a. A new classical theory of electrons. II. *Proc. Roy. Soc. London* A212:330-339.

Dirac, P.A.M. 1952b. Is there an aether ? *Nature* 169:146 (reply to H. Bondi and T. Gold); *ibid.* 169:702 (reply to L. Infeld).

Dirac, P.A.M. 1954a. A new classical theory of electrons. III. *Proc. Roy. Soc. London* A223:438–45.

Dirac, P.A.M. 1954b. Quantum mechanics and the aether. *Scientific Monthly* 78:142–146.

Dirac, P.A.M. 1957. The vacuum in quantum electrodynamics (suppl.). *Nuovo Cimento* (1)6:322–339.

Dirac, P.A.M. 1963a. Interview with T. S. Kuhn 1963. Archives for the History of Quantum Physics, Niels Bohr Library, AIP, New York.

Dirac, P.A.M. 1963b. The evolution of the physicist's picture of nature. *Scientific American* 208(5):45–53.

Dirac, P.A.M. 1968a. Methods in theoretical physics. Second evening lecture in the series "From a Life in Physics" at the International Symposium on Contemporary Physics, Trieste, 1968. Special suppl., *IAEA Bulletin*, Vienna, 1969.

Dirac, P.A.M. 1968b. The physical interpretation of quantum electrodyanamics. *Pontifica Academia Scientia, Commentarii* 2(13):1–12.

Dirac, P.A.M. 1969. Hopes and fear. *Eureka* 32:2–4.

Dirac, P.A.M. 1970. The mathematical foundations of quantum theory. In A. Royal Marlow, ed., *Mathematical Foundations of Quantum Theory*, pp. 1–8. New York: Academic Press.

Dirac, P.A.M. 1971. *The Development of Quantum Mechanics*. New York: Gordon and Breach.

Dirac, P.A.M. 1973a. Relativity and quantum mechanics. In C. G. Sudarshan and Y. Neéman, eds., *The Past Decades in Particle Theory*, pp. 741–772. New York: Gordon and Breach.

Dirac, P.A.M. 1973b. Development of the physicist's conception of nature. In J. Mehra, ed., *The Physicist's Conception of Nature*, pp. 1–14. Dordrecht, Holland: Reidel.

Dirac, P.A.M. 1977a. The relativistic electron wave equation. *Europhysics News* 8:1–5.

Dirac, P.A.M. 1977b. The relativistic electron wave equation. Hungarian Academy of Science, Central Research Institute of Physics Publication KFKI.

Dirac, P.A.M. 1977c. Recollections of an exciting era. In Weiner 1977.

Dirac, P.A.M. 1978a. *Directions in Physics*. New York: Wiley.

Dirac, P.A.M. 1978b. The prediction of antimatter. The first H. R. Crane Lecture. University of Michigan, Ann Arbor.

Dirac, P.A.M. 1980. A little 'prehistory.' *The Old Cothamian 1980*, p. 9.

Dirac, P.A.M. 1981. Does renormalization make sense? In D. W. Duke and J. F. Owens, eds., *Perturbative Quantum Chromodynamics*. AIP Conf. Proc. no. 74:129–130. New York: American Institute of Physics.

Dirac, P.A.M. 1982a. Pretty mathematics. *Intl. J. Theor. Phys.* 21:603–605.

Dirac, P.A.M. 1982b. The early years of relativity. In G. Holton and Y. Elkana, eds., *Albert Einstein, Historical and Cultural Perspectives: The Centenial Symposium in Jerusalem*, pp. 79–90. Princeton N.J.: Princeton University Press.

Dirac, P.A.M. 1983. The origin of quantum field theory. In Brown and Hoddeson 1983, pp. 39–55.

Dirac, P.A.M. 1984a. The future of atomic physics. *Intl. J. Theor. Phys.* 23(8):677–681.

Dirac, P.A.M. 1984b. The requirements of fundamental physical theory. *Eur. J. Phys.* 5: 65–67.

Dirac, P.A.M. 1984c. Blackett and the positron. In J. Hendry, ed., *Cambridge Physics in the Thirties*, pp. 61–62. Bristol, U.K.: Adam Hilger.

Dirac, P.A.M. 1987. The inadequacies of quantum field theory. In Kursunoglu and Wigner 1987.

Dirac, P.A.M., and Harding, J. W. 1932. Photo-electric absorption in hydrogen-like atoms. *Proc. Cambridge. Phil. Soc.* 28:209–218.

Dirac, P.A.M., and Kapitza, P. 1933. The reflection of electrons from standing light waves. *Proc. Cambridge Phil. Soc.* 29:297–300.

Dirac, P.A.M.; Fock, V. A.; and Podolsky, B. 1932a. On Quantum Electrodynamics. *Phys. Zeits. der Sowjetunion* 2:468–479.

Dirac, P.A.M.; Fock, V. A.; and Podolsky, B. 1932b. On Quantum Electrodynamics. *Phys. Zeits. der Sowjetunion* 3:64–72. (Reprinted in Schwinger 1958, pp. 29–40.)

Dirac, P.A.M.; Peierls, R.; and Pryce, M.H.L. 1942. On Lorentz invariance in the quantum theory. *Proc. Cambridge Phil. Soc.* 38:193–200.

Dresden, M. 1987. *H. A. Kramers. Between Tradition and Revolution.* New York: Springer-Verlag.

Drinkwater, J. W.; Richardson, O.; and Williams, W. E. 1940. Determinations of the Rydberg constants, e/m and the fine structure of H_α and D_α by means of a reflexion echelon. *Proc. Roy. Soc. London* A174:164–188.

DuBridge, L. 1947. The responsibility of the scientist. *Cal. Inst. Forum* 1:1–15.

DuBridge, L. 1949. The birth of two miracles. *Cal. Inst. Forum* 6:1–15.

Dupree, A. H. 1965a. Science in America—A historian's view. *Cahiers d' histoire mondiale* 8:666–681.

Dupree, A. H. 1965b. The structure of the government—university partnerships after World War II. *Bull. Hist. Med.* 39:245–251.

Dupree, A. H. 1972. *The Great Instauration of 1940*, In G. Holton, ed., *The Twentieth Century Sciences.* New York: Norton.

Dupree, A. H. 1986. National security and the post-war science establishment in the United States. *Nature* 323:213–216.

Durell, C. V. 1960. *Readable Relativity.* With foreword by F. J. Dyson. New York: Harper.

Dyson, F. J. 1943a. Three identities in combinatory analysis. *J. London Math. Soc.* 18: 35–39.

Dyson, F. J. 1943b. On the order of magnitude of the partial quotients of a continued fraction. *J. London Math. Soc.* 18:40–43.

Dyson, F. J. 1943c. A note on Kurtosis. *J. Roy. Stat. Soc.* 106 (part 4) :360–361.

Dyson, F. J. 1944a. A proof that every equation has a root. *Eureka* (Cambridge) 8:3–4.

Dyson, F. J. 1944b. Some guesses in the theory of partitions. *Eureka* (Cambridge) 8:10–15.

Dyson, F. J. 1944c. Note on the comparison of loss rates. Report to Operational Research Section Headquarters, R.A.F. Bomber Command, 1943–1945.

Dyson, F. J. 1945. A theorem on the densities of sets of integers. *J. London Math. Soc.* 20:8–14.

Dyson, F. J. 1946a. The problem of the pennies. *Mathematical Gazette London* 30 (29): 231–234.

Dyson, F. J. 1946b. Two problems in the theory of numbers. Fellowship dissertation. Trinity College.

Dyson, F. J. 1947a. On simultaneous diophantine approximations. *Proc. London Math. Soc.,* series 2, 49:409–420.

Dyson, F. J. 1947b. The approximation to algebraic numbers by rationals. *Acta Mathematica* (Uppsala) 89:225–240.

Dyson, F. J. 1948a. A theorem in algebraic topology. *Ann. Math.* 49(1):75–81.

Dyson, F. J. 1948b. On the product of four non-homogeneous linear forms. *Ann. Math.* 49(1):82–109.

Dyson, F. J. 1948c. The electromagnetic shift of energy levels. *Phys. Rev.* 73:617–626.

Dyson, F. J. 1948d. Interactions of nucleons with meson fields (L). *Phys. Rev.* 73:929–930.

Dyson, F. J. 1949a. The radiation theories of Tomonaga, Schwinger and Feynman. *Phys. Rev.* 75:486–502. (Reprinted in Schwinger 1958, pp. 275–291.)

Dyson, F. J. 1949b. The S-matrix in quantum electrodynamics. *Phys. Rev.* 75:1736–1755. (Reprinted in Schwinger 1958, pp. 292–311.)

Dyson, F. J. 1949c. Recent developments in the quantum theory of fields. Invited lecture, APS meeting, Washington, D.C., April 30, 1949. *Phys. Rev.* 76:162.

Dyson, F. J. 1950a. Longitudinal photons in quantum electrodynamics. *Phys. Rev.* 77: 428–439.

Dyson, F. J. 1950b. The radiation theory of Feynman. *Helv. Phys. Acta* 23 (supp. 3): 240–242.

Dyson, F. J. 1950c. Lectures on quantum electrodynamics. Michigan Symposium, Ann Arbor. Mimeographed.

Dyson, F. J. 1951a. Heisenberg operators in quantum electrodynamics. I. *Phys. Rev.* 82: 428–439.

Dyson, F. J. 1951b. Heisenberg operators in quantum electrodynamics. II. *Phys. Rev.* 82:608–627.

Dyson, F. J. 1951c. The Schrödinger equation in quantum electrodynamics. *Phys. Rev.* 83:1207–1216.

Dyson, F. J. 1951d. The renormalization method in quantum electrodynamics. *Proc. Roy. Soc. London* A207:395–401.

Dyson, F. J. 1951e. Continuous functions defined on spheres. *Ann. Math.* 54:534–536.

Dyson, F.J. 1951f. Review of Feynman (1950). *Math. Rev.* 12:889.

Dyson, F. J. 1952a. Quantum electrodynamics. *Phys. Today* 5:6–9.

Dyson, F. J. 1952b. Divergence of perturbation theory in quantum electrodynamics. *Phys. Rev.* 85:631–632.

Dyson, F. J. 1952c. Lecture notes on advanced quantum mechanics. Cornell Laboratory of Nuclear Studies, Cornell University, Ithaca, N.Y.

Dyson, F. J. 1953a. Field theory. *Scientific American* 188:57–64.

Dyson, F.J. 1953b. The use of the Tamm-Dancoff method in field theory. *Phys. Rev.* 90:994.

Dyson, F. J. 1953c. Mass-renormalization with the Tamm-Dancoff method. *Phys. Rev.* 91:421–422.

Dyson, F. J. 1953d. Fourier transforms of distribution functions. *Can. Journal Math.* 5: 554–558.

Dyson, F. J. 1954. Renormalization. Lecture notes edited by J. Lascoux and J. Mandelbrojt. Supplement to *Advanced Quantum Mechanics*, 2d ed. Les Houches.

Dyson, F. J. 1962. Statistical theory of the energy levels of complex systems. I. II. III. *J. Math. Phys.* 3:140–156, 157–165, 166–175.

Dyson, F. J. 1964. Mathematics in the physical sciences. *Scientific American* 211(9): 129–146.

Dyson, F. J. 1965a. Old and new fashions in field theory. *Phys. Today* 18(6):21–24.

Dyson, F. J. 1965b. Tomonaga, Schwinger, and Feynman awarded Nobel Prize for physics. *Science* 150:588–589.

Dyson, F. J. 1969. A new symmetry of partitions. *J. Comb. Theory* 7:56–61.

Dyson, F. J. 1972. Missed oportunities. *Bull. Am. Math. Soc.* 78:635–652.

Dyson, F. J. 1977. Letter to W. E. Lamb, December 2, 1977. In *Willis E. Lamb, Jr.: A Festschrift on the Occasion of His 65th Birthday*. Physics Reports Reprint Book Series, vol. 3. North-Holland, 1978.

Dyson, F. J. 1979. *Disturbing the Universe*. New York: Harper and Row.

Dyson, F. J. 1983. Review of Yu. I. Manin, *Mathematics and Physics* (Boston: Birkhäuser). *Math. Intelligencer* 5/2:54–57. (Reprinted in Dyson 1992.)

Dyson, F. J. 1987. A walk through Ramanujan's garden. In A. Ramanujan and G. E. Andrews, eds., *Ramanujan Revisited: Proceedings of the Centenary Conference*. Boston: Academic Press, 1988.

Dyson, F. J. 1992. *From Eros to Gaia*. New York: Pantheon Books.

Dyson, F. J.; Baranger, M.; and Salpeter, E. E. 1952. Fourth-order vacuum polarization. *Phys. Rev.* 88:680.

Dyson, G. 1924. *The New Music.* 1st ed. Oxford: Oxford University Press. (A second edition appeared in 1926.)

Einstein, A. 1909. Entwicklung unserer Anschauungen über das Wesen und die Konstitution der Strahlung. *Phys. Zeits.* 10:817–825.

Einstein, A. 1925. Bemerkung zu P. Jordans Abhandlung: Theorie der Quantenstrahlung. *Zeits. Phys.* 31:784–785.

Einstein, A., and Ehrenfest, P. 1923. Zur Quantentheorie des Strahlungsgleichgewichts. *Zeits. Phys.* 19:301–306.

Einstein, A.; Podolsky, B.; and Rosen, N. 1935. Can quantum-mechanical description of physical reality be considered complete? *Phys. Rev.* 47:777–780.

Eisenstein, J.; Feshbach, H.; and Schwinger, J. 1948. On tensor forces and the variation-iteration method. *Phys. Rev.* 74:1223.

Eliezer, C. Jayaratman. 1947. The interaction of an electron and an electromagnetic field. *Rev. Mod. Phys.* 19:147–184.

Elsasser, W. 1978. *Memoirs of a Physicist in the Atomic Age.* New York and Bristol, U.K.: Science History Publications.

Enz, C. P. 1973. W. Pauli's scientific work. In Mehra 1973, pp. 766–799.

Enz, C. P. 1986. Obituary for E.C.G. Stueckelberg. *Phys. Today* 39/119–121.

Epstein, S. T. 1948. Remarks on H. W. Lewis' paper "On reactive terms in quantum electrodynamics." *Phys. Rev.* 73:177(L) and 630(L).

Epstein, S. T. 1981. The Hellmann-Feynman theorem. In B. M. Deb, ed., *The Force Concept in Chemistry,* pp. 1–38. New York: Van Nostrand-Reinhold.

Ermolaev, A. M. 1978. Quantum electrodynamical effects in atomic spectra. In Hanle and Kleinpoppen 1978, pp. 149–181.

Estermann, I. 1946. Molecular beam techniques. *Rev. Mod. Phys.* 18:301–323.

Euler, H., and Heisenberg, W. 1938. Theorie den Höhestrahlen Phenomena. *Erg. exacten Naturwissenschaften* 17:1–69.

Euler, H., and Kockel,B. 1935. Über die Streuung von Licht an Licht nach der Diracschen Theorie. *Naturwissenschaften* 23:246–247.

Fairbank, J. 1974. *Japan: The Story of a Nation.* New York: Alfred Knopf.

Fairbank, J.; Reischauer, E.; and Craig, A. 1965. *East Asia: The Modern Transformation.* Boston: Houghton Mifflin.

Feldman, D. 1949. On realistic field theories and the polarization of the vacuum. *Phys. Rev.* 76:1369–1375.

Feldman, D., and Schwinger, J. 1949a. Radiative correction to the Klein-Nishina formula. *Phys. Rev.* 75:358.

Fermi, E. 1929. Sopra l'elettrodinamica quantistica. I. *Rend. Lincei* 9:881–887.

Fermi, E. 1930. Sopra l'elettrodinamica quantistica. II. *Rend. Lincei* 12:431–435.

Fermi, E. 1931. La masse elettromagnetiche nella elettrodinamica quantistica. *Nuovo Cimento* 8:121–132.

Fermi, E. 1932. Quantum theory of radiation. *Rev. Mod. Phys.* 4:87–132.

Fermi, E. 1933. Tentativo di una teoria dell'emissione dei raggi β. *Ric. Scientifica* 4(2):491–495.

Fermi, E. 1934a. Tentativo di una teoria dei raggi β. *Nuovo Cimento* 2:1–19.

Fermi, E. 1934b. Versuch einer Theorie der β-Strahlen. I. *Zeits. Phys.* 88:161–71.

Fermi, E. 1946. The development of the first chain reacting pile. *Proc. Amer. Phil. Soc.* 90:20–24.

Fermi, E. 1962. *Collected Papers.* Vol. 1. Chicago: University of Chicago Press.

Fermi, E., and Teller, E. 1947. The capture of negative mesotrons in matter. *Phys. Rev.* 72:399–408.

Fermi, E., and Uhlenbeck, G. E. 1933. On the recombination of electrons and positrons. *Phys. Rev.* 44:510.

Fermi, E.; Teller, E.; and Weisskopf, V. 1947. The decay of negative mesotrons in matter. *Phys. Rev.* 71:314–315.

Fermi, L. 1971. *Illustrious Immigrants: The Intellectual Migration from Europe, 1930–41.* Chicago: University of Chicago Press.

Feshbach, H. 1950. The new quantum electrodynamics. In *Electromagnetic Theory.* Vol. 2, pp. 1–19. Proceedings of Symposia in Applied Mathematics. New York : American Mathematical Society.

Feshbach, H. 1979. Schwinger and nuclear physics. In *Themes in Contemporary Physics.* Essays in honor of Julian Schwinger's 60th birthday. *Physica* 96A(1+2):17–26.

Feshbach, H., and Schwinger, J. 1951. On a phenomenological neutron-proton interaction. *Phys. Rev.* 84:194.

Feynman, R. P. 1939. Forces in molecules. *Phys. Rev.* 56:340–343.

Feynman, R. P. 1942. The principle of least action in quantum mechanics. Ph.D. diss., Princeton University, Princeton, N.J. (Ann Arbor: University Microfilms Publication No. 2948.)

Feynman, R. P. 1948a. A relativistic cut-off for classical electrodynamics. *Phys. Rev.* 74:939-946.

Feynman, R. P. 1948b. Relativistic cut-off for quantum electrodynamics. *Phys. Rev.* 74:1430–1438.

Feynman, R. P. 1948c. Space-time approach to non-relativistic quantum mechanics. *Rev. Mod. Phys.* 20:367–387. (Reprinted in Schwinger 1958, pp. 321–341.)

Feynman, R. P. 1948d. Pocono Conference. *Phys. Today* 1/2:8–10.

Feynman, R. P. 1949a. The theory of positrons. *Phys. Rev.* 76:749–759. (Reprinted in Schwinger 1958, 225–235.)

Feynman, R. P. 1949b. Space-time approach to quantum electrodynamics. *Phys. Rev.* 76:769–789. (Reprinted in Schwinger 1958, pp. 236–256.)

Feynman, R. P. 1950. Mathematical formulation of the quantum theory of electromagnetic interaction. *Phys. Rev.* 80:440–457.

Feynman, R. P. 1951a. An operator calculus having application in quantum electrodynamics. *Phys. Rev.* 84:108–128.

Feynman, R. P. 1951b. The concept of probability in quantum mechanics. In *Second Berkeley Symposium on Mathematical Statistics and Probability,* pp. 533–541. Berkeley: University of California.

Feynman, R. P. 1954. The present situation in fundamental theoretical physics. *Ann. Acad. Brasileira de Ciencias* 26/1:51–58.

Feynman, R. P. 1956a. The value of science. *Engineering and Science* 19:15–18. (Public address for National Academy of Science at California Institute of Technology, Pasadena.)

Feynman, R. P. 1956b. The relation of science and religion. *Engineering and Science* 19:20–23. (Public address, California Institute of Technology, Pasadena.)

Feynman, R. P. 1965. *The Character of Physical Law.* Cambridge, Mass.: MIT Press.

Feynman, R. P. 1966a. The development of the space-time view of quantum field theory. *Science* 153:699–708. Nobel lecture.

Feynman, R. P. 1966b. Nobel lecture, Stockholm. (Reprinted in *Phys. Today*, August 1966, pp. 31–44.)

Feynman, R. P. 1976. Los Alamos from below. Reminiscences of 1943–1945. *Engineering and Science* 39:11–30. (A slightly more polished version is in Feynman 1980.)

Feynman, R. P. 1980. Los Alamos from below. In Badash et al. 1980, pp. 105–132.

Feynman, R. P. 1987. *Elementary Particles and the Laws of Physics: The 1986 Dirac Memorial Lectures.* Cambridge (U.K.) and New York: Cambridge University Press.

Feynman, R. P., and Brown, L. M. 1952. Radiative correction to Compton scattering. *Phys. Rev.* 85:231–244.

Feynman, R. P., and R. G. Hibbs. 1965. *Quantum Mechanics and Path Integrals.* New York: McGraw-Hill.

Feynman, R. P., with Leighton, R. 1985. *"Surely You're Joking, Mr. Feynman!" : Adventures of a Curious Character.* Edited by E. Hutchings. New York: Norton.

Feynman, R. P., with Leighton, R. 1988. *What Do You Care What Other People Think?: Further Adventures of a Curious Character.* New York: Norton.

Feynman, R. P.; Leighton, R. B.; and Sands, M. 1964. *The Feynman Lectures on Physics.* Vol. 2. Reading, Mass.: Addison-Wesley.

Feynman, R. P.; Metropolis, N.; and Teller, E. 1949. Equations of state of elements based on the generalized Fermi-Thomas theory. *Phys. Rev.* 75:1561–1573.

Fierz, M. 1939. Über die relativistische Theorie kräfterfreier Teilchen mit beleibigen Spin. *Helv. Phys. Acta* 12:3.

Fierz, M. 1980. Physik in den dreissiger Jahren—ein Rückblick. *Phys. Blätter* 36:133–136.

Fierz, M., and Pauli, W. 1939. On the relativistic wave equations for particles of arbitrary spin in an electromagnetic field. *Proc. Roy. Soc. London* A173:211–232.

Fierz, M., and Weisskopf, V. F., eds. 1960. *Theoretical Physics in the Twentieth Century: A Memorial Volume to Wolfgang Pauli.* New York: Interscience.

Fleming, D. 1965. American science and the world scientific community. *Cahiers d'Histoire Mondiale* 8:666–681.

Fleming, D., and Bailyn, B., eds. 1969. *The Intellectual Migration: Europe and America, 1930–1960.* Cambridge, Mass: Belknap Press of Harvard University Press.

Flexner, A. 1930. *Universities: American, English, German.* New York: Oxford University Press.

Flexner, A. 1940. *I Remember: The Autobiography of Abraham Flexner.* New York: Simon and Schuster.

Flower, E., and Murphy, M. G. 1977. *A History of Philosophy in America.* 2 vols. New York: Capricorn Books.

Fock, V. 1933. Zur Theorie des Positrons. *Dokl. Akad. Nauk. USSR* 1:267–278.

Fock, V. 1934. Zur Quantenelektrodynamik. *Phys. Zeits. Sowjetunion* 6:425–469.

Fokker, A. D. 1929a. Ein invarianter Variationssatz für die Bewegung mehrerer elektrischen Massenteilchen. *Zeits. Phys.* 58:386–393.

Fokker, A. D. 1929b. Wederkeerigheit in der Werking van Geladen deeltjes. *Physica* 9: 33–42.

Fokker, A. D. 1932. Théorie relativiste de l'intéraction de deux particules chargées. *Physica* 12:145–152.

Foldy, L. 1952. Electron-neutron interaction. *Phys. Rev.* 87:693–696.

Forman, P. 1971. Weimar culture, causality and quantum theory, 1918–1927: Adaptation by German physicists and mathematicians to a hostile intellectual environment. *Hist. Stud. Phys. Sci.* 3:1–115.

Forman, P. 1979. The reception of an acausal quantum mechanics in Germany and Britain. In S. Mauskopf, ed., *The Reception of Unconventional Science.* AAAS Selected Symposium 25. Boulder, Colo.: Westview Press.

Forman, P. 1984. Kausalität, Anschaulichkeit, and Individualität, or how cultural values prescribed the character and the lessons ascribed to quantum mechanics. In N. Stehr and V. Meja, eds., *Society and Knowledge.* New Brunswick, N.J.: Transaction Books.

Forman, P.; Heilbron, J. L.; and Weart, S. R. 1975. Physics circa 1900: Personnel, funding and productivity of the academic establishments. *Hist. Stud. Phys. Sci.* 5:1–185.

Frampton, P. 1987. *Gauge Field Theories.* Menlo Park, Calif.: Benjamin/Cummings.

French, J. B. 1948. The shift of energy levels due to radiative coupling. Ph.D. diss., Massachusetts Institute of Technology, Cambridge.

French, B., and Weisskopf, V. 1949. The electromagnetic shift of energy levels. *Phys. Rev.* 75:1240–1248.

Frenkel, J. 1925. Zur elektrodynamik Punktförmiger Elektronen. *Zeits. Phys.* 32: 518–534.

Freund, E. O.; Goebel, C.; and Nambu, Y., eds. 1970. *Quanta: Essays in Honor of G. Wentzel.* Chicago: University of Chicago Press.

Frisch, O. 1979. *What Little I Remember.* Cambridge, U.K.: Cambridge University Press.

Fröhlich, H.; Heitler, W.; and Kahn, B. 1939a. Deviation from the Coulomb law for a proton. *Proc. Roy. Soc. London* A171:269–280.

Fröhlich, H.; Heitler, W.; and Kahn, B. 1939b. Deviation from the Coulomb law for a proton. *Phys. Rev.* 56:961–962.

Fröhlich, H.; Heitler, W.; and Kemmer, N. 1938. On the nuclear forces and the magnetic moment of the neutron and the proton. *Proc. Roy. Soc. London* A166:269–280.

Fukuda, H.; Miyamoto, Y.; and Tomonaga, S. 1949. A self-consistent subtraction method in the quantum field theory. II. *Prog. Theor. Phys.* 4:47–59, 121–129.

Furry, W. H. 1937. A symmetry theorem in the positron theory. *Phys. Rev.* 51:125.

Furry, W. H. 1947. Review of W. Pauli, *Meson Theory of Nuclear Forces* (Pauli 1946b). N.p.

Furry, W. H. 1951. On bound states and scattering in positron theory. *Phys. Rev.* 81: 115–124.

Furry, W., and Oppenheimer, J. R. 1934. On the theory of the electron and positron. *Phys. Rev.* 45:245–262.

Fussell, L. 1937. Production and absorption of cosmic ray showers. *Phys. Rev.* 51: 1005–1006.

Gabrielse, G., and Dehmelt, H. G. 1981. Geonium without a magnetic bottle–a new generation. In B. N. Taylor and Phillips, eds., *Precision Measurements and Fundamental Constants,* pp. 219–223. Washington, D.C.: National Bureau of Standards, U.S. Special Publication 617.

Galison, P. 1983. The discovery of the muon and the failed revolution against quantum electrodynamics. *Centaurus* 26:262–316.

Galison, P. 1987. *How Experiments End.* Chicago: University of Chicago Press.

Gamow, G. 1928. Zur Quantentheorie des Atomkernes. *Zeits. Phys.* 51:204–212.

Gasiorowicz, S. G.; Yennie, P. R.; and Suura, H. 1959. Magnitude of renormalization constants. *Phys. Rev. Lett.* 2:513–516.

Gay, P. 1987. *A Godless Jew: Freud, Atheism, and the Making of Psychoanalysis.* New Haven, Conn.: Yale University Press.

Gebhard, L. A. 1979. *Evolution of Naval Radioelectronics and Contributions of the NRL.* Washington, D.C.: Naval Research Laboratory.

Gell-Mann, M. 1953. Lectures on quantum electrodynamics. Department of Physics, Columbia University, New York.

Gell-Mann, M. 1987. Particle theory from S-matrix to quark. In M. G. Doncel, A. Hermann, L. Michel, and A. Pais, eds., *Symmetries in Physics (1600–1980),* pp. 474–497. Barcelona: Bellaterra.

Gell-Mann, M. 1989. Progress in elementary particle theory, 1950–1964. In L. M. Brown, M. Dresden, and L. Hoddeson, eds., *Pions to Quarks,* pp. 694–711. Cambridge, U.K.: Cambridge University Press.

Gell-Mann, M., and Low, F. E. 1951. Bound states in quantum field theory. *Phys. Rev.* 84: 350–354.

Gell-Mann, M., and Low, F. E. 1954. Quantum electrodynamics at small distances. *Phys. Rev.* 95:1300–1312.

Gellner, E. 1981. Pragmatism and the importance of being earnest. In R. J. Mulvaney and P. M. Zeltner, eds. *Pragmatism: Its Sources and Prospects,* pp. 41–66. Columbia, SC.: University of South Carolina Press.

Gentner, W.; Maier-Leibnitz, H.; and Bothe, W. 1954. *An Atlas of Typical Expansion Chamber Photographs.* New York: Interscience.

Georgi, H. 1989a. Grand unified field theories. In Paul Davies, ed., *The New Physics,* pp. 425–445. Cambridge, U.K.: Cambridge University Press.

Georgi, H. 1989b. Effective quantum field theories. In Paul Davies, ed., *The New Physics,* pp. 446–457. Cambridge, U.K.: Cambridge University Press.

Gerjuoy, E., and Schwinger, J. 1941. The theory of light nuclei. *Phys. Rev.* 60:158.

Gerjuoy, E., and Schwinger, J. 1942. On tensor forces and the theory of light nuclei. *Phys. Rev.* 61:138.

Gerstein, M. 1987. *Purcell's Co-Discovery of NMR: Contingency vs. Inevitability.* Dept. of History of Science, Harvard University, Cambridge, Mass.

Glashow, S. L. 1980. Toward a unified theory: Threads in a tapestry. *Rev. Mod. Phys.* 52:539–543.

Gleick, J. 1992. *Genius. The Life and Science of Richard Feynman.* New York: Pantheon.

Glick, T. F., ed. 1987. *The Comparative Reception of Relativity.* Dordrecht, Holland: Reidel.

Gold, T., ed. 1967. *The Nature of Time.* Ithaca, N.Y.: Cornell University Press.

Goldberg, S. 1984. Being operational is operationism: Bridgman on relativity. *Rivista di Storia della Scienza* 1:333–354.

Goldberg, S. 1987. Putting new wine in old bottles. In Glick 1987, pp. 1–26.

Goldhaber, M. 1979. The nuclear photoelectric effect and remarks on higher multipole transitions: A personal history. In R. H. Stuewer, ed., *Nuclear Physics in Retrospect,* pp. 83–110. Minneapolis: University of Minnesota Press.

Goldstein, J. S., moderator, with Rabi, I. I.; Schwinger, J.; Ramsey, N.; Millman, S.; and Zacharias, J. Reminiscences of the thirties. Videotaped at Brandeis University, March 29, 1984.

Goldstone, J. 1961. Field theories with "superconductor" solutions. *Nuovo Cimento* 19:154–164.

Goldstone, J.; Salam, A.; and Weinberg, S. 1962. Broken symmetries. *Phys. Rev.* 127: 965–970.

Goudsmit, S. A. 1961. The Michigan symposium in theoretical physics. *Michigan Alumnus Quart. Rev.* 67:178–182.

Gowing, M. 1964. *Britain and Atomic Energy: 1939–1945.* New York: St. Martin's Press.

Gray, L. H. 1939. The absorption of penetrating radiation. *Proc. Roy. Soc. London* A122:647.

Greber, L. 1987. The unholy trinity: Physics, gender, and the military. B.S. diss., Massachusetts Institute of Technology, Cambridge.

Green, M. B. 1985. Unification of forces and particles in superstring theories. *Nature* 314:409–414.

Green, M. B.; Schwarz, J. H.; and Witten, E. 1987. *Superstring Theory.* Cambridge, U.K.: Cambridge University Press.

Greenberg, D. S. 1967. *The Politics of Pure Science.* New York: New American Library.

Grodzins, M., and Rabinowitch, E., eds. 1963. *The Atomic Age, Scientists in National and World Affairs: Articles from the* Bulletin of the Atomic Scientists, *1945–1962.* New York: Basic Books.

Gross, D. 1985. Beyond quantum field theory. In J. Ambjorn, B. J. Durhuus, J. L. Petersen, eds., *Recent Developments in Quantum Field Theory,* pp. 151–168. New York: Elsevier.

Groueff, S. 1967. *Manhattan Project: The Untold Story of the Making of the Atomic Bomb.* Boston: Little, Brown and Co.

Grythe, I. 1982. Some remarks on the early S-matrix. *Centaurus* 26:198–203.

Guerlac, H. 1945. *The history of the Radiation Laboratory.* Mimeographed notes, MIT Archives, Cambridge, Mass.

Guerlac, H. 1987. *Radar in World War II.* Los Angeles: Tomash Publishers.

Gupta, S. N. 1950. Theory of longitudinal photons in quantum electrodynamics. *Proc. Phys. Soc. London* A63:681–691.

Gupta, S. N. 1951. On the elimination of divergences from quantum electrodynamics. *Proc. Phys. Soc. London* A64:426–427.

Haase, T. 1935. Über die Absorption decimeter Wellen in ionisierten Gasen. *Ann. der Phys.* 23:657–676.

Hacking, I. 1983. *Representing and Intervening.* Cambridge, U.K.: Cambridge University Press.

Hacking, I. 1985. Styles of Scientific Reasoning. In J. Rajchman and C. West, eds., *Post-Analytic Philosophy.* New York: Columbia University Press.

Halpern, O., and Schwinger, J. 1935. On the polarization of electrons by double scattering. *Phys. Rev.* 48:109.

Hamermesh, M. 1962. *Group Theory.* Reading, Mass.: Addison-Wesley.

Hamermesh, M., and Schwinger, J. 1939. The scattering of neutrons by hydrogen and deuterium molecules. *Phys. Rev.* 55:679.

Hamermesh, M., and Schwinger, J. 1946. Polarization of neutrons by resonance scattering in helium. *Phys. Rev.* 69:681.

Hamermesh, M., and Schwinger, J. 1947. Neutron scattering in ortho- and parahydrogen. *Phys. Rev.* 71:678.

Handlin, O. 1970. *The American University as an Instrument of Republican Culture.* Leicester, U.K.: Leicester University Press.

Handlin, O., and Handlin, M. F. 1970. *The American College and American Culture.* New York: McGraw-Hill.

Hanle, P. 1975. Erwin Schrödinger's statistical mechanics. Ph.D. diss., Yale University, New Haven, Conn.

Hanle, P. 1977. The coming of age of Erwin Schrödinger. *Arch. Hist. Exact Sci.* 17: 165–192.

Hanle, P. 1979. The Schrödinger-Einstein correspondence and the sources of wave mechanics. *Amer. J. Phys.* 47:644–648.

Hanle, P. 1982. *Bringing Aerodynamics to America.* Cambridge, Mass: MIT Press.

Hanle, W., and Kleinpoppen, H., eds. 1978. *Progress in Atomic Spectroscopy.* Part A. New York and London: Plenum Press.

Hanson, N. R. 1963. *The Concept of the Positron.* Cambridge, U.K.: Cambridge University Press.

Hardy, G. H. 1968. *A Mathematician's Apology.* With a foreword by C. P. Snow. Cambridge, U.K.: Cambridge University Press.

Hardy, G. H., and Wright, E. M. 1938. *An Introduction to the Theory of Numbers.* Oxford: The Clarendon Press.

Havas, P. 1944. On the interaction of radiation and two electrons. *Phys. Rev.* 66:69–76.

Havas, P. 1945. On the interaction of radiation and matter. *Phys. Rev.* 68:214–226.

Hawking, S. 1980. Is the end in sight for theoretical physics? Lecture on assuming the Lucasian Chair, Cambridge University.

Hawkins, D. 1983. *Project Y: The Los Alamos Story.* Part I, *Toward Trinity.* Los Angeles and San Francisco: Tomash Press.

Hayakawa, S. 1988a. Sakata model and activities in Nagoya. In Brown, et al. 1988, pp. 11–13.

Hayakawa, S. 1988b. Sin-Itiro Tomonaga and his contributions to quantum electrodynamics and high energy physics. In Brown et al. 1988, pp. 43–60.

Hayakawa, S.; Miyamoto, Y.; and Tomonaga, S. 1947a. On the elimination of the auxiliary condition in the quantum electrodynamics. *J. Phys. Soc. Japan* 2:172–198.

Hayakawa, S.; Miyamoto, Y.; and Tomonaga, S. 1947b. On the elimination of the auxiliary condition in the quantum electrodynamics. *J. Phys. Soc. Japan* 2:199–217.

Heilbron, J. 1985. The earliest missionaries of the Copenhagen spirit. *Rev. Hist. Sci.* 38/3–4:195–230.

Heilbron, J. L.; Seidel, R. W.; and Wheaton, B. R. 1981. *Lawrence and His Laboratory: Nuclear Science at Berkeley, 1931–1961.* Berkeley: Office for History of Science and Technology, University of California.

Heims, S. J. 1980. *John von Neumann and Norbert Weiner: From Mathematics to the Technologies of War and Death.* Cambridge, Mass.: MIT Press.

Heims, S. J. 1991. *The Cybernetics Group.* Cambridge, Mass.: MIT Press.

Heisenberg, W. 1925. Über quantentheoretische Umdeutung kinematischer und mechanischer Beziehung. *Zeits. Phys.* 33:879–883.

Heisenberg, W. 1930a. *The Physical Principles of the Quantum Theory.* Chicago: University of Chicago Press. (Reprinted by Dover, New York.)

Heisenberg, W. 1930b. Die Selbstenergie des Elektrons. *Zeits. Phys.* 65:4–13. (Reprinted in Heisenberg 1989.)

Heisenberg, W. 1931a. Bemerkung zur Strahlungstheorie. *Ann. der Phys.* 9:338–346. (Reprinted in Heisenberg 1989.)

Heisenberg, W. 1931b. Über Energieschwankungen in einem Strahlungsfeld. *Ber. Sächs. Akad. Wiss.* (Leipzig) 83:3–9. (Reprinted in Heisenberg 1989.)

Heisenberg, W. 1932. Über den Bau der Atomkerne. *Zeits. Phys.* 77:1–11, 78:156–64; and (1933) 80:587–96. (Reprinted in Heisenberg 1989.)

Heisenberg, W. 1934a. Über die mit der Entstehung von Materie aus Strahlung verknüpten Ladungsschwankungen. *Ber. Säch. Akad. Wiss.* (Leipzig) 86:317–322. (Reprinted in Heisenberg 1989.)

Heisenberg, W. 1934b. Bemerkung zur Diracschen Theorie des Positrons. *Zeits. Phys.* 90:209–231. (Reprinted in Heisenberg 1989.)

Heisenberg, W. 1934c. Berichtigung zu der Arbeit: "Bemerkung zur Diracschen Theorie des Positrons." *Zeits. Phys.* 92:692. (Reprinted in Heisenberg 1989.)

Heisenberg, W. 1938a. Über die in der Theorie der Elementarteilchen auftretende universelle Länge. *Ann. der Phys.* 32:20–33. (Reprinted in Heisenberg 1989.)

Heisenberg, W. 1938b. Report to the Congress of the Institute of Intellectual Cooperation held in Warsaw 1938 on "New Theories in Physics," Institute of Intellectual Cooperation, Paris, 1939.

Heisenberg, W. 1938c. Die Grenzen der Anwendbarkeit der bisherigen Quantentheorie. *Zeits. Phys.* 110:251–260. (Reprinted in Heisenberg 1989.)

Heisenberg, W. 1939a. Zur Theorie der explosionsartigen Schauer in der kosmische Strahlung. *Zeits. Phys.* 113:61–86.

Heisenberg, W. 1939b. Bericht über die allgemeinen Eigenschaften der Elementarteilchen. In Heisenberg 1989, series B, pp. 346–358.

Heisenberg, W. 1943a. Die "beobachtbaren Grössen" in der Theorie der Elementarteilchen. I. *Zeits. Phys.* 120:513–538. (Reprinted in Heisenberg 1989.)

Heisenberg, W. 1943b. Die "beobachtbaren Grössen" in der Theorie der Elementarteilchen. II. *Zeits. Phys.* 120:673–702. (Reprinted in Heisenberg 1989.)

Heisenberg, W. 1944. Die "beobachtbaren Grössen" in der Theorie der Elementarteilchen. III. *Zeits. Phys.* 123:93-112. (Reprinted in Heisenberg 1989.)

Heisenberg, W., ed. 1946. *Cosmic Radiation.* Trans. T. H. Johnson. New York: Dover.

Heisenberg, W. 1949. *Two Lectures.* Cambridge, U.K.: Cambridge University Press.

Heisenberg, W. 1963. Interview with T. S. Kuhn. Archives for the History of Quantum Physics, Niels Bohr Library, AIP, New York.

Heisenberg, W. 1969. Methods in theoretical physics. Evening lecture in the series "From a Life in Physics" at the International Centre for Theoretical Physics, Trieste, Italy. Special Suppl., *IAEA Bulletin,* Vienna.

Heisenberg, W. 1971. *Physics and Beyond: Encounters and Conversations.* New York: Harper and Row.

Heisenberg, W. 1989. *Collected Works.* Series A, part 2., ed. W. Blum, H. P. Dürr, and H. Rechenberg. Berlin: Springer-Verlag.

Heisenberg, W., and Euler, H. 1936. Folgerungen aus der Diracschen Theorie der Positrons. *Zeits. Phys.* 98:714–732. (Reprinted in Heisenberg 1989.)

Heisenberg, W., and Pauli, W. 1929. Zur Quantenelektrodynamik der Wellenfelder. I. *Zeits. Phys.* 56:1–61. (Reprinted in Heisenberg 1989.)

Heisenberg, W., and Pauli, W. 1930. Zur Quantenelektrodynamik der Wellenfelder. II. *Zeits. Phys.* 59:168–190. (Reprinted in Heisenberg 1989.)

Heitler, W. 1936. *The Quantum Theory of Radiation.* 1st ed. Oxford: Clarendon Press.

Heitler, W. 1940. Scattering of mesons and the magnetic moments of proton and neutron. *Nature* 145:29–30.

Heitler, W. 1941. The influence of radiation damping on the scattering of light and mesons by free particles. *Proc. Camb. Phil. Soc.* 37:291–300.

Heitler, W. 1945. *The Quantum Theory of Radiation.* 2d ed. Oxford: Clarendon Press.

Heitler, W. 1947. The quantum theory of damping as a proposal for Heisenberg's S-matrix. In *Report of an International Conference on Fundamental Particles and Low Temperature* held at the Cavendish Laboratory, Cambridge, on July 22–27, 1946. London: The Physical Society.

Heitler, W. 1961. Divergences in quantum field theory. In *The Quantum Theory of Fields.* Proceedings of the Twelfth Conference on Physics at the University of Brussels, October 1961, Instituts Solvay. New York: Interscience, 1962.

Heitler, W., and Peng, H. W. 1942. The influence of radiation damping on the scattering of mesons. *Proc. Cambridge Phil. Soc.* 38:296–312.

Hendry, J. 1980. Weimar culture and causality. *History of Science* 18:155–180.

Hendry, J. 1984. *The Creation of Quantum Mechanics and the Bohr-Pauli Dialogue.* Dordrecht, Holland: Reidel.

Hepp, K. 1966. Proof of the Bogoliubov-Parasiuk theorem on renormalization. *Comm. Math. Phys.* 2:301–326.

Herzberg, G. 1937. *Atomic Spectra and Atomic Stucture.* New York: Prentice Hall.

Hewlett, R. G., and Anderson, O. E., Jr. 1962. *The New World, 1939–1946.* Vol. 1, *A History of the United States Atomic Energy Commission.* College Station: Pennsylvania State University Press.

Hewlett, R. G., and Duncan, F. 1969. *Atomic Shield, 1947–1952.* Vol. 2, *A History of the United States Atomic Energy Commission.* College Station: Pennsylvania State University Press.

Hillis, W. D. 1989. Richard Feynman and the connection machine. *Phys. Today* 42/2: 91–100.

Hirosige, T. 1963. Social conditions for pre-war Japanese research in nuclear physics. *Jap. Stud. Hist. Sci.* 2:80–93. Also in Nakayama et al. 1974, pp. 202–220.

Hoch, P. 1983. The reception of Central European refugee physicists of the 1930's: USSR, UK, USA. *Annals of Science* 40:206–246.

Hoch, P. 1987. Migration and the generation of new scientific ideas. *Minerva* 25/3:209–237.

Hoch, P. K., and Yoxen, E. J. 1987. Schrödinger at Oxford: A hypothetical national cultural synthesis which failed. *Annals of Science* 44:593–616.

Hoddeson, L. H. 1983. Establishing KEK in Japan and Fermilab in the U.S.: Internationalism, Nationalism, and high energy accelerators. *Soc. Stud. Sci.* 13:1–48.

Hoddeson, L. H., and Baym, G. 1979. Interview with Richard Feynman. Los Alamos National Laboatory Record Center/Archives, T–79–0004.

Hoddeson, L. H., and Baym, G. 1980. The development of the quantum mechanical electron theory of metals: 1900–1928. *Proc. Roy. Soc. London* A371:3–23.

Holborn, H. 1951. *The Political Collapse of Europe.* New York: Alfred Knopf.

Holton, G. 1973. *Thematic Origins of Scientific Thought: Kepler to Einstein.* Cambridge, Mass.: Harvard University Press.

Holton, G. 1974. Striking gold in science: Fermi's group and the recapture of Italy's place in physics. *Minerva* 12:169–198.

Holton, G. 1981. The formation of the American physics community in the 1920's and the coming of Albert Einstein. *Minerva* 19:569–581.

Houriet, A., and Kind, A. 1949. Classification invariante des termes de la matrice. *S. Helv. Phys. Acta* 22:319–330.

Houston, W. V. 1937. A new method of analysis of the structure of H_α and D_α. *Phys. Rev.* 51:446–449.

Houston, W. V., and Hsieh, Y. M. 1934. The fine structure of the Balmer lines. *Phys. Rev.* 45:263–272.

Hurst, C. A. 1952. The enumeration of graphs in the Feynman-Dyson technique. *Proc. Roy. Soc. London* A214:44–61.

L'Institut International de Coopération Intellectuelle, 1925–1946. Paris: Institut International de Coopération Intellectuelle.

Irving, D. 1946. *The Virus House.* London: Kimber.

Itakura, K., and Yagi, E. 1974. The Japanese research system and the establishment of the Institute of Physical and Chemical Research. In Nakayama et al. 1974.

Ito, D. 1988. My positive and negative contribution to the development of Tomonaga's theory of renormalization. In Brown et al. 1988, pp. 61–62.

Ito, D.; Koba, Z.; and Tomonaga, S. 1947. Corrections due to the reaction of "cohesive force field" for the elastic scattering of an electron. *Prog. Theor. Phys.* 2:216. Errata 2:217.

Ito, D.; Koba, Z.; and Tomonaga, S. 1948a. Corrections due to the reaction of "cohesive force field" to the elastic scattering of an electron. *Prog. Theor. Phys.* 3:276–289.

Ito, D.; Koba, Z.; and Tomonaga, S. 1948b. On radiation reactions in collision processes. I. *Prog. Theor. Phys.* 3:325–337.

Iwanenko, D. 1934. Interaction of neutrons and protons. *Nature* 133:981–982.

Jackiw, R. 1972. Field investigations in current algebra. In S. B. Treiman, R. Jackiw, and D. J. Gross, eds. *Lectures on Current Algebra and Its Applications,* pp. 97–254. Princeton, N.J.: Princeton University Press.

Jackiw, R.; Khuri, N.; Weinberg, S.; and Witten, E., eds. 1985. *Shelter Island II.* Cambridge, Mass.: MIT Press.

Jackman, J. C., and Borden, C. M., eds. 1983. *The Muses Flee Hitler: Cultural Transfer and Adaptation, 1930–1945.* Washington D.C.: Smithsonian Institution Press.

Jaffe, A. 1965. Divergence of perturbation theory for bosons. *Comm. Math. Phys.* 1: 127–149.

Jammer, M. 1974. *The Philosophy of Quantum Mechanics: The Interpretation of Quantum Mechanics in Historical Perspective.* New York: McGraw-Hill.

Jauch, J. M., and Rohrlich, F. 1955. *The Theory of Photons and Electrons.* Reading, Mass.: Addison-Wesley.

Jewett, F. B. 1947. The future of scientific research in the post-war world. In G. A. Baitsell, ed., *Science in Progress,* 5th ser., pp. 3–23. New Haven, Conn.: Yale University Press.

Jordan, P. 1925. Über das thermische Gleichgewicht zwischen Quantenatomen und Hohlraumstrahlung. *Zeits. Phys.* 33:649–655.

Jordan, P. 1926. Bermerkungen über einen Zusammenhang zwischen Duanes Quantentheorie der Interferenz und den de Broglieschen Wellen. *Zeits. Phys.* 37:376–382.

Jordan, P. 1927a. Über quantenmechanische Darstellung von quanten Sprüngen. *Zeits. Phys.* 40:661–666.

Jordan, P. 1927b. Zur statistischen Deutung der Quantenmechanik. *Zeits. Phys.* 41: 797–800.

Jordan, P. 1927c. Kausalität und Statistik in der modernen Physik. *Naturwissenschaften* 15:105–110.

Jordan, P. 1927d. Die Entwicklung der neuen Quantenmechanik. *Naturwissenschaften* 15:614–623, 636–649.

Jordan, P. 1927e. Über eine neue Begründung der Quantenmechanik. II. *Zeits. Phys.* 44: 1–25.

Jordan, P. 1927f. Zur Quantenmechanik des Gasentartung. *Zeits. Phys.* 44:473–480.

Jordan, P. 1927g. Über Wellen und Korpuskeln in der Quantenmechanik. *Zeits. Phys.* 45:765–775.

Jordan, P. 1927h. Philosophical foundations of quantum theory. *Nature* 119:566–569, 779.

Jordan, P. 1929. Der gegenwärtige Stand der Quantenelectrodynamik. *Phys. Zeits.* 30:700–712.

Jordan, P. 1932. Zur Methode der zweiten Quantelung. *Zeits. Phys.* 75:648–653.

Jordan, P. 1933. *Statistische Mechanik auf quantentheoretischer Grundlage.* Braunschweig: F. Vieweg.

Jordan, P. 1935. Das Neutrinotheorie des Lichtes. *Zeits. Phys.* 93:464–472.

Jordan, P. 1936. *Anschaulische Quantentheorie. Eine Einführung in die moderne Auffassung der Quantenerscheinungen.* Berlin: Springer-Verlag.

Jordan, P. 1944. *Physics of the 20th Century.* Trans. E. Osky. New York: Philosophical Library.

Jordan, P. 1963. Interview with T. S. Kuhn. Archives for the History of Quantum Physics, Niels Bohr Library, AIP, New York.

Jordan, P. 1971. *Begegnungen.* Hamburg: Stalling-Verlag.

Jordan, P. 1973. Early years of quantum mechanics: Some reminiscences. In Mehra 1973, pp. 294–300.

Jordan, P., and Fock, V. 1930. Über Stimmung des elektromagnetischem Feldes. *Zeits. Phys.* 66:206–209.

Jordan, P., and Klein, O. 1927. Zum Mehrkörperproblem der Quantentheorie. *Zeits. Phys.* 45:751–765.

Jordan, P., and Pauli, W. 1928. Zur Quantenelektrodynamik ladungfreier Felder. *Zeits. Phys.* 47:151–173.

Jordan, P., and Wigner, E. P. 1928. Über das Paulische Äquivalenzverbot. *Zeits. Phys.* 47:631–651. (Reprinted in Schwinger 1958.)

Jordan, P.; Neumann, J. V.; and Wigner, E. 1934. On the algebraic generalization of the quantum mechanical formalism. *Annals of Math.* 35:29–64.

Jost, R. 1946. Bemerkung zur vorstehenden Arbeit. *Physica* 12:509–510.

Jost, R. 1947. Über die falschen Nullstellen der Eigenwerte der *S*-matrix. *Helv. Phys. Acta* 20:256–66.

Jost, R. 1960. Das Pauli-Prinzip und die Lorentz-Gruppe. In Fierz and Weisskopf 1960, pp. 107–136.

Jost, R. 1965. *The General Theory of Quantum Fields.* Providence, R.I.: American Mathematical Society.

Jost, R. 1972. Foundations of quantum field theory. In Salam and Wigner 1972, pp. 61–77.

Jost, R. 1984. Erinnerungen: Erlesenes und Erlebtes. *Physik. Blätter* 40:178–181.

Jost, R. and Luttinger, J. M. 1950. Vacuumpolarisation und e^4-Ladungrenormalisation für Elektronen. *Helv. Phys. Acta* 23:201–213.

Jost, R., and Rayski, J. 1949. Remarks on the problem of vacuum polarization and the photon self-energy. *Helv. Phys. Acta* 22:456–466.

Jungnickel, C., and McCormmach, R. 1986. *Intellectual Mastery of Nature: Theoretical Physics from Ohm to Einstein.* 2 vols. Chicago: University of Chicago Press.

Kahn, B. 1941. Remarks on deviations from the fine structure formula. *Physica* 8:58.

Kalckar, J. 1971. Measurability problems in the quantum theory of fields. In *Fundamenti di meccanica quantistica*, Rendiconti della Scuola Internazaionale di Fisica "Enrico Fermi." New York: Academic Press.

Källén, G. 1949. Higher approximations in the external field for the problem of vacuum polarization. *Helv. Phys. Acta* 22:636–653.

Källén, G. 1952. On the definition of the renormalization constants in quantum electrodynamics. *Helv. Phys. Acta* 25:417–434.

Källén, G. 1953. On the magnitude of the renormalization constants in quantum electrodynamics. *Dan. Vidensk Selskab. Mat.-Fys. Medd.* 27:1–18.

Källén, G. 1966. Review of consistency problems in quantum electrodynamics. *Acta Phys. Austr.* (suppl.)2:133–161.

Kaluza, Th. 1921. Zur Unitätsproblem der Physik. *Sitzungsber. der Berliner Akad. Wiss.* 54:966.

Kamefuchi, S. 1951. Note on the direct interaction between spinor fields. *Prog. Theor. Phys.* 6:175–181.

Kamefuchi, S. 1986. Quantum field theory in postwar Japan. In F. Mancini, ed., *Quantum Field Theory*, pp. 1–18. Amsterdam and New York: North-Holland.

Kamen, M. 1985. *Radiant Science, Dark Politics.* Berkeley: University of California Press.

Kaneseki, Y. 1974. The elementary particle theory group (1951). In Nakayama et al. 1974, pp. 221–252.

Kargon, R. H. 1981. Birth cries of the elements: Theory and experiment along Millikan route to cosmic rays. In H. Wolf, ed., *The Analytic Spirit.* Ithaca, N.Y.: Cornell University Press.

Kargon, R. H. 1982. *The Rise of Robert Millikan.* Ithaca, N.Y.: Cornell University Press.

Karplus, R., and Klein, A. 1952. Electrodynamic displacements of atomic energy levels. III. The hyperfine structure of positronium. *Phys. Rev.* 87:848–858.

Karplus, R., and Kroll, N. 1949. Fourth-order corrections in quantum electrodynamics and the magnetic moment of the electron. *Phys. Rev.* 76:846(L), 77:536–549.

Karplus, R., and Schwinger, J. 1948. A note on saturation in microwave spectroscopy. *Phys. Rev.* 73:1020–1026.

Karplus, R.; Klein, A.; and Schwinger, J., 1951. Electrodynamic displacement of atomic energy levels. *Phys. Rev.* 84:597.

Karplus, R.; Klein, A.; and Schwinger, J. 1952. Electrodynamic displacement of atomic energy levels. II. The Lamb shift. *Phys. Rev.* 86:288–301.

Kawabe, R. 1988. Chronological table for development of quantum field theory in 1940's Japan. In Brown et al. 1988, pp. 78–81.

Keith, S. T., and Hoch, P. K. 1986. Formation of a research school: Theoretical solid state physics at Bristol, 1930–54. *Brit. J. Hist. Sci.* 19:19–44.

Kellogg, J.B.M., and Millman, S. 1946. The molecular beam magnetic resonance method. The radiofrequency spectra of atoms and molecules. *Rev. Mod. Phys.* 18:323–352.

Kellogg, J.B.M.; Rabi, I. I.; Ramsay, N.; and Zacharias, J. 1939a. An electric quadrupole moment of the deuteron. *Phys. Rev.* 55:318–319.

Kellogg, J.B.M.; Rabi, I. I.; Ramsay, N.; and Zacharias, J. 1939b. Magnetic moment of the proton and the deuteron. *Phys. Rev.* 56:728–743.

Kellogg, J.B.M.; Rabi, I. I.; Ramsay, N.; and Zacharias, J. 1940. An electric quadrupole moment of the deuteron. *Phys. Rev.* 57:677–695.

Kemble, E. C. 1937. *The Fundamental Principles of Quantum Mechanics.* New York: McGraw-Hill.

Kemmer, N. 1938. The charge-dependence of nuclear forces. *Proc. Cambridge Phil. Soc.* 34:354–364.

Kemmer, N. 1965. The impact of Yukawa's meson theory on workers in Europe—A Reminiscence. *Suppl., Progress of Theoretical Phys.,* pp. 602–608. Commemoration issue for the 30th anniversary of the meson theory of Dr. H. Yukawa.

Kemmer, N. 1982. Isospin. Colloque international sur l'histoire de la physique des particules. *J. de Physique,* colloque C–8, suppl. au n° 12, 43:359–393.

Kemmer, N. 1983. Die Anfänge der Mesontheorie und des verallgemeinerten Isospins. *Phys. Blätter* 7:170–175.

Kemmer, N., and Weisskopf, V. F. 1936. Deviations from the Maxwell equations resulting from the theory of the positron. *Nature* 137:659.

Kevles, D. 1978. *The Physicists: The History of a Scientific Community in Modern America.* New York: Alfred Knopf.

Killian, J. R. 1895. *The Education of a College President: A Memoir.* Cambridge, Mass.: MIT Press.

Kinoshita, T. 1950. A note on the C meson hypothesis. *Prog. Theor. Phys.* 5:535–536.

Kinoshita, T. 1985. Theory of lepton anomalous moments, 1947–1983. In Jackiw et al. 1985, pp. 278–300.

Kinoshita, T. 1988. Personal recollections, 1944–1952. In Brown et al. 1988, pp. 7–11.

Kinoshita, T., and Saperstein, J. 1984. New developments in QED. In R. S. Van Dyck, Jr., and E. N. Fortson, eds., *Atomic Physics 9,* pp. 38–52. Singapore: World Scientific.

Kinster, L. E., and Houston, W. V. 1934. The value of e/m from the Zeeman effect. *Phys. Rev.* 45:104–108.

Kiyonubu, I., and Yagi, E. 1974. The Japanese research system and the establishment of the Institute of Physical and Chemical Research. In Nakayama et al. 1974, pp. 158–201.

Klein, O. 1926. Quantentheorie und fünfdimensionale Relativitätstheorie. *Zeits. Phys.* 37:895–906.

Klein, O. 1938. *Entretiens sur les idées fondamentalles de la physique moderne.* Paris: Hermann.

Klein, O., and Nishina, Y. 1929. Über die Streuung von Strahlung durch freie Elektronen nach der neuen relativistischen Quantenelektrodynamik von Dirac. *Zeits. Phys.* 52: 853–868.

Koba, Z., and Takeda, G. 1948. Radiation reactions in collision process. II. Radiative corrections for Compton scattering. *Prog. Theor. Phys.* 3/4:407–422.

Koba, Z., and Takeda, G. 1949. Radiation reactions in collision processes. III. *Prog. Theor. Phys.* 4(1):60–70.

Koba, Z., and Tomonaga, S. 1948. On radiation reactions in collision processes. I. Application of the "self-consistent" subtraction method to the elastic scattering of an electron. *Prog. Theor. Phys.* 3/3:290–303.

Koba, Z.; Tati, T.; and Tomonaga, S. 1947a. On a relativistically invariant formulation of quantum theory of wave fields. II. *Prog. Theor. Phys.* 2/3:101–116.

Koba, Z.; Tati, T.; and Tomonaga, S. 1947b. On a relativistically invariant formulation of quantum theory of wave fields. III. *Prog. Theor. Phys.* 2/4:198–208.

Koizumi, K. 1975. The emergence of Japan's first physicists: 1868–1900. In R. McCormmach, ed., *Historical Studies in the Physical Sciences,* vol. 6, pp. 3–108. Princeton, N.J.: Princeton University Press.

Kojevnikov, A. B. 1990. Dirac's quantum electrodynamics. In *Paul Dirac and the Physics of the XXth Century.* Moscow.

Komons, N. A. 1966. *Science and the Air Force: A History of the Air Force Office of Scientific Research.* Arlington, Va.: Office of Aerospace Research.

Kragh, H. 1981. The genesis of Dirac's relativistic theory of electrons. *Arch. Hist. Exact Sci.* 20:31–67.

Kragh, H. 1985. The fine structure of hydrogen and the gross structure of the physics community, 1916–26. *Hist. Stud. Phys. Sci.* 15:67–126.

Kragh, H. 1990. *Dirac: A Scientific Biography.* New York: Cambridge University Press.

Kramers, H. 1938a. Quantentheorie des Elektrons und der Strahlung. In *Hand und Jahrbuch der Chemische Physik.* I, part 2. Leipzig.

Kramers, H. 1938b. Die Wechselwirkung zwischen geladenen Teilchen und Strahlungsfeld. *Nuovo Cimento* 15:108–114 (reprinted in Kramers 1956).

Kramers, H. 1944. Fundamental difficulties of a theory of particles. *Ned. T. Nat.* 11: 134–147. (Reprinted in Kramers 1956.)

Kramers, H. 1948. Nonrelativistic quantum electrodynamics and correspondence principle. In *Rapports et discussions du 8e Congrès Solvay.* Brussels: R. Stoop, 1950.

Kramers, H. 1956. *Collected Scientific Papers.* Amsterdam: North-Holland.

Kroll, N. 1948. The unstrapped resonant system (chap. 2), and The rising sun system (chap. 3). In Collins 1948, pp. 49–117.

Kroll, N. 1987. Unpublished interview with Dr. Joan Bromberg, History of the Laser Project.

Kroll, N., and Lamb, W. E., Jr. 1949. On the self-energy of a bound electron. *Phys. Rev.* 75:388–398. (Reprinted in Schwinger 1958, pp. 414–424.)

Kronig, R. 1946. A supplementary condition in Heisenberg's theory of elementary particles. *Physica* 12:543–544.

Kuhn, T. S. 1962. *The Structure of Scientific Revolutions.* Chicago: University of Chicago Press.

Kuhn, T. S. 1970. *The Structure of Scientific Revolutions.* 2d ed. Chicago: University of Chicago Press.

Kuhn, T. S. 1978. *Blackbody Theory and the Quantum Discontinuity, 1894–1912.* Oxford and New York: Oxford University Press.

Kursunoglu, B.; Mintz, S. L.; and Perlmutter, A., eds. 1985. *High Energy Physics.* New York: Plenum Press.

Kursunoglu, B. N., and Wigner, E. P. 1987. *Reminiscences about a Great Physicist: Paul Adrien Maurice Dirac.* Cambridge (U.K.) and New York: Cambridge University Press.

Kusch, P. 1956. The magnetic moment of the electron. In *Nobel Lectures in Physics,* pp. 298–310. New York: Elsevier, 1964.

Kusch, P. Undated. The magnetic moment of the electron. A case history in the history of science. Mimeographed notes of lectures delivered at the University of Washington.

Kusch, P., and Foley, H. M. 1947. Precision measurement of the ratio of the atomic "g values" in the $^2P_{1/2}$ and $^2P_{3/2}$ state of gallium. *Phys. Rev.* 72:1256(L).

Kusch, P., and Foley, H. M. 1948a. On the intrinsic moment of the electron. *Phys. Rev.* 73:412 (L).

Kusch, P., and Foley, H. M. 1948b. The magnetic moment of the electron. *Phys. Rev.* 74:250–263.

Kusch, P., and Hughes, V. 1959. Atomic and molecular spectroscopy. In S. Flugge, ed. *Handbuch der Physik*. Vol. 37. Berlin: Springer-Verlag.

Lakatos, I. 1970. Falsification and the methodology of scientific research programmes. In I. Lakatos and A. Musgrave, eds., *Criticism and the Growth of Knowledge*. Cambridge, U.K.: Cambridge University Press.

Lamb, W. E., Jr. 1937. A note on the capture of slow neutrons in hydrogeneous substances. *Phys. Rev.* 51:187–190.

Lamb, W. E., Jr. 1939. Capture of neutrons by atoms in a crystal. *Phys. Rev.* 55:190–197.

Lamb, W. E., Jr. 1939. Deviation from the Coulomb law for a proton. *Phys. Rev.* 56: 384–385.

Lamb, W. E., Jr. 1940. Deviation from the Coulomb law for a proton. *Phys. Rev.* 57:458.

Lamb, W. E., Jr. 1946. Theory of a microwave spectroscope. *Phys. Rev.* 70:308–317.

Lamb, W. E., Jr. 1951. Anomalous fine structure of H and He$^+$. *Reports on Prog. in Phys.* 14:19–63.

Lamb, W. E., Jr. 1952. Fine structure of the hydrogen atom. III. *Phys. Rev.* 85:259–276.

Lamb, W. E., Jr. 1956. Fine structure of the hydrogen atom. *Science* 123:439–442. (Also in *Nobel Lectures, Physics*, pp. 283–297. New York: Elsevier, 1964.)

Lamb, W. E., Jr. 1960. The inverse square law in physics. *The Colorado Quarterly* 7(3):258–270.

Lamb, W. E., Jr. 1976. Some history of the hydrogen fine structure experiment. *Trans. N.Y. Acad. of Sciences*, series 2, 38:82–86.

Lamb, W. E., Jr. 1983. Fine structure of hydrogen. In Brown and Hoddeson 1983, pp. 311–328.

Lamb, W. E., Jr., and Retherford, R. C. 1946. Experiment to determine the fine structure of the hydrogen atom. *Columbia Univ. Rad. Lab. Rept.*, 1946, pp. 18–26. (See also Lamb 1976.)

Lamb, W. E., Jr., and Retherford, R. C. 1947. Fine structure of the hydrogen atom by microwave method. *Phys. Rev.* 72:241–243. (Reprinted in Schwinger 1958, 136–138.)

Lamb, W. E., Jr., and Retherford, R. C. 1950. Fine structure of the hydrogen atom. I and II. *Phys. Rev.* 79:549–572 and 81:222–232.

Lamb, W. E., Jr., and Retherford, R. C. 1952. Fine structure of the hydrogen atom. IV. *Phys. Rev.* 86:1114–1122.

Lamb, W. E., Jr., and Schiff, L. I. 1938. On the electromagnetic properties of nuclear systems. *Phys. Rev.* 53:651–661.

Landau, L. D. 1955. On the quantum theory of fields. In Landau and Pomeranchuck 1955, pp. 52–69.

Landau, L. D., and Peierls, R. 1931. Erweiterung des Unbestimmtheitsprinzips für die relativistische Quantentheorie. *Zeits. Phys.* 69:56–67. (Trans. in ter Haar 1965, pp. 40–51.)

Landau, L. D., and Pomeranchuck, I. 1955. On point interactions in quantum electrodynamics. *Dokl. Akad. Nauk. USSR* 102:489–491.

Landau, L. D.; Abrikosov, A. A.; and Khalatnikov, I. M. 1954a. The removal of infinities in quantum electrodynamics. *Dokl. Akad. Nauk. USSR* 95:497–499.

Landau, L. D.; Abrikosov, A. A.; and Khalatnikov, I. M. 1954b. An asymptotic expression for the electron Green function in quantum electrodynamics. *Dokl. Akad. Nauk. USSR* 95:773–776.

Landau, L. D.; Abrikosov, A. A.; and Khalatnikov, I. M. 1954c. An asymptotic expression for the photon Green function in quantum electrodynamics. *Dokl. Akad. Nauk. USSR* 95:1117–1120.

Landau, L. D.; Abrikosov, A. A.; and Khalatnikov, I. M. 1954d. The electron mass in quantum electrodynamics. *Dokl. Akad. Nauk. USSR* 96:261–263.

Landau, L. D.; Abrikosov, A. A.; and Khalatnikov, I. M. 1956. On the quantum theory of fields. *Nuovo Cimento* (suppl.) 3:80–104.

Lattes, C.M.G.; Muirhead, H.; Occhialini, G.P.S.; and Powell, C. F. 1947. Processes involving charged mesons. *Nature* 159:694–697.

Laudan, L. 1977. *Progress and Its Problems.* Berkeley: University of California Press.

Lawson, J. L., and Uhlenbeck, G., eds. 1949. *Threshold Signals.* Vol. 24 of MIT Radiation Laboratory Series. New York: McGraw-Hill.

Lee, B. W., and Zinn-Justin, J. 1972. Spontaneously broken gauge symmetries. *Phys. Rev.* D5:3121–3160.

Lehmann, H.; Symanzik, K.; and Zimmerman, W. 1955. Zur Formulierung quantisierter Feldtheorie. *Nuovo Cimento* 10/(1):205–225.

Lehmann, H.; Symanzik, K.; and Zimmerman, W. 1957. On the formulation of quantized field theories. *Nuovo Cimento* 10/(6):319–333.

Levine, H., and Schwinger, J. 1947. On the radiation of sound from an unflanged circular pipe. *Phys. Rev.* 72:742.

Levine, H., and Schwinger, J. 1948a. On the radiation of sound from an unflanged circular pipe. *Phys. Rev.* 73:383.

Levine, H., and Schwinger, J. 1948b. On the theory of diffraction by an aperture in an infinite plane screen. I. *Phys. Rev.* 74:958.

Levine, H., and Schwinger, J. 1948c. Variational principles for diffraction problems. *Phys. Rev.* 74:1212.

Levine, H., and Schwinger, J. 1949a. On the theory of diffraction by an aperture in an infinite plane screen. II. *Phys. Rev.* 75:1423.

Levine, H., and Schwinger, J. 1949b. On the transmission coefficient of a circular aperture. *Phys. Rev.* 75:1608.

Levine, H., and Schwinger, J. 1950. On the theory of electromagnetic wave diffraction by an aperture in an infinite plane conducting screen. *Comm. Pure Appl. Math. III* 4:355.

Lewis, H. W. 1948. On the reactive terms in quantum electrodynamics. *Phys. Rev.* 73: 173–176.

Lewis, H. W.; Oppenheimer, J. R.; and Wouthuysen, S. A. 1948. The multiple production of mesons. *Phys. Rev.* 73:127–140.

Lippman, B., and Schwinger, J. 1950. Variational principles for scattering processes. I. *Phys. Rev.* 79:469.

Livingston, M. S. 1969. *Particle Accelerators: A Brief History.* Cambridge, Mass.: Harvard University Press.

Longworth, L. G., and Shedlovshy, T. 1970. Duncan Arthur MacInnes, In *Biographical Memoirs of the National Academy of Sciences of the USA*, pp. 294–319. New York: Columbia University Press.

Lorentz, H. A. 1904a. Maxwells elektromagnetische Theorie. In *Encyc. Mat. Wiss.* 5/2: 63–144.

Lorentz, H. A. 1904b. Weiterbildung der Maxwellschen Theorie: Elektronentheorie. In *Encyc. Mat. Wiss.* 5/2:145–280.

Low, M. F. 1984. The impact of social and political factors upon Japanese meson physics research before and during World War II. M.S. diss., Griffith University, Australia.

Low, M. F. 1988. Accounting for the Sakata model. In Brown et al. 1988, pp. 112–122.

Lundeen, S. R., and Pipkin, F. M. 1986. Separated oscillatory field measurement of the Lamb shift in H, $n = 2$. *Metrologia* 22:9–54.

Ma, S. T. 1943. A relativistic formula for the scattering of mesons under the influence of radiation damping. *Proc. Cambridge Phil. Soc.* 39:168–172.

Mack, J. E., and Austern, N. 1947. Newly observed structure in He II λ4686. *Phys. Rev.* 72:972.

Mack, J. E., and Barkofski, E. C. 1942. Atomic beam apparatus for studying the atomic spectra of gases, especially hydrogen. *Rev. Mod. Phys.* 14:82–93.

MacKinnon, E. 1977. Heisenberg, models and the rise of matrix mechanics. *Hist. Stud. Phys. Sci.* 8:137–188.

McMillan, E. M. 1979. Early history of particle accelerators. In Stuewer 1979, pp. 113–156.

Majorana, E. 1937. Teoria simmetrica dell'electrone e del positrone. *Nuovo Cimento* 14:171–178.

Maki, Z. 1988. Tomonaga and the meson theory. In Brown et al. 1988, pp. 69–77.

Manley, J.; Goldsmith, H.; and Schwinger, J. 1937. Neutron energy levels. *Phys. Rev.* 51:1002.

Manley, J.; Goldsmith, H.; and Schwinger, J. 1939a. The widths of nuclear energy levels. *Phys. Rev.* 55:39.

Manley, J.; Goldsmith, H.; and Schwinger, J. 1939b. The resonance absorption of slow neutrons in indium. *Phys. Rev.* 55:107.

Manoukian, E. B. 1983. *Renormalization.* New York: Academic Press.

Manuel, F. 1979. Review of F. Dyson's *Disturbing the Universe. The New Republic,* August 18, 1979.

Marcuvitz, N., and Schwinger, J. 1951. On the representation of the electric and magnetic fields produced by currents and discontinuities in wave guides. I. *J. Appl. Phys.* 22:806.

Marshak, R. E. 1952. *Meson Physics.* New York: McGraw-Hill.

Marshak, R. E. 1970. The Rochester conferences. *Bull. Atom. Sci.* 26:92.

Marshak, R. E. 1983. Particle physics in rapid transition: 1947–1952. In Brown and Hoddeson 1983, pp. 376–401.

Marshak, R. E. 1985. Origin of the two-meson theory. In Jackiw et al. 1985.

Marshak, R. E., and Bethe, H. A. 1947. On the two-meson hypothesis. *Phys. Rev.* 72: 506–509.

Martin, P. 1979. Schwinger and statistical physics: A spin-off success story and some challenging sequels. In *Themes in Contemporary Physics: Essays in Honor of Julian Schwinger's 60th Birthday. Physica* 96A(1 + 2):70–88.

Matthews, P. T., and Salam, A. 1951. Renormalization. *Rev. Mod. Phys.* 23:311–314.

Mayer, A. J. 1967. *Politics and Diplomacy of Peacemaking: Containment and Counterrevolution at Versailles, 1918–1919.* New York: Alfred Knopf.

Mayoux, J. J., ed. 1947. *L'Institut International de Coopération Intellectuelle, 1925–1946.* Paris: Institut International de Coopération Intellectuelle.

Mehra, J., ed. 1973. *The Physicist's Conception of Nature*. Dordrecht, Holland: Reidel.

Mehra, J., 1975. *The Solvay Conferences on Physics*. Dordrecht, Holland: Reidel.

Mehra, J., and Rechenberg, H. 1978–1987. *The Historical Development of Quantum Theory*. Vols. 1–5. New York: Springer-Verlag.

Mehta, M. L. 1967. *Random Matrices*. New York: Academic Press.

Mie, G., 1912. Grundlagen einer Theorie der Materie. *Ann. der Phys.* 37:511–534.

Miller, A. 1984. *Imagery and Scientific Thought: Creating 20th Century Physics*. Boston: Birkhäuser.

Millikan, R. A. 1935. *Electrons*. Cambridge, U.K.: Cambridge University Press.

Millman, S. 1977. Recollections of a Rabi student of the early years in the molecular beam laboratory. *Trans. N.Y. Acad. Sci.,* series 2, vol. 38, no. 4.

Millman, S., and Kusch, P. 1940. The precision measurement of nuclear magnetic moments. *Phys. Rev.* 57:438–457.

Mills, R. L., and Yang, C. N. 1966. Treatment of overlapping divergences in the photon self-energy function. *Prog. Theor. Phys.* (suppl.) 37 and 38:507–511.

Misa, T. J. 1985. Military needs, commercial realities, and the transistor, 1948–1958. In M. R. Smith, ed., *Military Enterprise and Technological Change: Perspective on the American Experience*, pp. 253–288. Cambridge, Mass.: MIT Press.

Missner, M. 1985. Why Einstein became famous in America. *Soc. Stud. Sci.* 15:267–291.

Miyamiza, T., and Tomonaga, S. 1942. Zur Theorie des Mesons. II. *Sci. Papers IPCR* 40:21–67.

Miyamiza, T., and Tomonaga, S. 1943. On the mesotron theory of nuclear forces. *Sci. Papers IPCR* 40:274–310.

Miyamiza, T.; Tati, T.; and Tomonaga, S. 1948. On Wentzel's method in meson theory. *Prog. Theor. Phys.* 3:26–37.

Miyamoto, Y. 1988. Personal recollections. In Brown et al. 1988, pp. 63–67.

Molecular beams: Selected reprints. 1965. Vol. 1, *Experiments with Molecular Beams*. Vol. 2, *Atomic and Molecular Beam Spectroscopy*. New York: American Institute of Physics.

Møller, C. 1932. Zur Theorie des Durchgangs schneller Elektronen durch Materie. *Ann. der Phys.* 14:531–585.

Møller, C. 1945. General properties of the characteristic matrix in the theory of elementary particles. I. *Det. Kgl. Danske Videnskab. Selskab. Mat.-Phys. Medd.* 22:1–48.

Møller, C. 1946a. General properties of the characteristic matrix in the theory of elementary particles. II. *Det. Kgl. Danske Videnskab. Selskab. Mat.-Phys. Medd.* 23/19:1–46.

Møller, C. 1946b. New developments in relativistic quantum theory. *Nature* 158:403–406.

Møller, C. 1946c. Recent developments in relativistic quantum theory. Lectures given at the H. H. Wills Physical Laboratory, University of Bristol, U.K., spring term 1946. Notes by I. N. Sneddon.

Møller, C. 1947. On the theory of the characteristic matrix. In *Report of an International Conference on Fundamental Particles and Low Temperature Held at the Cavendish Laboratory on 22–27 July 1946*. London: The Physical Society.

Møller, C., and Rosenfeld, L. 1939. Theory of meson and nuclear forces. *Nature* 143: 241–242.

Møller, C., and Rosenfeld, L. 1940. On the field theory of nuclear forces. *Det. Konig. Danske Vid. Selskabs. Mat.-Phys. Medd.* 17/8:1–72.

Morette, C. 1951. *Particules élémentaires*. Acualités scientifiques et industrielles 1131. Paris: Hermann.

Morette, C.; Tiomno, J.; and Wheeler, J. A. 1951. Guide to literature of elementary particle physics. *Acualités scientifiques et industrielles* 1131. Paris: Hermann.

Morrison, M. 1986. More on the relationship between technically good and conceptually important experiments. *Brit. J. Phil. Sci.*. 37:101–22.

Morrison, P. 1939. Energy fluctuations in the electromagnetic field. *Phys. Rev.* 56:937–940.

Morrison, P. 1947. Physics in 1946. *J. Appl. Phys.* 18:133–152.

Morrison, P. 1948. Physics in 1947. *J. Appl. Phys.* 19:311–331.

Morse, P. M., and Young, L. A. 1933. Variational atomic wave functions. *Phys. Rev.* 43:501(A).

Morse, P. M.; Young, L. A.; and Haurwitz, E. A. 1935. Tables for determining atomic wave functions and energies. *Phys. Rev.* 48:948–957

Morse, P. M. 1977. *In at the Beginnings: A Physicist's Life.* Cambridge, Mass.: MIT Press.

Morse, R. W. 1966. Basic research and long-range national goals. *Naval Research Review* 19:1–7.

Mott, N. F. 1931. Influence of radiative forces on the scattering of electrons. *Proc. Cambridge Phil. Soc.* 27:255–267.

Mott, N. F., and Peierls, R. 1977. Werner Heisenberg, 1901–1976. *Biog. Mem. Fellows Roy. Soc.* 23:213–251.

Motz, L., and Schwinger, J. 1935. On the β-radioactivity of neutrons. *Phys. Rev.* 48:704.

Motz, L., and Schwinger, J. 1940a. Neutron-deuteron scattering cross section. *Phys. Rev.* 57:161.

Motz, L., and Schwinger, J. 1940b. The scattering of thermal neutrons by deuterons. *Phys. Rev.* 58:26.

Moyer, D. F. 1981a,b,c. Origins of Dirac's electron, 1925–28. *Amer. J. Phys.* 49:944–949; Evaluations of Dirac's electron, 1928–1932. *Amer. J. Phys.* 49:1055–1062; Vindications of Dirac's electron, 1932–1934. *Amer. J. Phys.* 49:1120–1125.

Mukherji, V. 1974. A history of the meson theory of nuclear forces from 1935 to 1952. *Arch. Hist. Exact Sci.* 13:27–102.

Musher, J. 1966. Comment on some theorems of quantum chemistry. *Am. J. Phys.* 34:267–268.

Nafe, J. E.; Nelson, E. B.; and Rabi, I. I. 1947. Hyperfine structure of atomic hydrogen and deuterium. *Phys. Rev.* 71:914–915.

Najita, T. 1974. *Japan.* Englewood Cliffs, N.J.: Prentice Hall.

Nakayama, S. 1974. A history of universities: An overview from the viewpoint of science history (1971). In Nakayama et al. 1974, pp. 72–80.

Nakayama, S. 1977. *Characteristics of Scientific Development in Japan.* New Delhi.

Nakayama, S.; Swain, D. L.; and Yagi, E., eds. 1974. *Science and Society in Modern Japan.* Cambridge, Mass.: MIT Press.

Nambu, Y. 1949. The level shift and the anomalous magnetic moment of the electron. *Prog. Theor. Phys.* 4/1:82–94.

Nambu, Y. 1950. Force potential in quantum field theory. *Prog. Theor. Phys.* 5:614–633.

Nambu, Y. 1960. A "superconductor" model of elementary particles and its consequences. In *Proceeedings of the Midwest Conference on Theoretical Physics,* Purdue University, West Lafayette, Indiana.

Nambu, Y. 1988. Summary of personal recollections of the Tokyo group. In Brown et al. 1988, pp. 3–7.

Nambu, Y., and Jona-Lasinio, G. 1961. Dynamical model of elementary particles based on an analogy with superconductivity. *Phys. Rev.* 122:345–358 and 124:246–254.

Needell, A. A. 1983. Nuclear reactors and the founding of Brookhaven National Laboratory. *Hist. Stud. Phys. Sci.* 14:93–122.

New Theories in Physics. 1939. Reports to the Congress of the Institute of Intellectual Cooperation held in Warsaw, September 1938. Paris: Institute of Intellectual Cooperation.

Nishina, Y., and Tomonaga, S. 1933. On the creation of positive and negative electrons. *Sci. Papers IPCR* 15:248–249.

Nishina, Y., and Tomonaga, S. 1934. On the negative-energy electrons. *Jap. J. Phys.* 9: 35–46.

Nishina, Y., and Tomonaga, S. 1936. A note on the interactions of the neutron and the proton. *Sci. Papers IPCR* 30:61–69.

Nishina, Y.; Sakata, S.; and Tomonaga, S. 1934. On the photo-electric creation of positive and negative electrons. *Sci. Papers IPCR* 24 (suppl. 17) :1–5.

Nishina, Y.; Takeuchi, M.; and Ichimiya, T. 1937. On the nature of cosmic-ray particles. *Phys. Rev.* 52:1198–1199.

Nishina, Y.; Tamaki, H.; and Tomonaga, S. 1934. On the annihilation of electrons and positrons. *Sci. Papers IPCR* 18:7–12.

Noble, D. F. 1984. *Forces of Production.* New York: Alfred Knopf.

Noble, D. F. 1985. Command performance: A perspective on military enterprise and technological change. In M. R. Smith 1985.

Nordheim, L., and Webb, N. 1939. On the production of the hard components in the cosmic radiation. *Phys. Rev.* 56:494–501.

Occhialini, G.P.S. 1933. La ricenti ricerche intorno all'ectrone positivo. *La Ricerca Scientifica*, part 1.

Old, B. S. 1961. The evolution of the Office of Naval Research. *Phys. Today* 14:30–35.

Oliner, A. A. 1984. Historical perspectives on microwave field theory. *IEEE Transactions on Microwave Theory and Techniques.* MTT 32:1022–1045.

Opechowski, W. 1941. Sur la quantification du système de l'electron et du rayonnement. *Physica* 8:161–176.

Oppenheimer, J. R. 1927a. Bemerkung zur Zerstreuung der α-Teilchen. *Zeits. Phys.* 43:413–415.

Oppenheimer, J. R. 1927b. Zur Quantenmechanik der Richtungsentartung. *Zeits. Phys.* 43:27–43.

Oppenheimer, J. R. 1927c. Zur Quantentheorie kontinuierlicher Spektren. *Zeits. Phys.* 41:268–293.

Oppenheimer, J. R. 1928. Three notes on the quantum theory of aperiodic effects. *Phys. Rev.* 31:66–81.

Oppenheimer, J. R. 1930a. Note on the theory of the interaction of field and matter. *Phys. Rev.* 35:461–477.

Oppenheimer, J. R. 1930b. On the theory of electrons and protons. *Phys. Rev.* 35:562–563.

Oppenheimer, J. R. 1930c. Two notes on the probability of radiative transitions. *Phys. Rev.* 35:939–947.

Oppenheimer, J . R. 1935. Are the formulae for the absorption of high energy radiations valid? *Phys. Rev.* 47:44–52.

Oppenheimer, J. R. 1941. The mesotron and the quantum theory of fields. In *Nuclear Physics: University of Pennsylvania Bicentennial Conference*. Philadelphia: University of Pennsylvania Press.

Oppenheimer, J. R. 1948. Electron theory. In *Rapport du 8ᵉ Conseil de Physique Solvay*, September 27–October 2, 1948. Brussels: R. Stoop, 1950. (Reprinted in Schwinger 1958, pp. 145–155.)

Oppenheimer, J. R. 1953. *Science and the Common Understanding*. New York: Simon and Schuster.

Oppenheimer, J. R. 1955. *Electron Theory: Description and Analogy* Deptartment of Physics, Iowa State University, Ames.

Oppenheimer, J. R. 1966. Thirty years of mesons. *Phys. Today* 19/11:51–58.

Oppenheimer, J. R. 1980. *Letters and Recollections*. Ed. A. K. Smith and C. Weiner. Cambridge, Mass.: Harvard University Press.

Oppenheimer, J. R., and Plesset, M. S. 1933. The production of the positive electron. *Phys. Rev.* 44:53–55.

Oppenheimer, J. R., and Schwinger, J. 1939. On pair emission in the proton bombardment of fluorine. *Phys. Rev.* 56:1066.

Oppenheimer, J. R., and Schwinger, J. 1941. On the interaction of mesotrons and nuclei. *Phys. Rev.* 60:150–152.

Oppenheimer, J. R., and Serber, R. 1937. Note on the nature of the cosmic-ray particles. *Phys. Rev.* 51:1113.

Osgood, C. 1951. *Lights in Nassau Hall—A Book of the Bicentennial: Princeton, 1746–1946*. Princeton, N.J.: Princeton University Press.

Osgood, T. H. 1940. Physics in 1939. *J. Appl. Phys.* 11:2–17.

Osgood, T. H. 1941. Physics in 1940. *J. Appl. Phys.* 12:84–99.

Osgood, T. H. 1942. Physics in 1941. *J. Appl. Phys.* 13:3–21.

Osgood, T. H. 1943. Physics in 1942. *J. Appl. Phys.* 14:53–68.

Osgood, T. H. 1944. Physics in 1943. *J. Appl. Phys.* 15:89–107.

Osgood, T. H. 1945. Physics in 1944. *J. Appl. Phys.* 16:61–76.

Pais, A. 1945. On the theory of the electron and of the nucleon. *Phys. Rev.* 68:227–228.

Pais, A. 1946. On the theory of elementary particles. *Verh. Kon. Ned. Akad. Nat.* 19/1:1–91.

Pais, A. 1947. Theory of the electron. Unpublished galley proofs.

Pais, A. 1948. *Developments in the Theory of the Electron*. Princeton, N.J.: Institute for Advanced Study and Princeton University.

Pais, A. 1972. Theory of the electron, 1897–1947. In Salam and Wigner 1972.

Pais, A. 1982. *'Subtle is the Lord . . .' The Science and the Life of Albert Einstein*. Oxford and New York: Oxford University Press.

Pais, A. 1986. *Inward Bound*. Oxford and New York: Oxford University Press.

Pais, A., and Uhlenbeck, G. E. 1950. On field theories with non-localized action. *Phys. Rev.* 79:145–165.

Les Particules Elémentaires. Rapports et Discussions du Huitième Conseil de Physique de l'Institut Solvay. 1950. Brussels: R. Stoop.

Palmer, C. 1984. *George Dyson—A Centenary Appreciation*. Seven Oaks, Kent, U.K.: Novello.

Pasternack, S. 1938. Note on the fine structure of H_α and D_α. *Phys. Rev.* 54:1113.

Pauli, W. 1927a. Über Gasentartung und Paramagnetismus. *Zeits. Phys.* 41:81–102.

Pauli, W. 1927b. Zur Quantenmechanik des magnetischen Elektrons. *Zeits. Phys.* 43: 603–623.

Pauli, W. 1933. Die allgemeinen Prinzipen der Wellenmechanik. In H. Geiger and K. Scheel, eds., *Handbuch der Physik.* 2d ed., vol. 24, part 1, pp. 82–272. Berlin: Springer-Verlag.

Pauli, W. 1935–1936. The theory of the positron and related topics. Notes by B. Hoffman. Institute for Advanced Study, Princeton, N.J.

Pauli, W. 1936. Théorie quantique relativistique des particules obéissant à la statistique de Einstein-Bose. *Ann. Inst. Henri Poincaré* 6:137–152.

Pauli, W. 1940. The connection between spin and statistics. *Phys. Rev.* 58:716–722.

Pauli, W. 1941. Relativistic field theories of elementary particles. *Rev. Mod. Phys.* 13: 203–232.

Pauli, W. 1943. On Dirac's new method of field quantization. *Rev. Mod. Phys.* 15:175–207.

Pauli, W. 1946a. Remarks on the history of the exclusion principle. *Science* 103:213–215.

Pauli, W. 1946b. *Meson Theory of Nuclear Forces.* New York: Interscience.

Pauli, W. 1947. Difficulties of field theories and of field quantization. In *Report of an International Conference on Fundamental Particles and Low Temperature Held at the Cavendish Laboratory, Cambridge, on July 22–27, 1946,* pp. 5–10. London: The Physical Society.

Pauli, W. 1948. Exclusion principle and quantum mechanics. In *Prix Nobel 1946.* Stockholm: Nobel Institute.

Pauli, W. 1950. One the connection between spin and statistics. *Prog. Theor. Phys.* 5: 526–543.

Pauli, W. 1955a. Exclusion principle, Lorentz group and reflection of space-time and charge. In Pauli 1955b, pp. 30–51.

Pauli, W. 1955b. *Niels Bohr and the Development of Physics.* Oxford: Pergamon.

Pauli, W. 1964. *Collected Scientific Papers.* Ed. R. Kronig and V. F. Weisskopf. 2 vols. New York: Interscience.

Pauli, W. 1979. *Wissenschaftlicher Briefwechsel.* Vol. 1, 1919–1929. Ed. A. Hermann, K. von Meyenn, and V. F. Weisskopf. New York: Springer-Verlag.

Pauli, W. 1985. *Wissenschaftlicher Briefwechsel.* Vol. 2, 1930–1939. Ed. K. von Meyenn. New York: Springer-Verlag.

Pauli, W., and Belifante, F. J. 1940. On the statistical behavior of known and unknown elementary particles. *Physica* 7:177–192.

Pauli, W., and Fierz, M. 1938. Zur Theorie der Emission langwelliger Lichtquanten. *Nuovo Cimento* 15:167–188.

Pauli, W., and Rose, M. E. 1936. Remarks on the polarization effects in the positron theory. *Phys. Rev.* 49:462–469.

Pauli, W., and Villars, F. 1949. On the invariant regularization in relativistic quantum theory. *Rev. Mod. Phys.* 21:434–444.

Pauli, W., and Weisskopf, V. 1934. Ueber die Quantisierung der skalaren relativistischen Wellengleichung. *Helv. Phys. Acta* 7:709–731.

Pauling, L., and Wilson, E. B. 1935. *Introduction to Quantum Mechanics.* New York: McGraw-Hill.

Peierls, R. E. 1934. The vacuum in Dirac's theory of the positive electron. *Proc. Roy. Soc. London* A146:420–421.

Peierls, R. E. 1948. Self-energy problems. In *Rapport du 8ᵉ Conseil de Physique Solvay,* September 27–October 2, 1948, pp. 291–317. Brussels: R. Stoop, 1950.

Peierls, R. E. 1959. Wolfgang Ernst Pauli. *Biog. Mem. Fellows Roy. Soc.* 5:175–192.

Peierls, R. E. 1963. Field theory since Maxwell. In C. Domb, ed., *Clerk Maxwell and Modern Science.* London: Athlone Press.

Peierls, R. E. 1973. The development of quantum field theory. In Mehra 1973, pp. 370–379.

Peierls, R. E. 1985. The glorious days of physics. In A. Zichichi, ed., *Lepton and Hadron Structure*. New York: Academic Press.

Peirce, C. S. 1931–1938. The fixation of belief. In *Collected Papers*. Vol. 5, pp. 358–387. Cambridge, Mass.: Harvard University Press.

Penick, J. L., Jr.; Pursell, C. W., Jr.; Sherwood, M.; and Swain, D. C. 1972. *The Politics of American Science: 1939 to the Present*. Cambridge, Mass.: MIT Press.

Pestre, D. 1984. *Physique et physiciens en France, 1918–1940*. Paris.

Peterman, A. 1953a. Divergence of perturbation expression. *Phys. Rev.* 89:1160–1161.

Peterman, A. 1953b. Renormalisation dans les séries divergentes. *Helv. Phys. Acta* 26: 291–299.

Petley, B. W. 1985. *The Fundamental Physical Constants and the Frontier of Measurement*. Bristol, U.K.: Adam Hilger.

Phillips, M. 1941. On the electronic *g* factor for alkali atoms. *Phys. Rev.* 60:100–101.

Piore, E. R. 1954. ONR research policy. *Naval Research Review* 7:6–11.

Poincaré, H. 1906. Sur la dynamique de l'électron. *Rend. Circ. Mat. Palermo* 21: 129–175.

Pollard, E. C. 1982. *Radiation: One Story of the MIT Radiation Laboratory*. Durham, N.C.: Woodburn Press.

Porter, L. S. 1988. From intellectual sanctuary to social responsibility. Ph.D. diss., Princeton University. Ann Arbor: University Microfilms.

Powell, C. F.; Fowler, P. H.; and Perkins, D. H. 1959. *The Study of Elementary Particles by the Photographic Method*. New York: Pergamon.

Proca, A. 1936. Sur la théorie ondulatoire des électrons positifs et négatifs. *J. Phys. Radium* 7:347–353.

Proca, A. 1938. Théorie nonrelativiste des particules à spin entier. *J. Phys. Radium* 9: 61–66.

Pursell, C. 1979. Science agencies in World War II: The OSRD and its challenges. In Nathan Reingold, ed., *The Sciences in the American Context: New Perspectives*. Washington, D.C.: Smithsonian Institution Press.

Rabi, I. I. 1937. Space quantization in a gyrating magnetic field. *Phys. Rev.* 51: 652–658.

Rabi, I. I., and Schwinger, J. 1937. Depolarization by neutron-proton scattering. *Phys. Rev.* 51:1003.

Racah, G. 1937. Sulla simmetrica tra particelle e anti-particelle. *Nuovo Cimento* 14: 322–329.

Racah, G. 1946. On the self-energy of the electron. *Phys. Rev.* 70:407–412.

Radder, H. 1983. Kramers and the Forman thesis. *History of Science* 21:165–182.

Rainwater, L. J.; Rabi, I. I.; and Havens, W. W. 1949. Interaction of neutrons with electrons in bismuth. *Phys. Rev.* 75:1295.

Rajkumari, W., ed. 1987. *The Making of Physicists*. Bristol, U.K.: Adam Hilger.

Raman, V. V., and Forman, P. 1969. Why was it Schrödinger who developed de Broglie's Ideas? *Hist. Stud. Phys. Sci.* 1:291–314.

Ramsey, N. F. 1966. Early history of Associated Universities and Brookhaven National Laboratory. Brookhaven Lecture Series No. 55, March 30, 1966.

Ramsey, N. F. 1985. Early history of magnetic resonance. *Bull. Magnetic Reson.* 7:94–99.

Rarita, W., and Schwinger, J. 1941a. The photodisintegration of the deuteron. *Phys. Rev.* 59:215.

Rarita, W., and Schwinger, J. 1941b. On the neutron-proton interaction. *Phys. Rev.* 59: 436.

Rarita, W., and Schwinger, J. 1941c. On the exchange properties of the neutron-proton interaction. *Phys. Rev.* 59:556.

Rarita, W., and Schwinger, J. 1941d. On a theory of particles with half-integral spin. *Phys. Rev.* 60:61.

Rarita, W.; Schwinger, J.; and Nye, H. 1941. The photodisintegration of the deuteron. *Phys. Rev.* 59:209.

Rayski, J. 1948. On simultaneous interaction of several fields and the self-energy problem. *Acta Phys. Polonica* 9:129–140.

Rechenberg, H. 1989. The early *S*-matrix and propagation (1942–1952). In Brown et al. 1989.

Redmond, K. C., and Smith, T. M. 1980. *Project Whirlwind—The History of a Pioneer Computer.* Maynard, Mass.: Digital Press.

Regis, E. 1987. *Who Got Einstein's Office? Eccentricity and Genius at the Institute for Advanced Study.* Reading, Mass.: Addision-Wesley.

Reid, C. 1970. *Hilbert.* New York: Springer-Verlag.

Reid, C. 1976. *Courant in Göttingen and New York: The Story of an Improbable Mathematician.* New York: Springer-Verlag.

Reingold, N., ed. 1976. *Science in America since 1820.* New York: Science History Publications.

Reischauer, E., and Fairbank, J. 1960. *East Asia: The Great Tradition.* Boston: Houghton Mifflin.

Reischauer, E., and Craig, A. 1973. *Japan: Tradition and Transformation.* Boston: Houghton Mifflin.

Report of an International Conference on Fundamental Particles and Low Temperature Held at the Cavendish Laboratory, Cambridge, on 22–27 July 1946. Vol. 1, *Fundamental Particles.* London: The Physical Society.

Rich, A., and Wesley, J. C. 1972. The current status of lepton *g* factors. *Rev. Mod. Phys.* 44:250–283.

Ridenour, L. N. 1947. *Radar System Engineering.* (Reprinted 1963 by Boston Technical Lithographers.)

Rider, R. E. 1985. Alarm and opportunity: Emigration of mathematicians and physicists to Britain and the United States. *Hist. Stud. Phys. Sci.* 15:107–176.

Rigden, J. S. 1983. Molecular beam experiments on hydrogens during the 1930s. *Hist. Stud. Phys. Sci.* 13:335–363.

Rigden, J. S. 1985. The birth of the magnetic resonance method. In P. Achenstein and O. Hannaway, eds., *Observation, Experiment, and Hypothesis in Modern Physical Science,* pp. 205–232. Cambridge, Mass.: MIT Press.

Rigden, J. S. 1986. Quantum states and precession: The two discoveries of NMR. *Rev. Mod. Phys.* 58:433–448.

Rigden, J. S. 1987. *Rabi: Scientist and Citizen.* New York: Basic Books.

Ritz, W. 1908. Recherches critiques sur l'electrodynamique générale. *Ann. Chimie et de Physique* 13:145–275.

710

Rivier, D. 1949. *Une méthode d'élimination des infinités en théorie des champs quantifiés. Application au moment magnétique du neutron.* Thèse presentée à la faculté des sciences de l'université de Lausanne au 13 Juillet 1948. Basel: Birkhäuser.

Rivier, D., and Stueckelberg, E.C.G. 1948. A convergent expression for the magnetic moment of the neutron. *Phys. Rev.* 74:218. Erratum 74:986.

Rivier, D., and Stueckelberg, E.C.G. 1949. Sur le problème de l'élimination des divergences dans la théorie des champs quantifiés. *Helv. Phys. Acta* 22:300.

Rochester, G. D., and Butler, C. C. 1947. Evidence for the existence of new unstable particles. *Nature* 160:855–857.

Rochester, G . D., and Wilson, J. G. 1952. *Cloud Chamber Photographs of the Cosmic Radiation.* New York: Academic Press.

Rohrlich, F. 1973. The electron: Development of the first elementary particle theory. In 'Mehra 1973, pp. 331–369.

Rohrlich, F. 1985. The art of doing physics in Dirac's way. In Kursunoglu et al. 1985, pp. 17–29.

Rojanski, V. B. 1938. *Introductory Quantum Mechanics.* New York: Prentice Hall.

Roland, A. 1985. Technology and war: A bibliographic essay. In M. R. Smith 1985.

Rosenfeld, L. 1929. Über die longitudinalen Eigenlösung Heisenberg-Paulischen electromagnetischen Gleichungen. *Zeits. Phys.* 58:540–555.

Rosenfeld, L. 1930a. Bemerkung über die Invarianz der kanonischen Vertauschungsrelation. *Zeits. Phys.* 63:574–575.

Rosenfeld, L. 1930b. Zur Quantelung der Wellenfelder. *Ann. der Physik* (5)5:113–152.

Rosenfeld, L. 1931. Zur Kritik der Diracschen Strahlung Theorie. *Zeits. Phys.* 70: 454–462.

Rosenfeld, L. 1932. Über eine mögliche Fassung des Diracschen Programms zur Quantenelectrodynamik und deren formalen Zusammenhang mit der Heisenberg-Paulischen Theorie. *Zeits. Phys.* 76:729–734.

Rosenfeld, L. 1951. *Theory of Electrons.* Amsterdam: North-Holland.

Rosenfeld, L. 1955. On quantum electrodynamics. In Pauli 1955b, pp. 70–85.

Rosenfeld, L., and Solomon, J. 1931a. Sur la théorie quantique du rayonnement. *J. Phys. et Rad.* (7)2:139–147.

Rossi, B. 1983. The decay of "mesotrons" (1939–1943). Experimental particle physics in the age of innocence. In Shapiro 1983, pp. 383–400.

Ruark, A. E., and Urey, H. C. 1930. *Atoms, Molecules and Quanta.* New York: McGraw-Hill.

Rueger, A. 1990. Independence from future theories: A research strategy in quantum theory. *PSA* 1:203–211. East Lansing: Philosophy of Science Association.

Rueger, A. 1991. Attitudes toward infinities. Response to anomalies in quantum electrodynamics, 1927–1947. *Hist. Stud. Phys. Bio. Sci.* 22(3):309–338.

Sabben-Clare, J. 1981. *Winchester College.* Southampton, U.K.: Paul Cave Publications Ltd.

Sachs, R., and Schwinger, J. 1942. On the magnetic moments of H^3 and He^3. *Phys. Rev.* 61:732.

Sachs, R., and Schwinger, J. 1946. The magnetic moments of H^3 and He^3. *Phys. Rev.* 70:41.

Sakata, S. 1947. The theory of the interaction of elementary particles. *Prog. Theor. Phys.* 2:145–147.

Sakata, S. 1950. On the direction of the theory of elementary particles. English trans. in *Prog. Theor. Phys.* (suppl.) 50 (1971):155–158.

Sakata, S. 1956. On a composite model for the new particles. *Prog. Theor. Phys.* 16:686–688.

Sakata, S., and Hara, O. 1947. The self-energy of the electron and the mass differences of nucleons. *Prog. Theor. Phys.* 2:30–31.

Sakata, S., and Tanikawa, Y. 1940. The spontaneous disintegration of the neutral mesotron (neutretto). *Phys. Rev.* 57:548.

Sakata, S., and Umezawa, H. 1950. On the applicability of the method of mixed fields in the theory of the elementary particles. *Prog. Theor. Phys.* 5:682–691.

Sakata, S.; Umezawa, H.; and Kamefuchi, S. 1952. On the structure of the interaction of the elementary particles. *Prog. Theor. Phys.* 7:377–390.

Sakita, B. 1988. Life as a graduate student in Japan and Rochester. In Brown et al. 1988, pp. 21–23.

Salam, A. 1950. Differential identities in three-field renormalization problem. *Phys. Rev.* 79:910–911.

Salam, A. 1951a. Overlapping divergence and the *S*-matrix. *Phys. Rev.* 82:217–227.

Salam, A. 1951b. Divergent integrals in renormalizable field theories. *Phys. Rev.* 84: 426–431.

Salam, A. 1968. Weak and electromagnetic interactions. In *Proceedings of Nobel Conference VIII*, pp. 367–377.

Salam, A., ed. 1969. Evening lectures in the series "From a Life of Physics" at the International Centre for Theoretical Physics, Trieste, Italy. Special suppl., *IAEA Bulletin*, Vienna.

Salam, A. 1973. Progress in renormalization theory since 1949. In Mehra 1973, pp. 430–446.

Salam, A. 1980. Gauge unification of fundamental forces. *Rev. Mod. Phys.* 52:525–538.

Salam, A. 1987. Physics and the excellences of the life it brings. In A. Salam, *Ideals and Realities: Selected Essays of Abdus Salam*, ed. C. H. Lai. 2d ed., pp. 291–303. Singapore: World Publishing.

Salam, A., and Strathdee, J. 1970. Quantum gravity and infinities in quantum electrodynamics. *Lett. Nuovo Cimento* 4:101–108.

Salam, A., and Wigner, E. P., eds. 1972. *Aspects of Quantum Theory.* Cambridge, U.K.: Cambridge University Press.

Salaman, E., and Salaman, M. 1986. Remembering Paul Dirac. *Encounter.* May, pp. 66–70.

Salkowitz, E. I. 1976. *Science, Technology and the Modern Navy: Thirtieth Anniversary, 1946–1976.* Arlington, Va.: Office of Naval Research.

Salpeter, E., and Bethe, H. A. 1951. A relativistic equation for bound-state problems. *Phys. Rev.* 84:1232–1242.

Sapolsky, H. M. 1979. Academic science and the military: The years since the second world war. In N. Reingold, ed., *The Sciences in the American Context: New Perspectives.* Washington, D.C.: Smithsonian Institution Press.

Saxon, D., and Schwinger, J. 1946. Electron orbits in the synchrotron. *Phys. Rev.* 69:702.

Saxon, D., and Schwinger, J. 1968. *Discontinuities in Wave Guides.* New York: Gordon and Breach.

Schiff, L. 1949. *Quantum Mechanics.* 1st ed. New York: McGraw-Hill.

Schwarzschild, K. 1903. Zur Elektrodynamik. I, II. *Göttingen Nachr. Math. Phys. Klasse,* pp. 126–142.

Schweber, S. S. 1961. *An Introduction to Relativistic Quantum Field Theory*. New York: Harper and Row.

Schweber, S. S. 1984. Some chapters for a history of quantum field theory, 1938–1952. In B. S. DeWitt and R. Stora, eds., *Relativity, Groups, and Topology*, pp. 40–220. Amsterdam and New York: North-Holland.

Schweber, S. S. 1985. A short history of Shelter Island I. In Jackiw 1985.

Schweber, S. S. 1986a. The empiricist temper regnant: Theoretical physics in the United States. *Hist. Studies Phys. Sci.* 17/1:55–98.

Schweber, S. S. 1986b. The Shelter Island, Pocono and Oldstone Conferences: The birth of American quantum electrodynamics after World War II. *Osiris* 2:265–302.

Schweber, S. S. 1986c. Feynman and the visualization of space-time processes. *Rev. Mod. Phys.* 58/2:449–509.

Schweber, S. S. 1989a. Some reflections on the history of particle physics in the 1950s. In Brown et al. 1989, pp. 668–693.

Schweber, S. S. 1989b. Molecular beam experiments, the Lamb shift, and the relation between experiment and theory. *Am. J. Phys.* 57/4:299–308.

Schweber, S. S.; Bethe, H. A.; and Hoffmann, F. de. 1955. *Mesons and Fields*. Vol. 1. New York: Row, Peterson and Co.

Schwinger, J. 1934. On the interaction of several electrons. Unpublished manuscript.

Schwinger, J. 1937a. On the magnetic scattering of neutrons. *Phys. Rev.* 51:544.

Schwinger, J. 1937b. On non-adiabatic processes in inhomogeneous fields. *Phys. Rev.* 51:648.

Schwinger, J. 1937c. On the spin of the neutron. *Phys. Rev.* 52:1250.

Schwinger, J. 1939. On the neutron-proton interaction. *Phys. Rev.* 55:235(A).

Schwinger, J. 1940. Neutron scattering in ortho- and parahydrogen and the range of nuclear forces. *Phys. Rev.* 58:1004.

Schwinger, J. 1941a. On the charged scalar mesotron field. *Phys. Rev.* 60:159.

Schwinger, J. 1941b. The quadrupole moment of the deuteron and the range of nuclear forces. *Phys. Rev.* 60:164.

Schwinger, J. 1942. On a field theory of nuclear forces. *Phys. Rev.* 61:387.

Schwinger, J. 1946. Electron radiation in high energy accelerators. *Phys. Rev.* 70:798.

Schwinger, J. 1947. A variational principle for scattering problems. *Phys. Rev.* 72:742.

Schwinger, J. 1948a. On the polarization of fast neutrons. *Phys. Rev.* 73:407.

Schwinger, J. 1948b. On quantum electrodynamics and the magnetic moment of the electron. *Phys. Rev.* 73:416–417. (Reprinted in Schwinger 1958, 142.)

Schwinger, J. 1948c. An invariant quantum electrodynamics. *Phys. Rev.* 74:1212.

Schwinger, J. 1948d. Quantum electrodynamics. I. A covariant formulation. *Phys. Rev.* 74:1439.

Schwinger, J. 1948e. Recent developments in quantum electrodynamics. Invited lecture to the American Physical Society, January 30, 1948, New York. *Phys. Rev.* 73:1263.

Schwinger, J. 1948f. Recent developments in quantum electrodynamics. Mimeographed notes on the lectures of Dr. Julian Schwinger. Chapter I by Dr. Schwinger. Chapter II by Mr. D. Park. Summer Symposium, University of Michigan, 1948.

Schwinger, J. 1949a. Quantum electrodynamics. II. Vacuum polarization and self energy. *Phys. Rev.* 75:651–679.

Schwinger, J. 1949b. On radiative corrections to electron scattering. *Phys. Rev.* 75:898–899.

Schwinger, J. 1949c. On the classical radiation of accelerated electrons. *Phys. Rev.* 75:1912.

Schwinger, J. 1949d. Quantum electrodynamics. III. The electromagnetic properties of the electron-radiative corrections to scattering. *Phys. Rev.* 76:790–817.

Schwinger, J. 1950. On the charge independence of nuclear forces. *Phys. Rev.* 78:135.

Schwinger, J. 1951a. On gauge invariance and vacuum polarization. *Phys. Rev.* 82: 664–679.

Schwinger, J. 1951b. The theory of quantized fields. I. *Phys. Rev.* 82:914–927.

Schwinger, J. 1951c. On the Green's functions of quantized fields. I, II. *Proc. Natl. Acad. Sci. USA* 37:452–459.

Schwinger, J. 1951d. Quantum dynamics. *Science* 113:479.

Schwinger, J., ed. 1958. *Selected Papers on Quantum Electrodynamics*. New York: Dover.

Schwinger, J. 1966. Relativistic quantum field theory. Nobel lecture. *Science* 153:949–953. (See also Schwinger 1972.)

Schwinger, J. 1970a. Charged scalar mesotron field. In Wentzel 1970, pp. 101–138.

Schwinger, J. 1970b. *Particles, Sources, and Fields*. Reading, Mass.: Addision-Wesley.

Schwinger, J. 1970c. *Quantum Kinematics and Dynamics*. New York: Benjamin.

Schwinger, J. 1972. Relativistic Quantum Field Theory. Nobel lecture. In *Nobel Lectures— Physics, 1963–1970*. Amsterdam: Elsevier.

Schwinger, J. 1973. *Particles, Sources, and Fields*. Vol. 2. Reading, Mass.: Addison-Wesley.

Schwinger, J. 1978. Physics in the future. In *Celebration of the Fiftieth Anniversary of the Pupin Laboratories*, pp. 118–123. Department of Physics, Columbia University, New York.

Schwinger, J. 1979. *Selected Papers*. Ed. M. Flato, C. Fronsdal, and K. A. Milton. Dordrecht, Holland: Reidel.

Schwinger, J. 1983a. Renormalization theory of quantum electrodynamics. In Brown and Hoddeson 1983, pp. 329–353.

Schwinger, J. 1983b. Two shakers of physics: Memorial lecture for Sin-itiro Tomonaga. In Brown and Hoddeson 1983, pp. 354–375.

Schwinger, J. 1989. A path to quantum electrodynamics. *Phys. Today* 42(2):42–49.

Schwinger, J., and Teller, E. 1937a. The scattering of neutrons by ortho- and parahydrogen. *Phys. Rev.* 51:775.

Schwinger, J., and Teller, E. 1937b. The scattering of neutrons by ortho- and parahydrogen. *Phys. Rev.* 52:286–295.

Schwinger, J., and Weisskopf, V. F. 1948. On the electromagnetic shift of energy levels. *Phys. Rev.* 73:1272A.

Seidel, R. W. 1976. The origins of academic physics in California. *J. College Sci. Teaching* 6:10–23.

Segrè, E. 1970. *Enrico Fermi: Physicist*. Chicago: University of Chicago Press.

Segrè, E. 1980. *From X-rays to Quarks*. Berkeley: University of California Press.

Serber, R. 1935. Linear modification in the Maxwell field equations. *Phys. Rev.* 48: 49–54.

Serber, R. 1936. A note on positron theory and proper energies. *Phys. Rev.* 49:545–550.

Serber, R. 1966. Interview by C. Weiner and G. Lubkin. Center for History of Physics, American Institute of Physics, New York.

Serber, R. 1967. The early years. *Phys. Today* 20/10:35–39.

Serber, R. 1983. Particle physics in the 1930s: A view from Berkeley. In Brown and Hoddeson 1983, pp. 206–221.

Serpe, J. 1940. Sur le problème de la largeur naturelle et du déplacement des raies spectrales. *Physica* 7:133–144.

Serpe, J. 1941. Sur la quantification du problème de l'intéraction d'un oscillateur linéaire harmonique. *Physica* 8:226–232.

Serwer, D. 1977. *Unmechanischer Zwang*: Pauli, Heisenberg and the rejection of the mechanical atom, 1923–1925. *Hist. Stud. Phys. Sci.* 8:189–256.

Shapere, D. 1984. *Reason and the Search for Knowledge.* Dordrecht, Holland: Reidel.

Shapiro, M. M., ed. 1983. *Composition and Origin of Cosmic Rays.* Dordrecht, Holland: Reidel.

Sherry, M. S. 1977. *Preparing for the Next War: American Plans for Post-War Defense, 1941–45.* New Haven, Conn.: Yale University Press.

Shimao, E. 1989. Some aspects of Japanese science, 1868–1945. *Annals of Science* 46: 69–91.

Siegert, A.J.F. 1937. Note on the interation between nuclei and electromagnetic radiation. *Phys. Rev.* 52:787–791.

Sigurdsson, S. 1991. Herman Weyl, mathematics and physics, 1900–1927. Ph.D. diss., Dept. of History of Science, Harvard University, Cambridge, Mass.

Slater, J. C. 1946. Microwave electronics. *Rev. Mod. Phys.* 18:441–512.

Slater, J. C. 1967. Quantum physics in America between the wars. *Int. J. Quant. Chem.* 1:1–23. (Also in *Physics Today* 21(1):43–51.)

Slater, J. C. 1975. *Solid-State and Molecular Theory: A Scientific Biography.* New York: Wiley.

Slater, J. C., and Frank, N. H. 1933. *Introduction to Theoretical Physics.* New York: McGraw-Hill.

Slotnick, M. 1949. The neutron-electron interaction. *Phys. Rev.* 75:1295.

Slotnick, M., and Heitler, W. 1949. The charge density and magnetic moment of the nucleons. *Phys. Rev.* 75:1645–1663.

Smith, A. K. 1965. *A Peril and Hope: The Scientists' Movement in America, 1945–1947.* Chicago: University of Chicago Press.

Smith, A. K. 1971. *A Peril and Hope: the Scientists' Movement in America, 1945–1947.* Cambridge, Mass.: MIT Press.

Smith, A. K., and Weiner, C., eds. 1980. *Robert Oppenheimer: Letters and Recollections.* Cambridge, Mass.: Harvard University Press.

Smith, M. R. 1985. *Military Enterprise and Technological Change.* Cambridge, Mass.: MIT Press.

Smythe, H. D. 1945. *Atomic Energy for Military Purposes.* Princeton, N.J.: Princeton University Press.

Snyder, H., and Weinberg, J. 1940. Stationary states of scalar and vector fields. *Phys. Rev.* 57:307–314.

Sommerfeld, A. 1931. Über die Beugung und Bremsung der Elektronen. *Ann. der Phys.* (Leipzig) 11:257–278.

Sommerfeld, A. 1939. *Atombau und Spektrallinien.* Part 2. Braunschweig: F. Vieweg.

Sommerfeld, A. 1941. Zur Theorie der Feinstruktur des Wasserstoffs. *Zeits. Phys.* 118: 295–311.

Spedding, F.; Shane, C. D.; and Graci, N. 1933. Fine structure of H_α^2. *Phys. Rev.* 44:58.

Spedding, F.; Shane, C. D.; and Graci, N. 1935. The fine structure of H_α. *Phys. Rev.* 47: 38–44.

Spradley, J. L. 1985. Particle physics in prewar Japan. *American Scientist* 73:263–269.

Steinberger, J.; Panofsky, W.K.H.; and Steller, J. 1950. Evidence for the production of neutral mesons by photons. *Phys. Rev.* 78:802–805.

Stone, R. S. 1946. Health protection activities of the plutonium project. *Proc. Am. Phil. Soc.* 90:11–90.

Streater, R. F., and Wightman, A.S. 1964. *PCT, Spin and Statistics, and All That.* New York: Benjamin. (Enlarged edition: Benjamin/Cummins, 1978.)

Street, J. C., and Stevenson, E. C. 1937. Penetrating corpuscular component of the cosmic radiation. *Phys. Rev.* 51:1005 (abstract).

Stueckelberg, E.C.G. 1934a. Bemerkung zur Intensität der Streustrahlung bewegter freier Elektronen. *Helv. Phys. Acta* 8:197–204.

Stueckelberg, E.C.G. 1934b. Relativistisch invariante Störungstheorie des Diracschen Elektrons. *Ann. der Phys.* 21/4:367–389.

Stueckelberg, E.C.G. 1935a. Remarque à propos des temps multiples dans la théorie d'intéraction des charges entre elles. *Compte Rendu SPHN (Genève)* 52:98–101.

Stueckelberg, E.C.G. 1935b. Remarque sur la production des paires d'electron. *Helv. Phys. Acta* 8:325–326.

Stueckelberg, E.C.G. 1936. Invariante Störungstheorie des Electron-Neutrino-Teilchens unter dem Einfluss von elektromagnetischem Feld und Kernkraftfeld (Feldtheorie der Materie II). *Helv. Phys. Acta* 9:533–554.

Stueckelberg, E.C.G. 1937a. La correspondance entre les potentiels retardés de la physique classique et de la physique quantique. *Compte Rendu SPHN (Genève)* 54:44–47.

Stueckelberg, E.C.G. 1937b. Etablissement de la formule des potentiels retardés dans la physique quantique. *Compte Rendu SPHN (Genève)* 54:47–50.

Stueckelberg, E.C.G. 1937c. On the existence of heavy electrons. *Phys. Rev.* 52:41–42.

Stueckelberg, E.C.G. 1938a. Die Wechselwirkungskräfte in der Elektrodynamik und in der Feldtheorie der Kern Kräfte. Parts 1, 2, and 3. *Helv. Phys. Acta* 11:225–244, 11: 299–328.

Stueckelberg, E.C.G. 1938b. A propos de l'intéraction entre les particules élementaires. *Comptes Rendus* 207:387–390.

Stueckelberg, E.C.G. 1938c. Rigorous theory of interaction between nuclear particles. *Phys. Rev.* 54:889–892.

Stueckelberg, E.C.G. 1939a. Theory of mesons and nuclear forces. *Nature* 143:560.

Stueckelberg, E.C.G. 1939b. Sur l'intégration de l'équation $(\Box - l^2)Q = -\rho$ en utilisant la méthode de Sommerfeld. *Comptes Rendu. SPHN (Genève)* 56:43–45.

Stueckelberg, E.C.G. 1940a. Die Schwierigkeiten in der Feldtheorie der Austauschkräfte. *Zeits. für Tech. Physik* 21/12:275–276.

Stueckelberg, E.C.G. 1940b. Influence du champ pseudo-scalaire sur la théorie classique des forces d'échange. *Helv. Phys. Acta* 13:347–354.

Stueckelberg, E.C.G. 1941a. La signification du temps propre en mécanique ondulatoire. *Helv. Phys. Acta* 14:321–322.

Stueckelberg, E.C.G. 1941b. Remarque à propos de la création de paires de particule en théorie de relativité. *Helv. Phys. Acta* 14:588–594.

Stueckelberg, E.C.G. 1942a. Solution invariantes $D_{\kappa^2}(x, y)$ de l'équation $(\Box - \kappa^2)D = 0$ dans l'éspace pseudo-euclidien. *Compte Rendu. SPHN (Genève)* 59:49–52.

Stueckelberg, E.C.G. 1942b. Solutions invariantes $D_{\kappa^2}(x, y)$ de l'équation de Schrödinger relativistique. *Compte Rendu SPHN (Genève)* 59:53–55.

Stueckelberg, E.C.G. 1942c. La mécanique du point matériel en théorie de relativité et en théorie des quanta. *Helv. Phys. Acta* 15:23–37.

Stueckelberg, E.C.G. 1942d. Remarques à propos de la relation entre spin et statistique. *Helv. Phys. Acta* 15:327–329.

Stueckelberg, E.C.G. 1942e. Le role de l'invariance spinorielle et l'invariance de jauge dans un nouveau principe fondamental. *Helv. Phys. Acta* 15:513–514.

Stueckelberg, E.C.G. 1943a. Une méthode nouvelle de la quantification des champs. *Arch. des Sci. Phys. et Nat. (Genève)* 25:1–70.

Stueckelberg, E.C.G. 1943b. Un principe qui relie la théorie de relativité et la théorie des quanta. *Helv. Phys. Acta* 16:173–202.

Stueckelberg, E.C.G. 1944a. An unambiguous method of avoiding divergence difficulties in quantum theory. *Nature* 153:143–148.

Stueckelberg, E.C.G. 1944b. Principe de correspondance d'une mécanique asymptotique classique. *Suppl. Arch. des Sci. Phys. et Nat. (Genève)* 61:155–158.

Stueckelberg, E.C.G. 1944c. Principe de correspondance d'une mécanique asymptotique quantifié. *Suppl. Arch. des Sci. Phys. et Nat. (Genève)* 61:159–161.

Stueckelberg, E.C.G. 1945. Mécanique foncionelle. *Helv. Phys. Acta* 28:195–220.

Stueckelberg, E.C.G. 1946. Une propriété de l'operateur S en mécanique asymptotique. *Helv. Phys. Acta* 19:242–243.

Stueckelberg, E.C.G. 1947. The present state of the S-operator theory. In *Report of an International Conference on Fundamental Particles and Low Temperature Held at the Cavendish Laboratory, Cambridge, on July 22–27, 1946*, pp. 5–10. London: The Physical Society.

Stueckelberg, E.C.G., and Bouvier, P. 1944. Le freinage de radiation de l'électron de Dirac en mécanique asymptotique. *Suppl. Arch. des Sci. Phys. et Nat. (Genève)* 61:162–165.

Stueckelberg, E.C.G., and Peterman, A. 1953. La normalisation des constantes dans la théorie des quanta. *Helv. Phys. Acta* 26:499–520.

Stueckelberg, E.C.G., and Rivier, D. 1950a. Causalité et structure de matrice. *S. Helv. Phys. Acta* 23:215–222.

Stueckelberg, E.C.G., and Rivier, D. 1950b. A propos des divergences en théorie des champs quantifiés. *Helv. Phys. Acta* 23 (supp. 3):236–239.

Stuewer, R. H. 1975. *The Compton Effect.* New York: Science History Publications.

Stuewer, R. H., ed. 1979. *Nuclear Physics in Retrospect: Proceedings of a Symposium on the 1930s.* Minneapolis: University of Minnesota Press.

Stuewer, R. H. 1984. Nuclear physicists in a new world: The emigrés of the 1930s in America. *Berichte zur Wiss. Gesch.* 1:23–40.

Swords, S. S. 1986. *Technical History of the Beginnings of RADAR.* London: Peter Peregrinus.

Synnott, M. G. 1979. *The Half-Opened Door.* Westport, Conn.: Greenwood Press.

Takabayasi, T. 1983. Some characteristic aspects of early elementary particle theory in Japan. In Brown and Hoddeson 1983, pp. 294–303.

Takeda, G. 1952. On the renormalization theory of the interaction of electrons and photons. *Prog. Theor. Phys.* 7:359–366.

Taketani, M. 1968. *Atomic Energy and Scientists.* Tokyo: Keiso-shobo.

Taketani, M. 1974. Methodological approaches in the development of the meson theory of Yukawa in Japan (1951). In Nakayama et al. 1974, pp. 24–38.

Tamm, I. 1930. Über die Wechselwirkung der freien Elektronen mit Strahlung nach der Diracschen Theorie des Elektrons und nach der Quantenelektrodynamik. *Zeits. Phys.* 62:545–568.

Tamm, I. 1934. Exchange forces between neutrons and protons, and Fermi's theory. *Nature* 133:981.

Tarrant, G.T.P. 1930. The absorption of hard monochromatic γ-radiation. *Proc. Roy. Soc. London* A128:345–356.

Tati, T., and Tomonaga, S. 1948. A self-consistent subtraction method in the quantum field theory. I. *Prog. Theor. Phys.* 3:391–406.

Teller, E.; Gamow, G.; and Fleming, J. A. 1941. The 7th Annual Washington Conference of Theoretical Physics. *Science* 94:92–94.

ter Haar, D. 1946. On the redundant zeros in the theory of the Heisenberg matrix. *Physica* 12:501–508.

ter Haar, D. 1965. *Collected Papers of Landau.* New York: Gordon and Breach.

ter Haar, D., and Scully, M. O. 1978. *Willis E. Lamb Jr.: A Festschrift on the Occasion of His 65th Birthday.* Amsterdam: North-Holland.

Tetrode, H. 1922. Über den Wirkungszusammenhang der Welt. Eine Erweiterung der klassischen Dynamik. *Zeits. Phys.* 10:317–326.

Thibaud, J. 1934. Positions: Focusing of beams, measurement of charge to mass ratio. *Phys. Rev.* 45:781–787.

Thirring, W. 1953. On the divergence of perturbation theory for quantum fields. *Helv. Phys. Acta* 16:33–52.

Thomas, L. H. 1985. Leon Nicolas Brillouin. *Biographical Memoirs of the National Academy of Sciences of the USA.* Vol. 55, pp. 69–92. New York: Columbia University Press.

Thomson, J. J. 1881. On the electric and magnetic effects produced by the motion of electrified bodies. *Phi. Mag.* 11:227–249.

't Hooft, G. 1971a. Renormalization of massless Yang-Mills fields. *Nuclear Phys.* B33: 173–199.

't Hooft, G. 1971b. Renormalizable Lagrangians for massive Yang-Mills fields. *Nuclear Phys.* B35:167–188.

't Hooft, G., and Veltman, M. 1972. Renormalization and regularization of gauge fields. *Nuclear Phys.* B44:189–213.

Tomonaga, S. 1937. Bemerkungen über die kinetische Kernenergie im Hartree-Fock Modell. *Sci. Papers IPCR* 32:229–232.

Tomonaga, S. 1938. Innere Reibung und Wärmeleitfähigkeit der Kernmaterie. *Zeits. Phys.* 110:573–575.

Tomonaga, S. 1941. Zur Theorie des Mesons. I. *Sci. Papers IPCR* 39:247–266.

Tomonaga, S. 1943a. Bemerkung über die Streuung des Mesonen aus Kernteilchen. *Sci. Papers IPCR* 40:274–310.

Tomonaga, S. 1943b. On a relativistic reformulation of quantum field theory. *Bull. IPCR (Rikeniko)* 22:545–557. (In Japanese; the English translation appeared in 1946.)

Tomonaga, S. 1946. On a relativistically invariant formulation of the quantum theory of wave fields. I. *Prog. Theor. Phys.* 1/2:1–13.

Tomonaga, S. 1948. On infinite reactions in quantum field theory. *Phys. Rev.* 74:224–225.

Tomonaga, S. 1949. Development of elementary particle physics. Discussions centering on the divergence difficulties. *Kakagu* 19:2 (in Japanese). (English trans. in Tomonaga 1971.)

Tomonaga, S. 1962. *Quantum Mechanics.* Trans. from the Japanese by Koshiba. Amsterdam: North-Holland.

Tomonaga, S. 1971. *Scientific Papers.* Vol. 1. Ed. T. Miyazima, Tokyo: Misuzu Shobo.

Tomonaga, S. 1973. Development of quantum electrodynamics. Personal Recollections. Nobel lecture, May 6, 1966. In Mehra 1973, pp. 404–412.

Tomonaga, S. 1976. *Scientific Papers.* Vol. 2. Ed. T. Miyazima. Tokyo: Misuzu Shobo.

Tomonaga, S. 1978. Nuclear Research at Riken. In Brown et al. 1988, pp. 1–25.

Tomonaga, S., and Koba, Z. 1947. Application of the "self-consistent" subtraction method to the elastic scattering of an electron. *Prog. Theor. Phys.* 2:318.

Tomonaga, S., and Koba, Z. 1948. On radiation reactions in collision processes. I. *Prog. Theor. Phys.* 3:290–303.

Toró, T. 1978. Alexandru Proca and the vector mesons in hadron physics. *Noesis* (Romania) 4:87–90.

Trigg, G. L. 1975. *Landmark Experiments in Twentieth Century Physics.* New York: Crane, Russak. Chapter 6: Precision values for nuclear magnetic moments, pp. 85–97. Chapter 7: Fine structure in the spectrum of hydrogen, pp. 99–134.

Tuan, S. F. 1980. Dirac, Heisenberg and the University of Hawaii. Mimeographed preprint.

Uehling, E. A. 1935. Polarization effects in the positron theory. *Phys. Rev.* 48:55–63.

Umezawa, H., and Kamafuchi, R. 1951. The vacuum in quantum field theory. *Prog. Theor. Phys.* 64:543–558.

Umezawa, H., and Kawabe, R. 1949a. Some general formulae relating to vacuum polarization. *Prog. Theor. Phys.* 4:423–442.

Umezawa, H., and Kawabe, R. 1949b. Vacuum polarization due to various charged particles. *Prog. Theor. Phys.* 4:443–460.

Umezawa, H.; Yukawa, J.; and Yamada, E. 1948. The problem of vacuum polarization. *Prog. Theor. Phys.* 3:317–318.

Vallarta, M. S., and Feynman, R. P. 1939. The scattering of cosmic rays by the stars of a galaxy. *Phys. Rev.* 55:505–507.

van der Waerden, B. L. 1967. *Sources of Quantum Mechanics.* Amsterdam: North-Holland.

Van Vleck, J. H. 1932. *The Theory of Electric and Magnetic Susceptibilities.* Oxford: The Clarendon Press.

Van Vleck, J. H. 1964. American physics comes of age. *Phys. Today* 17(6):21–26.

Van Vleck, J. H. 1972. Travels with Dirac. In Salam and Wigner 1972.

Velo, G. and Wightman, A. S., eds. 1973. *Constructive Quantum Field Theory.* Berlin and New York: Springer-Verlag.

Velo, G., and Wightman, A. S., eds. 1976. *Renormalization Theory.* Dordrecht and Boston: Reidel.

Veysey, L. R. 1965. *The Emergence of the American University.* Chicago: University of Chicago Press.

Villars, F. 1960. Regularization and non-singular interactions in quantum field theory. In Fierz and Weisskopf 1960, pp. 78–106.

Viner, J. 1946. The implications of the atomic bomb for international relations. *Proc. Am. Phil. Soc.* 90:53–58.

von Laue, M. 1934. Über Heisenbergs Ungenauigkeitsbeziehung und ihre erkenntnistheoretische Bedeutung. *Naturwissenschaften* 22:439–440.

von Neumann, J. 1932. *Mathematische Grundlagen der Quantenmechanik.* Berlin: Springer-Verlag. (Reprinted by Dover Publications, New York, 1943. Trans. into English by R. T. Beyer, *Mathematical Foundations of Quantum Mechanics.* Princeton, N. J.: Princeton University Press, 1955.)

von Weizsäcker, C. F. 1934. Ausstrahlung bei Stössen sehr schneller Elektronen. *Zeits. Phys.* 88:612–625.

Waller, I. 1930a. Die Streuung von Strahlung durch gebundene und freie Elektronen nach der Diracschen relativistischen Mechanik. *Zeits. Phys.* 61:837–851.

Waller, I. 1930b. Bemerkungen über die Rolle der Eigenenergie des Elektrons in der Quantentheorie der Strahlung. *Zeits. Phys.* 61:721–730.

Waller, I. 1930c. Bemerkungen über die Rolle der Eigenenergie des Elektrons in der Quantentheorie der Strahlung. *Zeits. Phys.* 62:673–676.

Walter, M.L.K. 1985. Science and cultural crisis: An intellectual biography of Percy Williams Bridgman. Ph.D. diss., Harvard University, Cambridge, Mass.

Walters, F. P. 1952. *A History of the League of Nations.* 2 vols. London: Oxford University Press.

Ward, J. 1950a. The scattering of light by light. *Phys. Rev.* 77:293.

Ward, J. 1950b. An identity in quantum electrodynamics. *Phys. Rev.* 78:182.

Ward, J. C. 1951. On the renormalization of quantum electrodynamics. *Proc. Phys. Soc. London* A64:54–56.

Weart, S. R. 1976. Scientists with a secret. *Phys. Today* 29:23–30.

Weart, S. R. 1977. Secrecy, simultaneous discovery, and the theory of nuclear reactors. *Am. J. Phys.* 45:1049–1060.

Weart, S. R. 1979a. The physics business in America, 1919–1940: A statistical reconnaissance. In N. Reingold ed., *Science in the American Context,* pp. 295–358. Washington, D.C.: Smithsonian Institution Press.

Weart, S. R. 1979b. *Scientists in Power.* Cambridge, Mass.: Harvard University Press.

Weart, S. R., and Philips, M., eds. 1985. *History of Physics.* New York: American Institute of Physics.

Weinberg, S. 1960. High energy behavior in quantum field theory. *Phys. Rev.* 118:838–849.

Weinberg, S. 1977. The search for unity: Notes for a history of quantum field theory. *Daedalus,* fall 1977. Vol. 2 of Discoveries and Interpretations in Contemporary Scholarship.

Weinberg, S. 1979. Phenomenological Lagrangian. *Physica* 96A:327–340.

Weinberg, S. 1980a. Conceptual foundations of the unified theory of weak and electromagnetic interactions. *Rev. Mod. Phys.* 52:515–523.

Weinberg, S. 1980b. Effective gauge theories. *Phys. Lett.* 91B:51–55.

Weinberg, S. 1983. Why the renormalization group is a good thing. In A. H. Guth, K. Huang, and R. L. Jaffee, eds., *Asymptotic Realms of Physics: Essays in Honor of Francis E. Low.* Cambridge, Mass.: MIT Press.

Weinberg, S. 1985a. The ultimate structure of matter. In C. DeTar, J. Finkelstein, and C. I. Tan, eds., *A Passion for Physics: Essays in Honor of Geoffrey Chew.* Singapore: World Scientific.

Weinberg, S. 1985b. Calculation of fine structure constants. In Jackiw et al. 1985.

Weinberg, S. 1986a. Particle physics: Past and future. *Int. J. Mod. Phys.* A1/1:135–145.

Weinberg, S. 1986b. Towards the final laws of physics. In *Elementary Particles and the Laws of Physics: The 1986 Dirac Memorial Lectures*. Cambridge, U.K.: Cambridge University Press.

Weinberg, S. 1986c. Particles, fields, and now strings. In de Boer et al. 1986.

Weinberg, S. 1987. Newtonianism, reductionism and the art of congressional testimony. Talk presented at the Tercentenary Celebration of Newton's *Principia*, University of Cambridge, June 30, 1987.

Weiner, C. 1969. A new site for the seminar: The refugees and American physics in the thirties. In Fleming and Baylin 1969, pp. 190–234.

Weiner, C. 1970. Physics in the Great Depression. *Phys. Today* 23:31–38. (Reprinted in Weart and Philips 1985, pp. 115–121.)

Weiner, C. 1972a. 1932–Moving into the new physics. *Phys. Today* 25:40–49.

Weiner, C. 1972b. *Exploring the History of Nuclear Physics*. New York: American Institute of Physics.

Weiner, C. 1973. Physics today and the spirit of the forties. *Phys. Today* 26:23–28.

Weiner, C. 1974. Institutional settings for scientific change: Episodes from the history of nuclear physics. In A. Thackray and E. Mendelsohn, eds., *Science and Values*, pp. 187–212. New York: Humanities Press.

Weiner, C. 1975. Cyclotrons and internationalism: Japan, Denmark and the United States, 1935–1945. In *Proceedings of the Fourteenth International Congress of the History of Science, 1974*, no. 2, pp. 353–365. Tokyo: Science Council of Japan.

Weiner, C., ed. 1977. *History of Twentieth Century Physics: Proceedings of the International School of Physics "Enrico Fermi,"* course 57, Varenna, Italy. New York: Academic Press.

Weisskopf, V. F. 1931. Zur Theorie der Resonanz Fluoreszenz. *Ann. der Phys.* 9:23–66.

Weisskopf, V. F. 1934. Über die Selbstenergie des Elektrons. *Zeits. Phys.* 89:27–39, and Berichtung, *Zeits. Phys.* 90:817–818.

Weisskopf, V. F. 1936. Über die Elektrodynamik des Vakuums auf Grund der Quantentheorie des Elektrons. *Det. Kgl. Danske Videnskab. Selskab. Mat.-Fys. Medd.* 14:1–39.

Weisskopf, V. F. 1939. On the self-energy of the electromagnetic field of the electron. *Phys. Rev.* 56:72–85.

Weisskopf, V. F. 1947. On the production process of mesons. *Phys. Rev.* 72:510.

Weisskopf, V. F. 1949. Recent developments in the theory of the electron. *Rev. Mod. Phys.* 21:305–328.

Weisskopf, V. F. 1972. *Physics in the Twentieth Century*. Cambridge, Mass.: MIT Press.

Weisskopf, V. F. 1983. Growing up with field theory: The development of quantum electrodynamics. In Brown and Hoddeson 1983, pp. 56–81.

Weisskopf, V. F., and Wigner, E.P. 1930. Berechnung der natürlichen Linienbreite auf Grund der Diracschen Lichttheorie. *Zeits. Phys.* 63:54–73.

Welton, T. A. 1948. Some observable effects of the quantum-mechanical fluctuations of the electromagnetic field. *Phys. Rev.* 74:1157–1167.

Welton, T. A. 1984. Memories. Recollections of T. A. Welton's friendship with R. P. Feynman, 1935–1950. Niels Bohr Library, American Institute of Physics, New York.

Wentzel, G. 1933a. Über die Eigenkräfte der Elementarteilchen. I. *Zeits. Phys.* 86:479–494.

Wentzel, G. 1933b. Über die Eigenkräfte der Elementarteilchen. II. *Zeits. Phys.* 86: 635–645.

Wentzel, G. 1934. Über die Eigenkräfte der Elementarteilchen. III. *Zeits. Phys.* 87: 726–733.

Wentzel, G. 1940. Zum Problem des statischen Mesonfeldes. *Helv. Phys. Acta* 13:269–308; Nachtag, ibid. 14:633–635.

Wentzel, G. 1941. Zur Hypothese der höheren Proton-Isobaren. *Helv. Phys. Acta* 14:3–20.

Wentzel, G. 1943. *Einführung in der Quantentheorie der Wellenfelder.* Vienna: Franz Deuticke. (Reprinted by Edwards Brothers Inc., Ann Arbor, Mich., 1946. Trans. into English as *Quantum Theory of Fields.* New York: Interscience, 1949.)

Wentzel, G. 1947. Recent research in meson theory. *Rev. Mod. Phys.* 19:1–18.

Wentzel, G. 1948. New aspects of the photon self-energy problem. *Phys. Rev.* 74: 1070–1075.

Wentzel, G. 1960. Quantum theory of fields (until 1947). In Fierz and Weisskopf 1960, pp. 44–67.

Wentzel, G. 1970. *Quanta: Essays in Theoretical Physics Dedicated to Gregor Wentzel.* Ed. P. Freund, C. J. Goebel, and Y. Nambu. Chicago: University of Chicago Press.

Wentzel, G. 1973. Quantum theory of fields (until 1947). In Mehra 1973, pp. 380–403.

Wessels, L. 1977. Schrödinger's route to wave mechanics. *Stud. Hist. Phil. Sci.* 10:311–340.

Wessels, L. 1983. Erwin Schrödinger and the descriptive tradition. In Aris et al. 1983, pp. 254–278.

Weyl, H. 1929. Elektron und Gravitation. I. *Zeits. Phys.* 56:332–353.

Weyl, H. 1930. *Gruppentheorie und Quantenmechanik.* 2d ed. (Trans. by H.P. Robertson as *The Theory of Groups and Quantum Mechanics.* New York: E.P. Dutton, 1931.)

Wheaton, B. 1983. *The Tiger and the Shark.* Cambridge, U.K.: Cambridge University Press.

Wheeler, J. A. 1937. On the mathematical description of light nuclei by the method of resonating group structure. *Phys. Rev.* 52:107–122.

Wheeler, J. A. 1940. Nuclear physics. Dittoed notes by R. P. Feynman. Department of Physics, Princeton University, Princeton, N.J.

Wheeler, J. A. 1946. Problems and prospects in elementary particle research. *Proc. Amer. Phil. Soc.* 90:36–52.

Wheeler, J. A. 1947. Correspondence principle analysis of double-photon emission. *J. Opt. Soc. America* 37:813–817.

Wheeler, J. A. 1953. Hendrik Kramers. *Yearbook of the Amer. Phil. Soc.*, pp. 355–360.

Wheeler, J. A. 1979. Some men and moments in nuclear physics. In Stuewer 1979, pp. 213–324.

Wheeler, J. A. 1989. The young Feynman. *Phys. Today* 42/2:24–31.

Wheeler, J. A., and Bohr, N. 1939. The mechanism of nuclear fission. *Phys. Rev.* 56: 426–450.

Wheeler, J. A., and Feynman, R. P. 1945. Interaction with the absorber as the mechanism of radiation. *Rev. Mod. Phys.* 17:157–181.

Wheeler, J. A., and Feynman, R. P. 1949. Classical electrodynamics in terms of direct interparticle action. *Rev. Mod. Phys.* 21:235–433.

Whipplesnaith (pseud.). 1937. *The Night Climbers of Cambridge.* London: Chatto and Windus.

Wick, G. C. 1950. The evaluation of the collision matrix. *Phys. Rev.* 80:268–272.

Wightman, A. S. 1947. Pais' *f*-field and the mirror nuclei. *Phys. Rev.* 71:447–448.

Wightman, A. S. 1956. Quantum field theories in terms of expecation values. *Phys. Rev.* 101:860–866.

Wightman, A. S. 1972. The Dirac equation. In Salam and Wigner 1972, pp. 95–115.

Wightman, A. S. 1978. Field theory: Axiomatic. In *Encyclopedia of Physics*, pp. 318–321. New York: McGraw-Hill.

Wightman, A. S. 1979. Should we believe in quantum field theory? In A. Zichichi, ed., *The Whys of Subnuclear Physics*. Lectures at "Ettore Majorna" International School of Subnuclear Physics, Erice, 1977. New York: Plenum Press.

Wightman, A. S. 1981a. Looking back at quantum field theory. *Physica Scripta* 24: 813–816.

Wightman, A. S. 1981b. The choice of test functions in quantum field theory. *Adv. Math. Suppl. Studies* 7B:769–790.

Wightman, A. S. 1986. Some lessons of renormalization theory. In deBoer et al. 1986, pp. 201–226.

Wightman, A. S. 1989. The general theory of quantized fields in the 1950s. In Brown et al. 1989, pp. 608–629.

Wigner, E. P. 1939. On the unitary representations of the Lorentz group. *Ann. Math.* 40: 140–204.

Wigner, E. P. 1946. Theoretical physics in the metallurgical laboratory of Chicago. *J. Appl. Phys.* 17:857–863.

Wigner, E. P. 1947. *Physical Science and Human Values.* Symposium volume with a fore-word by E. P. Wigner. Princeton, N.J.: Princeton University Press.

Wigner, E. P. 1963. Interview with T. Kuhn. Archives for the History of Quantum Physics, Niels Bohr Library, American Institute of Physics, New York.

Wigner, E. P. 1973. Relativistic equations in quantum mechanics. In Mehra 1973, pp. 320–330.

Williams, E. J. 1933. Applications of the method of impact parameter in collisions. *Proc. Roy. Soc. London* A139:163–186.

Williams, E. J. 1934. Nature of the high energy particles of penetrating radiation formulae. *Phys. Rev.* 45:729–730.

Williams, E. J. 1935. Correlation of certain collision problems with radiation theory. *Det. Kgl. Danske Videnskab. Selskab. Mat.-Fys. Medd.* (no.4) 13:1–50.

Williams, R. C. 1938. The fine structure of H_α and D_α under varying discharge conditions. *Phys. Rev.* 54:558–567.

Williams, R. C., and Gibbs, R. C. 1934. The fine structure analysis of H_α^1 and H_α^2. *Phys. Rev.* 45:475–479.

Williams, W. E. 1942. Attempts at obtaining excitation of an atomic beam of monoatomic hydrogen. *Rev. Mod. Phys.* 14:94–95.

Wilson, A. 1984. Theoretical physics in the late 1920s and early 1930s. In J. Hendry, ed., *Cambridge Physics in the Thirties*, pp. 174–175. Bristol, U.K.: Adam Hilger.

Wilson, A. H. 1941. The quantum theory of radiation damping. *Proc. Cambridge Phil. Soc.* 37:301–316.

Wilson, J., ed. 1975. *All in Our Time: The Reminiscences of Twelve Nuclear Pioneers.* Chicago: Bulletin of the Atomic Scientists.

Wilson, K. G. 1965. Model Hamiltonians for local quantum field theory. *Phys. Rev.* B140:445–457.

Wilson, K. G. 1969. Non-Lagrangian models of current algebra. *Phys. Rev.* B179: 1499–1512.

Wilson, K. G. 1970a. Operator-product expansions and anomalous dimensions in the Thirring model. *Phys. Rev.* D2:1473–1477.

Wilson, K. G. 1970b. Anomalous dimensions and the breakdown of scale invariance in perturbation theory. *Phys. Rev.* D2:1478–1493.

Wilson, K. G. 1975. The renormalization group: Critical phenomena and the Kondo problem. *Rev. Mod. Phys.* 47:773–840.

Wilson, K. G. 1983. The renormalization group and critical phenomena. *Rev. Mod. Phys.* 55:583–600.

Wilson, K. G., and Kogut, J. 1974. The renormalization group and the ϵ expansion. *Physics Reports* 12c:75–199.

Wilson, R. R. 1970. My fight against team research. *Daedalus*.99:4, 1076–1087. (Also G. Holton, ed., *The Twentieth-Century Sciences. Studies in the Biography of Ideas,* pp. 468–479. New York: Norton, 1972.

Wilson, R. R. 1977. Interview by S. Weart, May 19, 1977. American Institute of Physics, New York.

Wilson, R. R. 1979. Introduction of J. A. Wheeler. In Stuewer 1979, p. 213.

Yang, C. N. 1961. *Elementary Particles: A Short History of Some Discoveries in Atomic Physics:* The Vanuxem Lectures, 1959. Princeton, N.J.: Princeton University Press.

Yang, C. N. 1983. *Selected Papers, 1945–1980, with Commentary.* San Francisco: W. H. Freeman.

Yang, C. N., and Mills, R. L. 1954a. Isotopic spin conservation and a generalized gauge invariance. *Phys. Rev.* 95:631.

Yang, C. N., and Mills, R. L. 1954b. Conservation of isotopic spin and isotopic gauge invariance. *Phys. Rev.* 96:191–195.

Yukawa, H. 1935. On the interaction of elementary particles. I. *Proc. Phys.-Math. Soc. Japan* 17:48–57.

Yukawa, H. 1937. On a possible interpretation of the penetrating component of the cosmic rays. *Proc. Phys.-Math. Soc. Japan* 19:712–713.

Yukawa, H. 1942. On the direction of theoretical physics. *Kagaku-chishiki* (in Japanese). (Included in *Sonzai no Riho.* Tokyo: Iwanami-shoten Publishers, 1943.)

Yukawa, H. 1947. On the theory of elementary particles. *Prog. Theor. Phys.* 2(4):209–215.

Yukawa, H. 1950. Quantum theory of non-local fields. I, II. *Phys. Rev.* 77:219–226, and *Phys. Rev.* 80:1047–1052.

Yukawa, H. 1974. Hundred years of science in Japan from a physicist's point of view. *ICHS* 14. *Proceedings* 2:3–15; *Proceedings* 3:227–230.

Yukawa, H. 1982. *Tabibito* (The Traveller). Singapore: World Scientific.

Zirzel, P. R.; Blair, J. S.; and Powell, J. P. 1947. Effect of "*f*-interaction" of Pais on symmetry of nuclear Hamiltonian. *Phys. Rev.* 72:225–228.

INDEX